HYDROGEN FUEL

Production, Transport, and Storage

HYDROGEN FUEL

Production, Transport, and Storage

Edited by
Ram B. Gupta

CRC Press
Taylor & Francis Group
Boca Raton London New York

CRC Press is an imprint of the
Taylor & Francis Group, an **informa** business

CRC Press
Taylor & Francis Group
6000 Broken Sound Parkway NW, Suite 300
Boca Raton, FL 33487-2742

Printed in the United States of America on acid-free paper
10 9 8 7 6 5 4 3

International Standard Book Number-13: 978-1-4200-4575-8 (Hardcover)

Library of Congress Cataloging-in-Publication Data

Hydrogen fuel : production, transport, and storage / Ram B. Gupta, editor.
p. cm.
Includes bibliographical references and index.
ISBN 978-1-4200-4575-8 (hardcover : acid-free paper)
1. Hydrogen as fuel. 2. Fuel cells. I. Gupta, Ram B. II. Title.

TP359.H8H89 2008
665.8'1--dc22 2008000265

Visit the Taylor & Francis Web site at
http://www.taylorandfrancis.com

and the CRC Press Web site at
http://www.crcpress.com

Contents

Preface

The two most important environmental hazards faced by humankind today are air pollution and global warming. Both have a direct link with our current overdependence on fossil fuels. Pollutants produced from combustion of hydrocarbons now cause even more health problems due to the urbanization of world population. The net increase in environmental carbon dioxide from combustion is a suspect cause for global warming, which is endangering the Earth—the only known place to support human life. In addition, the import of expensive hydrocarbon fuel has become a heavy burden on many countries, causing political and economic unrest.

If we look at the past 2000 years' history of fuels, usage has consistently moved in the direction of a cleaner fuel: wood → coal → petroleum → propane → methane as shown on the next page.

With time, the fuel molecule has become smaller, leaner in carbon, and richer in hydrogen. The last major move was to methane, which is a much cleaner burn than gasoline. Our future move is expected to be to hydrogen, which has the potential to solve both the environmental hazards faced by humankind. Through its reaction with oxygen, hydrogen intensely releases energy in combustion engines or quietly releases it in fuel cells to produce water as its only by-product. There is no emission of smoke, CO, CO_2, NO_x, SO_x, or O_3. In fact, the health costs for urban populations can be reduced by switching to hydrogen automobiles. Hydrogen can be produced from water using a variety of energy sources including solar, wind, nuclear, biomass, petroleum, natural gas, and coal. Since renewable energy sources (solar, wind, and/or biomass) are available in all parts of the world, all countries will have access to hydrogen fuel. Hence, a greater democratization of energy resources will occur. Also the use of solar, wind, or biomass in producing hydrogen does not add to environmental CO_2. Before widescale use of hydrogen fuel can be accomplished, key technological challenges need to be resolved, including cost-effective production and storage of hydrogen. During the early adoption of hydrogen fuel, government incentives will be needed, which may be recovered from savings in the health care expenditures and carbon credits.

This book is organized into three sections: Chapters 1 through 8 deal with production and use aspects; Chapters 9 through 13 cover transportation and storage aspects, and Chapters 14 through 17 discuss safety and environmental aspects of hydrogen fuel.

The hydrogen molecule is the smallest and lightest of all the fuel molecules, with unique properties and uses (Chapter 1). Hydrogen can be produced from a variety of primary energies including hydrocarbons (Chapter 2), coal (Chapter 3), nuclear (Chapter 4), wind (Chapter 5), biomass (Chapter 6), and solar (Chapter 7). Wind, solar, and nuclear electrolyses can produce pure hydrogen ready for use in fuel cells or in internal combustion engines. However, hydrogen derived from the other energy sources will require separation and purification (Chapter 8).

A major technical challenge with hydrogen fuel is its transportation and storage. The U.S. Department of Energy has specified technical targets for storage (Chapter 9). Hydrogen can be transported using pipelines and tankers (Chapter 10) and stored using compressed tanks (Chapter 10), as metal hydrides (Chapter 11), adsorbed on carbons (Chapter 12), and as chemical hydrides (Chapter 13).

Proper codes and standards need to be adopted for effective utilization of hydrogen fuel (Chapter 14). Fuel and safety properties of hydrogen are different from conventional

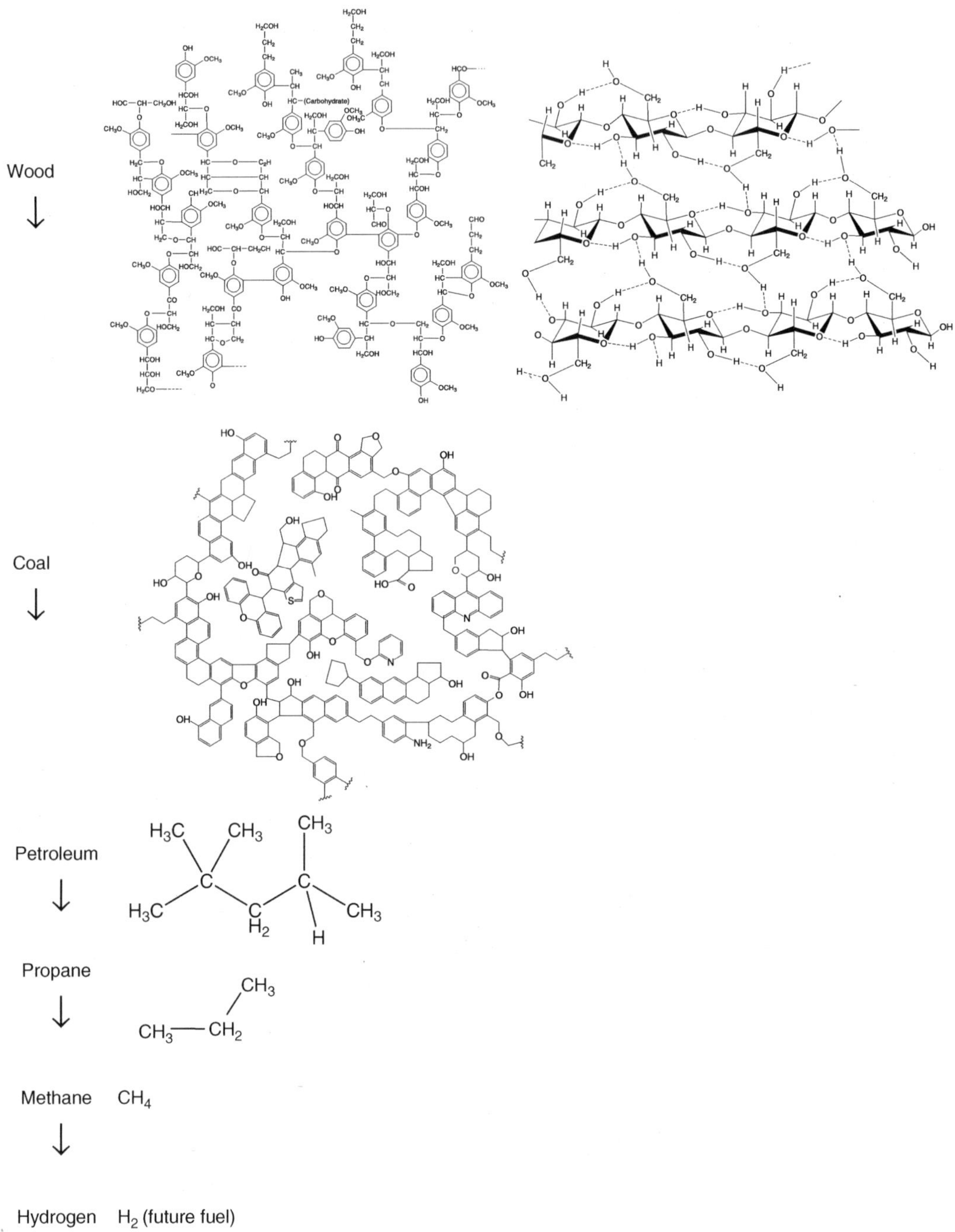

fuels. Hence, proper monitoring (Chapter 15) and safety designs need to be incorporated (Chapter 16). Finally, if hydrogen is produced from fossil fuels, the by-product CO_2 needs to be sequestered (Chapter 17).

Preparation of this book would not have been possible without the valuable contributions from various experts in the field. The timely contributions and support from the Alabama Center for Paper and Bioresource Engineering, Auburn University and the Consortium for Fossil Fuel Science are deeply appreciated.

Editor

Ram B. Gupta is an alumni (chair) professor of chemical engineering at Auburn University. He has published numerous research papers and holds several patents on hydrogen fuel and supercritical fluid technology, and is the recipient of the Distinguished Graduate Faculty Lectureship Award (2007) from Auburn University, the Science and Engineering Award (2002–2004) from DuPont, the Junior and Senior Research Awards (1998, 2002) from the Auburn Alumni Engineering Council, the James A. Shannon Director's Award (1998) from the National Institutes of Health, and the Young Faculty Career Enhancement Award (1997) from Alabama NSF-EPSCoR.

Dr. Gupta is a consultant to several energy companies. He received his BE (1987) from the Indian Institute of Technology, Roorkee; an MS (1989) from the University of Calgary, Canada; and his PhD (1993) from the University of Texas at Austin, in chemical engineering. He joined Auburn University in 1995, after two-year postdoctoral work at the University of California, Berkeley. His recent books are *Nanoparticle Technology for Drug Delivery* (2006, Taylor & Francis), *Solubility in Supercritical Carbon Dioxide* (2007, CRC Press), and *Hydrogen Fuel: Production, Transport, and Storage* (2008, CRC Press).

Contributors

Dimitrios A. Bechrakis
Hellenic Transmission System
Athens, Greece

Rajani S. Bennur
Department of Biochemistry
Karnataka University
Dharwad, India

Ashok Damle
Techverse, Inc.
Cary, North Carolina

Wiebren de Jong
Department of Process and Energy
Delft University of Technology
Delft, the Netherlands

Neelkanth G. Dhere
Florida Solar Energy Center
University of Central Florida
Cocoa Beach, Florida

Liang-Shih Fan
Department of Chemical and Biomolecular Engineering
The Ohio State University
Columbus, Ohio

Ming Gao
Blade Energy Partners
Houston, Texas

Ram B. Gupta
Department of Chemical Engineering
Auburn University
Auburn, Alabama

Ryutaro Hino
Japan Atomic Energy Agency
Ibaraki-Ken, Japan

Shinya Hodoshima
Department of Industrial Chemistry
Tokyo University of Science
Tokyo, Japan

Ravi Krishnamurthy
Blade Energy Partners
Houston, Texas

Klaus S. Lackner
Department of Earth and Environmental Engineering
Columbia University
New York, New York

Shi-Ying Lin
Japan Coal Energy Center
Tokyo, Japan

Nazim Z. Muradov
Florida Solar Energy Center
University of Central Florida
Cocoa Beach, Florida

James M. Ohi
Hydrogen Technologies and Systems
National Renewable Energy Laboratory
Golden, Colorado

K. K. Pant
Department of Chemical Engineering
Indian Institute of Technology
Delhi, India

Ah-Hyung Alissa Park
Department of Earth and Environmental Engineering
Columbia University
New York, New York

Fotis Rigas
School of Chemical Engineering
National Technical University of Athens
Athens, Greece

Yasukazu Saito
Department of Industrial Chemistry
Tokyo University of Science
Tokyo, Japan

Sunita Satyapal
Office of Hydrogen, Fuel Cells, and Infrastructure Technologies
U.S. Department of Energy
Washington, DC

Frank Schweighardt
Process Analytical Technology Consultant
Allentown, Pennsylvania

Spyros Sklavounos
School of Chemical Engineering
National Technical University of Athens
Athens, Greece

Prabhu Soundarrajan
H2scan Corporation
Valencia, California

George J. Thomas
Office of Hydrogen, Fuel Cells, and Infrastructure Technologies
U.S. Department of Energy
Washington, DC

Elli Varkaraki
Centre for Renewable Energy Sources
Attiki, Greece

Xing L. Yan
Japan Atomic Energy Agency
Ibaraki-Ken, Japan

Section I

Production and Use of Hydrogen

1

Fundamentals and Use of Hydrogen as a Fuel

K. K. Pant and Ram B. Gupta

CONTENTS

1.1 Introduction

Owing to an increasing world population and demands for higher standards of living and better air quality, the future energy demand is expected to increase significantly. To meet this demand poses great challenges. Currently, most of the world energy requirement for transportation and heating (which is two-third of the primary energy demand) is derived from petroleum or natural gas. These two fuels are generally favored due to the ease of transport of liquid or gaseous forms. Unfortunately, the combustion of hydrocarbon fuels for transportation and heating contributes over half of all greenhouse gas emissions and a large fraction of air pollutant emissions. Hence, today's world is facing an urgency in developing alternative fuels. Among various alternatives, hydrogen fuel offers the highest potential benefits in terms of diversified supply and reduced emissions of pollutants and greenhouse gases. For the past 40 years, environmentalists and several industrial organizations have promoted hydrogen fuel as the solution to the problems of air pollution and global warming. The key criteria for an ideal fuel are inexhaustibility, cleanliness, convenience, and independence from foreign control. Hydrogen possesses all these properties, and is being evaluated and promoted worldwide as an environmentally benign replacement for gasoline, heating oil, natural gas, and other fuels in both transportation and nontransportation applications. A number of reports are now available on several aspects of hydrogen [1–25].

Similar to electricity, hydrogen is a high-quality energy carrier, which can be used with a high efficiency and zero or near-zero emissions at the point of use. It has been technically demonstrated that hydrogen can be used for transportation, heating, and power generation, and could replace current fuels in all their present uses [2–6]. Hydrogen can be produced using a variety of starting materials, derived from both renewable and nonrenewable sources, through many different process routes. At present, two basic process technologies—(1) reformation of natural gas and (2) electrolysis of water—are widely used.

In the advent of hydrogen economy, the principal focus of hydrogen technology has shifted to the safe and affordable utilization of hydrogen as an alternative fuel based on seamless integration of generation, distribution, and storage technologies. Inaccuracies, inconsistencies, and contradictions abound in the seemingly persuasive arguments targeting the general public and politicians regarding the merits of the hydrogen case. These inaccuracies tend to create the global perception that hydrogen will become an active source for our energy needs, replacing today's relatively less-efficient machines with clean fuel cells, which will efficiently power cars, trucks, homes, and businesses, ending global warming and air pollution. The key assertions of the initiative for hydrogen production and utilization are based on the premise that the fuel cell is a proven technology and hydrogen is in abundant supply on Earth [10–12], but unfortunately, most of the hydrogen

TABLE 1.1

United States and World Hydrogen Consumptions by End-Use Category

	United States		World Total		
Captive Users	**Billion m³**	**Share (%)**	**Billion m³**	**Share (%)**	**U.S. Share of World Total (%)**
Ammonia producers	33.7	38	273.7	61	12
Oil refiners	32.9	37	105.4	23	31
Methanol producers	8.5	10	40.5	9	21
Other	3.4	4	13.6	3	25
Merchant users	10.8	12	16.1	4	67
Total	89.3	100	449.3	100	20

Source: Adapted from SRI Consulting Inc., *Chemical Economics Handbook 2001*, Menlo Park, CA, July 2001; Wee, J.H., *Renewable Sustainable Energy Rev.*, 11, 1720–1738, 2007.

on Earth is in the fully oxidized form as H_2O, which has no fuel value, and there are no natural sources of desirable molecular hydrogen (H_2).

At present, hydrogen production is a large and growing industry. Globally, some 50 million t of hydrogen, equivalent to about 170 million t of petroleum, were produced in 2004. And the production is increasing by about 10% every year. As of 2005, the economic value of all hydrogen produced worldwide was about $135 billion per year [3]. The current global hydrogen production is 48% from natural gas, 30% from petroleum, 18% from coal, and 4% from electrolysis [4]. Major end users of the hydrogen are listed in Table 1.1. Hydrogen is primarily consumed in two nonfuel uses: (1) about 60% to produce NH_3 by the Haber process for subsequent use in fertilizer manufacturing [14] and (2) about 40% in refinery, chemicals, and petrochemical sectors. If nonconvenentional resources, such as wind, solar, or nuclear power for hydrogen production were available, the use of hydrogen for hydrocarbon synfuel production could expand by 5- to 10-fold [4]. It is estimated that 37.7 million t per year of hydrogen would be sufficient to convert enough domestic coal to liquid fuels to end U.S. dependence on foreign oil imports, and less than half this figure to end dependence on Middle East oil. Figure 1.1 shows various application areas of hydrogen energy, out of which the use of hydrogen energy for vehicular application is of current focus [26].

1.2 Physical Properties

Hydrogen atom is the lightest element, with its most common isotope consisting of only one proton and one electron. Hydrogen atoms readily form H_2 molecules, which are smaller in size when compared to most other molecules. The molecular form, simply referred to as hydrogen is colorless, odorless, and tasteless and is about 14 times lighter than air, and diffuses faster than any other gas. On cooling, hydrogen condenses to liquid at −253°C and to solid at −259°C. The physical properties of hydrogen are summarized in Table 1.2. Ordinary hydrogen has a density of 0.09 kg/m³. Hence, it is the lightest substance known with a buoyancy in air of 1.2 kg/m³. Solid metallic hydrogen has a greater electrical conductivity than any other solid elements. Also, the gaseous hydrogen has one of the highest heat capacity (14.4 kJ/kg K).

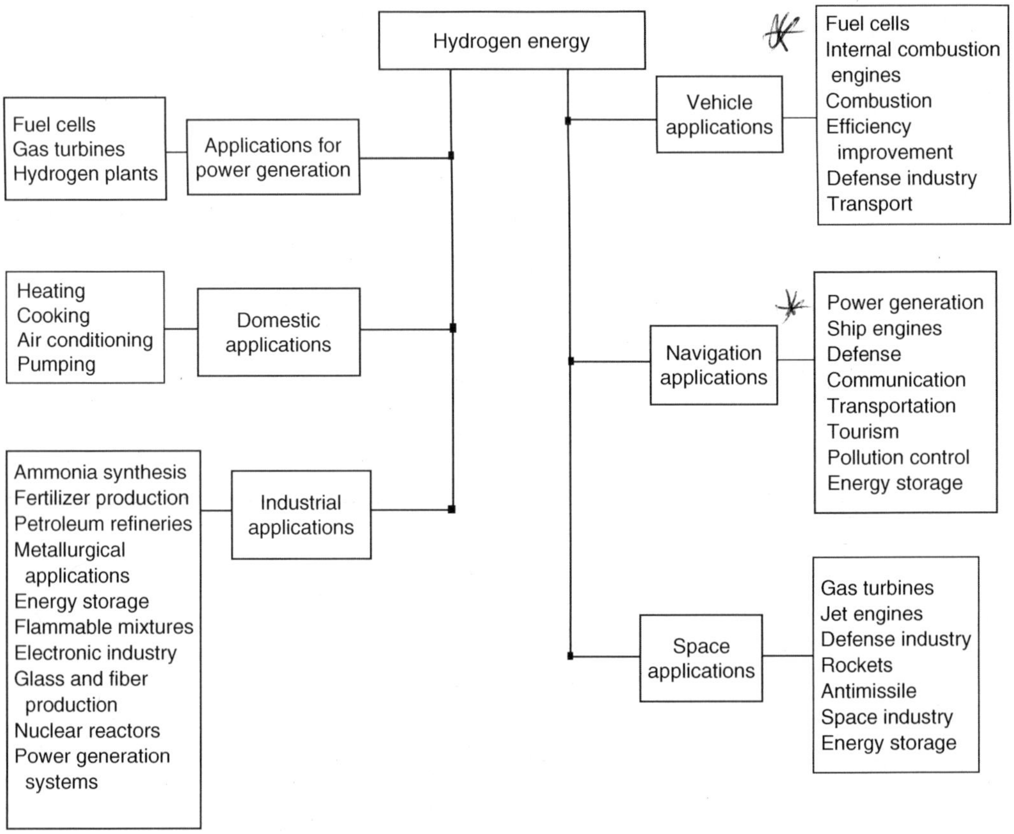

FIGURE 1.1
Application areas for hydrogen energy. (Reproduced with permission from Elsevier; Midilli, A., Dincer, I., and Rosen, M.A., *Renewable Sustainable Energy Rev.*, 9(3), 255–271, 2005.)

TABLE 1.2
Properties of Hydrogen

Property	Value
Molecular weight	2.01594
Density of gas at 0°C and 1 atm.	0.08987 kg/m^3
Density of solid at −259°C	858 kg/m^3
Density of liquid at −253°C	708 kg/m^3
Melting temperature	−259°C
Boiling temperature at 1 atm.	−253°C
Critical temperature	−240°C
Critical pressure	12.8 atm.
Critical density	31.2 kg/m^3
Heat of fusion at −259°C	58 kJ/kg
Heat of vaporization at −253°C	447 kJ/kg
Thermal conductivity at 25°C	0.019 kJ/(ms°C)
Viscosity at 25°C	0.00892 centipoise
Heat capacity (Cp) of gas at 25°C	14.3 kJ/(kg°C)
Heat capacity (Cp) of liquid at −256°C	8.1 kJ/(kg°C)
Heat capacity (Cp) of solid at −259.8°C	2.63 kJ/(kg°C)

Source: Adapted from *Kirk-Othmer Encyclopedia of Chemical Technology. Fundamentals and Use of Hydrogen as a Fuel*. 3rd ed., Vol. 4, Wiley, New York, 1992, 631p.

The hydrogen atom (H) consists of a nucleus of unit positive charge and a single electron. It has an atomic number of 1 and an atomic weight of 1.00797. This element is a major constituent of water and all organic matters, and is widely distributed not only on the earth but also throughout the Universe. There are three isotopes of hydrogen: (1) protium—mass 1, makes up 99.98% of the natural element; (2) deuterium—mass 2, makes up about 0.02%; and (3) tritium—mass 3, occurs in extremely small amounts in nature, but may be produced artificially by various nuclear reactions. The ionization potential of hydrogen atom is 13.54 V [7].

Hydrogen is a mixture of ortho- and para-hydrogen in equilibrium, distinguished by the relative rotation of the nuclear spin of the individual atoms in the molecule. Molecules with spins in the same direction (parallel) are termed ortho-hydrogen and those in the opposite direction as para-hydrogen. These two molecular forms have slightly different physical properties but have equivalent chemical properties. At an ambient temperature, the normal hydrogen contains 75% ortho-hydrogen and 25% para-hydrogen. The ortho-to-para conversion is associated with the release of heat. For example, at 20 K, a heat of 703 kJ/kg is released for ortho-to-para conversion. The conversion is slow but occurs at a finite rate (taking several days to complete) and continues even in the solid state. Catalysts can be used to accelerate the conversion for the production of liquid hydrogen, which is more than 95% para-hydrogen. The vapor pressure of liquid normal hydrogen is given by

$$P\,(\text{Pa}) = 10^{\left[-\frac{44.9569}{T\,(\text{K})}+6.79177+0.0205377\,(\text{K})\right]}$$

Hydrogen has a low solubility in solvents; for example, at ambient conditions, only 0.018 and 0.078 mL of gaseous H_2 dissolves into each milliliter of water and ethanol, respectively. However, the solubility is much more pronounced in metals. Palladium is particularly notable in this respect, which dissolves about 1000 times its volume of the gas. The adsorption of hydrogen in steel may cause "hydrogen embrittlement," which sometimes leads to the failure of chemical processing equipment [4].

1.3 Chemical Properties

At ordinary temperatures, hydrogen is comparatively nonreactive unless it has been activated in some manner. On the contrary, hydrogen atom is chemically very reactive, and that is why it is not found chemically free in nature. In fact, very high temperatures are needed to dissociate molecular hydrogen into atomic hydrogen. For example, even at 5000 K, about 5% of the hydrogen remains undissociated. In nature, mostly the hydrogen is bound to either oxygen or carbon atoms. Hence, to obtain hydrogen from natural compounds, energy expenditure is needed. Therefore, hydrogen must be considered as an energy carrier—a means to store and transmit energy derived from a primary energy source.

Atomic hydrogen is a powerful reducing agent, even at room temperature. For example, it reacts with the oxides and chlorides of many metals, including silver, copper, lead, bismuth, and mercury, to produce the free metals. It reduces some salts, such as nitrates, nitrites, and cyanides of sodium and potassium, to the metallic state. It reacts with a number of elements, both metals and nonmetals, to yield hydrides such as NH_3, NaH, KH, and PH_3. Sulfur forms a number of hydrides; the simplest is H_2S. Combining with oxygen, atomic hydrogen yields hydrogen peroxide, H_2O_2. With organic compounds, atomic hydrogen reacts to produce a complex mixture of products; for example, on reacting with ethylene, atomic hydrogen produces C_2H_6 and C_4H_{10}. Hydrogen reacts violently with oxidizers like nitrous oxide, halogens (especially with fluorine and chlorine), and unsaturated hydrocarbons (e.g., acetylene) with intense exothermic heat. When hydrogen reacts with oxygen in either a combustion or electrochemical conversion process to generate energy, the resulting reaction product is water vapor. At room temperature this reaction is immeasurably slow, but is accelerated by catalysts, such as platinum, or by an electric spark.

From the safety point of view, the following are the most important properties of hydrogen when compared to other conventional fuels:

- *Diffusion*. Hydrogen diffuses through air much more rapidly than other gaseous fuels. With a diffusion coefficient in air of .61 cm^2/s, the rapid dispersion rate of hydrogen is its greatest safety asset.
- *Buoyancy*. Hydrogen would rise more rapidly than methane (density at standard condition is 1.32 kg/m^3), propane (4.23 kg/m^3), or gasoline vapor (5.82 kg/m^3).
- *Color, odor, taste, and toxicity*. Hydrogen is colorless, odorless, tasteless, and nontoxic; similar to methane.
- *Flammability*. Flammability of hydrogen is a function of concentration level and is much greater than that of methane or other fuels. Hydrogen burns with a low-visibility flame. The flammability limits of mixtures of hydrogen with air, oxygen, or other oxidizers depend on the ignition energy, temperature, pressure, presence of diluents, and size and configuration of the equipment, facility, or apparatus. Such a mixture may be diluted with either of its constituents until its concentration shifts below the lower flammability limit (LFL) or above the upper flammability limit (UFL). The limit of flammability of hydrogen in air at ambient condition is 4–75%, methane in air is 4.3–15 vol%, and gasoline in air is 1.4–7.6 vol%.
- *Ignition energy*. When its concentration is in the flammability range, hydrogen can be ignited by a very small amount of energy due to its low ignition energy of 0.02 mJ as compared to 0.24 mJ for gasoline and 0.28 mJ for methane, at stoichiometry.
- *Detonation level*. Hydrogen is detonable over a wide range of concentrations when confined. However, it is difficult to detonate if unconfined, similar to other conventional fuels.
- *Flame velocity*. Hydrogen has a faster flame velocity (1.85 m/s) than other fuels (gasoline vapor—0.42 m/s; methane—0.38 m/s).
- *Flame temperature*. The hydrogen–air flame is hotter than methane–air flame and cooler than gasoline at stoichiometric conditions (2207°C compared to 1917°C for methane and 2307°C for gasoline).

Safety aspects of hydrogen are covered in more detail in Chapter 16.

1.4 Fuel Properties

Hydrogen is highly flammable over a wide range of temperature and concentration. Although its combustion efficiency is truly outstanding and welcomed as a fuel of the choice for the future, it inevitably renders several nontrivial technological challenges, such as

Hydrogen Challenges ① production ② storage ③ Transportation ④ Distribution ⑤ End Use

safety in production, storage, and transportation. On reacting with oxygen, hydrogen releases energy explosively in combustion engines or quietly in fuel cells to produce water as its only by-product. Unlike ready for fuel use coal or hydrocarbons, hydrogen is not available on the earth. It is, however, available as chemical compounds of oxygen and carbon. For example, hydrogen is present in water; fossil hydrocarbons such as coal, petroleum, natural gas; and biomass such as carbohydrates, protein, and cellulose. Hydrogen has both similarities and differences when compared to the conventional fuels such as methane (natural gas), liquefied petroleum gases (LPG), and liquid fuels such as gasoline. The technical and economic challenges of implementing a "hydrogen economy" require a solution to the fundamental problem of renewable energy production. There are many concerns to be addressed before hydrogen can serve as a universal energy medium, which includes difficulties with hydrogen production, transportation, storage, distribution, and end use [8–22].

1.4.1 Energy Content

Hydrogen has the highest energy content per unit mass of any fuel. For example, on a weight basis, hydrogen has nearly three times the energy content of gasoline (140.4 MJ/kg versus 48.6 MJ/kg). However, on a volume basis the situation is reversed: 8,491 MJ/m^3 for liquid hydrogen versus 31,150 MJ/m^3 for gasoline. The low volumetric density of hydrogen results in storage problem, especially for automotive applications. A large container is needed to store enough hydrogen for an adequate driving range. The energy density of hydrogen is also affected by the physical nature of the fuel, whether the fuel is stored as a liquid or as a gas; and if a gas, at what pressure. Energy-related properties of hydrogen are compared with other fuels in Tables 1.3 through 1.5.

One of the important and attractive features of hydrogen is its electrochemical property, which can be utilized in a fuel cell. At present, H_2/O_2 fuel cells are available operating at an efficiency of 50–60% with a lifetime of up to 3000 h. The current output range from 440 to 1720 A/m^2 of the electrode surface, which can give a power output ranging from 50 to 2500 W.

1.4.2 Combustibility Properties

Owing to the high diffusivity, low viscosity, and unique chemical nature, combustibility of hydrogen is somewhat different than the other fuels. Various combustibility properties are described in the following:

TABLE 1.3

Comparison of Hydrogen with Other Fuels

Fuel	LHV (MJ/kg)	HHV (MJ/kg)	Stoichiometric Air/Fuel Ratio (kg)	Combustible Range (%)	Flame Temperature (°C)	Min. Ignition Energy (MJ)	AutoIgnition Temperature (°C)
Methane	50.0	55.5	17.2	5–15	1914	0.30	540–630
Propane	45.6	50.3	15.6	2.1–9.5	1925	0.30	450
Octane	47.9	15.1	0.31	0.95–6.0	1980	0.26	415
Methanol	18.0	22.7	6.5	6.7–36.0	1870	0.14	460
Hydrogen	119.9	141.6	34.3	4.0–75.0	2207	0.017	585
Gasoline	44.5	47.3	14.6	1.3–7.1	2307	0.29	260–460
Diesel	42.5	44.8	14.5	0.6–5.5	2327		180–320

Source: Adapted from *Hydrogen Fuel Cell Engines and Related Technologies*, College of the Desert, Palm Desert, CA, 2001.

TABLE 1.4

Properties of Conventional and Alternative Fuels

Property	Gasoline	No. 2 Diesel	Methanol	Ethanol	Propane	CNG	Hydrogen
Chemical formula	C_4–C_{12}	C_9–C_{25}	CH_3OH	C_2H_5OH	C_3H_8	CH_4	H_2
Physical state	Liquid	Liquid	Liquid	Liquid	Compressed gas	Compressed gas	Compressed gas or liquid
Molecular weight	100–105	200–300	32	46	44	16	2
Composition (wt%)							
Carbon	85–88	84–87	39.5	52.2	82	75	0
Hydrogen	12–15	13–16	12.6	13.1	18	25	100
Oxygen	0	0	49.9	34.7	NA	NA	0
Specific gravity (15.5°C/15.5°C)	0.72–0.78	0.81–0.89	0.796	0.796	0.504	0.424	0.07
Boiling temperature (°C)	27–225	190–345	68	78	−42	−161	−252
Freezing temperature (°C)	−40	−34	−97.5	−114	−187.5	−183	−260
Reid vapor pressure (psi)	8–15	0.2	4.6	2.3	208	2400	NA

Source: Adapted from Alternative Fuels Data Center, Properties of Fuel, DOE Report, August 2005, available at www.afdc.doe.gov/fuel_comp.html, April 2007.

TABLE 1.5

LHV Energy Densities of Fuels

Fuel	Energy Density (MJ/m³ at 1 atm., 15°C)	Energy Density (MJ/m³ at 200 atm., 15°C)	Energy Density (MJ/m³ at 690 atm., 15°C)	Energy Density (MJ/m³ of Liquid)	Gravimetric Energy Density (MJ/kg)
Hydrogen	10.0	1,825	4,500	8,491	140.4
Methane	32.6	6,860		20,920	43.6
Propane	86.7			23,488	28.3
Gasoline				31,150	48.6
Diesel				31,435	33.8
Methanol				15,800	20.1

Source: Adapted from *Hydrogen Fuel Cell Engines and Related Technologies*, College of the Desert, Palm Desert, CA, 2001.

1.4.2.1 Wide Range of Flammability

In ambient air, hydrogen is flammable in 4–75% concentrations (which is much broader than gasoline range, 1–7.6%) and is explosive in 15–59% concentration range [9,13]. However, for internal combustion engines, it is more meaningful to define flammability range in terms of equivalence ratio (φ), defined as the mass ratio of actual fuel/air ratio to the stoichiometric fuel/air ratio. Then, the flammability range for hydrogen is $0.1 < \varphi < 7.1$, and that for gasoline is $0.7 < \varphi < 4$, which indicates that H_2 internal combustion engine is amenable to stable operation even under highly dilute conditions. In fact, the wider range gives additional control over the engine operation for emissions and fuel metering [25]. The engine operation at hydrogen-lean mixture (i.e., hydrogen amount less than the theoretical or stoichiometric amount needed for combustion with a given amount of air) allows an ease of start. Also, due to the complete combustion, the fuel economy is good. In addition, the final combustion temperature is generally lower with hydrogen fuel than with gasoline, reducing the amount of pollutants, such as nitrogen oxides, emitted in the exhaust.

1.4.2.2 *Low Ignition Energy*

The amount of energy needed to ignite hydrogen is 0.02 mJ, which is about 10-fold less than that required for gasoline (0.24 mJ). The low ignition energy enables hydrogen engines to ensure prompt ignition even for lean mixtures. Unfortunately, the low ignition energy means that hot gases and hot spots on the cylinder can serve as sources of ignition, creating problems of premature ignition and flashback. Prevention of hot spots is one of the challenges associated with running an engine on hydrogen, which is further exacerbated due to the wide flammability range.

1.4.2.3 *Small Quenching Distance*

Hydrogen has a smaller (0.64 mm) quenching distance than that for gasoline (~2 mm). Consequently, hydrogen flames travel closer to the cylinder wall than other fuels before extinguishing. Thus, it is more difficult to quench a hydrogen flame than a gasoline flame. The smaller quenching distance can also increase the tendency for backfire since the flame from a hydrogen–air mixture can more readily pass a nearly closed intake valve, than a hydrocarbon–air flame.

1.4.2.4 *Autoignition Temperature*

The autoignition temperature is the minimum temperature required to initiate self-sustained combustion in a combustible fuel mixture in the absence of an external ignition. For hydrogen, the autoignition temperature is relatively high—585°C. This makes it difficult to ignite a hydrogen–air mixture on the basis of heat alone without some additional ignition source. The autoignition temperatures of various fuels are shown in Table 1.3. This temperature has important implications when a hydrogen–air mixture is compressed. In fact, the autoignition temperature is an important factor in determining what maximum compression ratio an engine can use, since the temperature rise during compression is related to the compression ratio. The temperature should not exceed the autoignition temperature of hydrogen to avoid premature ignition. Thus, the absolute final temperature limits the compression ratio. The high autoignition temperature of hydrogen facilitates higher compression ratios than those in hydrocarbon engines. The higher compression ratio is important, since it is related to the thermal efficiency of the system. However, the drawback of a high autoignition temperature is that hydrogen is difficult to ignite in a compression ignition or diesel engine because the temperatures needed for these types of ignition are relatively high.

1.4.2.5 *High Flame Speed*

At stoichiometric ratio, hydrogen flame speed (3.46 m/s) is nearly an order of magnitude higher (faster) than that of gasoline (0.42 m/s). Hence, due to the high flame speed, hydrogen engines can more closely approach the thermodynamic engine cycle. However, at leaner mixtures, the flame velocity decreases significantly.

1.4.2.6 Hydrogen Embrittlement

Constant exposure to hydrogen causes hydrogen embrittlement in many materials, which can lead to leakage or catastrophic failures in both metal and nonmetallic components. Factors known to influence the rate and severity of hydrogen embrittlement include hydrogen concentration, purity, pressure, temperature, type of impurity, stress level, stress rate, metal composition, metal tensile strength, grain size, microstructure, and heat treatment history. Additionally, moisture content in the hydrogen gas may lead to metal embrittlement through the acceleration of the formation of fatigue cracks. Chapters 10 and 16 discuss various embrittlement aspects in detail.

1.4.2.7 Hydrogen Leakage

Owing to the low density and high diffusivity, hydrogen dispersion in air is considerably faster than that of gasoline, which is advantageous for two main reasons. First, high dispersion facilitates the formation of a uniform mixture of fuel and air. Second, if a hydrogen leak develops, then hydrogen disperses out rapidly. Thus, unsafe conditions can either be avoided or minimized. However, the high dispersibility makes hydrogen more difficult to contain than other gases. Leaks of liquid hydrogen evaporate very quickly since the boiling point of liquid hydrogen is extremely low. Hydrogen leaks are dangerous in that they pose a risk of fire where they mix with air. However, the small molecular size that increases the likelihood of a leak also results in very high buoyancy and diffusivity; therefore, leaked hydrogen rises and becomes diluted quickly, especially outdoors.

1.4.2.8 Air/Fuel Ratio

The stoichiometric air/fuel (A/F) mass ratio for the complete combustion of hydrogen in air is about 34:1, which is much higher than 15:1 A/F required for gasoline. Because hydrogen is a gaseous fuel at ambient conditions, it displaces more of the combustion chamber than a liquid fuel. Consequently, less of the combustion chamber can be occupied by air. At stoichiometric conditions, hydrogen displaces about 30% of the combustion chamber, compared to about 1–2% for gasoline. Because of hydrogen's wide range of flammability, hydrogen engines can run on A/F anywhere from 34:1 (stoichiometric) to 180:1. The lower volumetric energy density of gaseous hydrogen fuel leads to a 20% reduction in power compared to gasoline because a stoichiometric hydrogen air mixture contains 20% less energy than the same volume of gasoline(vapor)–air mixture.

1.5 Hydrogen Internal Combustion Engine

Hydrogen can be used as a fuel directly in an internal combustion engine, almost similar to a spark-ignited (SI) gasoline engine. Owing to low spark-energy requirement and wide flammability range, hydrogen is an excellent candidate for use in SI engines [25,27,28]. Owing to its high autoignition temperature, finite ignition delay, and the high flame velocity, hydrogen internal combustion engine (HICE) vehicles have less knocking tendency compared to gasoline engines. Hence, HICE have a higher research octane number (>120) than gasoline engines (91–99). HICE also offers CO_2 and hydrocarbon-free combustion and lean operation, resulting in lower NO_x emissions. Hydrogen cannot be used directly in a diesel

FIGURE 1.2
Hydrogen 7 combustion engine by BMW, which can operate on both gasoline and hydrogen fuels. (Reproduced with permission from BMW; Hydrogen 7 combustion engine by BMW, which can operate on both gasoline and hydrogen. Clean Energy BMW Group, BMW, March 22, 2007, NHA 2007.)

or "compression ignition" engine since hydrogen's autoignition temperature is too high. Thus, diesel engines must be outfitted with spark plugs or use a small amount of diesel fuel to ignite the gas known as pilot ignition.

In HICE, gaseous hydrogen is injected into the engine, which then burns the hydrogen fuel similar to a gasoline engine, and mostly designed to run at lean A/F of ≥30:1. Hydrogen being gaseous displaces the oxygen in the cylinders, and a supercharger is often needed to achieve the required power output. HICE vehicles can either be run as conventionally driven HICE vehicles or as hybrid HICE vehicles. In conventionally driven HICE vehicles, the hydrogen-burning engine mechanically drives the vehicle's wheels, similar to gasoline engine, whereas in hybrid HICE vehicles, the hydrogen engine is used to run an electric generator, similar to series hybrid drive systems operating on other fuels. Power from the electric generator is used to drive the vehicle's wheels, and is generally augmented by power from a battery or ultracapacitor pack. For illustration, a bifueled HICE is shown in Figure 1.2. This 6 L, 12-cylinder engine can operate on either hydrogen or gasoline, and provides a maximum output of 260 Hp (191 kW) at 5100 rpm [29].

1.5.1 Premature Ignition and Knock

Owing to hydrogen's lower ignition energy, wider flammability range, and shorter quenching distance, premature ignition is the major problem in HICEs when compared to gasoline internal combustion engines. Preignition is usually caused by hot spots in the combustion chamber, such as on a spark plug or exhaust valve, or on carbon deposits [30]. The well-examined external mixing of hydrogen with intake air causes backfire and knock, especially at higher engine loads. In addition, low heating value per unit of volume of hydrogen limits the maximum output power. Owing to low ignition energies of hydrogen–air mixtures, the HICE vehicles are predisposed toward the limiting effect of preignition. Premature ignition occurs when the fuel mixture in the combustion chamber becomes ignited before ignition by the spark plug, and results in an inefficient, rough running engine. The limiting effect of preignition is that this will produce an increased chemical heat release rate, which results in a rapid pressure rise, higher peak cylinder pressure, acoustic oscillations, and higher heat rejections [30].

Backfire conditions can also develop if the premature ignition occurs near the fuel intake valve, and the resultant flame travels back into the induction system. Spark knock is defined as autoignition of the hydrogen/air end gas ahead of the flame front that has oriented from the spark. Owing to superior fuel properties, knocking is less prevalent in HICE vehicles compared to gasoline vehicles. Preignition can be avoided through proper engine design, but knock is an inherent limit on the maximum compression ratio that can be used with a fuel. Preignition can be minimized by identifying the preignition sources such as in-cylinder hot spots, oil contaminants, combustion in crevice volumes, and residual energy in the ignition systems. These include use of cold-rated spark plugs, low coolant temperature, and optimized fuel injection [25,27,30].

1.5.2 Fuel Delivery Systems

Premature ignition can be reduced or eliminated by redesigning the fuel delivery system, which can be categorized into three types: central-, port-, and direct injection. Central and port fuel injections form the fuel–air mixture during the intake stroke. In the case of central injection (or a carburetor), the injection is at the inlet of the air intake manifold. In the case of port injection, fuel is injected at the inlet port. Direct injection is technologically sophisticated and involves forming the fuel–air mixture inside the combustion cylinder after the air intake valve has closed [27–36].

1.5.2.1 Central Injection

The simplest method of delivering hydrogen to a HICE is by way of a carburetor or central injection system. This method has several advantages. First, central injection does not require the hydrogen supply pressure to be as high as other methods. Second, central injection (or carburetors) is already used in gasoline ICE; hence, the conversion of a standard gasoline engine to a hydrogen or a gasoline/hydrogen engine is easy. The disadvantage of central injection is that it is more susceptible to irregular combustion due to preignition and backfires. Also, an increase in the amount of hydrogen–air mixture within the intake manifold can cause preignition [31,32].

1.5.2.2 Port Injection

In port injection, fuel is injected directly into the intake manifold at each intake port, rather than drawing fuel from a central point. Typically, hydrogen is injected into the manifold after the beginning of the intake stroke. At this point, conditions are much less severe and the probability for premature ignition is reduced. In port injection, the air is injected separately at the beginning of the intake stroke to dilute the hot residual gases, which cools any hot spots. Because less gas (hydrogen or air) is in the manifold at any one time, any preignition is less severe. The inlet supply pressure for port injection tends to be higher than for central injection, but less than for direct injection.

1.5.2.3 Direct Injection

More sophisticated hydrogen engines use direct injection into the combustion cylinder during the compression stroke. While injecting, the intake valve is closed when the fuel is injected; thus completely avoiding premature ignition during the intake stroke. Consequently, the engine does not backfire into the intake manifold. Typically, the power output of a direct injected HICE is about 42% more than a HICE using a central injection, and about 20% more than for a gasoline ICE [25,34–38]. Although direct injection solves the problem of preignition in the intake manifold, it does not necessarily prevent preignition within the combustion chamber. In addition, due to the reduced mixing time of the air and fuel in a direct injection engine, the air–fuel mixture can be nonhomogeneous. Studies have suggested that this can lead to higher NO_x emissions than the nondirect injection systems. Direct injection requires a higher fuel rail pressure than the other methods. The direct injection HICE operation requires hydrogen and air mixing within a very short time. For example, the maximum available mixing times range from approximately only 20–4 ms across the speed range 1000–5000 rpm [25].

1.5.3 Ignition Systems

Owing to low ignition energy, hydrogen can be easily ignited and gasoline ignition systems can be used. However, for very lean A/F ratios (130:1–180:1) the flame velocity is considerably low, which requires the use of a dual spark plug system. Spark plugs for a hydrogen engine should have cold rating and nonplatinum tips to reduce the chances of the spark plug tip igniting the A/F charge. A cold-rated spark plug transfers heat from the plug tip to the cylinder head quicker than a hot-rated spark plug.

1.5.4 Crankcase Ventilation

Crankcase ventilation is more important for HICE than for gasoline ICE. As with gasoline engines, unburnt fuel can seep by the piston rings and enter the crankcase. Because hydrogen has a lower ignition energy than gasoline, any unburnt hydrogen entering the crankcase has a greater chance of igniting. Hence, hydrogen should be prevented from accumulating through ventilation. Ignition within the crankcase can be just a startling noise or result in engine fire. When hydrogen ignites within the crankcase, a sudden pressure rise occurs, which needs to be relieved by using a pressure relief valve.

1.5.5 Power Output

The theoretical maximum power output from a HICE depends on the A/F ratio and the fuel injection method used, but is affected by volumetric efficiency, fuel energy density, and preignition. The stoichiometric A/F ratio for hydrogen is 34:1. At this A/F ratio, hydrogen will displace 29% of the combustion chamber leaving only 71% for the air. As a result, the energy content of the mixture is less than that for gasoline. Since the fuel and air, before entering the combustion chamber, are mixed through the central and port injection methods, these systems limit the maximum theoretical power output to approximately 85% of that of gasoline engines. However, in direct injection systems, which mix the fuel with the air after the intake valve has closed (and, thus, the combustion chamber has 100% air), the maximum output of the engine can be approximately 15% higher than that for gasoline engines.

However, at a stoichiometric A/F ratio, the combustion temperature is very high that causes the formation of a large amount of nitrogen oxides (NO_x), which is a criteria pollutant. Because one of the reasons for using hydrogen is low exhaust emissions, HICE are not normally designed to run at a stoichiometric A/F ratio. Instead, twice as much air is used, which reduces NO_x formation to near zero [35]. Unfortunately, doubling the air amount reduces the power output to about half that of a similarly sized gasoline engine. Hence, to make up for the power loss, HICEs are usually larger than gasoline engines or are equipped with turbochargers or superchargers. Overall, a hydrogen-fueled car has an approximate efficiency of 45%, which is much better than 25% efficiency for a standard gasoline car. Owing to a relatively higher flame speed, hydrogen also offers a possibility to increase the power output with the existing engine size.

For direct injection of hydrogen, the power density is roughly 120% that of an equivalent gasoline engine. Because of the easy combustion property, researchers are experimenting with a multiple injection approach, where hydrogen is injected directly into the cylinder once or twice during each combustion cycle [25,28,36].

1.5.6 Hydrogen Gas Mixtures

Hydrogen can be advantageously used in ICE as an additive to hydrocarbon fuels. For example, hydrogen and methane can be mixed and stored in the same tank. For blending with liquid fuels, hydrogen is stored separately and mixed in the gaseous state immediately before the injection. Hydrogen mixture–powered ICEs have many operating advantages. They perform well under all weather conditions, require no warm-up, have no cold-start issues even at subzero temperatures, and are highly efficient (up to 25% better than conventional spark-ignition engines). A commercially available gas mixture known as Hythane contains 20% hydrogen and 80% natural gas. At this ratio, no modifications are required to a natural gas engine, and studies have shown that the emissions are reduced by more than 20%. Mixtures of more than 20% hydrogen with natural gas can reduce emissions further but some engine modifications are required. Addition of hydrogen to methane reduces hydrocarbon, CO, and CO_2 emissions, although having a tendency to increase NO_x emissions. However, since hydrogen enrichment enables operating with leaner mixtures, lean operation results in NO_x reduction without scarifying engine output or thermal efficiency. Moreover, due to the high flame speed of hydrogen, retarded ignition timing is also possible without lowering thermal efficiency, which reduces flame temperature and NO_x levels consequently. Therefore, significant reductions in NO_x emissions are also obtained with hydrogen addition [30]. In gasoline engines, lean operation reduces emissions of CO and unburned hydrocarbons, as extra oxygen is available to combust the fuel and oxidize CO to CO_2. However, the drawback is reduction in the power output. On addition of hydrogen, hydrogen/carbon ratio increases, which improves the power output. The low ignition energy and high burning speed of hydrogen makes hydrogen/hydrocarbon mixture easier to ignite, reducing misfire, and thereby improving emissions, performance, and fuel economy [25,27,28].

1.5.7 Current Status

Several models of HICE vehicles have been demonstrated and few are commercially available [25,28,33,38]. However, hydrogen-powered vehicles will not be available to common public until there is an adequate refueling infrastructure and trained technicians to repair and maintain these vehicles. The design of each hydrogen-powered vehicle may vary from manufacturer to manufacturer and model to model. One model may be simple in design and operation, for example, a lean-burning fuel metering strategy using no emission control systems such as catalytic converter and evaporate fuel canister. Another model may be very sophisticated in design and operation, for example, using a fuel metering strategy with a catalytic converter and multiple spark plugs. One of the concerns for utilizing hydrogen for fuel is to modify and redesign internal combustion engines. Among the many factors related to the operation and performance of the engines, remarkable attention has been devoted to the introduction of hydrogen to combustion chambers. Direct hydrogen injection improves the efficiency, increases the power output, and significantly helps to eliminate abnormal combustion phenomena such as preignition and knocking [25–28]. In view of this, researchers usually have implemented direct combustion techniques using spring-loaded valves driven mechanically or electromagnetically.

1.6 Hydrogen Fuel Cells

Fuel cells convert the chemical energy of hydrogen directly into electrical and thermal energies. A fuel cell consists of two electrodes: the cathode (positive) and the anode (negative) connected by an electrolyte (Figure 1.3) [39]. Hydrogen and oxygen flow to the anode and cathode, respectively, giving an overall electrochemical reaction

$$H_2 + \tfrac{1}{2}O_2 \rightarrow H_2O$$

with a theoretical electrochemical potential of 1.23 V (0.40 $V_{hydrogen}$ + 0.83 V_{oxygen}). The electrodes serve two roles: (1) provide electron conduction and (2) provide the necessary surface for the initial deposition of the molecules into atomic species (e.g., electrocatalysts that reduce activation energy) before electron transfer. To get higher voltage, the individual fuel cells are combined into a fuel cell "stack," which is done efficiently by connecting each cell to the next in a way that avoids the current being taken off the edge of the electrode, but over the whole surface on the electrode. A bipolar plate is used to interconnect the cell as shown in Figure 1.4 [40]. The continuous operation of the stack requires effective heat, air, hydrogen, and water management, enabled by auxiliary equipment such as pumps, blowers, and controls.

1.6.1 Types of Fuel Cells

There are six different types of fuel cells (Table 1.6): (1) alkaline fuel cell (AFC), (2) direct methanol fuel cell (DMFC), (3) molten carbonate fuel cell (MCFC), (4) phosphoric acid fuel cell (PAFC), (5) proton exchange membrane fuel cell (PEMFC), and (6) the solid oxide fuel cell (SOFC). They all differ in applications, operating temperatures, cost, and efficiency.

Proton exchange membrane fuel cell is most suited for powering automobiles, due to its relatively low temperature (about 80°C) operation, high power density, rapid change in power on demand, and quick start-up. These features make PEMFCs the most promising and attractive candidate for a wide variety of power applications ranging from portable/micropower and transportation to large-scale stationary power systems for buildings and distributed generation [22]. The membrane is made of a thin poly(perfluorosulfonic) acid sheet, which acts as an electrolyte and allows the passage of hydrogen ions only. The membrane is coated on both sides with highly dispersed metal alloy particles

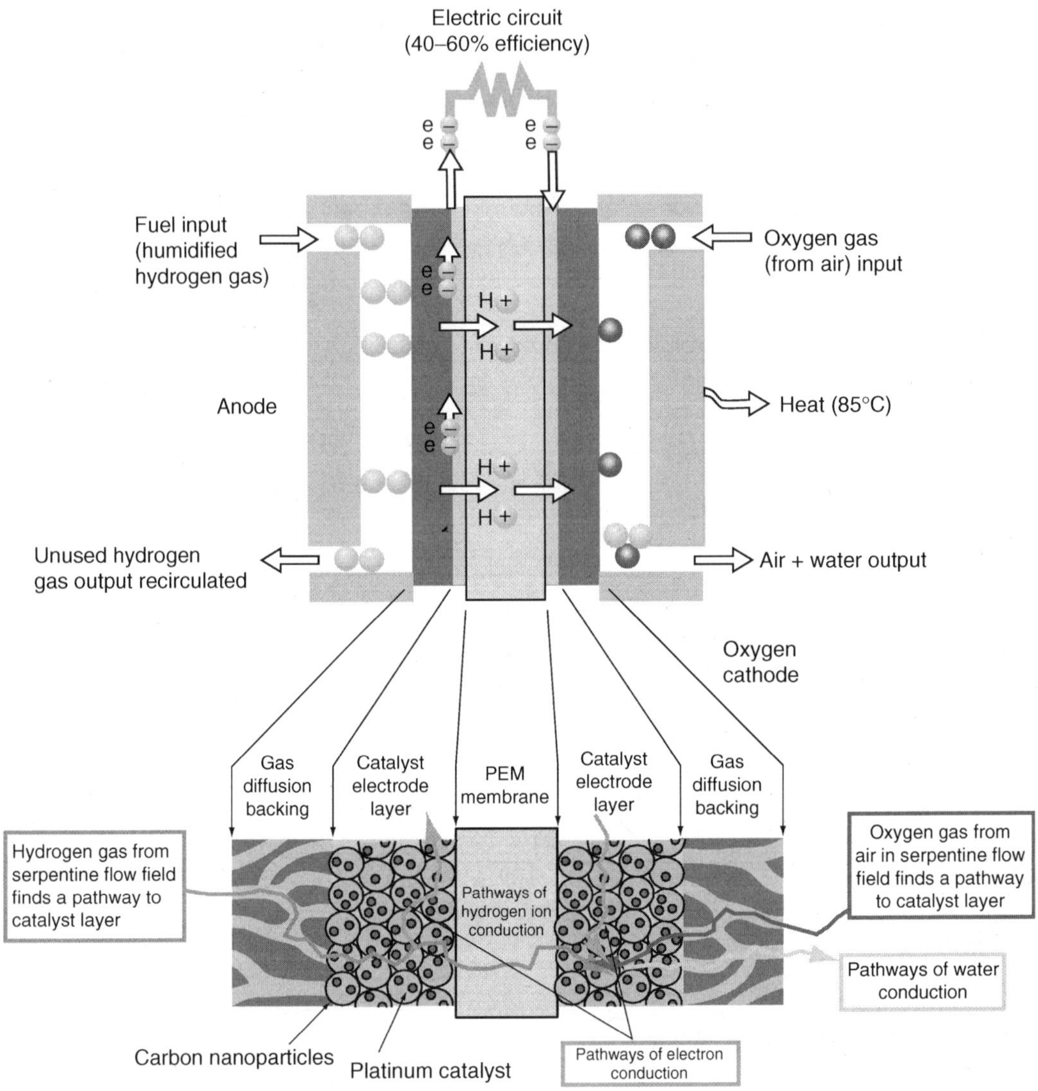

FIGURE 1.3
Schematic working of PEMFC. (Reprinted from Jacobson, D.L., http://physics.nist.gov/MajResFac/NIF/pemFuelCells.html, September 7, 2007.)

(mostly platinum) that act as catalysts. The DMFC is similar to the PEM cell in that it uses a polymer membrane as an electrolyte. However, a catalyst on the DMFC anode draws hydrogen from liquid methanol, eliminating the need for a fuel reformer. Therefore, pure methanol can be used as fuel. The MCFC uses a molten carbonate salt as the electrolyte. It has the potential to be fueled with coal-derived fuel gases, methane, or natural gas. These fuel cells can work at up to 60% efficiency with the possibility of increasing up to 80% when the waste heat is utilized. PAFC consists of an anode and a cathode made of a finely dispersed platinum catalyst on carbon and a silicon carbide structure that holds the phosphoric acid electrolyte. This is the most commercially developed type of fuel cell and is being used to power many commercial premises. The PAFC can also be used in large vehicles such as buses. SOFCs work at even higher temperatures (800–1000°C) than MCFCs and utilize a

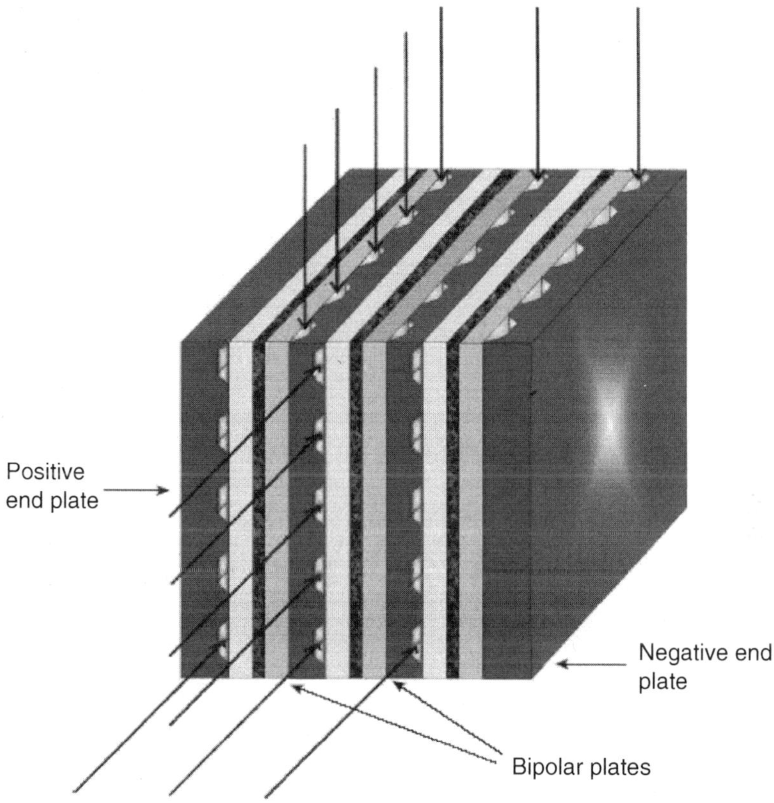

FIGURE 1.4
Bipolar plates for connecting fuel cells in a series. (Reproduced with permission from Wiley; *Kirk-Othmer Encyclopedia of Chemical Technology*, Vol. 12, Wiley, New York, 2002.)

TABLE 1.6
Different Types of Fuel Cells

	AFC	DMFC	MCFC	PAFC	PEMFC	SOFC
Electrolyte	Potassium hydroxide	Polymer membrane	Immobilized liquid molten carbonate	Immobilized liquid phos phoric acid	Ion exchange membrane	Ceramic
Operating temperature (°C)	60–90	60–130	650	200	80	1000
Efficiency (%)	45–60	40	45–60	35–40	40–60	50–65
Typical electrical power	Up to 20 kW	<10 kW	>1 MW	>50 kW	Up to 250 kW	>200 kW
Possible applications	Submarines, spacecraft	Portable applications	Power stations	Power stations	Vehicles, small stationary	Power stations

Source: Adapted from *Kirk-Othmer Encyclopedia of Chemical Technology*, Vol. 12, Wiley, New York, 2002.

solid ceramic electrolyte, such as zirconium oxide stabilized with yttrium oxide, instead of a liquid. These cells can reach efficiencies of about 60% and are expected to be used for generating electricity and heat in industry and potentially for providing auxiliary power in vehicles.

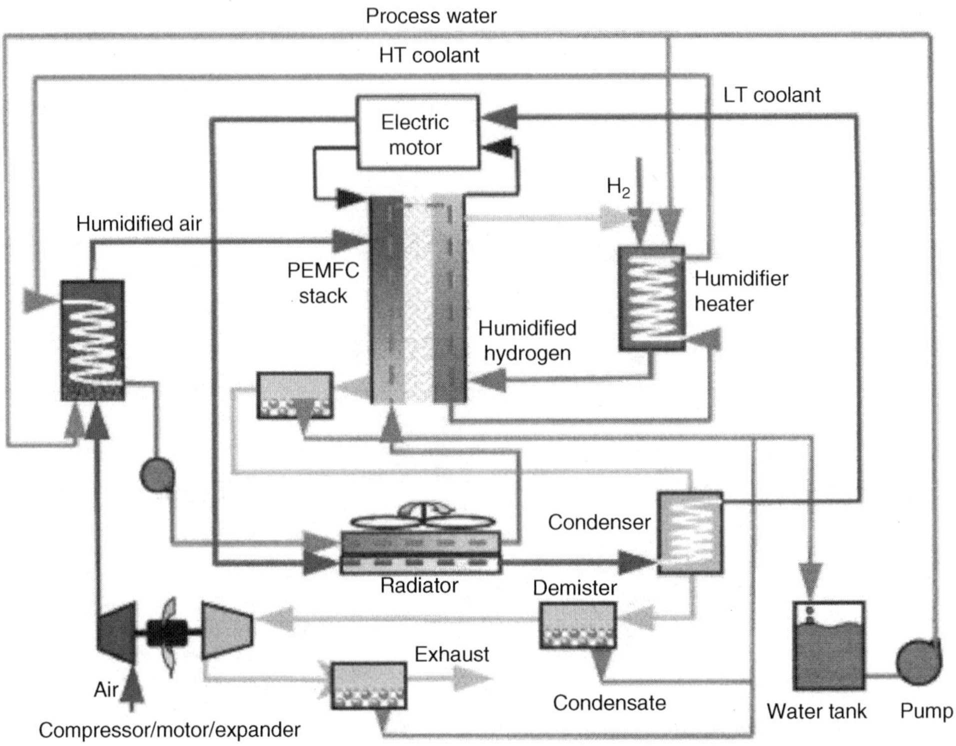

FIGURE 1.5
Schematic diagram of a hydrogen-fueled, PEMFC system for automotive applications. (Reproduced with permission from Elsevier; Ahluwalia, R.K., and Wang, X., *J. Power Sources*, 139(1–2), 152–164, 2005.)

A fuel cell system for automobile application is shown in Figure 1.5 [41]. At the rated power, the PEMFC stack operates at 2.5 atm. and 80°C to yield an overall system efficiency of 50% (based on lower heating value of hydrogen). Compressed hydrogen and air are humidified to 90% relative humidity at the stack temperature using process water and heat from the stack coolant. A lower system pressure is at part load and is determined by the operating map of the compressor–expander module. Process water is recovered from spent air in an inertial separator just downstream of the stack in a condenser and a demister at the turbine exhaust.

1.6.2 Major Challenges

The two major challenges for fuel cells are cost and durability. When compared to the cost for automotive internal combustion engines (about \$25–35/kW), current fuel cell systems are estimated to cost fivefold, even when cost savings for high-volume manufacturing are applied. Major contributors to the cost are platinum electrocatalyst, membrane, and bipolar plates. Automotive fuel cell systems will also need to be as durable and reliable as current automotive engines (5,000 h lifespan or 150,000 mi. drive range) under heavy load cycling. The variations in cell potential and relative humidity levels accelerate the degradation of both the catalyst and the membrane. Also, fuel cells need to be able to function over the full range of vehicle operating conditions (−40 to +40°C). Efforts are underway to reduce the cost. For example, the recent results have indicated substantial progress in reducing the platinum content in the catalyst.

1.7 Supply of Hydrogen

The important aspect of the hydrogen economy is the production of hydrogen and the total energy consumed and CO_2 emitted in the process. Current world hydrogen production is approximately 50 million t per year, which is equivalent to only 2% of world energy demand. Hydrogen can be produced from a diversity of energy resources using a variety of process technologies as described in Chapters 2 through 7. A brief summary is provided in Figure 1.6.

The worldwide consumption of the energy is divided as 38.1% in electricity, 44.3% in heating and industries, and 17.6% in transport, excluding electricity vehicles. About 10% of the electricity generated is lost during distribution, which represents about 4.2% loss in the total primary energy [42]. The worldwide primary energy during the year 2004 was 11.7 gigatons of oil equivalent (Gtoe) or 125,000 T Wh, which is equivalent to 496 quad. The consumption is expected to increase to more than 25 Gtoe/year by 2050. Considering the linear extrapolations of the rate of growth of oil consumption and the rate of increase of known oil reserves, it can be deduced that the end of the petroleum supply will probably take place around 2050 [42]. Hydrogen-based energy supply can be envisioned to meet the extra demand. A proposed management of energy supply and transformation is indicated in Figure 1.7. According to this scheme, proposed by Marban and Valdes-Solis [42], the traditional electricity network will be partially fed with natural gas and coal as it is done nowadays, although their percentage contribution will decrease. These fuels will be transformed in cogeneration thermal plants to produce H_2 and electricity with CO_2 sequestration, for instance, using integrated gasification in combined cycle (IGCC) plants provided with CO_2 separation systems (sorbents, membranes, etc.). The concept of high-capacity power plants based on coal will be maintained since this fuel is not appropriate for energy generation (electricity or hydrogen) at a smaller scale. These power plants will also be suitable for the processing of energetic biomass, either alone or in combination with coal. This biomass will be mainly made up of the short-rotation crops and organic wastes that are not destined to be employed in the reformers or biorefineries for the production of hydrogen and biofuels (Figure 1.7). To supply hydrogen to areas far from the general network it will be necessary to build refueling stations. Most of the supply will be provided by a network of refueling stations in which hydrogen will be supplied by a piping system connected to large-scale production plants. These H_2 production plants will use a mix of the primary energy sources most suited to each region [42,43].

1.7.1 Cost of Hydrogen Production

Hydrogen can be produced in a number of ways depending on the feedstock as described earlier. In addition, the design of a hydrogen energy system is site specific, depending on the type of demand, the local energy prices (for natural gas, coal, electricity, etc.), and the availability of primary energy resources. A typical cost analysis for hydrogen production and distribution from different feedstocks is given in Table 1.7. The cost estimation is based on the fact that the energy content of a gallon of gasoline and a kilogram of hydrogen are approximately equal on a lower heating value basis. Thus, a kilogram of hydrogen is approximately equal to a gallon of gasoline equivalent (gge) on an energy content basis [44,45]. The cost of producing hydrogen varies significantly by the type of technology and distribution channel used. According to an analysis in the year 2004 [45], the total cost of hydrogen ranged from \$1.91 to 6.58/kg for hydrogen made from coal and shipped by pipeline and for hydrogen made on-site from electrolysis.

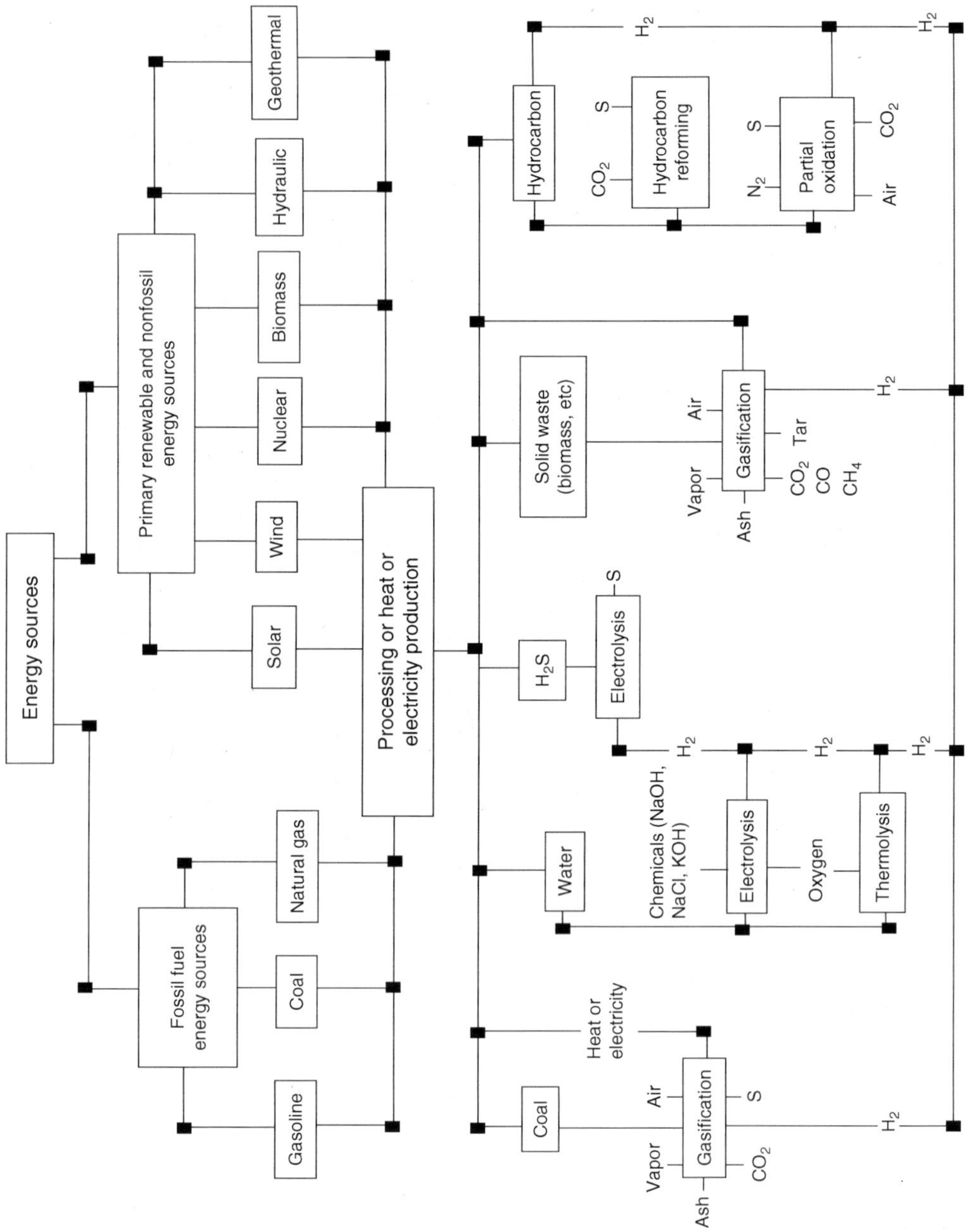

FIGURE 1.6
Hydrogen production techniques. (Reproduced with permission from Midilli, A., Dincer, I., and Rosen, M.A., *Renewable Sustainable Energy Rev.*, 9(3), 255–271, 2005.)

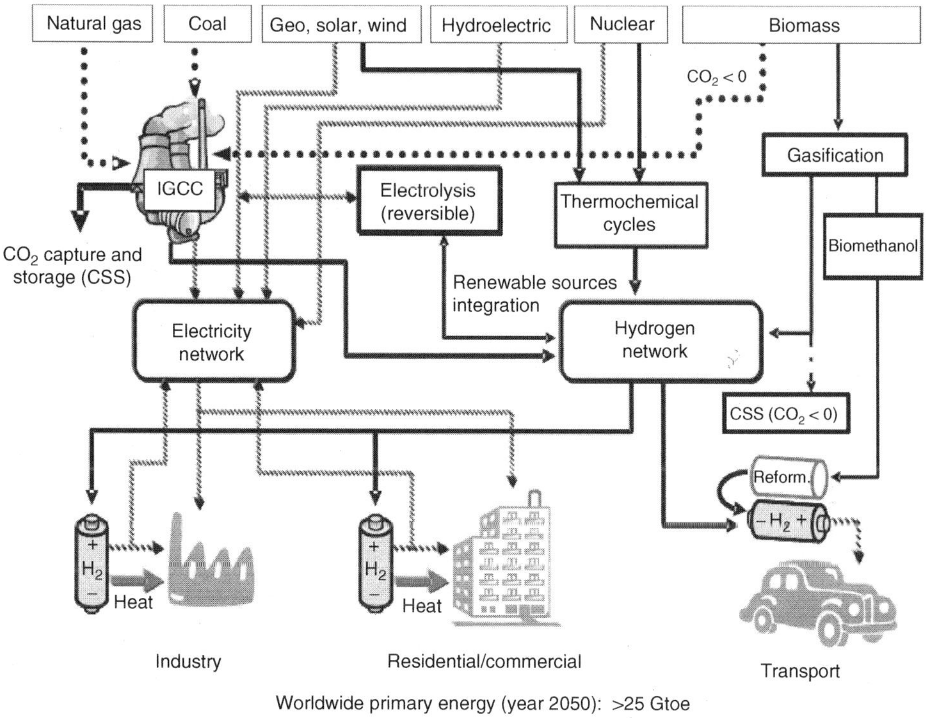

FIGURE 1.7
Predicted worldwide primary energy in the year 2050 (>25 Gtoe). (Reproduced with permission from Marban, G., and Valdes-Solis, T., *Int. J. Hydrogen Energy*, 32(12), 1625–1637, 2007.)

TABLE 1.7
Estimated Cost of Hydrogen Production Transportation and Distribution

Primary Energy Source	Production Cost ($/kg)($: Based on the Year 2003)	Distribution Cost via Pipeline ($/kg)[a]	Dispensing Cost ($/kg)[b]	Total Costs ($/kg H_2 or $/gge)
Natural gas reforming	1.03	0.42	0.54	1.99
Natural gas + CO_2 capture	1.22	0.42	0.54	2.17
Coal gasification	0.96	0.42	0.54	1.91
Coal + CO_2 capture	1.03	0.42	0.54	1.99
Wind electrolysis	6.64	0.42	0.54	7.60
Biomass gasification	4.63	1.80[a]	0.62[b]	7.04
Biomass pyrolysis	3.80	1.80[a]	0.62[b]	6.22
Nuclear thermal splitting of water	1.63	0.42	0.54	2.33
Gasoline (for reference)	$0.93/gal. refined	$0.19	—	$1.12/gal.

Note: Energy content of 1 kg hydrogen approximately equals the energy content of 1 gal. of gasoline.
[a] Liquid hydrogen via tanker.
[b] Liquid hydrogen fueling station.

Source: Adapted from Hydrogen Fuel cell, http://www1.eere.energy.gov/hydrogenandfuelcells/mypp/pdfs/production.pdf, May 2000; U.S. Department of Energy, Office of Basic Energy Sciences, Committee on Alternatives and Strategies for Future Hydrogen Production and Use, *The Hydrogen Economy: Opportunities, Costs, Barriers, and R&D Needs*, National Research Council, National Academies Press, Washington, 2004, available at http://www.nap.edu/catalog/10922.html, accessed May 2007.

The future objective is to reduce the cost of distributed production of hydrogen from natural gas to \$2.50/gge (delivered) at the pump by 2010 and \$2.00/gge (delivered) by 2015. From biomass gasification, the target is to reduce the cost of hydrogen produced to \$1.60/gge at the plant gate (<\$3.30/gge delivered) by 2012 and \$1.10/gge at the plant gate (\$2.10/gge delivered) by 2017 [44]. High-temperature, solar-driven, thermochemical hydrogen production using water-splitting chemical cycles is in an early stage of research. Research is also needed to couple the thermochemical cycles with advanced solar energy technology. If these efforts are cost effective and successful, high-temperature thermochemical processes may provide a clean, efficient, and sustainable route for producing hydrogen from water. For a hydrogen-based vehicle to be competitive with gasoline ICE or hybrid ICE, the estimated hydrogen cost needed is \$2–3/gge (untaxed) at the dispenser. The ultimate target for all of the production technologies being researched is a hydrogen cost that will be competitive for transportation on a well-to-wheels basis, regardless of the production method.

It is estimated that by the year 2040, approximately 150 million t of hydrogen will be required annually. To meet this demand solely from natural gas, approximately 0.43 million m^3 of natural gas per year of 777,000 facilities would be required, which would cost roughly 1 trillion dollars. If the demand is solely met by nuclear energy, it would require 240,000 t of unenriched uranium and about 2000, 600 MW power plants, which would cost \$840 billion. In the case of solar energy, 113 million, 40 kW system with 2500 kWh of sun/m^2 will be required, which will cost \$22 trillion. In the case of wind energy, 1 million, 2 MW turbines using an average wind speed of 7 m/s would be required, which will cost \$3 trillion. In the case of biomass, 1.5 billion t of dry biomass would be required, which will need 113.4 million acres of farm land, and then to produce hydrogen approximately 3300 plants will be needed, which will cost \$565 billion to build. Similarly, coal scenario will demand 1 billion ton per year of coal, and approximately 1000, 275 MW coal gasification/steam reformation plants, which will cost about \$500 billion [44,45].

According to the estimates of Ogden [11], the delivered costs of hydrogen varies from \$11 to 25/GJ (equivalent to \$1.6–3.8/gge on an energy basis) depending on the technology, which is more than untaxed gasoline. Because the hydrogen can be used more efficiently than gasoline in a fuel cell car, the fuel cost per kilometer is comparable. The costs of supplying hydrogen may further reduce if current research and development efforts become successful. Economically, producing hydrogen from coal is cheapest, but it produces a large amount of CO_2 as a by-product (see Chapter 17). Hence, research and development activities focused on CO_2 sequestration are needed [45]. The estimated CO_2 disposal cost is approximately \$10/t of CO_2.

1.7.2 Environmental Aspects

The combustion of fossil fuels accounts for a majority of anthropogenic greenhouse gas emissions. Currently, fossil fuel combustion produces 7 Gtn/year carbon emissions [6], and in a no-change scenario (the base scenario of the International Energy Agency [IEA]) CO_2 emissions in 2050 can be expected to reach 14 Gtn/year of carbon. There is a growing recognition that steps to reduce greenhouse gases are needed, and many countries are adopting policies to limit CO_2 emissions. Energy and transportation companies, many of which have multinational operations, are actively evaluating alternative sources of energy.

Hydrogen, when produced from reforming of natural gas, petroleum or coal, generates CO_2 as a by-product. In the current processes, CO_2 has been released into the atmosphere.

For each ton of hydrogen produced from hydrocarbons, approximately 2.5 t of carbon is vented to the atmosphere [44–47]. However, for each ton of hydrogen produced from current coal technology, approximately 5 t of carbon is emitted to the atmosphere. Principally, CO_2 capture and sequestration is a precondition for use of these fossil fuels. However, the sequestration necessity varies, because the relative atomic hydrogen-to-carbon ratios are 1:2:4 for coal:oil:natural gas. There are two basic approaches to CO_2 sequestration: either at the point of emission (*in situ* capture) or from the air (direct capture). In either case, CO_2 must be disposed off safely and permanently. With the capture and sequestration of CO_2, hydrogen is one path for coal, oil, and natural gas to remain viable energy resources [46]. Carbon sequestration technologies are discussed in detail in Chapter 17.

Recently, there have also been some concerns over possible problems related to hydrogen gas leakage; as the molecular hydrogen leaks from most containment vessels. It has been hypothesized that if significant amounts of H_2 escape to stratosphere, H^* free radicals can be formed due to ultraviolet radiation, which in turn can enhance the ozone depletion. However, the effect of these leakage problems may not be significant as the amount of hydrogen that leaks presently is much lower (by a factor of 10–100) than the hypothesized 10 20%.

1.7.3 Hydrogen Storage

One of the most critical factors in inducting hydrogen economy is transportation and on-vehicle storage of hydrogen [48–61]. Storing hydrogen that flexibly links its production and final use are key element of the hydrogen fuel utilization. The major contribution to the problem is from low gas density of hydrogen. For example, to store energy equivalent to one gasoline tank, an ambient pressure hydrogen gas tank would be more than 3000-fold the volume of the gasoline tank. Various storage options are discussed in Chapters 9 through 13, and briefly described in the following. Table 1.8 summarizes the weight requirements for various on-board hydrogen storage options; here 5 gal. of the gasoline has been taken as a reference, which is sufficient for a vehicle to drive up to a distance of 300 mi.

1.7.3.1 Compressed Hydrogen

Considering both storage and refueling technologies, the most promising short-term alternative is probably compressed gas storage [53,55]. Prototype hydrogen-powered vehicles

TABLE 1.8

Comparison of On-Board Hydrogen Storage

Fuel	Total Energy (MJ)	Fuel Weight (kg)	Tank Weight (kg)	Total Fuel System Weight (kg)	Volume (gal.)
5 gal. gasoline	662	14	6.4	20.4	5
Liquid hydrogen (20 K)	662	4.7	18.6	23.3	47
1.2 wt% H_2 stored in FeTi metal hydride	662	4.7	549.3	554	50
Compressed hydrogen (207–690 bar)	662	4.7	63.6–86.3	68.3–91	108–60

Source: Adapted from Kukkone, C.A. and Shelef, M., *Alternate Fuels*, National Research Council and National Academy of Engineering, National Academies Press, Washington, DC, 1992.

are using carbon fiber–reinforced polymer composite tanks for 350 and 700 bar compressed hydrogen (Chapter 10).

1.7.3.2 Liquid Hydrogen

Storing of hydrogen in liquid form at cryogenic condition is attractive in that it offers low weight and volume per unit energy when compared to compressed hydrogen.

Main issues with liquid hydrogen storage tanks are hydrogen boil-off, the energy required for hydrogen liquefaction, and tank cost. The rate of hydrogen boil-off mainly depends on the effectiveness of the thermal insulation, ambient conditions, geometry of the vessel, and length of time between driving [56]. For example, BMW's cryogenic tank in Hydrogen 7 will boil-off all hydrogen in <2 weeks. Liquid hydrogen needs to be stored at 20 K (–253°C): The storage tank has to be insulated, to preserve temperature, and needs to be reinforced to store the liquid hydrogen under some pressure (Chapter 10). The total fuel cycle energy efficiency is significantly lower for liquid hydrogen than for gaseous hydrogen because of the large amount of energy required for liquefaction [57,62]. Total theoretical amount of energy required for liquefaction is approximately 3.4 MJ/kg, whereas the actual amount of energy required for liquefaction is much higher, approximately 50.4 MJ/kg, with a refrigeration efficiency of 7.2% [62]. Overall, the liquefaction results in a loss of about 30% of the energy stored in liquid hydrogen.

1.7.3.3 Metal Hydrides

Metal hydrides are specific combinations of metallic alloys, which possess the unique ability to absorb hydrogen and release it later, either at room temperature or on heating (Chapter 11). The total amount of hydrogen absorbed is generally 1–2 wt% of the total weight of the tank. However, some metal hydrides are capable of storing 5–7 wt% [63]. The percentage of gas absorbed to volume of the metal is still relatively low, but hydrides offer a valuable solution to hydrogen storage [16]. The volume of this storage device is only two fold greater than the equivalent gasoline tank, but unfortunately it is 20-fold heavier. The life of a metal hydride storage tank is directly related to the purity of the hydrogen it is storing. The alloys act as a sponge, which absorbs hydrogen, but it also absorbs any impurities introduced into the tank by the hydrogen. Thus, the hydrogen released from the tank is highly pure, but the tank's lifetime and ability to store hydrogen is reduced as the impurities are deposited in the metal pores.

1.7.3.4 Organic Chemical Hydrides

By utilizing hydrogen and dehydrogenation cycles of organic compounds, one can achieve storage of hydrogen. Examples include organic chemical hydrides consisting of reversible catalysis pairs such as decalin dehydrogenation/naphthalene hydrogenation, methylcyclohexane dehydrogenation/toluene hydrogenation, and tetralin dehydrogenation/naphthalene hydrogenation. For example, decalin can store hydrogen with high capacities (7.3 wt%, 64.8 kg H_2/m^3). Decalin and naphthalene are accepted socially as safe commodity chemicals since they have been utilized as a solvent and an insect killer at home, respectively, for long periods. However, the energy and equipment requirements for the dehydrogenation reaction are too expensive for on-board use. Organic chemical hydrides are more suited for hydrogen production at the fueling station (Chapter 13).

1.7.3.5 Carbon Materials

Various forms of carbon such as graphite, fullerene, nanotubes, and activated carbon with high surface area may be utilized for the storage of hydrogen. Single-walled carbon nanotubes can store 2.5–3 wt% hydrogen [50,51,63]. Research on this technology has focused on the areas of improving manufacturing techniques and reducing costs as carbon nanotubes move toward commercialization [51]. Others have proposed storing hydrogen in fullerenes [64] or in activated carbon at low temperatures [64–68]. Details of carbon materials are presented in Chapter 12.

1.7.3.6 Silica Microspheres

Hydrogen can be stored in hollow silica microspheres. At high temperatures (e.g., 500°C), the wall of these microspheres is permeable to hydrogen, and at ambient temperatures the wall is impermeable to hydrogen. High-pressure hydrogen can be filled at high temperature and then locked in by cooling. When needed, the trapped hydrogen can be easily released by heating. Silica microspheres have the potential to be safe and resist contamination. Various researchers have studied high-pressure hydrogen storage in silica microspheres [69–71], which can be easily transported in bulk without the need for an external pressure vessel.

1.8 Current Challenges

The main obstacle in the utilization of hydrogen fuel in the automobiles is due to its low density. Even when the fuel is stored as a liquid in a cryogenic tank or in a pressurized tank as a gas, the amount of energy that can be stored in the space available is limited, and hydrogen cars, therefore, have a limited range compared to their conventional counterparts. Hence, storing hydrogen on board for 500 km driving range is a major challenge while meeting the performance (weight, volume, kinetics, etc.), safety, and cost requirements without compromising on passenger or cargo space. The fuel storage systems in today's gasoline vehicles have an energy density of about 6 kWh/L. With the improved fuel economy of a hydrogen fuel-based vehicle and a conformable hydrogen storage system, the requirement for a fuel cell vehicle is 2.7 kWh/L, which is higher than liquid hydrogen (20 K, 1 bar). The second major problem is the high cost of making reliable fuel cells. A combination of the two problems is hindering the development of hydrogen automobiles. Also, the existing hydrogen dispensers are not adequate to support commercial fueling. Cost, reliability, and safety must be significantly improved to allow commercial fueling.

Fuel cells have operating advantages for both stationary and mobile applications in that they are quiet and typically have high efficiency at partial loads. The direct conversion of the energy stored in the fuel to electricity in a fuel cell can be achieved at high efficiencies, avoiding limitations of standard heat-to-power cycles used in combustion engines and turbines. The higher efficiency of fuel cells (approximately 60%) compared to gasoline (22%) and diesel (45%) internal combustion engines would substantially help in the future energy needs. However, fundamental performance problem remain to be solved before hydrogen fuel cells can compete with gasoline. A primary factor limiting PEMFC

Slow reaction reduces voltage output

performance is the slow kinetics of the oxygen reduction reaction at the cathode. Even with the most effective platinum-based catalysts, the slow reaction reduces the voltage output of the fuel cell from the ideal 1.23–0.8 V or less when practical currents are drawn [52,72]. The solutions for speeding up the reaction are hidden in the complex reaction pathways and intermediate steps of the oxygen reduction reaction. The role of different key features as the atomic configuration of catalysts and their supports, and the electronic structure of surface reconstructed atoms and adsorbed intermediate species, is within reach of fundamental understanding.

The PEMFCs require expensive polymer membrane (e.g., Nafion®), and operate at a low temperature (e.g., 80°C). Although low temperature reduced the cost of material, the heat generated at low temperatures is more difficult to remove. Alternate proton conducting membranes (e.g., inorganic polymer composites) that will operate at a high temperature (e.g., 200°C) are required. The expensive platinum catalyst used for electrochemical reactions can be poisoned by even trace amounts of carbon monoxide in the hydrogen fuel stream. Hence, a more tolerant catalyst material needs to be developed.

1.9 Future Outlook

The hydrogen economy has both societal and technical appeals as a potential solution to provide abundant supply with minimal environmental impact. The success of a hydrogen economy will depend on the technological solution of the remaining storage and production issues. The investment requirements in research and development are high, and the outcome for specific, promising approaches is uncertain. The acceptance of hydrogen fuel depends not only on its practical and commercial appeal, but also on its record of safety in widespread use. The flammability, buoyancy, and permeability of hydrogen may present challenges to its safe handling that are different from, but not too difficult than those of other energy carriers. Also, a key to public acceptance of hydrogen is the development of safety standards and practices that are well known and routinely used [73–76]. The success of hydrogen economy will definitely depend on new energy sources and new energy converters. Bringing hydrogen and its energy conversion partner (e.g., fuel cells) to the hydrocarbon energy system level is a fascinating challenge, especially with the development of infrastructure.

Today, the technologies for producing, storing, and distributing hydrogen are well known and widely used in the chemical industry. Also, the technical building blocks for future hydrogen fuel system are well documented. Hydrogen end-use technologies such as fuel cells, hydrogen vehicles, power, and heating devices are undergoing rapid development. Still the costs and time factors inherent in changing the present energy system mean that building a large-scale hydrogen fuel infrastructure will take a long time. The development of hydrogen end-use systems, such as fuel cells, and their penetration into transportation or power markets will determine the pace of introducing hydrogen fuel. Because hydrogen can be produced from a variety of sources, a future hydrogen fuel infrastructure could evolve in a variety of ways. As demand increases and becomes more geographically dense, centralized production with local pipeline distribution would become more economically attractive. Once a large, geographically concentrated demand evolves, carbon sequestration can be incorporated. A variety of low or zero net-carbon emitting sources of hydrogen may be utilized at this time [77]. The development of such a market will depend on the political will to move toward a zero-emission energy system and on the relative economics of hydrogen versus other low-polluting alternative fuels.

1.10 Conclusions

Hydrogen has the potential to be an attractive alternative energy carrier for future fuel needs. The use of hydrogen as a fuel requires development in several industry segments, including production, delivery, storage, and end use. Fundamental to the creation of a hydrogen economy is a viable, safe, and affordable hydrogen energy system. The energy end-use applications of hydrogen include stationary, transportation, and portable devices. Hydrogen can be produced from a variety of feedstocks including natural gas, coal, biomass, wastes, solar, wind, or nuclear. Hence, the participation in hydrogen economy will be much more open than the present energy industry limited to countries with petroleum or coal reserves. In addition, the production of hydrogen from renewable sources, or from fossil fuel with carbon sequestration, does not emit greenhouse gases.

Owing to the differences in the properties, risks associated with hydrogen are different from conventional fuels. Hydrogen is a colorless, odorless, tasteless, and nontoxic gas. Hydrogen, when compared to gasoline is more flammable, explosive, buoyant, and diffusive. Hydrogen is the lightest of the elements, with a very low density per unit volume that is approximately 14-fold less than air. Owing to its lightness, there are problems with storage and transport of hydrogen. Various hydride-based storage, carbon materials, and compressed gas storage for automotive applications are being developed.

References

1. Kreith, F. and R. West, Fallacies of a hydrogen economy: A critical analysis of hydrogen production and utilization. *J. Energy Resour. Technol.*, 126(4), 249–257, 2004.
2. Hussain, M.M., I. Dincer and X. Li, A preliminary life cycle assessment of PEM fuel cell powered automobiles. *Appl. Thermal Eng.*, 27, 2294–2299, 2007.
3. Hydrogen Economy, Wikipedia Encyclopedia, available at http://en.wikipedia.org/wiki/Hydrogen_economy, March 2007.
4. Air Products and Chemicals, available at http://www.airproducts.com/Products/LiquidBulkGases/HydrogenEnergy/FuelCells/, 2003.
5. Rifkin, J., *The Hydrogen Economy*. Penguin Putnam Inc., New York, 2002.
6. The Hydrogen Economy: Opportunities, Cost, Barriers and R&D Needs, in *National Research Council and National Academy of Engineering*, National Academies Press, Washington, 2004.
7. *Kirk-Othmer Encyclopedia of Chemical Technology*, 3rd ed., Vol. 4, Wiley, New York, 1992, 631p.
8. Romm, J.J., *The Hype about Hydrogen, Fact and Fiction in the Race to Save the Climate.* Island Press, available at http://en.wikipedia.org/wiki/The Hype about Hydrogen, Washington, D. C., 2004.
9. A National Vision of America's Transition to a Hydrogen Economy—To 2030 and Beyond. National Hydrogen Vision Meeting, DOE Report, Washington, November 15–16, 2001.
10. Ogden, J.M., E. Dennis, M. Steinbugler and J. Strohbehn, Hydrogen Energy Systems Studies, Final Report to U.S. DOE for Contract No. XR-11265-2, January 18, 1995.
11. Ogden, J.M., Prospects for building a hydrogen energy infrastructure. *Annu. Rev. Energy Environ.*, 24, 227–279, 1999.

12. Ogden, J.M., T. Kreutz, S. Kartha and L. Iwan, Assessment of Technologies for Producing Hydrogen from Natural Gas at Small Scale, Center for Energy and Environmental Studies, University of California, Davis, CA, Draft Report, November 26, 1996.
13. Szklo, A. and R. Schaeffer, Fuel specification, energy consumption and CO_2 emission in oil refineries. *Energy*, 32, 1075–1092, 2007.
14. Transportation Energy Data Book, Centre for Transportation Analysis. 23rd ed., Oak Ridge National Laboratory, Oak Ridge, TN, 2003.
15. SRI Consulting Inc., *Chemical Economics Handbook 2001*, Menlo Park, CA, 2001.
16. Scholz, W., Processes for industrial production of hydrogen and associated environmental effect. *Gas Sep. Purif.*, 7, 131, 1993.
17. Kermode, R., Hydrogen from fossil fuels, in *Hydrogen: Its Technology and Applications*, Cox, K. and Williamson, K., Eds., CRC Press, Boca Raton, FL, 1977.
18. Spath, P. and M. Mann, Life Cycle Assessment of Hydrogen Production via Natural Gas Steam Reforming, Technical Report, NREL, NREL/TP-570-27637, 2000.
19. Yasuda, I. et al., Development of membrane reformer for highly-efficient hydrogen production from natural gas, *Proceeding of XV World Hydrogen Energy Conference*, Yokohama, Japan, 2004.
20. U.S. DOE-EERE 2003, available at www.eere.energy.gov/hydrogenand fuelcells (accessed March 2007).
21. DOE 2003. Alternative fuel data center, Properties of Fuels, available at www.afdc.doe.gov/pdfs/fueltable.pdf, 2007.
22. Wee, J.H., Applications of proton exchange membrane fuel cell systems. *Renewable Sustainable Energy Rev.*, 11, 1720–1738, 2007.
23. *Hydrogen Fuel Cell Engines and Related Technologies*, College of the Desert, Palm Desert, CA, 2001.
24. Alternative Fuels Data Center, Properties of Fuel, DOE Report, August 2005 available at www.afdc.doe.gov, April 2007.
25. White, C.M., R.R. Steeper and A.E. Lutz, The hydrogen fueled internal combustion engine: A technical review. *Int. J. Hydrogen Energy*, 31, 1292–1305, 2006.
26. Midilli, A., I. Dincer and M.A. Rosen, On hydrogen and hydrogen energy strategies. I: Current status and needs. *Renewable Sustainable Energy Rev.* 9(3), 255–271, 2005.
27. Das, L.M., Exhaust emission characterization of hydrogen operated design systems: Nature of pollutants and their control techniques. *Int. J. Hydrogen Energy*, 16, 765–775, 1991.
28. Das, L.M., Hydrogen engines: A view of the past and a look into the future. *Int. J. Hydrogen Energy*, 15, 425–443, 1990.
29. Hydrogen 7 combustion engine by BMW, which can operate both on gasoline and hydrogen. Clean Energy BMW Group, March 22, 2007, NHA 2007.
30. Mohammadia, A., M. Shiojib, Y. Nakaib, W. Ishikurab and E. Taboc, Performance and combustion characteristics of a direct injection SI hydrogen engine. *Int. J. Hydrogen Energy*, 32, 296–304, 2007.
31. Akansu, S.O., Z. Dulger, N. Kahraman and T.N. Veziroglu, Internal combustion engines fueled by natural gas-hydrogen mixtures. *Int. J. Hydrogen Energy*, 29, 1527–1539, 2004.
32. Toshio, S. and Y. Hiroyuki, Hydrogen as an ignition-controlling agent for HCCI combustion engine by suppressing the low-temperature oxidation, 32(14), 3066–3072, 2007.
33. Verhelst, S. and R. Sierens, Hydrogen engine-specific properties. *Int. J. Hydrogen Energy*, 26(9), 987–990, 2001.
34. Peschika, W., Hydrogen: The future cryofuel in internal combustion engines. *Int. J. Hydrogen Energy*, 23(1), 27–43, 1998.
35. AL-Garni, M., A simple and reliable approach for the direct injection of hydrogen in internal combustion engines at low and medium pressures. *Int. J. Hydrogen Energy*, 20(9), 723, 126, 1995.
36. Eichlseder, H., T. Wallner, R. Freyman and J. Ringler, The potential of hydrogen internal combustion engines in a future mobility scenario. SAE Paper 2003; Warrendale, PA, 2003-01-2267.
37. Galloni, E. and M. Minutillo, Performance of a spark ignition engine fuelled with reformate gas produced on-board vehicle. *Int. J. Hydrogen Energy*, 32(13), 2532–2538, 2007.

38. Schafer, A., J.B. Heywood, A. Malcolm, Weiss, Future fuel cell and internal combustion engine automobile technologies: A 25-year life cycle and fleet impact assessment. *Energy,* 31, 2064–2087, 2006.
39. Jacobson, D.L., Available at http://physics.nist.gov/MajResFac/NIF/pemFuelCells.html, September 7, 2007.
40. *Kirk-Othmer Encyclopedia of Chemical Technology*. Vol. 12, Fuel Cells, Wiley, New York, 2002.
41. Ahluwalia, R.K. and X. Wang, Direct hydrogen fuel cell systems for hybrid vehicles. *J. Power Sources,* 139(1–2), 152–164, 2005.
42. Marban, G. and T. Valdes-Solis, Towards the hydrogen economy? *Int. J. Hydrogen Energy,* 32(12), 1625–1637, 2007.
43. Chang, H.P., C.L. Chou, Y.S. Chen, T.L. Hou and B. Weng, The design and cost analysis of a portable PEMFC UPS system. *Int. J. Hydrogen Energy,* 32(3), 316–322, 2007.
44. Hydrogen Fuel Cell, http://www1.eere.energy.gov/hydrogenandfuelcells/mypp/pdfs/production.pdf, May 2000.
45. U.S. Department of Energy, Office of Basic Energy Sciences, Committee on Alternatives and Strategies for Future Hydrogen Production and Use, *The Hydrogen Economy: Opportunities, Costs, Barriers, and R&D Needs*, National Research Council, National Academies Press, Washington, available at http://www.nap.edu/catalog/10922.html (accessed May 2007), 2004.
46. EIA Annual Energy Outlook with Projections to 2025, Washington, 2005.
47. Blazek, C.F., R.T. Biederman, S.E. Foh and W. Jasionowski, Underground storage and transmission of hydrogen, *Proceedings of the Third Annual US Hydrogen Meeting*, March 18–20, Washington, 4–221, 1992.
48. Carpetis, C., Estimation of storage costs for large hydrogen storage facilities. *Int. J. Hydrogen Energy,* 7(2), 191–203, 1981.
49. Chambers, A., C. Park, T.K. Baker and N.M. Rodriguez, Hydrogen storage in graphite Nanofibers. *J. Phys. Chem. B,* 102(22), 4253–4256, 1998.
50. Dillon, A.C., P.A. Parilla, Y. Zhao, Y.-H. Kim, T. Gennett, C. Curtis, J.L. Blackburn, K.E.H. Gilbert, J.L. Alleman, K.M. Jones, S.B. Zhang and M.J. Heben, The 2005 U.S. Department of Energy Hydrogen Program Review, available at http://www.hydrogen.energy.gov/pdfs/review05/stp 63 heben.pdf (accessed March 2007), May 23–26, 2005.
51. Dillon, A.C., K.M. Jones, T.A. Bekkedahl, C.H. Kiang, D.S. Bethune and M.J. Heben, Storage of hydrogen in single wall carbon nanotubes. *Nature,* 386, 377–379, 1997.
52. Kukkone, C.A. and M. Shelef, Hydrogen as an alternative automotive fuel, in *Alternate Fuels*, Bata, R.M., Ed., National Research Council and National Academy of Engineering, National Academies Press, Washington, 1992.
53. Thomas, G., *Hydrogen Storage: State of the Art Workshop on Basic Energy Needs for the Hydrogen Economy*. Rockville, MD, 2003.
54. DeLuchi, M., Hydrogen vehicles: An evaluation of fuel storage, performance, safety, environmental impacts and cost. *Int. J. Hydrogen Energy,* 14(2), 88–130, 1989.
55. James, B.D., G.N. Baum, F.D. Lomax, C.E. Thomas and I.F. Kuhn, Comparison of Onboard Hydrogen Storage for Fuel Cell Vehicles, Directed Technologies, Inc., prepared for Ford Motor Company, under prime contract DE-AC02-94CE50389 to the United States Department of Energy, May 1996.
56. Satyapal, S., J. Petrovic, C. Read, G. Thomas and G. Ordaz, The U.S. department of energy's national hydrogen storage project: Progress towards meeting hydrogen-powered vehicle requirements. *Catal. Today,* 120, 246–256, 2007.
57. Michel, F., H. Fieseler, G. Meyer and F. Theiben, On-Board equipment for liquid hydrogen vehicles. *Int. J. Hydrogen Energy,* 23(3), 191–199, 1998.
58. Mitlitsky, F., A.H. Weisberg and B. Myers, Vehicular Hydrogen Storage Using Lightweight Tanks, U.S. DOE Hydrogen Program, Annual Review Meeting, Lakewood, CO, May 4–6, 1999.
59. Newson, E., T.H. Haletter, P. Hottinger, F. von Roth, W.H. Scherrer and T.H. Schucan, Seasonal storage of hydrogen in stationary systems with liquid organic hydrides. *Int. J. Hydrogen Energy,* 23(10), 905–909, 1998.

60. Verbetsky, V.N., S.P. Malyshenko, S. Mitrokhin, V.W. Solovei and Y. Shmalko, Metal hydrides: Properties and practical applications: 18 Review of works in the CIS countries. *Int. J. Hydrogen Energy*, 23(12), 1165–1178, 1998.
61. The Hydrogen Economy, *Physics Today*, available at http://www.aip.org/pt/vol-57/iss-12/p39.html, November 2006.
62. Wetzel, F., Improved handling of liquid hydrogen at filling stations: Review of six years of experience. *Int. J. Hydrogen Energy*, 23(5), 339–348, 1998.
63. Schlapbach, L. and A. Zuttel, Hydrogen-storage materials for mobile applications. *Nature*, 414, 353–358, 2001.
64. Review of DOE's vision 21, *Research and Development Program*. The National Academies Press, NRC, Washington, 2003.
65. Hirano, S., K.S. Young, A. Kwabena and J.A. Schwarz, The high surface area activated carbon hydrogen storage system. *Frontiers Sci. Ser.*, 7 (New Energy Systems and Conversions), 67–72, 1993.
66. Schwarz, J.A., Hydrogen program review, *Proc. DOE (NREL)*, 89–102, 1993.
67. Chen, J. and F. Wu, Review of hydrogen storage in inorganic fullerene-like nanotubes. *Appl. Phys. A*, 78, 989–994, 2004.
68. Strobel, R.J., J. Garche, P.T. Moseley, L. Jorissen and G. Wolf, Hydrogen storage by carbon material. *J. Power Sources*, 159, 781–801, 2006.
69. Rambach, G.G., and C. Hendricks, Hydrogen transport and storage in engineered microspheres, *Proceedings of the U.S. DOE Hydrogen Program Review*, Miami, May 1–2, 2, 765–772, 1996.
70. Rapp, D.B. and J.E. Shelby, Photo-induced hydrogen outgassing of glass. *J. Non-Crystalline Solids*, 349, 254–259, 2004.
71. Schmitt, M.L., J.E. Shelby and M.M. Hall, Preparation of hollow glass microspheres from sol–gel derived glass for application in hydrogen gas storage. *J. Non-Crystalline Solids*, 352, 626–631, 2006.
72. Hijikata, T., Research and development of international clean energy network using hydrogen energy (WE-NET). *Res. Dev. Int. Clean. Int. J. Hydrogen Energy*, 27, 115–129, 2002.
73. Hord, J., Is hydrogen a safe fuel? *Int. J. Hydrogen Energy*, 3, 157–176, 1976.
74. Ogden, J., T. Kreutz and M. Steinbugler, Fuels for Fuel Cell Vehicles: Vehicle Design and Infrastructure Issues, Society of Automotive Engineers Technical Paper No. 982500, Presented at the SAE Fall Fuels and Lubricants Meeting and Exposition, San Francisco, CA, October 19–22, 1998.
75. Thomas, C.E., B.D. James, F.D. Lomax and I.F. Kuhn, Societal Impacts of Fuel Options for Fuel Cell Vehicles, Society of Automotive Engineers Technical Paper No. 982496, Fuels and Lubricants Meeting and Exposition, San Francisco, CA, October 19–22, 1998.
76. Herzog, H., E. Drake and E. Adams, CO_2 Capture, Re-use and Storage Technologies for Mitigating Global Climate Change, Final Report for U.S. DOE Contract No. DE-AF22-96PC01257, January 1997.
77. *Hydrogen Posture Plan: An integrated Research*, Development and Demonstration Plan, U.S. Department of Energy, July 2003.

2

Production of Hydrogen from Hydrocarbons

Nazim Z. Muradov

CONTENTS

2.1 Introduction

2.1.1 Background

Hydrogen is an important raw material in chemical, petroleum, metallurgical, pharmaceutical, electronic, and food industries. Currently, the largest consumers of industrial hydrogen in the United States are ammonia synthesis facilities (40.3%), oil refineries (37.3%), and methanol production plants (10.0%) (in the world: 62.4%, 24.3%, and 8.7%, respectively) [1]. The nonrefinery utilization of hydrogen in the United States was reported as follows (in percent): chemical processing—73.9, electronics—8.1, food processing—3.6, metal manufacturing—2.7, other markets—11.7 [2]. As far as the oil industry is concerned, there has been a dramatic shift in the supply and demand for hydrogen in refineries during the past two-to-three decades [3]. Previously, refineries were viewed as net producers of hydrogen, because large amounts of hydrogen were produced in refineries as a by-product and some portion of it was consumed internally in a variety of hydrotreatment processes. At present, with increasingly tougher regulations on the automobile emissions (especially, on sulfur and nitrogen emissions) and larger volumes of heavy high-sulfur crudes to be processed, refineries substantially increased the capacities of hydrodesulfurization and hydrodenitrogenation processes. Moreover, in the past, large volumes of hydrogen in refineries were produced as a by-product of the aromatization (or catalytic reforming) process (which produced benzene). Emission control measures call for the reduction in the aromatics content in "reformulated" gasoline, resulting in less hydrogen production in refineries. This coupled with the sharp increase in hydrogen demand for hydrotreatment processes caused modern refineries to become major net consumers of hydrogen and establish independent hydrogen production plants on their perimeter. The medium- and long-term forecast for hydrogen supply and demand in refineries is expected to be very tight. A growing demand for hydrogen is also forecast in all other sectors with estimates of 10–15% increase per year [2].

With the transition to environmentally sustainable energy systems and the potential of hydrogen to become one of the two (along with electricity) major energy carriers of the future, the demand for hydrogen could explode in the next couple of decades. Given the advantages inherent in hydrocarbon fuels, such as their availability, cost competitiveness, convenience of storage and distribution, relatively high H/C ratio, they are likely to play a major role in hydrogen production for the near- to medium-term future. There are a number of factors controlling the selection of hydrogen production process including feedstock

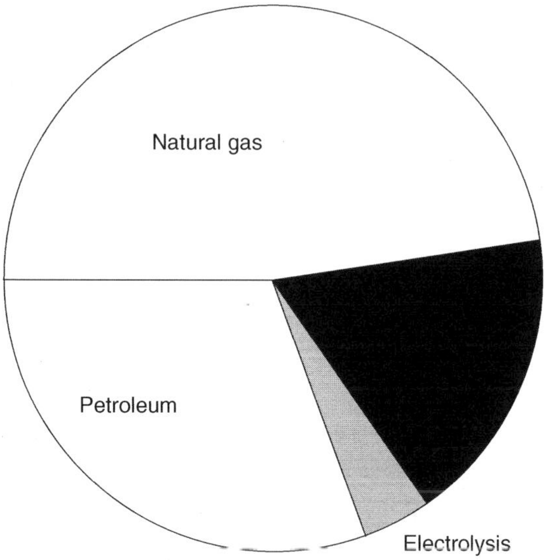

FIGURE 2.1
World hydrogen production structure. (Based on the data from The Hydrogen Economy: Opportunities, Costs, Barriers, and R&D Needs, The National Academic Press, Washington, 2004.)

availability and cost, capacity requirements, product purity, pressure needs, potential by-products, and cogeneration of steam or power [3]. Figure 2.1 depicts the current world's distribution of resources for hydrogen production (based on the data provided in [4 and 5]). It can be seen that hydrocarbons (natural gas (NG) and petroleum) account for 78% (48% and 30%, respectively) of the world's hydrogen production. It should be noted, however, that almost all hydrogen produced from petroleum-based feedstocks (e.g., refinery off-gases and residual oil) is consumed internally in refineries. A small percentage of hydrogen (about 4%) is produced by water electrolysis (it is noteworthy that a substantial portion of electricity is produced from hydrocarbon-based primary energy sources). Figure 2.2 provides a comparison of the theoretical energy consumption for producing hydrogen from various hydrocarbons, coal, and water (by electrolysis). It can be seen that the production of hydrogen from light hydrocarbons requires the least amount of energy, whereas hydrogen generation by water electrolysis is the most energy-intensive option. Figure 2.3 presents a comparative assessment of the theoretical yields of hydrogen produced by steam gasification of different hydrocarbon feedstocks and coal. One can note the following correlation: as the H/C ratio in the feedstock increases, the hydrogen yield also increases; the CO_2 by-product yields (not shown in the Figure 2.3) follow the reverse order. Thus, the preference of producing hydrogen from light hydrocarbons could be attributed to a large extent to the fact that the related processes are less energy intensive and have higher hydrogen yield and lower CO_2 emissions compared to alternative feedstocks (e.g., coal).

In this review, the author intends to provide an overview of the present day hydrocarbon-to-hydrogen technologies including well-established industrial processes as well as the latest technological developments that are still in the research and development (R&D) stage. Hydrogen production from petroleum coke, coal, and oxygenated fuels (e.g., methanol and ethanol) although closely related to the topic of discussion, will not be covered in this chapter (production from coal is covered in Chapter 3). The objective of this review is to give the reader a concise introduction to the field and provide an insight into some scientific, technological, and environmental aspects of hydrocarbon-to-hydrogen technologies not often

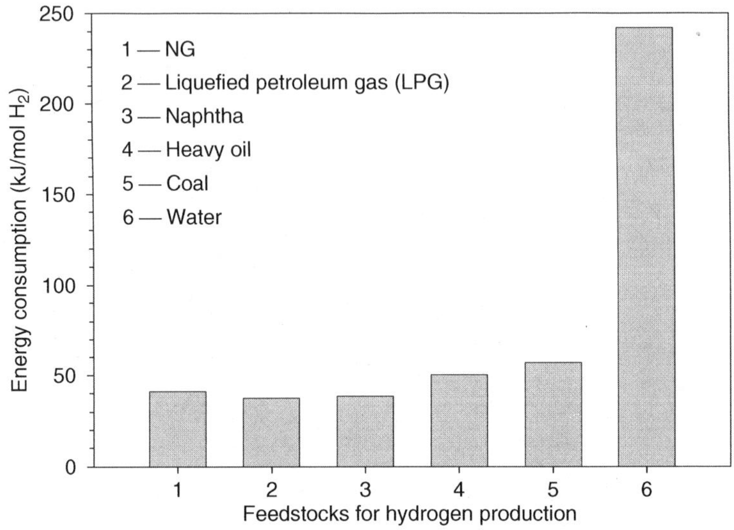

FIGURE 2.2
Theoretical energy consumption for hydrogen production from different feedstocks.

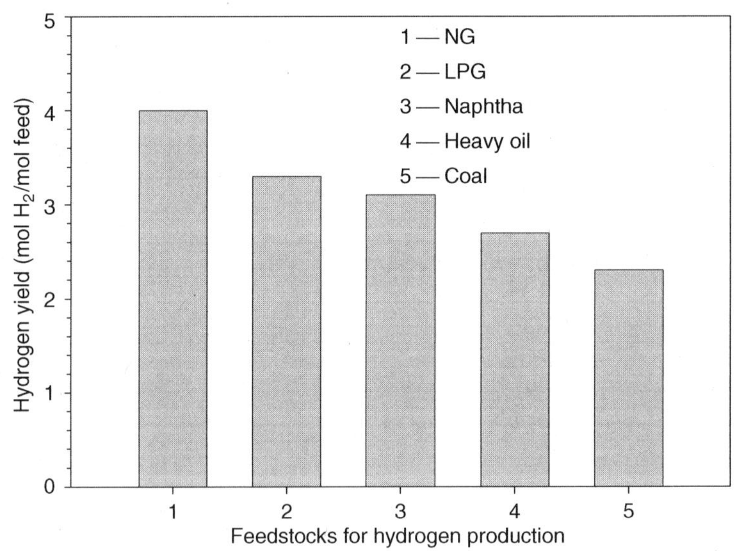

FIGURE 2.3
Maximum theoretical yield of hydrogen produced by steam reforming (gasification) of different feedstocks.

discussed in the literature. For in-depth background information on industrial hydrogen production technologies, the author would recommend a number of excellent review papers including Rostrup-Nielsen's [6] and Armor's [3] reviews, and an encyclopedic review [7].

2.1.2 Brief Overview of Hydrocarbon-to-Hydrogen Technologies

Before going into a detailed discussion of processes for the conversion of hydrocarbons to hydrogen, it would be useful to give a general classification of hydrocarbon-to-hydrogen technologies. There are several ways to categorize these technologies, for example, from a thermodynamic viewpoint (endothermic versus exothermic), by the use of catalysts (catalytic versus noncatalytic) or by the role of oxidants (oxidative versus nonoxidative). The last-mentioned classification (i.e., oxidative versus nonoxidative hydrocarbon-to-hydrogen processes) is used in this review because it best reveals the differences in process chemistry, the type of energy input, the role of catalysts, and the environmental implications of the technologies. Figure 2.4 provides the general classification of all the major technologies and methods for producing hydrogen from hydrocarbon-based feedstocks.

Oxidative processing of hydrocarbons to hydrogen occurs at elevated temperatures (typically, above 700°C) in the presence of oxidants such as oxygen (air), steam, CO_2, or a combination thereof. In general, the processes for oxidative conversion of hydrocarbons to hydrogen are well-established technologies, and most of the industrial hydrogen production processes (e.g., steam methane reforming [SMR], partial oxidation [POx], autothermal reforming [ATR], steam–iron process [SIP]) belong in this category. The oxidative conversion

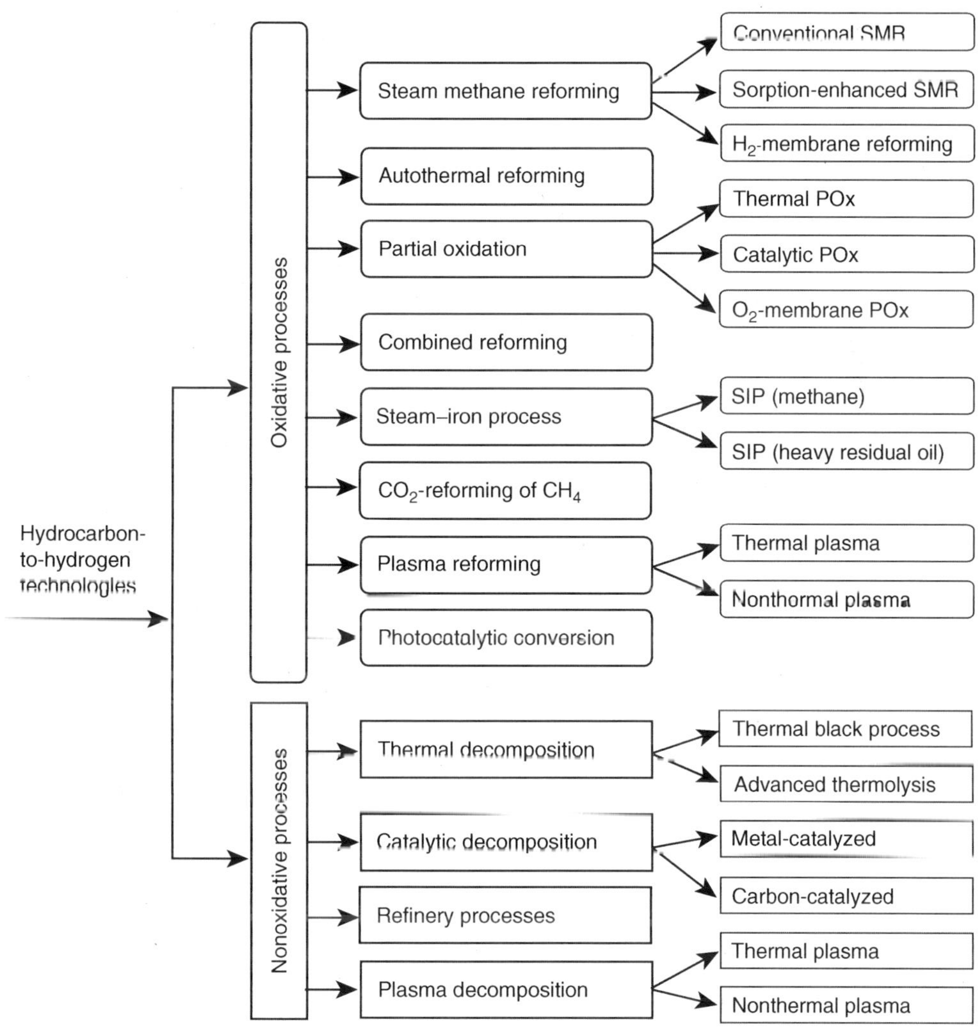

FIGURE 2.4
General classification of hydrocarbon-to-hydrogen technologies.

of hydrocarbons to hydrogen can be represented by the following generic chemical equation:

$$C_nH_m + [Ox] \rightarrow xH_2 + yCO + zCO_2 \tag{2.1}$$

where C_nH_m is a hydrocarbon ($n \geq 1, m \geq n$) and [Ox] is an oxidant such as O_2, H_2O, and CO_2.

Depending on the nature of the oxidant [Ox] the oxidative process could be exothermic (e.g., when [Ox] = O_2), endothermic (when [Ox] = H_2O, CO_2, or H_2O–CO_2 mixture), or near thermo-neutral (when [Ox] = O_2–H_2O or O_2–CO_2, or O_2–H_2O–CO_2 mixture with the suitable molar ratio of the reagents). High operational temperatures of the oxidative conversion processes are due to the relative "inertness" of methane and other saturated hydrocarbons (or alkanes) that make up most of the hydrocarbon feedstock for hydrogen production. However, water and CO_2 are also very inert compounds requiring high temperatures for their activation and interaction with other chemical compounds. Thus, the direct thermal (i.e., noncatalytic) interaction of methane with steam or CO_2 would require extremely high temperatures (in excess of 1000°C), therefore, catalysts are widely used to accomplish these processes at the practical range of temperatures (750–950°C).

Nonoxidative conversion of hydrocarbon feedstocks to hydrogen generally occurs by the splitting of C–H bonds in hydrocarbons in response to an energy input (heat, plasma, radiation, etc.), and it does not require the presence of oxidizing agents. Among the nonoxidative hydrocarbon-to-H_2 conversion processes are thermal, catalytic, and plasma hydrocarbon decomposition processes. The following generic chemical reaction describes the nonoxidative transformation of hydrocarbons to hydrogen:

$$C_nH_m + [\text{energy}] \rightarrow xH_2 + yC + zC_pH_q \tag{2.2}$$

where C_nH_m is the original hydrocarbon feedstock ($n \geq 1, m \geq n$), C_pH_q represents relatively stable products of the feedstock cracking ($z \geq 0, p \geq 1, q \geq p$; in most cases C_pH_q is CH_4 or C_2H_2), and [energy] is an energy input, for example, thermal energy or electrical energy (e.g., plasma) or radiation energy.

In most cases, the nonoxidative conversion processes are endothermic and require some form of an energy input to accomplish the decomposition reaction. In general, these processes require elevated temperatures (>500°C), especially when methane and other light alkanes are used as a feedstock. Sections 2.2 and 2.3 discuss oxidative and nonoxidative hydrocarbon-to-hydrogen technologies, respectively. Commercial hydrogen manufacturing processes will receive more extensive coverage compared to technologies that still are in the R&D stage. Small-scale hydrocarbon reformers for vehicular (onboard) and distributed hydrogen production applications will be included in Section 2.2. Finally, some environmental aspects of hydrogen production from hydrocarbon feedstocks will be discussed in Section 2.4, followed by concluding remarks.

2.2 Oxidative Processing of Hydrocarbons

Most industrial hydrogen is manufactured by the following hydrocarbon-based oxidative processes: steam reforming of light hydrocarbons (e.g., NG and naphtha), POx of heavy oil fractions, and ATR. Each of these technological approaches has numerous modifications depending on the type of feedstock, reactor design, heat input options, by-product treatment,

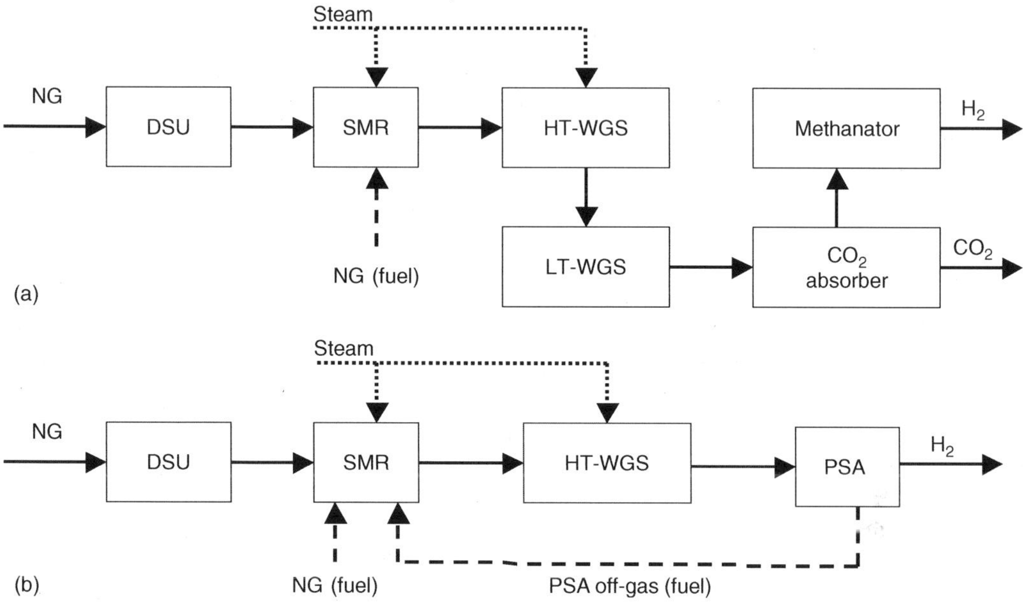

FIGURE 2.5
Simplified schematics of hydrogen production by SMR. (a) SMR with solvent removal of CO_2 and a methanation unit. (b) SMR with a PSA unit. HT- and LT-WGS: high- and low-temperature WGS reactors, respectively.

hydrogen purity, etc. Most typical process configurations will be discussed in this review, and, wherever possible the reference on other modifications to the process will be provided.

2.2.1 Steam Methane Reforming

The SMR is by far the most important and widely used process for the industrial manufacture of hydrogen, amounting to about 40% of the total world production [7]. The technology is well developed and commercially available at a wide capacity range, from <1 t/h H_2 for small decentralized units to about 100 t/h H_2 for large ammonia manufacturing plants [8].

2.2.1.1 Steam Methane Reforming Process Description

Figure 2.5 depicts the simplified block diagram of the SMR plant in two major technological variations differing from one another mostly by the final treatment of the product gas: (i) SMR including solvent CO_2 removal and a methanator (Figure 2.5a) and (ii) SMR equipped with a pressure-swing adsorption (PSA) system (Figure 2.5b). The main stages (or operating units) of an SMR plant include NG feedstock desulfurization, catalytic reforming, CO conversion (or water–gas shift [WGS]), and gas separation/H_2 purification. Owing to the high sensitivity of the reforming and WGS catalysts to sulfur poisoning, a high degree of feedstock desulfurization is required, and it is accomplished in the *desulfurization unit* (*DSU*). Sulfur–organic compounds (e.g., thiols) are first converted into H_2S by catalytic hydrogenation reaction (Co–Mo catalyst, 290–370°C [3]). A small splitstream of the H_2 product is used as a hydrogenating agent in this step. This is followed by H_2S scrubbing by a ZnO bed (at 340–390°C) according to the following reaction:

$$H_2S + ZnO \rightarrow ZnS + H_2O \quad (2.3)$$

In some cases, traces of halides (e.g., chlorides) are also present in the NG feedstock, and these are removed by an alumina guard bed.

After the DSU, NG is fed to a catalytic reforming Unit 2 where it reacts with steam to produce synthesis gas. Depending on the content of higher (C_2+) hydrocarbons in NG, there may be an additional step, called *prereforming*, which is designed to remove these objectionable higher (C_2+) hydrocarbons. (Higher hydrocarbons are more reactive than methane and are readily decomposed over the reforming catalyst leading to coke deposition on its surface, and consequently, catalyst deactivation.) The content of C_2+ hydrocarbons in NG could reach up to 10 vol% and higher [9]. The prereforming process converts C_2+ hydrocarbons in the feed to CH_4, CO_x, and H_2 in an adiabatic reactor at 300–525°C using alumina-supported high Ni with promoters [3]. The advantage of using prereforming units at SMR plants is twofold: (1) the process allows H_2 manufacturers to operate with a variety of feedstocks (i.e., fuel flexibility) and (2) the removal of C_2+ hydrocarbons allows for reducing the SMR unit's overall steam/carbon ratio (due to lesser coke formation), and, as a result, it increases the plant's efficiency.

Pretreated NG feedstock is mixed with steam (2.6 MPa), the resulting mixture is preheated to 500°C and introduced to the *catalytic reforming reactor*. In the reforming reactor, the steam–methane mixture is passed through externally heated reformer tubes filled with Ni catalyst, where it is converted to CO and H_2 at 850–900°C according to the following equation:

$$CH_4 + H_2O \rightarrow 3H_2 + CO \qquad \Delta H^\circ = 206 \text{ kJ/mol} \tag{2.4}$$

The reaction is highly endothermic and favored by low pressures (since 4 mol of gas are produced from 2 mol of gas). However, in most industrial application areas, hydrogen is required at the pressure of at least 2.0 MPa, therefore, the reformers are operated at elevated pressures (usually, 2.0–2.6 MPa). High pressures allow for a more compact reactor design, thus increasing the reactor throughput and reducing the cost of materials. According to reaction 2.4 stoichiometry, the molar ratio of steam to methane is H_2O:CH_4 = 1:1, however, in practice: an excess of steam (commonly, steam/methane ratio of 2.5:3) is used to prevent carbon (coke) deposition on the catalyst surface. Two reactions responsible for carbon deposition on the reforming catalyst surface are (1) methane decomposition to H_2 and carbon and (2) CO disproportionation

$$2CO \rightleftharpoons C + CO_2 \qquad \Delta H^\circ = -172 \text{ kJ/mol} \tag{2.5}$$

An addition of the supplemental steam shifts the reforming reaction equilibrium away from carbon formation.

To supply heat for the endothermic SMR reaction, the catalyst is loaded into a bundle of reactor tubes (about 15 m long and 12 cm inside diameter) made out of heat-resistant Ni alloy. These catalyst-containing tubes are placed inside a rectangular furnace box with multiple burners mounted along the inside walls of the box. The typical inlet temperatures are 450–650°C, with the product gas leaving the reformer at 700–950°C, depending on the applications [6]. In a tubular reformer furnace, about 50% of the heat produced by combustion in the burners is transferred through the reformer tube walls and utilized by the process. The other half of the fired duty is recovered for preheat duties and steam production. Owing to the high endothermicity of the reforming reaction 2.4, the process may suffer from heat transfer limitations. Therefore, a great deal of effort is spent to optimize the heat flux to the catalyst, the tube thickness, the catalyst load, etc. to ensure a good reactor performance.

The gaseous mixture containing H_2, CO, and steam (and usually about 4% of unconverted methane) leaves the reformer at about 800–900°C. It is cooled rapidly to about 350°C (thereby generating steam) and is fed to *WGS reactors*, where CO reacts with steam over a catalyst bed producing H_2 and CO_2:

$$CO + H_2O \rightarrow H_2 + CO_2 \qquad \Delta H^\circ = -41.2\ \text{kJ/mol} \tag{2.6}$$

Reaction 2.6 indicates that it is a moderately exothermic reaction. Figure 2.6 depicts the equilibrium concentrations of CO and CO_2 in the CO–H_2O–H_2–CO_2 mixture as a function of temperature at different H_2O/CO ratios (based on the data reported in Ref. 10). The increase in the catalyst temperature, while favorably increasing the rate of reaction, tends to shift the thermodynamic equilibrium to the left. In the practical system, a compromise is made between a high rate and unfavorable equilibrium and a lower rate and more favorable equilibrium [10]. To achieve high CO conversion, two reactors are commonly used in series: HT- and LT-WGS reactors, respectively (see Figure 2.5a). The HT-WGS reactor operates at the inlet temperatures of 340–360°C and uses an iron–chromium-based catalyst (90–95% magnetite iron oxide, stabilized with 5–10% of chromia, Cr_2O_3) [3]. A typical LT-WGS catalyst consists of CuO: 15–30%, ZnO: 30–60%, and Al_2O_3: balance, and it promotes a favorable reaction rate between 200 and 300°C [10]. The combination of HT- and LT-WGS reactors allows converting 92% of the CO in the reformate gas into H_2 [11] and lowering the CO content in the gas to about 0.1 vol%.

The H_2 is separated from CO_2 and purified at the final stage of the process. Older variations of the SMR process (Figure 2.5a) used solvents to remove the acid gas (CO_2) from the gaseous stream after WGS reactors. Solvents commercially used for CO_2 removal in the *gas separation unit* include monoethanolamine (most preferred and widely used solvent), water, ammonia solutions, potassium carbonate solutions, and methanol. This operation allows the reduction of CO_2 concentration in the process gas to about 100 ppm. The remaining

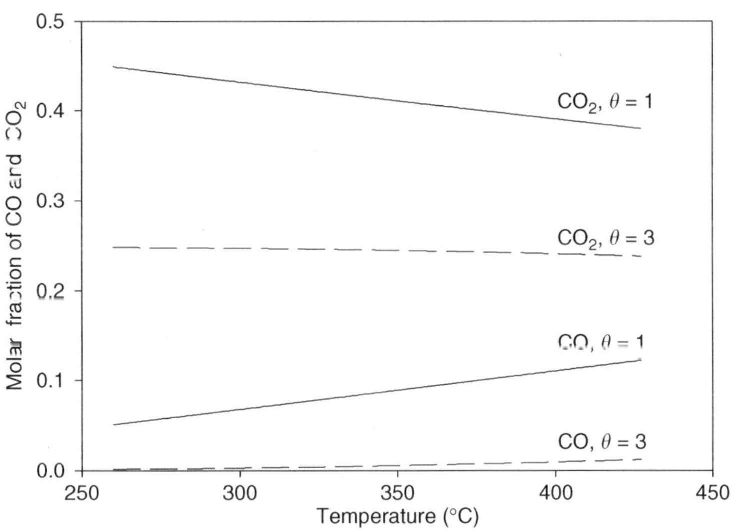

FIGURE 2.6
Equilibrium mole fraction of CO and CO_2 in CO–H_2O–H_2–CO_2 mixture as a function of temperature. θ is the initial H_2O/CO molar ratio.

residual CO_2 and CO are removed in the *methanation reactor* where they are converted to CH_4 in the presence of H_2 (320°C, catalyst: Ni or Ru on oxide support).

Modern SMR plants (Figure 2.5b) incorporate a PSA unit for purifying hydrogen from CO_2, CO, and CH_4 impurities (moisture is preliminarily removed from the process gas). The PSA unit consists of multiple (parallel) adsorption beds, most commonly filled with molecular sieves of suitable pore size; it operates at the pressure of about 20 atm. The PSA off-gas is composed of (mol%) CO_2–55, H_2–27, CH_4–14, CO–3, N_2–0.4, and some water vapor [11] and is burned as a fuel in the primary reformer furnace. Generally, SMR plants with PSA need only a HT-WGS stage, which may somewhat simplify the process.

The hydrogen plant efficiency is defined as the total energy produced by the hydrogen plant divided by the total energy consumed by the plant, determined by the following formula [11]:

$$\eta = \frac{E_{H_2} + E_{steam,\,4.8\,MPa}}{E_{NG} + \text{electricity} + E_{steam,\,2.6\,MPa}} \tag{2.7}$$

where η is the energy efficiency, E_{H_2} the energy in H_2 product, $E_{steam,\,4.8\,MPa}$ the 4.8 MPa steam energy (exported), E_{NG} the NG energy, and $E_{steam,\,2.6\,MPa}$ the 2.6 MPa steam (required). The authors estimated that for the hydrogen plant with the capacity of 1.5×10^6 Nm^3/day, the energy efficiency was 89% (on high heating value [HHV] basis). If the steam were not included in the equation (e.g., 4.8 MPa steam could not be used by another source), the energy efficiency would decrease to 79.2%. The hydrogen plant efficiency drops to 69.1% if 2.6 MPa steam is considered as an energy input (i.e., it is not produced internally). Scholz [4] estimated the energy efficiency of SMR plant at 81.2%.

2.2.1.2 Catalysts for Steam Methane Reforming Process

Because of a great practical importance of SMR as a major industrial process for manufacturing H_2, the development of efficient steam reforming catalysts is a very active area of research. Nickel and noble metals are known to be catalytically active metals in the SMR process. The relative catalytic activity of metals in the SMR reaction (at 550°C, 0.1 MPa and steam/carbon ratio of 4) is as follows [12]:

$$\text{Ru} > \text{Rh} > \text{Ir} > \text{Ni} > \text{Pt} > \text{Pd}$$

Although Ni is less active than some noble metals and more prone to deactivation (e.g., by coking), it is the most widely used catalyst for the SMR process due to its relatively low cost. The activity of a catalyst is related to the surface concentration of active sites, which implies that, generally, the catalytic activity increases with the increase in dispersion of metal particles over the support surface. The typical size of metal particles in the SMR catalyst is in the range 20–50 nm [13]. Although the Ni surface area is increased with higher metal loadings, there is an optimum (about 15–20 wt%) beyond which an increase in Ni loading does not result in an increase in catalytic activity.

The catalyst most commonly used in the reforming reaction is the high-content Ni catalyst (~12–20% Ni as NiO) supported on a refractory material (e.g., α-Al_2O_3) containing a variety of promoters [3]. Key promoters include potassium or calcium alkali ions designed to suppress carbon deposition on the catalyst surface. The Ni catalyst is manufactured in a variety of shapes to ensure high surface to volume ratio, optimal heat and mass transfer, low pressure drop, high strength, etc. (e.g., commonly the catalyst is extruded in the shape of multichannel wheels). There are several stringent requirements to the reforming catalyst performance, which include long-term stability and high tolerance of the extreme operating conditions (e.g., very high temperature); robustness to withstand the stress of start-up and transient operational conditions; nonuniformity of the feedstock, which may expose the catalyst to poisons (e.g., sulfur); and excessive coke deposition. The industrial reforming catalysts are supposed to perform in excess of 50,000 h (or 5 years) of continuous operation before their replacement [3].

The role of the support (or carrier) is to provide support for the catalytically active metal to achieve a stable and high surface area. The influence of the support on the activity of catalysts in the SMR reaction can hardly be overestimated. It not only determines the dispersion of the catalytically active metal particles, but it also affects the catalyst's reactivity, resistance to sintering, and coke deposition, and may even participate in the catalytic action itself [14]. From this viewpoint, the support is an integral part of the catalyst and cannot be considered separately. Among the most common supports for SMR catalysts are α- and γ-Al_2O_3, MgO, $MgAl_2O_4$, SiO_2, ZrO_2, and TiO_2. These supports have relatively high surface area and porosity and suitable pore structure and surface morphology, which are conducive to better contact between the reactants and the catalyst. Furthermore, due to the nature of the chemical bonding between the support and the metal particles, the electronic properties of the metal, and hence, its catalytic activities are affected. For example, the supports with pronounced acidic properties are known to facilitate decomposition of methane. Generally, a strong interaction between a metal and a support makes the catalyst more resistant to sintering and coking, thus resulting in an enhanced long-term stability of catalysts [15].

2.2.1.3 Reaction Kinetics and Mechanism

The reaction mechanism of the SMR reaction strongly depends on the nature of the catalytically active metal and the support (the detailed discussion is provided in the review [14]). The kinetics and mechanism of the SMR reaction over Ni-based catalysts have been extensively studied by several research groups worldwide. For example, Xu and Froment [16] investigated the intrinsic kinetics of the reforming reaction over Ni/$MgAl_2O_4$ catalyst. They arrived at the reaction model based on the Langmuir–Hinshelwood reaction mechanism, which includes several reaction steps as follows:

$$H_2O + * \leftrightharpoons O{-}* + H_2 \tag{2.8}$$

$$CH_4 + * \leftrightharpoons CH_4{-}* \tag{2.9}$$

$$CH_4{-}* + * \leftrightharpoons CH_3{-}* + H{-}* \tag{2.10}$$

$$CH_3{-}* + * \leftrightharpoons CH_2{-}* + H{-}* \tag{2.11}$$

$$CH_2{-}* + O{-}* \leftrightharpoons CH_2O{-}* + * \tag{2.12}$$

$$CH_2O{-}* + * \leftrightharpoons CHO{-}* + H{-}* \tag{2.13}$$

$$CHO{-}* + * \leftrightharpoons CO{-}* + H{-}* \tag{2.14}$$

$$CO{-}* + O{-}* \leftrightharpoons CO_2{-}* + * \tag{2.15}$$

$$\text{CHO—}* + \text{O—}* \leftrightarrows \text{CO}_2\text{—}* + \text{H—}* \quad (2.16)$$

$$\text{CO—}* \leftrightarrows \text{CO} + * \quad (2.17)$$

$$\text{CO}_2\text{—}* \leftrightarrows \text{CO}_2 + * \quad (2.18)$$

$$2\text{H—}* \leftrightarrows \text{H}_2\text{—}* + * \quad (2.19)$$

$$\text{H}_2\text{—}* \leftrightarrows \text{H}_2 + * \quad (2.20)$$

where $*$ denotes an active surface site.

The rate equations of the reaction between steam and methane (reaction 2.4) can be written as

$$r_1 = k_1\left(\frac{P_{CH_4}P_{H_2O}}{P_{H_2}^{2.5}} - \frac{P_{H_2}^{0.5}P_{CO}}{K_1}\right)\Big/Z^2 \quad (2.21)$$

where r_1 is the reaction rate, k_1 the rate constant, P_i the partial pressure, and K_1 the equilibrium constant. Z is a function of P_i and adsorption constants C_i as follows:

$$Z = 1 + C_{CO}P_{CO} + C_{H_2}P_{H_2} + C_{CH_4}P_{CH_4} + C_{H_2O}\frac{P_{H_2O}}{P_{H_2}} \quad (2.22)$$

In a number of publications, Rostrup-Nielsen discusses different mechanism of methane steam reforming over Ni catalysts [17]. The proposed simplified reaction sequence for reforming of methane is as follows:

$$\text{CH}_4 + * \xrightarrow{k_1} \text{CH}_x* \quad (2.23)$$

$$\text{CH}_x* \xrightarrow{k_2} \text{C}* \longleftrightarrow [\text{C,Ni}]_{\text{bulk}} \longrightarrow \text{carbon} \quad (2.24)$$

$$\text{CH}_x* + \text{OH}_y* \xrightarrow{k_3} \text{gas} \quad (2.25)$$

$$\text{C}* + \text{OH}_y* \xrightarrow{k_4} \text{gas} \quad (2.26)$$

where $*$ represents a nickel active site.

On the surface of the Ni catalyst, carbon is normally produced in a whisker (or filamentous) form. According to Rostrup-Nielsen, carbon formation is avoided when the concentration of carbon dissolved in Ni crystal is smaller than that at the equilibrium. The steady-state activity is proportional to [C*], which can be expressed by the following equation:

$$a_c^s \sim [\text{C}*] \sim \frac{k_1k_2}{k_3k_4} \cdot \frac{1}{\left[\text{OH}_y*\right]^2} \quad (2.27)$$

where a_c^s is a steady-state activity for carbon whisker.

The carbon whisker mechanism can be blocked by the use of noble metal catalysts because these metals do not dissolve the carbon.

2.2.1.4 *Steam Reforming of Naphtha*

At the beginning of the 1960s, light naphtha was available in large quantities for the production of syngas, town gas, ammonia, and methanol. Generally, direct reforming of naphtha is not different from NG reforming. Like NG, naphtha reforming is carried out in externally heated tubes over Ni catalyst, and it produces a mixture consisting of H_2, CO, CO_2, CH_4, and steam. Owing to a higher carbon content in the process feed, naphtha reforming produces a gas with increased CO and CO_2 content compared to the NG feed. Naphtha reforming plants are distinguished from those based on NG by the following features [7]:

1. More complex desulfurization system
2. Use of a special catalyst in the tubular reformer, and a special start-up system
3. Fewer reformer tubes per quantity of hydrogen and CO produced at equal heat loads per unit area
4. Larger CO_2 washing system

Currently, naphtha reforming is of minor importance. Some hydrogen and syngas production based on naphtha still takes place at a few locations with no access to NG.

2.2.1.5 *Advanced Steam Reforming Systems*

Although SMR is a well-developed technology, there is room for further technological improvement, in particular, with regard to energy efficiency, gas separation, and H_2 purification stages.

2.2.1.5.1 *Sorption-Enhanced Reforming*

In the sorption-enhanced reforming (SER) process, one of the gaseous reaction products (CO_2) of the catalytic reforming reaction is separated from the reaction zone by sorption. As a result, the equilibrium of the reaction is shifted toward products according to the Le Chatelier's principle. Balasubramanian et al. [18] studied the SMR reaction in the presence of CaO as a CO_2 acceptor. Thus, in addition to reactions 2.4 and 2.6, the reaction of CO_2 with the CO_2 acceptor (CaO) takes place in the reaction zone:

$$CaO\,(s) + CO_2\,(g) \rightarrow CaCO_3\,(s) \qquad (2.28)$$

The advantages are fourfold: (1) fewer processing steps, (2) improved energy efficiency, (3) elimination of the need for shift catalysts, and (4) reduction in the temperature of the primary reforming reactor by 150–200°C. Figure 2.7 depicts the simplified diagram of the SER of methane (based on the process description provided in Ref. 18). In this process, the three simultaneous reactions (i.e., reactions 2.4, 2.6, and 2.28) occur in an adiabatic fluidized bed reactor (FBR) containing a mixture of the reforming catalyst and CO_2 acceptor at 725°C. The heat released by the exothermic shift and carbonation (Equation 2.28) reactions balances the heat input required by the endothermic reforming reaction (thus, no supplemental fuel is required in the reforming reactor). About 88% conversion of methane

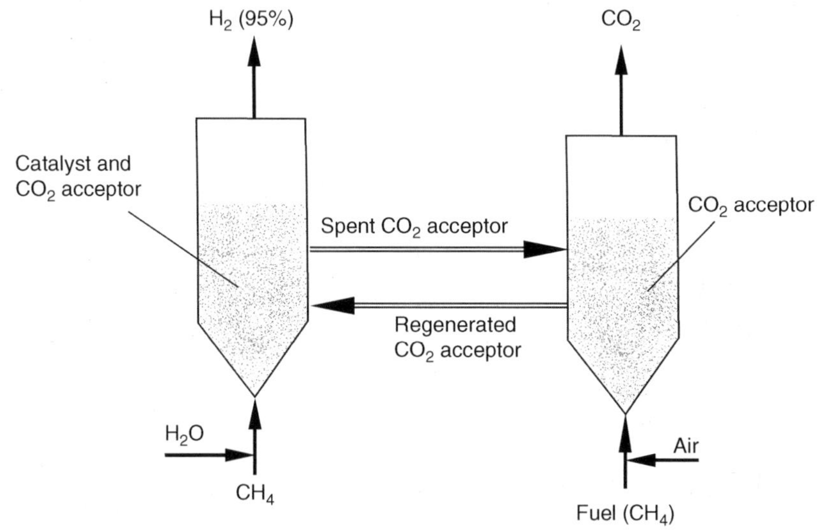

FIGURE 2.7
Simplified schematics of SER of methane.

is thermodynamically achievable, and the product gas contains 95 vol% H_2. The regeneration of the spent acceptor ($CaCO_3$) is accomplished in the adiabatic FBR regenerator at about 975°C with the CO_2 acceptor continuously recirculated between the reforming and the regenerator reactors. Key unanswered questions related to this technology include continuous separation of the reforming catalyst and the CO_2 acceptor, and the durability of the acceptor for multiple cycle operations. In the Air Products and Chemicals, Inc. version of the SER process, CO_2 was extracted from the reaction zone by K-promoted hydrotalcites (which are layered double hydroxides) [19]. As a result, large conversion of methane (90%+) could be achieved at relatively low temperatures (300–500°C).

2.2.1.5.2 *Hydrogen Membrane Reactor*

The concept of the hydrogen membrane reactor (HMR) is based on a similar application of Le Chatelier's principle in that the hydrogen produced in the reforming reaction selectively permeates through a membrane and exits the reaction zone. Typically, the membranes are made of Pd or Pd/Ag or other Pd-based alloys several microns thick. Figure 2.8 illustrates one of the conceptual designs of the HMR, which includes a reforming catalyst bed, and a H_2-permeable membrane. The main advantages of an HMR are as follows: (a) the H_2 producing reactions are free from the limitations of chemical equilibrium (i.e., equilibrium is shifted toward products), (b) high methane conversions are reached at lower temperatures (compared to a conventional reactor), (c) the process produces separate H_2 and CO_2 flows, (d) there is no need for additional CO-shift converters, (e) the reactor has a more simple and compact configuration, and (f) overall efficiency is higher. Owing to the relatively low-temperature resistance of the Pd-based membranes, HMRs operate at temperatures of 400–600°C (compared to 800–950°C typical of conventional reformers). As a result, the catalysts for an HMR must be very active in the low-temperature range.

Yasuda et al. [20] reported on the development and testing of an HMR equipped with Pd-based alloy modules with the total capacity of 20 Nm^3/h. The unit operated at the temperature of 540–560°C and produced hydrogen with purity of 99.999% at the average

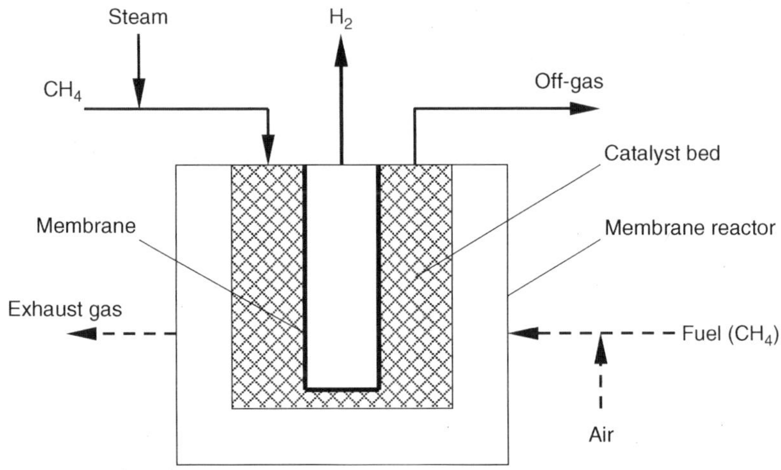

FIGURE 2.8
Conceptual design of a HMR for steam reforming of methane.

hydrogen recovery yield of 93% and energy efficiency of 70%. The system efficiency (η) was defined as follows:

$$\eta\,(\%) = \frac{F(H_2) \times Q(H_2)}{F(NG) \times Q(NG) + W(Aux)} \times 100 \tag{2.29}$$

where $F(H_2)$ and $Q(H_2)$ are the production rate and heat value of H_2, F(NG) and Q(NG) are consumption rate and heat value of NG, and W(Aux) is the electric power consumed by the auxiliary equipment.

In this study, the long-term performance test of the reformer with 35 start-up and shut-down cycles had to be terminated after 2071 h due to the failure of hydrogen separation modules. Tong et al. [21] reported experimental studies of steam reforming of methane in a thin Pd-based membrane reactor. A high hydrogen permeation flux of 0.26 mol/(m^2 s) and complete hydrogen selectivity were obtained at 500°C and a pressure difference of 100 kPa using a thin (6 μm) defect-free Pd film supported on a macroporous stainless-steel (MPSS) membrane tube. The catalytic membrane reactor for SMR was constructed from Pd/MPSS composite membrane and a commercial Ni/Al_2O_3 reforming catalyst. The authors demonstrated a dramatic improvement in the membrane reformer performance compared to the reformer made out of a dense stainless-steel tube. A SMR membrane reactor for pure hydrogen production was studied by Barbiery et al. [22]. The membrane reactor consisted of two tubular membranes, one Pd-based and another made from porous alumina. The reactor operated at 350–500°C with no sweep gas, and the steam/methane molar ratio varied in the range 3.5–5.9. The use of the membrane allowed a 7% increase in methane conversion over its thermodynamic equilibrium value.

2.2.1.6 Steam Methane Reforming Using Alternative Energy Sources

Because SMR is a highly endothermic process, the use of alternative (nonfossil) energy sources would result in a dramatic conservation of NG or other hydrocarbon fuels. From this viewpoint, the possibility of using high-temperature nuclear and solar heat sources has long attracted the interest of researchers.

2.2.1.6.1 *Steam Methane Reforming with Nuclear Heat Input*

Steam reforming of NG has the greatest potential for near-term development into a nuclear process heat system [23]. According to a study conducted by General Atomic Co. researchers, the efficiency of the nuclear-heated reformer system is considerably higher than that of the conventional one (85% versus 74%) [24]. In high-temperature gas-cooled nuclear reactors (HTGR), recycled helium is heated to temperatures up to 950°C, which is suitable for carrying out the SMR reaction. Hot helium is circulated in indirectly heated heat exchangers countercurrent to methane and steam flowing through the reformer tube, releasing its sensible heat and being cooled from 950°C to 600°C. The preferred reformer tube design has an inner helical tube through which the reformed gas is discharged to heat the catalyst-filled tube. Thermal energy is supplied to a helium heat carrier in the core of a high-temperature nuclear reactor. Such reactors, have been under testing and pilot-scale operation since 1971 and are considered suitable for commercial syngas production [7]. The reformed gas can be used to produce basic chemicals (e.g., H_2, NH_3, and CH_3OH) or, in conjunction with methanation, to transfer heat for long distances (the latter option is termed ADAM-EVA system). A simulation model for a steam reformer of the heat exchanger type was reported by Hiroshi [25]. The steam reformer is intended to produce reducing gas for direct steel making where heat is supplied by a high-temperature He carrier for a nuclear reactor. The basic reaction data on steam reforming were obtained from the experiments with a microreactor. The developed simulation model was used extensively in the design and development of the pilot-scale steam reformer and its control.

2.2.1.6.2 *Steam Methane Reforming with Solar Heat Input*

The state-of-the-art solar concentrators can provide solar flux concentrations in the following ranges, depending on the type of the concentrator [26]:

- Trough concentrators: 30–100 suns
- Tower systems: 500–5,000 suns
- Dish systems: 1,000–10,000 suns

For a solar concentration of 5000, the optimum temperature of the solar receiver is about 1270°C, giving a maximum theoretical efficiency of 75% (i.e., the portion of solar energy that can be converted to the chemical energy of fuels). This temperature is adequate to conduct high-temperature endothermic SMR or CO_2 reforming of methane processes. Solar chemical reactors for highly concentrated solar systems typically utilize a cavity receiver-type configuration, that is, a well-insulated enclosure with a small opening (the aperture) to let in concentrated solar radiation.

Solar reforming of methane (NG) has been extensively studied in solar furnaces as well as in solar simulators using different reactor configurations and catalysts [27–29]. In his review paper, Steinfeld discussed an indirectly-irradiated solar reforming reactor consisting of a pentagonal cavity receiver insulated with ceramic fibers and containing a set of Inconel tubes (the solar reactor was developed at the Weizmann Institute of Science, Israel). The tubes were filled with a packed bed of Rh(2%)/Al_2O_3 catalyst [26]. Berman et al. [30] reported the experimental results on the development of a high-temperature steam reforming catalyst for "DIAPR-Kippod" volumetric-type reformer. The absorber in this reformer consisted of an array of ceramic pins. The authors have developed and tested alumina-supported Ru catalysts promoted with La and Mn oxides. The activity of the catalysts in SMR and CO_2-methane reforming reactions was measured in the temperature range 500–1100°C. The catalysts showed a stable operation at 1100°C for 100 h.

A series of studies on solar-driven SMR has been conducted at the Boreskov Institute of Catalysis in Russia. SMR was conducted under the direct illumination of a catalyst by concentrated light in the reactor–receiver with a transparent wall [31]. In this reactor, the specific rate of H_2 production and the specific power loading of the solar energy conversion device appeared to increase considerably compared to a conventional stainless-steel reactor, reaching 130 Ndm^3/h per 1 g of catalyst and 50–100 W/cm^3, respectively. It was proposed that the increase in the reaction rate is caused by a significant intensification of energy input into the catalyst bed due to the absorption of light directly by the catalyst granules. Yokota et al. [32] reported steam reforming of methane over Ni/Al_2O_3 catalyst using a solar simulator (Xe lamp). The reaction was conducted at $H_2O/CH_4 = 1/1$ ratio at the range of temperatures 650–950°C. At 850°C and molar ratio of $H_2O/CH_4 = 1/1$, methane conversion was in excess of 85% under atmospheric pressure.

2.2.2 Partial Oxidation of Hydrocarbons

The POx of hydrocarbons is another major route to hydrogen production on a commercial scale. In POx process, a fuel and oxygen (or air) are combined in proportions such that a fuel is converted into a mixture of H_2 and CO. There are several modifications of the POx process, depending on the composition of the process feed and the type of the reactor used. The overall process is exothermic due to a sufficient amount of oxygen added to a reagent stream. The POx process can be carried out catalytically or noncatalytically. The noncatalytic POx process operates at high temperatures (1100–1500°C), and it can utilize any possible carbonaceous feedstock including heavy residual oils (HROs) and coal. The catalytic process is carried out at a significantly lower range of temperatures (600–900°C) and, generally, uses light hydrocarbon fuels as a feedstock, for example, NG and naphtha. If pure oxygen is used in the process, it has to be produced and stored, which significantly adds to the cost of the system. In contrast, if the POx process uses air as an oxidizer, the effluent gas would be heavily diluted with nitrogen resulting in larger WGS reactors and gas purification units.

2.2.2.1 Partial Oxidation (Noncatalytic) of Heavy Residual Oil

A key advantage of a noncatalytic POx process is that it can utilize all kinds of petroleum-based feedstocks from light hydrocarbons to HROs and even petroleum coke. Heavy residues from refineries are the preferred feedstocks for the production of hydrogen for the following two reasons: (1) residual oils high in sulfur and heavy metals (e.g., Ni and V) are very difficult and costly to upgrade (e.g., by hydrogenation) and (2) there are environmental restrictions on their use as fuels (due to heavy SO_x and NO_x emissions). Hydrogen production by POx of heavy residues is an economically viable process, and has been commercially practiced for decades by Texaco and Shell [7]. Although the principal steps for both Texaco and Shell POx processes are similar, there are some differences in the burner construction, the reactor cooling, the waste heat boiler, and the soot recycle. Occasionally, gaseous hydrocarbons are also processed to hydrogen in POx processes. Typically, this is done if these gases cannot be used as a feedstock for the catalytic steam reforming process due to the high content of olefins and sulfur compounds, or if the process must operate with a variety of feedstocks from NG to oil fractions.

The major reaction during POx of sulfurous heavy oil fractions can be presented by the following generic chemical reaction:

$$C_mH_nS_p + m/2O_2 \rightarrow mCO + (n/2 - p)H_2 + pH_2S \quad \text{(exothermic)} \qquad (2.30)$$

Besides this exothermic reaction, a number of other (both exothermic and endothermic) reactions also occur in the POx reactor including the cracking reaction:

$$C_mH_nS_p \rightarrow (m - x)C + (n - y)H_2 + (m - z)CH_4 + C_kH_l + pH_2S \quad \text{(endothermic)} \quad (2.31)$$

incomplete carbon (coke) combustion:

$$C + 1/2O_2 \rightarrow CO \quad \text{(exothermic)} \quad (2.32)$$

and coke gasification:

$$C + H_2O \rightarrow H_2 + CO \quad \text{(endothermic)} \quad (2.33)$$

as well as WGS (reaction 2.6) and methane reforming (reaction 2.4) reactions.

A typical POx plant processes about 44 t/h vacuum residue with a sulfur content of up to 5 wt%, resulting in 116,000 m^3/h of 96 vol% pure hydrogen [7]. Figure 2.9 depicts the simplified block diagram of POx of high-sulfur content vacuum residue. Feed oil is preheated, atomized with steam, and partially burnt with oxygen in a special burner. Pressurized oxygen (>95 vol%), supplied by an air separation unit (ASU), is used in the POx process, which occurs primarily in the flame in an empty brick-lined reactor. The typical temperature and pressure ranges are 1250–1500°C and 3–12 MPa, respectively (the pressure is often adjusted to accommodate subsequent processes) [7]. At these operational conditions, the reactions kinetics are very fast, therefore, no catalyst is used in the process. About 2% of the carbon feed exits the reactor in the form of soot. Soot is extracted from the wash quench water streams and is fed back to the reactor with the feed oil. After quenching and scrubbing, the process gas contains a sufficient amount of water vapor for the catalytic conversion of CO to H_2 and CO_2 over a sulfur-proof Co–Mo catalyst. Nonreacted water is mostly removed from the stream in a condenser, and the condensate is recycled to the process. At this point, the process gas contains 30 vol% CO_2, about 1 vol% H_2S, and up to 0.5 vol% CO (the balance—hydrogen). (Owing to the net reducing conditions within the reactor, most of the feedstock sulfur is converted into H_2S.) The acid gases: CO_2 and H_2S are removed from the stream using a selective absorption (wash) process, in most cases, Rectisol, which operates with cooled methanol as a solvent. The acid gases are removed in

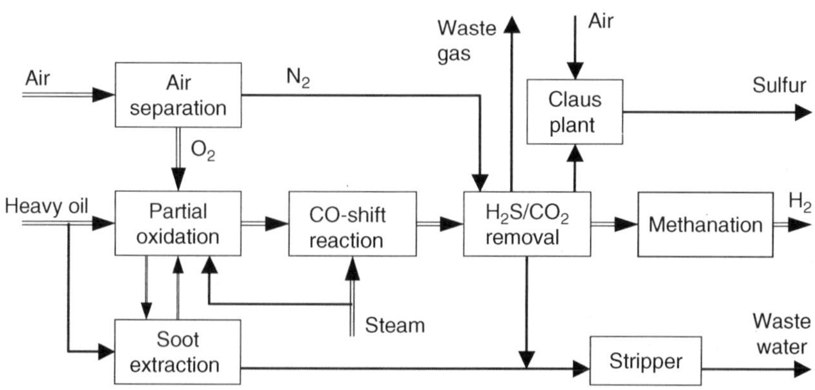

FIGURE 2.9
Simplified flow diagram of hydrogen production by POx of sulfurous HRO.

two steps: first, H_2S followed by CO_2 is washed out from the stream. The solvent is regenerated using nitrogen (from the ASU) as a stripping gas. H_2S-containing gas is treated at the Claus plant to produce elemental sulfur. At the final purification step, the residual CO (some tenths of a percent) is catalytically converted to CH_4 in the presence of H_2 by methanation reaction (which is the reverse of the methane reforming reaction):

$$CO + 3H_2 \rightarrow CH_4 + H_2O \tag{2.34}$$

The resulting water is removed by adsorption and the final product is dry hydrogen with the purity of about 98.6 vol% (the balance: CH_4, N_2, and Ar) at 5.0 MPa. The thermal efficiency of the process is 69.5% [4]. Major disadvantages of the POx process are the need for large quantities of pure oxygen (thus, requiring an expensive air separation plant), and the production of large volumes of CO_2 emissions (0.53–0.63 Nm^3 CO_2 per Nm^3 H_2 product).

2.2.2.2 *Catalytic Partial Oxidation*

The production of synthesis gas based on heterogeneous catalytic reactions using O_2 (air) as an oxidant is referred to as catalytic POx (CPO). Although the process is potentially able to process a wide range of hydrocarbon feedstocks, including heavy hydrocarbons, most of the information in the literature relates to the CPO of methane (or NG). The CPO of methane can be presented by the following equation:

$$CH_4 + 1/2O_2 \rightarrow CO + 2H_2 \qquad \Delta H^\circ = -38\ \text{kJ/mol} \tag{2.35}$$

Figure 2.10 provides a thermodynamic equilibrium molar fraction of the products of CPO of methane as a function of temperature. It is evident that at temperatures above 800°C, hydrogen and CO (in molar ratio of 2:1) are two major products of the reaction. The oxidant (oxygen or air) and the hydrocarbon feedstock (e.g., methane) are premixed in a mixer

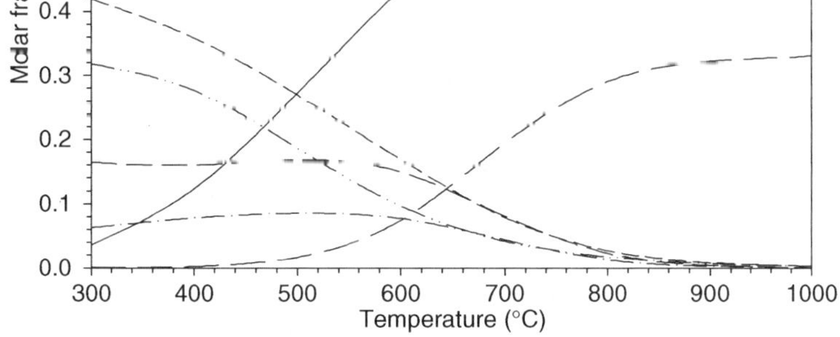

FIGURE 2.10
Thermodynamic equilibrium composition of POx products obtained from CH_4:O_2 = 2:1 (molar) mixture as a function of temperature. Pressure—atmospheric.

before the feed enters the catalyst bed. In the catalytic section, the hydrocarbons are oxidized by a number of heterogeneous reactions including partial and complete combustion, steam reforming, and WGS reactions. These reactions are typically at (or close to) equilibrium at the reactor exit. There may be a catalyst over-heating problem in the front layer of the catalyst bed due to the excessive combustion of the feedstock (the catalyst temperature could rise to 1000°C and higher). The reaction 2.35 should be conducted outside the explosive limits of the CH_4–O_2 mixture. For safety reasons, the inlet temperatures of the hydrocarbon and oxygen must be kept low. This, however, increases the oxygen consumption [33]. Both air and pure oxygen could be used in the CPO reaction. However, any practical implementation of air-fed CPO would require a complex unit for separation and purification of hydrogen.

2.2.2.2.1 *Catalysts*

The CPO of methane and other light hydrocarbons has been extensively researched for several decades. Most commonly used catalysts for the CPO reaction include refractory-supported Ni and noble metal–based (e.g., Rh, Pt, Pd, Ir, Ru, and Re) catalysts in the form of pellets, monoliths, and foams [34]. For example, Dissanayake et al. [35] studied 25 wt% Ni/Al_2O_3 catalysts in the temperature range 450–900°C. The authors observed CO selectivities approaching 95% at a nearly complete conversion of methane at temperatures above 700°C. Hickman and Schmidt [36] produced syngas with a H_2:CO molar ratio of 2:1 by the direct catalytic oxidation of methane over Pt or Rh catalyst surfaces in a monolith reactor with extremely short residence times (10 ms) using oxygen or air. The system's steady-state temperature was 850–1150°C (for inlet temperatures between 25°C and 460°C). It was shown that Rh exhibited higher selectivities to H_2 and higher CH_4 fractional conversions compared to Pt catalysts. The authors have developed a complete model that incorporates the adsorption, desorption, and surface reaction steps. In particular, it was assumed that the first step was dissociative adsorption of methane on the catalyst surface:

$$CH_4 \rightarrow C_s + 4H_s \tag{2.36}$$

Different barriers for $H + O \rightarrow OH$ reaction on Pt and Rh correlate well with the difference in H_2 selectivity.

The CPO of methane to syngas with high efficiency can be achieved at temperatures as low as 650°C using mixed perovskites with the general formula of $LaNi_{1-x}Fe_xO_3$ [37]. The authors demonstrated that for $x = 0.3$ and the reaction temperature of 800°C, CH_4 conversion and CO selectivity could reach 97.0% and 98.7%, respectively. In another work, CPO of methane was conducted in a spouted-bed reactor at 850–1000°C using the catalyst, 0.15 wt% Ni/1.7 wt% La/Al_2O_3, with a mean particle diameter of 150 and 670 μm [38]. CH_4 conversion of 80–90% and CO selectivity of 90%–95% were obtained at 950°C. A mathematical model has been developed to study the temperature profile along the catalyst bed in the reactor for CPO of methane [39]. Based on an elementary material balance and stoichiometry, the extent of each reaction for the selected set of methane oxidation reactions was calculated and used for the prediction of the overall reaction enthalpy in a volumetric reactor element according to the following equation:

$$Q = \sum_{i=1}^{N} \Delta H_r \frac{dF_i}{dz} = \sum_{i=1}^{N} \Delta H_r \Delta F_i \tag{2.37}$$

where Q is the heat generated at the specific axial element of the reactor, ΔH_r the standard heat of reaction for the reaction r, and ΔF_i the number of moles of component i reacted in the reaction r. The temperature gradient in the CPO reactor at the selected axial coordinate was calculated according to the following equation:

$$\frac{dT}{dz} = \frac{Q}{\sum_{i=1}^{N} F_i C_p^i} \tag{2.38}$$

where dT is the temperature change, dz the length of the differential element of the reactor, F_i the molar flow of the component i, and C_p^i its heat capacity in the stream. The authors concluded that the simulation and experimental results for CPO of methane are in good agreement. The model allows to perform sensitivity studies for a variety of reactor designs in the capacity range 1–200 Nm^3 syngas per hour.

CPO of hydrocarbons is a very complicated process, and mechanistic investigations of this reaction are still challenging. The complexity of the technology is compounded by the fact that, although the overall process is essentially adiabatic, it is characterized by high catalyst surface temperatures (hot spots), resulting in a lack of thermal equilibrium between the solid and the gaseous phases [40]. Because the process is conducted at high temperatures for extremely short contact times (i.e., within the domain of kinetic control), it is evident that heat and mass transfer play a decisive role in determining the process characteristics, temperature, and product concentration profiles, which may completely change the distribution of the reaction products. Owing to large temperature gradients in the reactors (hot spots), the process may be difficult to control in large-scale industrial operations. The problem can potentially be solved by using FBRs with a more uniform distribution of temperature in the catalyst bed. For a more detailed discussion of the CPO process, see Hu and Ruckenstein [34].

2.2.2.2.2 *Oxygen Membrane Reforming*

Although POx of hydrocarbons with air as an oxidant seems to be economically advantageous over the process using pure oxygen (because it avoids the very high capital cost of an ASU), the downstream process requirements negate the benefits of using air. This is due to the fact that in air-blown POx, syngas becomes heavily diluted with nitrogen, which requires a more complex and expensive gas separation unit. Also, the cost of compression of N_2-diluted syngas to pressures >2.0 MPa (which is typically required for a majority of downstream industrial processes) is high [34]. Thus, the main focus of recent research is aimed at the development of POx systems that can use air, but do not involve the dilution of syngas with N_2.

Recently, important advances have been made in syngas production technology using oxygen-permeable membranes (OPMs). The use of OPM allows for integrating the oxygen separation and POx processes in one reactor (thus avoiding a costly oxygen plant, which could reduce the cost of syngas production by 25–40% [34]). To be a viable membrane material for the POx process, it has to meet several stringent requirements, in particular, the membrane must be chemically and mechanically stable at elevated temperatures (up to 1000°C and higher) with one side of the membrane exposed to oxidizing atmosphere (air) and the other side to the reducing agents (H_2, CH_4, and CO).

It is well known that dense ceramic membranes made of the mixture of ionic and electron conductors are permeable to oxygen at elevated temperatures. For example, perovskite-type oxides (e.g., La–Sr–Fe–Co, Sr–Fe–Co, and Ba–Sr–Co–Fe-based mixed oxide systems) are good oxygen-permeable ceramics. Figure 2.11 depicts a conceptual design of an oxygen membrane reactor equipped with an OPM. A detail of the ceramic membrane wall

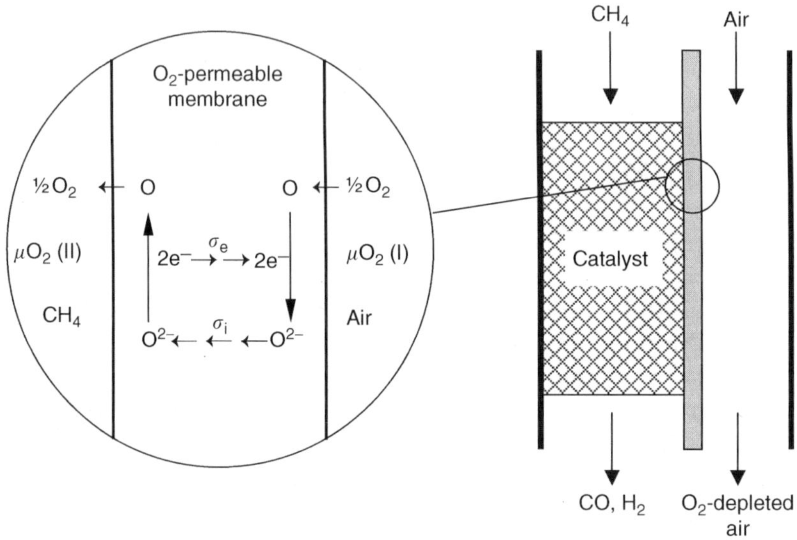

FIGURE 2.11
Schematics of an oxygen membrane reactor for catalytic POx of methane. A blown up section on the left-hand side shows the details of the ceramic membrane wall explaining the mechanism of oxygen permeation across the membrane. μ is the chemical potential of oxygen and σ_i and σ_e are the ionic and electronic components of the conductivity, respectively.

depicting the mechanism of oxygen permeation through the membrane is given on the left-hand side of Figure 2.11. The gradient of oxygen pressure (or its chemical potential, μ) is the driving force for oxygen transport across the ceramic membrane. Air is introduced at the right-hand side of the membrane, through which oxygen is transported in the form of O^{2-} ions to the other side of the ceramic membrane. At the left-hand side of the membrane, oxygen oxidizes methane to syngas in the presence of a catalyst (the catalyst could be in the form of powder, pellets, or immobilized onto the surface of the membrane). Advantageously, the O_2-permeable ceramic membranes operate efficiently at the same temperature range as the POx process (i.e., 800–900°C).

Extensive research has been carried out on the application of ceramic OPMs to POx and other processes requiring pure oxygen. For example, Balachandran et al. [41] studied the POx of methane in a tubular membrane reactor using La–Sr–Fe–Co–O OPM and Rh-based reforming catalyst. They reported methane conversion of >99% at 850°C with the tubes operating for over 1000 h. A membrane reactor containing O_2-permeable ceramic membrane made of $Ba_{0.5}Sr_{0.5}Co_{0.8}Fe_{0.2}O_{3-\delta}$ has been successfully applied to the POx of methane at 900°C [42].

Notwithstanding the extensive research and significant progress achieved in the O_2-membrane POx area, there are significant technical challenges to overcome [34]. Although the mixed conducting membranes demonstrate relatively high oxygen fluxes, more research is needed to improve their mechanical and chemical stability. Moreover, integrating the ceramic membranes into large-scale commercial units will be challenging because they easily break and it would be very difficult to interconnect them with other materials, for example, steel tubes. Furthermore, these materials are very difficult to manufacture and process without creating microcracks, voids, and fractures.

2.2.2.3 *Metal Oxide–Mediated Partial Oxidation*

Metal oxide–mediated oxidation of methane using air as a primary oxidant is an alternative way to produce N_2-free syngas. The concept is based on the oxidation of methane by transition metal oxides in high-oxidation state yielding syngas and corresponding metal oxide in a low-oxidation state:

$$M^{n+}O_x + CH_4 \rightarrow M^{(n-a)+}O_y + CO + 2H_2 \quad (2.39)$$

where $x, y, a \geq 1$.

In the second stage of the process, the metal oxide in a low-oxidation state is reoxidized by air to the original metal oxide in a high-oxidation state:

$$M^{(n-a)+}O_y + O_2/N_2 \rightarrow M^{n+}O_x + N_2 \quad (2.40)$$

Because both stages are carried out in separate reactors, the resulting syngas is not diluted with N_2. A number of transition metal oxides can be used to accomplish POx of methane according to Equation 2.39, most prominently, copper oxide. Lewis patented the process for the production of syngas by a controlled reaction between the hydrocarbons and the metal oxides [43]. It was demonstrated that the copper oxide, CuO, oxidizes methane to syngas with high selectivity and conversion (>90%) at about 1090–1300°C; the resulting reduced form of the oxide (Cu_2O) can be efficiently reoxidized to its original oxidized form by air. The process utilized CuO (3–30 wt%) supported on a silica gel. According to the patented process, continuous POx of methane can be carried out in two FBRs: an oxidation reactor (where methane is oxidized to syngas) and a regenerator (where reoxidation of a reduced form of the oxide takes place). CuO particles are circulating between these two apparatuses in a fluidized state. The composition of the reformer gas is as follows (vol%, on dry basis): H_2—58.3, CO—29.9, CO_2—3.5, CH_4—8.3. Despite some attractive features, the process has not been implemented on a large scale.

2.2.3 Autothermal Reforming

2.2.3.1 *Autothermal Reforming Reactor*

The ATR process is a combination of SMR and POx technologies in that the thermal energy for the production of syngas is provided by POx (combustion) of the hydrocarbon feedstock in an adiabatic reactor. The ATR process has been used to produce hydrogen and CO-rich synthesis gas for decades. In the 1950s and 1960s, autothermal reformers were used to produce syngas for ammonia and methanol production [33]. The autothermal reformer shown in Figure 2.12 consists of three zones: combustion, thermal, and catalytic zones (the schematics of the reformer is based on the description provided in Ref. 33). The feed is introduced to the combustion zone (which is, essentially, a turbulent diffusion flame with the temperature as high as 3000°C at the core) and mixed intensively with steam and a substoichiometric amount of oxygen or air. The resulting combustion reaction can be presented as follows:

$$CH_4 + 3/2O_2 \rightarrow CO + 2H_2O \quad \Delta H^\circ = -519 \text{ kJ/mol} \quad (2.41)$$

In the thermal zone, above the catalyst bed, further conversion occurs by homogeneous gas-phase reactions. The main reactions in the thermal zone are homogeneous gas-phase

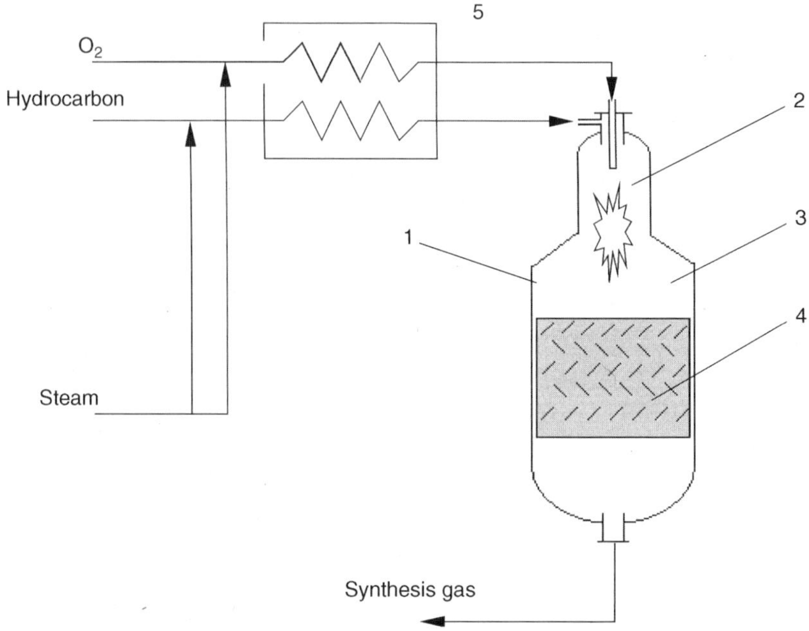

FIGURE 2.12
Schematic diagram of the reactor for ATR of hydrocarbons. 1 = Autothermal reformer, 2 = burner section, 3 = combustion chamber, 4 = catalyst, and 5 = heater.

SMR (reaction 2.4) and shift (reaction 2.6) reactions. Reactions between N_2 and hydrocarbon radicals leading to the formation of such by-products as NH_3 and HCN can also take place in the thermal zone. By proper adjustment of O_2/CH_4 and H_2O/CH_4 ratios, the partial combustion in the thermal zone provides the heat for the subsequent endothermic steam reforming reaction taking place in the catalytic zone [40]. Thus, simplistically, ATR of methane at temperature T can be represented as follows:

$$CH_4 + xO_2 + yH_2O \rightarrow \text{syngas} \qquad \Delta H_T \approx 0 \tag{2.42}$$

In the catalytic zone, a reforming catalyst bed 4 (typically, alumina-supported Ni catalyst) carries out steam reforming of hydrocarbons at an operating pressure in the range 1.8–6.7 MPa. Owing to the high endothermicity of the steam reforming reaction, the temperature decreases from typically 1200°C to 1300°C at the inlet of the catalyst bed to about 1000°C at the exit of the catalyst bed. Thus, the product gas composition is fixed thermodynamically through the pressure, exit temperature, and O_2/CH_4 and H_2O/CH_4 ratios. The process has a high CO-shift activity leading to a considerable amount of CO_2 in the product gas. Oxygen consumption is high due to the simultaneous adiabatic steam reforming reaction (usually, the O_2 to carbon ratio is in the order of 0.60–0.65 [44]). Because ATR uses less oxygen than POx (per unit of H_2 produced), the economics are less sensitive to the price of oxygen [3]. If the product hydrogen is intended for ammonia (NH_3) production, an air feed can be used instead of pure oxygen. ATR requires no external fuel while offering some flexibility in feedstock (although ATR does not lend itself to heavy hydrocarbon feedstocks). One of the advantages of the ATR process is that the pressure can be increased compared to conventional steam reforming, resulting in lower-energy requirement for syngas compression.

The key elements of the ATR technology are the burner and the catalyst. The burner provides proper mixing of the feed streams, and the fuel-rich combustion is taking place as a turbulent diffusion flame. Intensive mixing is essential to avoid soot formation (by C_2 radicals and polyaromatics as soot precursors) [33]. The catalyst equilibrates the synthesis gas and destroys soot precursors. The catalyst particle size and shape are optimized to achieve high activity and low pressure drop and minimize the reactor size. Generally, an ATR catalyst is robust with a high thermal stability, however, some deactivation may occur, mainly due to sintering, sulfur poisoning, or fouling. A careful design of the burner, the combustion chamber, and the catalyst bed is of utmost importance to ensure efficient and safe operation of the ATR unit. The design of the unit is facilitated by the reactor modeling using chemical process simulators and computational fluid dynamics (CFD) tools.

2.2.3.2 *Combined Reforming*

If the objective of the process is to control the H_2/CO ratio in the synthesis gas, or to increase the pressure and at the same time to reduce the consumption of oxygen, the combination of steam reforming and ATR reactors might be advantageous. In this case, the first reactor is a relatively small steam reformer from which the reformate gas goes to a secondary ATR reactor. Combined reforming (also called "secondary reforming") is the dominating process for the manufacture of synthesis gas for NH_3 production from NG and naphtha. The reforming section of the process is shown in Figure 2.13. In this process, NG is desulfurized, mixed with steam (steam/carbon ratio is 2.5:3.5) and passed to a fired tubular reformer (the "primary reformer") 1. The product gas from the primary reformer reacts with air in the secondary reformer 3 to produce the raw syngas, which is further processed by CO-shift conversion; removal of CO_2 and methanation produces a

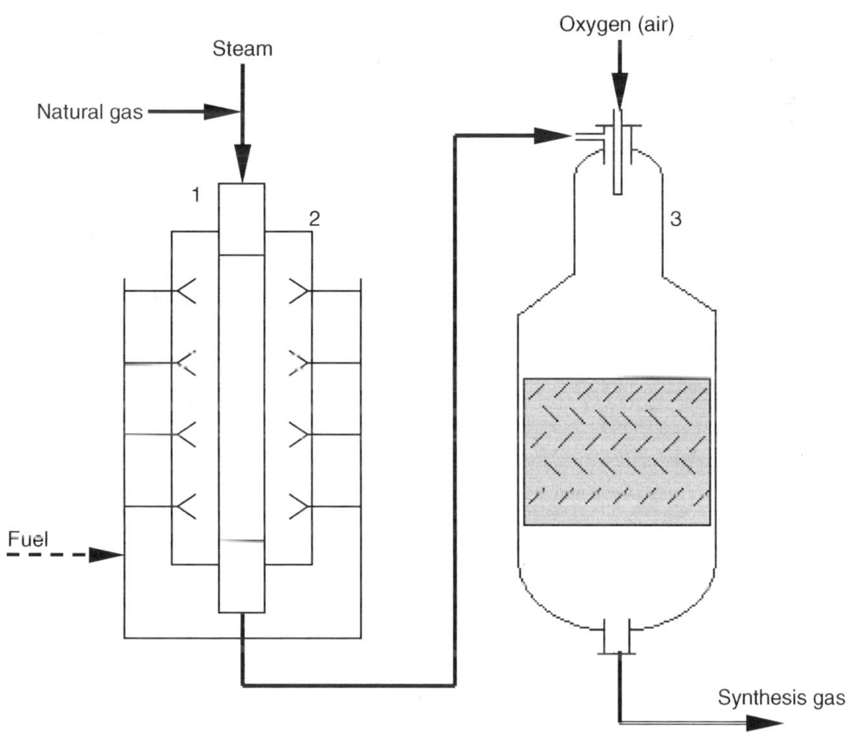

FIGURE 2.13
Schematics of the reformer section of combined reforming of NG. 1 = Fired tubular reformer, 2 = furnace, and 3 = autothermal reformer.

N_2–H_2 mixture (with small amounts of Ar and CH_4) [33]. The amount of air added to the secondary reformer is adjusted to give the desirable H_2/N_2 ratio (which is close to 3 for the NH_3 synthesis). The secondary reformer is similar to the autothermal reformer described in the previous section. The pressure at the outlet of the secondary reformer is in the range 2.5–3.5 MPa. The outlet temperatures from the primary and secondary reformers are 750–850°C and 950–1050°C, respectively.

When H_2 is not intended for ammonia synthesis (i.e., N_2 is not a desirable component of the syngas), pure oxygen may be used in the secondary reformer. This is the case when syngas is used for subsequent production of methanol, dimethylether, or is used in Fischer–Tropsch (FT) synthesis. A major advantage of combined reforming is that the pressure can be increased up to 3.5–4.5 MPa due to the relatively low exit temperature of the primary reformer. This would reduce the size of the compressor almost in half compared to the conventional steam reforming process. NO_x emissions from the combined reforming process would be significantly less than from a conventional SMR process (because H_2 is not a part of the fuel combusted in the furnace) [44].

2.2.4 Carbon Dioxide Reforming of Methane

The CO_2 reforming of methane is an alternative to SMR and POx processes, where CO_2 plays the role of an oxidant. Sometimes the process is also called *stoichiometric reforming*, but more often it is referred to as *dry reforming*. Like SMR, it is a highly endothermic process requiring high operational temperatures of 800–1000°C. Owing to the presence of CO_2 in the feedstock, the process produces synthesis gas with high CO/H_2 ratio (1:1) according to the following equation:

$$CH_4 + CO_2 \rightarrow 2CO + 2H_2 \qquad \Delta H^\circ = 247 \text{ kJ/mol} \tag{2.43}$$

Figure 2.14 depicts the thermodynamic equilibrium data related to CO_2 reforming of methane at atmospheric pressure. It is noteworthy that at temperatures below 800°C, elemental

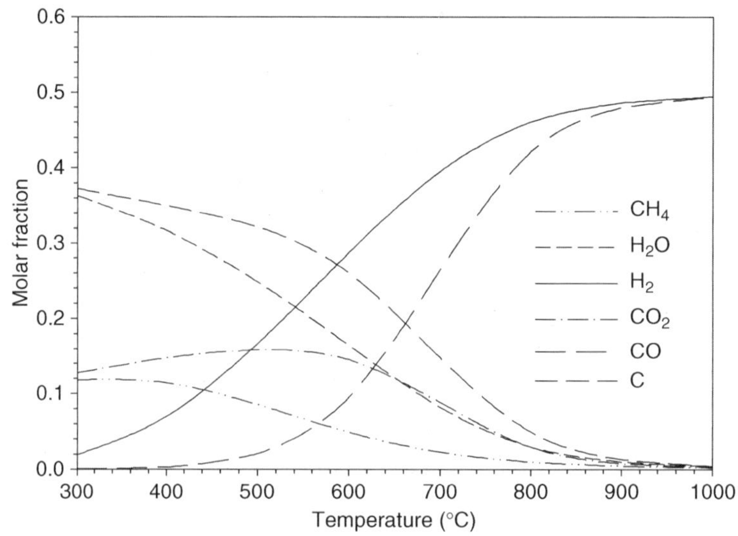

FIGURE 2.14
Temperature dependence of thermodynamic equilibrium composition of the products obtained from CH_4:CO_2 = 1:1 (molar) mixture at atmospheric pressure.

carbon is one of the major components of the equilibrium mix. At higher temperatures (>800°C), the carbon molar fraction in the mix dramatically drops, and the H_2 and CO molar fractions become predominant and reach the plateau.

With the growing concerns about negative environmental impact of CO_2 (i.e., global warming), the CO_2 reforming technology is getting more attention increasingly. It should be noted, however, that if the objective were to produce H_2 only, this process would not result in the overall reduction of CO_2 emission compared to SMR (because CO_2 is produced in the WGS stage of the process). However, if the process targets the production of syngas with relatively high content of CO (e.g., for FT synthesis), then this approach can be conducive to increasing CO/H_2 ratio in the syngas. In this case, CO_2 from the feedstock will be sequestered in the form of synthetic fuels (e.g., FT gasoline or diesel) or oxygenated compounds (alcohols, esters, etc.). Another advantage of this approach is related to the fact that pure CO_2 is produced as a coproduct of the methane reforming process [3].

Practical implementation of CO_2 reforming of methane faces several key challenges, technically and economically. At the preferred (from the economical viewpoint) pressure of the syngas plant (2.0–4.0 MPa), CO_2 reforming will result in the noncomplete conversion of methane due to thermodynamic limitations [17]. Furthermore, the process economy strongly depends on the pressure and the cost of CO_2 available. However, the most serious problem hindering the practical application of CO_2 reforming is the deactivation of metal catalysts due to the deposition of carbon (or coke). Carbon formation can be attributed to two reactions: methane decomposition and CO disproportionation (Equation 2.5). CO disproportionation is an exothermic reaction; it is favored by temperatures below 700°C and high pressures. From a practical viewpoint, it is preferable to operate CO_2 reforming of methane at moderate temperatures and with the CH_4:CO_2 ratio close to unity, which would require a catalyst that kinetically inhibits the carbon formation under conditions that are thermodynamically favorable for carbon deposition [34]. Iron-, cobalt-, and nickel-based catalysts are particularly active in methane decomposition and CO disproportionation reaction, and noticeable deposition of carbon on the surface of these catalysts would occur at temperatures as low as 350°C. The form of carbon deposited on metal surfaces is controlled by the reaction temperature: in the lower temperature range 350–600°C, amorphous and filamentous carbons are the predominant form of carbon, whereas an ordered graphitic structure dominates at the temperatures above 700°C [34].

Most of the reported research on CO_2 reforming of methane relates to Ni-based catalysts, because Ni exhibits high catalytic activity (comparable to that of noble metals) at lesser cost. However, Ni catalysts are prone to carbon deposition and deactivation as discussed earlier. Therefore, much research has been conducted to improve the resistance of Ni catalysts to deactivation and eliminate carbon deposition during the process. Comprehensive reviews on the topic of CO_2 reforming of methane using Ni-based catalysts and other nonprecious metal catalysts were recently published by Hu and Ruckenstein [34] and Bradford and Vannice [45]. The surface crystallographic structure and the surface acidity are two main factors affecting the carbon deposition. It was found, for example, that Ni(100) and Ni(110) surfaces are more catalytically active in methane decomposition and, hence, prone to carbon deposition than the Ni(111) surface [46]. By controlling the size of the ensembles of metal atoms on the catalyst surface (since larger ensembles favor carbon formation), it would be possible to inhibit carbon deposition. Coke deposition on the commercial Ni catalyst (used in SMR) can be minimized by selectively passivating catalytically active sites (e.g., by sulfiding the catalyst). The suppression of carbon deposition by sulfur passivation is attributed to the strong adsorption of sulfur that controls the size of active metal ensembles (i.e., it eliminates larger ensembles that favor carbon deposition). Haldor Topsoe has been operating its SPARG process, which used a partially sulfided Ni catalyst [12]. However, the addition of sulfur to the catalyst reduces the catalyst activity and overall process throughput [3].

Another important factor affecting carbon deposition is the catalyst surface basicity. In particular, it was demonstrated that carbon formation can be diminished or even suppressed when the metal is supported on a metal oxide carrier with a strong Lewis basicity [47]. This effect can be attributed to the fact that high Lewis basicity of the support enhances the CO_2 chemisorption on the catalyst surface resulting in the removal of carbon (by surface gasification reactions). According to Rostrup-Nielsen and Hansen [12], the amount of carbon deposited on the metal catalysts decreases in the following order:

$$\mathrm{Ni} \gg \mathrm{Rh} > \mathrm{Ir} \approx \mathrm{Ru} > \mathrm{Pt} \approx \mathrm{Pd} \qquad (\text{at } 500°\text{C})$$

$$\mathrm{Ni} > \mathrm{Pd} \approx \mathrm{Rh} > \mathrm{Ir} > \mathrm{Pt} \gg \mathrm{Ru} \qquad (\text{at } 650°\text{C})$$

The catalysts supports and promoters have a significant effect on the rate of carbon deposition. In particular, ZrO_2 has been widely used as a support for Pt because of the lower rate of carbon deposition compared to other supports [48]. The authors demonstrated the following order of the carbon formation rate:

$$\mathrm{Pt/Al_2O_3} \gg \mathrm{Pt/TiO_2} > \mathrm{Pt/ZrO_2}$$

It was shown that vanadium oxide had a promoting effect on the Rh/SiO_2 catalyst by decreasing the rate of carbon deposition, while enhancing its catalytic activity [45]. The effect was attributed to the formation of VO_x overlayer on the Rh surface, decreasing the size of the ensembles of Rh atoms and, thus, hindering coke formation. It was found that SiO_2- and ZrO_2-supported bimetallic Pt–Au, Pt–Sn catalysts were less prone to carbon deposition during CO_2 reforming of methane than the respective monometallic Pt catalysts [49]. Although noble metal-based catalysts are less sensitive to deactivation by carbon deposition compared to nonprecious metal catalysts, their high cost hinders their large-scale application. A relatively simple mechanism for CO_2 reforming of methane has been suggested by Lercher et al. [50]:

$$\mathrm{CH_4} + * \rightarrow \mathrm{C}* + 2\mathrm{H_2} \tag{2.44}$$

$$\mathrm{CO_2} + 2* \leftrightarrows \mathrm{CO}* + \mathrm{O}* \tag{2.45}$$

$$\mathrm{C}* + \mathrm{O}* \leftrightarrows \mathrm{CO}* + * \tag{2.46}$$

$$2\mathrm{CO}* \leftrightarrows 2\mathrm{CO} + 2* \tag{2.47}$$

where $*$ is an active site.

This sequence of steps was derived from a series of pulsed adsorption experiments, in which CH_4 was decomposed stoichiometrically to carbon and H_2, whereas CO_2 was shown to react stoichiometrically with the surface carbon yielding CO.

The addition of steam to the CH_4/CO_2 feedstock to avoid excessive carbon formation is a widely used technique in practical systems [3]. The resulting CO_2-steam gasification of methane process can be described by the following chemical equation:

$$2\mathrm{CH_4} + \mathrm{CO_2} + \mathrm{H_2O} \rightarrow 3\mathrm{CO} + 5\mathrm{H_2} \tag{2.48}$$

H_2:CO ratio in the resulting syngas is about 1.7. However, due to the relatively high content of CO in the syngas, carbon deposition may still be a problem, especially for Ni-based catalysts widely used for steam reforming.

2.2.5 Steam–Iron Process

The production of hydrogen by the SIP is one of the oldest commercial methods of H_2 manufacturing. SIP was practiced from the early 1900s well into 1930s for supplying small quantities (less than about 1000 m^3/h) of pure hydrogen to some industries (e.g., oil hardening) [10]. Later, SIP was supplanted by the more efficient and economical SMR process, which produced hydrogen at larger volumes and higher pressure. The SIP produces high-purity hydrogen by separating the hydrogen production and feedstock oxidation steps using iron oxide reduction–oxidation regenerative system. Thus, it does not require WGS and CO_2 removal stages, which significantly simplifies the process. Recently, there has been a renewed interest in the process, particularly for small-scale applications due to its simplicity, the purity of hydrogen obtained, feedstock flexibility, and other factors.

Simplistically, SIP is based on two subsequent reactions: in the first step, iron oxide (magnetite, Fe_3O_4) is reduced to wustite, FeO, or even to metallic iron, by a reducing gas (e.g., hydrocarbons and syngas), and in the second step, wustite (or metallic iron) reacts with steam producing hydrogen gas and the original form of iron oxide (magnetite). In reality, the process is much more complex due to the presence of various reduced forms of iron oxides and the simultaneous occurrence of a great number of feedstock decomposition/oxidation reactions.

2.2.5.1 Steam–Iron Process Using Methane as Feedstock

Hydrogen production by SIP can be accomplished through direct and indirect employment of hydrocarbon feedstocks (e.g., NG). In the *direct employment method,* iron oxide directly reacts with methane or other hydrocarbons to produce the reduced form of iron oxide and methane oxidation products, according to the following generic reaction:

$$Fe_3O_4 + CH_4 \rightarrow 3/(1-y)Fe_{1-y}O\ (3Fe) + CO_x,\ H_2O \tag{2.49}$$

where "y" relates to the cation vacancy in wustite, ranging from 0.05 to 0.17 [10].

The direct reduction of magnetite with methane has been reported in a number of papers [51,52]. Reaction 2.49 is complicated by the fact that the dissociation of CH_4 to carbon and hydrogen is thermodynamically and kinetically favorable (because Fe catalyzes methane decomposition) at temperatures above about 600°C. Carbon deposited on the iron oxide surface can act as a reducing agent or can directly react with iron oxide and form iron carbide (cementite, Fe_3C). Further, complications arise from the possibility of formation of C_2 hydrocarbons (by oxidative coupling reaction) and simultaneous occurrence of a number of competing reactions involving CO_2, methane, carbon, and hydrogen.

Recently, Takenaka et al. [53] studied hydrogen production through direct reaction of methane with transition metal–doped iron oxides as a way of producing pure hydrogen by SIP. The authors showed that the reduction of metal-doped Fe_3O_4 with methane at a temperature of 750°C resulted in the formation of metallic iron and a gaseous mixture consisting of H_2, CO, and CO_2. It was reported that the Ni–Cr-doped iron oxides showed excellent resistance to sintering and exhibited enhanced performance in the reaction with methane. The Ni–Cr–FeO_x redox system repeatedly produced pure H_2 with high reproducibility by the reduction with methane and the subsequent oxidation with steam. Interestingly, the

Ni species in Ni–Cr–FeO$_x$ were present as Ni–Fe alloys after the reduction with methane and as Ni metal crystallites after oxidation with steam. Kodama et al. [54] demonstrated the production of syngas by oxidizing methane with $Ni_{0.39}Fe_{2.61}O_4/ZrO_2$ at 800–900°C as follows:

$$CH_4 + Ni_{0.39}Fe_{2.61}O_4 \rightarrow 1.2H_2 + 0.6CO + 0.4CO_2 + 0.8H_2O + \text{(reduced Ni–Fe alloy)} \quad (2.50)$$

Reduced Ni–Fe alloy was oxidized back to the original ferrite in the presence of steam.

Indirect employment of hydrocarbons involves a preliminary conversion (e.g., by steam reforming) of hydrocarbons to syngas followed by the reduction of iron oxides with H_2 and CO components of the syngas. The following reactions occur during the reduction of magnetite by H_2 and CO (heat of the reactions relates to a mole of iron oxide):

$$Fe_3O_4 + 0.83CO \rightleftharpoons 3.17Fe_{0.95}O + 0.83CO_2 \qquad \Delta H_{1100\text{ K}} = -22.4 \text{ kJ/mol} \quad (2.51)$$

$$Fe_{0.95}O + CO \rightleftharpoons 0.95Fe + CO_2 \qquad \Delta H_{1100\text{ K}} = -17.4 \text{ kJ/mol} \quad (2.52)$$

$$Fe_3O_4 + 0.83H_2 \rightleftharpoons 3.17Fe_{0.95}O + 0.83(H_2O)_g \qquad \Delta H_{1100\text{ K}} = 46.1 \text{ kJ/mol} \quad (2.53)$$

$$Fe_{0.95}O + H_2 \rightleftharpoons 0.95Fe + (H_2O)_g \qquad \Delta H_{1100\text{ K}} = 16.3 \text{ kJ/mol} \quad (2.54)$$

The subsequent reaction of the reduced iron oxide with steam regenerates the original oxidized form of iron oxide (i.e., magnetite) and yields pure hydrogen (mixed with unreacted steam) as follows:

$$0.95Fe + (H_2O)_g \rightleftharpoons Fe_{0.95}O + H_2 \qquad \Delta H_{1100\text{ K}} = -16.3 \text{ kJ/mol} \quad (2.55)$$

$$3.17Fe_{0.95}O + 0.83(H_2O)_g \rightleftharpoons Fe_3O_4 + 0.83H_2 \qquad \Delta H_{1100\text{ K}} = -46.1 \text{ kJ/mol} \quad (2.56)$$

It should be noted that the choice of $Fe_{0.95}O$ (wustite) rather than FeO in the preceding reactions is not arbitrary [10]. The steam–iron reaction would produce very little hydrogen at these temperatures if magnetite were reduced to FeO instead of $Fe_{0.95}O$. Hacker et al. [55] determined that the activation energy of magnetite reduction with H_2 and CO is equal to 95 and 98 kJ/mol, respectively. The energy of activation of wustite oxidation with steam was found to be 29 kJ/mol.

Figure 2.15a and 2.15b depicts the cyclic and continuous reactors, respectively, for hydrogen production by SIP through the indirect employment of a hydrocarbon feedstock (the sketches are based on the process description presented in Ref. 10). In the cyclic reactor (Figure 2.15a), syngas enters the bottom of the reactor where it gains additional heat by contacting a hot refractory material, and subsequently, releases this thermal energy to provide the heat input needed for the magnetite reduction reactions. Air is then mixed with the remaining H_2–CO, raising the temperature of the gas as it leaves the iron oxide bed, and heating the upper refractory material. The flow of the syngas is then stopped, and after the bed is purged, steam is introduced into the top of the reactor. The upper (hot) refractory material superheats the steam, which then oxidizes the $Fe_{0.95}O$ and Fe (reactions 2.55 and 2.56). The hot steam–H_2 mixture then reheats the bottom refractory material. The use of several beds coupled with a gasholder allows for a continuous flow of hydrogen.

Figure 2.15b depicts the continuous flow reactor (designed by the Institute of Gas Technology), consisting of four fluidized beds of iron oxide particles within a single pressurized

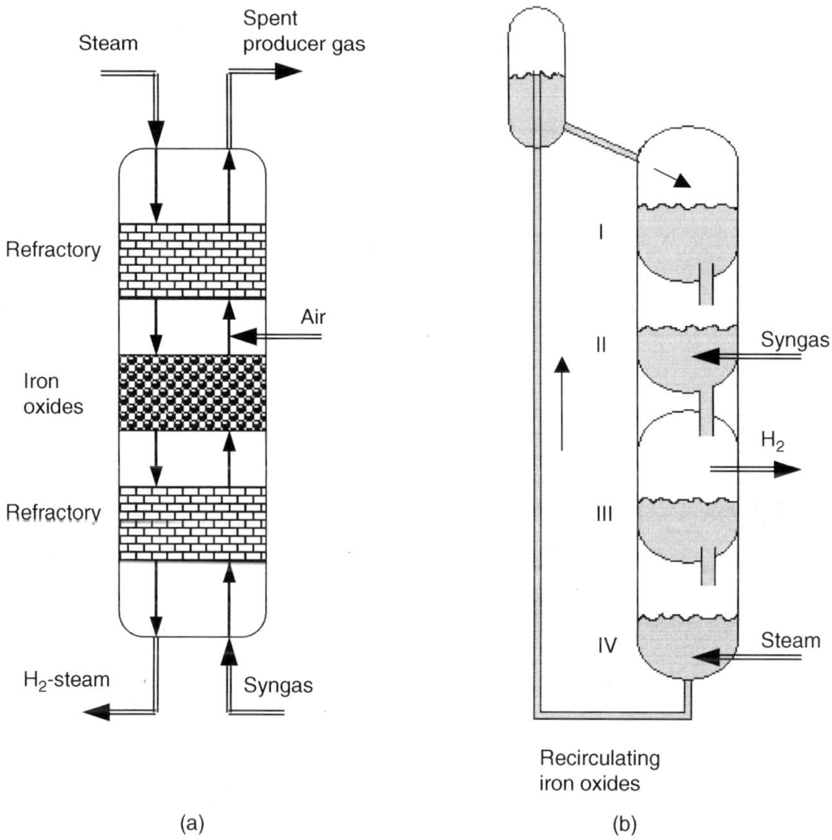

FIGURE 2.15
Schematics of cyclic (a) and continuous (b) reactor for hydrogen production by steam–iron process. I–IV denote the reactor zones.

vessel operating at a pressure of 6.7 MPa and a temperature of about 770°C [10]. Beds I and II are the upper and lower reducers, respectively, where the reducing reactions 2.51 through 2.54 occur. Beds III and IV are the two oxidizers, where H_2 is produced by the reactions of steam with reduced iron oxides (Equations 2.55 and 2.56). At the preceding operational conditions, a steam–H_2 mixture with the ratio of 60:40 can be produced. The advantages of SIP are as follows: (a) H_2 with the purity higher that 99% is produced after the water is condensed, (b) because N_2-diluted syngas can be used as a reducer for iron oxides, air rather than oxygen can be used (thus, eliminating the need for an expensive O_2 plant), (c) the H_2–steam mixture contains considerable thermal energy, thus electric power can be generated (e.g., through a turbine), and (d) feedstock flexibility (since practically any carbon-containing feedstock can be used).

Recently, SIP was modified for hydrogen storage and fuel cell (FC) applications [51]. The authors [55] proposed to store the energy of syngas in the form of sponge iron and produce hydrogen on demand by reacting sponge iron with steam. The energy density of the system was estimated at 575 kJ(H_2)/kg (sponge iron pellet). The sponge iron is oxidized in a reactor to provide high-purity hydrogen to a FC, whereas already depleted beds are regenerated (reduced) using syngas. Otsuka et al. [56] proposed a method for the storage and production of hydrogen from methane mediated by indium and iron oxides. First, methane is decomposed to H_2 and carbon over Ni catalysts, and H_2 reduces metal oxides to

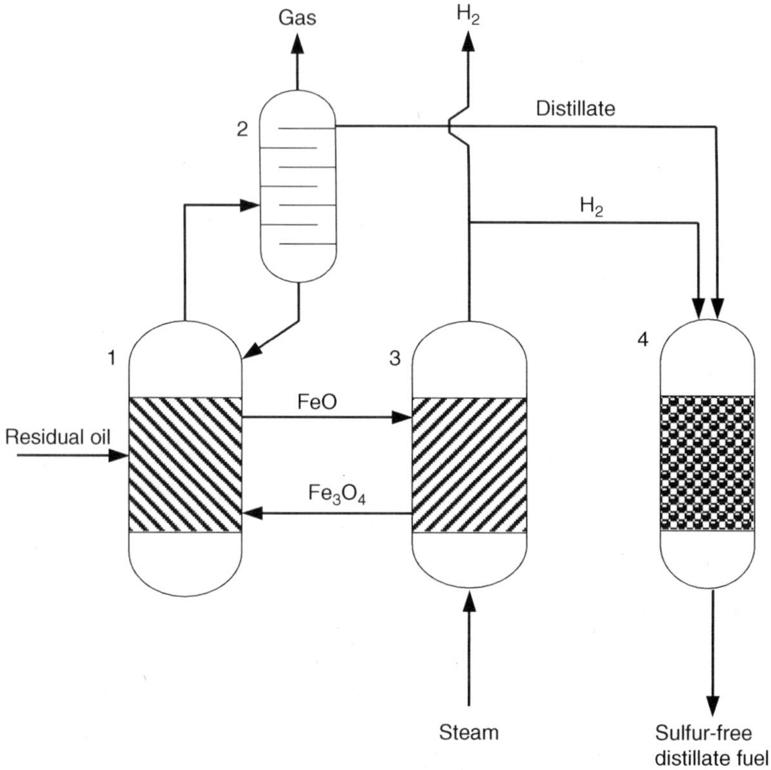

FIGURE 2.16
Flow diagram of the process for hydrogen and distillate fuel production from residual oil using iron oxides and steam. 1 = Cracking reactor, 2 = distillation column, 3 = hydrogen generator, and 4 = hydrodesulfurization reactor.

reduced oxides. Then, H_2 is produced at 400°C by the reaction of reduced In and Fe oxides with steam.

2.2.5.2 Hydrogen Production from Residual Oil Using Steam–Iron System

One of the major advantages of the steam–iron process relates to high feedstock flexibility: it can be applied not only to gaseous, but also to solid (coal, biomass) and liquid hydrocarbon feedstocks. In this section, we will consider the application of SIP to HRO. Although HRO could be converted to hydrogen through a noncatalytic POx process (as discussed in Section 2.2.2.1), the application of the SIP concept to HRO processing offers certain advantages over POx technology, namely, it produces distillate fuels along with H_2. Figure 2.16 illustrates the concept by providing the simplified schematics of HRO cracking with the simultaneous production of H_2 by SIP. HRO is fed to a cracking reactor 1 where it is cracked over the magnetite form of the catalyst with the production of gaseous and liquid products and coke deposited on the catalyst surface. Simultaneously, the magnetite form of the catalyst is reduced to wustite (and possibly metallic iron) by gaseous products of the cracking and coke as follows:

$$Fe_3O_4 + C_nH_m \rightarrow 3FeO\ (Fe) + xH_2 + yCO + (HC)_{gas} + (HC)_{liquid} + C_{coke} \quad (2.57)$$

$$Fe_3O_4 + C_{coke} \rightarrow 3FeO\ (Fe) + CO\ (CO_2) \quad (2.58)$$

where C_nH_m is HRO, HC is a mix of hydrocarbons (gaseous or liquid), and C_{coke} is the petroleum coke.

The reduced form of the catalyst is directed to a hydrogen generator 3, where it is oxidized by steam producing pure hydrogen and the original magnetite form of the catalyst, which is recycled back to the cracking reactor. Liquid products of the cracking enter the distillation column 2, where light and middle distillate fuels are recovered from the liquid products of HRO cracking. Because HRO typically contains a large amount of sulfurous compounds (up to 5 wt% and even higher), a certain percentage of the sulfur ends up in the distillate fuels. Hydrodesulfurization of the distillate fuels takes place in the hydrogenation reactor 4, which utilizes the hydrogen produced in the hydrogen generator 3.

In practice, the system is very complex because iron oxides are sulfided in the presence of sulfurous HRO. Furthermore, a certain amount of coke remains on the catalyst surface when it enters the hydrogen generator, thus, contaminating hydrogen with carbon oxides. In the early 1980s, the research groups in Japan and U.S.S.R. studied this process as a means of increasing the yield of distillate fuels from heavy petroleum feedstocks [57,58]. In particular, the authors reported [58] on the development of a process consisting of the following steps: (1) cracking of HRO (vacuum residual oil) at 450–600°C over magnetite-based catalyst with the formation of cracking products (gas, liquid, and coke), (2) partial combustion of coke to CO, which reduces magnetite to the wustite form of the catalyst at 800–850°C (this also provides heat input for the process), and (3) generation of hydrogen by the reaction of steam with wustite form of the catalyst at 650°C [59]. During HRO cracking, sulfur compounds partially sulfide the iron oxide catalyst to FeS. In the presence of steam in the hydrogen generator, FeS is converted to magnetite and releases H_2 and H_2S as follows:

$$3FeS + 4H_2O \rightarrow Fe_3O_4 + 3H_2S + H_2 \tag{2.59}$$

H_2S can be scrubbed from hydrogen gas by off-the-shelf technologies. The product distribution of HRO cracking at 600°C was as follows (wt%): gas–25.6, fraction 60–200°C–8.1, fraction 200–350°C–6.1, fraction >350°C–41.5, coke–18.7. The yield of the hydrogen gas with the purity of 98% reached up to 400 L/kg HRO.

Fukase and Suzuka [60] reported on the development of the process involving HRO cracking with generation of H_2 in the presence of iron oxides. The process consists of three major steps taking place in FBRs:

1. In the cracker, HRO is cracked over iron oxide catalyst at 540°C in the presence of steam with the production of gas, liquid products, coke (12–16%), and hydrogen (by reaction of wustite with steam)
2. In the regenerator (830°C), the coke deposited on the catalyst is partially oxidized to CO, thus reducing magnetite to wustite. Sulfur dioxide produced by POx of sulfurous coke is fixed into the catalyst as follows:

$$FeO + SO_2 + 3CO \rightarrow FeS + 3CO_2 \tag{2.60}$$

3. In the desulfurizer, FeS is converted to magnetite by roasting with air:

$$3FeS + 5O_2 \rightarrow Fe_3O_4 + 3SO_2 \tag{2.61}$$

It was found that the hydrogen-producing activity of catalysts declined during consecutive reduction and oxidation cycles. The authors concluded that the catalyst deactivation could be prevented by careful balancing of the stoichiometry of the reduction oxidation reactions. The amount of H_2 produced was estimated at 210 Nm^3/kL of vacuum HRO [57].

This amount of hydrogen is about twice the amount required for the desulfurization of the cracked oil obtained at the cracking stage of the process.

2.2.6 Plasma Reforming

There is a growing interest in electricity-assisted generation of syngas and hydrogen. In these processes, electricity alone or a mixed source of energy (i.e., electrical and chemical) can be used to provide the syngas generation process with the required energy input. Use of electricity allows a better control and useful modularity of the syngas generation equipment [61]. Hydro Quebec's technology research laboratory (LTE) developed an approach based on the joint use of electron transport and catalysis. A kinetic model was developed to analyze gaseous, homogeneous, and heterogeneous complex reactive systems. The model allows the computation of reaction rates in the search for a material with desired catalytic properties in electricity-assisted syngas production. Depending on the type of electric arc used and the chemical environment in the reformer reactor, electricity-assisted systems for hydrogen production can be categorized as follows: thermal versus nonthermal plasma and oxidative versus oxidant-free plasma systems. In this section, we will consider oxidative plasma (both thermal and nonthermal) systems only, leaving oxidant-free plasma systems for the discussion in Section 2.3.3.

2.2.6.1 Thermal Plasma Reforming

Thermal plasma consists of an electric arc (with the temperatures exceeding 5000°C) through which a gaseous feedstock diffuses at a high velocity, generating ionized species. Figure 2.17 depicts the simplified sketch of a thermal plasma reformer. It consists of two electrodes (an anode and a cathode) spatially arranged within a so-called plasmatron. Thermal plasmas operate at very high power densities and they can catalyze chemical reactions through the intermediate formation of active radicals and ionized species. Thus, thermal plasma reformers use less hydrocarbon fuel since reactant heating is provided by the electric plasma torch. Other advantages of thermal plasma reformers include high conversion efficiencies, a rapid response, compactness, fuel flexibility, and no need for the use of catalysts (thus, no catalyst deactivation problem). The heat generation is independent

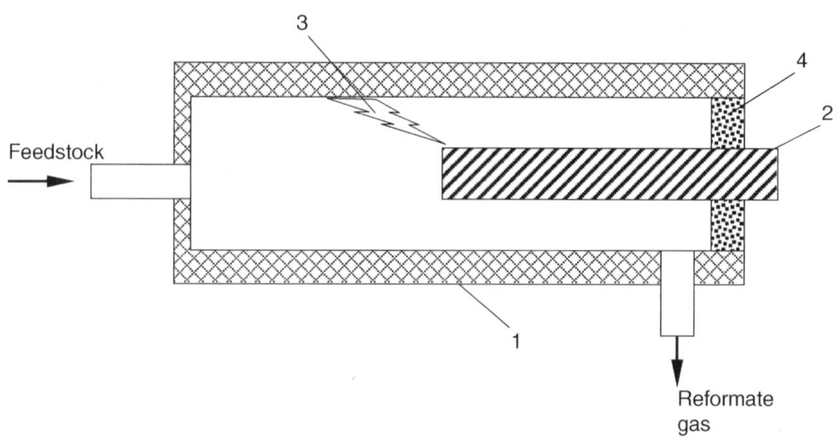

FIGURE 2.17
Simplified schematics of a thermal plasma reformer for the production of synthesis gas from hydrocarbons. 1 = Anode, 2 = cathode, 3 = discharge, and 4 = insulator.

of reaction chemistry, and optimum operating conditions can be maintained over a wide range of feed rates and gas composition [62]. Disadvantages of plasma reformers are as follows: the difficulty of a high-pressure operation, the need for cooling electrodes (to reduce their thermal erosion), high-energy consumption, and dependence on electrical energy. Owing to the high-energy intensity, the process energetics may be less favorable than that of purely thermal processes, especially, endothermic reactions such as steam reforming.

Bromberg et al. [62,63] studied thermal plasma-assisted POx and steam reforming of methane. The process involved a combination of air and steam as an oxidizer. The authors demonstrated that hydrogen-rich gas (50–75% H_2 and 25–50% CO, for steam reforming mode) could be efficiently produced in compact plasma reformers (2–3 kW). For the POx regime, it was determined that the specific energy consumption in the plasma reformers is 40 MJ/kg H_2. Plasma-catalytic reforming of methane by the air–water mixture was also conducted. In this case, the reactor was filled with NiO/Al_2O_3 catalyst, and the composition of reformate gas was (vol%): H_2—40, N_2—38, CH_4—3.4, CO—3.4, CO_2—13.5. The comparison of homogeneous plasma and plasma-catalytic modes of operation showed the significant advantage of the latter regime. The hydrogen yields increased two to three times at significantly lower (1/3) specific power consumption.

2.2.6.2 Nonthermal Plasma Reforming

Nonthermal plasma (also referred to as "cold" or nonequilibrium plasma) systems operate under nonequilibrium thermal conditions (i.e., electrons are at much higher temperature than the ions, radicals, and neutral molecules, which are at near room temperature). An electric discharge produces chemically active species, for example, electrons, ions, atoms, free radicals, excited-state molecules, and photons, which can catalyze chemical reactions involving hydrocarbons and oxidants (e.g., oxygen, steam, and CO_2). Nonthermal plasma reformers operate at much reduced electric currents and consume less electrical energy compared to thermal plasma systems (since energy is not consumed in heating the bulk of the gas). Other advantages of using nonthermal plasma reactors relate to lower temperature and insignificant electrode erosion (thus, cooling of electrodes is not necessary), compactness, etc.

Several types of nonthermal plasma systems have been reported in the literature for reforming of hydrocarbons to hydrogen-rich gas:

1. A gliding arc technology [64,65]
2. A corona discharge [66,67]
3. A microwave plasma [68]
4. A dielectric barrier discharge [69]

Figure 2.18 depicts the simplified representation of the *gliding arc plasma reactor.* In the gliding arc reactor, a gaseous feedstock (e.g., a mixture of methane and air or steam) is injected into the plasma region and flows between the electrodes at high velocity. The electric discharge glides along two diverging electrodes starting from the point where the distance between the electrodes is the shortest, then progressing in the direction of flow toward the position where the distance is greatest. When the arc reaches the farthest position it dies out, and a new arc is generated at the opposite end. Massachusetts Institute of technology (MIT) researchers investigated nonthermal plasma-assisted POx of methane and liquid fuels [70]. The plasmatron featured the following parameters: power, 50–300 W; current, 15–120 mA; volume, 2 L; weight, 3 kg; and H_2 flow rate, 30–50 L/min. Most of the heating

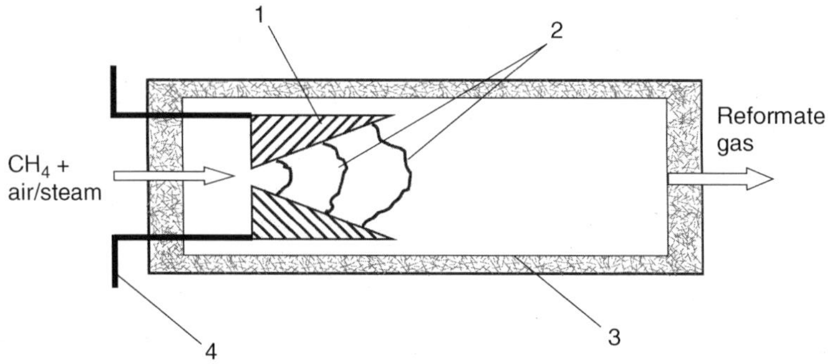

FIGURE 2.18
Simplified schematics of a gliding arc-type plasma reformer. 1 = Electrodes, 2 = discharges, 3 = vessel with insulation, and 4 = electrode connectors.

was provided by the exothermicity of the POx reaction. The plasmatron was also capable of reforming gasoline and diesel fuel into the hydrogen-rich gas. Typical power conversion efficiencies in the POx mode were 60–85%.

Paulmier and Fulcheri [64] reported on the application of the gliding arc technology for onboard reforming of gasoline. The plasma reformer was designed to operate in autothermal or steam reforming regimes at pressures up to 3.105 Pa and temperatures up to 500°C. The feedstock used was a nonsulfurous synthetic gasoline with the average formula of $C_7H_{15.2}$. The plasma reformer efficiency was defined as follows:

$$\eta_{\text{ref}} = \frac{(n_{\text{CO}} + n_{\text{H}_2})\Delta H_{\text{H}_2}}{n_{\text{fuel}}\Delta H_{\text{fuel}} + P_{\text{elec}}} \tag{2.62}$$

where n_{CO} is CO molar flux, n_{H_2} the H_2 molar flux, ΔH_{H_2} the lower heating value of H_2, ΔH_{fuel} the lower heating value of gasoline, and P_{elec} the electric power provided to the reactants by the plasma discharge.

The plasma reformer efficiency reached 12.3% and 26% in gasoline autothermal and steam reforming regimes, respectively. The typical composition of the effluent gas from the reformer operating in steam reforming mode was (vol%) H_2—28.7, CO—15, CO_2—3, and CH_4—40.

A pulsed corona discharge reactor was used by Sobacchi et al. [66] as a source of nonthermal plasma for conducting ATR of hydrocarbons (e.g., isooctane). Application of the voltage in the form of fast rising pulses enabled creation of a highly nonequilibrium discharge with a small total discharge power. Power to the plasma source (1–20 W) is supplied by a thyratron-based power supply that generated pulses of about 100 ns duration and 10 ns rise time. The central part of the wire cylinder reactor was placed inside a high-temperature furnace and maintained at 400°C. In the combined plasma-catalytic reforming mode, the effluent gas from the plasma reactor entered a catalytic reactor operating at 630–800°C. It was found that conversion to hydrogen increased by a factor of 2.5 when plasma processing was employed jointly with the catalytic system. Huang et al. [71] studied CO_2 reforming of methane using atmospheric pressure alternating current discharge plasmas. The experiments were carried out in a Y-type plasma reactor. Plasmas were generated between the inner- (stainless-steel rods) and the outer electrodes (aluminum foil wrapped around walls of quartz tubes). The authors found that the glow discharge plasma was effective in converting CH_4 and CO_2 into syngas. The reactions' products included H_2, CO, and small amounts of C_2+ hydrocarbons. With the increase in CO_2/CH_4 ratio, the selectivity to CO increased and less coke is formed. The mechanism proposed for this system included excited species as follows:

$$CH_4 + * \rightarrow CH_4* \rightarrow CH_3* \rightarrow CH_2* \rightarrow CH* \rightarrow C* \quad (2.63)$$

$$CO_2 + * \rightarrow CO_2* \rightarrow CO + O* \quad (2.64)$$

$$CH_x* + O* \rightarrow CO + H_2 + 2* \quad (2.65)$$

$$C* + O* \rightarrow CO + 2* \quad (2.66)$$

$$CH_x* \rightarrow C_2H_6, C_2H_4, C_2H_2, C_3H_6 \quad (2.67)$$

$$CH_x* + CO_2 \rightarrow CO + x/2H_2 + * \quad (2.68)$$

$$CH_4 + O* \rightarrow CO + 2H_2 + * \quad (2.69)$$

where * denotes excited species.

Micro-arc formation between excited methane and excited CO_2 increased the conversions of CH_4 and CO_2 and favored the production of CO. The energy efficiency of the reaction reached a maximum at $CH_4/CO_2 = 1$.

2.2.7 Onboard Reforming of Hydrocarbons to Hydrogen

The objective of onboard reforming is to convert liquid fuels into a hydrogen-rich gas to be used in an internal combustion engine (ICE) or a FC. In many respects, liquid fuels (e.g., gasoline, diesel, and jet fuels) represent a more attractive means of carrying hydrogen than compressed H_2 itself, promising greater vehicle range, shorter refueling times, increased safety, and perhaps most importantly, utilization of the current fuel distribution infrastructure [72]. The drawbacks of onboard reformers include their inherent complexity, weight, high cost, the need for an elaborate purification of hydrogen from impurities (e.g., CO and H_2S), which could degrade FC performance, etc. Hydrogen purity requirements for the ICE applications are much less stringent than those for FC. According to U.S. DOE Hydrogen Quality Standard (as of January 2006), the upper limits on the selected impurities in the hydrogen gas for proton exchange membrane (PEM) FC are as follows (μmol/mol): CO—0.2, total sulfur—0.004, NH_3—0.1, formaldehyde—0.01, and total halogenates—0.05 [73].

Onboard reforming of liquid fuels to H_2 for use in PEM FC electric vehicles is a very active area of research [74–76]. Three main liquid fuel reforming strategies for an onboard H_2 production are SMR, POx, and ATR. POx and ATR have a number of advantages over SMR, namely, a shorter start-up time and better transient behavior, and less weight; they can reform a wide range of fuels including gasoline and diesel fuel [72]. However, the POx and ATR reformers suffer from some disadvantages with respect to SMR: (i) the reformate gases from POx and ATR are diluted with N_2 (since air is used in the process), which results in a lower FC performance compared to SMR, (ii) H_2 in FC anode exhaust is not readily integrated in the POx or ATR system, and (iii) due to high exothermicity of POx process, the reformer may be subject to greater thermal loses, etc. It was estimated that the energy efficiency (based on HHV) of POx reactors could reach up to 97% for gasoline and 74% for diesel and jet fuels.

Different fuels lay different constraints on the reformer design, the catalysts, and on the operating conditions. For example, it is much more difficult to reform diesel fuel than gasoline due to a number of factors, for example, the former has a lower H/C ratio (thus, soot formation is favored), higher energy consumption due to diesel fuel evaporation, higher temperatures in the reformer, and higher sulfur content. The problems associated with the development of multifuel reformers are discussed by Pettersson and Westerholm [77] in an excellent review paper. The authors concluded that at present it does not seem feasible to develop an efficient and reliable multifuel reformer for automotive applications. The dedicated fuel reformers are to be used and petroleum-based fuels should be specially designed for the potential use in FC vehicles. The effect of the fuel composition on the fuel processor performance was investigated by Borup et al. [78]. It was demonstrated that short-chained aliphatic hydrocarbons tend to have favorable reforming characteristics for catalytic ATR compared to longer-chained and aromatic components.

The Argonne National Laboratory researchers investigated the reactor characteristics and the efficiency of a kilowatt-scale catalytic autothermal reformer using surrogates of diesel fuel (dodecane and hexadecane) as feedstock [79]. The catalyst used was 1 wt% Pt supported on cerium and gadolinium oxides. The reforming of these hydrocarbons was examined at a variety of oxygen-to-carbon and steam-to-carbon ratios at gas space velocities ranging from 10,000 to 100,000 per hour. H_2 selectivity reached up to 86%. The application of the exhaust gas–assisted fuel reforming in diesel engines was investigated by Tsolakis et al. [80]. The process involves hydrogen generation by direct catalytic interaction of diesel fuel with an engine exhaust gas. The exhaust gas fuel reforming is a combination of several basic reforming reactions including steam reforming, POx, ATR, and CO_2 reforming. Two representative chemical equations for ATR and CO_2 reforming of diesel fuel with the average formula of $C_{12.31}H_{22.17}$ are shown as follows:

$$C_{12.31}H_{22.17} + 5.65O_2 + H_2O \rightarrow 12.31CO + 12.08H_2 \tag{2.70}$$

$$C_{12.31}H_{22.17} + 12.31CO_2 \rightarrow 24.62CO + 11.08H_2 \tag{2.71}$$

The authors concluded that for the production of hydrogen-rich gas, diesel fuel could be reformed by exhaust gas at temperatures typical of exhaust gas temperatures (from 200 to 700°C).

2.2.8 Photoproduction of Hydrogen from Hydrocarbons

Photocatalytic production of hydrogen is a potentially attractive approach in converting solar photon energy to chemical energy of hydrogen. Owing to the high dissociation energy of CH_3–H bond (4.48 eV), methane absorbs irradiation in vacuum ultraviolet (UV) region. The absorption spectrum of methane is continuous in the region from 1100 to 1600 Å (absorption coefficient $k = 500$/atm/cm) [81]. Unfortunately, the wavelengths shorter than $\lambda = 160$ nm are present neither in the solar spectrum, nor in the output of most UV lamps. Therefore, the production of hydrogen and other products by direct photolysis of methane does not seem to be practical. However, the use of special photocatalysts allows activating and converting hydrocarbons to H_2 under the exposure to the wavelengths extending well into near-UV area (300–360 nm) that are present in solar spectrum (about 4–5% of the total spectrum).

Hashimoto et al. [82] reported the photocatalytic production of hydrogen from aliphatic and aromatic hydrocarbons using a powdered Pt/TiO_2 photocatalyst suspended in deaerated water. The hydrocarbon–water–photocatalyst system was exposed to UV irradiation

($\lambda > 320$ nm) from 500 W Xe lamp at 40–50°C. Under these conditions *n*-hexadecane ($C_{16}H_{34}$) was converted to a mixture of H_2 and CO_2 as follows:

$$C_{16}H_{34} + 32H_2O + \text{light} \rightarrow 49H_2 + 16CO_2 \qquad \Delta G^\circ = 1232 \text{ kJ/mol} \tag{2.72}$$

No other products were detected in the gas phase. The amount of H_2 produced from 85 μmol of *n*-$C_{16}H_{34}$ was 4.14 mmol, which is close to the stoichiometric value. One can note that reaction 2.72 stoichiometry resembles that of steam reforming of hexadecane. The authors proposed the following mechanism, which involves the initial generation of active species: holes (p^+) in the valence band and electrons (e^-) in the conduction band of the semiconductor TiO_2, as follows:

$$H_2O + p^+ \rightarrow {}^\bullet OH + H^+ \tag{2.73}$$

$$H^+ + e^- \rightarrow 1/2H_2 \tag{2.74}$$

$$RCH_2CH_3 + 2^\bullet OH \rightarrow RCH_2CH_2OH + H_2O \tag{2.75}$$

$$RCH_2CH_2OH \rightarrow RCH_2CHO + H_2 \tag{2.76}$$

$$RCH_2CHO + H_2O \rightarrow RCH_2COOH + 3H_2 \tag{2.77}$$

$$RCH_2COOH \rightarrow RCH_3 + CO_2 \tag{2.78}$$

$$\text{Overall: } RCH_2CH_3 + 2H_2O + \text{light} \rightarrow RCH_3 + 3H_2 + CO_2 \tag{2.79}$$

where R is an alkyl group in an aliphatic hydrocarbon.

Renneke and Hill [83] reported a new system for the homogeneous photochemical activation and functionalization of alkanes with simultaneous production of hydrogen in the presence of polyoxometalates. The chemistry involves the irradiation ($\lambda > 260$ nm) of heteropolytungstic acids (e.g., $H_3PW_{12}O_{40} \cdot 6H_2O$) and alkanes in acetonitrile solution. For example, in case of cyclohexane, the reaction produces cyclohexylmethyl ketone and hydrogen with the selectivity of >90%. Hydrogen is produced from the reduced form of the heteropolyacid through a thermodynamically favorable process catalyzed by Pt(0). The photocatalytic system based on other subclasses of polyoxometalates, namely, isopolytungstates (IPT) and light alkanes (C_1–C_6) dissolved in aqueous solutions was reported by Muradov and Rustamov [84]. IPT is formed by self-assembling of tungstate anions $(WO_4)^{2-}$ in acidic aqueous solutions, which can be described by the following general equation:

$$pH^+ + qWO_4^{2-} \rightarrow [W_qO_y]^{n-} + (4q - y)\, H_2O \tag{2.80}$$

where $p = 8q - 2y$.

In aqueous solutions, IPT exists as an equilibrium mixture of several polymeric anions, for example, $[W_7O_{24}]^{6-}$, $[W_{10}O_{32}]^{4-}$, and $[W_6O_{19}]^{2-}$. The extent of polymerization is a function of the $H^+/(WO_4)^{2-}$ ratio, ionic strength, and temperature. The rate of hydrogen production from IPT/organic aqueous solutions was found to be a function of several parameters: light intensity, pH, tungstate and organic concentrations, and temperature and catalytic additives. Hydrogen is the main component of the gaseous product of the photoreaction. Alkanes produced corresponding alcohols, ketones, and in some cases, products of dimerization (e.g., propane yielded *n*-propanol, isopropanol, acetone, and hexane).

The mechanism of IPT-catalyzed conversion of alkanes (RH) involves the photoinduced charge transfer in the photoactive $W^{VI} = O$ group in the octahedral moiety (**A**) leading to the formation of a reactive electron-deficient radicallike specie (**B**) capable of abstracting H atom from organic substrates (* denotes an excited state):

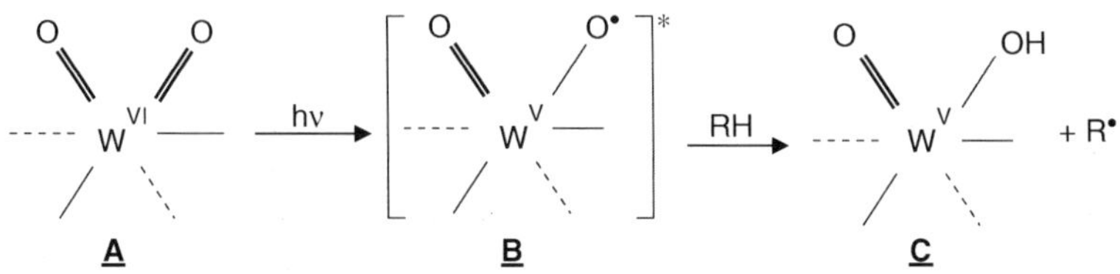

(2.81)

The intermediately formed alkyl radical could be either further oxidized in presence of IPT and water molecules yielding corresponding alcohols or recombine with other radical (dimerization reaction):

$$\underline{\mathbf{B}} + R^{\bullet} + H_2O \rightarrow \underline{\mathbf{C}} + ROH \tag{2.82}$$

$$2R^{\bullet} \rightarrow R\text{–}R \tag{2.83}$$

Photoreduced (**C**) species undergo a number of dark redox transformations (including disproportionation: $2W^{V} \rightarrow W^{VI} + W^{IV}$) leading to their original oxidized form (**A**) with the release of hydrogen. In a simplified form, hydrogen production can be written as follows:

$$2\underline{\mathbf{C}} \rightarrow 2\underline{\mathbf{A}} + H_2 \tag{2.84}$$

The presence of colloidal Pt(0) dramatically accelerates this process.

Recent German patent by Guemter et al. [85] describes a photocatalytic composite element for the cleavage of hydrogen-containing compounds, especially, for hydrocarbon decomposition. The composite includes a porous support on which a photoactive semiconducting material is deposited (e.g., Al_2O_3 and SiO_2). Photoactive materials are prepared from a number of metals or their oxides, for example, Cd, Cu, Ti, and Zn. The authors claim that such a photocatalytic system can photolytically decompose methane to hydrogen and carbon, and can be used for decomposition of organic contaminants in wastewater.

2.3 Nonoxidative Processing of Hydrocarbons

2.3.1 Thermal Decomposition of Methane

When hydrocarbons are heated to a high temperature, they thermally decompose into their constituent elements: hydrogen and carbon,

$$C_nH_m \rightarrow nC + m/2H_2 \tag{2.85}$$

The amount of energy required to carry out this process depends on the nature of the hydrocarbon: it is the highest for saturated hydrocarbons (alkanes, cycloalkanes) and low for unsaturated and aromatic hydrocarbons (in fact, decomposition of acetylene and benzene are exothermic reactions). Methane is one of the most thermally stable organic molecules.

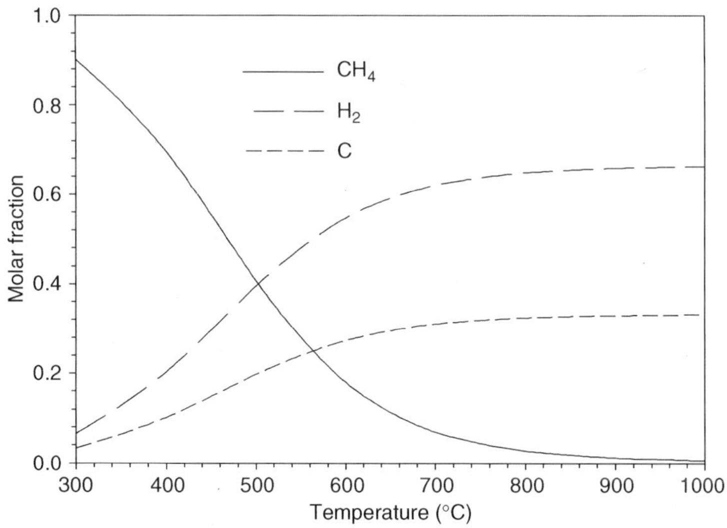

FIGURE 2.19
Thermodynamic equilibrium data for methane decomposition reaction at atmospheric pressure.

The dissociation energy for C–H bond in methane (E = 436 kJ/mol) is one of the highest among all organic compounds. Its electronic structure (i.e., the lack of π- and n-electrons), lack of polarity, and any functional group makes it extremely difficult to thermally decompose the methane molecule into its constituent elements.

Methane decomposition reaction is a moderately endothermic process:

$$CH_4 \rightarrow C + 2H_2 \qquad \Delta H^\circ = 75.6 \text{ kJ/mol} \tag{2.86}$$

Figure 2.19 provides the thermodynamic equilibrium data for methane decomposition reaction. At temperatures above 800°C, molar fractions of hydrogen and carbon products approach their maximum equilibrium value. The effect of pressure on the molar fraction of H_2 at different temperatures is shown in Figure 2.20. It is evident that the H_2 production yield is favored by low pressure. The energy requirement per mole of hydrogen produced (37.8 kJ/mol H_2) is significantly less than that for the SMR reaction (68.7 kJ/mol H_2). Owing to a relatively low endothermicity of the process, <10% of the heat of methane combustion is needed to drive the process. In addition to hydrogen as a major product, the process produces a very important by-product: clean carbon. Because no CO is formed in the reaction, there is no need for the WGS reaction and energy-intensive gas separation stages.

Since the 1950s, thermal decomposition of NG has been practiced for the production of carbon black (CB) with hydrogen being a by-product and supplementary fuel for the process [86]. The process (called thermal black process) was operated in a semicontinuous mode using two tandem reactors running at high operational temperatures. After one of the reactors was heated to a pyrolysis temperature of approximately 1400°C by fuel–air flame, the air was cut off and hydrocarbon was pyrolyzed over the heated contact (firebrick) to hydrogen and CB particles. Simultaneously, another reactor was heated to pyrolysis temperature, followed by the reversing flow of hydrocarbon feedstock from the pyrolysis reactor to the heated reactor, and the process continued in a cyclic mode. The CB produced has been mainly used in the tire and pigment industries. Currently, the thermal black process, as a source of CB, has a very limited application, having been

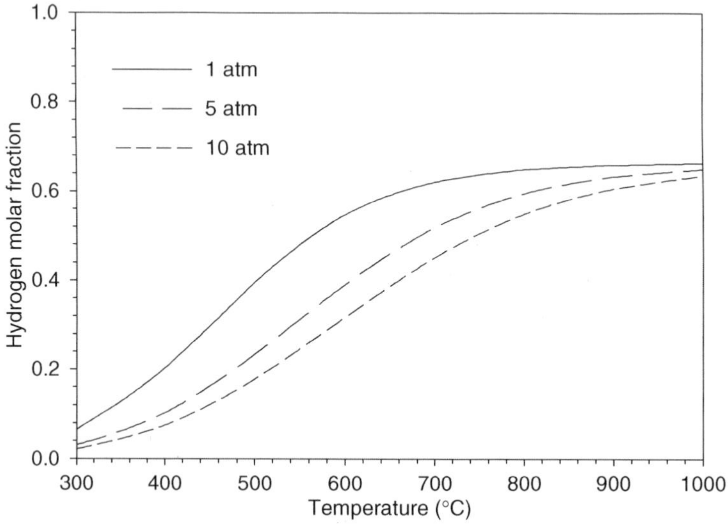

FIGURE 2.20
Thermodynamic equilibrium molar fraction of hydrogen at different pressures as a function of temperature.

supplanted by a more efficient, continuous oil furnace process based on POx of heavy petroleum feedstocks.

Recently, several new processes for methane thermal decomposition were reported in the literature. Steinberg [87] proposed a methane decomposition reactor consisting of a molten metal bath. Methane bubbles through molten tin or copper bath at high temperatures (900°C and higher). The advantages of this system are the efficient heat transfer to a methane gas stream and the ease of carbon separation from the liquid metal surface by density difference. In another work [88], methane decomposition was carried out in a continuous process using a metallic tubular reactor in the temperature range 700–900°C and pressure 2.8–5.6 MPa. It was shown that at 900°C, 5.6 MPa, and sufficiently high residence time (>100 s), the concentration of methane in the effluent gas appeared to be approaching equilibrium values. The determined reaction activation energy of $E_a = 131.1$ kJ/mol was substantially lower than the E_a reported in the literature for homogeneous methane decomposition (272.4 kJ/mol), which pointed to a significant contribution of the heterogeneous processes caused by the submicron-sized carbon particles adhered to the reactor surface. A high-temperature regenerative gas heater (HTRGH) reactor for hydrogen and carbon production from NG has been developed in Russia [89]. In this process, the thermal decomposition of NG to hydrogen-rich gas and CB was conducted in a "free volume" of HTRGH using a carrier gas (N_2 or H_2), which was preheated up to 1627–1727°C in the matrix of the regenerative gas heater. The reactor was combined with a steam turbine to increase the overall efficiency of the system.

The mechanism of thermal decomposition (pyrolysis) of methane has been extensively studied [90,91]. Because C–H bonds in the methane molecule are significantly stronger than C–H and C–C bonds of the products, the secondary and tertiary reactions contribute at the very early stages of the reaction, which obscure the initial processes. According to Holmen et al. [92], the overall methane thermal decomposition reaction at high temperatures can be described as a stepwise dehydrogenation as follows:

$$2CH_4 \rightarrow C_2H_6 \rightarrow C_2H_4 \rightarrow C_2H_2 \rightarrow 2C \quad (2.87)$$

The formation of products is explained by the free radical mechanism, where the initiation step and the primary formation of ethane are described by the following equations:

$$CH_4 \rightarrow CH_3^{\bullet} + H^{\bullet} \tag{2.88}$$

$$CH_4 + H^{\bullet} \rightarrow CH_3^{\bullet} + H_2 \tag{2.89}$$

$$2CH_3^{\bullet} \rightarrow C_2H_6 \tag{2.90}$$

It has been shown [90] that the homogeneous dissociation of methane is the only primary source of free radicals and it controls the rate of the overall process. This reaction is followed by a series of consecutive and parallel reactions with much lower activation energies. After the formation of acetylene (C_2H_2), a sequence of very fast reactions occurs, leading to the production of higher unsaturated and aromatic hydrocarbons and finally carbon:

$$C_2H_2 \rightarrow \text{unsaturated hydrocarbons} \rightarrow \text{aromatics} \rightarrow \text{polynuclear aromatics} \rightarrow \text{carbon} \tag{2.91}$$

This involves simultaneous decomposition and polymerization processes and phase changes from gas to liquid to solid. A detailed mechanism of the final transformations to carbon is rather complex and is not well understood.

2.3.2 Catalytic Methane Decomposition

Because methane decomposition reaction requires high temperatures, there have been attempts to use catalysts to reduce the temperature of thermal decomposition of methane. Figure 2.21 summarizes reported literature data on different catalysts for methane decomposition and the preferred temperature range. It can be seen that transition metals

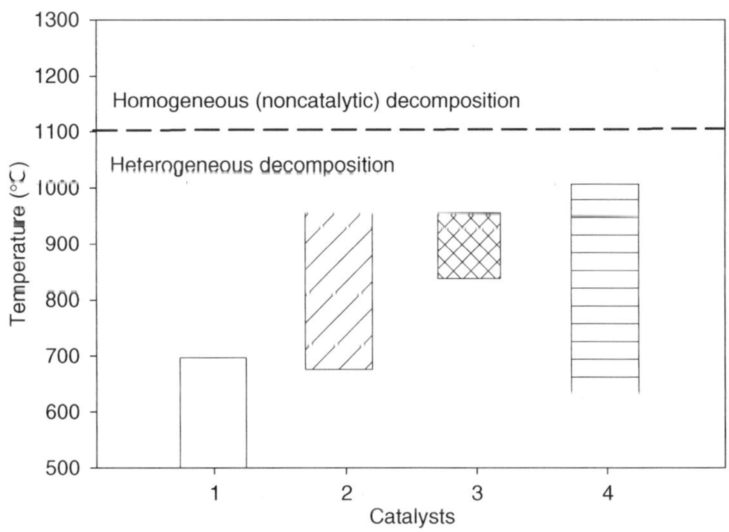

FIGURE 2.21
Summary of literature data on methane decomposition catalysts and preferred temperature range. Catalysts: 1 = nickel, 2 = iron, 3 = carbon, and 4 = other transition metals (Co, Pd, Pt, Cr, Ru, Mo, W). The dotted line arbitrarily separates heterogeneous (catalytic) and homogeneous (noncatalytic, gas phase) temperature regimes of the methane decomposition reaction.

catalyze methane decomposition in a wide range of temperatures (500–1000°C), whereas carbon-based catalysts are efficient at a somewhat higher temperature range (850–950°C). At temperatures above 1000–1100°C, the contribution of gas-phase (or homogeneous) free radical reactions leading to decomposition of methane dramatically increases.

2.3.2.1 Metal-Catalyzed Decomposition of Methane

It has long been known that certain transition metals (d-metals), most prominently, Fe, Co, and Ni, exhibited high catalytic activity toward the decomposition of methane and other hydrocarbons to hydrogen and carbon at moderate temperatures. In the 1960s, Universal Oil Products Co. (UOP) developed the HYPRO™ process for the continuous production of hydrogen by catalytic decomposition of gaseous hydrocarbon streams [93]. Methane decomposition was carried out in a fluidized-bed catalytic reactor in the temperature range from 815°C to 1093°C. Supported Ni, Fe, and Co catalysts (preferably, Ni/Al_2O_3) were used in the process. The coked catalyst was continuously removed from the reactor to the regeneration section where carbon was burned off by air, and the regenerated catalyst was recycled to the reactor. However, the system with two fluidized bed reactors and the solids-circulation system was too complex and expensive and could not compete with the SMR process.

NASA conducted studies on the development of the catalysts for methane decomposition process for space life-support systems [94]. A special catalytic reactor with a rotating magnetic field to support Co catalyst at 850°C was designed. In the 1970s, a U.S. Army researcher M. Callahan [95] developed a fuel processor to catalytically convert different hydrocarbon fuels to hydrogen, which was used to feed a 1.5 kW FC. He screened a number of metals for the catalytic activity in the methane decomposition reaction including Ni, Co, Fe, Pt, and Cr. Alumina-supported Ni catalyst was selected as the most suitable for the process. The following rate equation for methane decomposition was reported:

$$-\frac{d[CH_4]}{dt} = k'S(1 - \Phi)[CH_4] \tag{2.92}$$

where k' is an intrinsic rate constant, Φ is the fraction of the sites covered by carbon, and S is the surface area.

In the Callahan's system, a stream of the gaseous fuel entered one of two reactor beds, where hydrocarbon decomposition to hydrogen took place at 870–980°C and carbon was deposited on the Ni catalyst. Simultaneously, air entered the second reactor where the catalyst regeneration by burning coke off the catalyst surface occurred. The streams of the fuel and air to the reactors were then reversed for another cycle of decomposition–regeneration. The reported fuel processor did not require WGS and gas separation stages, which was a significant advantage. However, the energy efficiency of the processor was relatively low (<60%) and it produced CO_2 by-product in quantities comparable to that of the SMR and POx processes. The kinetics of the deposition of laminar graphite on Ni by pyrolysis of methane, ethane, and ethylene at temperatures of 700–1000°C was reported by Derbishire and Trimm [96]. The authors determined that at the lower range of temperatures, the reaction was of zero order, and the apparent activation energy was E_a = 134 kJ/mol, which is close to the E_a value for the diffusion of Ni in carbon. At higher temperatures, the rates were first order with regard to hydrocarbon. Catalytic decomposition of methane over a number of supported transition metal catalysts including Pt, Mo, Fe, Ni, and Co using pulse and continuous flow reactors was reported by Muradov [97–99] in a number of publications. Fe- and Ni-based catalysts were found to be most active in the methane decomposition reaction; hydrogen was detected in the effluent gas of methane decomposition over alumina-supported Ni catalyst at temperatures as low as 200°C. It was shown that the kinetic curve of methane decomposition over the alumina-supported (10 wt%) Fe_2O_3 at 850°C could be broken down to four zones (A–D) [99]. The zone A corresponds to the reduction of iron oxide to metallic Fe accompanied by the formation of CO_x. In the zone B, quasi-steady process of methane decomposition (lasted about 20–30 min) occurred with the peak H_2 concentration in the effluent gas approaching 98 vol%. This was followed by the drastic reduction in the catalytic activity of iron catalyst in the zone C and, finally, the steady-state methane decomposition (although at low methane conversion yield) in the zone D. The author demonstrated the Fe/alumina-catalyzed decomposition of liquid hydrocarbons including gasoline at 850°C with the maximum hydrogen concentration in the product gas of 92 vol% (the balance, predominantly, methane) and the yield of about 900 mL of gas per milliliter of gasoline [98].

Alstrup and Tavares [100] studied the kinetics of carbon formation from CH_4–H_2 over silica-supported Ni and Ni–Cu catalysts at the range of temperatures 450–590°C. The data were used to test various kinetic models suggested in the literature. The authors demonstrated that the Grabke-type model based on a rate-limiting dissociative chemisorption step and stepwise dehydrogenation of the chemisorbed methyl group is in a good agreement with the experimental data for the Ni/SiO_2 catalyst. The proposed Grabke-type model for carbon deposition from methane on the Ni catalyst surface is shown as follows:

$$H_2 + 2* \leftrightarrows 2H* \tag{2.93}$$

$$CH_4 + 2* \leftrightarrows CH_3* + H* \tag{2.94}$$

$$CH_3* + * \leftrightarrows CH_2* + H* \tag{2.95}$$

$$CH_2* + * \leftrightarrows CH* + H* \tag{2.96}$$

$$CH* + * \leftrightarrows C* + H* \tag{2.97}$$

$$C* + ss1 \leftrightarrows C_{b1} + * \tag{2.98}$$

$$C_{b1} + ss2 \leftrightarrows C_{b2} + ss1 \tag{2.99}$$

$$C_{b2} + ws \leftrightarrows C_w + ss2 \tag{2.100}$$

where * denotes a surface site and CH_x (with $x = 0$–3) are the chemisorbed species. C_{b1} is a carbon atom in the bulk phase of Ni crystal at a subsurface site ss1. C_{b2} is a carbon atom at a subsurface site ss2 just below the interface between the Ni particle and the carbon segregating out of the Ni particle. C_w is a carbon atom at a site ws in the final carbon phase (which is most likely to be a carbon filament). If the second step is rate limiting, then the rate expression is as follows:

$$r_2 = k_2\left(P_{CH_4}\theta*^2 - \frac{1}{K_2}\theta_{CH_3}\theta_H\right) \tag{2.101}$$

where

$$\theta_{CH_3} = \frac{a_H^3}{K_3K_4K_5}\theta_C \quad \text{and} \quad a_H = \sqrt{K_1P_{H_2}}$$

where K_n, $n = 1$–5, are equilibrium constants of the steps 1–5, K_2 is a rate constant, and $\theta*$ is the concentration of unoccupied sites.

The authors showed that the Grabke-type kinetic model can explain the results at a low carbon activity for Ni–Cu catalysts, but that at higher carbon activities, the rates for the $Ni_{0.9}Cu_{0.1}$ catalysts are higher than the model-predicted rates. Low-temperature decomposition of methane over the silica-supported Ni catalyst has been reported by Kuijpers et al. [101]. It was demonstrated that at temperatures as low as 175°C, methane adsorbed on the Ni catalysts dissociates completely into adsorbed carbon atoms and hydrogen.

$$CH_4(g) \leftrightarrows CH_{(4-x)}(ads) + xH(ads) \quad \text{with } 0 < x < 4 \tag{2.102}$$

The methane conversion is thus determined not only by the temperature, but also by the hydrogen surface coverage. Matsukata et al. [102] compared catalytic activities of different transition metals in the methane decomposition reaction. They found that of the group of transition metals including Cu, Fe, Ni, and W, only Ni was catalytically active in methane decomposition at 600–650°C. Pourier and Sapundzhiev [103] screened a number of supported transition metal catalysts including Co, Ni, Rh, Fe, Pd, and Ru. The authors concluded that Pd is by far the most active catalyst. According to the data presented in Ref. 104, Co catalyst demonstrated the highest activity in methane decomposition reaction. Koerts et al. [105] demonstrated that the rate of methane activation in the presence of transition metals followed the order:

$$\text{Co, Ru, Ni, Rh} > \text{Pt, Re, Ir} > \text{Pd, Cu, W, Fe, Mo}$$

Decomposition of methane to H_2 and carbon over Ni/SiO_2 was carried out in a membrane reactor (membrane: 90Pd–10Ag) [106]. The use of the membrane reactor allowed increasing the H_2 yield by shifting the reaction equilibrium toward the products. An excellent review of the literature data on nonoxidative methane activation over the surface of transition metals was recently published by Choudhary et al. [107].

2.3.2.2 Simultaneous Production of Hydrogen and Filamentous Carbon

Owing to a high value and practical importance of carbon filaments, catalytic decomposition of methane and other hydrocarbons as a means of production of different types of filamentous carbon (carbon nanotubes [CNTs], carbon nanofibers, carbon whiskers, etc.) has been a very active area of research for several decades. Carbon filaments with their unique mechanical and electrical properties have important practical applications in areas such as composite materials, electronics, catalysis, space, and military). NG is an abundant feedstock for the production of hydrogen and filamentous carbon. In fact, the earliest reference to vapor-grown carbon fibers relates to the 1889 U.S. Patent by Hughes and Chambers [108] who observed the growth of relatively thick carbon filaments from the CH_4–H_2 mixture in an iron crucible. G. Tibbetts, R.T.K. Baker, M. Endo, T. Koyama, A. Oberlin, and other researchers have made significant contributions to the understanding of the role of catalytic particles in producing the carbon filaments (for an excellent book see Ref. 109). According to a widely accepted mechanism of the carbon filament formation, the reaction product—carbon dissolves into the metal particle, diffuses through it, and precipitates at the rear of the metal crystallite with the formation of a carbon filament. The growth of the carbon filament is inhibited when the catalyst particle is encapsulated in carbon layers,

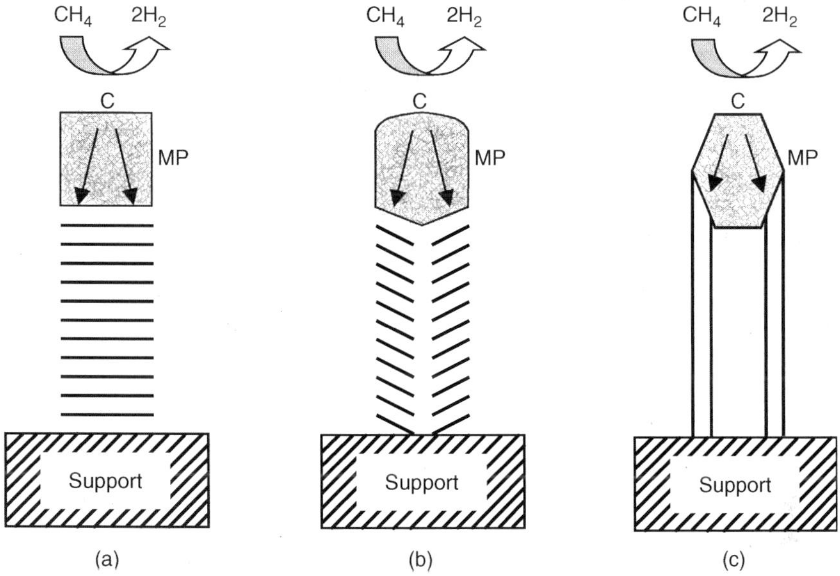

FIGURE 2.22
Schematic representation of carbon filaments of different structure produced by metal-catalyzed decomposition of methane. (a) Platelet structure, (b) "herringbone" structure, and (c) ribbon structure. MP denotes a nano-sized metal particle.

thus preventing further hydrocarbon decomposition. Figure 2.22 depicts the schematic representation of carbon filaments with a metal particle on the growing tip of the filaments. Depending on the carbon filament growth conditions, they may possess different structures, namely, "platelet," "herringbone," and "ribbon" (Figure 2.22a through 2.22c, respectively) (the sketches are based on the description provided in Ref. 110). It should be noted that most of the early works were concerned with the production of carbon filaments (or CNTs, carbon nanofibers) only, and not much consideration was given to hydrogen as a reaction product.

Recently, several groups of researchers reported on the development of the processes for simultaneous production of hydrogen and filamentous carbon by catalytic decomposition of methane and other light hydrocarbons. The objective of these efforts is to simultaneously produce two valuable products: pure hydrogen and filamentous carbon, which has a much higher value than CB (soot) or graphitelike carbon. For example, researchers at the Florida Solar Energy Center (FSEC) studied the production of hydrogen and filamentous carbon from pipeline quality NG using alumina-supported iron catalysts. Figure 2.23a through 2.23c shows the transmission electron microscopic (TEM) images of carbon filaments produced by the NG decomposition over Fe (10 wt%)/Al_2O_3 at 850°C (unpublished data, courtesy of FSEC). The dark spots on the image (Figure 2.23a) are nanosized iron particles embedded in the carbon filaments. A high-resolution TEM image in the inset (Figure 2.23b) clearly shows the "ribbon"-type structure of the carbon filament. The TEM image (Figure 2.23c) depicts an iron nanoparticle encapsulated in the several carbon layers at the tip of a carbon filament.

Li et al. [111] reported the simultaneous production of hydrogen and nanocarbon by decomposition of methane on Ni and Ni–Cu catalysts. The authors demonstrated the production of hydrogen with a purity of 80 vol% over 10 h; simultaneously, 180 g of nanotubes

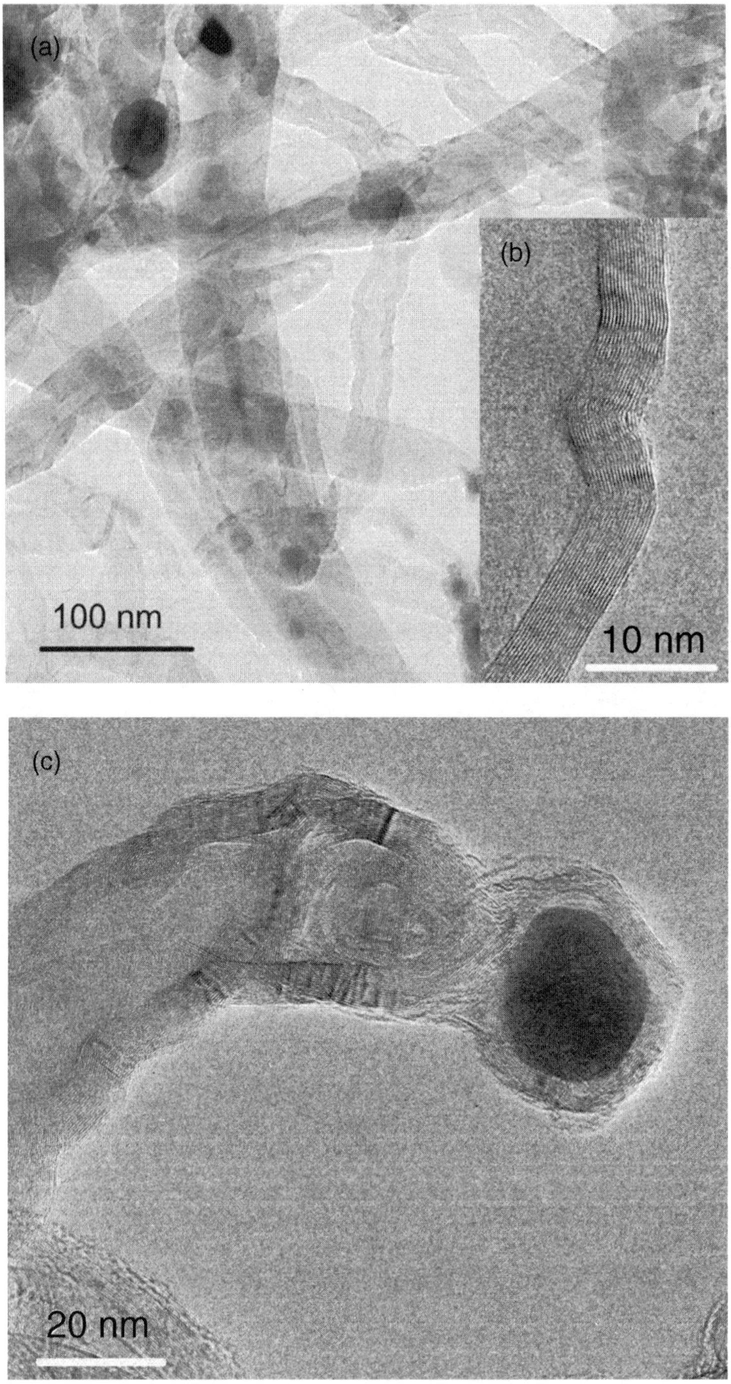

FIGURE 2.23
TEM images of carbon filaments produced by decomposition of NG over Fe(10 wt%)/Al_2O_3 catalyst at 850°C. (a) Carbon filaments with embedded iron nanoparticles, (inset b) high-resolution TEM image of the wall of a carbon filament, and (c) = an iron nanoparticle encapsulated in carbon layers at the tip of a carbon filament.

was produced per 1 g of catalyst. The effect of the addition of different metals (e.g., Cu, Rh, Pd, Ir, and Pt) to the supported Ni catalysts on their catalytic activity for methane decomposition into H_2 and carbon was examined by Takenaka et al. [112]. It was observed that the addition of Pd resulted in a considerable increase in the catalytic activity and stability, whereas modification with other metals decreased the yields of H_2 and carbon compared to the nonmodified Ni/SiO_2 catalyst. Methane decomposition to H_2 and CNTs over an alumina-supported Ni aerogel catalyst at 450–700°C was reported by Piao et al. [113]. TEM images showed that the CNTs were curved, with the diameters in the range 10–20 nm. Production of hydrogen and filamentous carbon by methane decomposition over silica- and zeolite-supported Ni catalysts was reported by Choudhary et al. [114] and Aiello et al. [115]. In particular, Choudhary and coworkers [114] showed the production of filamentous carbon using HY (zeolite)- and silica-supported Ni catalysts at 450–600°C. Two forms of carbon (carbidic and graphitic) were observed on the Ni/SiO_2 catalyst after methane decomposition at 450°C, whereas only graphitic carbon was observed at 550°C.

A series of kinetic studies on the carbon filament formation by methane decomposition over Ni catalysts was reported by Snoeck et al. [116]. The authors derived a rigorous kinetic model for the formation of the filamentous carbon and hydrogen by methane cracking. The model includes the following steps:

a. Surface reactions:

$$CH_4 + * \leftrightarrows CH_4{-}* \qquad K_{CH_4} \tag{2.103}$$

$$CH_4{-}* + * \leftrightarrows CH_3{-}* + H{-}* \qquad (\text{rds})\ k_M^+ \text{ and } k_M^- \tag{2.104}$$

$$CH_3{-}* + * \leftrightarrows CH_2{-}* + H{-}* \qquad K_3 \tag{2.105}$$

$$CH_2{-}* + * \leftrightarrows CH{-}* + H{-}* \qquad K_4 \tag{2.106}$$

$$CH{-}* + * \leftrightarrows C{-}* + H{-}* \qquad K_5 \tag{2.107}$$

$$2H{-}* \leftrightarrows H_2 + 2* \qquad 1/K_H \tag{2.108}$$

b. Carbon dissolution/segregation:

$$C{-}1 \leftrightarrows C_{Ni,f} + 1 \tag{2.109}$$

c. Diffusion of carbon through Ni:

$$C_{Ni,f} \leftrightarrows C_{Ni,r} \tag{2.110}$$

d. Precipitation of carbon:

$$C_{Ni,r} \leftrightarrows C_W \tag{2.111}$$

where "*" is an active site (please note that in the original paper [116], the active site was designated as "1") and rds is a rate-limiting step.

The rate-determining step is the abstraction of the first hydrogen atom from molecularly adsorbed methane with the formation of an adsorbed methyl group. Based on this model, the rate equation for the carbon filament formation by methane cracking can be written as

$$r_{C,M} = \frac{k_M^+ \cdot K_{CH_4} \cdot \left(p_{CH_4} - \frac{1}{K_M^*} \cdot p_{H_2}^2\right)}{\left(1 + \frac{1}{K_r''} \cdot p_{H_2}^{3/2} + K_{CH_4} \cdot p_{CH_4}\right)^2} \tag{2.112}$$

where $r_{C,M}$ is the rate of carbon filament formation, k_M^+ is the rate coefficient for the forward reaction of the rate-determining step, K_{CH_4} is the equilibrium coefficient, p_{CH_4} and p_{H_2} are the partial pressures of methane and hydrogen, respectively, K_M^* is the experimentally determined threshold constant for methane cracking, and K_r'' is an equilibrium coefficient derived from the equilibrium coefficients of the surface reactions and the dissolution/segregation step.

The rigorous kinetic modeling with the incorporation of the diffusion step allows explaining the deactivation of the carbon filament growth and the influence of the affinity for carbon formation on the nucleation of the filamentous carbon.

Carbon capacious Ni–Cu–Al_2O_3 catalysts for methane decomposition was reported by Ismagilov et al. from the Boreskov Institute of Catalysis [117]. The authors demonstrated that the steady and efficient decomposition of methane could be achieved at 625–675°C over Cu-promoted (8–15 wt% Cu) nickel catalysts. The catalyst allowed increasing the yield of filamentous carbon and controlling the microstructural and textural properties of the filaments. The yield of carbon filaments under the optimal condition was 700 g/g Ni. Ermakova and Ermakov [118] studied Ni/SiO_2 and Fe/SiO_2 catalysts for the production of hydrogen and filamentous carbon by methane decomposition. High-loaded silica-supported Ni and Fe (85–90 wt% of the metal) was used in the methane decomposition experiment at 550–700°C. The authors compared the yields of filamentous carbon using silica-supported Ni and Fe catalysts. The carbon yield of 384 g C/g Ni was observed with 90% Ni–10% SiO_2 catalyst, whereas Fe-based catalyst yielded 30 g C/g of Fe.

2.3.2.3 Carbon-Catalyzed Decomposition of Methane

Although transition metal-based catalysts enjoy a high catalytic activity at moderate temperatures (500–750°C), there are some remaining technical challenges, particularly with regard to (1) the catalyst deactivation (due to either carbon deposition or the catalyst particle encapsulation), (2) sulfur poisoning, and (3) separation of the carbon product from the metal catalyst. There have been suggestions to combust carbon off the catalyst surface to regenerate its original activity [115]. In the HYPRO process developed by UOP, the carbon deposited on the catalyst surface (after methane decomposition) was burned in a heater-regenerator apparatus to provide heat input for the process and regenerate the catalyst [93]. This measure, however, would eliminate the production of a valuable by-product—carbon, contaminate H_2 with CO_x and produce large amounts of CO_2 emissions.

The use of carbon-based catalysts offers certain advantages over metal catalysts due to their availability, durability, and low cost. In contrast to the metal-based catalysts, carbon catalysts are sulfur resistant and can withstand much higher temperatures. Muradov [98,99] screened a variety of carbon materials and demonstrated that the efficient catalytic methane decomposition can be accomplished over high surface area carbons at temperatures typical of an SMR process (800–900°C). Particularly, in the presence of activated carbon (AC) (coconut), methane was decomposed at 850°C to produce the effluent gas with the initial hydrogen concentration of 77 vol%, with the balance being unconverted methane. It was reported in the preceding papers that the catalytic activity of the carbon catalysts tested was not sustainable. In the later publications, the range of carbon-based catalysts was significantly expanded and included ACs of different origin and activation methods, a variety of CBs with a wide range of surface areas, microcrystalline graphites, glassy carbon, diamond powder, and nanostructured carbons (i.e., fullerenes and nanotubes; [119,120]). The experimental results indicated that ACs had the highest initial catalytic activity among all the carbon materials tested. However, the initial reaction rate of ACs could not be sustained and the rate of an AC-catalyzed methane decomposition gradually dropped to a quasi-steady value (in most cases, over a period of 2–3 h). The methane decomposition rates over different grades of CB and acetylene black were somewhat lower than that of the AC samples. Structurally ordered carbons (i.e., natural and polycrystalline graphites and diamond powder) demonstrated negligible catalytic activity. Nanostructured carbons (e.g., CNTs and fullerenes) also showed relatively low catalytic activity, whereas fullerene soot was especially active for methane decomposition. Generally, there is a correlation between the initial rate of methane decomposition and the carbon surface area, which points to the heterogeneous nature of the methane decomposition over carbon materials.

The apparent reaction order of carbon-catalyzed methane decomposition reaction was determined to be 0.6 ± 0.1 for AC (lignite) and 0.5 ± 0.1 for CB (BP2000) catalysts. Thus, the rate equation for carbon-catalyzed decomposition of methane can be written as follows:

$$-r_{CH_4} = kP_{CH_4}^{\ 0.5} \tag{2.113}$$

The apparent activation energies for carbon-catalyzed methane decomposition varied in a wide range not only within different types of carbon (e.g., AC versus CB), but also within the family of carbons. For example, among the AC-based catalysts tested, the apparent activation energies (E_a) varied in a range 160–201 kJ/mol (at 600–900°C). For CB-based catalysts, the measured E_a were somewhat higher (205–236 kJ/mol, at the same conditions). It is interesting to note that the values of activation energies for carbon-catalyzed methane decomposition lie between the E_a for noncatalytic and metal-catalyzed reactions. For reference, the activation energies for noncatalytic methane decomposition were reported in the range 370–433 kJ/mol [121], whereas E_a for transition metal catalyzed reactions were reported in the range 60 kJ/mol and below [105].

X-ray diffraction (XRD) studies of carbon samples indicated that after exposure to hydrocarbons, the ordering in the "columnar" or stacking direction has evolved. The distance between graphene layers (d spacing) was found to be d = 3.4948 Å, which is somewhat larger than the d spacing of pure crystalline graphite (d = 3.3480 Å). Thus, carbons produced by decomposition of methane or propane feature a more ordered structure compared to amorphous carbons, but they are less structurally ordered than graphite (which is the characteristic of turbostratic carbon). The surface concentration of high energy sites (which are presumed to be catalytic active sites) is the most important factor governing the activity of carbons. No definite conclusion has yet been made on the mechanism for carbon-catalyzed methane decomposition. Most probably, the reaction starts with the dissociative adsorption of methane molecule on the surface active sites:

$$(CH_4)_g \rightarrow (CH_3)_a + (H)_a \tag{2.114}$$

This is followed by a series of surface stepwise dissociation reactions leading to elemental carbon and hydrogen (the stepwise mechanism was suggested for Ni-catalyzed methane decomposition in Refs 100 and 122):

$$(CH_{3-x})_a \rightarrow (CH_{2-x})_a + (H)_a \tag{2.115}$$

$$(C)_a \rightarrow 1/n(C_n)_c \qquad \text{(carbon crystallite growth)} \tag{2.116}$$

$$2(H)_a \rightarrow (H_2)_g \tag{2.117}$$

where $0 < x < 2$; subscripts a, c, and g denote adsorbed, crystalline, and gaseous species, respectively.

Kim et al. [123] conducted the kinetic study of methane catalytic decomposition over ACs. Several domestic (South Korea) ACs made out of coconut shell and coal were tested as catalysts for methane decomposition at the range of temperatures 750–900°C using a fixed-bed reactor. The authors reported that no significant difference in kinetic behavior of different AC samples was observed despite the differences in their surface area and method of activation. The reaction order was 0.5 for all the AC samples tested and their activation energies were also very close (about 200 kJ/mol) regardless of the origin. The ashes derived from AC and coal did not show appreciable catalytic effect on methane decomposition.

Moliner et al. [124] investigated the effect of textural properties and surface chemistry of AC on the thermocatalytic decomposition of methane. It was demonstrated that the surface chemistry and the pore size distribution play an important role in the initial stage of the methane decomposition process. Microporous carbons with the high content of oxygenated surface groups exhibit a high initial activity, but they rapidly deactivate. In contrast, mesoporous carbons with a high surface area showed better stability. The production of hydrogen by thermal decomposition of methane in a FBR was reported by Dunker et al. [125]. The authors studied the effect of a catalyst, temperature, and residence time on the rate of methane decomposition over a fluidized bed of CB particles at the range of temperatures of 810–980°C and space velocities of 95–210/h. Under optimum conditions, the process produced an effluent gas with hydrogen concentration ≥40 vol%. There was an initial rapid decrease in the catalyst activity during the first 50 min, followed by a slower decrease beginning at about 1000 min after the start of the experiment. The reduction in the hydrogen production yield is explained by a loss of micropores and filling of internal cavities in the catalyst by deposited carbon. Kushch et al. have found that fullerene black is a very active catalyst for the dehydrogenation of methane. Hydrogen and pyrocarbon were formed during the first hours of methane pyrolysis at 1000°C [126]. Methane conversion decreased with time due to the formation of carbon. The reaction also produced the hydrodimerization products (C_2+) such as ethylene, propane, and propylene. During pyrolysis of CH_4–Ar (50–50) mixture, a stable methane conversion (4.2%) was observed with the selectivity to ethylene 89–94%. The kinetics of methane decomposition over ACs was reported by Bai et al. [127]. Methane decomposition over ACs was carried out in a fixed-bed flow reactor made out of a quartz tube. A reaction order of 0.5 and activation energies in the range from 117 to 185 kJ/mol were determined for the methane decomposition reaction. Changes in surface properties of carbons before and after the reaction were investigated. The pore size change in the course of methane decomposition over ACs indicated that the catalytic reaction occurs mainly in the micropores.

2.3.2.4 Catalytic Decomposition of Methane for Fuel Cell Applications

Recently, catalytic decomposition of methane has attracted a great deal of interest as a simple (i.e., one step, and practically no side reactions) and relatively efficient route to produce CO_x-free hydrogen suitable for FC applications. Some types of FCs do not tolerate CO_x impurities in hydrogen gas even at very low levels. For example, the CO concentration in the hydrogen fuel for a PEMFC is restricted to <10 ppmv; an alkaline FC does not tolerate CO_2 in hydrogen because it reacts with the electrolyte. Conventional hydrocarbon reformers (e.g., the reformers based on SMR or POx processes) can potentially deliver hydrogen gas with the required level of the impurities, but they are very complex and bulky systems because they involve several gas conditioning and purification steps (see Section 2.2.7). In contrast, methane decomposition systems do not involve oxidants, and thus, do not produce CO and CO_2. In practical systems, however, it would be very difficult to completely avoid CO_x because they may originate from the traces of moisture or air in the feedstock or from water and air entrained in the catalyst pores. In this case, the addition of a methanation unit could reduce the concentration of CO_x in hydrogen gas down to parts per million by volume levels.

The work by Callahan [95] on the reactor development for metal-catalyzed decomposition of hydrocarbons to carbon and hydrogen to be used in PEMFC has already been mentioned in Section 2.3.2.1. Pourier and Sapundzhiev [103] described a concept of a fuel processor for PEMFC applications based on the catalytic decomposition of NG. In this concept, NG is decomposed over a ceramic-supported Pd (0.1 wt%) catalyst; hydrogen-rich gas is produced and carbon is deposited on the catalyst surface. Once the catalytic bed is filled with carbon, it is regenerated by burning the carbon by air. A hydrogen gas stream with the purity of 95 vol% and trace amounts of CO and CO_2 could be produced. The residual CO was removed by the methanation reaction. Catalytic decomposition of methane over oxidized diamond-supported Ni and Pd catalysts at 600°C yielded nanostructured carbon and hydrogen for FC as reported by Nakagawa et al. [128]. Whisker-type carbon nanomaterials were produced using Ni- and Pd-loaded oxidized diamond. It was concluded that high catalytic activity could be attributed to the chemical interaction between the metal and the support surface.

Shah et al. [129] investigated catalytic decomposition of undiluted methane into hydrogen and carbon over alumina-supported nanostructured binary Fe–M catalysts as a means of producing H_2 for FC applications. The authors observed that all the Fe–M (where M = Pd, Mo, or Ni) exhibited significantly higher catalytic activity than Fe or any secondary metals alone. At the reaction temperatures of 700–800°C and space velocities of 600 mL/g/h, the product stream is composed of over 80 vol% of H_2 (balance being unconverted methane). Multiwalled CNTs were the dominant form of carbon produced. No CO_x or C_2 hydrocarbons were observed in the hydrogen gas, which makes it potentially suitable for PEMFC. Choudhary et al. [114] reported hydrogen production by catalytic decomposition of methane over supported Ni catalysts. One of the objectives was to produce hydrogen gas suitable for FC applications. The CO content in the hydrogen stream was 50 and 250 ppm for Ni/SiO_2 and Ni/HY zeolite, respectively. The low levels of CO coupled with the stability of the catalysts for methane decomposition make this system particularly attractive for an FC. Direct cracking of methane over Ni/SiO_2 as a potential route to CO-free hydrogen production was investigated by Aiello et al. [115]. The authors focused on the catalyst deactivation and regeneration issues. It was shown that deactivated Ni (15 wt%)/SiO_2 catalyst could be fully regenerated at 650°C with steam for up to 10 successive cracking–regeneration cycles without any significant loss of catalytic activity. Initial methane conversion ranged from 24% to 34%.

The main objective of the work reported in the paper [130] was to develop an efficient, compact, and emission-free fuel reformer for mobile and portable FC applications. The concept is based on single-stage catalytic pyrolysis of hydrocarbon fuels into hydrogen-rich gas and carbon products. A hydrocarbon fuel (preferably, propane, or gasoline vapor) enters the pyrolytic reformer where it is decomposed over the carbon catalyst at 850–950°C producing a stream of the hydrogen-rich gas with the average $[H_2] = 70$–80 vol%, the remaining being predominantly methane. Because commercial hydrocarbon fuels contain sulfurous compounds and moisture (the latter would result in formation of CO_x), a sulfur trap and a methanator are added to the scheme to prevent rapid deactivation of PEMFC by the reactive impurities (e.g., CO and H_2S). The estimated specific energy of a propane-fueled pyrolytic reformer is in excess of 1 $kW_{th}h/kg$. The carbon product in the form of fine particulates remains within the reactor for the duration of a power generating cycle and is dislodged from the reactor during a refueling operation (e.g., by blowing it with nitrogen into a special container). Otsuka et al. [131] reported on the development of a new technology for the catalytic decomposition of gasoline-range alkanes to carbon and hydrogen for PEMFC applications. Decomposition of C_1–C_8 alkanes was carried out over Ni/fumed silica catalysts at 500°C. Main products were hydrogen with a low concentration of CH_4 and carbon fibers. C/Ni values (i.e., the number of carbon atoms per one Ni atom) after complete deactivation of the catalyst was in the range 750–1000 depending on the alkane used.

2.3.2.5 Methane Decomposition Using Nuclear and Solar Energy Input

Decomposition of methane is a high-temperature endothermic process. The thermal energy for the process is usually provided by combusting some portion (about 10–15%) of methane. Alternatively, the heat input to the process could be provided by nonfossil energy sources such as high-temperature nuclear reactors and concentrated solar radiation. The advantages are twofold: CO_2 emissions will be avoided and the calorific value of the products will be upgraded. Serban et al. [132] proposed the concept of utilizing the "waste" heat from the Generation IV high-temperature nuclear reactors to produce hydrogen and carbon from methane or NG by direct contact pyrolysis. The Generation IV nuclear reactors use heavy liquid metal coolant (Pb or Sn). According to the concept, high-temperature gas (e.g., He) is used for the heat transfer from the nuclear reactor to the thermal cracking unit. NG was bubbled through a bed of low melting point metals (Pb or Sn) at temperatures ranging from 600°C to 900°C. No other gaseous products beside hydrogen were observed in the effluent stream. The main advantage of the proposed system is the ease of separation of the generated carbon by-product from the heat transfer media (by buoyancy due to difference in densities). At 750°C, methane conversion was about 9%. Two types of carbon were produced: soot and pyrocarbon. These experiments lay the groundwork for developing technical expertise in producing pure H_2 cost effectively by utilizing heat energy contained in the liquid metal coolant in the Generation IV nuclear reactors.

Steinfeld et al. [133] demonstrated the technical feasibility of solar decomposition of methane using a reactor with a fluidized bed of catalyst particulates. Experimentation was conducted at the Paul Scherrer Institute (PSI, Switzerland) solar furnace delivering up to 15 kW with a peak concentration ratio of 3500 sun. A quartz reactor (diameter 2 cm) with a fluidized bed of Ni (90%)/Al_2O_3 catalyst and alumina grains was positioned in the focus of the solar furnace. The direct irradiation of the catalyst provided effective heat transfer to the reaction zone. The temperature was maintained below 577°C to prevent rapid deactivation of the catalyst. The outlet gas composition corresponded to 40% conversion of methane to H_2 in a single pass. Concentrated solar radiation was used as a source of high-temperature process heat for the production of hydrogen and filamentous carbon by the decomposition of gaseous hydrocarbons, CH_4 and C_4H_{10}, in the presence of catalyst particles as reported by Meier et al. [134]. Depending on the nature of the catalyst, two types of filamentous carbon were produced in the solar experiments: nanotubes were formed using Co/MgO catalyst and nanofibers were formed over Ni/Al_2O_3. The conversion of CH_4 to H_2 was approximately 30% as determined from the outlet gas composition during the solar experiments.

Kogan et al. [135] investigated production of hydrogen and carbon by solar-powered decomposition of methane using a volumetric solar receiver. The authors demonstrated that the volumetric absorption of radiation in a solar reactor could be achieved by seeding the gas in the reaction chamber with a cloud of fine particles (CB). The experiments indicated that the loading of 0.1 g of CB particles per 1 m^3 of the irradiated gas would be sufficient for the attainment of adequate radiation absorption. Solar thermal dissociation of methane in a fluid-wall aerosol flow reactor was reported by Dahl et al. [136]. A fluid-wall aerosol flow reactor was heated to temperatures in excess of 1700°C by concentrated sunlight, and it was used to thermally decompose methane to CB and hydrogen. Approximately, 90% conversion of methane to products was obtained at the reactor wall temperature of 1860°C and the average residence time of 0.01 s. The CB produced had primary particle sizes of 20–40 nm. A nozzle-type solar reactor for thermal methane splitting has been reported by Abanades and Flamant [137]. In the tested reactor, a graphite nozzle absorbs concentrated solar radiation provided by a solar furnace. The heat is then transferred to the reagent flow. The CB product was recovered in the carbon trap. With a 2 m diameter solar concentrator, methane-to-H_2 conversion was in the range 30–90%.

2.3.3 Plasma-Assisted Decomposition of Hydrocarbons

The objective of plasma-assisted decomposition of hydrocarbons is to produce hydrogen and carbon in an oxidant-free environment (as opposed to plasma-assisted POx and steam reforming that produce hydrogen and CO_2), according to the following generic reaction:

$$C_nH_m + \text{Energy} \rightarrow nC + m/2H_2 \quad (2.118)$$

The plasma decomposition process is applicable to any hydrocarbon fuel, from methane to heavy hydrocarbons. Similar to oxidative plasma reforming, plasma decomposition processes fall into two major categories: thermal and nonthermal plasma systems.

2.3.3.1 Thermal Plasma Systems

In thermal plasma reactors, the energy required to accomplish an endothermic hydrocarbon decomposition process is provided by high-temperature plasma. Figure 2.24 illustrates the simplified schematics of a thermal plasma reformer for the decomposition of methane to hydrogen and carbon. In the 1990s, Kværner company of Norway developed the Kværner CB&H process for the production of CB and hydrogen by thermal plasma decomposition of NG [138]. In the Kværner's reactor, a plasma torch supplies the necessary energy to pyrolyze methane (the electrodes are made of graphite). The plasma gas is hydrogen, which is recirculated from the process. The characteristics of the Kværner CB&H process are as follows: maximum output per generator—8 MW, thermal efficiency of the process—in excess of 90%, hydrogen purity—98%, energy consumption—1.25 kW/Nm^3 H_2, hourly capacity—270 kg of CB, and 1000 Nm^3 H_2 [139]. The advantages of the CB&H process include (a) feedstock flexibility, (b) nearly 100% conversion of the feedstock, (c) no need for the catalyst, and (d) easy scale-up due to modular concept. The CB product

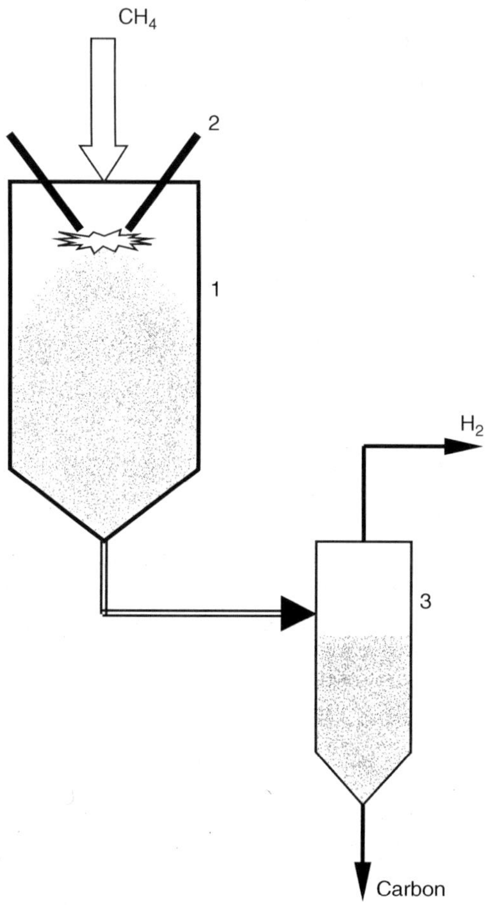

FIGURE 2.24
Schematics of thermal plasma reformer for decomposition of methane to hydrogen and carbon. 1 = Thermal plasma reactor, 2 = graphite electrodes, and 3 = hydrogen–carbon separation unit (cyclone).

of the Kværner CB&H process was evaluated and found to be suitable in rubber, plastic, and metallurgical industries.

Fulcheri and Schwob [140] analyzed the thermal plasma-assisted decomposition of methane to hydrogen and CB. The authors estimated that, assuming the electrothermic efficiency of modern plasma arc devices (80%), the electric energy consumption of the plasma pyrolyzer would be 4–7 kWh/kg of CB and 1–1.9 kWh/m^3 of H_2. Bromberg et al. [62] studied plasma pyrolysis of methane using a water-cooled plasmatron with a cathode with a zirconium tip and a copper tubular anode (plasma-forming gas: N_2). The average specific power consumption was in the range 200–250 MJ/kg H_2. The hydrogen yield, defined as a ratio of measured hydrogen concentration to that of complete conversion, increased from 30% to 70% when the specific power input increased from 250 to 650 kJ/mol CH_4. The product gas composition in the case of the highest hydrogen yield was (vol%) H_2–33.3, N_2–54.4, CH_4–6.8, C_2H_2–3.2, and C_2H_4–1.2.

Thermal plasma-assisted decomposition (cracking) of heavy hydrocarbons is reported in Ref. 141. Decomposition of paraffinic hydrocarbons including $C_{16}H_{34}$ by Ar–H_2 plasma was carried out by continuous injection of the hydrocarbon in the spouted-bed reactor. Controlling temperature, residence time, and H_2 concentration allows reaching the hydrocarbon decomposition yield of about 95%. The authors demonstrated that excitation and dissociation of molecular hydrogen and methane into atomic hydrogen and methyl radicals in plasma plays an important role in the hydrocarbon cracking reactions. Thermodynamic analysis indicated that hydrogen atoms appear in significant quantities at about 3200–3700°C, and methane dissociation occurs at about 1230–2730°C according to the following equations:

$$H_2 \rightarrow 2H^{\bullet} \qquad 436\ \text{kJ} \tag{2.119}$$

$$CH_4 \rightarrow H^{\bullet} + CH_3^{\bullet} \qquad 419\ \text{kJ} \tag{2.120}$$

The experimental results obtained from the plasma fluidized-bed cracking device were in an agreement with the theoretical predictions taking into account the role of reactive species produced by the plasma.

2.3.3.2 Nonthermal Plasma Systems

Czernichowski et al. [142] studied nonthermal plasma decomposition (pyrolysis) of NG in a gliding arc reactor (similar to the one described in Section 2.2.6.2). The methane decom position reaction was carried out in a steel tube reactor equipped with two or six symmetrically located steel electrodes. The gliding arc plasma reactor was powered by the voltage in the range 0.5–10 kV and the currents from 0.1 to 5 A (per discharge). The gaseous feed input was in the range 0.13–2.3 Nm^3/h. The electric power dissipated in the gliding arc plasma reactor reached 1.6 kW. No chemical corrosion or thermal erosion of the electrodes was observed during the plasma reactor operation (hence, no cooling of the electrodes were required). At these conditions, up to 34% of methane was converted into soot, hydrogen, and acetylene. The energy efficiency of the process was estimated at about 29%, which corresponded to an electrical energy consumption of 4.1 kWh to produce 1.0 Nm^3 H_2 and 0.22 Nm^3 C_2H_2.

Nozaki et al. [143] reported on the development of a glow plasma-enhanced chemical vapor deposition (CVD) system for simultaneous production of hydrogen and CNTs through direct decomposition of methane. A typical barrier discharge reactor operating at atmospheric pressure was used in this work. A He/H_2/CH_4 (90:10:1 by volume) mixture was introduced into the reacting chamber maintained at 600°C and containing Ni-coated quartz substrate (as a catalyst for CNT growth). Scanning electron microscopy (SEM) confirmed the formation of CNT with the diameter of 20–30 nm and the density of 10^9–10^{10}/cm^2. Babaritskiy et al. [144] studied microwave plasma-catalyzed decomposition of methane at relatively low temperatures. The experimental setup consisted of two main blocks: a gas heater and a plasma block. The plasma block was exposed to a pulse microwave discharge with the frequency of 9.04 MHz and pulse duration of 300–1000 ns. The products of microwave plasma-assisted decomposition of methane were H_2, C_2H_6, C_2H_4, C_3H_8, C_3H_6, C_2H_2, and CB. The specific energy input (E) was determined as follows:

$$E = \frac{[(H_h - H_o) + W]}{Q} \tag{2.121}$$

where H_h and H_o are the gas enthalpies at the inlet and outlet of the heater, W is the plasma energy, and Q is the methane flow rate.

The electric energy consumption was about 0.2 eV/mol H_2. The possible role of chain-radical reactions involving H, CH_3, and CH_2 radicals generated by the plasma is discussed in the paper. The most probable mechanism of the methane decomposition reaction catalyzed by the microwave plasma impulses is similar to ion–molecular Winchester mechanism as follows:

Initiation:

$$CH_4 + e^- \rightarrow (CH_3)^- + H^\bullet \quad (2.122)$$

Cluster growth:

$$\begin{cases} (CH_3)^- + CH_4 \rightarrow (C_2H_5)^- + H_2 & (2.123) \\ (C_nH_{2n+1})^- + CH_4 \rightarrow (C_{n+1}H_{2(n-q)+3})^- + (1-q)H_2, \quad \text{where } q = 0.1 & (2.124) \end{cases}$$

Termination:

$$(CH_4)^+ + (C_nH_m)^- \rightarrow C_{n+1}H_{4+m-x} + xH \quad (2.125)$$

The authors demonstrated that this ion–molecular mechanism explains the experimental results from the viewpoints of the process energetics and the reaction kinetics.

2.3.4 Refinery Processes

The issues related to hydrogen production and consumption in refineries were briefly discussed in Section 2.1.1. In refineries, hydrogen is produced by steam reforming of hydrocarbons (e.g., SMR) or by its recovery from hydrogen-rich gases and refinery off-gases (that were formerly used as gaseous fuels for heating purposes). With the increasing quality demands for transportation fuels and other refinery products and with the increasing use of heavy sulfurous feedstocks, the transfer of hydrogen within the refinery is rising [7]. The following refinery processes produce gaseous streams from which the recovery of hydrogen could be economically viable:

1. Thermal cracking processes (e.g., cokers and visbreakers)
2. Catalytic cracking (CC)
3. Catalytic reforming
4. Olefins production
5. Acetylene production

Thermal cracking processes. In cokers and visbreakers, heavy crude and residues are converted to petroleum coke, oil, light fractions (benzene, naphtha, LPG), and gas. Depending on the particular process, hydrogen is present in the process gases in a wide range of concentrations. The PSA process is best suited for the recovery of H_2 from these gases because they feature a high-purity H_2, and its off-gas is suitable as a fuel for the process-heating purposes. If more H_2 is required in the refinery, it can be produced by steam reforming of light fractions (LPG and naphtha), or even by the gasification of petroleum coke.

Catalytic crackers. The CC is designed to produce light products from gas oil and increasingly from vacuum gas oil and residues. In the CC process, the molecular mass of the main fraction of the feedstock is lowered, whereas the remaining part is converted to coke, which is deposited on the catalyst. The catalyst is regenerated by burning the coke off with air, which also provides the heat input for the endothermic cracking process. During cracking processes (1) paraffins and naphthens are cracked to olefins and lighter alkanes, (2) monoaromatics are dealkylated without ring cleavage, and (3) polyaromatics are dealkylated and converted to coke [7]. Hydrogen is produced only in the last-mentioned process, whereas the first two processes produce light hydrocarbons, and therefore, need a certain amount of hydrogen. Increase in coke production results in an increase in the overall hydrogen yield.

Catalytic reformers. Catalytic reforming is an important step to improve the quality of gasoline. During the reforming process, naphthens are dehydrogenated to aromatics. As a representative example, hydrogen is produced by cyclohexane dehydrogenation to benzene as follows:

$$C_6H_{12} \rightarrow C_6H_6 + 3H_2 \qquad \Delta H° = 206.2 \text{ kJ/mol} \tag{2.126}$$

This reaction is endothermic and is favored by low pressure. In practice, however, the process is conducted at a pressure of 1–3 MPa (because of a concurrent hydrocracking reaction) and a temperature of 300–450°C using Pt-based catalysts [7]. The feedstock for the reforming process must be carefully purified from S- and N-compounds (below 1 ppm), which may use up a significant portion of hydrogen produced. The typical composition of the off-gas from the catalytic reforming of naphtha is as follows (vol%): H_2–82, CH_4–7, C_2–5, C_3–4, and C_4–2 [7].

Olefin production. During the production of ethylene and propylene by the steam cracking process, large amounts of hydrogen are formed as a by-product. The concentration of H_2 in crude gases from steam cracking of ethane, naphtha, and hydrogenated vacuum gas oil is (mol%) 33, 16, and 10, respectively. Because ethane, ethylene, propane, and propylene are fractionated at low temperatures, hydrogen is usually separated from methane by condensation, cryogenically. The resulting gas has the following composition (mol%): H_2—80–95, CO—0.5–1, CH_4—4.5–19, and C_2—0.1 [7].

Acetylene production. Hydrogen is produced as a by-product in the production of acetylene by high-temperature thermal cracking in a plasma arc process. In a simplified form,

$$2CH_4 \rightarrow C_2H_2 + 3H_2 \tag{2.127}$$

Owing to extremely high temperatures in the reaction zone, the hydrogen content in the cracking gas is particularly high. For example, the Hüls electric arc process generates about 3000 Nm^3 of hydrogen per ton of acetylene produced [7]. Acetylene is separated from the raw gas by absorption and pure H_2 is recovered from the remaining gas by cryogenic processes, sometimes in combination with adsorption processes.

2.4 Some Environmental Aspects of Hydrogen Production

Hydrocarbons (predominantly, NG) are likely to play a major role in hydrogen production in the near- to medium-term future. However, hydrocarbons significantly contribute to anthropogenic CO_2 emissions into the atmosphere. Spath and Mann [11] estimated that the global warming potential (GWP) of hydrogen production by the SMR process is 13.7 kg CO_2 (equiv.) per kg of net hydrogen produced. GWP is defined as a combination of CO_2, CH_4, and N_2O emissions expressed as CO_2 equivalence for a 100-year time frame.

CO_2 accounts for 77.6% of the system's GWP. According to Scholz, CO_2 emissions from the SMR process is 0.44 Nm^3 CO_2/Nm^3 H_2 [4]. Thus, a typical SMR hydrogen plant with the capacity of 1 million m^3 of hydrogen per day produces over 0.4 million standard m^3 of CO_2 per day, which is normally vented into the atmosphere.

For the near- to medium-term, three options to mitigate the CO_2 emissions problem associated with hydrogen production are being actively discussed in the literature: (1) the use of nuclear reactors, (2) sequestration of CO_2 produced by a SMR hydrogen plant, and (3) thermal dissociation of hydrocarbons into hydrogen and carbon. Nuclear power systems, a relatively clean and abundant energy source, have a great potential to contribute to the hydrogen economy. The advanced high-temperature reactor (AHTR) concept, proposed for the U.S. Department of Energy's Generation IV nuclear plant development program, is specifically designed for hydrogen production [145] (see the discussion in Section 2.2.1.6.1). Economically sound hydrogen production utilizing nuclear reactors would require major technological developments in the years ahead. Also, a major shift away from the negative public perception of nuclear energy *per se* would be necessary to base a long-term energy scenario on the nuclear-hydrogen option.

2.4.1 Hydrogen Production by Steam Methane Reforming with CO_2 Sequestration

The main objective of CO_2 sequestration is to prevent anthropogenic CO_2 emissions from reaching the atmosphere by capturing and securely storing CO_2 in sinks such as geologic formations (e.g., deep coal seams, depleted oil and gas reservoirs, and salt domes), the ocean, aquifers, and terrestrial ecosystems. Of particular practical interest is the sequestration of CO_2 produced by the conventional hydrogen production processes, for example, SMR. Generally, CO_2 sequestration involves several steps: the capture, pressurization, transportation, and injection of liquid CO_2 (L-CO_2) under the ocean or into geologic formations. Figure 2.25 illustrates the concept of H_2 production by the SMR hydrogen plant coupled with ocean sequestration of CO_2 using two injection strategies: (a) droplet plume technique, where L-CO_2 is injected below 1000 m forming a rising droplet plume (Figure 2.25a) and (b) CO_2 lake option, where L-CO_2 is introduced to a sea floor depression forming a stable "deep L-CO_2 lake" at a depth below 3000 m (Figure 2.25b). At typical pressures and temperatures that exist in the ocean, CO_2 would exist in a gaseous form above 500 m and in a liquid form below that depth. Owing to the relatively low density, the L-CO_2 would be positively buoyant (i.e., it will rise) in seawater at depths down to 3000 m, but it will sink below that depth [146]. The droplet plume is considered a viable short-term option since it relies on commercially available technology, but it is also more prone to the CO_2 leakage to the atmosphere. The CO_2 lake concept would require more advanced technology and higher cost, but it could potentially minimize the leakage to the atmosphere and exposure to biota.

There have been some estimates reported in the literature on the economics of CO_2 sequestration associated with hydrogen production from fossil fuels. Audus et al. [147] analyzed the concept of hydrogen production by SMR process with CO_2 capture and disposal in a saline aquifer. The authors estimated that the CO_2 capture (80–85% of CO_2 captured from the concentrated streams after PSA) and sequestration would add about 25–30% to the cost of hydrogen production. The costs of avoiding CO_2 emissions are about U.S.$20/t of CO_2 avoided. It was concluded that, for several decades, hydrogen derived from fossil fuels would be cheaper than that from renewable sources, even taking into account the costs of separating and sequestering the associated CO_2. The net cost of CO_2 disposal, however, could be significantly reduced if CO_2 sequestration is accompanied by enhanced oil recovery, or the hydrogen plant is located near depleted NG wells where recovered CO_2 could be reinjected [148]. It was estimated that a carbon tax of $100–300/t would be necessary to

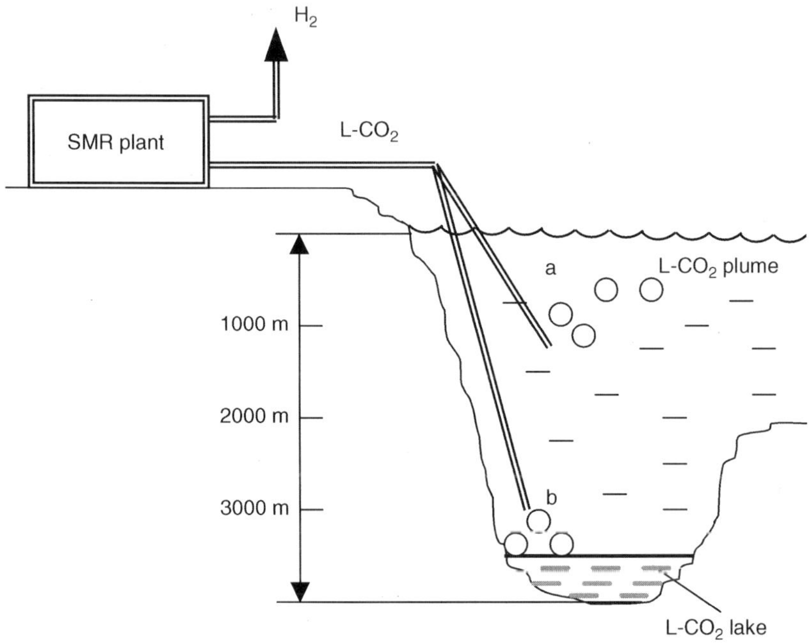

FIGURE 2.25
Schematic diagram of hydrogen production by SMR process coupled with ocean CO_2 sequestration. CO_2 sequestration techniques: (a) = L-CO_2 droplet plume option and (b) = liquid CO_2 lake option.

make CO_2 sequestration viable with the current technology [149]. The technical feasibility of CO_2 sequestration has already been proven on large-scale systems, for example, deep saline aquifer disposal of CO_2 at Sleipner West field in the North Sea by Statoil [149]. Over 28 million t/year of CO_2 is currently being sequestered for enhanced oil recovery in Texas and New Mexico [150].

Two major challenges remain for CO_2 sequestration: (1) bringing its cost down and (2) understanding the reservoir options (e.g., size, permanence, and, most importantly, environmental effect). The key risk factor associated with the CO_2 sequestration stems from uncertain long-term ecological consequences of this approach. Sequestration in oceans is actively debated because of possible adverse impact on the aquatic environment (by the reduction of seawater pH near the disposal site). Although CO_2 injection into geologic formations is being practiced today by the petroleum industry for enhanced oil recovery, it is not yet possible to predict with confidence storage volumes, formation integrity, and permanence over long time periods. It is realized that real ecological responses to CO_2 sequestration cannot be adequately examined or predicted, given the current levels of understanding, and more research and field demonstrations are needed to confirm practical considerations such as economics, safety, permanence, and public acceptance.

2.4.2 Methane Decomposition as Fuel Decarbonization Strategy

In a number of publications, Steinberg [151,152] discussed fossil fuel decarbonization technologies for mitigating global warming. He critically assessed the concept of H_2 production by SMR with CO_2 sequestration. Steinberg estimated that assuming 15% energy loss due to the sequestering operation, about 40% of energy would be lost in SMR process avoiding CO_2 emission. An alternative process is the thermal decomposition of methane (or NG) to hydrogen and carbon. The carbon can be either sequestered or sold on the market or used as a fuel at a later date under less severe CO_2 restraints. The energy sequestered in the carbon amounts to about 42% of the energy of NG resource, which is stored and not destroyed. The author compares well-developed SMR process with the less-developed methane decomposition process including technological status, efficiency, carbon management, and cost. In particular, he points out that the overall efficiencies of SMR and methane decomposition processes would be fairly close (after accounting for CO_2 sequestration) although the energy value of carbon in the latter process is not taken into consideration. It is much easier to sequester carbon as a solid stable product compared to CO_2. It is possible to reduce the cost of H_2 production by methane decomposition by marketing the carbon coproduct.

Methane decomposition as a means of decarbonization of hydrocarbon fuels was discussed by Muradov [153,154]. The author analyzed different technological approaches to catalytic methane decomposition, for example, process heat input options and conceptual designs for the reactor with continuous withdrawal of carbon product. Among the reactors considered were tubular, fluid wall, spouted bed, and FBRs. The latter reactor was selected as the most promising for the large-scale production of hydrogen and carbon. Of particular importance is the issue of finding sufficiently large markets for the carbon by-product. It was estimated that several areas of carbon utilization could potentially absorb tens of millions of tons of carbon: building and construction materials, soil amendment, and environmental remediation.

Hirsch et al. [155] discussed the perspectives of solar thermal decarbonization of NG. The authors analyzed endothermic methane decomposition into carbon and H_2 using concentrated solar energy as a model reaction for conducting a Second Law analysis of a solar decarbonization process, in which carbon is removed from fossil fuels before their use for power generation. It was estimated that the theoretical maximum exergy efficiency of the system could be as high as 35% for a blackbody solar reactor operating at 1223°C under a mean solar flux concentration ratio of 1000. Four routes were examined for extracting power from the products of solar methane decomposition: (1) H_2 is used as a fuel, and carbon is sequestered, (2) carbon is used to fuel a Rankine cycle, and H_2 is used in the FC, (3) carbon is gasified to syngas, and it is further processed to H_2 to be used in the FC, and (4) carbon is used as a reducing agent for ZnO reduction to Zn in water-splitting cycle to produce H_2 to be used in the FC. The analysis indicated that for the third and fourth power generation routes, the energy efficiency exceeded 65%. For the first route, the energy penalty for avoiding CO_2 emissions amounts to 30% of the electrical output.

2.5 Summary and Conclusion

With the transition to environmentally sustainable energy systems based on hydrogen and electricity as two major energy carriers of the future, it is forecasted that the demand for hydrogen would explode in the next one-to-two decades. Given the advantages inherent in hydrocarbon fuels, such as their availability, cost competitiveness, and convenience of storage and distribution, they are likely to continue to play a major role in hydrogen production for the near-to-medium-term future. The author provides an overview of the present day hydrocarbon-to-hydrogen technologies including modern industrial large-scale processes as well as the latest technological developments that still are in R&D stage (totaling about 30 different technologies). All the existing processes for hydrogen production from hydrocarbon feedstocks are classified into two major categories: oxidative and nonoxidative processes. Oxidative processing of hydrocarbons to hydrogen occurs in the presence of oxidants such as steam, oxygen (air), CO_2, or combination thereof. Generally, the processes related to oxidative conversion of hydrocarbons to hydrogen are well-established technologies, and most of the industrial hydrogen production processes (e.g., SMR, POx, and ATR) belong to this category. Nonoxidative conversion of hydrocarbons to hydrogen occurs by splitting of C–H bonds in hydrocarbons in response to an energy input (heat, plasma, radiation, etc.), and it does not require the presence of oxidizing agents. These technologies do not (or minimally) produce CO_2 emissions, which is certainly a great advantage over the oxidative processes (it should be noted, however, that in some cases, CO_2 emissions could be produced indirectly, for example, through the consumption of fossil-derived electricity).

As long as NG and petroleum-based hydrocarbons will remain relatively low-cost feedstock options, in all likelihood, SMR (or other advanced hydrocarbon reforming technologies) will continue to be a cost-effective approach to make hydrogen. Although SMR is a mature technology, there is room for further technological developments and improvements. For example, additional research is needed to identify more durable reforming and shift catalysts (especially, at high pressures >2.0 MPa), improve reforming efficiency, and develop advanced H_2 purification technologies. It should be noted that hydrocarbon reforming technologies can also be applied to some renewable methane-containing feedstocks such as landfill gas, biogas, and digester gas (methane content in these gases reach up to 70 vol%, the balance predominantly CO_2).

It is widely acknowledged that the environmentally sustainable production of hydrogen from hydrocarbons by conventional reforming technologies (e.g., SMR, POx, and ATR) would not be possible without energy-efficient and cost-effective capture and sequestration of the CO_2 by-product. Although the technical feasibility of CO_2 sequestration has already been technically proven on a large scale (e.g., in an enhanced oil recovery, disposal of CO_2 in the North Sea by Statoil Co.), it is not yet possible to predict with confidence storage volumes, formation integrity, and permanence over long time periods. Two major challenges remain for CO_2 sequestration: (1) bringing its cost down and (2) understanding the reservoir options (e.g., size, permanence, and, most importantly, long-term environmental effect). It is realized that the real ecological responses to CO_2 sequestration cannot be adequately examined or predicted, given the current levels of understanding, and more research and field demonstrations are needed.

One alternative to conventional reforming technologies coupled with CO_2 sequestration is the thermal or catalytic decomposition of methane to hydrogen and carbon. This technical approach is not as well developed as SMR, POx, or other reforming technologies, and it would require substantial process development efforts in terms of improving process energy efficiency, the catalyst activity and long-term stability, optimal design of the reactor with continuous withdrawal of carbon, increasing hydrogen yield, etc. The cost of producing hydrogen by methane decomposition is a function of carbon selling price, thus, this process would be able to compete with SMR if sufficiently large markets for the by-product carbon will be found. Because of lack of CO_2 emissions, methane decomposition technology may potentially serve as a link between present-day hydrocarbon reforming technologies and nonfossil-based hydrogen production technologies of the future. Although a number of novel promising technological approaches to hydrogen production from hydrocarbons emerged recently, the main challenge remains with the renewable hydrogen technologies where major breakthroughs are needed. Most likely, in the future, hydrogen will be produced from a diverse mix of primary energy resources and feedstocks including renewable, nuclear, and some fossil sources by energy efficient and environmentally acceptable processes.

Acknowledgments

The support of the U.S. Department of Energy, Hydrogen Program, and National Aeronautics and Space Administration (NASA) (Glenn Research Center) for much of the author's work in hydrogen area is gratefully acknowledged. The author thanks Drs Ali T-Raissi and James Fenton of FSEC for the fruitful discussions related to the topic of the review. Special thanks to Franklyn Smith, Zia Ur-Rahman, Adrienne Henzmann, and Janice Matley for the technical assistance.

References

1. SRI International, Chemical Economics Handbook Product Review. Hydrogen. Industrial Gases, 743.5000A-743.5006E, Menlo Park, CA, 1998.
2. Hairston, D., Hail hydrogen, *Chem. Eng.*, 103, 59, 1996.
3. Armor, J., The multiple roles for catalysis in the production of H_2, *Appl. Catal. A: General*, 176, 159, 1999.
4. Scholz, W., Processes for industrial production of hydrogen and associated environmental effect, *Gas Sep. Purif.*, 7, 131, 1993.
5. Committe on Alternatives and Strategies for Future Hydrogen Production and Use. The Hydrogen Economy: Opportunities, Costs, Barriers, and R&D Needs. The National Academic Press, Washington, 2004.
6. Rostrup-Nielsen, J., Production of synthesis gas, *Catal. Today*, 18, 305, 1993.
7. Haussinger, P., Lohmuller, R., and Watson, A., *Ullmann's Encyclopedia of Industrial Chemistry*, 7th ed., Wiley-VCH Verlag GmbH & Co., Weinheim, Germany, http://www.wiley-vch.de. Online edition, 2002.
8. Ibsen, K., Equipment design and cost estimation for small modular biomass systems, synthesis gas cleanup, and oxygen separation equipment, Contract Report NREL/SR-510-39943, NREL Technical Monitor, Section 2, May 2006.
9. Green, D., Ed., *Perry's Chemical Engineer's Handbook*, 6th ed., Chap. 9, McGraw-Hill, New York, 1984.
10. Kermode, R., Hydrogen from fossil fuels, in *Hydrogen: Its Technology and Applications*, Chap. 3, Cox, K. and Williamson, K., Eds., CRC Press, Boca Raton, FL, 1977.
11. Spath, P. and Mann, M., Life cycle assessment of hydrogen production via natural gas steam reforming, Technical Report, NREL, NREL/TP-570-27637, 2000.
12. Rostrup-Nielsen, J. and Bak Hansen, J., CO_2-reforming of methane over transition metals, *J. Catal.*, 144, 38, 1993.
13. Rostrup-Nielsen, J., Catalytic steam reforming, in *Catalysis: Science and Technology*, Anderson, J. and Boudart, M., Eds., Springer, Berlin, 1984.
14. van Beurden, P., On the catalytic aspects of steam-methane reforming, Technical Report ECN-I-04-003, ECN Energy Innovation, Netherlands, December, 2004.
15. Wang, S. and Lu, G., CO_2 reforming of methane on Ni catalysts: Effect of the support phase and preparation technique, *Appl. Catal. B*, 16, 269, 1998.
16. Xu, Z. and Froment, G., Methane steam reforming, methanation and water-gas shift: I. Intrinsic kinetics, *AIChE J.*, 35, 88, 1989.
17. Rostrup-Nielsen, J., New aspects of syngas production and use, *Catal. Today*, 63, 159, 2000.
18. Balasubramanian, U. et al., Hydrogen from methane in a single-step process, *Chem. Eng. Sci.*, 54, 3543, 1999.
19. Hufton, J. et al., Sorption enhanced reaction process for the production of hydrogen, *Proc. U.S. DOE Hydrogen Prog. Rev.*, Herndon, VA, 179, 1997.

20. Yasuda, I. et al., Development of membrane reformer for highly-efficient hydrogen production from natural gas, *Proc. XV World Hydrogen Energy Conf.*, Yokohama, Japan, 2004.
21. Tong, J. et al., Experimental studies of steam reforming of methane in a thin Pd-based membrane reactor, *Ind. Eng. Chem. Res.*, 44, 1454, 2005.
22. Barbiery, G. et al., Hydrogen production using membrane reactor, *Korean Membrane J.*, 5, 68, 2003.
23. Jiacoletti, R., Hydrogen from nuclear energy, in *Hydrogen: Its Technology and Applications*, Chap. 4, Cox, K. and Williamson, K., Eds., CRC Press, Boca Raton, FL, 1977.
24. Peterma, D. et al., Studies of the use of high temperature nuclear heat from an HTGR for hydrogen production, General Atomic Company Report GA-A11391, San Diego, CA, 1975.
25. Hiroshi, Y., *Proc. Summer Comput. Simul. Conf.*, Yokohama, Japan, 286, 1981.
26. Steinfeld, A. and Meier, A., Solar fuels and materials, *Encyclopedia of Energy*, vol. 5, 623, 2004.
27. De Maria, G. et al., Thermochemical conversion of solar energy by steam reforming of methane, *Energy*, 11, 805, 1986.
28. Tamme, R. et al., Solar upgrading of fuels for generation of electricity, *J. Solar Energy Eng.*, 123, 160, 2001.
29. Boehmer, M. et al., Solar steam reforming of methane, *Proc. Bienn. Congr. Int. Sol. Energy Soc.*, Pergamon, Oxford, 823, 1992.
30. Berman, A., Karn, R., and Epstein, M., Steam reforming of methane on Ru-catalysts for solar hydrogen production, *Abstracts of 232 ACS National Meeting*, Paper: FUEL-140, Amer. Chem. Soc. San Francisco, CA, 2006.
31. Aristov, Yu., Fedoseev, V., and Parmon, V., High-density conversion of light energy via direct illumination of catalyst, *Int. J. Hydrogen Energ.*, 22, 869, 1997.
32. Yokota, O. et al., Steam reforming of methane by using a solar simulator controlled by $H_2O/CH_4 = 1/1$, *Appl. Organomet. Chem.*, 14, 867, 2000.
33. Aasberg-Petersen, K., Synthesis gas production for FT synthesis, in *Fischer–Tropsch Technology*, Chap. 4, Steynberg, A. and Dry, M., Eds., Elsevier, Amsterdam, 2004.
34. Hu, Y. and Ruckenstein, E., Catalytic conversion of methane to synthesis gas by partial oxidation and CO_2 reforming, *Adv. Catal.*, 48, 298, 2004.
35. Dissanayake, M. et al., Partial oxidation of methane to carbon monoxide and hydrogen over Ni/Al_2O_3 catalyst, *J. Catal.*, 132, 117, 1991.
36. Hickman, D. and Schmidt, L., Production of syngas by direct catalytic oxidation of methane, *Science*, 259, 343, 1993.
37. Petiti, C. et al., Hydrogen production by catalytic oxidation of methane, *Proc. 9th World Hydrogen Conf.*, Paris, 53, 1992.
38. Voutetakis, S. et al., Catalytic partial oxidation of methane in a spouted bed reactor, in *Natural Gas Conversion V, Studies in Surface Science and Catalysis*, vol. 119, Parmaliana, A. et al., Eds., Elsevier, Amsterdam, 807, 1998.
39. Maggio, G. et al., Modeling studies to proper size a hydrogen generator for fuel cell, *Proc. 1996 Fuel Cell Seminar*, Orlando, FL, 738, 1996.
40. Joensen, F. and Rostrup-Nielsen, R., Conversion of hydrocarbons and alcohols for fuel cells, *J. Power Sources*, 105, 195, 2002.
41. Balachandran, U. et al., Dense ceramic membranes for partial oxidation of methane to syngas, *Appl. Catal.*, 133, 19, 1995.
42. Chen, C. et al., Conversion of methane to syngas by a membrane-based oxidation reforming process, *Angew. Chem. Int. Ed.*, 42, 5196, 2003.
43. Lewis, W., Control method in reactions between hydrocarbons and metal oxides, U.S. Patent 2, 603, 608, 1952.
44. Solbaken, A., Synthesis gas production, in *Natural Gas Conversion*, Holmen, A. et al., Eds., Elsevier, Amsterdam, 1991.
45. Bradford, M. and Vannice, M.A., CO_2 reforming of CH_4, *Catal. Rev.*, 41, 1, 1999.
46. Schouten, S.C., Gijzeman, O.L., and Bootsma, G.A., Reaction of methane with nickel single crystal surfaces and the stability of surface nickel carbides, *Bull. Soc. Chim. Belg.*, 88, 541, 1979.
47. Zhang, Z.L. and Verykios, X.E., Carbon dioxide reforming of methane to synthesis gas over supported Ni catalysts, *Catal. Today*, 21, 589, 1994.

48. Bitter, J.H., Seshan, K., and Lercher, J.A., Deactivation and coke accumulation during CO_2/CH_4 reforming over Pt catalysts, *J. Catal.*, 183, 336, 1999.
49. Stagg, S.M. and Resaco, D.E., Effect of promoters on supported Pt catalysts for CO_2 reforming of CH_4, *Stud. Surf. Sci. Catal.*, 119, 813, 1998.
50. Lercher, J. et al., Design of stable catalysts for methane carbon dioxide reforming, *Stud. Surf. Sci. Catal.*, 101, 463, 1996.
51. Lehrhofer, J. et al., Integrated analysis of the "sponge iron reactor" and fuel cell system, *Proc. 1996 Fuel Cell Seminar*, Orlando, FL, 710, 1996.
52. Steinfeld, A. and Kuhn, P., High-temperature solar thermochemistry: Production of iron and synthesis gas by Fe_3O_4 reduction with methane, *Energy*, 18, 239, 1993.
53. Takenaka, S. et al., Production of pure hydrogen from methane mediated by the redox of Ni- and Cr-added iron oxides, *J. Catal.*, 228, 405, 2004.
54. Kodama, T. et al., Stepwise production of CO-rich syngas and hydrogen via solar methane reforming by using Ni(II)-ferrite redox system, *Sol. Energy*, 73, 363, 2002.
55. Hacker, V. et al., Usage of biomass gas for fuel cells by the SIR process, *J. Power Sourc.*, 71, 226, 1998.
56. Otsuka, K. et al., Production of hydrogen from methane without CO_2-emission mediated by indium oxide and iron oxide, *Int. J. Hydrogen Energy*, 26, 191, 2001.
57. Suzuka, T. et al., Residual oil cracking with generation of hydrogen, *J. Jpn. Petrol. Inst.*, 26, 174, 1983.
58. Muradov, N., Rustamov, M., and Guseinova, A., Hydrogen production by thermooxidative processing of residual oil, *Nuclear-Hydrogen Energy Technol. Ser.*, 3(19), 34, 1984 (in Russian).
59. Muradov, N. et al., Environmentally compatible processing of residual oil with simultaneous hydrogen production, in *Energy and Environmental Progress*, Vol. D, Veziroglu, N., Ed., Nova Science, New York, 105, 1991.
60. Fukase, S. and Suzuka, T., Residual oil cracking with generation of hydrogen: Deactivation of iron oxide catalyst in the steam-iron reaction, *Appl. Catal. A: General*, 100, 1, 1993.
61. Labrecque, R. and Laflamme, C., Electricity-assisted syngas generation, Report Hydro Quebec Institut de Recherche, June 2002.
62. Bromberg, L. et al., Plasma reforming of methane, *Energy Fuels*, 12, 11, 1998.
63. Bromberg, L. et al., Plasma catalytic reforming of methane, *Int. J. Hydrogen Energ.*, 24, 1131, 1999.
64. Paulmier, T. and Fulcheri, L., Use of non-thermal plasma for hydrocarbon reforming, *Chem. Engin. J.*, 106, 59, 2005.
65. Lesuer, H., Czernichowski, A., and Chapelle, J., Electrically assisted partial oxidation of methane, *Int. J. Hydrogen Energ.*, 19, 139, 1994.
66. Sobacchi, M. et al., Experimental assessment of combined plasma/catalytic system for hydrogen production via partial oxidation of hydrocarbon fuels, *Int. J. Hydrogen Energ.*, 27, 635, 2002.
67. Rusu, I. and Cormier, J.-M., On a possible mechanism of the methane steam reforming in a gliding arc reactor, *Chem. Eng. J.*, 91, 23, 2003.
68. Seguichi, H. and Mori, Y., Steam plasma reforming using microwave discharge, *Thin Solid Films*, 435, 44, 2003.
69. Jiang, T. et al., Plasma methane conversion using dielectric-barrier discharges with zeolite, *Catal. Today*, 72, 229, 2002.
70. Bromberg, L. et al., Emissions reductions using hydrogen from plasmatron fuel converters, *Int. J. Hydrogen Energ.*, 26, 1115, 2001.
71. Huang, A. et al., CO_2 reforming of CH_4 by atmospheric pressure as discharge plasmas, *J. Catal.*, 189, 349, 2000.
72. Kreutz, T., Steinbugler, M., and Ogden, J., Onboard fuel reformers for fuel cell vehicles: Equilibrium, kinetic and system modeling, *Proc. 1996 Fuel Cell Seminar*, Orlando, FL, 714, 1996.
73. U.S. DOE-EERE web site: www.eere.energy.gov/hydrogenandfuelcells.
74. Thomas, C. et al., Fuel options for the fuel cell vehicle: Hydrogen, methanol or gasoline, *Int. J. Hydrogen Energ.*, 25, 551, 2000.

75. Avchi, A., Onsan, I., and Trimm, D., On-board fuel conversion for hydrogen fuel cells: Comparison of different fuels by computer simulations, *Appl. Catal. A: General*, 216, 243, 2001.
76. Ahmed, S. et al., Catalytic partial oxidation reforming of hydrocarbon fuels, *Proc. of 1998 Fuel Cell Seminar*, Palm Springs, CA, 242, 1998.
77. Pettersson, L. and Westerholm, R., State of the art of multi-fuel reformers for fuel cell vehicles: Problem identification and research needs, *Int. J. Hydrogen Energ.*, 26, 243, 2001.
78. Borup, R. et al., Fuel composition effects on transportation fuel cell reforming, *Catal. Today*, 99, 263, 2005.
79. Liu, D. et al., Characterization of kilowatt-scale autothermal reformer for production of hydrogen from heavy hydrocarbons, *Int. J. Hydrogen Energ.*, 29, 1035, 2004.
80. Tsolakis, A., Low temperature exhaust fuel reforming of diesel fuel, *Fuel*, 83, 1837, 2004.
81. Okabe, H., *Photochemistry of Small Molecules*, Chap. 7, Wiley, New York, 1978.
82. Hashimoto, K., Kawai, T., and Sakata, T., Photocatalytic reactions of hydrocarbons and fossil fuels with water. Hydrogen production and oxidation, *J. Phys. Chem.*, 88, 4083, 1984.
83. Renneke, R. and Hill, C., Homogeneous catalytic photochemical functionalization of alkanes by polyoxometalates, *J. Am. Chem. Soc.*, 108, 3528, 1986.
84. Muradov, N. and Rustamov, M., Photocatalytic activation of low alkanes in aqueous solutions of polyoxocomplexes of tungsten, *Acc. USSR Acad. Sci.*, 303, 656, 1988 (in Russian).
85. Guemter, S. et al., Photocatalytic composite element for cleavage of hydrogen containing compounds, especially, for hydrocarbon decomposition and wastewater treatment, German Patent No. DE 10210465, 2003.
86. *Kirk-Othmer Encyclopedia of Chemical Technology*, 3rd ed., Vol. 4, Wiley, New York, 631, 1992.
87. Steinberg, M., Fossil fuel decarbonization technology for mitigating global warming, *Int. J. Hydrogen Energ.*, 24, 771, 1999.
88. Kobayashi, A. and Steinberg, M., The thermal decomposition of methane in tubular reactor, Technical Report BNL-47159, Brookhaven National Laboratory, Upton, NY, 1992.
89. Shpilrain, E., Shterenberg, V., and Zaichenko, V., Comparative analysis of different natural gas pyrolysis methods, *Int. J. Hydrogen Energ.*, 24, 613, 1999.
90. Chen, C., Back, M., and Back, R., Mechanism of the thermal decomposition of methane, in *Symp. Industrial Lab. Pyrolysis*, ACS Symp. Series, 1, 1976.
91. Dean, A., Detailed kinetic modeling of autocatalysis in methane pyrolysis, *J. Phys. Chem.*, 94, 1432, 1990.
92. Holmen, A., Olsvik, O., and Rockstad, O., Pyrolysis of natural gas: Chemistry and process concepts, *Fuel Proces. Technol.*, 42, 249, 1995.
93. Pohlenz, J. and Scott, N., Method for hydrogen production by catalytic decomposition of a gaseous hydrocarbon stream, U.S. Patent No. 3,284,161 (UOP), 1966.
94. Kim, B., Zupan, J., Hillebrand, L., and Clifford, J., Continuous catalytic decomposition of methane, NASA Contractor Report, NASA CR-1662, NASA, Washington, 1970.
95. Callahan, M., Hydrocarbon fuel conditioner for a 1.5 kW fuel cell power plant, *Proc. 26th Power Sources Symp.*, Red Bank, NJ, 181, 1974.
96. Derbishire, F. and Trimm, D., Kinetics of deposition of pyrolytic carbon on nickel, *Carbon*, 13, 189, 1975.
97. Muradov, N., How to produce hydrogen from fossil fuels without CO_2 emission, *Energy and Environmental Progress-I*, Vol. D, Veziroglu, N., Ed., Nova Science, New York, 93, 1991.
98. Muradov, N., Thermocatalytic decomposition of methane using fixed bed reactor, *Proc. 1996 U.S. DOE Hydrogen Prog. Rev.*, Miami, FL, 1996.
99. Muradov, N., CO_2-free production of hydrogen by catalytic pyrolysis of hydrocarbon fuel, *Energy Fuel*, 12, 41, 1998.
100. Alstrup, I. and Tavares, T., Kinetics of carbon formation from CH_4–H_2 on silica-supported nickel and Ni–Cu catalysts, *J. Catal.*, 139, 513, 1993.
101. Kuijpers, E. et al., The reversible decomposition of methane on Ni/SiO_2 catalyst, *J. Catal.*, 72, 75, 1981.
102. Matsukata, M., Matsushita, T., and Ueyama, K., A circulating fluidized bed CH_4 reformer: Performance of supported Ni catalysts, *Energy Fuels*, 9, 822, 1995.

103. Pourier, M. and Sapundzhiev, C., Catalytic decomposition of natural gas to hydrogen for fuel cell applications, *Int. J. Hydrogen Energ.*, 22, 429, 1997.
104. Pyatenko, A. et al., Investigation of the process of catalytic decomposition of methane on d-metals, *Khimiya Tverdogo Topliva*, 23, 682, 1989 (in Russian).
105. Koerts, T., Deelen, M., and van Santen, R., Hydrocarbon formation from methane by a low-temperature two-step reaction sequence, *J. Catal.*, 138, 101, 1992.
106. Ishihara, T. et al., Decomposition of methane over Ni/SiO_2 catalysts with membrane reactor for the production of hydrogen, *Chem. Lett.*, 93, 1995.
107. Choudhary, T., Aksoylu, E., and Goodman, D., Nonoxidative activation of methane, *Catal. Rev.*, 45, 151, 2003.
108. Hughes, T. and Chambers, C., Manufacture of carbon filaments, U.S. Patent No. 405, 480, 1889.
109. Dresselhaus, M., Dresselhaus, G., and Avouris, P., Eds., *Carbon Nanotubes. Synthesis, Structure, Properties and Applications*, Springer, Berlin, 2001.
110. Bessel, C. et al., Graphite nanofibers as an electrode for fuel cell applications, *J. Phys. Chem.*, 105, 1115, 2001.
111. Li, Y. et al., Simultaneous production of hydrogen and nanocarbon from decomposition of methane on nickel-based catalyst, *Energy Fuel*, 14, 1188, 2000.
112. Takenaka, S. et al., Methane decomposition into hydrogen and carbon nanofibers over supported Pd–Ni catalysts, *J. Catal.*, 220, 468, 2003.
113. Piao, L. et al., Methane decomposition to carbon nanotubes and hydrogen on an alumina supported nickel aerogel catalyst, *Catal. Today*, 74, 145, 2002.
114. Choudhary, T. et al., Hydrogen production via catalytic decomposition of methane, *J. Catal.*, 199, 9, 2001.
115. Aiello, R. et al., Hydrogen production via direct cracking of methane over Ni/SiO_2: Catalyst deactivation and regeneration, *Appl. Catal. A: General*, 192, 227, 2000.
116. Snoeck, J., Froment, G., and Fowles, M., Kinetic study of the carbon filament formation by methane cracking on a nickel catalyst, *J. Catal.*, 169, 250, 1997.
117. Reshetenko, T. et al., Carbon capacious $Ni\text{-}Cu\text{-}Al_2O_3$ catalysts for high-temperature methane decomposition, *Appl. Catal. A: General*, 247, 51, 2003.
118. Ermakova, M. and Ermakov, D., Ni/SiO_2 and Fe/SiO_2 catalysts for production of hydrogen and filamentous carbon via methane decomposition, *Catal. Today*, 77, 225, 2002.
119. Muradov, N., Catalysis of methane decomposition over elemental carbon, *Catal. Commun.*, 2, 89, 2001.
120. Muradov, N., Catalytic activity of carbon for methane decomposition reaction, *Catal. Today*, 102–103, 225, 2005.
121. Holmen, A., Rokstad, O., and Solbakken, A., High temperature pyrolysis of hydrocarbons. 1. Methane to acetylene, *Ind. Eng. Chem., Process. Des. Dev.*, 15, 439, 1976.
122. Solymosi, F. et al., Decomposition of CH_4 over supported Pd catalysts, *J. Catal.*, 147, 272, 1994.
123. Kim, M. et al., Hydrogen production by catalytic decomposition of methane over activated carbons: Kinetic study, *Int. J. Hydrogen Energ.*, 29, 187, 2004.
124. Moliner, R. et al., Thermocatalytic decomposition of methane over activated carbons: Influence of textural properties and surface chemistry, *Int. J. Hydrogen Energ.*, 30, 293, 2005.
125. Dunker, A., Kumar, S., and Mulawa, P., Production of hydrogen by thermal decomposition of methane in a fluidized bed reactor—effect of catalyst, temperature and residence time, *Int. J. Hydrogen Energ.*, 31, 473, 2006.
126. Kushch, S. et al., Methane activation by fullerene black, *Nephtekhimiya*, 37, 117, 1997 (in Russian).
127. Bai, Z. et al., Catalytic decomposition of methane over activated carbon, *J. Anal. Appl. Pyrolysis*, 73, 335, 2005.
128. Nakagawa, K. et al., Oxidized diamond as a simultaneous production medium of carbon nanomaterials and hydrogen for fuel cell, *Chem. Mater.*, 15, 4571, 2003.
129. Shah, N., Panjala, D., and Huffman, G., Hydrogen production by catalytic decomposition of methane, *Energy Fuel*, 15, 1528, 2001.

130. Muradov, N., Emission-free fuel reformers for mobile and portable fuel cell applications, *J. Power Sourc.*, 118, 320, 2003.
131. Otsuka, K., Shigeta, Y., and Takenaka, S., Production of hydrogen from gasoline range alkanes with reduced CO_2 emission, *Int. J. Hydrogen Energ.*, 27, 11, 2002.
132. Serban, M. et al., Hydrogen production by direct contact pyrolysis of natural gas, *Preprints of Symposia—Am. Chem. Soc., Div. Fuel Chem.*, 47, 746, 2002.
133. Steinfeld, A. et al., Production of filamentous carbon and hydrogen by solar thermal catalytic cracking of methane, *Chem. Eng. Sci.*, 52, 3599, 1997.
134. Meier, A. et al., Solar thermal decomposition of hydrocarbons and carbon monoxide for the production of catalytic filamentous carbon, *Chem. Eng. Sci.*, 54, 3341, 1999.
135. Kogan, A., Kogan, M., and Barak, S., Production of hydrogen and carbon by solar thermal methane splitting. III. Fluidization, entrainment and seeding powder particles into a volumetric solar receiver, *Int. J. Hydrogen Energ.*, 30, 35, 2005.
136. Dahl, J. et al., Solar thermal dissociation of methane in a fluid-wall aerosol flow reactor, *Int. J. Hydrogen Energ.*, 29, 725, 2004.
137. Abanades, S. and Flamant, G., Production of hydrogen by thermal methane splitting in a nozzle-type laboratory-scale solar reactor, *Int. J. Hydrogen Energ.*, 30, 843, 2005.
138. Lynum, S. and Gaudernack, B., Hydrogen from natural gas without release of CO_2 to the atmosphere, *Proc. 11th World Hydrogen Energy Conf.*, Stuttgart, Germany, 1996.
139. Lynum, S. et al., Kværner based technologies for environmentally friendly energy and hydrogen production, *Proc. 12th World Hydrogen Energy Conf.*, Buenos Aires, Argentina, 1998.
140. Fulcheri, L. and Schwob, Y., From methane to hydrogen, carbon black and water, *Int. J. Hydrogen Energ.*, 20, 197, 1995.
141. Nikravech, M. et al., Plasma-fluidized bed hydrocracking process of heavy hydrocarbons, *Proc. 9th Int. Symp. on Plasma Chemistry*, Pugnochiuso, Italy, 209, 1989.
142. Czernichowski, A., Czernichowski, P., and Ranaivosolarimanana, A., Plasma pyrolysis of natural gas in gliding arc reactor, *Proc. 11th World Hydrogen Energy Conf.*, Stuttgart, Germany, 1996.
143. Nozaki, T., Kimura, Y., and Okazaki, K., Carbon nanotubes and hydrogen co-production from methane using atmospheric pressure non-equilibrium plasma, *Proc. 16th ESCAMPIG and 5th ICRP Joint Conf.*, Grenoble, France, 2002.
144. Babaritskiy, A. et al., Plasma catalysis processes for hydrogen and syngas production from hydrocarbons, *Proc. 13th World Hydrogen Energy Conf.*, Beijing, China, 2000.
145. Zink, J., Trend favors nuclear-hydrogen economy, *Power Eng.*, 107, 20, 2003.
146. Herzog, H., Caldeira, K., and Adams, E., Carbon Sequestration via Direct Injection. Workshop on Carbon Sequestration Science, NETL Publication, Pittsburgh, PA, 2001.
147. Audus, H., Kaarstad, O., and Kowal, M., Decarbonization of fossil fuels: Hydrogen as an energy vector, *Proc. 11th World Hydrogen Energy Conf.*, Stuttgart, Germany, 1996.
148. Blok, K. et al., Hydrogen production from natural gas, sequestration of recovered CO_2 in depleted gas wells and enhanced natural gas recovery, *Energy*, 22, 161, 1997.
149. Herzog, H., CO_2 sequestration. *Proc. Greenhouse Gas Reduction Programs and Technologies Symp.*, Dedham, MA, 2002.
150. Stevens, S. and Gale, J., Geologic CO_2 sequestration, *Oil Gas J.*, May 15, 40, 2000.
151. Steinberg, M., Fossil fuel decarbonization technology for mitigating global warming, *Int. J. Hydrogen Energ.*, 24, 771, 1999.
152. Steinberg, M., Production of hydrogen and methanol from natural gas with reduced CO_2 emission, *Int. J. Hydrogen Energ.*, 23, 419, 1998.
153. Muradov, N., Hydrogen via methane decomposition: An application to decarbonization of fossil fuels, *Int. J. Hydrogen Energ.*, 26, 1165, 2001.
154. Muradov, N., From hydrocarbon to hydrogen-carbon to hydrogen economy, *Int. J. Hydrogen Energ.*, 30, 225, 2005.
155. Hirsch, D., Epstein, M., and Steinfeld, A., The solar thermal decarbonization of natural gas, *Int. J. Hydrogen Energ.*, 26, 1023, 2001.

3

Hydrogen Production from Coal

Shi-Ying Lin

CONTENTS

3.1 Introduction

3.1.1 Overview of Coal

Coal is the most abundant fossil fuel on Earth and is expected to continue to be an important source of energy over the next several hundred years. Coal deposits are widely distributed on Earth, and few areas do not have reasonable access to coal. Currently, about 23% of the world's primary energy comes from coal, and the International Energy Agency expects consumption of coal to continue to increase in the future. For example, about 7.5 billion t of coal is projected to be used in 2025. This amount, which includes coal used for production of hydrogen, will be nearly twice the amount consumed in 2004 (Table 3.1).

Coal is formed from plant materials by the process of coalification, which occurs underground over long periods of time. Coal is a complex mixture of organic chemical substances made up of carbon, hydrogen, oxygen, and smaller amounts of nitrogen and sulfur, as well as moisture and minerals. According to its degree of coalification, coal is classified into different ranks: lignite (brown coal), subbituminous coal, bituminous coal, and anthracite, each having a different heating value. Table 3.2 shows typical compositions of these types of coal.[3] Table 3.3 shows proximate- and ultimate analysis of some coals from around the world.

TABLE 3.1

Projected Worldwide Consumption of Coal

	Year					
	2002	2004	2010	2020	2025	
Coal demand (billion t)[1]	3.4	4.0			7.5	Proven coal reserves discovered till 2002—907 billion tons[2]
Shares in total primary energy supply (%)[2]	23		23	22		

TABLE 3.2

Contents and Heating Values of Different Types of Coal

Coal Rank	Natural Water (wt%)	Volatile Matter (Dry, Ash Free) (wt%)	Total Carbon (Dry, Ash Free) (wt%)	Heat of Combustion (MJ/kg)
Peat	70–75	60–62	60–64	~6.7
Lignite (brown coal)	35–40	~53	~67	18.8–19.2
Subbituminous	~10	45–50	~77	23.9–26.8
Bituminous	~3	10–14	91–92	~35.4

Source: Adapted from *Survey of Energy Resources*, BGR, WEC (World Energy Conference, 2004).

TABLE 3.3

Proximate and Ultimate Analyses of Coals

	Proximate Analysis (wt%)				Ultimate Analysis (wt%)				
Coals	W	A	VM	FC	C	H	O	N	S
Lignite; Wyoming (United States)	15.2	7.0	46.6	31.2	54.1	3.9	19.0	0.8	1.1
Subbituminous; Taiheiyo (Japan)	4.8	7.6	48.8	38.8	67.7	5.8	13.0	1.1	0.2
Bituminous; Datong (China)	4.6	6.4	24.3	64.6	72.3	4.2	11.6	0.9	0.4
Bituminous; Ebenezer (Australia)	1.73	12.9	36.5	48.9	69.7	5.2	9.2	1.3	0.5

Note: W—water, A—ash, VM—volatile matter, and FC—fixed carbon.

3.1.2 Hydrogen Production from Coal

Hydrogen is mainly produced from water (H_2O) by splitting the water molecule into H_2 and O. The key aspects of hydrogen production from water are supplying sufficient energy to split the water molecule and fixing the oxygen produced. One method for producing hydrogen from water involves using a reductant M in an oxidation–reduction reaction.

$$H_2O + M \xrightarrow{\text{Energy}} H_2 + MO \tag{3.1}$$

M may be a metal or carbon, C. Fortunately, carbon is readily available from abundant materials such as coal (CH_mO_n).

$$(2-n)H_2O + CH_mO_n \xrightarrow{\text{Energy}} (2+m/2)H_2 + CO_2 \tag{3.2}$$

Equation 3.2 offers a large-scale method for the production of hydrogen from water and coal, which are readily and inexpensively obtained to meet hydrogen energy needs worldwide.

Production of hydrogen from coal is a well-established technology, in which O_2 or steam is passed over coal to produce a mixture of H_2, CO, and CO_2 from which H_2 is separated (Figure 3.1).

During the past several decades, coal made hydrogen is mainly used in areas for the production of chemicals such as ammonia, methanol, methane, and Fischer–Tropsch products (Figure 3.2).

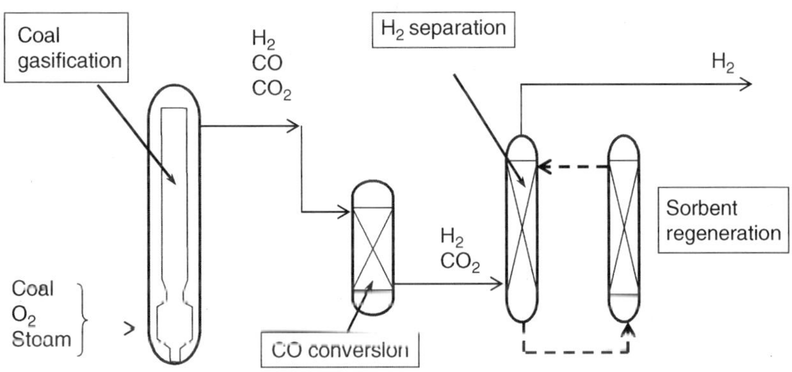

FIGURE 3.1
Production of hydrogen from coal gasification.

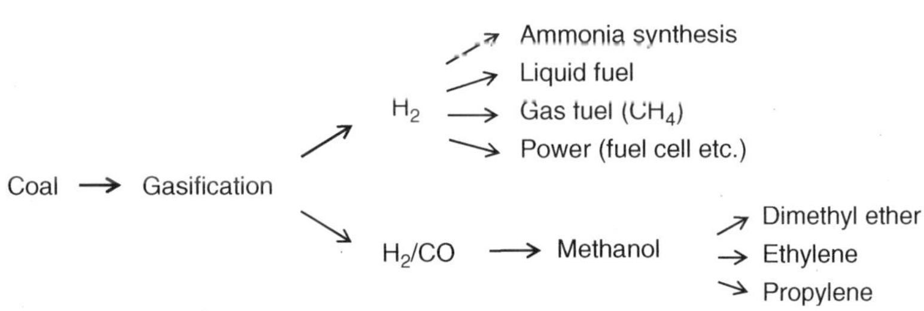

FIGURE 3.2
Current uses of hydrogen derived from coal.

Although, producing hydrogen from coal is not as cost effective as producing hydrogen from oil or natural gas, coal can be used where oil or natural gas is not readily available and where coal is abundant.

3.2 Projects Using Hydrogen Derived from Coal

3.2.1 Synthesis of Ammonia

Ammonia is synthesized from hydrogen and nitrogen.

$$N_2 + 3H_2 \rightarrow 2NH_3 \tag{3.3}$$

Most of the hydrogen for synthesis of ammonia is derived from oil or natural gas; however, a number of cost-effective coal gasification installations have been commissioned in Asia, North America, and Eastern Europe where oil or natural gas is not available and where coal is abundant.

3.2.2 Synthesis of Liquid Fuel

3.2.2.1 Direct Coal Liquefaction

Liquid fuel can be synthesized by the direct reaction between coal and hydrogen.

$$H_2 + \text{coal} \rightarrow \text{liquid fuel} \tag{3.4}$$

The best-known processes are the IGOR (Germany), HTI (America), and NEDOL (Japan). The New Energy and Industrial Technology Development Organization (NEDO) finished a 150 t coal per day coal liquefaction pilot plant in 1998. Shenhua Group Corporation is building a 4000 t coal per day (1 Mt oil/year) commercial plant in China.

3.2.2.2 Indirect Coal Liquefaction

Indirect coal liquefaction is a technology in which coal is first gasified to synthesis gas ("syngas," CO + H_2), which is used to synthesize liquid fuel by the Fischer–Tropsch process.

$$\text{Coal gasification} \rightarrow n\text{CO} + (n+1)H_2 \rightarrow (CH_2)_n + H_2O \tag{3.5}$$

This technology is primarily used by Sasol, South Africa.

3.2.3 Production of Methane from Coal

3.2.3.1 Direct Coal Methanation

Direct production of methane from coal is called coal hydrogasification. Several technologies have been developed to make city gas from coal. The main reaction in coal hydrogasification is

$$\text{C(coal)} + 2H_2 \rightarrow CH_4 \tag{3.6}$$

3.2.3.2 Indirect Coal Methanation

In this process, coal is first gasified to CO and H_2, which then react to form CH_4 as follows:

$$\text{Coal gasification} \rightarrow CO + 3H_2 \rightarrow CH_4 + H_2O \tag{3.7}$$

Laboratory and pilot plant projects have demonstrated the feasibility of producing high-calorie gas from coal.

3.2.4 Production of Methanol from Coal

Methanol can be synthesized from syngas derived from coal.

$$\text{Coal gasification} \rightarrow CO + 2H_2 \rightarrow CH_3OH \tag{3.8}$$

Although most methanol is made from natural gas, a number of cost-effective coal gasification installations have been commissioned in areas where natural gas and oil are not readily available and where coal is abundant.

Methanol is used directly in furnaces or mixed with gasoline to fuel cars, and also is an important material for making DME and plastics (methane-to-olefin [MTO] and methane-to-propylene [MTP] processes).

3.2.5 Hydrogen Power Generation

The integrated coal gasification fuel cell combined cycle (IGFC; Figure 3.3) is an efficient means of generating power from coal. In this cycle, hydrogen produced from coal is used to run a fuel cell, which in turn powers a gas turbine and steam turbine. The efficiency of IGFC for electricity generation is expected to be as high as 55%. An IGFC project called EAGLE (Energy Application for Gas, Liquid and Electricity) is currently under development in Japan.

Recently, the United States launched the FutureGen Project (Figure 3.4). FutureGen is an initiative to build the world's first integrated carbon dioxide sequestration–hydrogen

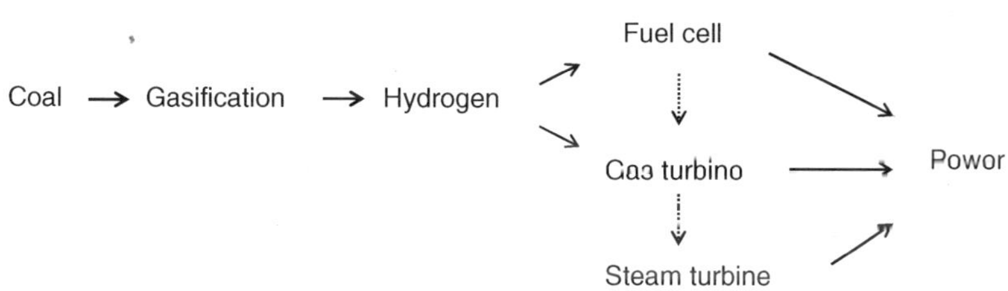

FIGURE 3.3
Integrated coal gasification fuel cell combined cycle.

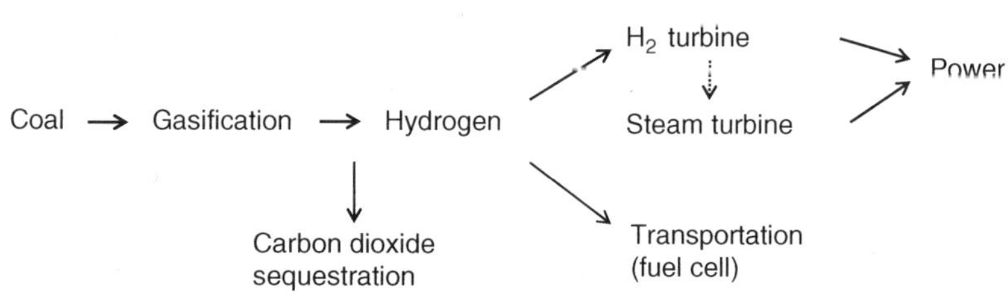

FIGURE 3.4
The FutureGen concept.

production research power plant. FutureGen will initiate operations around 2012. The project is intended to create the world's first zero-emission fossil fuel plant, which would be sized to generate approximately 270 MW of electricity.

Several other countries are also planning to build zero-emission power plants. The GreenGen Project, started in 2005 in China, will employ hydrogen to generate power and will include a CO_2-sequestering system.

Hydrogen derived from coal is a potential source of fuel for fuel cell vehicles of the future.

3.3 Technologies for Producing Hydrogen from Coal

The production of hydrogen-containing synthesis gas by the gasification of coal, gas clean up, the CO shift reaction, and methanation is illustrated in Figure 3.5. The gasification of coal is a well-established technology, but not yet economically competitive with steam reforming of natural gas (LPG) or naphtha for production of hydrogen. However, a number of cost-effective coal gasification installations have been commissioned in areas where natural gas and oil are not readily available and where coal is abundant. For example, Sasol (South Africa) uses coal to produce synthesis gas for the Fischer–Tropsch synthesis of gasoline, and coal is used to synthesize ammonia in China.

3.3.1 Process Components

Figure 3.5 shows the complete process for the production of hydrogen from coal. The process consists of the following components: coal preparation, air separation, coal gasification,

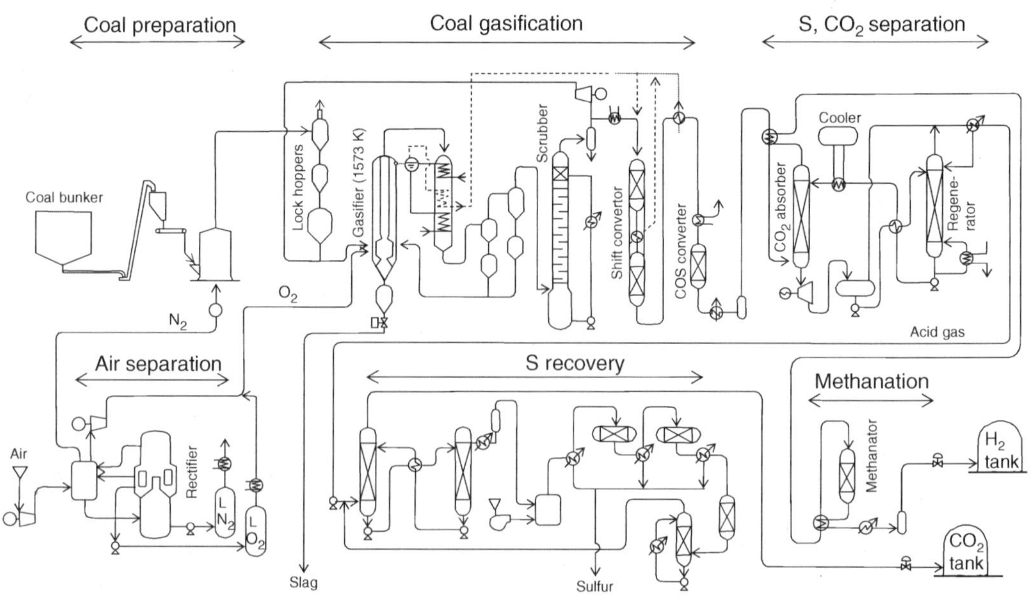

FIGURE 3.5
Production of hydrogen from coal.

syngas cooling, CO conversion to CO_2, CO_2/sulfur separation, sulfur recovery, and methanation. Here, we focus on current coal gasification technologies.

3.3.2 Coal Gasification

In a gasifier, coal (solid carbon) is converted to syngas by the following gasification reactions:

$$C + O_2 \rightarrow 2CO \quad \text{(oxygen gasification)} \tag{3.9}$$

$$C + H_2O \rightarrow H_2 + CO \quad \text{(steam gasification)} \tag{3.10}$$

$$C + 2H_2 \rightarrow CH_4 \quad \text{(hydrogasification)} \tag{3.11}$$

The gas-phase water–gas shift reaction is an important reaction that controls the equilibrium among CO, H_2, CO_2, and H_2O.

$$CO + H_2O \rightarrow H_2 + CO_2 \quad \text{(water–gas shift reaction)} \tag{3.12}$$

Figure 3.6 shows the equilibrium characteristics for the C–O_2–H_2O reaction system. To favor production of CO and H_2 from coal, reactions 3.9 and 3.10 should be carried out at a comparatively low pressure and low temperature. However, during actual production, synthesis of chemicals usually occur at high pressures of CO and H_2, and therefore, the gasifier should be operated at high pressure and high temperature to obtain high process efficiency.

A special concern in gasification is coal ash. The ash content of coal is approximately 10% or more. At high temperatures, ash undergoes a phase change, from solid to an intermediate soft stage to liquid. Because softened ash has a high viscosity and adheres to the gasifier wall and gas outline, resulting in production difficulties, coal gasification is usually performed at a temperature below or above the temperature range in which ash softens. A fluidized bed gasifier is employed for coal gasification below the ash-softening temperature, whereas an entrained bed gasifier is used above the ash-softening temperature.

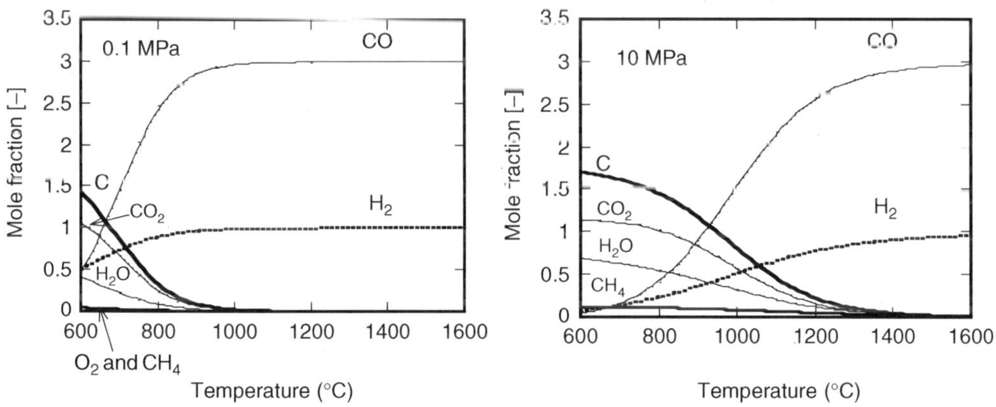

FIGURE 3.6
Equilibrium curves for the C–O_2–H_2O reaction system at 0.1 and 10 MPa total pressure (C/O_2/H_2O = 3/1/1 mol).

3.3.3 Entrained-Bed Gasification Technology

Coal contains several components that must be removed or substantially altered to obtain an acceptable fuel gases. Heat must be added and removed at various process stages. Finally, because coal is a solid, special techniques must be employed to feed it into the gasifier to ensure contact with reacting gases and to remove ash. There are three principal types of gasifiers: Fixed bed, fluidized bed, and entrained-bed (Shell, Texaco [General Electric, GE]) gasifiers. The following section focusses on the Shell and Texaco gasification technologies.

3.3.3.1 Shell Gasification Technology

The Shell gasification process is based on an entrained upflow, oxygen-blown, slagging gasifier using a dry-feed system, in which the pulverized coal is injected into the gasifier by nitrogen gas. The gasifier is equipped with multiple burners configured in pairs, and an inner membrane wall consisting of tubes through which water circulates. A separate air separation unit provides oxygen for gasification and nitrogen for coal transportation. The syngas exiting the gasifier is quenched by cold, recycled syngas and further cooled in a syngas cooler. The raw syngas is cleaned by filtering and water scrubbing to remove fly ash, and by acid gas treating including sulfur recovery.[4]

The main features of the Shell gasification process are illustrated in Figure 3.7.

The Shell process uses partial oxygen gasification. Because insufficient oxygen exists for complete combustion (20–30% of the oxygen required for complete combustion is used), only a fraction of carbon in the coal is oxidized completely to CO_2. The heat released from this combustion provides most of the energy needed for endothermic coal gasification reactions and raises the gasifier temperature. Some steam is usually added to prevent excessive

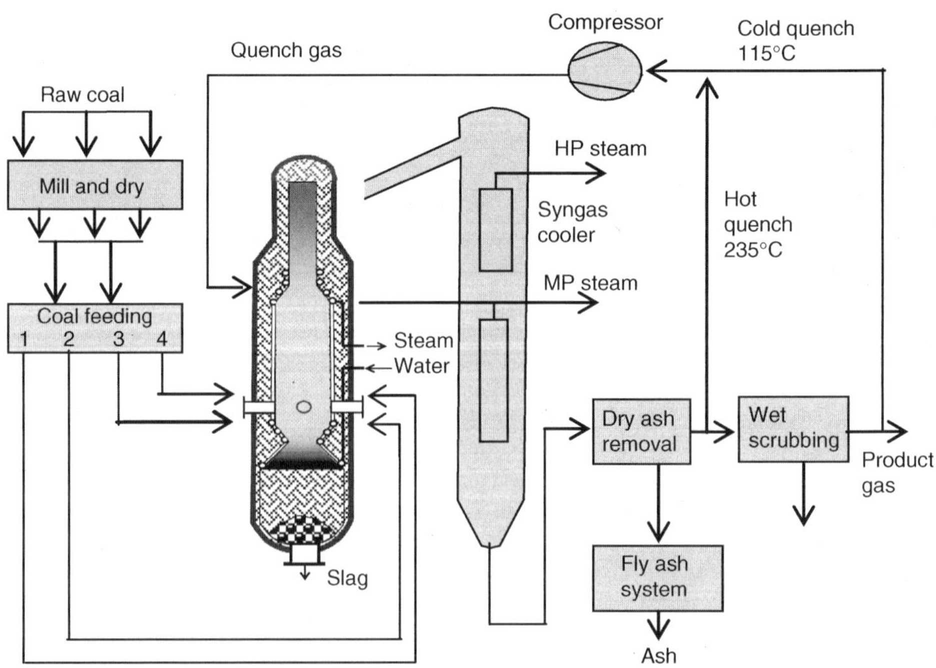

FIGURE 3.7
Main features of the Shell coal gasification process.

increase in temperature. The reactivity of the coal and caking tendency are important parameters affecting the choice of coal gasification technology.

The Shell syngas contains about 80–83% of the energy in the coal feed (the cold gas efficiency). This high efficiency is due to the high carbon conversion (>96%) during gasification (Tables 3.4 through 3.6).[5]

The development of the Shell gasification technology began in the early 1970s, being driven mainly by the high oil prices and the strategic importance to secure a reliable supply of feedstock over the long term. In 1987, a larger plant was put into operation at Shell Oil's Deer Park Manufacturing Complex near Houston. This plant was designed to gasify 250 t of high sulfur bituminous coal per day or 400 t of high moisture coal as lignite per day. In 1998, a 2000 t/day Shell gasifier was placed in operation at Demkolec for integrated gasification combined cycle (IGCC) power generation. Currently, Shell gasification

TABLE 3.4

Coal Used in the Shell Gasifier

Moisture	3.7–34%
Ash	0.5–30%
Sulfur	0.5–5.2%
High heating value	23–33 MJ/kg
Ash melting point	1090–1500°C

Source: Adapted from Hailong, X., The Shell Coal Gasification Process (SCGP), *International Hi-Tech Symposium on Coal Chemical Industry & Coal Conversion*, Oct. 30–31, Shanghai, China, 2004.

TABLE 3.5

Consumption of Feedstock and Utility during Shell Gasification

Coal (18 wt% ash and 28 MJ/kg high heating value; both moisture free)	0.59 kg/Nm^3 (CO + H_2)
Oxygen	0.48 kg/Nm^3 (CO + H_2)

Source: Adapted from Hailong, X., The Shell Coal Gasification Process (SCGP), *International Hi-Tech Symposium on Coal Chemical Industry & Coal Conversion*, Oct. 30–31, Shanghai, China, 2004.

TABLE 3.6

Gas Products from the Shell Gasifier

Gases	Volume (%)
H_2	26.7
CO	63.3
CO_2	1.5
H_2O	2.0
H_2S	1.3
N_2	4.1
Ar	1.1
CH_4	0.0

Source: Adapted from Hailong, X., The Shell Coal Gasification Process (SCGP), *International Hi-Tech Symposium on Coal Chemical Industry & Coal Conversion*, Oct. 30–31, Shanghai, China, 2004.

TABLE 3.7

Number of Commercial Shell Gasifiers (Including Those Under Construction)

The Netherlands	2
America	1
Australia	1
China	18

Source: Adapted from Bos, H. and Van Dongen, F. Shell Coal Gasification Process, *The 23rd International Pittsburgh Coal Conference*, Sep. 25–28, 2006.

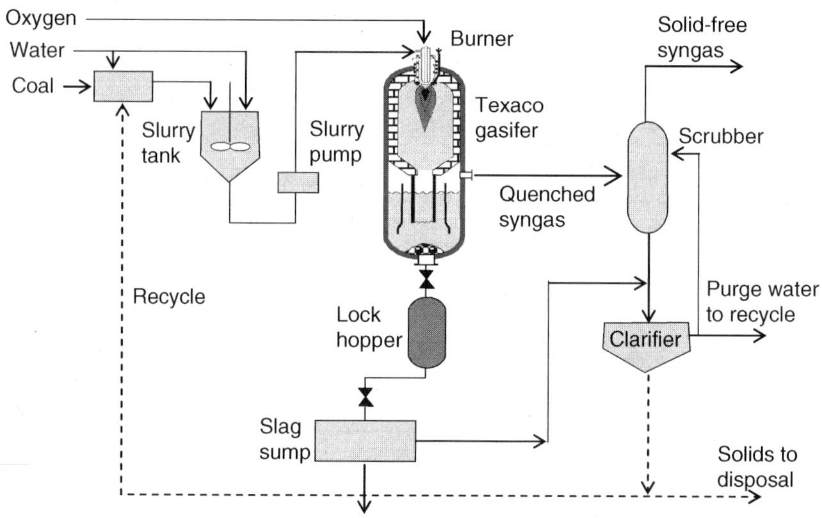

FIGURE 3.8
Texaco (GE) gasification process.

technology is mainly used in China to produce ammonia, methanol, and coal-to-liquid (CTL) products (Table 3.7).[6]

3.3.3.2 Texaco (GE) Gasification Technology

The Texaco (GE) gasification process employs an entrained downflow gasifier with a slurry feed based on a design similar to that of a natural gas or heavy residual fuel oil gasifier (Figure 3.8). A slurry of pulverized coal and water is fed into the gasifier from the top by a slurry pump. Oxygen is also introduced through the top of the gasifier, and steam is added to moderate the reaction temperature. The residence time in the gasifier is a few seconds. Because of the high reaction temperature (1573–1773 K), the product gas consists principally of CO and H_2 (Table 3.8).[7] Gas and unconverted solids from the bottom of the gasifier chamber are quenched in water. Slag is solidified and removed periodically from the bottom of the vessel through a lock hopper.

In 1956, a 100 t/day Texaco gasifier was built at the Morgantown Ordnance Works in West Virginia. In 1980, a 100 MW (1000 t/day) IGCC power generation plant using the Texaco gasification process was constructed near Dagget, California. In 2004, GE acquired the Texaco gasification technology. The number of commercial units in various countries are shown in Table 3.9.[4,8]

TABLE 3.8

Gas Products from the Texaco (GE) Gasifier

Gases	Volume (%)
H_2	34
CO	48
CO_2	17
N_2 + Ar	1

Source: Adapted from Bisio, A. and Boots, S., *Energy Technology and the Environment*, Wiley, New York, 1995.

TABLE 3.9

Number of Commercial Texaco (GE) Gasification Plants

United States	17	Sweden	1
Germany	7	China	24
France	2	Japan	4
Italy	5	Singapore	2
United Kingdom	2	South Korea	1
Spain	1	Australia	1

Source: Adapted from Qianlin, Z., GE Gasification, the best available Clean Coal Technology, *International Forum on C1 Chemical Industry and Clean Coal Technology*, Jun. 13–17, Shihezi, China, 2005.

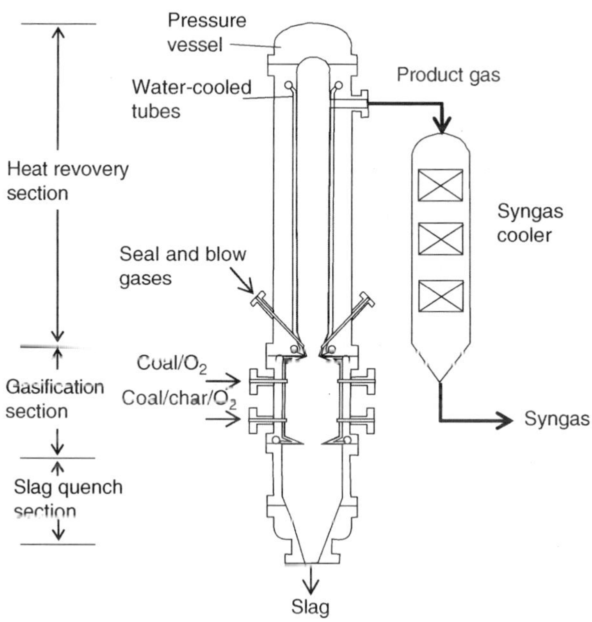

FIGURE 3.9
EAGLE gasifier.

3.3.3.3 EAGLE Gasification Technology

EAGLE is an entrained coal gasification technology (150 t/day). EAGLE gasification technology is widely applicable to the production of a variety of chemicals, hydrogen, synthetic liquid fuel, and electric power. The EAGLE gasifier (Figure 3.9) is an entrained upflow type with a two-stage swirling burner in a single chamber. High-efficiency power generation has been achieved by combining EAGLE gasification technology with gas turbines, steam turbines, and fuel cells.[9,10]

3.4 New Developments for Production of Hydrogen from Coal

Because coal is an important economical source for production of hydrogen, developing new technologies to improve the efficiency of hydrogen production is an important priority. Some new approaches for producing hydrogen from coal are discussed in the following section.

3.4.1 HyPr-RING Process, Direct Production of Hydrogen from Coal

3.4.1.1 Principle of HyPr-RING Method

HyPr-RING is an acronym for *hy*drogen *pr*oduction by *r*eaction *i*ntegrated *n*ovel *g*asification. It is a method trying to produce hydrogen directly from coal by using a single reactor.

In conventional production of hydrogen from coal, as described earlier, coal is converted to hydrogen and carbon monoxide (CO) through the water–carbon reaction as shown in reactions 3.9 through 3.11. CO is then converted to hydrogen and carbon dioxide by the water–gas shift reaction as shown in reaction 3.12.

Reaction 3.10 is endothermic, about 1273 K or higher temperature is necessary to obtain a sufficiently fast reaction rate. In contrast, reaction 3.12 is exothermic and does not require such high temperatures to obtain a higher conversion of the CO because it is governed by equilibrium ($K_C = P_{CO_2}P_{H_2}/P_{H_2O}P_{CO}$). In most of the conventional processes,[4] reactions 3.9 and 3.10 are performed in the first reactor with an operating temperature above 1273 K. The produced gases are then introduced into the second reactor, which is usually operated below 673 K to perform the water–gas shift reaction (reaction 3.12). However, the product of reaction 3.12 contains the CO_2, so that an extra separation process is required to obtain pure hydrogen (see Figure 3.5). Conventional process is very complicated with multiple steps and conditions.

If the operating temperature of reaction 3.10 can be reduced and that of reaction 3.12 can be raised, reactions 3.10 and 3.12 can occur in the same reactor, and as a result, the time and cost of the process can be reduced. To integrate the reactions in one reactor, HyPr-RING method introduced CO_2 absorption reaction into the reaction system as shown in Equations 3.13 through 3.15.[11,12]

$$C + H_2O \rightarrow CO + H_2 \qquad \Delta H^\circ_{298} = 132\ \text{kJ/mol} \tag{3.13}$$

$$CO + H_2O \rightarrow CO_2 + H_2 \qquad \Delta H^\circ_{298} = -41\ \text{kJ/mol} \tag{3.14}$$

$$CaO + H_2O \rightarrow Ca(OH)_2 \qquad \Delta H^\circ_{298} = -109\ \text{kJ/mol} \tag{3.15}$$

$$Ca(OH)_2 + CO_2 \rightarrow CaCO_3 + H_2O \qquad \Delta H^\circ_{298} = -69\ \text{kJ/mol} \tag{3.16}$$

The partial pressure ratio, $P_{CO_2}P_{H_2}/P_{H_2O}P_{CO}$, is decreased by the absorption of carbon dioxide. Thus, the equilibrium of the reaction 3.12 may be kept at a higher temperature simultaneously with the CO_2 absorption. The expression of the overall reaction is written as follows:

$$C + CaO + 2H_2O \rightarrow CaCO_3 + 2H_2 \qquad \Delta H^\circ_{298} = -88\ \text{kJ/mol} \tag{3.17}$$

CaO is also reported to absorb H_2S and catalyze NH_3 and tar decomposition.

$$CaO + H_2S \rightarrow CaS + H_2O \tag{3.18}$$

$$2NH_3 \rightarrow N_2 + 3H_2 \tag{3.19}$$

Equation 3.17 shows the new reaction system and also shows several possibilities: (1) producing high concentration hydrogen with a single gasifier (Figure 3.10), (2) no need of combustion in the gasifier, and (3) high cold gas efficiency.

Equilibrium compositions from C–H_2O–CaO and C–H_2O reaction systems are calculated by using HSC Chemistry 4.0 software. Gas compositions for the C–H_2O–CaO reaction system and for the C–H_2O reaction system are shown in Figure 3.11a and 3.11b, respectively. In the C–H_2O–CaO reaction system, CO, CO_2, and CH_4 are lower, and H_2 is higher than those in the C–H_2O reaction system. CO and CO_2 decreased with increasing pressure, and

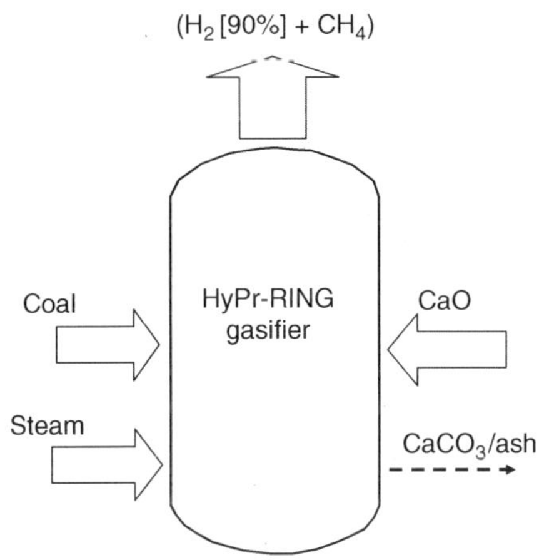

FIGURE 3.10
Direct hydrogen production from coal.

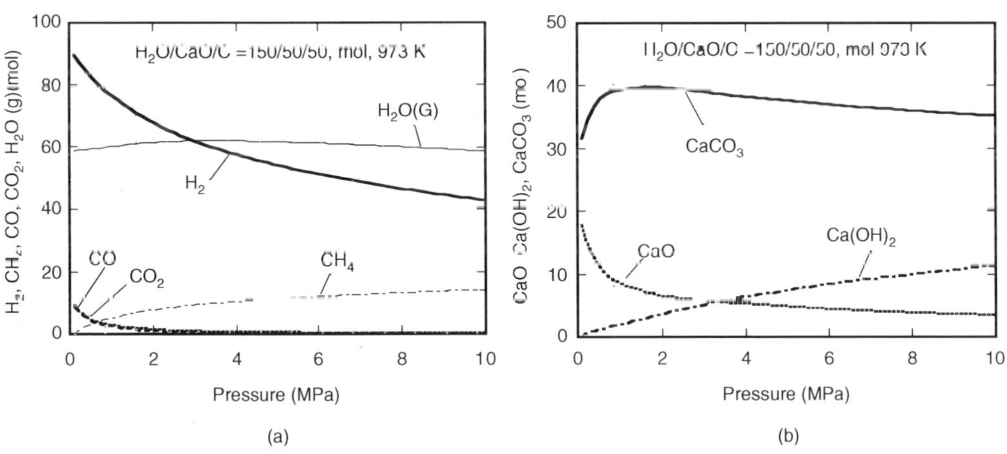

FIGURE 3.11
Equilibrium C–CaO–H_2O 923 K and 973 K, gas and calcium. (Adapted from Shiying, L., Michiaki, H., Yoshizo, S., and Hiroyuki, H., *Fuel*, 81, 2079–2085, 2002.)

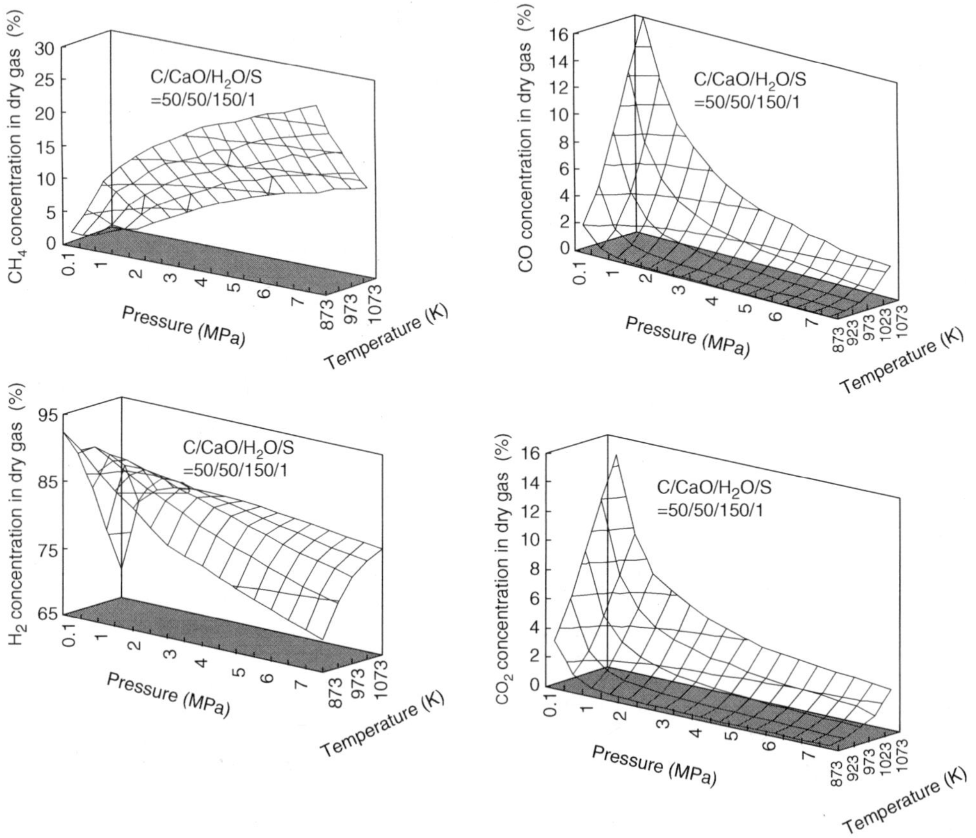

FIGURE 3.12
Equilibrium of gas products under various temperatures and pressures for C–CaO–H_2O system.

reached approximately zero at pressures above 4 MPa. CO_2 fixation by CaO is a significant factor in the reduction of CO and CH_4 and the increase of H_2.[13] Without CO and CO_2, H_2 is the primary resultant gas together with a small amount of CH_4. The ratio of H_2/CH_4 was about 6/1 at 4 Mpa (Figure 3.12).

The phase equilibriums between CaO, $Ca(OH)_2$, and $CaCO_3$ (Figure 3.13)[14] are also examined. The equilibrium constants for reactions 3.20 through 3.22 are defined as $K_{C19} = 1/P^*_{H_2O}$, $K_{C20} = P^*_{H_2O}/P^*_{CO_2}$, and $K_{C21} = 1/P^*_{CO_2}$, where $P^*_{H_2O}$ and $P^*_{CO_2}$ are the equilibrium partial pressures of H_2O and CO_2. By fixing $P^*_{H_2O}$, we can see how $P^*_{CO_2}$ changes when the equilibrium constant K_{C20} varies with temperature.

$$CaO + H_2O \rightleftharpoons Ca(OH)_2 \tag{3.20}$$

$$Ca(OH)_2 + CO_2 \rightleftharpoons CaCO_3 + H_2O \tag{3.21}$$

$$CaO + CO_2 \rightleftharpoons CaCO_3 \tag{3.22}$$

3.4.1.2 Development of HyPr-RING Process in Japan

HyPr-RING method was proposed in 1998; in 2000, a project started in Japan to develop this method to a commercial process. Figure 3.14 shows the concept of HyPr-RING process. Two chemical loops are included in this process. The first chemical loop is the water cycle.

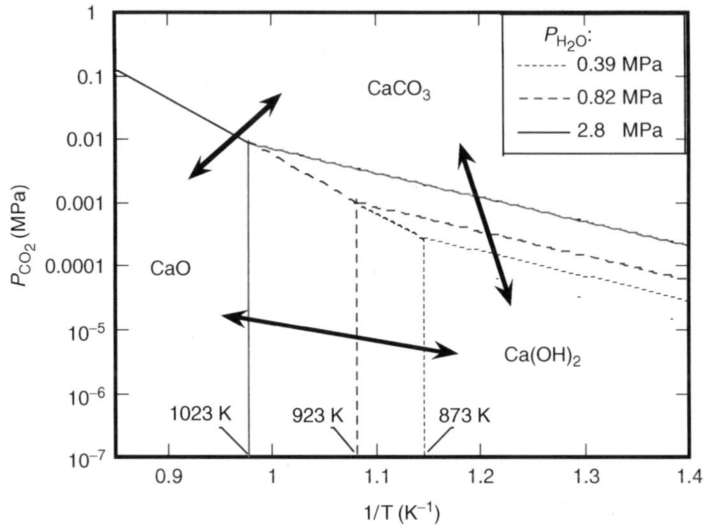

FIGURE 3.13
Phase equilibriums between CaO, $Ca(OH)_2$, and $CaCO_3$. (Adapted from Shiying, L., Michiaki, H., Yoshizo, S., and Hiroyuki, H., *Fuel*, 85, 1143–1150, 2006.)

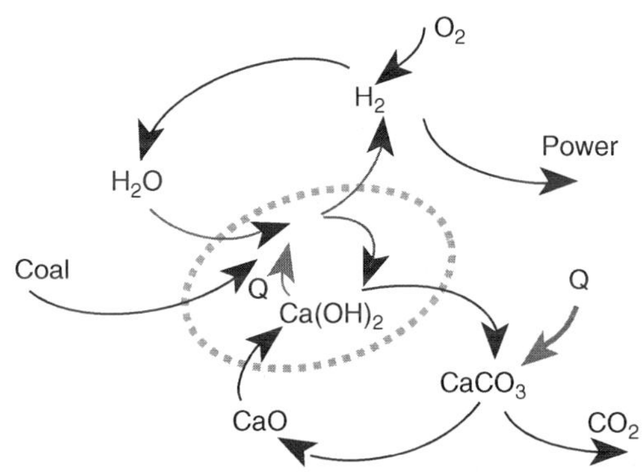

FIGURE 3.14
Concept of HyPr RING process. (Adapted from Shiying, L., Michiaki, H., Yoshizo, S., and Hiroyuki, H., *Energ. Convers. Manag.*, 46, 869–880, 2005.)

Water and coal react to produce hydrogen and CO_2, and absorb heat. The hydrogen reacts with O_2 and forms H_2O (H_2O–H_2–H_2O) to make power. The second is the calcium cycle. CaO absorbs CO_2 and this reaction releases heat to form $CaCO_3$, then $CaCO_3$ is regenerated as CaO to be recycled (CaO–$CaCO_3$–CaO) and CO_2 is released. Reactions of carbon with water and CaO with CO_2 would occur in the same reactor. For a complete HyPr-RING process, raw materials supplied are the hydrocarbons such as coal, water, and oxygen. Major products are hydrogen, power, and pure CO_2. The HyPr-RING process is the first attempt at complete absorption of CO_2 and simultaneous conversion of CO in one reactor.

3.4.1.2.1 *Process Analysis*

The feasibility of performing the water–carbon oxidation–reduction reaction in combination with the CO_2 absorption by CaO in a single reactor has been confirmed in experiments.[11,12] The underlying concept of the HyPr-RING process involves two reactors: a main reactor (the gasifier) and a regenerator. For the whole process, the raw materials supplied are the hydrocarbons, water, and CaO; the major products are hydrogen and pure CO_2. In the gasifier, hydrocarbons and H_2O react to produce H_2 and CO_2. CaO first reacts with high pressure H_2O to form reactive $Ca(OH)_2$, which then absorbs CO_2, giving $CaCO_3$ and releasing heat. In the regenerator, the CaO is regenerated from $CaCO_3$ with the release of CO_2.

Figure 3.15 shows the mass and energy flows calculated for a typical HyPr-RING process of 1000 t coal per day.[15] In the gasifier, 100% of H and O and about 53% of carbon in the coal are suggested to be gasified, and 47% of carbon (char) remains in solid residues as fuel for $CaCO_3$ calcination in the regenerator. Products of the gasifier and the regenerator are in equilibrium composition, calculated on the basis of the input materials. When the inputs consist of 1×10^6 kg of coal, 2.18×10^6 kg of CaO, and 2.03×10^6 kg of steam, the equilibrium gas products at 923 K and 3 MPa are 5.78×10^4 kmol of H_2, 0.58×10^4 kmol of CH_4, and 6.38×10^4 kmol of steam. The dry gas contains 90.6 vol% H_2 and 9 vol% CH_4; CO and CO_2 comprise <0.4 vol%. The equilibrium gas compositions are very sensitive to temperature and pressure, as shown in Figures 3.11 and 3.12. The proportion of H_2 increases with increasing temperature and decreasing pressure.

The solid products of the gasifier in equilibrium are 2.54×10^6 kg of $CaCO_3$, 0.70×10^6 kg of $Ca(OH)_2$, and 0.30×10^6 kg of CaO, with 0.32×10^6 kg unreacted carbon.

Input heats to the gasifier are coal 27.98×10^9 kJ (heat value), CaO 4.89×10^9 kJ (heat of reaction to form $Ca(OH)_2$ and $CaCO_3$), and the superheated steam 7.34×10^9 kJ (873 K, 3 MPa). To improve the heat balance of the gasifier, it is suggested that the temperatures of the input material and steam should be as high as possible. CaO is recycled at 573 K, and then mixed with coal (298 K; Figure 3.15). Steam is injected at 873 K. The steam is heated by heat recovered from the exhausts of the gasifier and the regenerator.

Output heats from the gasifier are in the form of fuel gases and unreacted carbon (reaction heat to form H_2O and CO_2) and sensible and latent heats in the gas and solid exhausts. The input heat is larger than the output for the gasifier, given that the allowable heat loss of the gasifier is 0.15×10^9 kJ, which is about 0.5% of the total heat generated actually in the gasifier.

Input materials for the regenerator are the solid exhausts from the gasifier (at 473 K after heat exchange), supplementary $CaCO_3$ (0.587×10^6 kg), and O_2 (40 vol%) with a dilution gas of CO_2 and H_2O (60 vol%). The dilution gas is the recycled gas from the regenerator. The regenerator is composed of two sections, one for preheating and the other for calcination. The calcination section is operated at 1373 K and 0.1 MPa and is heated by internal combustion of carbon contained in the solid exhaust from the gasifier. The preheating section is heated by exhaust gas (1373 K) from the calcination section. The solid exhaust of the gasifier is first preheated to 923 K in the preheating section, while $Ca(OH)_2$ decomposes. After preheating, the solid exhaust and the supplementary $CaCO_3$ are injected into the calcination section to calcine the $CaCO_3$. The calcination of $CaCO_3$ was presumed to occur under an average CO_2 partial pressure of 0.072 MPa at 1373 K. The equilibrium composition shows that, under these conditions, $CaCO_3$ is nearly completely (about 96 Ca mol%) decomposed into CaO. Accordingly, the solid output flows from the calcination section are 2.49×10^6 kg CaO, 0.18×10^6 kg $CaCO_3$, and 0.51×10^6 kg ash (1373 K).

Input heats for the regenerator are the heat of combustion of the carbon contained in the solid exhaust of the gasifier and the sensible and latent heats contained in the injection gases and solids. Output heats are the heat of reaction of CaO (calcination heat in the regenerator), recovered heat, and the heat contained in the gaseous and solid exhausts. The input heat was designed to be larger than the output heat to allow a heat loss in the regenerator of about 0.21×10^9 kJ.

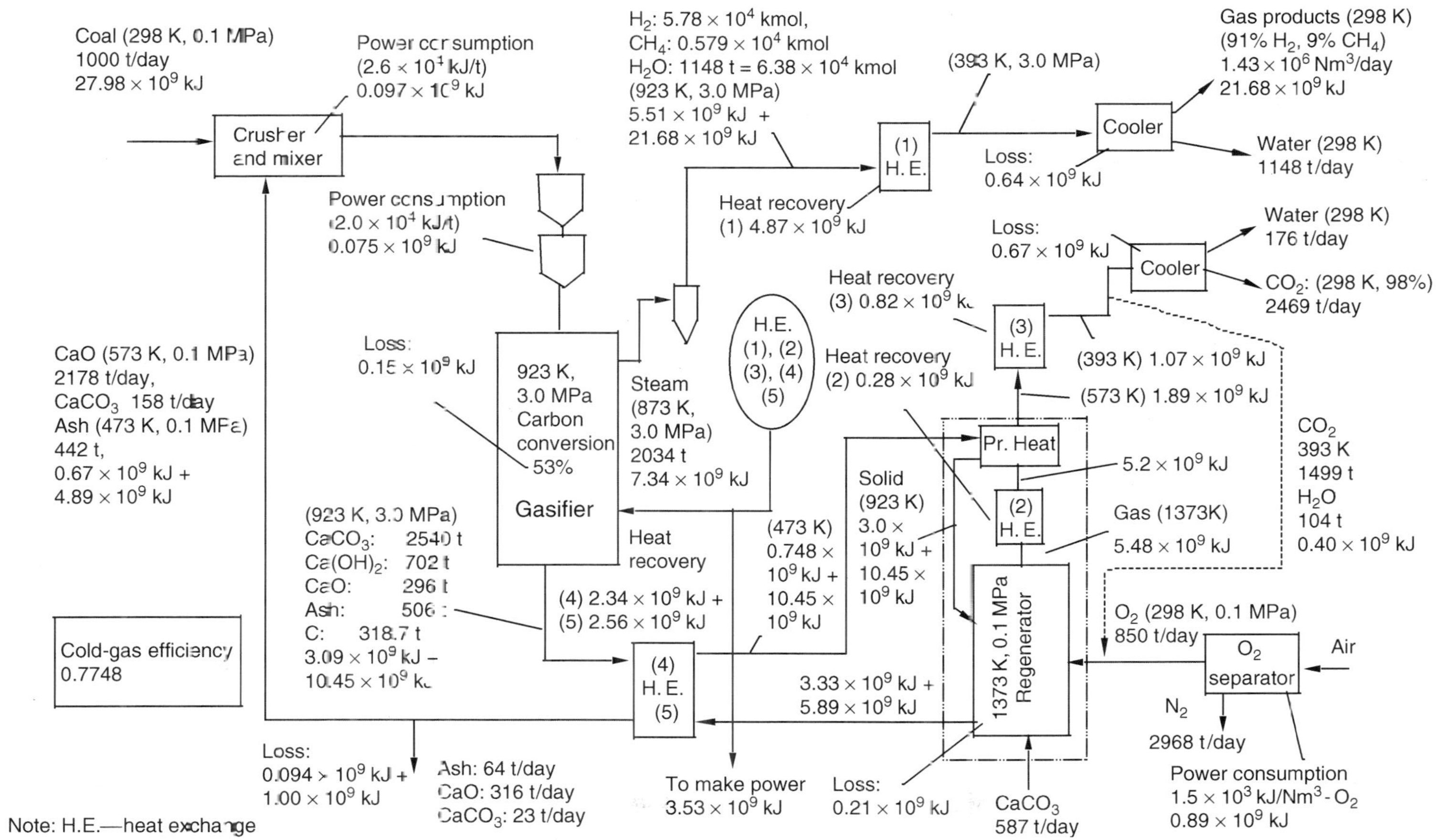

FIGURE 3.15
Mass-energy flowchart of the HyPr-RING process. (Adapted from Shiying, L., Michiaki, H., Yoshizo, S., and Hiroyuki, H., *Energ. Convers. Manag.*, 46, 869–880, 2005.)

Heat exchangers for heat recovery from the process are located mainly in the gaseous exhausts of the gasifier (1), the regenerator (2) and (3), and in the solid exhausts of the gasifier (4), and the regenerator (5) (in Figure 3.15). Recovered heat is used to superheat steam for the gasifier (7.34×10^9 kJ), and to generate power for oxygen separation and other power-consuming parts of the process (3.53×10^9 kJ).

The mass and heat balances of the process are satisfactory when coal is gasified to a 53% carbon conversion in the gasifier and the remaining 47% of carbon is sent to the regenerator. From the heat values of the produced fuel gas and the input coal, the cold-gas efficiency (91 vol% H_2 with 9 vol% CH_4; 298 K, 0.1 MPa) was calculated to be higher than 0.77.

$$\text{Cold-gas efficiency} = \frac{\text{heat value of fuel gas}}{\text{heat value of coal}} \tag{3.23}$$

The amount of steam introduced into the gasifier is an important factor in maintaining the steam partial pressure, not only for the gas–solid reactions, but also for the product gas equilibrium. As seen in Figure 3.16,[15] increasing the steam content would increase the H_2 content and decrease the CH_4 content of the product gas. For example, when the steam content is 50 vol%, the H_2/CH_4 ratio is about 10, whereas when the steam content is 60 vol%, the H_2/CH_4 ratio increases to 18; however, increasing the steam content in the product gas could cause an increase in heat loss. It can also be seen in Figure 3.16 that decreasing the operating pressure or increasing temperature also increases H_2/CH_4 ratio, but the effects are less than those achieved by increasing the steam content.

3.4.1.2.2 *Experimental Confirmation*

A continuously supplied reaction system was built and the integration of reactions was tested.[14,16] The coal/CaO mixture was quantitatively introduced into the reactor by a

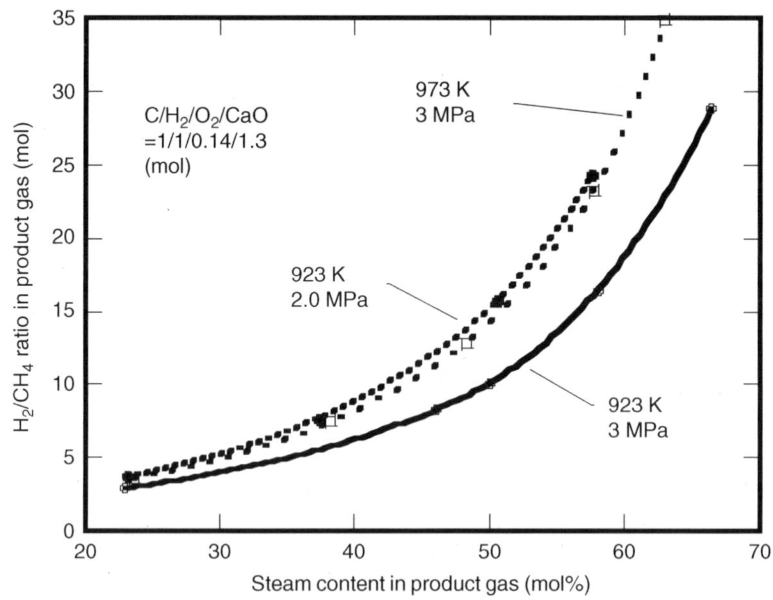

FIGURE 3.16
Effect of steam content on product gas composition. (Adapted from Shiying, L., Michiaki, H., Yoshizo, S., and Hiroyuki, H., *Energ. Convers. Manag.*, 46, 869–880, 2005.)

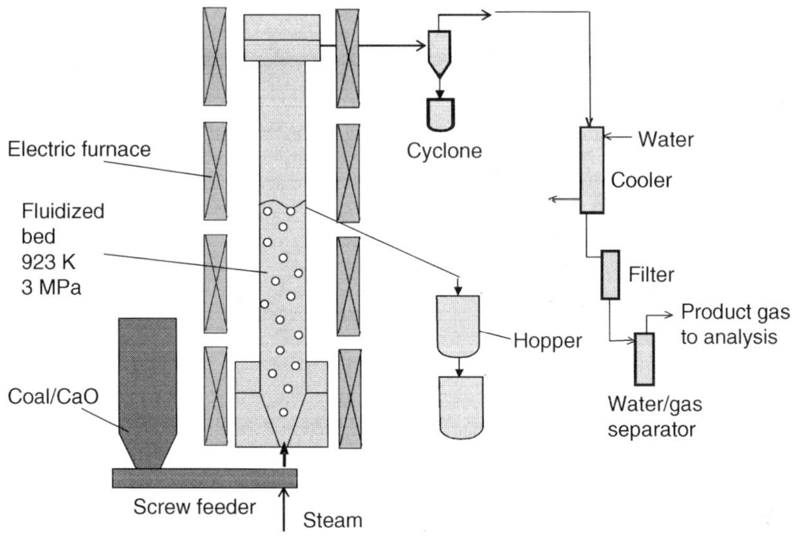

FIGURE 3.17
Schematic diagram of the continuous experiment high-pressure reactor. (Adapted from Shiying, L., Michiaki, H., Yoshizo, S., and Hiroyuki, H., Continuous Experiment Regarding Hydrogen Production by Coal/CaO Reaction with Steam (HyPr-RING), *21st Pittsburgh Coal Conference*, Osaka, Japan, Sep. 13–17, 2004.)

screw feeder (Figure 3.17).[17] High-pressure (0.1–10 MPa) and high-temperature steam was generated by forcing water through a Hastelloy coil (length 43 m, inner diameter 1.2 mm) housed in a fluidized hot sand bath (16 kW). The coal/CaO mixture and steam were injected into the reactor from the bottom by a nozzle. Product gases passed through the cyclone into a cooling section consisting of a stainless tube coil (length 6.4 m, inner diameter 6 mm) housed in a flowing water bath. In the cooling section, steam was condensed to water and was separated from the other gas products. After passing through a filter and a pressure regulator, the gas products were continuously analyzed.

The rate of gas production from the Taiheiyo coal/CaO–steam reaction at 923 K and 3 MPa increased rapidly during the first 40 min of reaction (Figure 3.18).[18] During this time, char and calcium particles were formed in the reactor. Gasification of the char caused the rapid increase in the rate of gas production. After the initial 40 min of the reaction, the rate of gas production stabilized, increasing only slightly with increasing reaction time (steady state). The increase in char with supply was roughly equal to the decrease in char by gasification. Gas analysis showed that, excluding H_2O and N_2, H_2 was the main gas; CH_4 and a few other gases were produced in smaller amounts. At point A in the initial stage of reaction (cf. Figure 3.18), the gas concentrations were 77% H_2, 14% CH_4, 1.9% C_2H_4, 1.3% C_2H_6, and about 6% $CO + CO_2$; at point B in the stable stage (steady state), the gas concentrations were 84% H_2, 10% CH_4, 1.2% C_2H_4, 0.8% C_2H_6, and about 2.7% $CO + CO_2$. Thus, the amount of H_2 increased significantly and the amount of CO_2 decreased somewhat over the course of the reaction. At the steady state, the product gas concentrations were approximately equal to the equilibrium gas concentration. However, the H_2/CH_4 ratio in the initial stage of the reaction was smaller than that in the stable stage because CH_4 as well as C_2H_6 and C_2H_4 produced by coal pyrolysis is proportionally more abundant in the initial stage than that in the stable stage of reaction.

In 2004, a 50 kg coal per day continuous run facility was built to confirm reactor (fluidized bed) operating condition, feeding method, product gas composition, sorbent recycle, etc.[18]

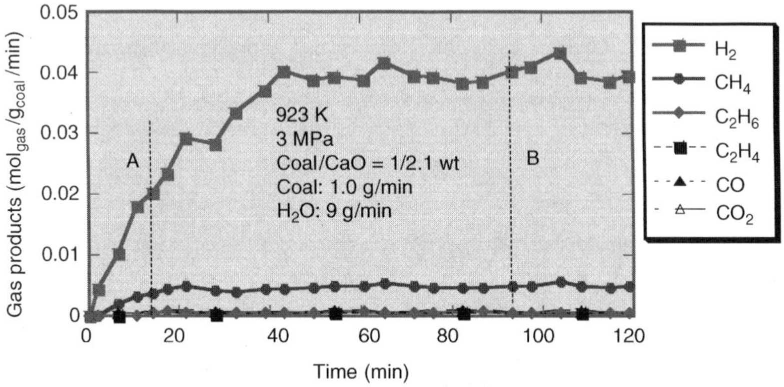

Vol (%)	H_2	CH_4	C_2H_6	C_2H_4	CO	CO_2	H_2S	NH_3
A	77	14	1.9	1.3	2.4	3.4		
B	84	10	1.2	0.8	1.2	1.5	3	0

FIGURE 3.18
Gas products from continuous supply reaction of coal/CaO mixture with steam. (Adapted from Shiying, L., Michiaki, H., Yoshizo, S., and Hiroyuki, H., Continuous Experiment Regarding Hydrogen Production by Coal/CaO Reaction with Steam (HyPr-RING), *21st Pittsburgh Coal Conference*, Osaka, Japan, Sep. 13–17, 2004.)

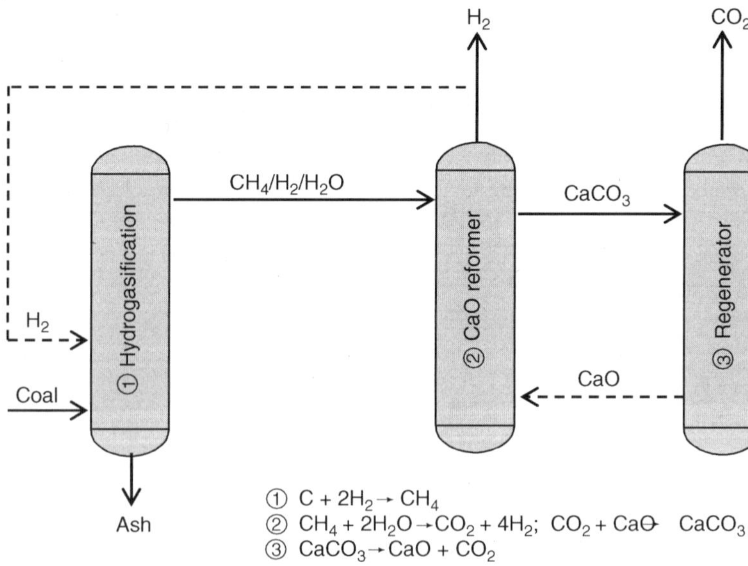

FIGURE 3.19
Concept of ZECA project for hydrogen production from coal.

3.4.2 ZECA Project

A new concept of hydrogen production from coal, called *Z*ero *E*mission *C*oal *A*lliance (ZECA) process, was proposed by the Los Alamos National Laboratory.[19] ZECA process is illustrated in Figure 3.19. The figure shows a schematic of the hydrogen production conception, which integrates coal gasification without combustion, hydrogen production with

CO_2 absorption, and sorbent regeneration. Figure 3.19 shows the concept of ZECA process for hydrogen production from coal. In the ZECA concept,[20] coal is first hydrogasified to produce CH_4 in reactor 1. The reactor 1 operates with recycled hydrogen gas at a temperature of 1088 K and pressure of 6.2 MPa. The gas leaving reactor 1 is composed mainly of CH_4 and H_2 as well as steam CO and CO_2. The produced gas from reactor 1 is then sent to a reformer (reactor 2) with the CaO bed material. Here, CH_4 is reformed by steam to CO_2 and H_2, and CO_2 is fixed by CaO to $CaCO_3$. The reactor 2 (reformer) may be operated under 1088 K and 3 MPa. The used CO_2 sorbent as $CaCO_3$ can then be sent to a regenerator, which operates under 1403 K with a pressure just above the atmospheric.[19]

3.4.3 Coal Gasification with CCR Process

The Ohio state university proposed a method to combine coal gasification with a carbonation/calcinations reaction (CCR) process to produce hydrogen. Sorbents based on CaO are reported to be very effective for CO_2 absorption at high temperatures. Coal gasification produces the synthesis gas containing CO, H_2, CO_2, and H_2O. The synthesis gas is then passed through a water–gas shift reactor with the injection of a CaO-based sorbent. In the shift reactor, the following reactions take place (Figure 3.20):

$$CO + H_2O \rightarrow CO_2 + H_2 \tag{3.24}$$

$$CO_2 + CaO \rightarrow CaCO_3 \tag{3.25}$$

The $CaCO_3$ produced in Equation 3.25 can then be converted back to CaO in a rotary calciner and reused.[21]

3.4.4 AGC Project

The Advanced Gasification Combustion (AGC) project was developed by the National Energy Technology Laboratory (NETL) and General Electric and Environmental Research Corporation (GE-EER), etc. The AGC process is considered to have three fluidized bed reactors. In reactor 1, coal is partly gasified with steam in the presence of a CO_2 sorbent. The remaining char together with the CO_2 sorbent is sent to reactor 2. In reactor 2, char is

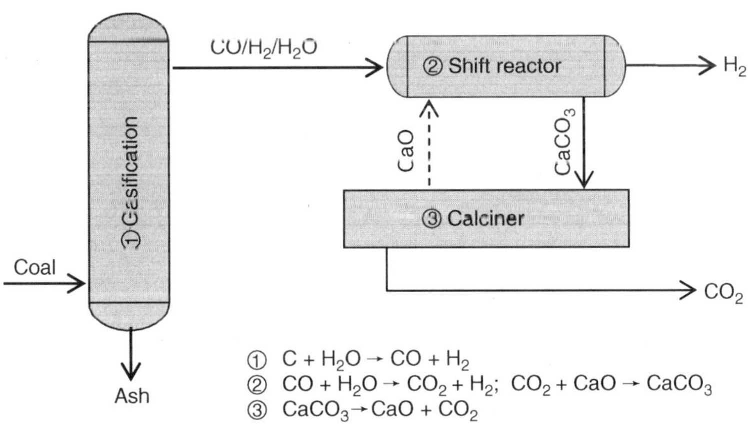

FIGURE 3.20
Hydrogen production by combining coal gasification with CCR process.

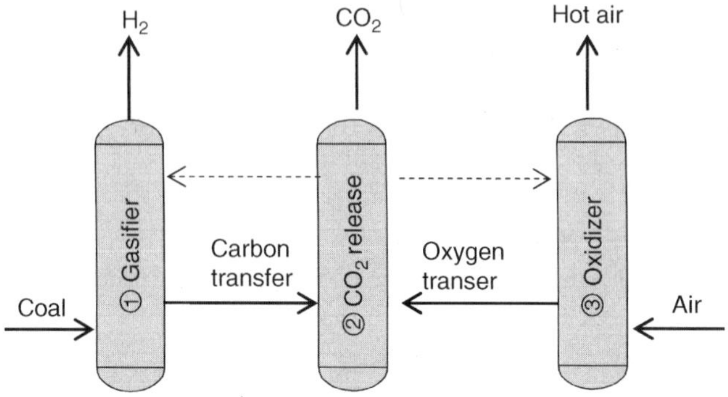

FIGURE 3.21
Concept of AGC project for production of hydrogen from coal.

oxidized and CO_2 sorbent is regenerated in a bed containing O_2 transfer material, which is transferred from reactor 3 at a temperature of 1223 K. The heat released by the combustion of the char is used to regenerate the CO_2 sorbent, which releases CO_2. The O_2 transfer material is then returned to reactor 3, operating at 1023 K to be generated under air (Figure 3.21). This project is now under pilot experiment.[22]

3.5 Conclusion

Production of pure hydrogen from coal involves coal gasification and CO_2 separation together with conversion of CO to H_2 by the water–gas shift reaction. Coal gasification is the most expensive part in the process of producing hydrogen from coal. Several entrained bed coal gasification technologies were introduced in this chapter.

Several new concepts under development for hydrogen production from coal were also described. The HyPr-RING method combines coal gasification, CO_2 separation, and water–gas shift reaction in a gasifier to produce hydrogen.

References

1. *Coal Information 2005*, International Energy Agency (IEA), Head of Publications Service, Paris, 2005.
2. *World Energy Outlook 2004*, International Energy Agency, 2004.
3. Survey of Energy Resources, BGR, WEC (World Energy Conference 1980).
4. Meyers R., *Coal Handbook*, Marcel Dekker, New York, 1981.
5. Hailong X., The Shell Coal Gasification Process (SCGP), *International Hi-Tech Symposium on Coal Chemical Industry & Coal Conversion*, Oct. 30–31, Shanghai, China, 2004.
6. Bos H. and Van Dongen F., Shell Coal Gasification Process, *The 23rd International Pittsburgh Coal Conference*, Pittsburgh, Sep. 25–28, 2006.
7. Bisio A. and Boots S., *Energy Technology and the Environment*, Wiley, New York, 1995.
8. Qianlin Z., GE Gasification, The Best Available Clean Coal Technology, *International Forum on C1 Chemical Industry and Clean Coal Technology*, Jun. 13–17, Shihezi, China, 2005.

9. Eiki S., The Status of the EAGLE Project, *Advanced Clean Coal Technology International Symposium 2006*, Sep. 5, Tokyo, Japan, 2006.
10. Multi-purpose Coal Gasification Technology Development, EAGLE, *Coal Energy Application for Gas, Liquid & Electricity*, New Energy and Industrial Technology Development Organization, Japan.
11. Shiying L., Yoshizo S., Hiroyuki H. and Michiaki H., A New Method (HyPr-RING) for Producing Hydrogen from Coals, *The 10th International Conference on Coal Science*, Sep. 12–17, Taiyuan, China, 1999.
12. Shiying L., Yoshizo S., Hiroyuki H. and Michiaki H., Producing Hydrogen from Coals by using a Method of Reaction Integrated Novel Gasification (HyPr-RING), *16th International Pittsburgh Coal Conference*, Pittsburgh, Oct. 11–15, 1999.
13. Shiying L., Michiaki H., Yoshizo S. and Hiroyuki H., Hydrogen Production from Coal by Separating Carbon Dioxide during Gasification, *Fuel*, 81, 2079–2085, 2002.
14. Shiying L., Michiaki H., Yoshizo S. and Hiroyuki H., Continuous Experiment Regarding Hydrogen Production by Coal/CaO Reaction with Steam (II)—Solid Formation, *Fuel*, 85, 1143–1150, 2006.
15. Shiying L., Michiaki H., Yoshizo S. and Hiroyuki H., Process Analysis for Hydrogen Production by Reaction Integrated Novel Gasification (HyPr-RING), *Energy Conversion and Management*, 46, 869–880, 2005.
16. Shiying L., Michiaki H., Yoshizo S. and Hiroyuki H., Continuous Experiment Regarding Hydrogen Production by Coal/CaO Reaction with Steam (I)—Gas Products, *Fuel*, 83, 869–874, 2004.
17. Shiying L., Michiaki H., Yoshizo S. and Hiroyuki H., Continuous Experiment Regarding Hydrogen Production by Coal/CaO Reaction with Steam (HyPr-RING), *21st Pittsburgh Coal Conference*, Osaka, Japan, Sep. 13–17, 2004.
18. Shiying L., Progress of HyPr-RING Process Development for Hydrogen Production from Fossil Fuels, *16th World Hydrogen Energy Conference*, Jun. 13–16, Lyon, France, 2006.
19. Hans-Joachim Z. and Klaus S.L., Overview of the ZECA (Zero Emission Coal Alliance) Technology, *4th Advanced Clean Coal Technology Int. Symposium*, Jan. 29–30, Tokyo, Japan, 2001.
20. Collot A.-G., Prospects for Hydrogen from Coal, IEA Clean Coal Center, London, 2004.
21. Iyer M.V., Gupta H., Sakadjian B.B. and Fan, L.-S., High Temperature CO_2 Capture and Hydrogen Production Using Calcium Oxide: Process Development and Economics for Combustion and Gasification Systems, *The 2005 Clearwater Coal Conference*, Apr. 17–21, Clearwater, FL, 2005.
22. Rizeq G., West J., Frydman A., Subia R., Zamansky V. and Das K., Advanced Gasification-Combustion Technology for Production of Hydrogen Power and Sequestration-ready CO_2, *Gasification Technologies 2003*, Oct. 12–15, San Francisco, CA, 2003.

4

Hydrogen Production from Nuclear Energy

Ryutaro Hino and Xing L. Yan

CONTENTS

4.1 Introduction

Today, we are alarmed by the depletion of fossil energy and the adverse effects of the rapid increase in fossil fuel burning on the environment, such as climate changes and acid rain, because our lives still depend heavily on fossil energy. It is, thus, widely recognized that hydrogen is one of the important future energy carriers because it is used without the emission of carbon dioxide greenhouse gas and atmospheric pollutants and that the demand to it is expected to increase greatly as the fuel cells are developed and applied widely in the near future. However, hydrogen hardly exists naturally. Hydrogen production by direct thermal decomposition of water molecules would require a high-temperature heat of several thousands of Kelvin. By using electrolysis of water or combining high-temperature endothermic chemical reactions and low-temperature exothermic chemical reactions, in which the net chemical change resulting from the sequence of component chemical reactions is water decomposition; it is possible, in principle, to split the water molecule with heat at a considerably lower temperature. And a practical answer for protecting the global environment would be to produce hydrogen economically and from energy sources without carbon emission and air pollution.

Hydrogen production from water using nuclear energy offers one of the most attractive zero-emission energy strategies and the only one that is practical on a substantial scale. Recently, strong interest is seen in hydrogen production using heat of a high-temperature gas-cooled reactor (HTGR), a.k.a., very high-temperature reactor (VHTR) in the form of very high-temperature advanced design. The HTGR is particularly attractive because it has unique capability, among potential future generation nuclear power options, to produce high-temperature heat ideally suited for nuclear-heated hydrogen production, in addition to being inherently and passively safe and near-term deployable. This chapter introduces representative HTGR hydrogen production methods, high-temperature electrolysis of steam (HTE) and thermochemical splitting of water in an iodine–sulfur (IS) process, and coupling these methods to the HTGR to enable economical and safe hydrogen production from nuclear energy.

4.2 High-Temperature Electrolysis of Steam

4.2.1 Reaction Scheme

The process of the HTE is a reverse reaction of the solid-oxide fuel cell (SOFC): an oxygen ionic conductor is usually used as a solid oxide electrolyte as shown in Figure 4.1. Steam (H_2O) is dissociated at the cathode with hydrogen molecules (H_2) forming on the cathode surface: $H_2O(g) + 2e^- \rightarrow H_2(g) + O^{2-}$, whereas oxygen ions migrate simultaneously through oxygen vacancies in the lattice of the electrolyte material. Oxygen molecules (O_2) form on the anode surface with the release of electrons: $O^{2-} \rightarrow 1/2O_2(g) + 2e^-$. The products, hydrogen and oxygen, are separated by the gas-tight electrolyte. Reactions on the two electrodes are summed up as

$$H_2O(g) \rightarrow H_2(g) + \tfrac{1}{2}O_2(g) \tag{4.1}$$

In reaction 4.1, theoretical energy demand (ΔH) for water and steam decomposition is the sum of the Gibbs energy (ΔG) and the heat energy ($T\Delta S$). Figure 4.2 shows an energy demand for water and steam electrolysis. The electrical energy demand, ΔG, decreases with increasing temperature as shown in the figure; the ratio of ΔG to ΔH is about 93% at 100°C and about 70% at 1000°C. Thus, the HTE demands less electricity to produce hydrogen than the conventional water electrolysis. This reaction can be expressed as follows:

$$\Delta G = \Delta G_o + RT \ln\left(\frac{a_{H_2} a_{O_2}^{1/2}}{a_{H_2O}}\right) \tag{4.2}$$

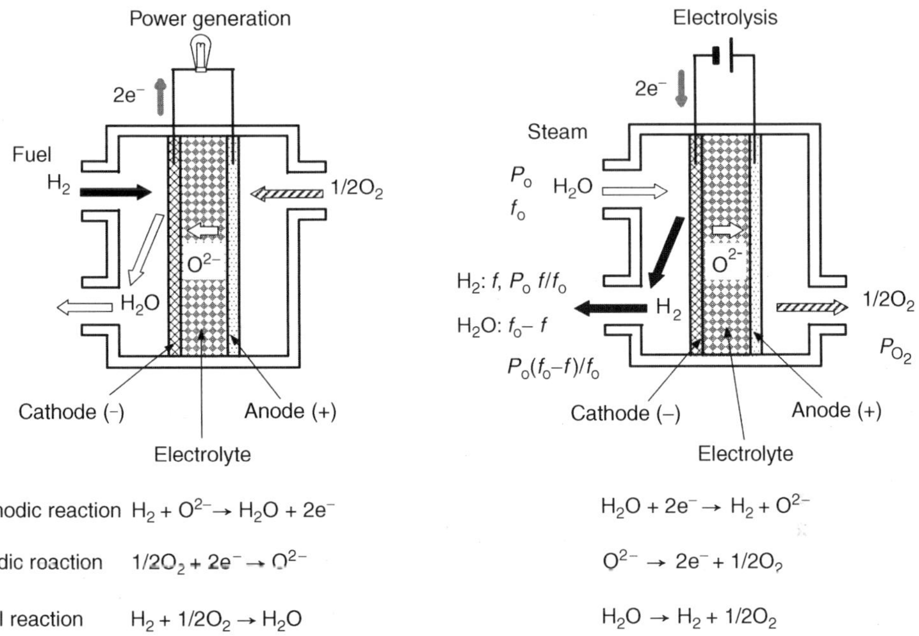

FIGURE 4.1
Principle of high-temperature electrolysis of steam (reverse reaction of SOFC).

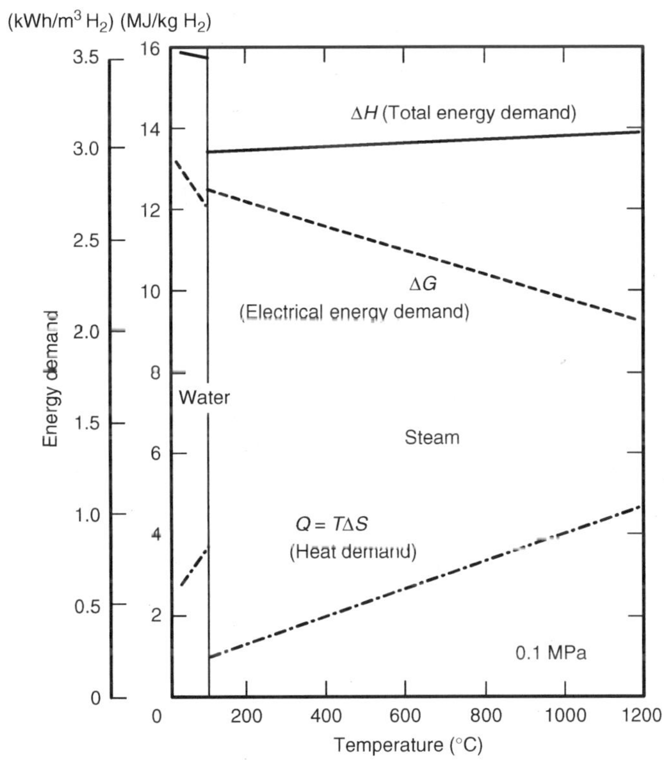

FIGURE 4.2
Energy demand for water and steam electrolysis.

where ΔG_o is the standard Gibbs free energy change (per mole) for the reaction 4.1 at a temperature of T, R the gas constant, and a_{H_2}, a_{O_2}, and a_{H_2O} the activities of H_2, O_2, and H_2O in the cell. Equation 4.2 can be rewritten using relations of $E = \Delta G/2F$ and $E_o = \Delta G_o/2F$ as

$$E = E_o + \left(\frac{RT}{2F}\right)\ln\left(\frac{a_{H_2}a_{O_2}^{1/2}}{a_{H_2O}}\right) \tag{4.3}$$

where F is the Faraday constant and E_o ($= \Delta G_o/2F$) the standard electromotive force (EMF) of the following reaction in the quasi-static state:

$$H_2(g) + \tfrac{1}{2}O_2(g) \rightarrow H_2O(g)$$

Equation 4.3 is equivalent to the Nernst equation for EMF of SOFC. Activities of reactants and products can be expressed as partial pressures in the cell because the reaction proceeds in the gas phase regardless of the electrolyte. The applied voltage increases with polarization in the operating cell. The overvoltage, η, is caused mainly by shortage of steam concentration at the cathode (concentration overvoltage) and electrical resistance including electrical leads, interconnections, and others (resistance overvoltage). Equation 4.3 can be rewritten as

$$E = E_o + \left(\frac{RT}{2F}\right)\ln\left(\frac{P_{H_2}P_{O_2}^{1/2}}{P_{H_2O}}\right) + \eta \tag{4.4}$$

where P_{H_2}, P_{O_2}, and P_{H_2O} are the partial pressures of H_2, O_2, and H_2O, respectively.

In the model cell shown in Figure 4.1, steam of molar flow rate, f_o, and pressure, P_o, is reduced to hydrogen gas of flow rate, $f_o - f$, and partial pressure, $P_o(f_o - f)/f_o$, in the cathode compartment: $(f_o - f)/f_o$ is a conversion ratio from steam (H_2O) to H_2 (steam conversion ratio). The partial pressure of O_2, P_{O_2}, is assumed to be unity. Using the steam conversion ratio, $X = (f_o - f)/f_o$, Equation 4.4 can be written as follows:

$$E - \eta = E_o + \left(\frac{RT}{2F}\right)\ln\left\{\frac{X}{1-X}\right\} \tag{4.5}$$

The value of $E - \eta$ is called the open-circuit voltage of the cell, which is related to the composition of the product. Note that the steam conversion ratio, X, depends on the open-circuit voltage, and is not affected by the pressure or flow rate of the reactant. Also, the open-circuit voltage decreases with increasing temperature because of the endothermic nature of the reaction. However, due to the temperature dependence of the logarithmic term in Equation 4.5, this effect decreases with the value of X.

Performance parameters of the electrolysis include the applied voltage, E (V), the applied current, I (A), and the hydrogen production rate, Q (Nℓ/h) at the reference condition of 0.1 MPa (1 bar) and 273 K (0°C). The Faraday efficiency, ε_f, expressed in Equation 4.6, is the ratio of ΔG to the applied power, $I * E$, that is, the ratio of the theoretical electric power needed for the electrolysis to the actually applied power of the cell. Thus, the Faraday efficiency is one useful measurement to judge electrolysis performance.

$$\varepsilon_f = \frac{\Delta G}{I * E} = \frac{2F*(E-\eta)*Q}{(I * E)} \tag{4.6}$$

The energy efficiency, ε_e, is the ratio of the produced hydrogen (H_2) energy and the applied power. Then, the produced H_2 energy is estimated with the combustion heat of produced H_2, by using the lower heating value, ΔH (kJ/mol), provided along the reaction ($H_2(g) + 1/2\ O_2(g) \rightarrow H_2O\ (g)$). The energy efficiency is useful for comparing the performances of hydrogen production systems.

$$\varepsilon_e = \Delta H * Q/22.4/(I * E) \tag{4.7}$$

where $Q/22.4$ is the molar rate of produced hydrogen.

4.2.2 Experimental Results Obtained at JAEA

Several experimental studies have been carried out using practical tubular electrolysis cells, a representative of which was the HOTELLY (high operating temperature electrolysis) program, which included research and developments (R&Ds) on solid-oxide cells and their assembly technologies as well as demonstration tests using a pilot plant. In the program, self-supporting electrolysis tubes were fabricated by connecting solid-oxide cells in a series. Basic hydrogen production data on HTE was obtained with a multicell electrolysis tube [1–3]. The cell consisted of a gas-tight cylindrical electrolyte and porous electrodes. The electrodes (cathode and the anode) were thin porous layers formed on the inner and outer surfaces of the electrolyte cylinder. Ten cells were connected in series by interconnections. The cell specifications were as follows:

Active cell length		10 mm
Electrolyte outer diameter		14 mm
Thickness	Electrolyte	0.3 mm
	Anode	Less than 0.25 mm
	Cathode	Less than 0.1 mm
Material	Electrolyte	Yttria (Y_2O_3)-stabilized zirconia (ZrO_2) (YSZ; Y_2O_3 containment 8–12 mol%)
	Cathode (air electrode)	Nickel containing YSZ (Ni cermet)
	Anode (fuel electrode)	Strontium-doped $LaMnO_3$ (LSM)

Using the 10-cell electrolysis tube, hydrogen was produced at the maximum rate of 6.78 Nℓ/h, 997°C of electrolysis temperature, 370 mA/cm^2 of current density, and a power of 21.7 W. Based on the HTE experimental results and operational experience, an HTE module consisting of 10 electrolysis tubes was integrated, and thereafter, 10 modules (including 1000 electrolysis cells) were assembled into a stack, in which hydrogen was produced at the maximum rate of 0.6 Nm^3/h [4]. However, the stability in a long-term operation and the reliability of the electrolysis tube and the module against thermal cycles were not published.

Maskalic had also carried out the HTE experiment using a single-cell electrolysis tube supported by a ceramic tube, which was developed for SOFC at Westinghouse Electric Co. The electrolyte and electrode layers were formed on a closed-one-end porous support tube made of calcia-stabilized ZrO_2 (CSZ), whose porosity was 35% approximately. Dimensions of the support tube were 12–13 mm in inner diameter, 1–1.5 mm in thickness, and 1000 mm in length. The support tube was overlaid with a porous LSM layer, 1.4 mm thick, working as the anode. A gas-tight electrolyte layer made of YSZ (containing 10 mol% of Y_2O_3), 0.04 mm thick, covered the anode along the cell's active length. The Ni cermet layer working as the cathode, 0.1 mm thick, covered the entire electrolyte surface. By using this type of cell, hydrogen was produced at a maximum rate of 17.6 Nℓ/h at a current density of 400 mA/cm^2, an applied power of 39.3 W, and 1000°C of electrolysis temperature [5].

However, the HTE data obtained below 950°C in the 1990s were not sufficient to support the data needed for the design of the HTGR heat utilization system. In addition, the HTE operational issues were not clarified. To improve the situation of HTE, the Japan Atomic Energy Research Institute (currently, the Japan Atomic Energy Agency [JAEA]) carried out the laboratory-scale experiments under HTGR operation temperatures from 850°C to 950°C, by using a practical electrolysis tube with 12 solid-oxide cells connected in series and a self-supporting planar cell of a practical size [6].

In 2003, R&D on HTE was started at the Idaho National Laboratory (INL) aiming to apply HTE to Generation IV HTGRs, a.k.a. VHTR. They fabricated refined self-supporting planar cells composed of YSZ (containing 8 mol% of Y_2O_3), Ni cermet, and LSM; and demonstrated a 1000-h continuous hydrogen production at a rate of about 160 Nℓ/h (H_2) by using a 25-plannar cell stack [7]. Private companies, such as Toshiba, are developing HTE for realizing a HTGR cogeneration system [8].

The remainder of the section introduces the HTE experimental results obtained at JAEA and describes the key process engineering issues [9].

4.2.2.1 Experimental Results with Tubular Cell

Figure 4.3 illustrates the structural drawing of an electrolysis tube with 12 cells, and Figure 4.4 shows the view of the tube. This type of electrolysis tube had been originally developed for SOFC [10]. The electrolysis tube was composed of 12 electrolysis cells of 19 mm in length. The length was determined so as to keep the current density uniform in the electrolysis region. These cells were connected in series. The electrolyte layer made of YSZ (containing 8 mol% of Y_2O_3) was sandwiched between the porous cathode and anode layers: Ni cermet (Ni + YSZ) for the cathode and $LaCoO_3$ for the anode. At the ends of the

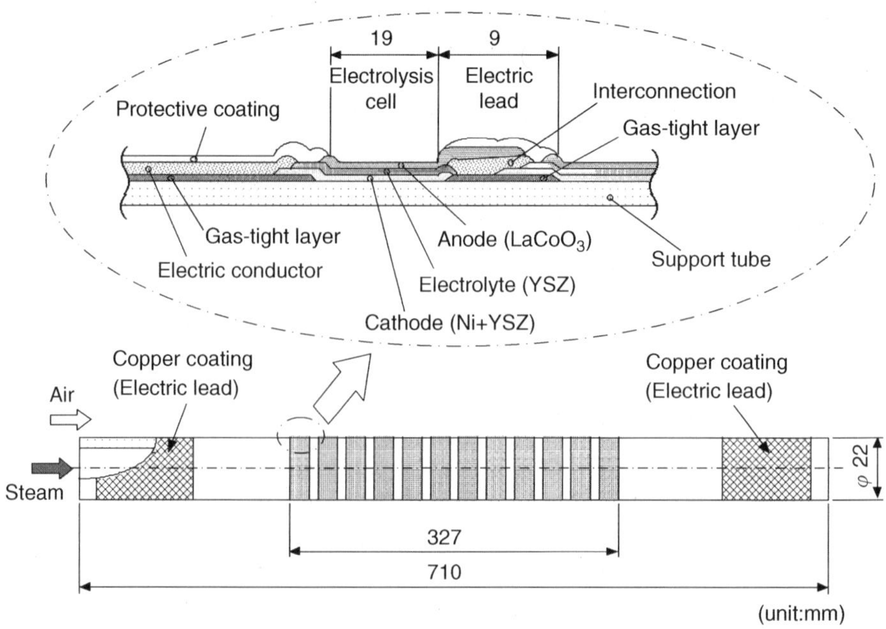

FIGURE 4.3
Structural drawing of an electrolysis tube with 12 cells (tubular type cell).

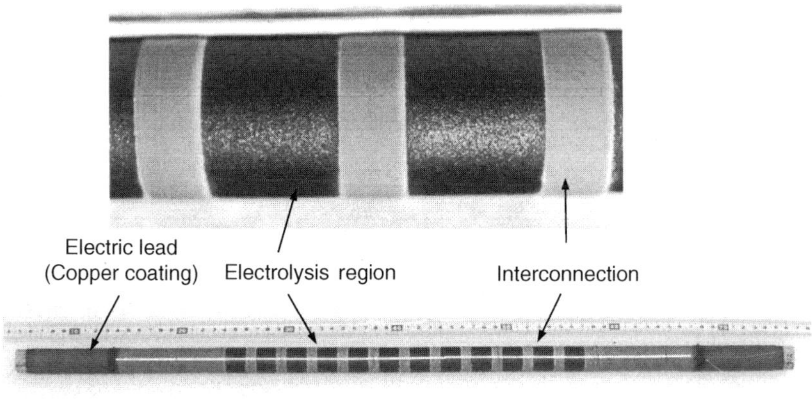

FIGURE 4.4
Outer view of an electrolysis tube with 12 cells.

electrolysis tube, platinum (Pt) wires working as electric leads were welded on copper-coating layers. The copper-coating layers were connected to thin layers of electric conductors. The outside of the electrolysis tube, except the cells and the copper coatings, was coated by thin alumina (Al_2O_3) gas-tight layers and YSZ protective layers. These layers were formed on a porous CSZ tube (support tube) of 22 mm in outer diameter, 3 mm in thickness, and about 38% of porosity, by using the plasma spraying. The thickness of each layer was in the range from 0.1 to 0.25 mm. The total length of the electrolysis tube was 710 mm.

Experiments were carried out at the electrolysis temperatures of 850°C, 900°C, and 950°C under the pressure of 0.11 MPa (abs). Steam (H_2O) was supplied with argon carrier gas at a rate of about 0.32 g/min. The heating and cooling rates were set to be <20°C/h during start-up and shut-down of the test section so as to avoid generating high interfacial stress from differential thermal expansion among the electrolysis tube components otherwise occurring under rapid heating or cooling. Before applying the electrolysis voltage, the cathode material, Ni cermet, was chemically reduced with hydrogen mixed with argon carrier gas for 1 h at least.

Figure 4.5 shows the relation between the hydrogen (H_2) production rate and the applied voltage. H_2 production rate increased with the applied voltage and the electrolysis temperature. The maximum production rate was 3.8 Nℓ/h at 15.7 V, 1.42 A, and 850°C, 4.3 Nℓ/h at 16.3 V, 1.38 A, and 900°C; and 6.9 Nℓ/h at 15.6 V, 1.72 A, and 950°C, respectively. These H_2 production rates were quoted to the reference condition of 273 K (0°C) and 0.1 MPa (1 bar) as mentioned earlier. Lines in the figure express the hydrogen production rate calculated with Equation 4.5. The differences between experimental values of the applied voltage and the corresponding theoretical lines indicate the extent of overvoltage, η. As seen in the figure, the overvoltage increases with the hydrogen production rate. The Faraday efficiency at 950°C and in a range of 0.74 to 0.85 V applied voltage with more than 6 W applied power, which results in current density of more than 45 mA/cm^2, is estimated to range from 0.27 to 0.56 in accordance with Equation 4.6. The experimental result obtained in the HOTELLY program, 6.78 Nℓ/h of the hydrogen production rate under the condition of the electrolysis temperature of 997°C and the applied power of 21.7 W [2], showed the Faraday efficiency of 0.75. Then the open-circuit voltage, $E - \eta$, was 1.006 V at 997°C. This low Faraday efficiency could have been caused mainly by lower oxygen ion conductivity than expected for 997°C. Also, other possible causes included rather high ohmic losses at interconnections and electric lead layers, and an increase in the concentration overvoltage because steam could not be

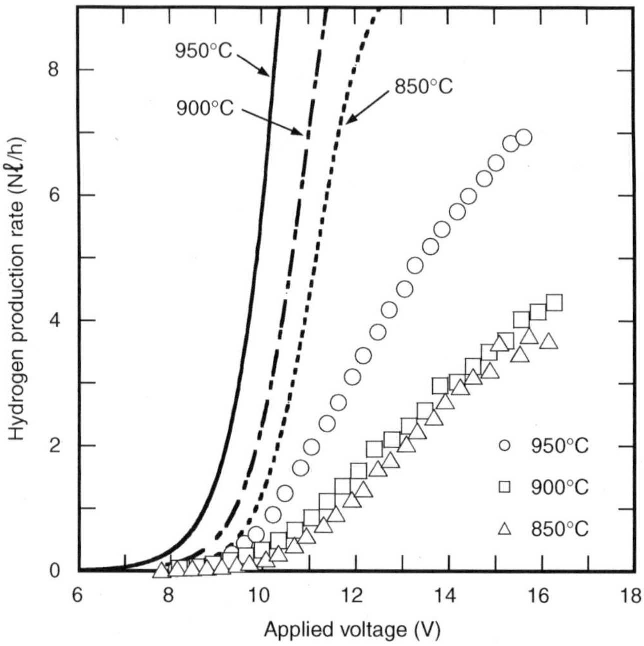

FIGURE 4.5
Applied voltage versus hydrogen production rate. Symbols: experimental results, lines: theoretical hydrogen production rates.

supplied effectively to the cathode through the support tube and the hydrogen produced at the cathode was not fully discharged into the main flow through the support tube.

In these experiments, the steam conversion ratio (steam utilization ratio) was <40%. The lower than expected steam conversion ratio might be caused by the low permeation rate of steam through the support tube to the cathode, although the support tube had a rather high porosity of about 38%. If steam could reach the cathode sufficiently, the hydrogen production rate would have increased, whereas the concentration overvoltage would have decreased. Turbulent flow generated by twist tapes to be inserted inside the tube might be effective. Otherwise, from the viewpoint of steam supplying performance to the cathode, the self-supporting cell structure would be better in the HTE than in the multicell structure supported by the porous ceramic tube.

Figure 4.6 shows the relation between the current density and the energy efficiency. The energy efficiency is a quotient of combustion heat of generated hydrogen/applied power, which could be derived with Equation 4.7 using the low heating value. As seen in the figure, the energy efficiency has a peak in the range of current density from 80 to 100 mA/cm^2. The maximum energy efficiency below the electrolysis temperature of 950°C was 86% at a current density of 98 mA/cm^2. This would be mainly due to a rather low oxygen ion conductivity at 950°C and a shortage of steam supply at the interface between the electrolysis and the electrode as mentioned earlier as the reason for the increase of the concentration overvoltage.

When connected to the HTGR, the HTE operating temperature will be up to 900°C. It is necessary to use other solid-oxide electrolytes working below 900°C and having considerably higher oxygen ion conductivity than that of YSZ. One candidate is ytterbia (Yb_2O_3)-stabilized ZrO_2 (YbSZ). It is also necessary to enhance the hydraulics of the steam supply to the interface between the electrolysis and the electrode, which might be realized by decreasing the fuel electrode layer thickness and by increasing its porosity.

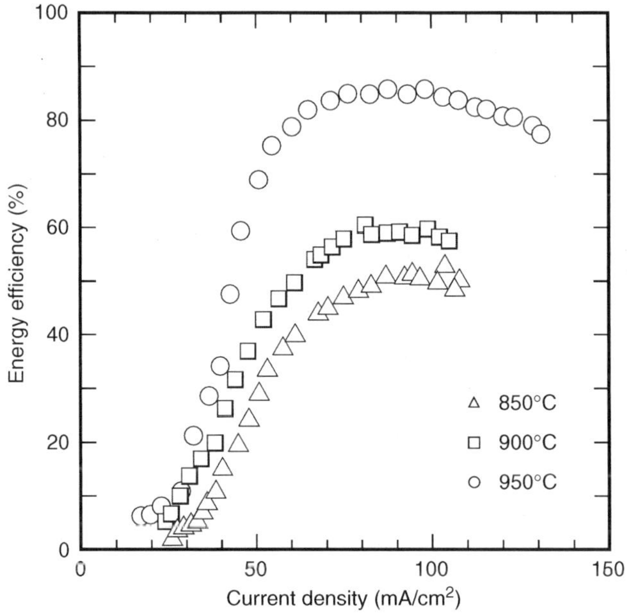

FIGURE 4.6
Current density versus energy efficiency.

After the experiments, it was observed that large parts of the anode (air electrode) layers had been separated from the electrolyte layers, although the electrolysis tube had served for one thermal cycle only. This indicates that the durability of the cell against thermal cycles, especially the durability of the anode layer, is one of the key issues of HTE. It could be met by raising the bonding force and keeping high residual compression stress of the anode layer to the electrolyte layer against large differential thermal expansion of these layers considering the relevant service conditions.

4.2.2.2 Preliminary Experiments with Self-Supporting Planar Cell

Because the tubular cell shown in Figure 4.3 is rather bulky and costly, preliminary HTE experiments were conducted with a planar cell, especially a self-supporting planar cell. Figure 4.7 shows a schematic drawing of the self-supporting planar cell and an electrolysis chamber [6,9]. The cell consisted of the electrolyte plate of YSZ and porous electrodes. The YSZ plate was a 100 mm square plate with a thickness of 0.3 mm, and the electrodes were baked on an area of 80 × 80 mm^2 with a thickness of <0.03 mm.The material of the cathode was Ni cermet (Ni + YSZ) and that of the anode was LSM. The cell was sandwiched with metal housings made of SUS310S. Each housing had a metal rod for the electric lead, an inlet and an outlet piping for H_2 and H_2O, and a Pt sheet of 0.1 mm thick was welded on the inner surface of each housing opposite to the electrodes of the cell. A corrugated electrical lead plate made of Pt mesh (0.1 mm thick) was installed in each electrode compartment to work as the electric lead. Then, the flat Pt mesh sheet was inserted between the corrugated electrical lead plate and the electrode to realize a uniform current density. Direct current (DC) power was supplied through the metal rods, housings, Pt sheets, and corrugated electrical lead plates to the cell. Compression seals were installed at the edge of the cell plate; the compression load was up to 20 kg. Then, an

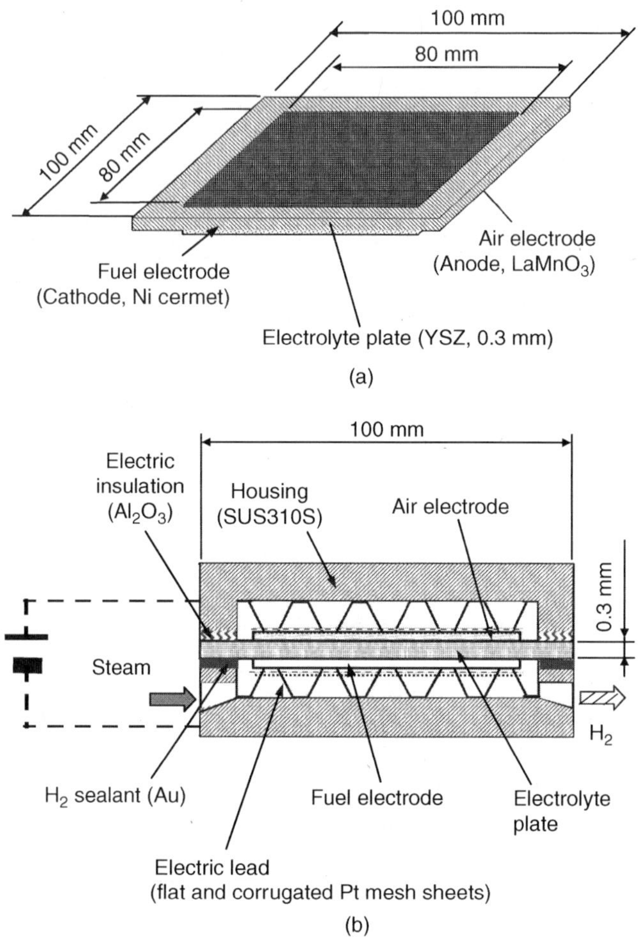

FIGURE 4.7
Schematic drawing of a self-supporting planar cell (a) and an electrolysis chamber (b).

alumina sheet of 0.3 mm thick was inserted between the cell plate and the housing in the anode side (air supplying side) to prevent current leak through the cell plate edge, and a gold (Au) sheet of 0.1 mm thick was inserted in the cathode side (steam supplying side) to prevent any H_2 leak.

Figure 4.8 shows the relation between current and H_2 production densities obtained at 850°C of electrolysis temperature. The maximum H_2 production density was 38 Ncm^3/cm^{2}h, which was higher than that of the electrolysis tube obtained at 950°C. The maximum H_2 production rate was 2.4 Nℓ/h at the applied power of 10 W: applied voltage and current were 2.68 and 3.72 A. Then the open-circuit voltage was 0.847 V. Hence, the Faraday efficiency and the energy efficiencies were 0.5 and around 0.73, respectively, which were almost the same values as those of the electrolysis tube obtained at 950°C.

After the experiment, a crack crossing the cell was observed, which could have been caused by the thermal expansion due to the Joule heating. In designing a module consisting of stacked planar cells, special relief structures against the thermal expansion should be installed, which also seem to be necessary for providing both seal and electric insulation performances. Recently, Herring et al. have been improving the self-supporting

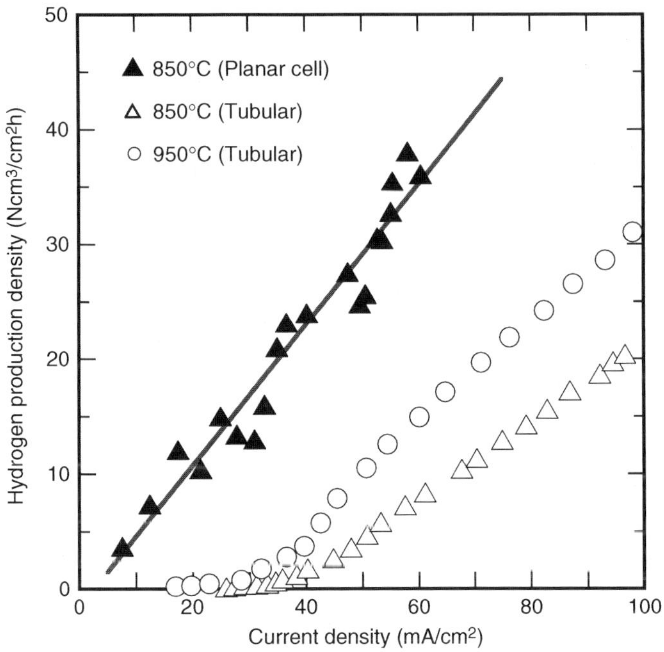

FIGURE 4.8
Relation between current and hydrogen production densities.

planar cell—by using update materials such as NbB_2-ScSZ and Strontium-doped $LaCrNiO_3$—and the cell stack structure based on numerical analysis [7].

4.3 Thermochemical Water Splitting by Iodine–Sulfur Cycle

Thermochemical water splitting cycle was first studied by Funk [11], and an actual example was proposed by researchers of CEC, JRC Ispra establishment, in early 1970s [12]. Since then, a number of thermochemical cycles have been proposed assuming HTGR as the heat source, which can supply heat with its maximum temperature close to 1000°C. After mid-1980s, activities in Europe and in North America slowed down in accordance with the slowing down of their HTGR projects. Recently, however, with the emerging interest in the "hydrogen energy system" in accordance with the progress of fuel cell technology, the thermochemical method attracts growing interest again. In United States, France, Korea, as well as Japan, R&D on thermochemical water splitting process, especially IS cycle, is ongoing [13–15].

4.3.1 Previous Studies on Iodine–Sulfur Cycle

Hundreds of cycles have been studied from the viewpoint of the feasibility of component chemical reactions in terms of conversion ratio or product separation, theoretical thermal efficiency of hydrogen production, etc. [16]. Among them, those that utilize thermal decomposition of sulfuric acid, which are categorized as "sulfur cycles," have been considered one of the most promising cycles.

Thermal decomposition of sulfuric acid, reaction 4.1, proceeds in the following two steps:

$$\begin{array}{ll} H_2SO_4(aq) = H_2O(g) + SO_3(g) & 300-500°C \\ SO_3(g) = SO_2(g) + 0.5O_2(g) & 800-900°C \\ \hline H_2SO_4 = H_2O + SO_2 + 0.5O_2 & \end{array} \tag{4.8}$$

Both steps are highly endothermic and proceed smoothly without side reactions and with a high conversion ratio at the temperature range indicated. The endothermic characteristics match well with the temperature distribution of the heat source—HTGR. The heat of HTGR is transferred to the chemical process through the sensible heat of helium gas, the temperature of which varies from 900°C to 400°C. Therefore, the reaction is quite suited as the high-temperature endothermic reaction for the thermochemical water splitting cycle [17].

IS cycle (or IS process) combines the following chemical reactions with the sulfuric acid decomposition reaction:

$$SO_2 + I_2 + 2H_2O = 2HI + H_2SO_4 \tag{4.9}$$

$$2HI = H_2 + I_2 \tag{4.10}$$

Here, reaction 4.9, known as "Bunsen reaction," is the low-temperature exothermic reaction, where the raw material, water, reacts with iodine and gaseous sulfur dioxide producing an aqueous solution of hydriodic acid and sulfuric acid. The acids are then separated and thermally decomposed to produce hydrogen and oxygen. The total reaction scheme is shown in Figure 4.9.

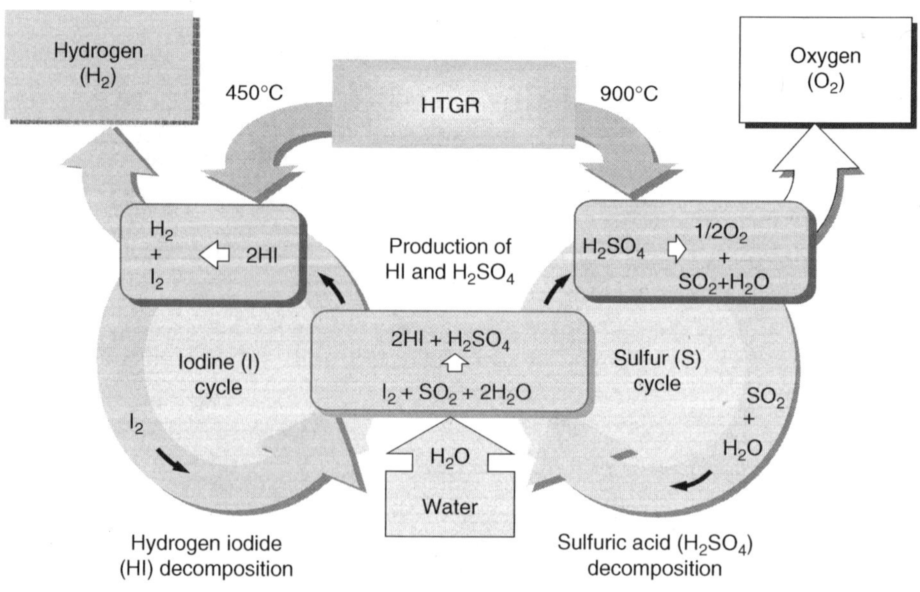

FIGURE 4.9
Schematic of the IS process.

The IS process has been studied in United States, Europe, and Japan since 1970s, and some important breakthroughs were attained by General Atomics (GA). So far, the research has been carried out in the following fields:

1. Study on the chemistry of each reaction section
2. Demonstration of the closed-cycle hydrogen production
3. Heat/mass balance analysis of the process flow sheet
4. Screening of corrosion-resistant materials

4.3.1.1 Study on the Chemistry of Each Reaction Section

There are two main issues concerning the chemistry of the reaction and the separation. One is how to separate the hydriodic acid and sulfuric acid produced by the Bunsen reaction. The other is how to carry out the hydrogen iodide (HI) decomposition section, where the presence of azeotrope in the vapor–liquid equilibrium of the hydriodic acid makes the energy-efficient separation of HI from its aqueous solution difficult, and also, the unfavorable reaction equilibrium limits the attainable conversion ratio of HI to a low level, around 20%.

As for the former problem, the researchers of GA found that the mixed acid solution produced by the Bunsen reaction separates spontaneously into two liquid phases in the presence of excess amount of iodine [17]. The heavier phase is mainly composed of HI, I_2, and H_2O, and is called "Hix" solution. The main components of the lighter phase are H_2SO_4 and H_2O. The phenomenon (liquid–liquid (LL)-phase separation) offered an easy way of separating the hydriodic acid and the sulfuric acid. As for the HI processing, some ideas have been proposed by GA [17], RWTH Aachen [18], and JAEA. JAEA studied the utilization of membrane technologies for concentrating the HIx solution to facilitate the HI separation and also for enhancing the one-pass conversion of HI decomposition [19,20].

4.3.1.2 Demonstration of the Closed-Cycle Hydrogen Production

One of the specific and important characteristics of thermochemical water splitting cycles is that the reactants except water are cyclically used in the process. The closed-cycle continuous hydrogen production by the IS process featuring the LL-phase separation has been examined at JAEA. Although the chemistry of sulfuric acid decomposition section and that of hydrogen iodide decomposition section are rather straightforward in terms of reaction and separation, in the Bunsen reaction section, occurrence of side reactions forming sulfur or hydrogen sulfide should be suppressed while maintaining the LL-phase separation. JAEA has devised a basic methodology for the closed-cycle continuous hydrogen production and also for the reaction control in the Bunsen reaction step. Feasibility of the methodology was verified in small-scale continuous hydrogen production experiments of 1 Nℓ/h (H_2) in 1997 by using a laboratory-scale experimental apparatus shown in Figure 4.10 [21]. Further tests have been conducted by using a bench-scale test apparatus made of glass as shown in Figure 4.11. In June 2004, continuous hydrogen production was successfully achieved with a hydrogen production rate of about 31 Nℓ/h for 1 week [22]. Figure 4.12 shows the experimental result, and the relationship between the operation time and the integrated amount of H_2 and O_2 [22]. In the experiments, newly developed automatic control system with ultrasonic detectors, etc. were applied to the control of the Bunsen reaction. These laboratory- and bench-scale tests were carried out under atmospheric pressure (0.1 MPa).

FIGURE 4.10
Outer view of laboratory-scale test apparatus.

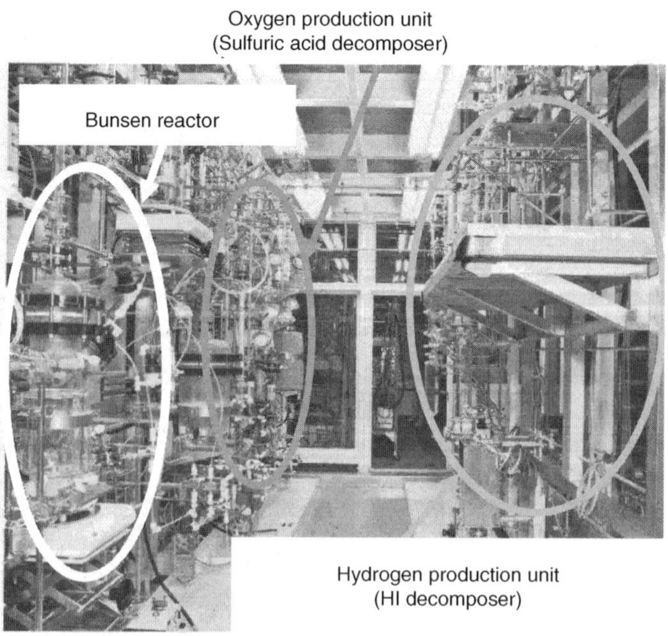

FIGURE 4.11
Outer view of the bench-scale test apparatus.

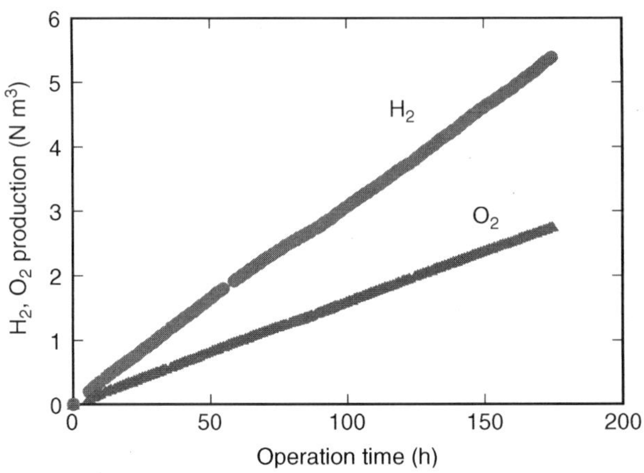

FIGURE 4.12
One week continuous hydrogen production with the bench-scale test apparatus—relationship between operation time and integrated amount of H_2 and O_2.

4.3.1.3 Heat and Mass Balance Analysis of the Process Flow Sheet

Preliminary flow sheeting studies have been carried out based on the LL-phase separation for the separation of the Bunsen reaction products. In 1982, GA published a result of the total flow sheet analysis, where extractive distillation using phosphoric acid was applied for the hydrogen iodide processing [23]. Assuming intensive heat recovery, a "process thermal efficiency" of 47% was estimated. Here, the thermal efficiency is defined as the ratio of the higher heating value (HHV) of hydrogen to the net energy input for the process. Later, a reactive distillation was proposed by RWTH Aachen [18] for the hydrogen iodide processing, where the separation of hydrogen iodide from the HIx solution and the decomposition of hydrogen iodide into hydrogen and iodine were carried out in one column. The concept was further studied by GA [24] and CEA [25] as well. The reactive distillation is quite interesting because of its simplicity, although the estimated process thermal efficiency varies from 35% to 50% at present. The application of the membrane systems for the hydrogen iodide processing proposed by JAEA also gives a similar estimation of the process thermal efficiency [26]. As for the sulfuric acid processing, main efforts in the flow sheeting study have been on how to recover the heat of concentration. A steam recompression method [23] and a multiple effect evaporator system [27] have been proposed. Ozturk et al. [28] proposed and analyzed an interesting idea to use oxygen gas as the heat transfer media so as to mitigate the corrosion issue. The precise thermodynamic data concerning the concentrated process solutions is desired for the much more accurate evaluation of the heat/mass balance.

4.3.1.4 Screening of Corrosion-Resistant Materials

Because sulfuric acid and halogen are very corrosive, selection of the structural materials is an important issue. Screening tests have been carried out using test pieces of commercially available materials at GA [29], JAEA [30,31], etc. As for the gas-phase environment of the H_2SO_4 decomposition step, some refractory alloys that have been used in conventional chemical plants showed good corrosion resistance. Figure 4.13 shows one of the experimental results of Alloys 800 and −600 obtained under gas-phase sulfuric acid decomposition environments at 850°C. Gas compositions in the upstream and downstream

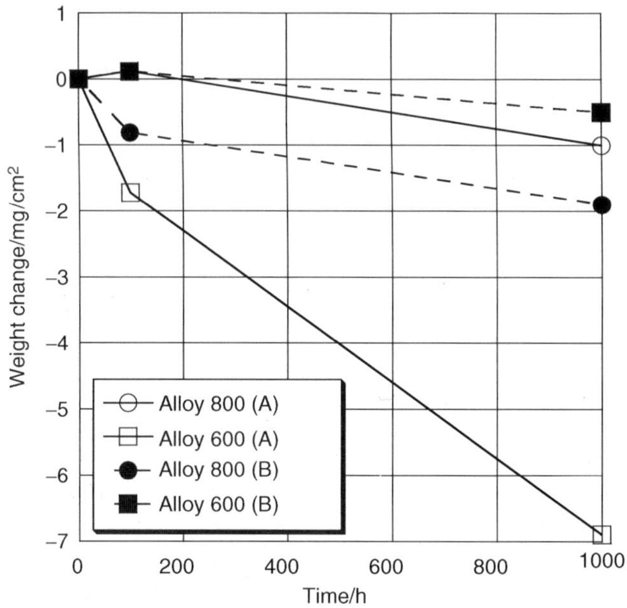

FIGURE 4.13
Weight loss of base materials in the gas-phase sulfuric acid decomposition environments at 850°C; (a) upstream of catalyst bed and (b) downstream of catalyst bed.

of the catalyst bed are a mixture of H_2O, SO_3, and H_2SO_4, and that of H_2O, SO_2, O_2, SO_3, and H_2SO_4, respectively.

In the gas-phase environment of the HI decomposition step, a Ni–Cr–Mo–Ta alloy was found to show good corrosion resistance under 450°C. As for the Bunsen reaction step operated at about 100°C, glass-lining materials, Ta, etc., showed good corrosion resistance. In the environment of the HIx distillation operated at about 230°C, Ta showed excellent corrosion resistance. The severest environment is the boiling condition of concentrated sulfuric acid under high pressure (e.g., 2 MPa) and temperature up to 400°C, where ceramic materials containing Si such as Si–SiC, SiC, and Si_3N_4 were the only materials that showed excellent corrosion resistance [31,32]. For example, SiC has a much better corrosion resistance than the stainless steel as shown in Figure 4.14, whose results were obtained under a boiling condition of 95 wt% sulfuric acid and temperature up to 320°C.

4.3.2 Pilot Test Program at JAEA

United States, France, South Korea, and others have proposed pilot-scale test programs of the IS process as an out-of-pile demonstration test for the VHTR system in the Generation IV International Forum (GIF). Laboratory-scale experiments are being conducted before the advanced stage of the pilot tests. Similarly, JAEA is conducting design works of an IS process pilot plant on the database and know-how accumulated from hydrogen production with the bench-scale test apparatus and other significant investigations such as corrosion and component tests for a H_2SO_4 decomposer and so on in the framework of the High Temperature Engineering Test Reactor (HTTR) program [15], which covers R&D on HTGR technology and the system integration technology to connect HTGR and the hydrogen production plant. The HTTR is the only Japanese HTGR with thermal power of 30 MW,

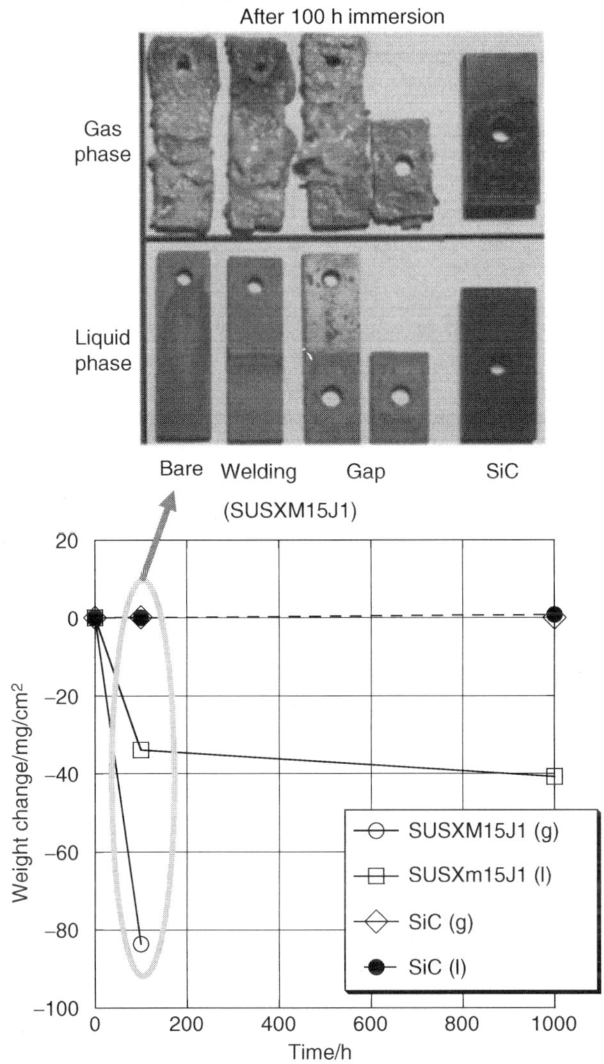

FIGURE 4.14
Weight loss of base materials immersed in a boiling condition of 95 wt% sulfuric acid at 320°C; (g) gas phase, and (l) liquid phase.

built and operated on the site of the Oarai Research and Development Center of JAEA. By using the IS process pilot plant, the following tests will be carried out:

1. Hydrogen production test using the test apparatus driven by electrically heated helium gas. Operation of the test plant will demonstrate the technical feasibility of IS cycle, and also, the test data will be used to verify the analytical codes to be developed.
2. Development of computer code system for analyzing the heat/mass balance, dynamic process simulation, supporting the component design works, etc.

Figure 4.15 shows a tentative flow diagram of the pilot plant [33]. Sensible heat of high-temperature (up to 880°C) and high-pressure (4 MPa) helium (He) gas, which is the same

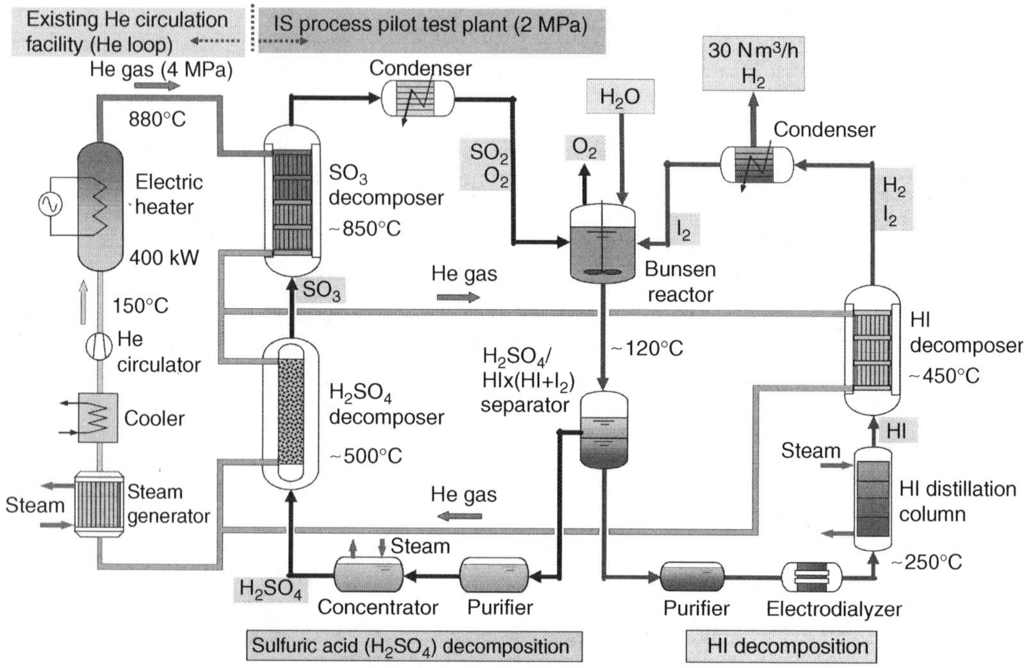

FIGURE 4.15
Flow diagram of the IS process pilot plant (tentative).

TABLE 4.1
Candidate Materials Applied for JAEA Pilot Plant

Equipments	Candidate Material
Bunsen reactor	Glass lining
Two-phase separator	
HI distillation column	Gas phase: Ni–Cr–Mo alloy
HI decomposer	Liquid phase: Glass lining
H_2SO_4 concentrator	SiC ceramics
H_2SO_4 decomposer	
SO_3 decomposer	Alloy 800 or SiC ceramics

as the HTTR secondary helium gas temperature and pressure, will drive continuous hydrogen production. The pilot plant will have the specification of the hydrogen production rate of up to 30 Nm^3/h at process pressures up to 2 MPa, with which the smallest components used in industrial chemical plants can be tested, such as valves and pumps. All components are fabricated with industrial materials such as SiC ceramics and glass-coated steels as shown in Table 4.1.

One of the key components to be used in the boiling H_2SO_4 environments is the sulfuric acid decomposer as shown in Figure 4.15, in which the H_2SO_4 solution with a concentration of more than 90 wt% is evaporated and, simultaneously, H_2SO_4 is decomposed into gaseous SO_3 and H_2O under high-temperature conditions of up to 500°C. Recently, JAEA proposed a concept of the sulfuric acid decomposer applying to the pilot plant. Figure 4.16 shows the

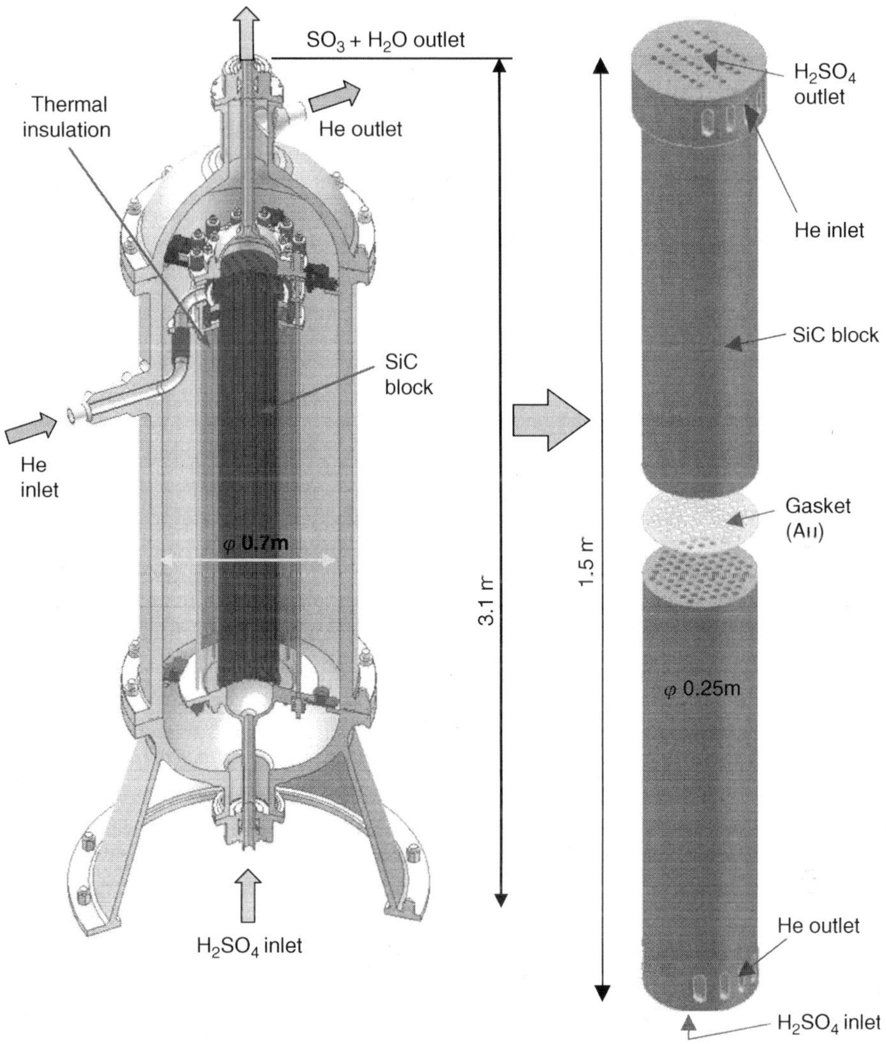

FIGURE 4.16
Concept of H_2SO_4 decomposer for JAEA pilot plant.

sulfuric acid decomposer featuring multihole heat exchanger blocks made of SiC ceramic. He gas flows through the inside channels of the SiC ceramic blocks, and exchanges the heat with countercurrent H_2SO_4 flow. The SiC ceramic blocks of 0.25 m in outer diameters and 0.75 m in height were piled vertically. The pure gold, which is excellent in the corrosion resistance to high-temperature H_2SO_4, has been currently chosen as the seal material between the SiC ceramic blocks. The mock-up model of the sulfuric acid decomposer was test-fabricated to confirm fabricability and structural integrity as shown in Figure 4.17 [34].

As per the present plan, the JAEA pilot test program precedes the immediate next stage of nuclear hydrogen demonstration using the HTTR, namely, the HTTR-IS test, which will produce hydrogen at a rate of up to 1000 Nm^3/h. Figure 4.18 shows the conceptual plan.

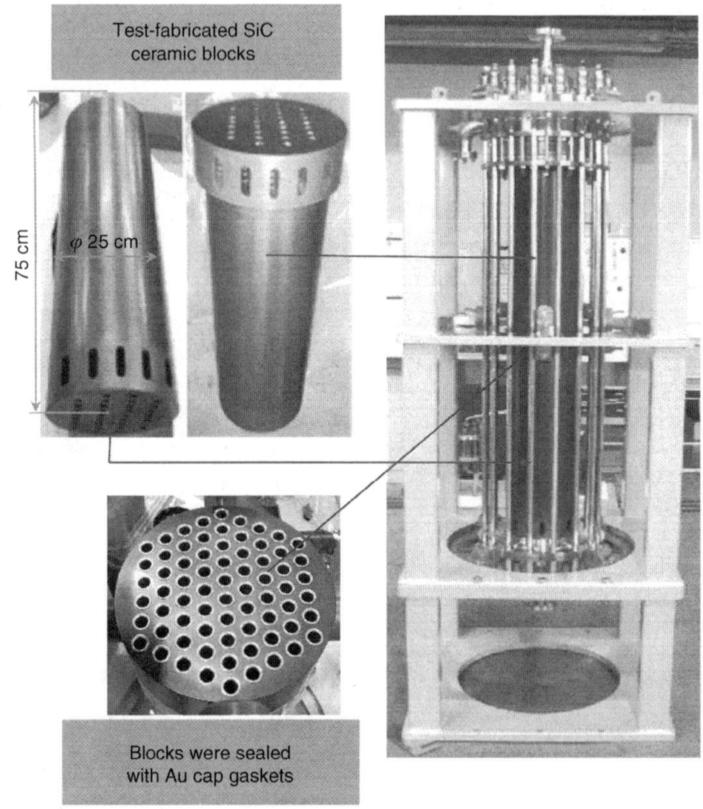

FIGURE 4.17
Mock-up model of the H_2SO_4 decomposer.

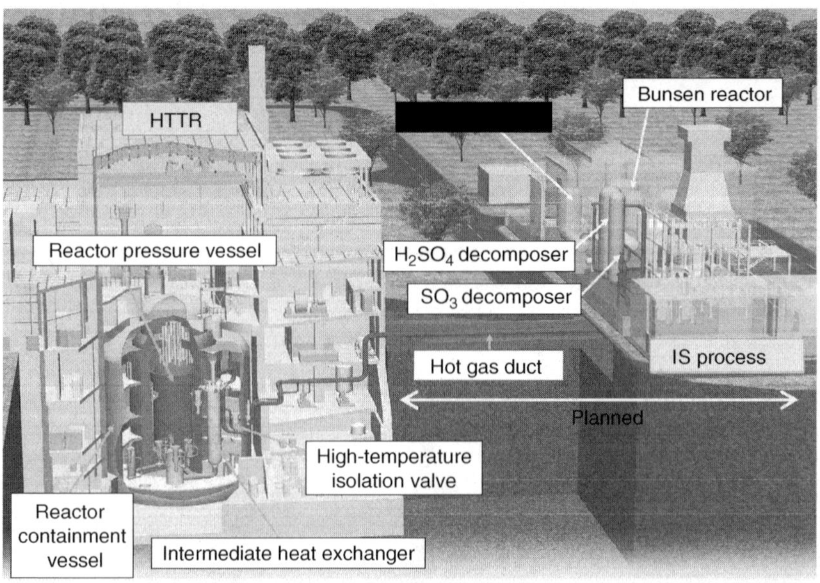

FIGURE 4.18
Bird's eye view of the HTTR-IS system.

4.4 Hydrogen Production with Very High-Temperature Reactor

4.4.1 Overview of the Very High-Temperature Reactor

The VHTR is well placed to take on the nuclear energy's traditional role of generating electricity. It has many advantages over other types of nuclear reactor, however. The chief advantage among them is its unique capability of producing a high-temperature heat of about 950°C, which increases the thermal efficiency of electricity generation to 50% by Brayton cycle from the present 34% in the light water reactor with Rankine cycle and expands the role of nuclear energy to various other fields including mass production of hydrogen to help fuel a sustainable world economical development.

National programs are underway to develop and deploy the reactor and applications in China, France, South Korea, and United States [35–38]. Many more countries are participating in R&D through the High Temperature Reactor Technology Network of the European Union [39] and the GIF including the 10 member countries [40]. Builds of full-scale commercial demonstration reactors are actively planned in several countries such as South Africa [41] and the United States [38].

In Japan, JAEA collaborated with domestic nuclear fuel, material, and heavy industries in the successful development and demonstration of the HTTR shown in Figure 4.19 [42–43]. A similar JAEA–industry collaboration has been developing commercial-scale plant system design for electricity and hydrogen production for nearly a decade. An initial feasibility study revealed economical potential and exposed critical issues to be resolved through design development. The follow-on conceptual design established the original first-system variant–the GTHTR300 [44]. The plant combines a 600 MW block reactor and a direct cycle gas turbine for electricity generation.

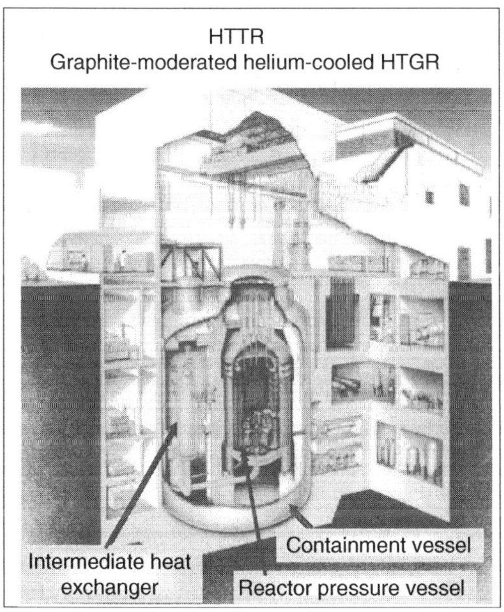

Major specification

Thermal power	30 MW
Fuel	Coated fuel particles in prismatic block
Core material	Graphite
Coolant	Helium
Inlet temperature	395°C
Outlet temperature	950°C (Max.)
Pressure	4 MPa

Operations History

- First criticality : 1998
- Full power operation : 2001
- Safety demonstration test : 2002
- High-temperature operation (950°C) : 2004

FIGURE 4.19
JAEA's HTTR.

The basic design for the GTHTR300, initiated in 2001, has involved detailed design and engineering to the extent verifiable by tests of new component and system technologies at an appropriate scale. Having confirmed economical and safe performance targets, the design evolved to add three more members including a growth system for electricity generation, GTHTR300+, a hydrogen cogeneration system, GTHTR300C (the letter "C" denotes *c*ogeneration), and most recently, a hydrogen production system, GTHTR300H ("H" denotes *h*ydrogen) [45]. The system variants that produce hydrogen may be based on the HTE or thermochemical IS process, described earlier in the chapter. Hydrogen is produced from water as the only material feed, with reactor-supplied heat and electricity and without greenhouse gas emission and air pollution. A more detailed description of the systems is given in Section 4.4.2, followed by a review of the technology requirement and status for nuclear hydrogen production in Section 4.4.3.

4.4.2 Nuclear Reactor Hydrogen Production Systems

The commercial design in JAEA aims at developing a family of plant options capable of producing competitive electricity, hydrogen, or a variable mix of both, and yet deployable in the near- to mid-term. To develop the multiple systems simultaneously does not necessarily suggest to have investment and risk multiplied. Rather, development has been minimized thanks to the three tiers of a SECO design philosophy: simplicity, economical competitiveness, and originality.

The design simplification is built on the premise that all plant variants share a common reactor system, an aerodynamically and mechanically similar line of helium gas turbines used for electricity production and the IS process system for hydrogen production in a common plant arrangement as shown in Figure 4.20. Note that the hydrogen plant along with

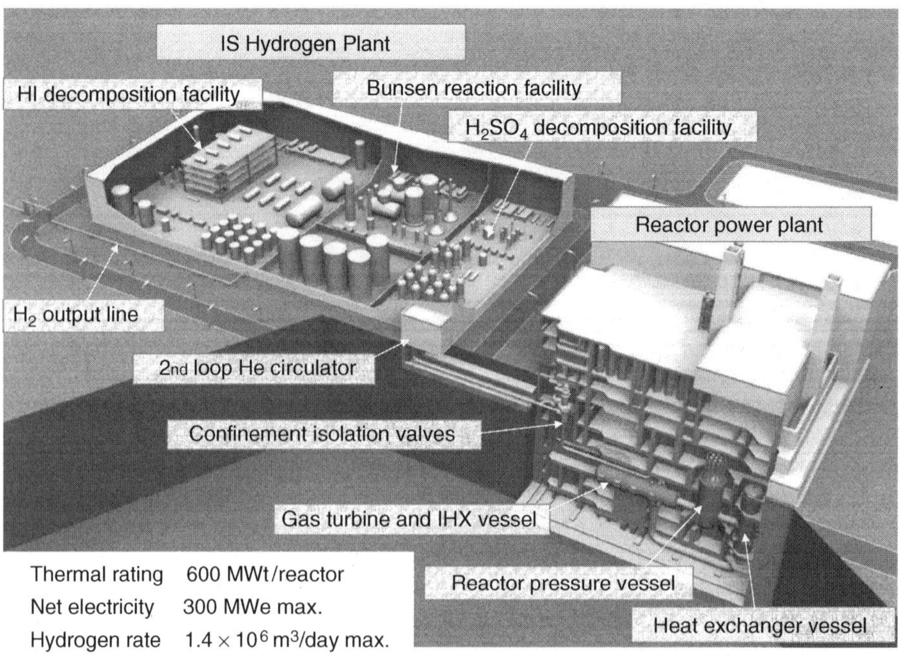

FIGURE 4.20
The GTHTR300 system variants produce electricity, hydrogen, or concurrently both.

the second heat transport loop is not present in the plant should electricity be the only form of product desired. To achieve competitive economics, the original design attributes that are less demanding on the system technologies are adopted such as conventional steel reactor pressure vessel, horizontal gas turbine, and distributed modular system arrangement among others. The pursuit of technology and design simplification has led to a focused development scope that comes with a low overall risk and investment. Furthermore, since the technologies to be developed are shared, the benefit of investing in any one development is increased.

Combination of a VHTR with a recuperated Brayton cycle forms the basic production process as shown in Figure 4.21. The direct cycle gas turbine generates electricity and circulates the reactor coolant, performing both tasks most efficiently relative to all other forms of process arrangement. Hydrogen cogeneration is enabled by adding an intermediate heat exchanger (IHX) in serial between the reactor and the gas turbine. The particular serial arrangement allows for large logarithmic mean temperature difference (LMTD) between the primary and the secondary fluids, creating a desired condition for a compact IHX. A secondary loop delivers hot helium from the IHX to the hydrogen plant over a sufficient distance that together with the isolation valves located in the secondary loop circuits provides a safe and environmental separation between the nuclear plant and the conventional-grade hydrogen plant.

Electricity is generated in a range 45–50% efficiency in the GTHTR300 and GTHTR300+ based on 850°C and 950°C reactor coolant temperature, respectively. In the GTHTR300C, a nominal 170 MWt of the total 600 MWt reactor thermal power is extracted by the IHX as 900°C process heat supply to the hydrogen process and the balance of reactor thermal power is used for electricity generation at 850°C gas turbine inlet. A fraction of the 200 MWe electricity generated supplies in-house operations, mainly the hydrogen plant operations, to power electrolyzers, circulators, pumps, and other utilities, while exporting a more significant fraction to grid. In the GTHTR300H, the process heat extracted from the IHX is increased to 371 MWt while electricity is generated at 730°C gas turbine inlet and

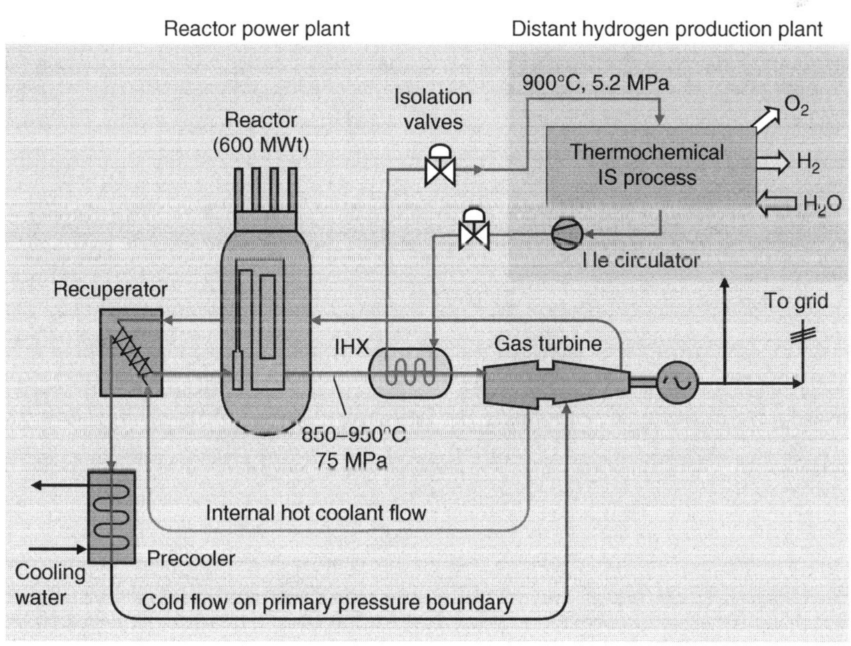

FIGURE 4.21
VHTR-based nuclear electricity and hydrogen generation process.

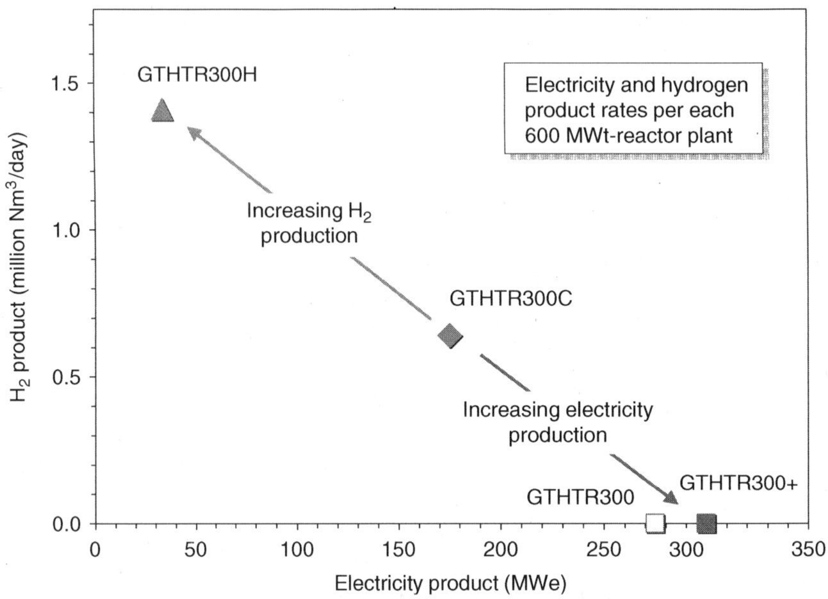

FIGURE 4.22
GTHTR300's variable production ratios of electricity and hydrogen.

38% efficiency mainly to meet the significant demand of electricity for hydrogen production. Figure 4.22 depicts the variable production range of electricity and hydrogen from the GTHTR300 plant family.

The cogeneration reactor pressure is lowered to 5 MPa from 7 MPa for the power-only systems due to two considerations. The first is to reduce pressure loads on a host of high-temperature heat exchangers including the IHX and chemical reactors to secure economical lifetime of these cost-intensive components. The second consideration is to maintain geometric similarity of the gas turbines to the baseline unit of the GTHTR300, as discussed in detail on the helium gas turbine in Section 4.4.3.2. Although the lower pressure increases the specific cost of gas turbine equipment, the benefits gained in the heat exchanger life cost saving and for gas turbine technology simplification offer more a compelling design advantage.

4.4.3 Technology Requirement and Status

4.4.3.1 The Reactor

The basic reactor technologies are developed with the test reactor HTTR [43,46]. The fuel is in the form of tri-isotropic (TRISO) ceramic-coated particles containing low-enriched UO_2 kernels as depicted in Figure 4.23. The particles secure containment for fission products at 1600°C or higher. The particles are dispersed in a graphite matrix and then sintered to form a fuel compact. A number of fuel compacts are encaged in a fuel rod for insertion into the graphite fuel block. Helium coolant flows in the annular space provided around the outer surface of a fuel rod. The HTTR's cylindrical active core, measuring 2.9 m in height and 2.3 m in diameter, contains 30 fuel columns, each comprising a stack of five fuel blocks high, and additional seven control rod guide columns. Dowels and mating sockets provided on the fuel block secure vertical alignment of a column. The active core is enclosed by replaceable reflector graphite blocks, which are laterally tightened by a graphite block–lined core restraint mechanism. The core structure is contained in a steel reactor pressure vessel.

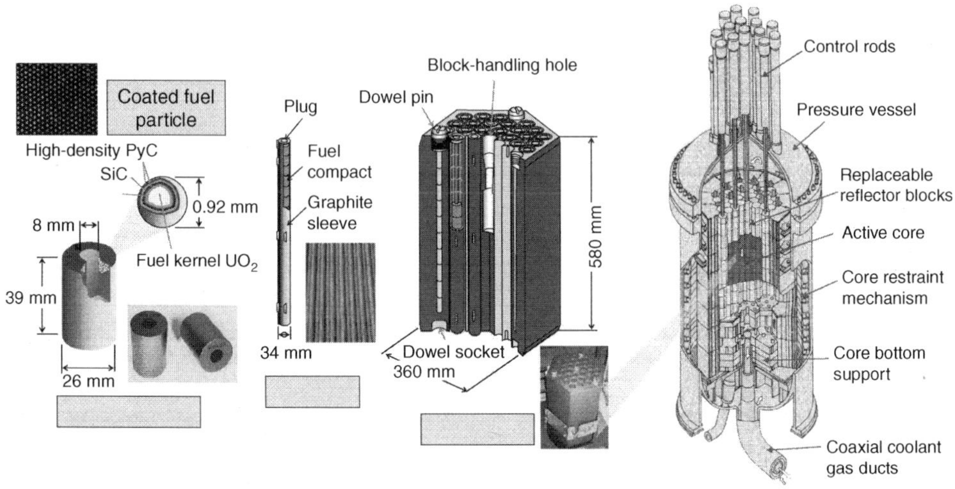

FIGURE 4.23
Prismatic block fuel element for the HTTR.

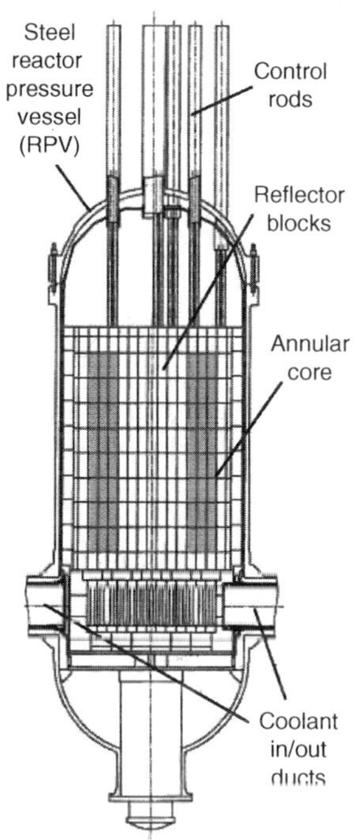

		GTHTR300 (electricity); baseline	GTHTR300+ (electricity); growth	GTHTR300C (cogeneration); GTHTR300H (hydrogen)
Operating parameters				
Reactor rating	MWt	600	600	600
Coolant outlet temperature	°C	850	950	950
Coolant inlet temperature	°C	587	663	594
Coolant inlet pressure	MPa	7.0	6.4	5.1
Coolant flow rate	kg/s	439	403	322
Max fuel operating temperature	°C	1108	1206	1244
Max fuel temperature	°C	1216	1314	1372
Max fuel LOCA temperature	°C	1546	1564	1535
Design parameters				
Reactor core type			Prismatic block annular core	
Reactor pressure vessel steel			SA533/508 (Mn–Mo)	
RPV inner diameter	m		7.6	
Fuel type			TRISO coated particle of low-enriched UO_2	
Fuel average enrichment	%		14	
Average fuel burnup	GW day/t		120	
Number of fuel block columns			90	
Inner reflector columns			73	
Outer reflector columns			48	
Annular core height	m		8.4	
Annular core inner/outer diameter	m		3.7/5.6	
Average core power density	W/cm^3		5.4	

FIGURE 4.24
Full-scale commercial reactor design of the GTHTR300.

The coolant is ducted into and away from the reactor by a concentric double pipe. The HTTR has demonstrated inherent and passive reactor safety.

Comprehensive experience and know-how in design, construction, operation, and maintenance acquired through three decades of the HTTR project are the basis for the full-scale commercial VHTR reactor design used by the GTHTR300 system variants as shown in Figure 4.24. To meet economical and safety requirements of the commercial reactor, several design details are necessarily modified from those of the HTTR [47]. For instance, a conventional carbon steel replaces the low alloy steel to lower reactor pressure vessel construction cost. An annular active core instead of the cylindrical one is required to raise reactor power without exceeding the limit of 1600°C for fuel during passive conduction cool down in accidents. Fuel burn up is extended to 120 GW day/t from the HTTR's maximum 33 GW day/t mainly by resizing kernel buffer and coating layers. R&D have, thus, been made to extend the technical base for higher burnup fuel [48]. More work is being done in this area. Carbon–carbon composites are being developed to replace the existing alloy cladding of control rods to improve the high-temperature structural integrity [49].

Alternative reactor types are possible for the VHTR. China's HTR-10 [35] and South Africa's pebble bed modular reactor (PBMR) [41] adopted major elements of pebble bed reactor design including fuel element from the past German experience. The fuel cycles might be thorium- or plutonium-based or potentially use mixed oxide (MOX) fuel.

4.4.3.2 Helium Gas Turbine

Helium gas turbine of high aerodynamic efficiency, reliability, and serviceability is required to qualify for nuclear service. The development of a helium gas turbine to meet these goals presents technical challenges, the extent of which depends heavily on design choices to be made. The baseline design shown in Figure 4.25 is a single-shaft unit consisting of 6 turbine stages and 20 compressor stages without compressor intercooling. Rated at 300 MWe and 3600 rpm, it drives a synchronous generator from the shaft cold end. The unit is horizontally installed to minimize bearing loads. These design features are aligned with established practice in industrial gas turbines. The design elements unique to the present application include the narrow compressor flowpath, which is the result of working in helium, and the magnetic bearings used to avoid large high-power shaft penetration to the primary pressure boundary or possibility of bearing lubricant contamination to reactor. Subscale tests are being conducted, as shown in Figure 4.26, for the compressor aerodynamics to validate the design and performance bases and for the magnetic bearing test rig to develop rotor clearance and dynamic control techniques to operate the heavy-duty flexible turbomachinery rotor.

The baseline design is for use in the GTHTR300. Geometric scaling from the baseline design is applied for units used in other plant variants and the resulting specifications are given in Table 4.2. The scaling is based on the principle that one can increase or decrease system pressure or rotor diameter while holding the speed constant to produce aerodynamically and mechanically similar gas turbines of larger or smaller unit capacity. Aerodynamic conditions, such as stage pressure ratio and efficiency are not changed; neither are the stresses in blades and discs. This makes the technologies developed for the baseline unit also applicable in other units.

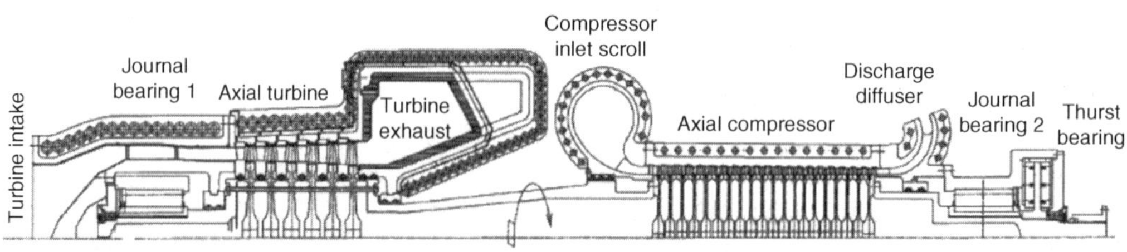

FIGURE 4.25
Baseline design of 300 MWe class helium gas turbine.

FIGURE 4.26
One-third scale development tests for helium compressor aerodynamics (on the left-hand side) and magnetic bearing suspended rotor dynamics.

Different choices for gas turbines are made by other designers based on specific project requirements. Relative features of the major systems were compared [38]. The gas turbine-modular helium reactor (GT-MHR) selects a bold vertical, asynchronous unit to facilitate a tight plant arrangement and compressor intercooling. The PBMR changes to a new single-shaft and horizontal unit using oil bearings from its earlier choice of a multishaft and vertical installation. The global nuclear industry supplier AREVA-NP prefers an indirect combined cycle of gas and steam turbines using a mixed nitrogen–helium gas to relate to more familiar air gas turbine conditions.

4.4.3.3 *Intermediate Heat Exchanger*

The IHX is a barrier of reactor pressure boundary. A proven nuclear design is the helical tube and shell heat exchanger for services at 950°C and 4.0 MPa in the HTTR as shown in Figure 4.27 [50]. Tubing is made of *Hastelloy-XR*, a corrosion and heat-resistant superalloy developed by JAEA. The IHX designs are compared for the GTHTR300C and HTTR. Similar operating conditions are observed and the same tubing material and similar stress limits are followed in both designs. Note that the GTHTR300C IHX is relatively compact owing to a large LMTD, such that it is integrated into an extended gas turbine horizontal pressure vessel. It is also possible to adapt to the HTTR's IHX installation practice.

New promising designs have been proposed such as the one under an extensive industrial study on plate IHX to develop fabrication, ISI methods and a design standard for gas reactors [51], and another plate-type design for the indirect cycle reactor plant [52]. These works demonstrate that a plate IHX would simplify integration in the primary system because of its drastically reduced size.

4.4.3.4 *Integration and Safety of Hydrogen Plant*

The JAEA selects the IS-process to be the basis for commercial development mainly because it is seen more suited to large-scale nuclear hydrogen production than HTE [9] and other alternatives. However, an available HTE-based plant can be connected to the reactor in the same manner as the IS process plant is connected. The HTE similarly requires a high-temperature process heat, and about 25% of its total energy input is heat and the balance electricity, which are fully and efficiently met in-house by the reactor heat and gas turbine power plant.

Hydrogen plant and gas turbine generator are each a partial load of the nuclear reactor during normal cogeneration operations. Transient or upset conditions in one system may

TABLE 4.2

Aerodynamically Scaled Parameters of Helium Gas Turbines

	Gas Turbine			Compressor Section				Turbine Section			
Unit	Rotation speed (rpm)	Pressure Ratio	Mass Flow (kg/s)	Inlet Temperature (°C)	Inlet Pressure (MPa)	Rim Speed (m/s)	Number of Stages	Inlet Temperature	Inlet Pressure (MPa)	Mean Speed (m/s)	Number of Stages
GTHTR300	3600	2.00	445	28.0	3.5	282	20	850	6.8	377	6
GTHTR300+	3600	2.00	408	28.0	3.2	282	20	950	6.2	377	6
GTHTR300C	3600	2.00	327	26.2	2.6	282	20	850	5.0	377	6
GTHTR300H	4215	1.47	327	26.2	3.5	282	11	730	5.0	377	3

	GTHTR300C IHX	HTTR IHX
Thermal rating (MWt)	168	10
Primary inlet/outlet temperature (°C)	950/850	950/389
Secondary inlet/outlet temperature (°C)	500/900	237/869
Primary side pressure (MPa)	5.02	4.06
Secondary side pressure (MPa)	5.15	4.21
LMTD (°C)	154	113
Heat transfer tubing	Hastelloy XR	
OD × *t* (mm)	45 × 5	31.8 × 3.5
Average length (m)	14	22
Tube number	724	96
Helical tube bundle sizing		
OD × *L* (m)	4.57 × 2.97	1.31 × 4.87
Weight (ton)	51	5.4
Design lifetime (y)	20	20

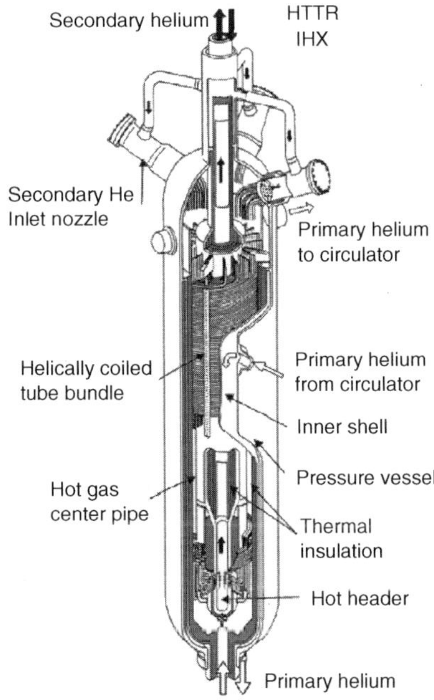

FIGURE 4.27
High-temperature IHX designs.

disturb or interrupt the operations of the other system or the reactor. The GTHTR300C is shown by simulation to operate with one production system in partial or full loss of the other, ensuring operational flexibility and stability in scheduled or forced outage [45]. Another control technology has been validated in the pilot test, which mitigates any major disturbance of the process plant to the reactor [53].

Prior experience to connect process heat applications, particularly, high-temperature processes, such as hydrogen plant, to nuclear reactor is limited. JAEA is studying to establish safety design rules for reactor protection and mitigation against the explosion of hydrogen and other combustibles. Another safety issue is tritium permeation from the reactor through the IHX and chemical reactors into product hydrogen. JAEA tested underlying tritium permeation characteristics. Design measures will be developed to limit tritium contamination to an insignificant level.

High-temperature gas isolation valves are safety components located in the secondary gas loop (see Figure 4.21). They protect the IHX from chemicals released from any failed hydrogen plant equipment and act as nuclear reactor confinement barriers to prevent radioactive releases from accidental rupture of the IHX tubing. Designs and tests for a scaled-down isolation valve, shown in Figure 4.28, at 900°C and 4 MPa helium confirmed the basic structural integrity, leak-tightness, and maintainability [54].

4.5 Concluding Remarks

Nuclear energy can produce hydrogen in several ways: (1) nuclear heated steam reforming of natural gas, (2) electrolysis of water using nuclear power, (3) HTE using minor heat and major electricity from nuclear reactor, and (4) thermochemical splitting of water using

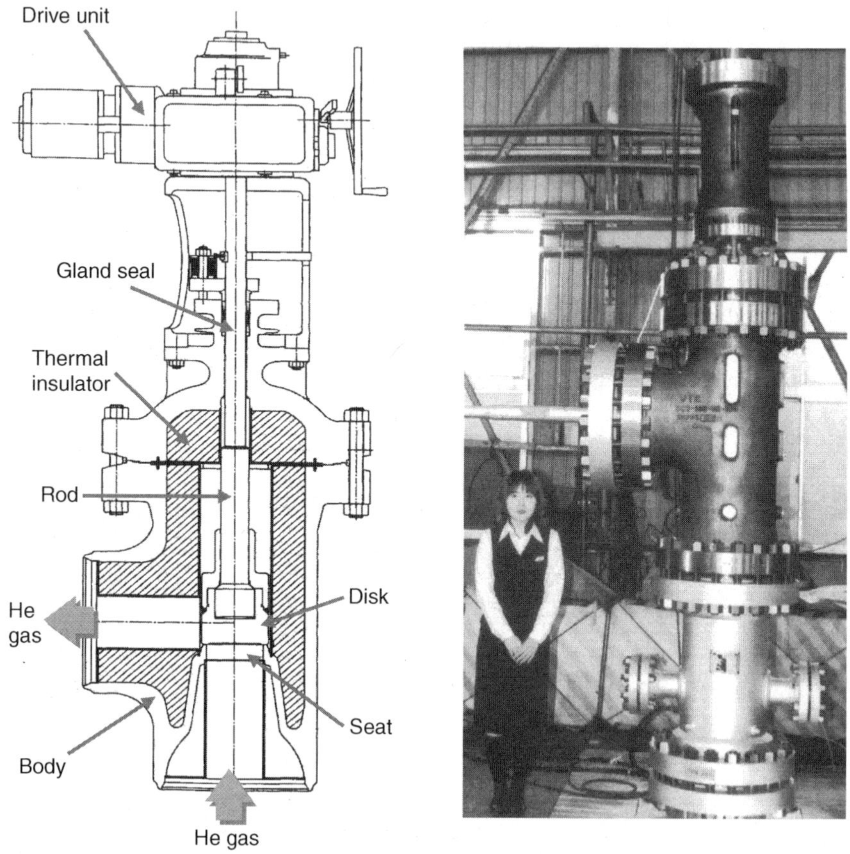

FIGURE 4.28
Development test of hot (900°C) helium gas isolation valve.

major heat and minor electricity from the nuclear reactor. The existing light water reactor is only applicable to electrolysis of water and at <30% efficiency.

The GIF identified six nuclear reactor systems for future development that can be licensed, constructed, and operated to provide economical and reliable energy products while satisfactorily addressing nuclear safety, waste, proliferation, and public acceptance [40]. Of the six technologies, the VHTR is uniquely suited for both electricity and hydrogen production and doing so at the highest potential efficiency of about 50%. It is also considered the nearest-term deployable reactor. Because of these important attributes the U.S. Department of Energy has chosen it to demonstrate the Next Generation Nuclear Plant (NGNP) including nuclear-heated hydrogen cogeneration.

Figure 4.29 summarizes the present efforts and future plans in JAEA to develop the VHTR electricity and hydrogen production. The next milestone toward nuclear hydrogen include the pilot test of IS process and the demonstration of 1000 Nm^3/h-scale hydrogen production with the HTTR. The ultimate goal is commercializing the family of the GTHTR300 plants that will produce electricity and hydrogen economically and yet flexibly to be able to adapt to future market demands. Although any one plant is suited to prototypical demonstration, the GTHTR300C being a substantial cogeneration system is best suited in this role because it employs the full spectrum of common technologies used in the family.

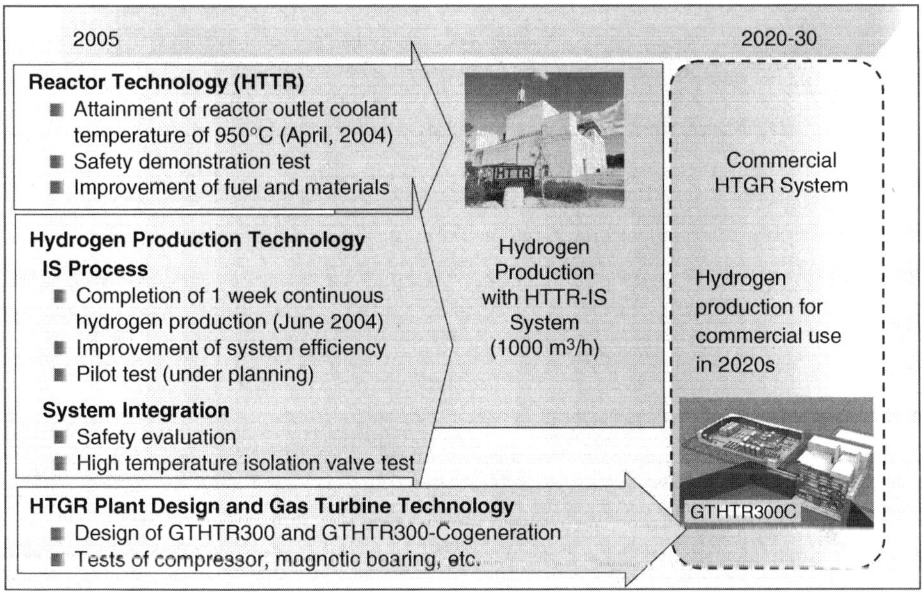

FIGURE 4.29
JAEA's plan for nuclear hydrogen production with the VHTR.

References

1. Doenitz, W. and Schmidberger, R., Concepts and design for scaling up high-temperature water vapor electrolysis, *Int. J. Hydrogen Energ.*, 7, 321, 1982.
2. Doenitz, W. and Eedle, E., High-temperature electrolysis of water vapor status of development and perspective for application, *Int. J. Hydrogen Energ.*, 10, 291, 1985.
3. Doenitz, W. et al., Electrochemical high temperature technology for hydrogen production or direct electricity generation, *Int. J. Hydrogen Energ.*, 13, 283, 1988.
4. Doenitz, W. et al., Recent advances in the development of high-temperature electrolysis technology in Germany, in *Proc. 7th World Hydrogen Energy Conf.*, Moscow, 65, 1988.
5. Maskalick, N.J., High temperature electrolysis cell performance characterization. *Int. J. Hydrogen Energ.*, 11, 563, 1986.
6. Hino, R. et al., Present status of R&D on hydrogen production by high temperature electrolysis of steam, Japan Atomic Energy Research Institute report, JAERI Research 95-057, 1995.
7. Herring, S. et al., Laboratory-scale high temperature electrolysis system, 2006 Annual Merit Review Proc., Hydrogen production and delivery, U. Nuclear Energy Initiative, http://www.hydrogen.energy.gov/annual_review06_delivery.html.
8. Matsunaga, K. et al., Hydrogen production system with high temperature electrolysis for nuclear power plant, Paper 6282 in *Proc. ICAPP'06*, Reno, NV, June 4–8, 2006.
9. Hino, R. et al., R&D on hydrogen production by high temperature electrolysis of steam, *Nucl. Eng. Des.*, 233, 363, 2004.
10. Nagata, S. et al., Fabrication of high temperature solid electrolyte fuel cell and power generation test (in Japanese), *J. High Temp. Soc.*, 7, 217, 1981.
11. Funk, J.E. and Reinstrom, R.M., Energy requirements in the production of hydrogen from water, *I&EC Process Design Dev.*, 5, 336, 1966.
12. Beghi, G.E., A decade of research on thermochemical hydrogen at the Joint Research Centre, ISPRA, *Int. J. Hydrogen Energ.*, 11, 761, 1986.

13. Pickard, P., Sulfur–iodine thermochemical cycle, 2006 Annual Merit Review Proc., Hydrogen Production and Delivery, D. Nuclear Energy Initiative, http://www.hydrogen.energy.gov/annual_review06_delivery.html.
14. Hong, J.H., Status and plans for nuclear power in Korea, in *WNFC Conf.*, Miami, Florida, April 5–7, 2006.
15. Shiozawa, S., Development status on hydrogen production technology using high-temperature gas-cooled reactor at JAEA, Japan, ICAPP'06, Reno, NV, June 4–8, 2006.
16. Funk, J.E., Thermochemical hydrogen production: past and present, *Int. J. Hydrogen Energ.*, 26, 185, 2001.
17. Norman, J.H. et al., Thermochemical water-splitting for hydrogen production, Gas Research Institute report GRI-80/0105, 1980.
18. Engels, H. et al., Direct dissociation of hydrogen iodide—an alternative to the General Atomic proposal, in *Proc. 6th World Hydrogen Energy Conf.*, 2, 657–662, Vienna, Austria, July 1986.
19. Onuki, K. et al., Electro-electrodialysis of hydriodic acid in the presence of iodine at elevated temperature, *J. Membr. Sci.*, 192, 193, 2001.
20. Hwang, G.-J. et al., Separation of hydrogen from a H_2–H_2O–HI gaseous mixture using a silica membrane, *AIChE J.*, 46, 92, 2000.
21. Nakajima, H. et al., A study on a closed-cycle hydrogen production by thermochemical water-splitting IS process, ICONE-7104, in *Proc. 7th Int. Conf. Nucl. Eng.*, Tokyo, Japan, April 19–23, 1999.
22. Kubo, S. et al., A demonstration study on a closed-cycle hydrogen production by thermochemical water-splitting Iodine–Sulfur process, *Nucl. Eng. Des.*, 233, 347, 2004.
23. Norman, J.H. et al., Thermochemical water-splitting cycle, bench-scale investigations and process engineering, GA-A 16713, 1982.
24. Brown, L.C. et al., High efficiency generation of hydrogen fuels using nuclear power, General Atomics report GA-A24285, 2003.
25. Goldstein, S. et al., Upper bound and best estimate of the efficiency of the iodine sulphur cycle, *Int. J. Hydrogen Energ.*, 30, 619, 2005.
26. Kasahara, S. et al., Effects of process parameters of the IS process on total thermal efficiency to produce hydrogen from water, *J. Chem. Eng. Jpn.*, 36, 887, 2003.
27. Knoche, K.F. et al., Second law and cost analysis of the oxygen generation step of the general atomic sulfur–iodine cycle, in *Proc. 5th World Hydrogen Energy Conf.*, Toronto, 2, 487, July, 1984.
28. Ozturk, I.T. et al., An improved process for H_2SO_4 decomposition of the sulfur-iodine cycle, *Energ. Convers. Manag.*, 36, 11, 1995.
29. Trester, P.W. and Staley, H.G., Assessment and investigation of containment materials for the sulfur–iodine thermochemical water-splitting process for hydrogen production, Gas Research Institute report GRI-80/0081, 1981.
30. Onuki, K. et al., IS process for thermochemical hydrogen production, JAERI-Review 94-006, 1994.
31. Kubo, S. et al., Corrosion test on structural materials for iodine–sulfur thermochemical water splitting cycle, in *Proc. 2nd Topical Conf. on Fuel Cell Tech.*, AIChE 2003 Spring National Meeting, New Orleans, March 30–April 3, 2003.
32. Tanaka, N. et al., Corrosion resistance of materials of construction for high temperature sulfuric acid service in Thermochemical IS Process (Alloy 800, Alloy 600, SUSXM15J1 and SiC) (in Japanese), *Zairyo-to-Kankyo*, 55, 320, 2006.
33. Terada, A. et al., Development program of hydrogen production by thermo-chemical water splitting IS process, ICONE-13-50183, in *Proc. 13th Int. Conf. Nucl. Eng.*, Beijing, China, May 16–20, 2005.
34. Ota, H. et al., Conceptual design study on sulfuric-acid decomposer for thermo-chemical iodine–sulfur process, ICONE-13-50494, in *Proc. 13th Int. Conf. Nucl. Eng.*, Beijing, China, May 16–20, 2005.
35. Xu, Y. and Zuo, K., Overview of the 10 MW high temperature gas cooled reactor—test module project, *Nucl. Eng. Des.* 218, 13, 2002.

36. Billot, P. and Barbier, D., Very High Temperature Reactor (VHTR) The French Atomic Energy Commission (CEA) R&D Program, in *Proc. of the 2nd International Topical Meeting on High Temperature Reactor Technology*, Beijing, September 22–24, 2004.
37. Chang, J., Overview of HTR development in Korea, in *Proc. of the HTR2006 Conference*, South Africa, October 2–4, 2006.
38. Next Generation Nuclear Plant Project—Preliminary Project Management Plan, INL/EXT-05-00952 Rev.1 Idaho National Laboratory, March 2006.
39. Hittner, D., The European programme of development of HTR/VHTR technology, in *Proc. of the 2nd International Topical Meeting on High Temperature Reactor Technology*, Beijing, September 22–24, 2004.
40. U.S. Department of Energy and the Generation IV International Forum, A Technology Roadmap for Generation IV Nuclear Energy Systems, 2002, http://gif.inel.gov/roadmap/pdfs/gen_iv_roadmap.pdf.
41. Matzner, D., PBMR project status and the way ahead, in *Proc. of the 2nd International Topical Meeting on High Temperature Reactor Technology*, Beijing, September 22–24, 2004.
42. Ogawa, M. and Nishihara, T., Present status of energy in Japan and HTTR project, *Nucl. Eng. Des.*, 233, 5, 2004.
43. Shiozawa, S. et al., Overview of HTTR design features, *Nucl. Eng. Des.*, 233, 11, 2004.
44. Yan, X. et al., GTHTR300 design and development, *Nucl. Eng. Des.*, 222, 247–262, 2003.
45. Yan, X. et al., GTHTR300 design variants for production of electricity, hydrogen or both, in *Proc. of the OECD/NEA 3rd Information Exchange Meeting on the Nuclear Production of Hydrogen*, OECD/NEA, Oarai, Japan, October 5–7, p. 121, 2005.
46. Saito, S. et al., Design of high temperature engineering test reactor (HTTR), JAERI, 1332, 1994.
47. Kunitomi, K. et al., Japan's future HTR—the GTHTR300, *Nucl. Eng. Des.*, 233, 309, 2004.
48. Sawa, K. et al., Investigation of irradiation behavior of SiC-coated fuel particle at extended burnup, *J. Nucl. Tech.*, 142, 250, 2002.
49. Sumita, J. et al., Anisotropic deformation effect on the fracture of core components made of two-dimensional C/C composite, *Key Engineering Materials*, Trans Tech Publications Ltd., Switzerland, 143, 2005.
50. Inagaki, M. et al., R&D on high temperature components, *Nucl. Eng. Des.*, 233, 211, 2004.
51. Tokunaga, K. et al., Technology development of high efficiency and high capacity gas to gas heat exchanger which is necessary to practical application of gas-cooled reactor, in *Proc. Global 2005*, Tsukuba, Japan, October 9–13, 2005.
52. Gauthier, J. et al., Antares: The HTR/VHTR project at Framatome ANP, in *Proc. of the 2nd International Topical Meeting on High Temperature Reactor Technology*, Beijing, September 22–24, 2004.
53. Nishihara, T. et al., Development of control technology for the HTGR hydrogen production system, New Orleans, in *Proc. of the Global 2003*, New Orleans, Louisiana, November 16–20, p. 320, 2003.
54. Nishihara, T. et al., Development of high temperature isolation valve for the HTTR hydrogen production system, *AESJ Transaction*, 3(4), 381, 2004.

5

Hydrogen Production from Wind Energy

Dimitrios A. Bechrakis and Elli Varkaraki

CONTENTS

5.1 Introduction

The process of producing hydrogen from wind energy is, more precisely, the hydrogen production by the electrolysis of water, where the main power input is electrical power produced from the conversion of wind energy. As a concept, it is a very promising way to produce hydrogen not only because both procedures are renewable and nonpolluting, but also because wind energy conversion systems can become more effective by incorporating an energy storage medium with such a versatile energy carrier.

Wind power penetration in the electrical grids has increased significantly and is still expanding even without the adoption of any appropriate energy storage technique, which would solve many problems associated with the unstable performance of wind energy systems. Wind farms (WFs) may use some part of their energy production for storage by balancing their load with respect to the current wind conditions and the energy demand of the grid for on-site hydrogen production and distribution.

5.2 Water Electrolysis

More than 400 industrial water electrolyzers were in operation by the beginning of the nineteenth century. In 1939, the first large water electrolysis plant of 10,000 Nm^3 H_2/h capacity went into operation and in 1948, Zdansky/Lonza built the first pressurized industrial electrolyzer [1].

The major part of the hydrogen produced worldwide is consumed in the treatment of petrochemicals, and in the production of ammonia and methanol. Electrolytic hydrogen is generally used in smaller markets and specialty applications, where higher purity is required. In food processing, hydrogen is used to increase the degree of saturation in fats and oils, increasing their melting point and resistance to oxidation. In the electronics industry, it is used as a reducing agent in the epitaxial growth of polysilicon and in the manufacture of integrated circuits. In the nuclear industry, it is used as an oxygen scavenger to remove traces of oxygen, which may cause stress corrosion cracking [2]. In float glass manufacture, it is used as a protective atmosphere. Small hydrogen quantities are consumed in the pharmaceutical and plasma industry. On account of its high thermal conductivity, it is also used as a coolant for the electrical generators in power stations.

The electrolysis process consists of electrochemically splitting water into its constituents, namely hydrogen and oxygen. According to the electrolyte used for electricity conduction in the process, four types of water electrolysis may be found. The first electrolysis units in the beginning of the past century were using an acid electrolyte, but later the acidic medium was abandoned due to severe corrosion problems. Polymer electrolyte membrane electrolyzers use a proton-conducting membrane as electrolyte, which presents the advantage of being oxygen impermeable, but there are no large electrolyzers actually available because it is a relatively new technology. Steam electrolysis at very high temperatures, using an oxygen ion conducting ceramic electrolyte may achieve very high efficiencies in the future, but is still under investigation at the laboratory scale. Today, most commercial water electrolyzers use an alkaline electrolyte (Figure 5.1). Three of the largest

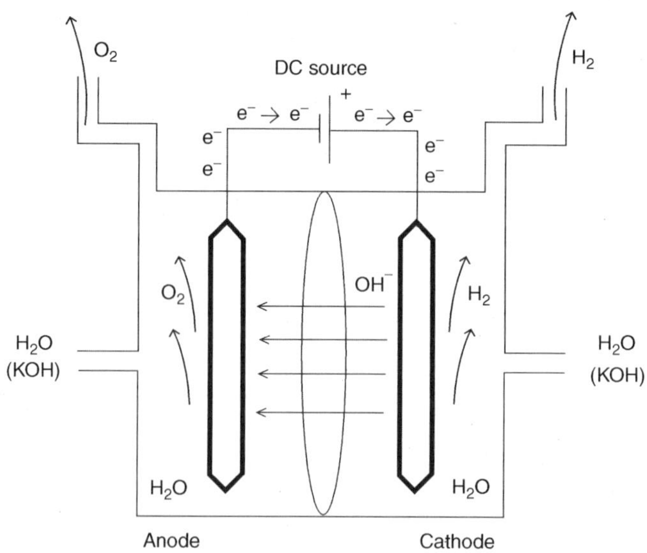

FIGURE 5.1
Schematic view of an alkaline electrolytic cell.

electrolyzers in the world were installed at the end of the 1940s, supplying hydrogen to ammonia plants for the fertilizer industry in India (De Nora), Norway (Norsk-Hydro), and Egypt (Demag) [3].

In an electrolysis cell, the following simplified reactions take place:

In the electrolyte	$2H_2O \rightarrow 2H^+ + 2OH^-$
At the cathode	$2H^+ + 2e^- \rightarrow H_2$
At the anode	$2OH^- \rightarrow \frac{1}{2}O_2 + H_2O + 2e^-$
Overall reaction	$H_2O \rightarrow \frac{1}{2}O_2 + H_2 \qquad \Delta H_r = 285.6$ kJ/mol

where ΔH_r is the enthalpy of the reaction.

In alkaline electrolyzers, hydrogen is obtained at the cathode with a purity of approximately 98 vol%, with oxygen and water vapor as the only impurities. Hydrogen may be further purified to almost 100% by the removal of oxygen in a catalytic deoxidizer and the subsequent removal of water vapor in a dryer. In the purification step, 5–10% of the produced hydrogen may be lost; therefore, the use of electrolytic hydrogen without purification should always be considered in priority for each application.

In bipolar electrolyzers, more than 100 cells may be mounted together in a filter press arrangement, to form a stack with common hydrogen and oxygen outlets. The cells are interconnected electrically in series and hydraulically in parallel. Current flows from one end of the stack to the other, and therefore is the same for each cell. The total voltage is equal to the sum of all the individual cell voltages in the stack. Each cell voltage mainly depends on the current density, that is, the current per unit electrode area, and the electrolysis temperature. In practice, the current–voltage (*I*–*V*) curve of an electrolyzer is rather unique, depending on the electrode activity, thermal efficiency, and several construction characteristics of the electrolyzer.

There are no electrolyzers developed specifically for operation with wind turbines. However, the rapid response of electrochemical systems to power variations makes them suitable "loads" for wind turbines. Industrial electrolyzers are designed for continuous operation, mainly because their elevated investment cost requires high-capacity factors for reasonable payback times, but they are subject to a considerable number of current interruptions through their lifetime due to occasional power interruptions, accidental trips of safety systems, and planned stops for maintenance. Current interruptions are more frequent in specialty applications, where electrolyzers supply hydrogen "on demand." Therefore, the discontinuous use of the equipment is not new, and most commercial electrolyzers may be used in intermittent operation although a significant performance decrease is expected with time. In fact, it is not power variation, but current interruptions that may cause severe corrosion problems to the electrodes, if the latter are not protected by the application of a polarization current when idle.

Research and development on water electrolysis presented an impressive boom in the 1970s, after the first oil crisis, but was subsequently reduced due to the lowering of oil prices. The efficiency of actual electrolyzers mainly depends on the developments of these years [4–8]. Conventional water electrolyzers operated in 25–35 wt% KOH solutions with nickel electrodes, requiring a cell voltage of 2.0–2.2 V at 2000 A/m^2 current density at 80°C [9]. Murray and Yaffe [10] studied alkaline electrolyzers and materials for operation at temperatures up to 150°C. At that time, the long-term instability of the asbestos separator used in the cell required limiting the operating temperature to <82°C. Several engineering thermoplastic polymers were therefore evaluated as electrode separators in alkaline electrolyzers, to replace asbestos, for stability rather than toxicity reasons. Polysulfone fiber was identified as a promising material for the application [10]. The relatively inexpensive nickel wire mesh electrodes showed reasonable electrochemical activity and low corrosion rates. A typical cell voltage of 1.9 V DC at 4000 A/m^2 current density was obtained with an asbestos separator at 100°C, and a linear thermal voltage behavior, slightly depending on the applied current density, was observed. This value corresponds to an average DC input of 50 kWh/kg H_2. A filter press, pressurized alkaline electrolyzer, with a 25 wt% KOH solution, generating 52 kg/h H_2 at 0.7 MPa, capable of operating from 32% to 145% of nominal capacity was subsequently designed [10].

Already in 1977, the solid polymer electrolyte (SPE) water electrolysis technology had demonstrated efficiencies over 90% at current densities up to 5380 A/m^2 [11]. Prototype systems had been developed to generate oxygen for life support in spacecraft, but a 62,500 standard cubic feet per hour (SCFH) (1770 m^3/h) hydrogen generation plant had also been studied, operating in the range 0–100% of rated capacity, to store off-peak electrical energy. Hydrogen would be stored in metal hydrides to fuel a 26 MW_e fuel cell [11]. Cell voltages of <1.6 V DC at 4,000 A/m^2 and 1.7 V DC at 10,000 A/m^2 current density were obtained at 80°C, and this performance is much higher than for any commercial electrolyzer available currently.

To address the high operating costs of water electrolysis, several studies have been performed during the past 30 years for the development of stable, efficient electrodes [12–14] with low overpotential, both for the hydrogen evolution [15–20] and the oxygen evolution [21–26] reactions at the cathode and anode, respectively. Less effort has been concentrated on the development of appropriate diaphragms for alkaline water electrolysis [27], with low ohmic resistance, low gas permeation rates, and high mechanical and chemical stability. Asbestos has been the material of choice for decades, but most commercial electrolyzers currently use asbestos-free diaphragms due to its high toxicity.

Since oxygen production from air is much simpler and cheaper, water electrolysis is not generally regarded as an oxygen production method, except for special applications such as "life support" in submarines. Oxygen recovery from water electrolyzers, however, is often considered as a means to increase the overall energetic and economical efficiency of the process. Saxe and Alvfors [28] recently evaluated the advantages of using oxygen recovered from a water electrolyzer in a pulp and paper industry.

A simplified process diagram of an industrial alkaline water electrolyzer is shown in Figure 5.2. Hydrogen and oxygen leave the electrolysis stack separately, where the electrochemical dissociation of water occurs, and pass through distinct vessels HV and OV, respectively, where they are separated from the electrolyte solution. The electrolyte and gases are cooled to remove the heat of reaction. From the vessels they go through additional filters HF and OF to remove traces of electrolyte. Oxygen is then stored or vented according to the application. Hydrogen may also be used directly after the filter. In the case presented in Figure 5.2, there is an additional purification section, composed of a deoxidizer where the oxygen content in hydrogen is burned to produce water vapor and a drier to remove the initial humidity of the saturated hydrogen gas and the additional water of reaction from the deoxidizer.

The utilization of oxygen by-product from the electrolytic hydrogen production may reduce the electricity consumption of oxygen production by air-separation technologies, enhancing the overall efficiency and improving the economics of water electrolysis. Kato et al. [29] compared the relative costs of hydrogen production by means of water electrolysis and steam methane reforming, taking into account the use of pure electrolytic oxygen for medical purposes.

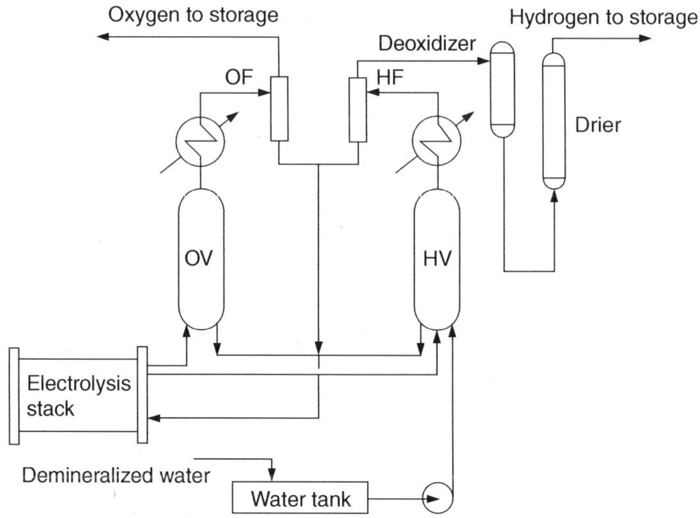

FIGURE 5.2
Simplified process diagram of an alkaline water electrolyzer.

5.2.1 Efficiency

The efficiency of a water electrolyzer is generally defined as the ratio of the electrical energy spent to produce a certain quantity of hydrogen by the energy content of the produced hydrogen. Here, the higher heating value (HHV) of the fuel is used because it reflects its true energy content, according to the energy conservation principle (first law of thermodynamics) [30].

Depending on the size, type, and condition of an electrolysis plant, the energy requirement to produce 1 Nm^3 of H_2 lies in the range 4–6 kWh. With a hydrogen HHV of 3.5 kWh/Nm^3, the efficiency of water electrolyzers lies in the range 58–87%.

The power consumption of the electrolyzer is generally calculated by multiplying the applied current by the stack voltage that are both continuously measured. However, this power consumption refers to DC current and therefore to the electrolysis stack. The AC power consumption is not usually measured, and most manufacturers assume a 5–10% energy loss for the AC/DC converter.

In Figure 5.3, some measurements of electrolyzer efficiency are reported with respect to both DC and AC (before conversion) power consumptions. The efficiency is clearly higher at the nominal operating temperature of 80°C than at a temperature of 44°C, which may be attained if the electrolyzer operates at partial load for long periods of time. The AC efficiency of the electrolyzer, as measured with the help of a power meter at the inlet of the power supply cabinet, is much lower than the DC efficiency. The DC efficiency of the electrolyzer increases at partial load due to the lower ohmic resistance in the cells. However, the AC efficiency remains almost constant through the whole operating power range, probably because of a higher AC/DC conversion efficiency at nominal power and a lower efficiency at low power output. A similar finding was reported for the Hydrogen and Renewables Integration (HARI) project, where significant losses were associated to the power conversion electronics, ranging from 11% at rated power to 28% at low power [31].

An example of an alkaline electrolyzer operating under variable power input is shown in Figure 5.4, where the power output of the wind turbine (P_{wt}) is depicted versus the power

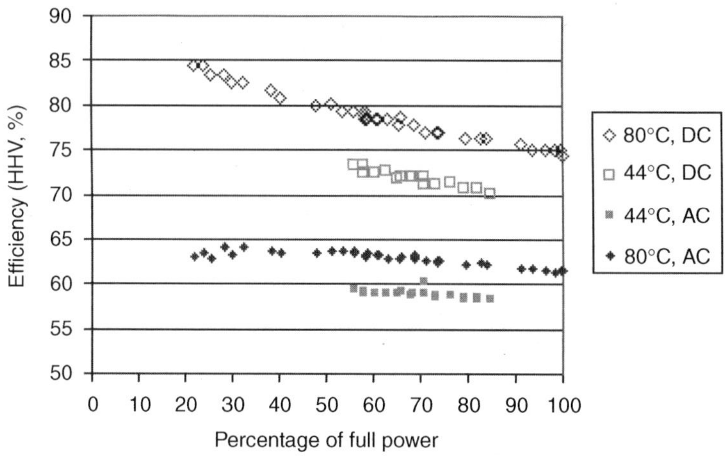

FIGURE 5.3
Efficiency range of an electrolyzer.

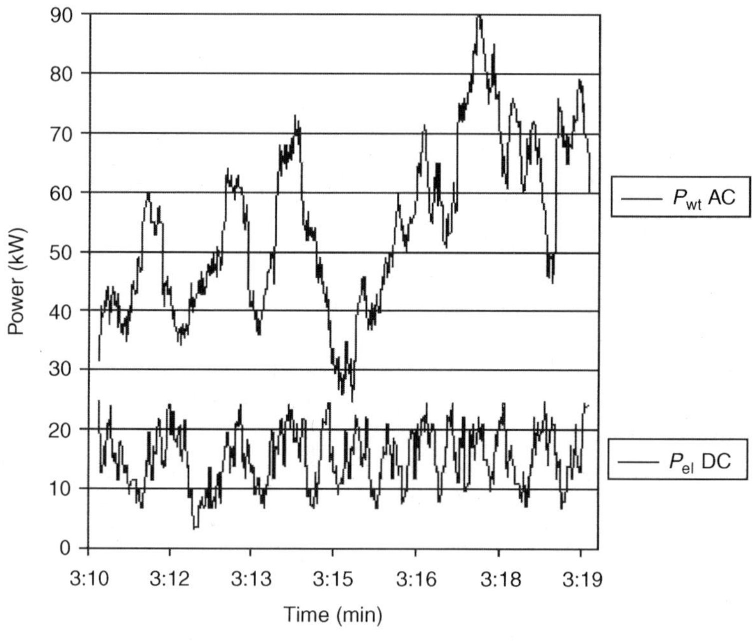

FIGURE 5.4
An example of an electrolyzer operation under variable power input.

consumption of the electrolyzer (P_{el}). The measurements were taken at the wind-hydrogen plant of the Centre for Renewable Energy Sources (CRES) in Greece during a day with low wind potential.

5.2.2 Development Goals for Water Electrolyzers

The development goals for water electrolyzers are oriented toward increasing the efficiency and reducing the cost. Most of the issues to be addressed are not specific to intermittent power operation. The efficiency of water electrolysis increases with the increasing temperature, but the operating temperature is generally limited to 80–90°C due to severe corrosion of the materials at higher temperatures. The electrolysis efficiency may exceed 90% if appropriate corrosion-resistant materials are developed for operation at higher temperatures and pressures.

Advanced alkaline electrolyzers may be subject to input power variations in the range 15–120% of their nominal power within 1 s and this feature makes them attractive for coupling with the wind turbines. However, during operation at a very low load, the purity of the gases decreases due to diffusion through the diaphragm. In addition, most electrolyzers require the application of a protective polarization current on the order of 40–100 A/m^2 when they are not in operation to avoid the corrosion of the electrodes. The protective polarization current affects the purity of the gases in a similar way as the operation at low load. For safety reasons, an emergency stop of the electrolyzer is generally foreseen if the oxygen-in-hydrogen or hydrogen-in-oxygen content exceeds a certain value below the limit of explosive mixture formation. In industrial electrolyzers with high safety standards, an emergency stop is accompanied by a pressure release (in the case of a pressurized operation) and subsequent purging of the system with an inert gas such as nitrogen resulting in hydrogen and energy losses. It may also take 10–30 min for the electrolyzer to restart the operation after an emergency stop; therefore, the wind-hydrogen system should be dimensioned and designed in a way to avoid the formation of impure gases to the extent possible.

It was recently reported that the application of a protective polarization may be avoided by using corrosion-resistant activated electrodes [32]. The polarization protection of the electrodes represents approximately 1–2% of the nominal electrolyzer power. Assuming the capacity factor of the electrolyzer cannot exceed the one of the wind turbine, the polarization protection should be applied during 60% of the electrolyzer's lifetime or higher depending on the relative sizes of the electrolyzer and wind turbine.

5.3 Wind Energy Conversion Systems

Wind is the motion of air masses caused by the different thermal conditions that occur over the earth's surface as a result of the transmission of solar radiation. Wind energy is defined as the kinetic energy of the wind converted into mechanical work. This mechanical work can be used to drive an electrical generator for the production of electricity. A machine that performs this conversion is called a wind turbine generator (WTG) and a group of these, including the auxiliary equipment, constitute a WF.

Electrical energy production from wind has gained great support by governments toward a policy of reducing greenhouse gas emissions. This support is not only expressed in terms of subsidies; in many countries, wind power stations as well as other renewable energy sources (RES) do have special privileges in dispatching and, generally, in the energy market competition [33]. For the past decade, wind energy was the fastest growing energy source in the world. It has evolved to such an extent that now it can be characterized as a mature technology for electricity production [34].

Wind energy exists in almost every part of the world. It is a safe, clean, and inexhaustible source of energy with a slight environmental impact caused mostly during the time of installation of its equipment [35]. In parallel, the demand for energy is increasing all over the world although the problems of using fossil fuels are already obvious. However, there are many hindrances that refrain the use of wind power systems from producing electrical energy at a very high percentage in a system.

Although wind energy is abundant on Earth, it is advantageously exploitable only in certain sites with high wind potential and appropriate topographical conditions where the investment of the whole installation and operation is viable. The stochastic nature of wind is responsible for the intermittent operation of the wind energy systems; an effect that does not allow the use of wind power exclusively to meet the energy needs of consumers. However, wind is never ceased on Earth and there is always a part of, at least, a continent where its potential is always significant.

There are wind energy systems that are connected to an energy transmission system or operate in an autonomous power plant, the "grid-connected" and the "stand-alone" systems, respectively. The first category includes the on- and off-shore WFs of various capacities, which feed the energy transmission systems at a continuous base. The stand-alone wind energy systems use wind-generated electricity for direct use or to charge a heavy-duty battery cluster when the load demand is lower than the power production.

5.3.1 Grid-Connected Wind Energy Systems

A schematic of a typical grid-connected WF is shown in Figure 5.5. The dashed line defines a single WTG, which includes its transformer to the medium voltage level. The area of the WF is defined by the dashed line, which includes the distribution lines and the necessary infrastructure to get connected to the transmission system. Each wind turbine gets connected to the transmission system as long as the wind resource at their hub height is sufficient and the grid conditions allow the initializing of the particular generator.

Integration of large-scale WFs in a power system has an impact on the transmission systems' stability as well as on power quality issues. This is due to the intermittent operation of the wind turbines and the technical constraints of the generators that work under these conditions. In a transmission system with large WFs, the capability to maintain constant values on frequency and voltage is limited. WFs must be able to remain connected to the grid during voltage dips caused by network faults (fault ride-through capability). They also have to provide voltage and reactive power control and to contribute in the system's frequency regulation. These requirements have been included in the grid codes

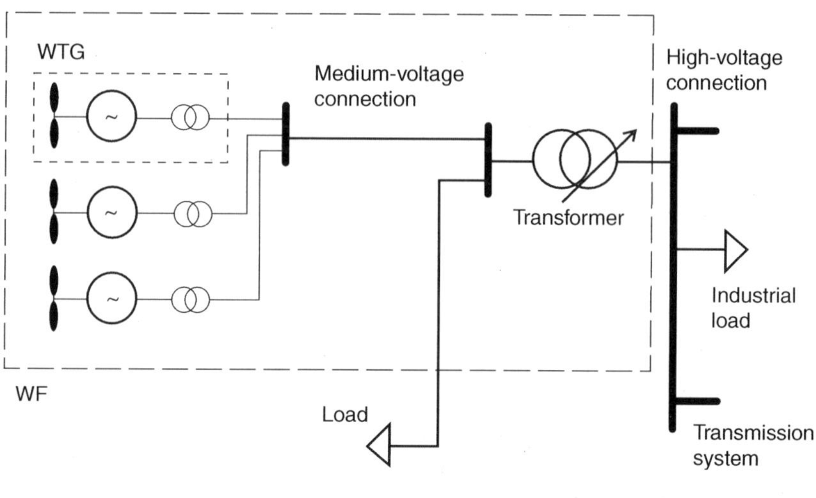

FIGURE 5.5
A schematic of a typical grid-connected wind energy conversion system.

published in countries with high wind power penetration [36–38]. Some of them set the WF's normal operation level at 90% of the available active power output. A transmission system operator (TSO) should also have the option to reduce the average power output of a wind plant through a Supervisory Control and Data Acquisition (SCADA) system in real time whenever it is required.

Another very important factor that confines the penetration of WFs in a transmission system is the congestion caused occasionally by the flow of high power in transmission lines of low capacity. Usually, WFs are installed on sites that are far away from inhabited areas and consequently, the existent grid infrastructure in these areas cannot accept large amounts of power and support the operation of large WFs. Also, in case that there are thermal power stations connected to these lines, the limitation is even stricter because these plants should always operate at a technical minimum stage of generation. To solve the problem of congestion, the operators have to curtail the power output of a WF that exceeds a specified set point at the high-voltage connection node [39].

Almost all the cases of problematic operation of a wind energy system are followed by a forced termination or curtailment resulting in wasting significant amounts of energy that could be stored or used in another application. This energy wasting practice is even more controversial when it concerns an energy source that is renewable and nonpolluting.

5.3.2 Stand-Alone Wind Energy Systems

Stand-alone wind energy systems are driven by wind turbines of smaller sizes and supply the loads of an isolated system. A power controller is responsible for the load management of the installation. In these systems, WTGs are supported by a storage medium, which also operates as a backup power supply unit. This medium is usually a rechargeable lead acid battery cluster with the necessary power conditioning equipment. The batteries are charged through the AC/DC converter, which is activated with respect to the load conditions. The backup unit is set on operation by the power controller when there is no sufficient wind potential or during a possible fault in the operation of the WTG.

Figure 5.6 depicts a typical stand-alone wind energy system, which serves AC loads. When the storage unit is full and the load demand is lower than the power production, the

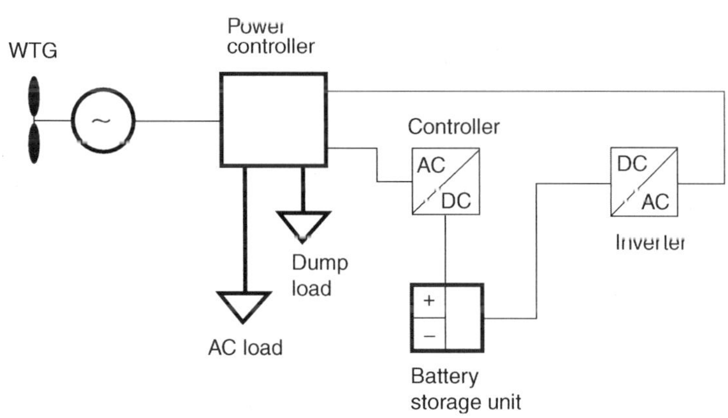

FIGURE 5.6
A typical stand-alone wind power system with a rechargeable battery storage unit.

surplus is converted into heat through a dump load. If there is access to an existing electricity network, a temporary energy surplus is fed to the utility instead of dumping it.

Wind power stand-alone systems are bound to keep their load consumption stable with respect to the wind conditions, avoiding temporary peaks. In stand-alone systems with critical loads, the main power unit must be a fully controlled diesel generator instead of a larger wind turbine. An alternative option for energy storage used currently in large stand-alone wind energy systems is pumping water in reservoirs to be utilized in a combined hydroelectric energy system. Generally, batteries are not considered as a reliable storage medium because after a specific number of charging–discharging cycles their efficiency decreases [40]. Also, battery disposal causes environmental hazards and it is still an issue that has not been solved thoroughly.

5.3.3 Noninterconnected Wind Energy Systems

An intermediate category of wind energy systems is the ones that are used in microgrids or in small grids, which are installed in places where the access to the interconnected system is infeasible (e.g., on islands). In these systems, for security reasons the primary power production often comes from diesel engines. Although the main power comes from the wind turbines, there is a need for a reactive power compensation device to maintain the voltage stability of the grid. Figure 5.7 gives the basic configuration of a noninterconnected wind energy system.

In these systems, the cost of energy is very high because of the pressing need to use diesel fuel at an almost daily basis. Concerning the use of the wind energy system, there is always the demand for balancing the load consumption with respect to the power production and the need for an auxiliary power supply with "black start" capability—the process of restoring power after a wide-area power outage.

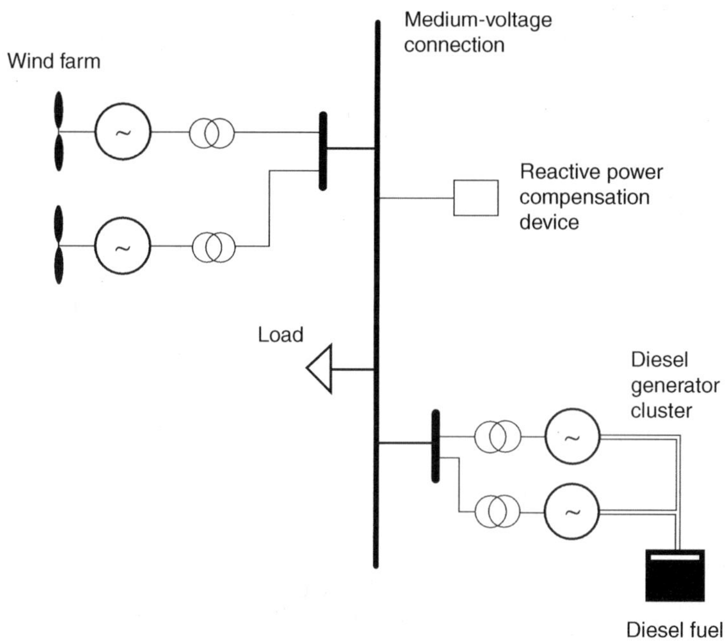

FIGURE 5.7
A wind energy system in a noninterconnected grid.

5.4 Wind Power for Electrolysis

To estimate the annual electrical energy yield from a wind turbine to be used for electrolysis, it is important to make an assessment of the wind potential of the particular site where the installation will take place. Wind speed distribution expressed into a histogram of the wind speed values under which a WTG produces power is the key factor through which the energy production is calculated. It is convenient to calculate these magnitudes in per hour units to obtain the results in commonly used units of energy. Equation 5.1 describes how the annual energy production from a wind turbine can be estimated at the generator output.

$$\mathrm{WE_Y} = \sum_{v=\mathrm{ci}}^{\mathrm{co}} (P_v \cdot H_v) \tag{5.1}$$

For each wind speed value (v) from the cut-in (ci) to the cut-out (co) phase of the WTG, the product of its corresponding power output (P_v) multiplied by the time H_v (in hours) during which value v appears in a year is calculated. The sum of these products gives the annual energy production ($\mathrm{WE_Y}$). Wind speed values are referred to the hub height of the turbine. The power output values, which compose the power curve of a WTG, are provided by the turbine manufacturer. H_v values are calculated from the annual distribution (or the histogram) of the wind speed values.

The theoretical power in the wind is proportional to the cube of wind speed. In fact, the power that can be harnessed by a wind turbine, converted to electrical, is delimited by the curve of Figure 5.8. The Figure 5.8 depicts the electrical power that can be derived from a doubly fed (wound-rotor) WTG—a type that prevailed in the WTG industry. The WTG starts to operate producing power at a wind speed threshold, which is called cut-in speed. Within a specific range of wind speed after this threshold, the power increases exponentially till it reaches the nominal value. At this point, the velocity of the rotor has to remain constant at all wind speed values till the cut-out verge where the turbine suspends its operation for safety reasons. Consequently, the wind speed input is desirable to fall within the nominal operational conditions of the WTG. Sites showing extreme gusts might be of lower-energy yield performance compared to a calmer one. Additionally, these sites are more hazardous for the mechanical parts of a WTG.

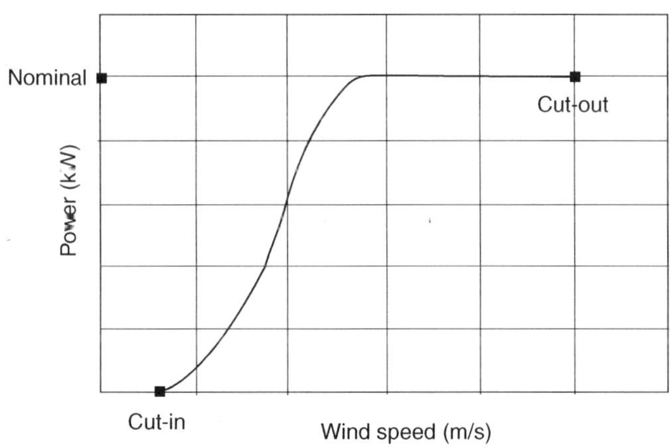

FIGURE 5.8
The power curve of a typical doubly fed induction WTG.

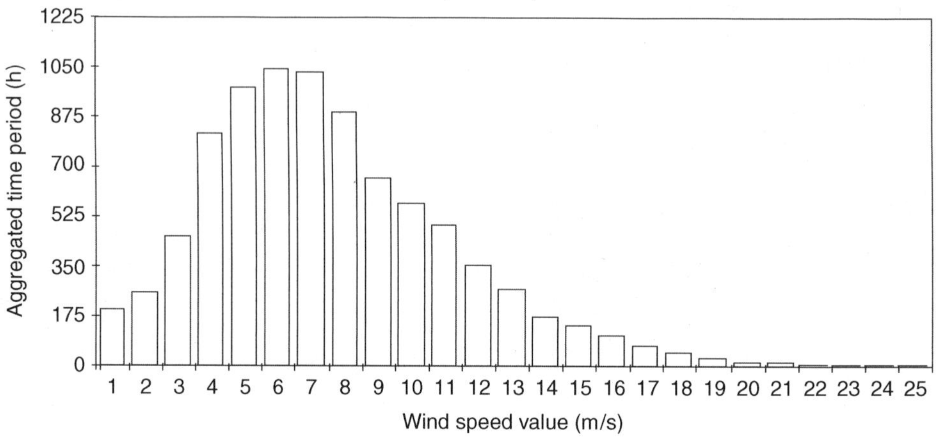

FIGURE 5.9
The annual distribution of wind speed at a site with sufficient potential.

Figure 5.9 gives the annual wind speed distribution of a site where the wind potential is exploitable. The wind speed values are sorted in the ascending order to be used in Equation 5.1. An alternative way to simulate the annual wind speed distribution is using the Weibull distribution with the proper pair of constants.

The ratio of the net electricity generated, for the time considered, to the energy that could have been generated at continuous full-power operation during the same period is the capacity factor (cf) of a WTG. This factor allows the comparison of a WTG with a conventional (or of other source) energy generator with respect to the energy yield capability. A site on which a wind power plant shows annual capacity factor over 30% can be considered as a site of high wind potential, appropriate for wind energy exploitation. The life cycle of a wind turbine is guaranteed for 20 years.

To calculate the amount of hydrogen produced by electrolysis powered from a wind energy conversion system within a year, the efficiency of the AC/DC (or DC–DC*) conversion (η_c) and the energy consumption of the electrolyzer (ec_{el}) per newton cubic meter of H_2 production need to be defined. The efficiency of a standard AC/DC converter ranges from 80% to 95% [41]. High values of η_c occur in the conversion of large amounts of power. Typical values of ec_{el} range from 5 to 6 kWh/Nm^3.

Provided that the efficiency of the electrolyzer is linearly proportional to the energy supplied and all the possible energy losses are included in these two factors, Equation 5.2 gives the amount of hydrogen (aH_w) in newton cubic meter that can be produced by a wind energy system in a year.

$$aH_w = \frac{WE_Y}{ec_{el}} \cdot \eta_c \tag{5.2}$$

This formula stands for other time periods as well, if the wind speed distribution values are ensured for the particular period. In most cases, the annual wind speed distribution is relatively the same over the years due to the seasonal characteristics of the weather in a site [42].

* For DC generators used or for alternators that include an AC/DC converter as a standard equipment.

For a safe estimation of the wind speed distribution, as well as for a preliminary WF designing, at least wind measurements of 1 year are required. A complete measurement of the wind regime of a site includes wind speed and wind direction measurements with statistical processing at the hub height of each WTG position.

The power that is fed to the electrolyzer should be over 15% of the rated value to maintain an acceptable purity level of the produced hydrogen. At lower power input, the electrolyzer produces hydrogen with a significant amount of impurities, mostly of oxygen.

5.4.1 Sizing of the Electrolyzer

For a given wind power installed, the sizing of the electrolyzer is not trivial. In the case of "stand-alone" systems, a one-to-one approach is often proposed, with the electrolyzer power input being equal to the nominal power output of the wind turbine. In this way, the electrolyzer should be able to retrieve all the wind power in the absence of load. In grid-connected systems, the same approach leads to the choice of an electrolyzer with a power supply equal to the power output of the wind turbine minus a "base load." However, the specific capital cost of the electrolyzer being almost equal to the cost of a wind turbine, it is important to take into account the capacity factor of the electrolyzer that will always be smaller than that of the wind turbine.

Electrolyzers are generally current-controlled, which means that a certain DC is imposed according to the desired hydrogen production. In a wind-hydrogen system, the wind turbine power available for the operation of the electrolyzer is generally known; therefore, the power input should be transformed to a current input. The voltage–current relation of an electrolyzer is not very simple because it depends on the temperature, pressure, and other construction characteristics. For a given electrolyzer, it is possible to experimentally establish the I–V curve at different temperatures and pressures, and deduce a temperature-dependent current–power curve.

The sizing of the wind turbine, electrolyzer, and auxiliaries is greatly affected by whether the wind-hydrogen system is grid connected or stand alone. The cells can be made in any size, and any number of cells can be stacked in a series depending on the desired output and design point selected. The design point, namely the current density for the cell, is case sensitive and depends on whether the system is operational cost critical or capital cost critical. The cell voltage and therefore the specific power consumption of the electrolyzer depend on the current density according to a *I–V* curve. If the capital cost is more important than the operating cost, which depends on electricity price, the optimum current density should be shifted toward the higher range with a lower efficiency. If the operating cost is more significant than the cost of money, the current density should be shifted toward the lower range, or expensive activated electrodes may be used to increase the efficiency. Module material costs usually represent between 44% and 56% of the total material costs [3].

Commercial water electrolyzers cover a wide range of hydrogen production rates from 0.001 to 750 Nm^3/h. Small hydrogen generators are intended for laboratory use, where hydrogen is often used as a carrier in analytical instruments, whereas large units are used in different fields of the chemical industry.

The actual capital cost of different electrolyzers operating at pressures between atmospheric pressure and 30 bars is presented in Figure 5.10 based on offers from manufacturers. There is a wide variation in the specific capital cost of small-size electrolyzers due to the different technical characteristics and options included. The specific capital cost of very small laboratory electrolyzers may exceed 30,000 €/kW, but has not been shown here for simplicity. There are no large variations in the specific cost of medium-to-large-size electrolyzers, that is, above 200 kW, mainly because there are very

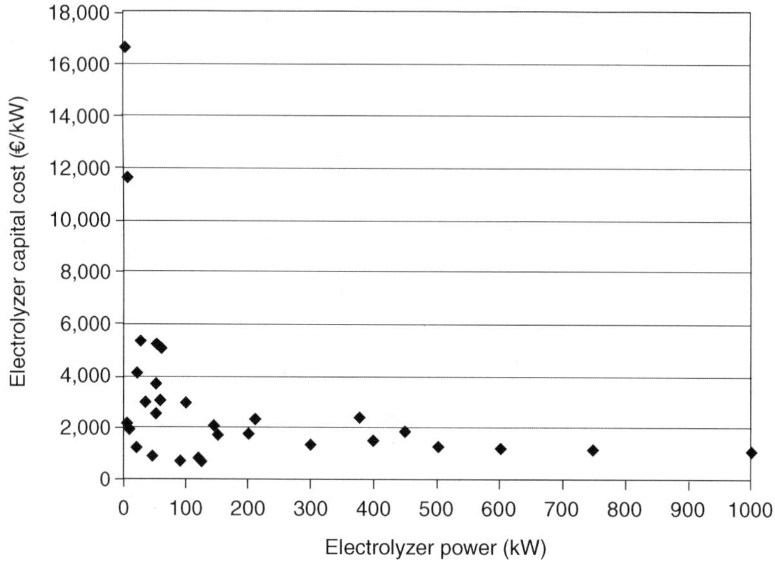

FIGURE 5.10
Electrolyzer cost.

few manufacturers around the world that are actually supplying equipment of such a size. The manufacturing process of electrolyzers is still intensely manual and the prices shown here refer to individual items without taking into account any volume effect. The specific capital cost of very small laboratory electrolyzers, producing <0.5 Nm^3/h H_2, has not been included in Figure 5.10 because it is difficult to compare it to industrial electrolyzers. These electrolyzers generally have a very high specific electrical consumption per newton cubic meter of hydrogen produced due to the fact that the power consumption of the auxiliaries is often higher than that of the electrolysis stack. However, they generally produce ultrapure hydrogen.

Generally, the electrolyzer cost increases with hydrogen purity and delivery pressure. In fact, a purification section for the removal of oxygen and the reduction of humidity to a concentration below 10 ppm may represent 30–40% of the total cost. Advanced electrolyzers actually supply hydrogen at pressures up to 30 bars and there are prototypes with a delivery pressure of up to 100 bars. They are more expensive than atmospheric electrolyzers, but their use decreases the cost of storage: hydrogen may be directly stored at the delivery pressure, and if a compressor is used for filling high-pressure cylinders, it only needs one stage instead of three or four.

5.5 Wind Energy Systems for Hydrogen Production

For each wind energy system category, there is a particular approach of how to apply an electrolyzing hydrogen production unit. The most common application studied currently is the stand-alone wind energy system combined with an electrolyzer. The latter is connected in the place of the battery cluster. The electrolyzer unit can be either a PEM or an alkaline type where there is a constant feed with water. The produced hydrogen is usually stored in a tank at the output pressure of the electrolyzer or compressed at a higher

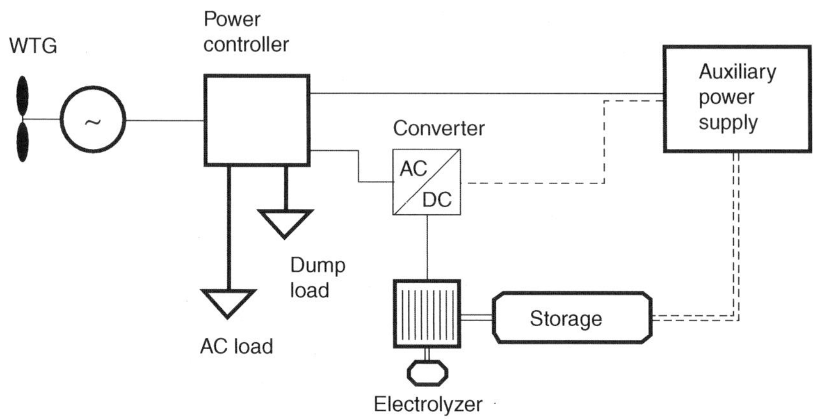

FIGURE 5.11
A stand-alone wind-hydrogen power station diagram including the basic modules.

pressure by a gas compressor. The storage medium for such an application can also be a metal hydride device in which the storing and releasing conditions do not aggravate the load consumption of the system.

For energy security reasons, the presence of an auxiliary power supply unit is necessary. This unit can be preferably either a hydrogen internal combustion engine (H_2 ICE) or a fuel cell of corresponding capacity to meet at least the minimum needs of the system. In this case, the system is an autonomous power plant. Figure 5.11 shows a stand-alone wind-hydrogen system that is autonomous. The dashed line in some parts of it implies that these connections may not exist as well. The DC/AC converter/controller should have the capability to operate vice versa and power up the lines through the power controller.

Alternatively, the auxiliary power device can be some other RES with the appropriate equipment or a diesel engine, which can be obtained in various sizes and specifications. It is presumed that in a stand-alone system the requirements for power quality are not the same as that for a grid-connected consumer [43].

The flowchart of the operation of an autonomous wind-hydrogen system where the auxiliary power supply is a PEM fuel-cell stack is given in Figure 5.12. The controller takes into account the power available to the electrolyzer (wind power minus the electric load) and the hydrogen levels in the tank. The electrolyzer operates when the power available for the electrolyzer is equal to the rated power or within its operation range and the tank is not full. Also, while the power available to the electrolyzer is greater than the rated power and the storage tank is not full, the electrolyzer is run at full capacity and the excess power is dumped. If the storage tank is full, the power available to the electrolyzer is also dumped. The dumped energy can be used in other facilities as water heating or water pumping. In the case that there is a shortage in the power supply, the use of a first fuel-cell stack is commenced. If the shortage remains, the second set is selected and so on. The simultaneous use of all the fuel-cell stacks is avoided to manage the consumption more effectively and to increase the lifetime of some of its components. On a constant basis, fuel-cell usage may also contribute to thermal energy production for domestic applications.

The determination of the size of the hydrogen equipment for these stations is very important to decrease the total cost and the amount of the dumped load due to technical constraints [44]. This determination can be achieved by simulating the operation of the whole system based on real wind data of the site and considering all the possible energy losses that might occur.

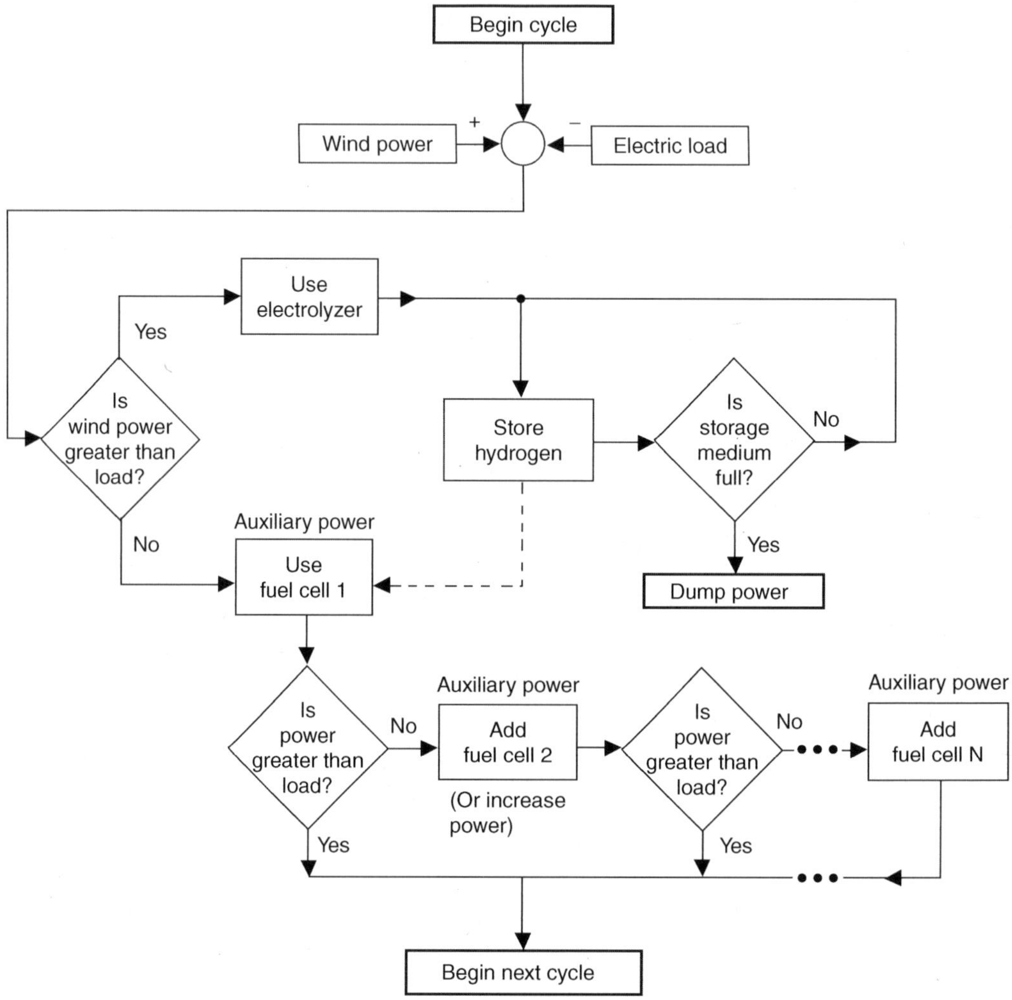

FIGURE 5.12
The flowchart of an autonomous wind-hydrogen energy system.

Hydrogen production from wind energy has not been implemented in large-scale WFs yet. The main reason for this, apart from the high cost, is that the present commercially available electrolyzers are designed to operate at lower capacities. An increase in the size of an electrolyzer is achieved by connecting electrolysis stacks in series.

Figure 5.13 shows a configuration for applying an electrolyzing system in a WF that is connected to a transmission system. The same configuration may be applied to a noninterconnected grid as well. The electrolyzer is connected to the medium-voltage bus bar, which assembles all, or a part, of the WTG of a WF. The electrolyzer is powered through an AC/DC controller along with a transformer that provides the proper voltage level in the device. In the output of the electrolyzer, there is a compressor that compresses the gas at higher pressure to be stored in steel tanks. From this point, hydrogen is available to be dispensed through a regulator to another storage medium or to be fed to an auxiliary power supply unit in the WF, which can be either a fuel cell or a H_2 ICE cluster. The topology of the hydrogen consumption equipment that can be added in a WF is drawn by the dashed line. By the time hydrogen technologies become competitive to the conventional

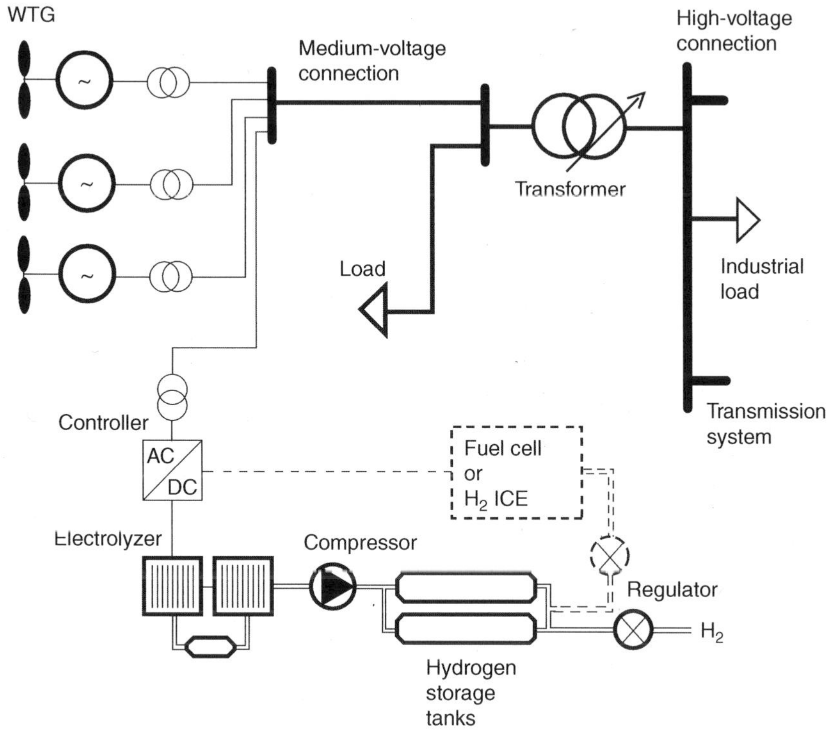

FIGURE 5.13
A simplified diagram of a hydrogen production unit inside a WF.

power supply systems, hydrogen energy produced from WFs may be used as an auxiliary service during peak load events.

This terminal point in hydrogen production can be used for refueling hydrogen-powered or hydrogen transport vehicles. There is also an option to canalize hydrogen in a natural gas (NG) pipeline provided that the distance from the WF is fairly short. Hydrogen injection in NG pipelines may improve NG energy properties [45].

Considering the grid code provisions for WFs that set a threshold in production around 90% of the maximum available active power output, an electrolyzing unit of power consumption equal to at least 1.5% of the WF's installed capacity can be always in operation as long as the WTG function. In weak grids with high wind power penetration, or in noninterconnected grids, the proportion of the wind energy that should be converted can be even higher. The preceding calculations concern the typical operation of a WF to feed the grid. Provided that hydrogen production will become a major part of the wind energy, dispatching these figures will increase by other criteria based on the optimal operation of a unified wind hydrogen power plant.

5.6 Case Studies

Poul la Cour (1846–1908), a Danish pioneer who constructed the first wind turbine for the generation of electricity, was also probably the first person to supply an electrolyzer with electrical energy from wind turbines to produce hydrogen for gas light [46].

Water electrolyzers have been designed in the past for continuous or discontinuous operation with grid current converted to DC current. During the past 20 years, the emergence of RES and their stochastic nature led to the first systematic approaches of water electrolysis under intermittent power operation. The major part of the first hydrogen production systems from RES focused on solar energy because of the fact that the DC output of the photovoltaic panels could be directly coupled to the electrolyzer stack by an appropriate matching of the voltages. Although the cost of wind electricity was clearly lower than that of solar electricity, the possibility to avoid the cost and efficiency loss of an AC/DC inverter was certainly an advantage. Most of the case studies reported in the literature refer to stand-alone systems. In fact, several of these stand-alone systems have part of their auxiliaries connected to the grid, or use the electrical grid as backup power.

A wind-hydrogen system of 20 kW installed power has been developed, constructed, and optimized at the Fachhochschule Wiesbaden in Germany since 1985 [47]. A 20 kW wind energy converter, designed for stand-alone operation, feeds DC to a pressurized alkaline electrolyzer of 20 kW, and the produced hydrogen is used in two gas motor generators of 8 and 4 kW electrical output. The overall energy efficiency from wind electricity to gas generator electricity is around 15% [48].

At the University of Stralsund in Germany, a 100 kW wind turbine and a 20 kW alkaline electrolyzer supplying hydrogen at 25 bars to a storage tank of 8 m^3 have been in operation for several years. According to the wind speed, the asynchronous wind generator can be operated at 1000 or 1500 rpm producing 20 or 100 kW of electricity, respectively. Both the static and the dynamic behaviors of the system were investigated, and an electrolyzer efficiency of approximately 65% with respect to the HHV has been reported [49].

The Hydrogen Research Institute in Canada has developed and tested a stand-alone renewable energy system composed of a 10 kW wind turbine, a 1 kW_{peak} photovoltaic array, a 5 kW alkaline electrolyzer, and a 5 kW PEM fuel cell. The components of the system are electrically integrated on a 48 V DC bus [50].

A small stand-alone, wind-powered hydrogen production plant was designed, constructed, and tested in Italy, at the Casaccia Research Centre of Ente per le Nuove tecnologie, l'Energia el'Ambiente (ENEA); National Agency for New Technologies, Energy and the Environment; [trans.]. The main aim of the project was to study the control of a wind turbine to produce a smooth power output, the tolerance of an electrolyzer to fluctuating power inputs, and the overall economics of a wind-hydrogen system [51]. The system was composed of a 5.2 kW wind turbine with a synchronous generator at variable speed, a 2.25 kW electrolyzer, and a 330 Ah battery bank. The variable frequency AC power output from the turbine was rectified and supplied to the electrolyzer. Hydrogen purity remained satisfactory even at low-capacity factors, with oxygen content in hydrogen in the range 0.15–0.35 vol%, although the efficiency of the electrolyzer stack (50%) was lower than state-of-the-art electrolyzers (70%) [51].

The intermittent operation of electrolyzers powered from a wind turbine is characterized by a dynamic power fluctuation with periods of varying overload, partial load, and shutdown. The operation of the electrolyzer below 10% of its nominal power remarkably reduced the current efficiency and the purity of the product gases, inducing the shutdown of the electrolyzer, which was programmed at a safety limit of 2 vol% hydrogen-in-oxygen [52].

A systematic investigation of an electrolyzer response to rapid load variation was performed by operating the electrolyzer according to 1 s power data from a wind turbine, then according to 10, 60, and 150 s averages as well as under the mean constant power for a time period of 24 h each [52]. The pressure and temperature values were not affected by the different modes of operation. The quantity of produced hydrogen was lower by about 2.5% for the 1 s regime with respect to constant power. However, the purity of the gases was greatly affected with the oxygen-in-hydrogen content passing from 0.24% for constant operation to 0.41% for the 1 s power regime [52]. The gas impurities could be reduced by 30% through replacement of the discontinuously controlled expansion valves supported by a two-position controller with continuously controlled valves combined with a proportional integral differential controller [52].

During 2004 and 2005, four major wind-hydrogen projects have been launched in Europe: the Utsira project in Norway; the HARI project at Loughborough; the promoting unst renewable energy project in Unst, UK; and the RES2H2 project in Greece and Spain.

On the Utsira Island, a wind-hydrogen system was installed to serve 10 households with a peak demand of 45 kW in an autonomous mode. The existing grid connection was kept for emergency situations to avoid the high costs of redundancy. The hydrogen plant consists of a 48 kW alkaline electrolyzer producing 10 Nm^3/h H_2, a hydrogen compressor for the filling of a 12 m^3 storage tank at 200 bars with a 2400 Nm^3 H_2 capacity, a hydrogen generation set of 55 kW, and a PEM fuel cell of 10 kW nominal power. A 600 kW wind turbine supplies a varying portion of its power to the system feeding the rest of its power to the grid. At maximum load, the electrolyzer and compressor need approximately 54 kW of electrical power [53].

In the "PURE" project on the Shetland Islands, the wind-hydrogen system is composed of two wind generators of 15 kW power each, a 15 kW advanced alkaline electrolyzer operating at 55 bars, a 16-cylinder stack of 44 Nm^3 H_2 capacity at the same pressure, and a 5 kW PEM fuel cell [54].

A hydrogen plant was integrated to an existing infrastructure of renewables, which included two 25 kW wind turbines, photovoltaics, and two microhydroelectric turbines in the HARI project. The wind turbines are two-bladed, stall-regulated, and pitch overspeed. The hydrogen plant is composed of a 36 kW alkaline electrolyzer, a hydrogen compressor, 48 pressurized hydrogen cylinders of 2856 Nm^3 capacity at 13.7 MPa, and two different fuel cells of 2 and 5 kW nominal power. The alkaline electrolyzer with a 46-cell stack produces 8 Nm^3/h H_2 at 2.5 MPa and operates in a 20–100% range of its nominal power. The 3.75 kW single-stage hydrogen compressor has a capacity of 11 Nm^3/h for an inlet pressure of 2.5 MPa and an 8:1 compression rate. To reduce the on/off switching cycles of the electrolyzer, which affect its long-term stability and performance, a 20 kWh battery was incorporated. The battery helped to moderate the power supply variations to the electrolyzer. The mean electrolyzer stack efficiency measured was 75%, whereas the average conversion efficiency of the electrolyzer–battery–compressor system was 49%, including the balance of plant losses [30]. The performance of the electrolyzer decreased gradually over 2 years of operation, with the stack power requirement increasing from 36 to 39 kW and the total power requirement, including the hydrogen purification section and auxiliaries, from a maximum of 43 to 45 kW [30].

In the frame of the RES2H2 project, a hydrogen system was installed and tested in a wind park near Athens, Greece. The hydrogen system consists of a 25 kW alkaline electrolyzer producing 5 Nm^3/h H_2, a hydrogen compressor for filling high-pressure cylinders at 220 bars and metal hydride tanks. The plant and all the auxiliaries are connected to a 500 kW wind turbine, which feeds the rest of its power to the grid. Under variable power input, the electrolyzer rarely reaches its nominal operating temperature of 80°C, and its efficiency varies from 70% to 85% for an electrolysis temperature in the range 45–70°C [55]. Figure 5.14 shows a view of the filling station at the wind-hydrogen plant of the RES2H2 project.

FIGURE 5.14
View of the wind-hydrogen plant of the RES2H2 project in Greece.

5.7 Critical Issues

Hydrogen production from electrolysis of water powered from a wind energy system depends on the wind potential of the site that the plant is installed. Planning of these stations should be accompanied by an elaborate analysis of the wind regime of the area to have a substantiated view of the amount of hydrogen that will be produced. In wind energy systems that are already in operation, it is easier to make this assessment and to conclude about the size of the electrolysis and storage equipment.

In addition to the assessment of the wind potential comes the need for a forecasting system that is essential for the optimization of the energy generation and management of a wind energy system. Energy damping in a wind-hydrogen system may be confined if there is certain information about the hourly wind conditions on a site in 1–3 days ahead time frame.

A very important issue related to the performance of the whole wind-hydrogen system is the quality of its electrical equipment. There is a strong dependence on the reliability of the power electronics used for the AC/DC conversion as well as for the sensitivity of the fault detection devices that often result in a lower overall efficiency rate.

There are also a few auxiliary independent circuits that support hydrogen production and storage processes that should also be considered. The most important are the cooling system of the electrolyzer, the compressor of the gaseous hydrogen, and the temperature adjustment of other storage devices. The purification units of water and hydrogen are also important with regard to the specifications of the equipment used.

Existence of water, especially when it concerns large hydrogen production stations, is another factor that affects the construction of an electrolysis unit. However, water consumption (0.9 L/Nm^3 H_2) for electrolysis is relatively low compared to other matters in an electrolysis station. Its transport is easy and it may be more convenient to acquire the water ready for electrolysis than to construct a secondary unit for water purification inside the plant.

The need of reinforcement of the grid in windy areas to support the operation of high-capacity WFs is a matter that also concerns a possible large-scale hydrogen production from wind energy. However, grid infrastructure development should be considered as a more general issue for high level of safety and reliability in the transmission systems.

5.8 Conclusions

Hydrogen production from electrolysis of water using wind power is a challenging process because it is environmental friendly and irrespective of fossil fuels. Although wind energy exploitation and water electrolysis have been used extensively in the past, their combination has not been applied at a commercial level yet. Hydrogen technologies applied in wind energy systems are still at a research and development phase, which is confined in small-scale applications. From the time that hydrogen becomes the primary energy carrier for transportation and other mobile applications, this method will be considered as one of the most competitive investments for disaggregated energy markets.

Use of hydrogen as a carrier for the storage of wind energy presents several advantages. First, the existing experience regarding the handling of hydrogen is already high, thanks to its wide use in the chemical sector. Hydrogen is well adapted for seasonal energy storage without energy loss over time. Water electrolyzers are able to handle power fluctuations, addressing the problems relative to the intermittent nature of wind power, and allowing wind parks to provide a guaranteed power output. Wind-hydrogen systems present a potential for high-density energy storage with low and predictable operating and maintenance costs.

The combination of hydrogen and wind energy systems may contribute to the further growth of both technologies while one serves the other for different reasons. Hydrogen production is an efficient and benign way to store energy from wind while wind power is available and not required. However, electrical energy that is derived from the movement of air masses is sustainable, does not produce polluting emissions during its generation, and is applicable in most of the energy conversion schemes. The increase of wind power penetration in the transmission systems may become higher under the support of hydrogen production as an energy storage technique.

References

1. Kreuter W., Hofmann H., Electrolysis: The important energy transformer in a world of sustainable energy, *Int. J. Hydrogen Energ.*, 23(8), 661–666, 1998.
2. Ramachandran R., Menon R.K., An overview of industrial uses of hydrogen, *Int. J. Hydrogen Energ.*, 23(7), 593–598, 1998.
3. Laskin J.B., Feldwick R.D., Recent development of large electrolytic hydrogen generators, *Int. J. Hydrogen Energ.*, 3, 311–320, 1978.
4. Appleby A.J., Crepy G., Jacquelin J., High efficiency water electrolysis in alkaline solution, *Int. J. Hydrogen Energ.*, 3(1), 21–37, 1978.

5. Vandenborre H., Leysen R., Baetsle L.H., Alkaline inorganic-membrane-electrolyte (IME) water electrolysis, *Int. J. Hydrogen Energ.*, 5(2), 165–171, 1980.
6. Nayar M.G., Ragunathan P., Mitra S.K., Development and operation of a high current density high pressure advanced electrolysis cell, *Int. J. Hydrogen Energ.*, 5(1), 65–74, 1980.
7. Divisek J., Mergel J., Schmitz H., Improvements of water electrolysis in alkaline media at intermediate temperatures, *Int. J. Hydrogen Energ.*, 7(9), 695–701, 1982.
8. Abe I., Fujimaki T., Matsubara M., Hydrogen production by high temperature, high pressure water electrolysis, results of test plant operation, *Int. J. Hydrogen Energ.*, 9(9), 753–758, 1984.
9. Srinivasan S., Salzano F.J., Prospects for hydrogen production by water electrolysis to be competitive with conventional methods, *Int. J. Hydrogen Energ.*, 2, 53–59, 1977.
10. Murray J.N., Yaffe M.R., Testing aqueous caustic electrolyzers at high temperatures, *Int. J. Hydrogen Energ.*, 4, 193–204, 1979.
11. Nuttall L.J., Conceptual design of large scale water electrolysis plant using solid polymer electrolyte technology, *Int. J. Hydrogen Energ.*, 2, 395–403, 1977.
12. Nidola A., Water electrolysis in alkaline solutions. New electrode materials, *Int. J. Hydrogen Energ.*, 9(5), 367–375, 1984.
13. Angelinetta C., Trasatti S., Atanososka Lj. D., Atanasoski R.T., Surface properties of RuO_2 + IrO_2 mixed oxide electrodes, *J. Electroanal. Chem.*, 214(1–2), 535–546, 1986.
14. Schiller G., Henne R., Mohr P., Peinecke V., High performance electrodes for an advanced intermittently operated 10-kW alkaline water electrolyzer, *Int. J. Hydrogen Energ.*, 23(9), 761–765, 1998.
15. Enyo M., Yamazaki T., Kai K., Suzuki K., Amorphous Pd–Zr alloys for water electrolysis cathode materials, *Electrochim. Acta*, 28(11), 1573–1579, 1983.
16. Dyer C.K., Improved cathodes for industrial electrolytic hydrogen production, *Int. J. Hydrogen Energ.*, 9(12), 993–995, 1984.
17. Chonglun F., Piron D.L., Paradis P., Hydrogen evolution on electrodeposited nickel-cobalt-molybdenum in alkaline water electrolysis, *Electrochim. Acta*, 39(18), 2715–2722, 1994.
18. Shafia Hoor F., Aravinda C.L., Ahmed M.F., Mayanna S.M., Fe–P and Fe–P–Pt co-deposits as hydrogen electrodes in alkaline solution, *J. Power Sources*, 103(1), 147–149, 2001.
19. Dos Santos A.M.H., Acioli M.L., Da Silva J.G. Jr., Pereira S.J.C., Vilar E.O., Tonholo J., Preliminary investigation of some commercial alloys for hydrogen evolution in alkaline water electrolysis, *Int. J. Hydrogen Energ.*, 29, 235–241, 2004.
20. Rosalbino F., Macciò D., Angelini E., Saccone A., Delfino S., Electrocatalytic properties of Fe–R (R = rare earth metal) crystalline alloys as hydrogen electrodes in alkaline water electrolysis, *J. Alloys Compd.*, 403(1–2), 275–282, 2005.
21. Tseung A.C.C., Jasem S., Oxygen evolution on semiconducting oxides, *Electrochim. Acta*, 22(1), 31–34, 1977.
22. Balej J., Electrocatalysts for oxygen evolution in advanced water electrolysis, *Int. J. Hydrogen Energ.*, 10(2), 89–99, 1985.
23. Potvin E., Brossard L., Electrocatalytic activity of Ni–Fe anodes for alkaline water electrolysis, *Mater. Chem. Phys.*, 31(4), 311–318, 1992.
24. Singh R.N., Pandey J.P., Anitha K.L., Preparation of electrodeposited thin films of nickel–iron alloys on mild steel for alkaline water electrolysis. Part I: Studies on oxygen evolution, *Int. J. Hydrogen Energ.*, 18(6), 467–473, 1993.
25. Singh S.P., Singh R.N., Poillearat G., Chartier P., Physicochemical and electrochemical characterization of active films of $LaNiO_3$ for use as anode in alkaline water electrolysis, *Int. J. Hydrogen Energ.*, 20(3), 203–210, 1995.
26. Cappadonia M., Divisek J., von der Heyden T., Stimming U., Oxygen evolution at nickel anodes in concentrated alkaline solution, *Electrochim. Acta*, 39(11–12), 1559–1564, 1994.
27. Wendt H., Hofmann H., Cermet diaphragms and integrated electrode-diaphragm units for advanced alkaline water electrolysis, *Int. J. Hydrogen Energ.*, 10(6), 375–381, 1985.
28. Saxe M., Alvfors P., Advantages of integration with industry for electrolytic hydrogen production, *Energy*, 32(1), 42–50, 2007.
29. Kato T., Kubota M., Kobayashi N., Suzuoki Y., Effective utilization of by-product oxygen from electrolysis hydrogen production, *Energy*, 30, 2580–2595, 2005.

30. Bossel U., Well-to-wheel studies, heating values, and the energy conservation principle, Report, 2003, http://www.efcf.com/reports/E10.pdf, 01/2007.
31. Gammon R., Roy A., Barton J., Little M., Case Study—Hydrogen and Renewables Integration (HARI), *IEA–Hydrogen Implementing Agreement*, 2006.
32. Varkaraki E., Lymberopoulos N., Zoulias E., Kalyvas E., Christodoulou C., Vionis P., Chaviaropoulos P., Integrated wind-hydrogen systems for wind parks, *Proceedings of the European Wind Energy Conference*, Feb 27–Mar 2, Athens, Greece, 2006.
33. Hatziargyriou N., Zervos A., Wind power development in Europe, *Proc. IEEE*, 89(12), 1765–1782, 2001.
34. Zervos A., Developing wind energy to meet the Kyoto targets in the European Union, *Wind Energ.*, 6(3), 309–319, 2003.
35. Ackermann T., Soder L., Wind energy technology and current status: A review. *Renew Sust. Energ. Rev.*, 4, 315–374, 2000.
36. CER/04/237, Wind Farm Transmission Grid Code Provisions, 2004.
37. SP Transmission and Distribution Scottish Hydro-Electric, Guidance Note for the Connection of Wind Farms, Issue No. 2.1.4, 2004.
38. Eek J., Gjengedal T., Influence of high wind power penetration on the Nordic power grid primary control response, 2006 CIGRE SESSION, Paris, France, August 27–September 01, 2006.
39. Jauch C., Sorensen P., Bak-Jensen B., International review of grid connection requirements for wind turbines, *Nordic Wind Power Conference*, Gothenburg, Sweden, Mar 1–2, 2004.
40. Barton J.P., Infield D.G., Energy storage and its use with intermittent renewable energy, *IEEE Trans. Energ. Convers.*, 19(2), 441–448, 2004.
41. Mohan N., Undeland T., Robins W., *Power Electronics: Converters, Applications and Design*, Wiley, New York, 2003.
42. Frost W., Aspliden C., Characteristics of the wind, *Fundamental Concepts of Wind Turbine Engineering*, D.A. Spera, Ed., ASME Press, New York, 1994, Chapter 8.
43. Tande J.O.G., Grid integration of wind farms, *Wind Energ.*, 6, 281–295, 2003.
44. Bechrakis D.A., McKeogh E.J., Gallagher P.D., Simulation and operational assessment for a small autonomous wind-hydrogen energy system, *Energ. Convers. Manag.*, 47, 46–59, 2006.
45. Hythane Company, LLC, http://www.hythane.com/, 2007.
46. DWIA (DanishWindIndustryAssociation), http://www.windpower.org/en/pictures/lacour.htm, 2007.
47. Schulien S., Sandstede G., Hahn H.W., Hydrogen and carbon dioxide as raw materials for an ecological energy-technology, *Int. J. Hydrogen Energ.*, 24, 299–303, 1999.
48. Altmann M., Gamallo F., Autarke Wind-Wasserstoff-Systeme, http://www.hyweb.de/Wissen/autarke.htm, 2007.
49. Lehmann J., Luschtinetz T., Menzl F., The wind hydrogen–fuel cell chain, *Proceedings of the 14th World Hydrogen Energy Conference*, Montreal, Canada, 2002.
50. Agbossou K., Kolhe M.L., Hamelin J., Bernier E., Bose T.K., Electrolytic hydrogen based renewable energy system with oxygen recovery and re-utilisation, *Ren. Energ.*, 29, 1305–1318, 2004.
51. Schucan T., Hydrogen generation from stand-alone wind-powered electrolysis systems, in Case Studies of Integrated Hydrogen Energy Systems, *Final Report of the IEA Task 11: Integrated Systems*, 129–145, Switzerland, 1998, Chapter 9.
52. Dutton A.G., Ruddell A.J., Falchetta M., Prischich D., Dienhart H., Hug W., Bleijs J.A.M., Hydrogen generation from stand-alone wind-powered electrolysis systems, Final report, EC contract no. JOU2-CT93-0413, Dec 1996.
53. Eide P.O., Hagen E., Kuhlmann M., Rohden R., Construction and commissioning of the Utsira wind/hydrogen stand-alone power system, *Proceedings of the 16th World Hydrogen Energy Conference*, Lyon, France, June 13–16, 2006.
54. AccaGen SA, PURE project: stand alone hydrogen energy system, Report, 2005.
55. Varkaraki E., Lymberopoulos N., Zoulias E., Kalyvas E., Christodoulou C., Karagiorgis G., Stolzenburg K., Integrated wind-hydrogen systems for wind parks, *Proceedings of the 16th World Hydrogen Energy Conference*, Lyon, France, June 13–16, 2006.

6

Sustainable Hydrogen Production by Thermochemical Biomass Processing

Wiebren de Jong

CONTENTS

6.1 Introduction

6.1.1 Current Production and Use of Hydrogen

On a global scale, the consumption of energy is growing due to the increasing world population and the increased level of wealth standards. Nowadays, energy is mainly produced from fossil fuels.[1] This situation leads to a further depletion of the limited cheap fossil reserves, increased dependency on fossil fuel imports and, furthermore, gives rise to an increasing concentration of greenhouse gases (mainly CO_2) in the atmosphere. There are increasing indications that this is a significantly contributing factor to global climate changes.[2] One of the ways to reduce the anthropogenic contribution to this problem is the application of a CO_2 neutral energy carrier instead of the net CO_2 producing fossil fuels, which are currently used. The role of primary renewable energy sources in the global energy economy could be greatly extended if they could be converted to an energy carrier that is easily stored and transported and this could serve fluid-fuel markets not readily served by electricity; see Ogden and Nitsch in Johansson et al.[3] Potentially, hydrogen is a carrier that can also reduce the exclusive dependency on fossil fuels; see, for example, Turner.[4]

Consumption of hydrogen these days mainly takes place in petrochemical processes (such as hydrotreating, desulfurization, dealkylation, and cracking); in chemical production (e.g., ammonia and methanol); metallurgical processing; electronics industry; and food processing (oil and fat hydrogenation) and not so much for energy production purposes.[5]

The diverse sources for hydrogen are depicted in Figure 6.1.[6] Today, the least expensive and the most widely applied technology to produce hydrogen is catalytic steam reforming

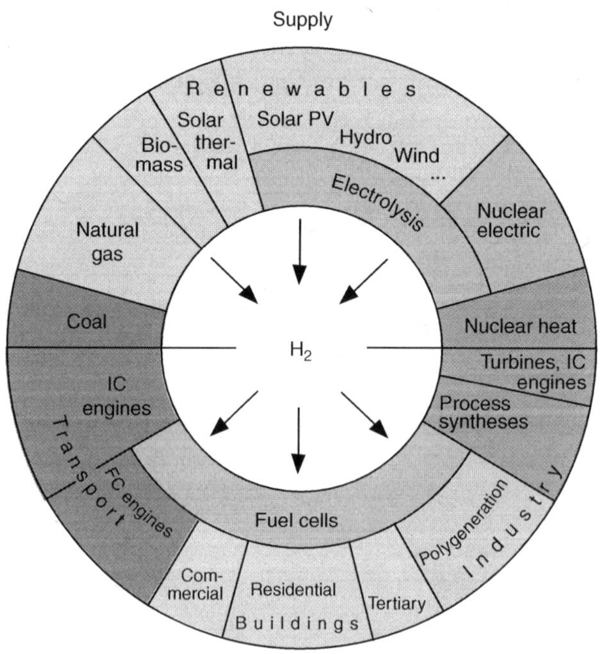

FIGURE 6.1
Hydrogen production and utilization. (Adapted from EC, *Hydrogen Energy and Fuel Cells—A Vision of Our Future.* European Commission—High level group on hydrogen and fuel cells, Brussels, 2003.)

of natural gas.[5,7] The global hydrogen production currently exceeds 1 billion m^3/day, of which 48% is produced from natural gas, 30% from oil (often on-site in refineries), 18% from coal, and the remaining (4%) by water electrolysis. The current production of hydrogen in the United States is 9 million t/year.[8] The production of hydrogen from fossil fuels, however, is accompanied by the production of CO_2, CH_4, and other polluting gases that contribute to the enhancement of the natural greenhouse effect.[9]

Hydrogen, however, can be produced using sustainable technologies that are nowadays still not applied at large scale: the most promising ones are water splitting by electrolysis using sustainably produced electricity (e.g., wind, hydro, and sun) and processes based on sustainable production and utilization of biomass.[3]

6.1.2 Biomass as Renewable Hydrogen Source

Biomass as the source of hydrogen production is considered to be a leading practical and viable option for the near- and mid-term future according to several studies,[5,10,11] although critical comments have been made regarding this route concerning costs and biomass supply.[12] Before the discovery of inexpensive fossil fuels, our society was highly dependent on plant biomass to meet its energy demands.

To enable the large-scale introduction of sustainably produced hydrogen, a number of technical and nontechnical issues have to be addressed, such as the configuration of the production technology chain, storage options, smooth integration into the existing energy system, and social acceptance. The transition from the present fossil fuel-based energy system into a sustainable, hydrogen-based energy system is expected to be fragmented and involves a diverse mixture of fossil and renewable energy sources and carriers. The use of hydrogen produced from fossil fuels will probably be an initial step toward the realization of a sustainable energy system based on hydrogen.

This chapter focuses on "biomass to hydrogen conversion technologies." A variety of biomass resources can be converted for energy supply. They can be divided into four general categories:[13]

- Energy crops
- Agricultural residues and wastes
- Forestry wastes and residues
- Industrial and municipal wastes

6.1.3 Biomass Availability

As such, biomass are abundantly available and compared to fossil fuels are more evenly distributed geographically on a world-scale. It is the solar energy stored in living material and also the oldest energy source known to humankind. Currently, its use is still mainly traditional, and often unhealthy and laborious practices have been established for combustion, generating heat for cooking in the developing countries. In modern societies, the use of biomass as an energy source was gradually abandoned after the discovery of huge amounts of cheap fossil fuels that were not only cheaper but also easier to process and use. Furthermore, the "new" fuels allowed applications that were not so easy to realize with biomass: transportation based on liquid fuels from crude oil and cooking on fossil gases. Biomass seemed to have no future anymore as an energy source. The confrontation with the limitations of the oil availability in the early 1970s due to political reasons (first oil crisis) and ecological concerns (see, e.g., Meadows et al.[14]) gave rise to a first renewed interest in biomass and with it virtually all forms of renewable energy. After this crisis, however, again the prices of fossil fuels decreased due to higher availability and this impeded the further development of bioenergy technology. In the 1980s and 1990s, there was a growing concern that global warming and the resulting climate change were enhanced by CO_2 emissions. This led to the Kyoto Protocol[15] aimed at a reduction of the emission of greenhouse gases. This development has again stimulated research in the area of renewables and, in particular, biomass.

Regional issues greatly affect biomass costs, which from cheapest to most expensive are typically lignocellulose < starches < vegetable oils < terpenes < algae.[16] Regarding the availability of biomass, current figures concerning the United States indicate a resource of lignocellulosic biomass of over 500 t/year, representing a potential annual production of 40–50 t hydrogen, enough to fuel 150 million fuel cell vehicles.[8,17] Hall et al. in Johansson et al.[3] indicated that by the middle of this century some 890 million ha or 10% of the world's land area now in croplands, forests, woodlands, and permanent pasture might be put into biomass production for energy recovery. If average crop yields of 15 t/ha were realized on the indicated area then the resulting global energy production rate would exceed 260 EJ, being more than 80% of worldwide commercial energy use in 1985. This is next to residue biomass (derived) sources that are expected to be more than 30 EJ, and which can already be exploited faster. This is in-line with studies of International Energy Agency (IEA) that present a range 150–450 EJ total worldwide biomass energy potential.[18] Based on different studies, a minimum of 50 EJ is indicated.[19] Even wider ranges have been proposed, considering also energy farming on surplus land and the worldwide biomass to energy contribution of 33–1130 EJ/year as prospects for 2050. By assessing the land use and productivity on global grid cell ($1° \times 1°$) level, and application of Intergovernmental Panel on Climate Change (IPCC) Special Reports on Emission Scenarios (SRES) scenarios, this range has been narrowed down to 310–660 EJ.[20] The requirement of species offering high crop yield that are soil and climate adapted; resistant to pests and diseases; grown in a sustainable, biodiverse way considering use of land, minerals, and water needed; and accompanied by good management practice, is challenging.

In different global energy demand scenarios, biomass plays a significant role. In a renewables-intensive global energy scenario presented by Johansson et al.,[3] the renewables contribution to fuels used directly can be as high as 25% by 2025 and 40% by 2050, with most of the contribution coming from biomass-derived fuels: methanol, ethanol, hydrogen, and biogas. These authors indicated that methanol and hydrogen may well prove to be *the* biofuels of choice because they are the energy carriers most easily used with a high efficiency in transportation.

6.1.4 Energy-Related Properties of Biomass

The nature of the diverse biomass feedstock is not discussed in great detail here. The biomass feedstock varies to a large extent with regard to chemical composition and physical appearance. Its structural components include cellulose, hemicelluloses, lignin, lipids, proteins, simple sugars, starches, moisture, hydrocarbons, and mineral matter ("ash"). Typical (dry) weight percentages for C, H, and O are 30–60%, 5–7%, and 30–45%, respectively. The carbon content is significantly lower compared to coals, whereas the oxygen content is much higher. As a consequence, the heating values of biomasses are generally lower than those of coals: typically lower heating values (LHVs) of 10–18 MJ/kg on an as received (a.r.) basis versus approximately 30 MJ/kg a.r. for black coal. The volatility of biomass, generally, is higher than that with as coal and due to the oxygen-rich structure the fuel is more reactive in nature.

Biomass has a relatively low hydrogen content as indicated earlier. Hydrogen yield not only depends on this bound hydrogen, but also on the chemical splitting of water in the steam reforming reaction; it is a disadvantage compared to, for example, natural gas as the hydrogen source, especially in view of the lower energy content per unit mass.

Widely varying moisture contents of biomass are encountered; they may range from 10% to 70%. This water content influences the thermochemical conversion and the volume of gases produced per energy unit. The heating value decreases with an increasing moisture content. Also, it can create ignition problems and it reduces the combustion temperature. This moisture presence influences the composition of volatiles released in the early stages of particle conversion.[21] Also, a broad spectrum of ash contents, in terms of composition and amount, is characteristic for biomass. Increasing ash contents also decreases the heating value of the fuel. The mineral constituents of biomass may give rise to problems of corrosion, agglomeration, and slagging/fouling of high-temperature equipment. This is especially the case with fuels that contain appreciable amounts of alkalis in combination with chlorine. Sulfur is mostly less problematic for biomass in comparison with coal. Properties of a large range of biomass species are reported in the Phyllis Web database provided by Energy Research Centre of the Netherlands (ECN).[22]

Biomass also usually has a low mass and energy density. For example, the bulk density of chopped straw and rice husks are 50–120[23] and 100–125 kg/m^3.[24] These ranges are low compared to bulk densities of coals, which are in the range 560–600 kg/m^3 for brown coals and between 800 and 900 kg/m^3 for bituminous coals. A number of disadvantages are attributed to the lower-bulk densities of biomass: relatively low heating values per unit volume, process control difficulties with regard to feeding, requirement of large storage facilities, expensive transportation, and limitation of fuel-conversion technologies applied.[23–25] Densification is well known to overcome these issues; it removes both the inter- and intraparticle voids. The three most commonly used techniques are baling, briquetting, and pelletization.

When implementing a biomass-based thermochemical conversion system, it is important to critically evaluate the feedstock characteristics such as cost, distribution, mass, and physical and chemical properties. The feedstock qualities must be considered when matching feedstocks with a proper conversion technology.

6.1.5 Biomass Conversion Processes

Biomass-based processes for the production of hydrogen can be either thermochemical or biological and can produce this clean carrier directly or through an intermediate, storable product. Also, the use of coproducts has to be addressed to improve the process economics and in view of the sustainability of using this natural resource. A recent overview of the status of the different technologies is given in an IEA/NREL review[3] and further by Czernik et al.[8]

Our focus will be on a large-scale application by means of *thermochemical biomass processing*. This technology class for biomass conversion has two attractive features: (1) it is omnivoric, that is, a very broad range of biomass feedstock can be completely converted and (2) it can be integrated with fossil-based infrastructure for large-scale production of synthesis gas that is existing and foreseen to be biobased in the medium- to long-term future as can be seen in Figure 6.2.[26–28]

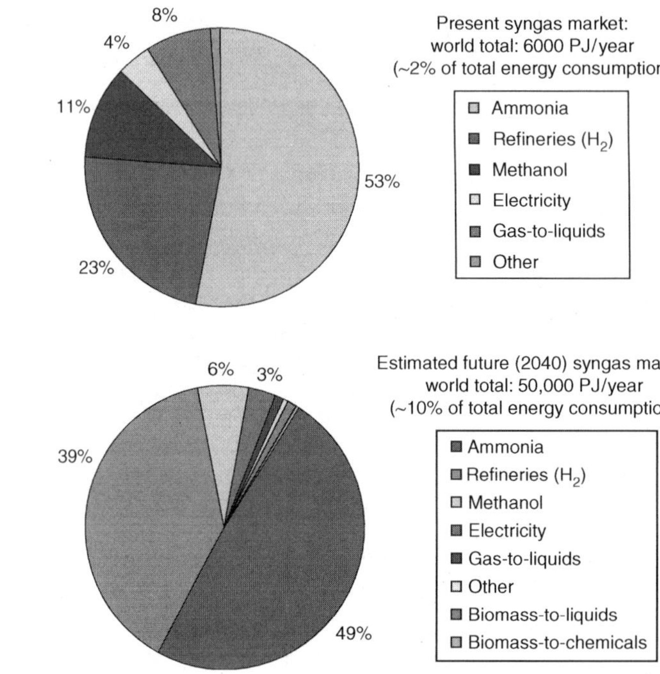

FIGURE 6.2
Foreseen syngas market in 2040 compared with the current one, differentiated toward utilization. (Reproduced from Boerrigter, H., and van der Drift, A., Bio Key Intermediate in Production of Renewable Transportation Fuel, Chemicals and Electricity, Optimum and Economic Properties of Fischer–Tropsch Plants, ETA, 2005. Copyright by ECN. With permission.)

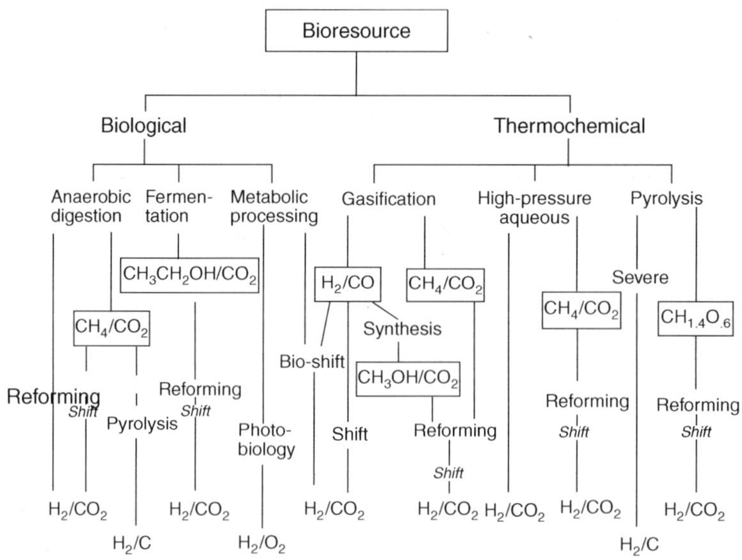

FIGURE 6.3
Pathways from biomass to hydrogen. (Adapted from Milne, T.A., Elam, C.C., and Evans, R.J., *Hydrogen from Biomass—State of the Art and Research Challenges*. IEA/H2/TR-02/001, A Report for the International Energy Agency Agreement on the Production and Utilization of Hydrogen Task 16, Hydrogen from Carbon-Containing Materials, Golden, CO USA, 2001.)

6.2 Overview of Biomass Conversion Routes to Hydrogen

Figure 6.3 depicts a schematic overview of the different routes of conversion of biomass into hydrogen. Two groups of conversion technologies are distinguished: biological and thermochemical. The main difference between these routes is the temperature level. The biological processes take place at ambient to slightly higher-temperature levels, whereas the thermochemical conversion routes take place at temperatures of several hundred degrees celsius.

This review limits itself to the treatment of high-temperature thermochemical biomass conversion technologies. There are very good overviews of biological conversion technologies for hydrogen production, for example, Ni et al.[13] and Zaborsky.[29]

6.3 Hydrogen from Thermochemical Gasification of Biomass

Thermochemical gasification by partial oxidation is already an old art (well over 150 years), that has been developed to substantially transfer the combustion value of a solid feedstock to a gaseous energy carrier in the form of chemical energy. This is done because a gas offers many advantages as a heat carrier.[30] Gasification of biomass generates the so-called biosyngas, which contains CO and H_2 in amounts depending on the applied process, oxidation medium, and fuel/oxidizer ratio. Comparatively, low-temperature (<1000°C) gasification processes yield a product gas that contains significant amounts of hydrocarbon compounds. Higher-temperature processes yield a biosyngas that contains virtually no hydrocarbons anymore.[31]

The main reactions occurring during gasification are outlined in Table 6.1, occurring together with moisture evaporation and devolatilization of the fuel in which gases, tars, and char are formed.

A comparatively simple way of representing the process of hydrogen formation from biomass by (steam/O_2-blown) gasification is by writing the following approximate reactions.[8] Here, an average typical elemental composition is assumed (not corresponding to an existing structure):

$$CH_{1.46}O_{0.67} + 0.16O_2 \rightarrow CO + 0.73H_2$$

$$\underline{CO + H_2O \leftrightarrow CO_2 + H_2}$$

$$CH_{1.46}O_{0.67} + 0.16O_2 + H_2O \rightarrow CO + 1.73H_2$$

TABLE 6.1

Important Gasification Reactions

	Reaction	ΔH_r (298.15 K) (kJ/mol)	ΔG_r (298.15 K) (kJ/mol)	Reaction Name
R1	$C + O_2 \rightarrow CO_2$	−393.5	−394.4	Complete carbon oxidation
R2	$C + 1/2O_2 \rightarrow CO$	−110.5	−137.2	Partial oxidation of carbon
R3	$C + H_2O \leftrightarrow H_2 + CO$	131.3	91.4	Water–gas reaction
R4	$C + CO_2 \leftrightarrow 2CO$	172.5	120.1	Boudouard reaction
R5	$C + 2H_2 \leftrightarrow CH_4$	−74.8	−50.8	Carbon hydrogenation
R6	$CO + H_2O \leftrightarrow H_2 + CO_2$	−41.2	−28.7	Water–gas shift reaction
R7	$CO + 3H_2 \leftrightarrow H_2O + CH_4$	−206.1	−142.2	Reverse methane reforming reaction

This shows that the maximum stoichiometric H_2 yield is of the order of 14.3 mass%. Practical yields are (significantly) lower because, in reality, part of the biomass is converted to char, tars, and hydrocarbons (all H-containing species) and CO_2, resulting in less CO available for production of hydrogen by the water–gas shift reaction; also, the equilibrium in the preceding equation was considered to be at the right-hand side of the equation, which is not the case in practical process conditions. An advantage of oxygen-based gasification is that it does not require any external source of energy. Higher yields of hydrogen can be obtained when gasifying with steam only (around 19.3 mass%, theoretically[32]), but this is a highly endothermic process, accompanied by the generation of hydrocarbons that must further be (catalytically) reformed before the water–gas shift reaction stage.

In the past 20–30 years, significant progress has been made in the field of thermochemical biomass gasification, mainly through demonstration projects for (combined) heat and power generation on a medium- to large-scale; see, for example, Beenackers and Van Swaaij,[30] Bridgwater,[33] and Morris et al. in Knoef.[1] The applied reactor technologies are often based on principles known from coal processing, although the chemicophysical properties of biomass are much different so that at least the adaptions in gas cleaning need to be made. In spite of the considerable efforts devoted to the commercialization of this conversion technology, it has so far remained in the development stage with rare exceptions. On the one hand, this is due to its relatively higher cost compared to conventional combustion in liberalized markets which makes it less attractive to investors. On the other hand, it is related to the still low reliability for long-term operation; see Maniatis.[34] However, for example, gasification of lignocellulosic biomass producing a hydrogen-rich synthesis gas and cleaning of the gas by steam reforming and upgrading promises higher efficiencies in the production of, for example, hydrogen for transportation use since the achievable yield of motor fuel from cellulosic biomass is higher through this route than the hydrolysis/fermentation process route.

Different gasification technologies are available[35] and those that are promising for large-scale production of hydrogen from biomass are described in the following text. Fixed- and moving-bed gasification technologies that have been reported in the 1980s[30] for medium calorific value syngas generation are not considered due to their relatively low thermal input scale. Also, some atmospheric large-scale technologies developed in that era in the European Union such as the $CaSO_4$-based "chemical looping" Oxygen Donor Process of John Brown/Wellman and the indirect gasification process of Italenergie/AGIP are not dealt with as these concepts have been abandoned later. According to the author's knowledge, no new initiatives in this area for hydrogen production have started that base the technology on this type of reactors.

The gasifiers considered in this overview can be differentiated into direct (autothermal) and indirect (allothermal) processes. In the indirect processes, the heat needed to drive the endothermic heterogeneous reactions is provided by inert (circulating) solid material or by heat exchangers. The processes are

- Autothermal fluidized-bed (FB) gasification, applying bubbling fluidized-bed (BFB) or circulating fluidized-bed (CFB) technology (steam/O_2 blown)
- Indirect FB technology (steam blown)
- Autothermal entrained-flow (EF) gasification, operated in a slagging or nonslagging mode

Figure 6.4[36] shows an example of a biosyngas production scheme for the trigeneration of methanol, hydrogen, and electricity. The process scheme typically consists of the following basic steps: biomass pretreatment, gasification, gas cleaning, reforming of higher hydrocarbons, shift to obtain suitable H_2:CO ratios, and gas separation for hydrogen production

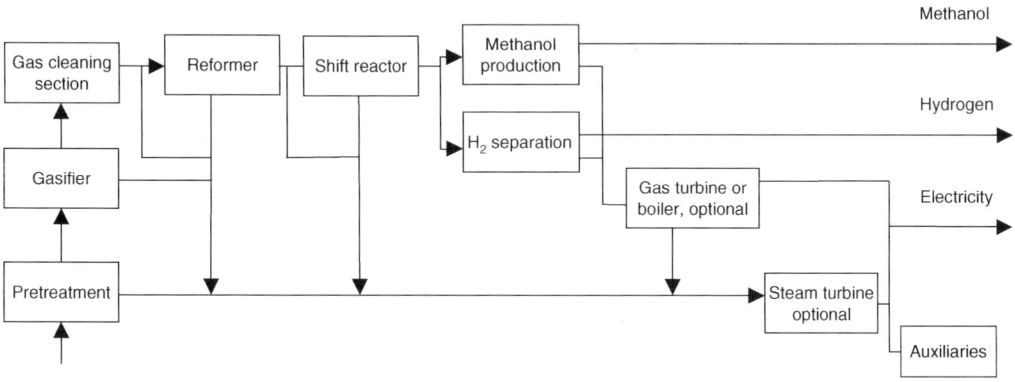

FIGURE 6.4
Process flow schematic for the production of fuels, including hydrogen, based on biomass gasification. (Reproduced from Hamelinck, C., and Faaij, A.P.C. *J. Power Sources*, 111(1), 1–22, 2002. With permission.)

and purification. Optional are a gas turbine or boiler to utilize the unconverted gas and a steam turbine to produce electricity.

6.3.1 Pretreatment

Pretreatment of biomass is of great importance for a reliable feeding system, and furthermore, has a big influence on storage, handling, and transportation characteristics. The pretreatment techniques may include drying, chipping, sizing, densification, and optional torrefaction. Leaching biomass to remove alkali salts that may cause downstream problems of slagging, fouling, or agglomeration in gasifiers can also be applied as pretreatment; see Arvelakis et al.[37] The extent of necessary drying depends on the applied gasification process, usually a maximal moisture content of 35 mass% is taken as feasible.[38] Size reduction is necessary especially for EF gasification proceses characterized by a low residence time of the solids in combination with a high process temperature. The costs associated with it are high when reducing the size almost equal to that of coal in pulverized fuel systems; milling wood to 100 μm particles, similar to coal, would consume as much as 0.08 $kW_e/kW_{th,wood}$ of electric energy relative to the heating value of the wood. Calculated to primary energy this means approximately 20%, which is unacceptable.[39] The comparatively novel process of torrefaction that is investigated for energy generating purposes today (see, e.g., Prins,[40] Zanzi et al.[41]) makes biomass fuel look like peat and is easier to crush economically. Integration of the drying process within the system (steam drying) and development of reliable feeding systems for pressurized gasifiers are areas that still need substantial R&D.

6.3.2 Gasification Reactors

As mentioned earlier, FB gasification together with EF gasification of solid biomass feedstock are technologies, which can in principle realize the required large production capacity for a hydrogen production facility benefitting from economy of scale with comparatively high thermal efficiencies.

6.3.2.1 *Fluidized-Bed Reactor Systems*

Figure 6.5 shows the process configuration of different FB gasification concepts, from single-reactor concepts to twin fluid-bed systems. The FB gasifiers are operated in a typical temperature window of 800–950°C. Producers of such FB gasifiers (also including air

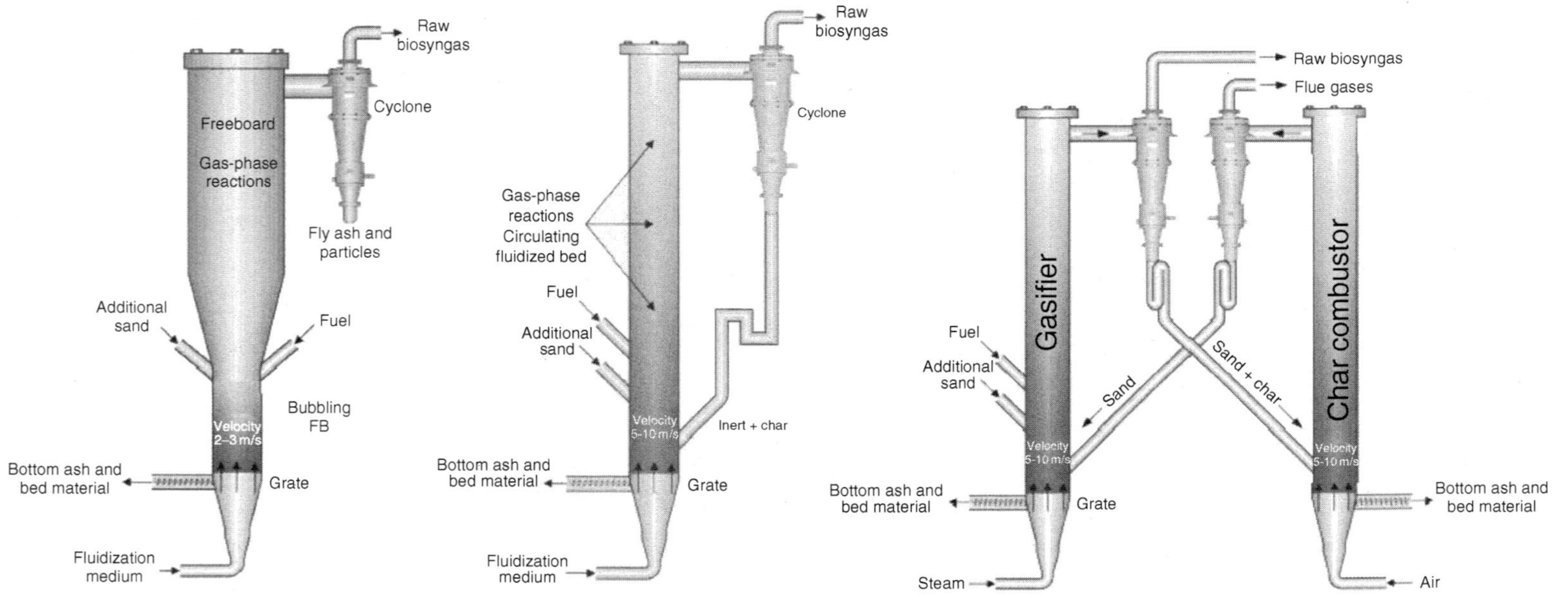

FIGURE 6.5
Examples of large-scale gasification reactor technology based on FB gasification. Fuel in these schematics can be widely differing in nature including biomass. (Reproduced from Olofsson, I., Nordin, A., and Söderlind, U., *Initial Review and Evaluation of Process Technologies and Systems Suitable for Cost-Efficient Medium-Scale Gasification for Biomass to Liquid Fuels,* ETPC Report 05-02, University of Umeå/Mid Sweden University, Umeå, 2005, p. 90. With permission.)

blown types) are, for example, Austrian Engineering (Austria), Carbona (United States), Ebara (Japan), EPI (United States), Foster Wheeler (United States), HoSt (The Netherlands), Lurgi (Germany), and TPS (Sweden).[42,43]

There are different process configuration concepts where FBs form the heart of the technology for the hydrogen-rich gas production. The following recently developed gasification reactor technology can be especially mentioned in this respect:[44]

- The fast internally circulating fluidized-bed (FICFB) gasifier (developed by Technical University (TU) Vienna, Austria)
- The SilvaGas gasifier (developed by Battelle Columbus lab., United States)
- The Institute for Gas Technology (IGT) gasifier (also called (Ren)U-Gas, developed at IGT, United States)
- The CFB gasifier at Värnamo, Sweden (modified Foster Wheeler technology)

The FICFB gasification process has been developed at TU Vienna[45] and is in an upscaled combined heat and power generation demonstration scale (fuel input 8 MW_{th}; heat generation 4.5 MW_{th} and 2 MW_e), currently operating in the Austrian village Güssing; see Hofbauer in Knoef. The manufacturer is AE/Repotec. It comprises an indirect gasifier in which the gasification zone—a bubbling bed (typically 800°C, 1.1 bar) and the combustion zone—a circulating bed (around 1050°C, 1.1. bar) are separated in different, closely integrated reactor compartments. Heat required to drive the endothermal reactions in the gasifier is supplied by hot bed material leaving the combustor, in which unburnt char is oxidized. Steam (400°C, 1.1. bar) is added as the gasification medium. Air is provided as the fluidization agent in the combustion zone. The producer gas is first cooled down to 160–180°C, particles are removed, and a scrubber operated at 40°C removes the soluble pollutants. To control the temperature in the combustion zone, around 20% of the producer gas leaving the scrubber is preheated by heat exchanging it with the flue gas, which has been cleaned of particles using a cyclone, and finally, this air is used to generate steam. Bottom air from the combustion section may leak into the gasification section. Owing to optimized geometry, this leakage is $<3\%$. Depending on the tar content of the producer gas, biodiesel is added. Another measure for tar control is the use of catalytic active bed material, in this case, olivine $(Mg, Fe)_2SiO_4$ is used. Olivine has shown a high attrition resistance and it can be catalytically very active in tar reduction; see Devi et al.[46] Excess heat is used to produce steam for the gasifier and heat the combustion air. A schematic of the CHP plant layout in Güssing is given in Figure 6.6.

The *SilvaGas* gasification process underwent initial development at Battelle's Colombus Laboratories since the late 1970s as a part of DOE's Biomass Power Program; see Paisley and Welch.[47] A demonstration scale facility was stationed at the Burlington Electric Department's McNeil power station—a 50 MW wood fired unit. The partners in the development of the process at this site are FERCO, Burlington Electric, Battelle, U.S. DOE, and the National Renewable Energy Laboratory (NREL).[48] It uses two CFB reactors that are interconnected: a steam-blown CFB gasifier (800–1000°C, 1.1. bar) and a combustor (1050°C, 1.2 bar). A schematic of the process is shown in Figure 6.7. Heat is supplied to the gasifier by the exothermal reactions taking place in the combustor, and thereby, heating the fluidization medium: sand. Steam at 150°C (2 bar) is introduced into the gasifier to enhance hydrogen production. A medium calorific value product gas is generated by this process, with a heating value of about 17–19 MJ/m_n^3.[48] The SilvaGas process has been tested successfully in the period 1999–2000 on a large scale with a biomass input of over 200 t/day. A wide range of fuels can be used in this type of gasifier, and it produces a gas with a relevant heating value that is not so strongly dependent on the feedstock's moisture or ash content.

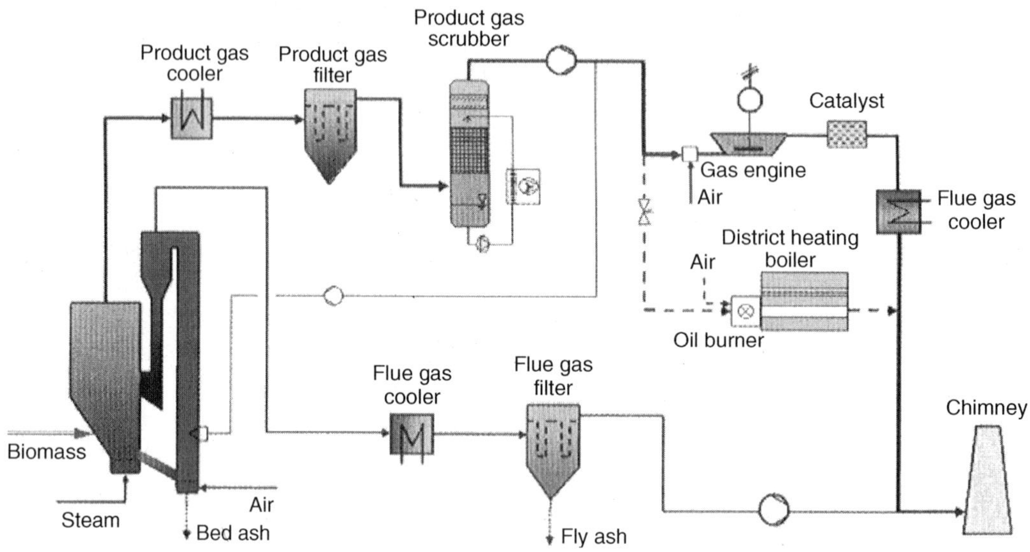

FIGURE 6.6
The FICFB process concept applied in the 8 MW_{th} Güssing plant (Austria). (Reproduced from Knoef, H., *Handbook Biomass Gasification*. BTG (Biomass Technology Group), Enschede, 2005. With permission.)

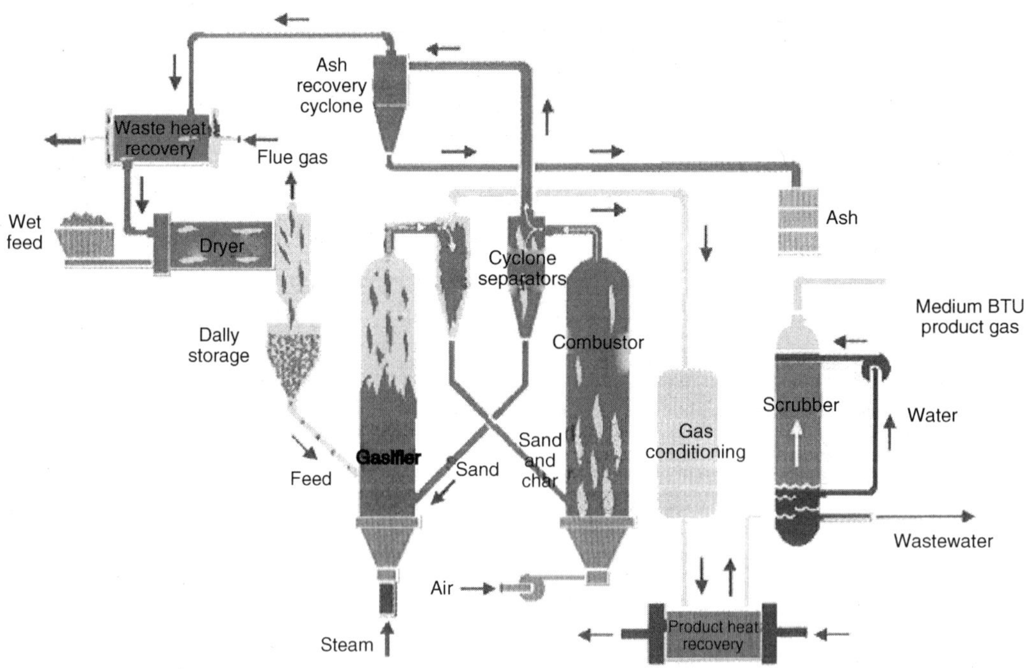

FIGURE 6.7
The SilvaGas process concept demonstrated at the McNeil station (Burlington, VT). (Reproduced from Knoef, H., *Handbook of Biomass Gasification*, BTG, Enschede, 2005. With permission.)

The IGT has developed a pressurized, oxygen-blown (air-blown optional) BFB gasifier. The process was named *U-Gas*®, originally developed for coal-based hydrogen production.[49] Pilot-scale experiments of steam gasification of charred cellulosic wastes were already performed in the 1980s; see Rabah and Eldighidy.[50] The pressure prevailing in the gasifier is approximately 25 bar. Alumina beads are used as bed material. The use of oxygen is advantageous as there is no N_2 to ballast the gas, although the oxygen needs to be generated, which also comprises expensive equipment. The process has been demonstrated at a larger scale. It produces a CO_2-rich gas and the hydrogen content can be increased and optimized by using a biomass ratio of 0.8 kg/kg dry biomass.[36] Pressurized gasification allows a large throughput per reactor volume and diminishes the need for pressurization downstream. In this way, less overall power is needed in this process.

Comparable to IGT technology for syngas generation is the high-temperature Winkler (HTW) gasification process. A commercial HTW unit was installed in 1988 at a Finnish ammonia synthesis plant and has operated successfully using peat as a feedstock.[51]

Within a recently started EU project (Framework 6, Integrated Project), CHRISGAS 18 partners from industry, institutes, and universities develop and optimize an energy- and cost-efficient method to produce hydrogen-rich gases from biomass, using *steam-oxygen-blown CFB* gasification technology; see, for example, Waldheim et al.[52] and the project's Website.[53] The produced gas can be upgraded to commercial quality hydrogen or synthesis gas for further upgrading to liquid fuels such as dimethyl ether (DME) and methanol or the Fischer-Tropsch diesel.

New equipment is developed, tested, and implemented in a demonstration facility to produce clean biosyngas for vehicle fuel production. This facility, the 18 MW_{th} Värnamo plant, is to be reconstructed so as to enable the production of around 7000 m_n^3/h biosyngas. The facility is depicted in Figure 6.8.

Also included in the project are studies related to the large-scale use of such plants and their impact on the environment and society. The project consists of tasks for which the objectives are

- Solid biofuel conversion into syngas by pressurized circulating fluidized bed (PCFB) gasification using steam and oxygen.
- Improved *in situ* desulfurization by means of dedicated additives.
- Cleaning of the generated gas from particulates in a high-temperature filter.
- Purification of the generated gas by catalytic autothermal steam reforming of a broad spectrum of hydrocarbons to generate a raw synthesis gas consisting mainly of CO and H_2 as energy carriers.
- Conditioning of the raw syngas, meeting the quality stipulated for manufacture of DME or other potential products. This is attained by catalytic water–gas shift to adjust the CO/H_2 ratio and catalytic hydrogenation of minor contaminants followed by the removal of CO_2 and acidic gases in a conventional wash system or in a novel, selective processes for deeper S-removal.
- Studies of fuel production from various biofuels at a scale and cost representative of biomass fuel chains in various regions in Europe, thereby rendering realistic data for future production potential on a European scale.
- Large-scale feeding system testing of a compact, high-capacity piston feeder with the advantage of much lower total energy consumption than using lock hoppers.

FIGURE 6.8
Pictures of the Värnamo 18 MW_{th} pressurized CFB demonstration facility. (Reproduced from www.chrisgas.com (accessed May 9, 2007). Copyright by VVBGC. With permission.)

6.3.2.2 *Other Processes Involving Fluidized-Bed Technology*

A novel development in hydrogen-rich gas generation from biomass using FB technology is the so-called absorption-enhanced reforming (AER) process concept,[54,55] which is demonstrated in a European project. It is an example of allothermal gasification. The process comprises an unpressurized steam-blown gasification in a primary reactor using *in situ* CO_2 absorbent bed material. The CO_2-loaded absorbent is regenerated in a second coupled FB combustor using some additional fuel (see Figure 6.9 for the working principle). Heat is transferred to the primary reactor (operating typically at 600–700°C) through the exchange of solids with the second reactor (with combustion taking place there at approximately 800°C). The process makes use of a shift of the homogeneous water–gas shift reaction toward the $CO_2 + H_2$ side, thereby maximizing H_2 yield by the CO_2 absorption that is taken away from

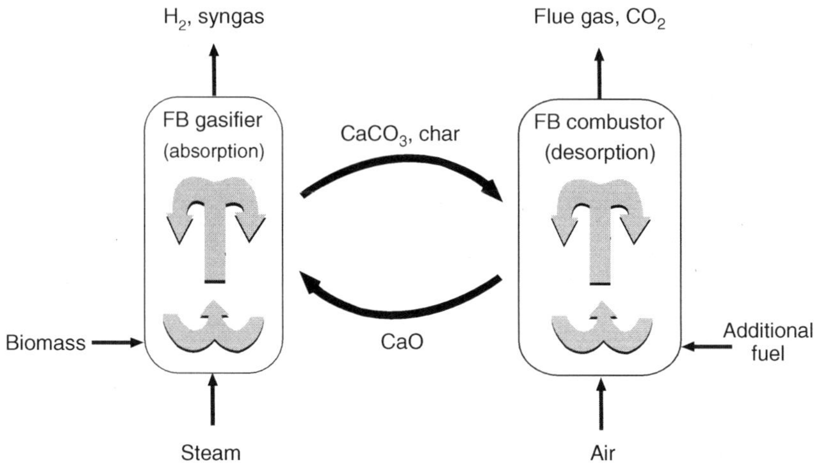

FIGURE 6.9
The AER process concept for hydrogen rich syngas production. (Adapted from Marquard Möllenstedt, T., Sichler, P., Specht, M., Michel, M., Berger, R., Hein, K.R.G., Höftberger, E., Rauch, R., and Hofbauer, H., New approach for biomass gasification to hydrogen. *2nd World Conference on Biomass for Energy, Industry and Climate Protection, Rome,* ETA-Florence, Rome, 2004.)

the gas phase. For coal in Japan a similar process was developed, the HyPr-RING process,[56] but research activities there are also directed toward the use of hydrogen from biomass.[57] There is some concern about the long-term regenerative behavior of the CO_2 absorbent, limestone, which tends to decrease with increasing number of absorption/desorption cycles.[58]

An example of a combined steam gasification of biomass and reforming process of the gases is the "Blaue Turm" (= Blue Tower) process (developed by DM2, Germany). It consists of three separate reactor sections. In the first section, pyrolysis takes place at temperatures in the range 500–600°C. Practically, all of the volatile content (about 80% of the biomass) is released within this temperature range. The generated pyrolysis gas, to increase its hydrogen content, is reformed at 950°C in the reforming section where steam is added to promote the hydrogen formation by the water–gas shift reaction. The solid material that remains in the pyrolyzer is burned at 1000°C together with a heat carrier in a combustor. The hot heat carrier is then, after being separated from the flue gas, used to provide the heat required for the reforming reactions. Because of the high-temperature prevailing in the reformer, tars are destroyed and partly converted to hydrogen. The advantage of this process is the simple construction, no support fuel is needed, and a hydrogen-rich gas is generated. The steam that is used in the reformer is preheated by using the hot producer gas. A schematic of the process is shown in Figure 6.10.[59]

Somewhat related is a process proposed and demonstrated on labscale by the University of Siegen (Germany). The process is called the (Herhof)-Integrierte Pyrolyse und Verbrennung (IPV) process and is decribed in detail by Hamel et al.[60] In this process, biomass is converted with high-temperature steam to pyrolysis gas in a fixed-bed reactor. The generated carbon from this reactor is led to a stationary FB combustor from which the hot ash is returned to the first-mentioned reactor. The ash works catalytically to reduce the tar content of the gas produced. The gas is further cleaned and conditioned using a scrubber and electrostatic filter from which the catch is returned to the FB combustor.

Another novel development in gasification using allothermal gasification based on FB reactor technology is the so-called heatpipe reformer. The technology has been developed at TU Munich[61] in recent years. It does not make use of circulating sand as the heat transfer

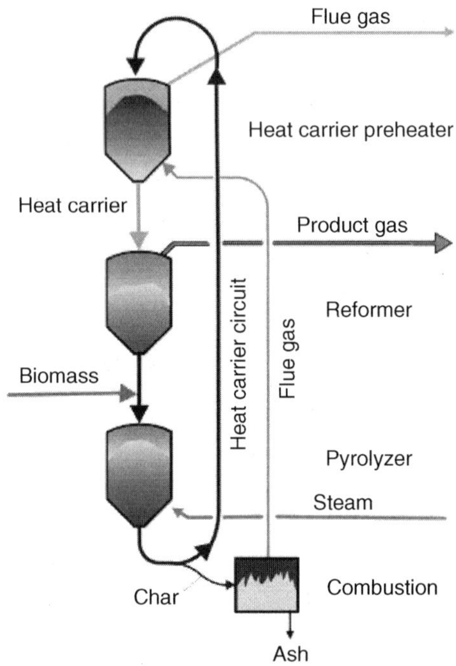

FIGURE 6.10
The "Blaue Turm" process. (Reproduced from www.dml-2.de/ (accessed May 8, 2007). Copyright by DM2 Verwertunstec, chnologien. With permission.)

medium, but uses heatpipes as the new feature of the technology. Owing to the low heat transfer coefficients of common heat exchangers, it is nearly impossible to integrate the necessary surface into the reactor for flue gas–driven heat exchangers even with atmospheric gasification. Higher pressures will even enhance that problem. The use of high-temperature heatpipes shifts the determining thermal resistance from the inner tube side to the outer heat exchanger surface into the FB and reduces the necessary surface significantly. Thus, the FB heat transfer determines the necessary heat exchanger surface and the necessary temperature difference between the heating medium and the gasifier bed. The high-temperature heatpipe is a pipe containing an alkali metal (Na, K, etc.). It has two zones, the evaporation zone, where heat is provided to the heatpipe and the condensation zone, where heat from the heatpipe is transferred to its environment. The biomass heatpipe reformer (BioHPR) is an innovative gasification concept, which uses the technology of liquid metal heatpipes to transfer the high-heat flux needed for the gasification reactions from a combustion chamber into the gasifier. It consists of three main parts and can be seen in Figure 6.11.[62]

The first part is a BFB gasifier, the so-called reformer. Biomass is fed into the reformer from the top through a specially designed lock hopper system. The gasification takes place either in atmospheric pressure or under a pressure of up to 5 bar. The product gas from the reformer is first driven to the sand filter, which is the second part of the BioHPR. This integrated sand filter captures char particles from the product gas to transfer them into the combustion chamber. After it passes through a cyclone and a ceramic or metallic filter, the product gas can be used in a gas turbine or a fuel cell. The third part of the BioHPR is the FB combustion chamber. The dust and char particles, which are filtered in the integrated filter, enter the combustion chamber with the help of a siphon system, which separates the gasifier and the filter from the combustor. Liquid metal heatpipes filled with sodium transfer the heat from the combustion chamber to the reformer.

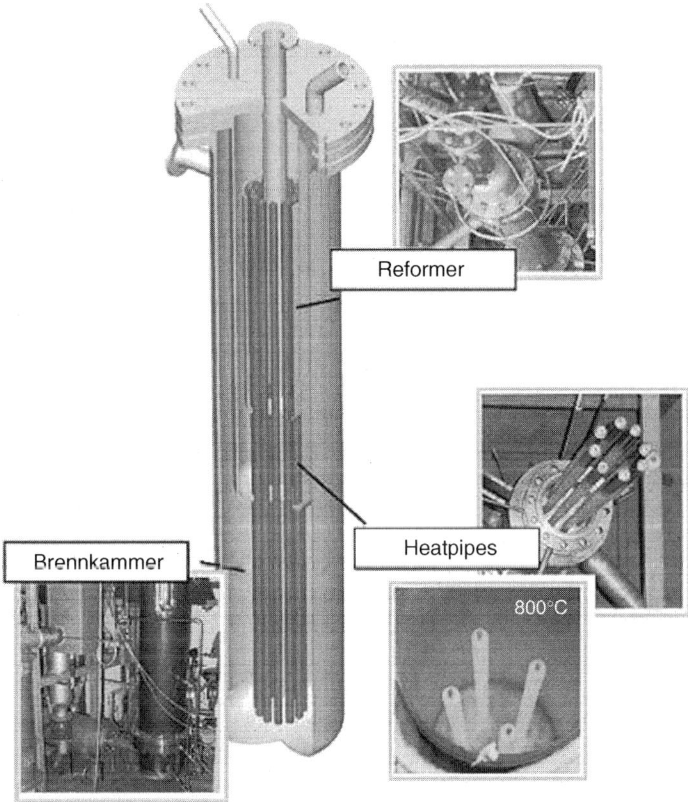

FIGURE 6.11
The heatpipe reformer process concept for hydrogen-rich syngas production. (Reproduced from Karellas, S., Metz, T., Kuhn, S., and Karl, J., Online analysis of the tar content of the product gas from biomass gasification. Application on the BIOHPR. *14th European Biomass Conference & Exhibition, Biomass for Energy, Industry and Climate Protection*, ETA-Renewable Energies, Paris, 2005. With permission.)

6.3.2.3 Entrained-Flow Reactor Systems

Figures 6.12[63,64] and 6.13[65] depict some current commerical EF gasification reactor technologies for the hydrogen-rich syngas generation. Typical maximal operating temperatures of such EF gasifiers are in the range 1200–1500°C. Most of the gasifiers have only shown operation on diverse coal types. Slagging EF gasifier manufacturers are Shell, Texaco, Krupp-Uhde, Future Energy (formerly Noell/Babcock Borsig Power), Conoco-Phillips (E-gas technology, formerly Destec/Dow), MHI (Mitsubishi Heavy Industries), and Hitachi and Choren (formerly UET, Germany).

Some of these gasifiers are described in the following text (a large part is adapted from Van der Burgt in Knoef[1]). The first EF gasifier was developed for coal gasification in the fifties of the past century by Koppers. The process comprised atmospheric gasification, using the so-called Koppers-Totzek slagging gasifiers for syngas production aimed at the production of ammonia. In the process, coal and oxygen are fed by pairs of diametrically opposed burners inside niches in the side of a more or less egg-shaped reactor. The product gas leaves at the top part of the reactor at a maximum temperature of about 1500°C and it is quenched using water to a temperature of about 900°C, which is cooled down further in a syngas cooler in which steam is produced. This quenching is necessary to prevent sticky slag droplets to contact the heat exchanger, which otherwise could be severely fouled.

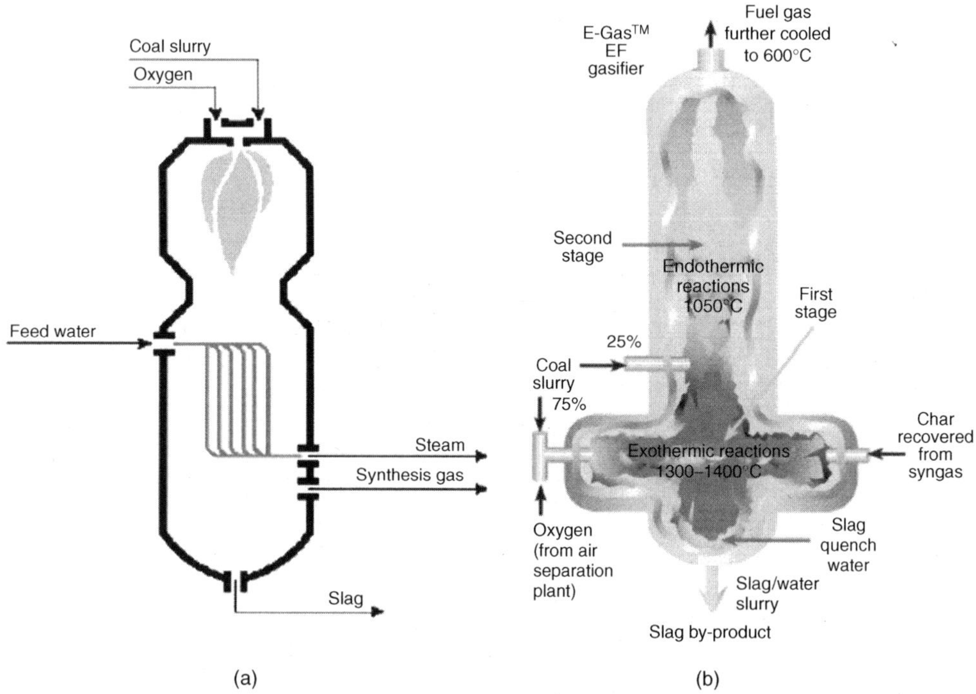

FIGURE 6.12
Large-scale gasification reactor technology based on EF gasification from (a) General Electric (GE; Texaco process) and (b) Conoco-Phillips (E-Gas). (Adapted from Meier, D. Faix, O. Fast pyrolysis: A route for energy and chemicals from wood—fluidized vs. ablative pyrolysis. In *Wood and Biomass Utilization for the Carbon Uptake*, Seoul National University, Seoul, 2005, pp. 55–68.)

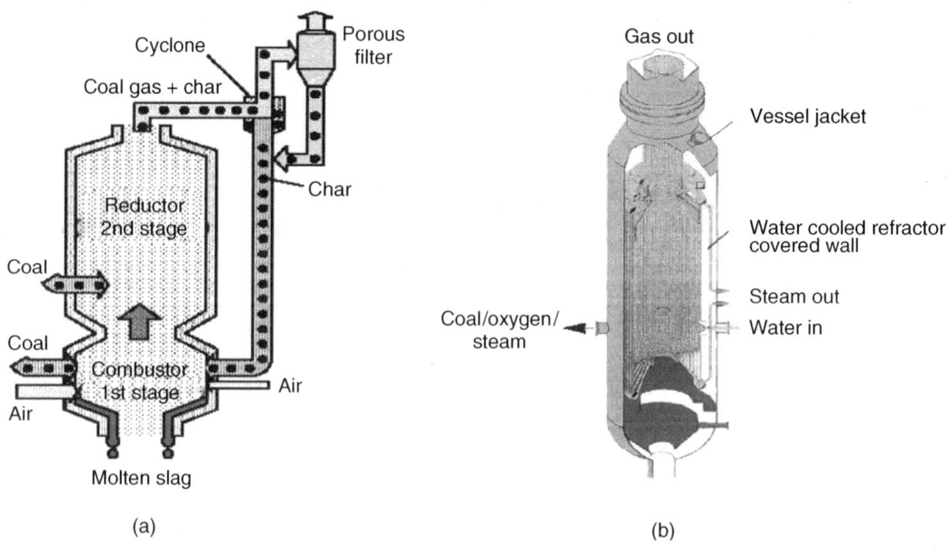

FIGURE 6.13
Large-scale gasification reactor technology based on EF gasification from MHI (a) and Shell (SCGP) (b). (Adapted from Hooper, M., Waste coal to ultra clean fuels applying the shell coal gasification process. *Gasification Technologies Conference*, www.gasification.org/Presentations/2003.htm (accessed May 11, 2007), San Francisco, CA, 12–15 October, 2003.)

The reactor wall is constructed of a steam jacket to protect the steel from high temperatures. The slag produced is mainly leaving the reactor through a bottom hole.

The Shell coal gasification process (SCGP) and the Prenflo process that were developed in the last decades of the twentieth century are of higher efficiency, pressurized versions (30–40 bar) of the Koppers-Totzek process maintaining the feature of diametrically opposed burners inside niches in the side of a vertical cylindrical reactor. The developments led to the erection of a 2000 t/day SCGP unit in Buggenum (The Netherlands) and a 3000 t/day Prenflo unit in Puertollano (Spain). Both gasifiers are integrated in an integrated gasification combined cycle (IGCC) unit for electricity production. Today, a gas quench has replaced the water quench. Also, membrane walls are applied instead of a steam jacket to protect the pressure shell from very high temperatures. These walls are used for the generation of high-pressure saturated steam for additional power production. At the Buggenum site (see Figure 6.14[66]), the power producing company NUON has demonstrated cofiring of coal with chicken manure upto about 30 wt% biomass share in the Shell process–based 253 MW_e plant for electricity production; see, for example, Van Dongen and Kanaar.[67] The concept and reactor technology can be extended to the polygeneration of heat, power, and syngas-based chemicals.

The Texaco gasification process, currently owned by GE, is also of the EF type. A main difference with the aforementioned technologies is that the fuel is fed as a water slurry, whereas Koppers-Totzek- and Shell-based coal gasifiers use some steam.

The company Noell in Germany, currently owned by Future Energy GmbH, has developed a down-fired entrained flow reactor (EFR). Instead of burners at the side of the reactor, this gasifier only consists of a single burner in the top part, which results in a simpler and lower cost reactor. Also, the control of this single burner is easier than the former types mentioned with a resulting lower capital and operational cost related.

The company Choren in Germany has developed a three-stage gasification process involving the following subprocesses (see Figure 6.15[68]):

1. Low-temperature pyrolysis/gasification
2. High-temperature gasification
3. Endothermic EF gasification

FIGURE 6.14
The Buggenum 253 MW_e IGCC plant based on EF gasification technology. (Adapted from Zuideveld, P., Overview of shell global solution's worldwide gasification developments. *Gasification Technologies Conference*, www.gasification.org/Presentations/2003.htm (accessed May 11, 2007), San Francisco, CA, October 12–15, 2003.)

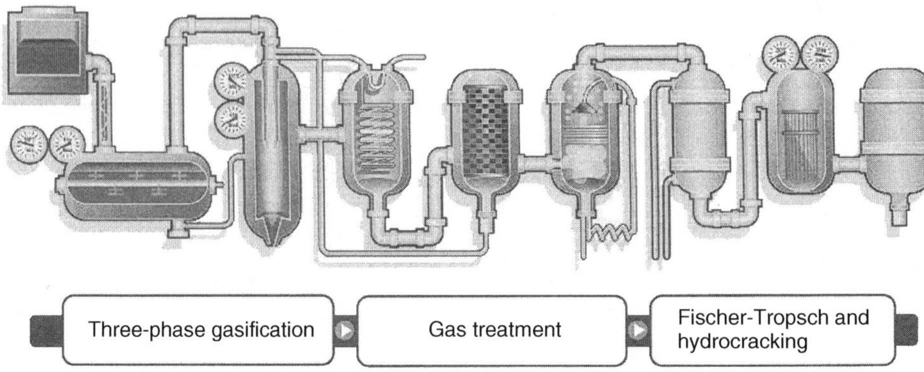

FIGURE 6.15
The Carbo-V™ process of CHOREN GmbH. (Reproduced from www.choren.com (accessed May 10, 2007). Copyright by CHOREN Industries GmbH. With permission.)

During the first process stage, relatively dry biomass (with moisture content of 15–20 wt%) is continually carbonized by means of partial oxidation with air/oxygen at mild temperatures of 400–500°C. Here volatiles and char are formed. The second stage comprises the understoichiometric combustion of the generated volatiles with oxygen or air in a combustion chamber that operates above the melting point of the fuel's ash to turn it into a hot gasification medium. The third stage, finally, consists of char grinding to a pulverized fuel that is blown into the hot gasification medium created in the EFR of the second stage (bottom part of the reactor). The pulverized fuel here reacts in an endothermic process that generates a raw syngas. Further treatment of cooling and cleaning follow. The syngas can be used for different purposes; nowadays, the focus is on diesel fuel production, called SunDiesel.

It is also possible to use liquid feedstock, and three commercial processes are currently available that are all based on the design of a single burner and an outlet for the gas and char/soot/ash near the bottom part of the reactor. The first oil gasifier was developed by Texaco in the late 1940s followed by Shell (Shell gasification process [SGP]) and finally by Lurgi (Lurgi multipurpose gasification [MPG] process). In all three processes, a refractory wall is being used. Very heavy oil residues can be processed, and it is expected that all three can process pyrolysis-derived bio-oil.

6.3.2.4 Evaluation of Gasification Technologies

The two basic reactor technologies for large-scale syngas generation, FBs and EFRs, have their own specific advantages and disadvantages. Table 6.2 gives an overview of these features. The greatest risks reported for EF gasification are fuel preparation costs as EF gasification needs very small particles ($\ll$1 mm) for good fuel conversion. This requires an extra process step of biomass milling + sizing or torrefaction + milling + sizing. Also, slagging phenomena in the gasifier are different for biomass when compared to more conventionally applied coals, and fouling in the syngas cooler is a risk. However, for CFB gasification, risks are mainly concentrated in the fouling of the tar cracker and syngas cooler and agglomeration tendencies, especially when using high alkali agricultural residues. The CFB process and also the (pressurized) BFB process, contrary to EF gasification, is more fuel flexible.

TABLE 6.2

Features of Different Gasifiers for Large-Scale Hydrogen-Rich Gas Production from Biomass

Reactor Type	Positive Aspects	Negative Aspects
BFB	Good temperature control	Intermediately high amount of solids in raw gas (of the order of some g/m_n^3)
	Good scale-up potential to 10–15 dry t/h	Tar content may be high in raw gas (2–10 g/m_n^3)
	Easy start-up and shutdown	Carbon loss with ash
	Good tolerance for fuel input particle size range	
	High ash fuels can be accepted	
	In-bed catalysts can be applied (tar cracking/gas upgrading)	
	Thermodynamic losses limited	
	Experience with biomass processing	
CFB	Good temperature control	High amount of solids in raw gas (up to several g/m_n^3)
	Very good scale-up potential to >15 dry t/h, for atmospheric, a max. value of 300 MW_{th} is mentioned[69]	Tar content generally high (order of magnitude 2–20 g/m_n^3)
	Easy start-up and shutdown	Moderate carbon loss with ash
	Good tolerance for fuel input particle size range	Use of twin configurations add complexity, but medium high calorific value (MHV) gas can be generated using just air
	High ash fuels can be accepted	
	In-bed catalysts can be applied (tar cracking/gas upgrading)	
	Thermodynamic losses limited	
	Experience with biomass processing	
EF	Very good scale-up to *huge sizes* (>1000 MW)	Fuel size range flexibility limited to very fine particle sizes, thus costly preparation, also limited for high ash fuels
	Simple construction	Application of biomass alone is limited, practically only in cogasification
	No tar generated	More release of trace metals into gas phase due to higher temperatures
		Higher thermodynamic losses due to inherent high temperature operation

The use of air alone leads to a relatively low calorific value product gas, of the order of 4–6 MJ/m_n^3 (LHV basis), which is not attractive for H_2 production in view of the large bulk of N_2 to be separated from it compared to a preseparation from the air. Only the application of hydrogen in ammonia production would need N_2 as cofeedstock. Therefore, in this context only steam- or oxygen-blown gasification concepts are dealt with. The raw product gas can, thus, be produced by an oxygen-blown or indirectly heated steam-blown processes.

6.3.3 Gas Cleaning

The different gasification processes produce a more or less dirty raw gas, which must be cleaned further and upgraded to the downstream application. Particulates, bound sulfur

TABLE 6.3

Main Gas Quality Requirements for Different Hydrogen Gas Applications

Synthesis	H_2 or Refinery	Ammonia	Methanol	Fischer–Tropsch	Oxo Alcohols
H_2	>98%	75%	71	60	60
CO	<10–50 ppmv	$CO + CO_2$	19	30	40
CO_2	<10–50 ppmv	<20 ppmv	4–8%		
N_2		25%			
Inert	N_2, Ar, CH_4 balance	Ar, CH_4 As low as possible	N_2, Ar, CH_4 As low as possible	CO_2, N_2, Ar, CH_4 Low	
H_2/N_2		~3			
H_2/CO				0.6–2	1–1.5
$H_2/(2 * CO + 3 * CO_2)$			1.3–1.4		
Process temperature (°C)		350–550	300–400	200–350	85–200
Process pressure (bar)	>50	100–250	50–300	25–60	15–350

Source: Reproduced from Knoef, H., *Handbook Biomass Gasification*. BTG (biomass technology group), Enschede, 2005. With permission.

TABLE 6.4

Overview of Allowance of Minor Species in the Hydrogen-Rich Product Gas in Gas Engines, Turbines, and Syngas Production for Methanol Production

Application	Tar (mg/Nm³)	Particles (mg/Nm³)	Alkalis (mg/Nm³)	Ammonia (mg/Nm³)	Chloride (mg/Nm³)	Sulfur (H_2S, COS) (mg/Nm³)
Gas engine[a]	<50	<50	<1	<50	<10	<100
Gas turbine[b]		<30	<0.25			
Syngas/Methanol[c]	<0.1	<0.02			Below	ppmv level

[a] If an engine exhaust catalyst is used, some specifications are more strict such as sulfur, chloride, alkalis, and heavy metals.

[b] Turbines are not sensitive to tar because gas temperature is usually high and tars are in vapor form. Alkalis (Na, K, P) are critical compounds for erosion.

[c] Specifications are very strict for the shift plant.

Source: Reproduced from Knoef, H., *Handbook Biomass Gasification*. BTG (biomass technology group), Enschede, 2005. With permission.

species (H_2S, COS, CS_2), halogens (mainly HCl), bound nitrogen species (HCN, NH_3), alkali (Na, K salts), and other trace species and tars have a potentially harmful effect on downstream equipment, especially catalysts.

In FBs, tars are generated, but these can be partially decomposed to gas using *in situ* catalysts such as dolomite[70] and treated olivine sand.[46] The tars can further be cracked or reformed downstream in separate beds yielding additional syngas, and thus, hydrogen (see Section 6.3.4). A new development is the combination of ceramic gas filtration and catalytic tar cracking; see Heidenreich and Nacken.[71]

Table 6.3 gives an overview of the main gas composition specifications for selected applications; see Knoef.[1]

For minor species in the produced syngas, the maximum acceptable levels are reported to be even more stringent as the catalysts used are intrinsically very sensitive to small amounts of impurities (see Tables 6.4 and 6.5). Generally, economic considerations govern

TABLE 6.5

Overview of Allowance of Minor Species in the Hydrogen-Rich Product Gas in Fuel Cells

Fuel Cell Type	Tar (mg/Nm^3)	Particles (mg/Nm^3)	Alkalis (mg/Nm^3)	Ammonia (mg/Nm^3)	Chloride (mg/Nm^3)	Sulfur (H_2S, COS) (mg/Nm^3)
Polymer electrolyte fuel cell (PEFC)	n.i.	n.i.	n.i.	n.i.	n.i.	n.i.
Phosphoric acid fuel cell (PAFC)	Poison <0.5% olefins	n.i.	n.i.	Poison: <0.2 mol% $(NH_4)_3PO_4$ in electrolyte	Poison: <4 ppm[73]	Poison: <4 ppm[73]
Molten carbonate fuel cell (MCFC)	Aromatics <0.5%[74] <2000 ppm (benzene)[75]	Anode blockage: 10 ppmw[74]	Electrolyte loss: 1–10 ppm	Fuel/Inert: <1 vol%[75]	Poison: <0.1 ppm[75]	Poison: <0.5 ppm[75] in fuel
						<10 ppm H_2S in fuel[a]
		100 ppm, dp>3 μm				
						<1 ppm SO_2 in oxidant
Solid oxide fuel cell (SOFC)	Aromatics <0.5%[74]	n.i.	n.i.	Fuel: <5000 ppm[75]	Poison: <1 ppm[75]	Poison (reversible): <0.1 ppm[75]
				Identified as fuel![76]		
						Tests ongoing on new materials to prove higher tolerances[77–79]

Note: When no further specification is given, data are from Dayton et al.,[80] n.i. = not indicated.

[a] Tolerance level depends on anode gas composition and partial pressure of H_2; these concentration limits increase when temperature increases, but they decrease at increasing pressures.[75]

the purity requirements. Thus, additional expenses for gas cleaning are weighed against the acceptance of lower production due to catalyst poisoning or higher maintenance frequency. This forms the background of the fact that there are practically no requirement criteria given in open literature sources for the gas impurities. For Fischer–Tropsch synthesis, although, some levels have been published.[72] They indicate that the sulfur species (H_2S, COS, CS_2) should be reduced to levels below 1 ppmv; nitrogen species (NH_3 and HCN) to below 1 ppmv; acid halogen species (HCl, HF, HBr) to 10 ppbv; alkali metals to 10 ppbv; organic compounds should be removed to below dewpoint, and no solids should virtually be allowed to be present. Further experimental in-depth investigation, although, is necessary toward achievable and tolerable levels of these minor gas constituents for the specific system components applied.

As gas cleaning options, a choice can be made for either low-temperature "wet" or high-temperature "dry" gas cleaning. The latter type of gas cleaning is advantageous for the overall energy balance but many process elements are not yet fully proven technology.

The ultimate selection depends on the properties of the produced gas and the end use. The sequence and temperature levels of the different steps require a careful analysis of input properties and output requirements.

6.3.4 Hydrogen Yield Maximization

The syngas can contain a considerable amount of methane and other light hydrocarbons, representing a significant part of the heating value of the gas. This is especially the case for gasification processes carried out at comparatively low temperatures; for large-scale syngas generation, there are the different FB processes as described in Section 6.3.2.

Steam reforming converts these hydrocarbons to CO and H_2 driven by steam addition. This can be either *thermal* reforming at high temperatures (typically well above 1100°C) or *catalytic* reforming (temperatures well above 650°C) over nickel-based catalysts. The pioneering work on the use of nickel catalysts for biomass gasification product upgrading was carried out in Sweden in the framework of the MINO process for methanol production from biomass (see Blackader and Rensfelt[81]) and at Pacific Northwest Laboratory (PNL) (see Baker et al.[82–83] and Mudge et al. [84–85]). Use of the catalyst in the gasifier was not so successful. The use of a steam reforming catalyst applied in a fixed-bed downstream of an FB gasifier, however, was reported to be working well and increased the H_2 content of the gas to 52–59 vol% on a dry basis.[86,87] The use of a guard bed for further reduction of tar and sulfur content has been reported.[88–90] However, it has been shown that comparable tar contents can be achieved (at maximum 2 g/m_n^3) when dolomite is used within the FB gasifier as bed material and additive,[70] thus dismissing the need for an additional guard bed unit in this application.[91] Autothermal reforming combines partial oxidation in the first part of the reactor with steam reforming in the second part, thereby optimally integrating the heat flows.

The syngas produced by, for example, the SilvaGas and IGT gasifiers, has a low H_2:CO ratio. The water–gas shift reaction is a common process operation to shift the energy value of CO to H_2. The equilibrium constant for the water–gas shift reaction increases as temperature decreases. Hence, to increase the production to H_2 from CO, it is desirable to conduct the reaction at lower temperatures, which is also preferred in view of the steam economy. However, to achieve necessary rates higher temperatures are required. The application of water–gas shift catalytic reactors for biomass gasification product gas is a quite novel application developed in the past decade.[92] The hydrogen content is reported to be significantly increased to values of above 60 vol% dry, N_2-free basis for FB steam-blown gasification[87]; even in the case of air-blown gasification, a value of 26.7 vol% H_2 was obtained.[89] In a further optimized system, it was shown by Aznar et al.[91] that a 73 vol% H_2-containing syngas could be realized using a combination of commercially available high-temperature CO-shift catalyst (350–480°C) and low-temperature CO-shift catalyst (200–260°C) applying a H_2O/CO ratio >2 for the gas downstream of the steam reforming catalyst. These authors gave an outlook to even higher obtainable H_2 concentrations based on a thermodynamic equilibrium consideration.

The combination of CO shift with membrane separation of H_2 and *in situ* CO_2 capture is mentioned as an advanced process to optimize the hydrogen yield; LHV-based cold hydrogen efficiencies of 72% are reported in such configurations.[93]

6.3.5 Purification and Separation

After reforming and shifting of raw syngas, hydrogen can be separated and compressed. Today, hydrogen plants use different process concepts. New hydrogen plants use pressure swing adsorption (PSA) for purification. The hydrogen quality is a major issue for its eventual automotive application. Specifically, CO is a strong poison to polymer electrolyte membrane (PEM) fuel cells. Studies indicate that levels as low as 1–2 ppb will deactivate the platinum anode material of such fuel cells.

Membrane technology is promising for gas separation due to its simple design and the ability of combining shift and separation in one reactor. Much R&D effort is put in decreasing the pore size, so that membranes can act as molecular sieves. Membranes are expected to perform better at high temperature, implying an important energy advantage in combination with hot gas cleaning, as no temperature drop would have to occur between the gasifier and the gas turbine. The membrane surface also catalyzes the water–gas shift reaction. This reaction is driven to hydrogen as it is removed by the selectively permeable membrane. Ceramic membranes have the advantage of a broad temperature and operating pressure range. Construction of membrane separation units is potentially very simple and cheap when compared to other technologies such as PSA. Moreover, membranes do not suffer the efficiency losses and high capital costs for heat exchangers, associated with the need to cool down the syngas.

6.3.6 Some Economic Prospects of Gasification-Based Hydrogen Production

Spath and Dayton[94] analyzed the process thermal efficiency (PTE), defined as the energy in the product fuel divided by the energy of the biomass feedstock. They also studied the economics of gasification routes to syngas-based fuels. Based on a biomass cost of \$33/dry t, they conclude that biomass to hydrogen is a comparatively attractive alternative. The price would vary between \$9 and \$17/GJ (LHV basis), with a resulting selling price of \$1.1–2.0/kg. This is even within the current price margin of about \$0.7–1.4/kg. Also, the PTE was in the higher ranges (36–73%) among the alternatives of methanol, ethanol, olefins, and Fischer-Tropsch diesel. Hamelinck and Faaij[95] made an economic analysis for hydrogen production based on some gasification processes, that is, the IGT FB gasifier and the SilvaGas gasifier with different cleaning and upgrading options. With the assumption of a feedstock price of \$2/GJ, feasible in biomass producing regions such as Latin America, and a reasonably large production capacity of 400 MW_{th}, they arrive at a production price of \$8–11/GJ, on the low side of the previously mentioned analysis but quite well in-line with it.

6.4 Hydrogen from Pyrolysis-Derived Bio-Oil

Pyrolysis of biomass is a thermochemical process carried out at typical temperatures in the range 650–800 K and atmospheric or a few bar pressure, in which the fuel is converted under supply of heat and principally with the exclusion of oxygen. The process produces a liquid product called bio-oil as the main product in a single-reactor configuration, next to gases and charcoal.[31] This is a complex mixture of a multitude of compounds including acids, alcohols, aldehydes, esters, ketones, and (oxygenated) aromatics, which can be further converted into different fuels, chemicals, and materials. Bio-oils are already used (practically) commercially as boiler fuels for stationary power and heat production, but need upgrading if they are to be used as transportation fuels.

The pyrolysis process can further be classified into slow- and fast-pyrolysis, referring to the applied heating rates in the process.

Slow pyrolysis, also called carbonization, is characterized by a high charcoal yield and is not considered for hydrogen production processes. The slow pyrolysis of wood (24 h typical residence time) was a common industrial technology to produce charcoal, acetic acid, methanol, and ethanol from wood until the early 1900s.

Fast pyrolysis is a high temperature, fast heating process in which vapor is produced, that is, subsequently condensed as a dark brownish liquid, bio-oil, with some coproduction from char and permanent gases. High yields of pyrolysis vapors/bio-oil can be achieved by a proper choice of reactor configuration, biomass particle sizes, and heat and mass transfer rates.

Research activities in the area of fast pyrolysis of biomass started toward the end of the 1970s of past century. A thorough overview of the basic principles and actual state of the art is given by Meier,[96] Bridgwater and Peacocke,[97] and Mohan et al.[98]

Pyrolysis processes have some advantages. The produced bio-oil is easier to transport than raw biomass, provided it is stabilized so as to give minimum losses. Also, to improve economics for larger-scale processing, bio-oil/char slurries can be transported.[99] A series of small-sized pyrolysis units could be constructed at sites where low-cost feedstocks, such as agricultural residues, are available. In this case, the bio-oil would be transported to a central steam reforming unit at an industrial site with hydrogen storage and distribution infrastructure. Also, higher value coproducts could be derived from the pyrolysis oil, although the oil is not uniquely defined in quality.

Different reactor types can be employed[100] as follows:

- Ablative reactors
- FBs (BFB, CFB)
- EFRs

Basically, nowadays there are five processes that are based on a production capacity of several tons per day; these are[101]

1. The BioTherm™ process of the company Dynamotive (Vancouver, Canada). The process is based on a stationary FB and the largest plant processes about 100 t/day of biomass. This unit is situated in Ontario (Canada) and its feedstock is woody waste. In this application, the obtained bio-oil is converted into electricity using an Orenda gas turbine (2.5 MW_e).
2. The rapid thermal processing (RTP)™ of the company Ensyn Group Inc., situated in Ottawa (Canada). Diverse working process plants for the production of biobased aromates are available of which the largest has a biomass processing capacity of 70 t/day. The heart of the process is a CFB reactor.
3. The Biomass Technology Group (BTG, Enschede, The Netherlands) process is based on a rotating cone reactor that serves mechanical fluidization of hot sand, which is recirculated in the reactor system. Since 2005, there is a plant in Malaysia with a throughput capacity of 50 t/day of palmoil residues.
4. The Biolig process of the research center Karlsruhe FZK, Germany. Here, flash pyrolysis, with emphasis on straw as feedstock, is tested to produce a bio-oil-char slurry. The pyrolysis reactor compares to the LR reactor (Lurgi-Ruhrgas) by which sand as heat carrier is mixed and transported together with biomass in a double (twin) screw feeder. A novel unit is constructed with a biomass processing capacity of 12 t/day.
5. The BTO process of the company PYTEC in Lüneburg, Germany.[102] The reactor principle is that of a hot rotating disc, which is in direct contact with biomass to let it pyrolyse; the so-called ablative pyrolysis is taking place in this process. So far, a working unit of 6 t/day is working on wood residues at a sawmill near Cuxhaven. The produced pyrolysis oil is combusted in an adapted diesel engine.

The respective flash pyrolysis technologies are depicted in Figure 6.16; see Meier and Faix.[103]

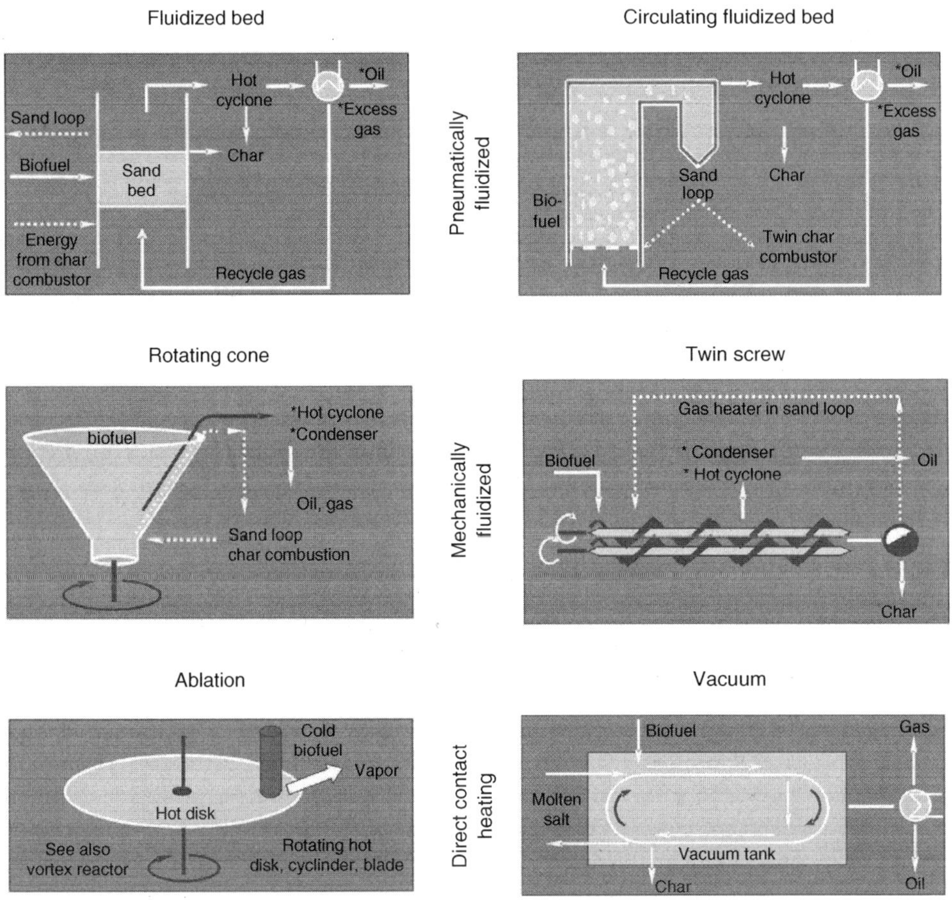

FIGURE 6.16
Schematic overview of flash pyrolysis reactor technologies. (Reproduced from Meier, D., and Faix, O., *Wood and Biomass Utilization for the Carbon Uptake*, Seoul National University, 2005. With permission.)

Typical bio-oil production prices are still relatively high: Peacocke et al.[104] mention values of €13.2–14.6/GJ for FB and rotating cone technology, respectively, based on the assumptions of biomass costs of €50/t dry and feeding rates of 10 dry t/h.

An emerging option for hydrogen production is a two-step process including fast pyrolysis of biomass followed by catalytic steam cracking/reforming of the pyrolysis vapors/liquids. Steam reforming of fossil fuels is a well-established technology,[105] and this process applied for pyrolysis-derived bio-oils is an extension of this technology. Typical temperatures for this process are 600–800°C. At NREL laboratory, work has been conducted as of the early 1990s and has been reviewed by Milne et al.[5] Ni-based catalysts were reported to be favorable, but enough CO remained in the product gas fraction to require the addition of a water–gas shift process step. Main concern reported is the deactivation of the catalyst by carbon deposition. Czernik et al.[106] developed a FB steam reformer for these bio-oils. Catalysts were more stable in this reactor type than in a fixed-bed reactor due to improved contact of the catalysts with steam. A point of concern is the attrition resistance of the catalyst. Commercial catalysts developed for steam reforming of natural gas and crude oil fractions proved to be more efficient for bio-oil than most research catalysts developed, mainly due to their higher water–gas shift activity.

The hydrogen producing potential of pyrolysis can be illustrated by the following overall reactions. Here, an average typical elemental composition is assumed (not corresponding to an existing structure):

$$\underset{\text{Biomass}}{CH_{1.46}O_{0.67}} \rightarrow \underset{\text{Bio-oil}}{0.71CH_{1.98}O_{0.76}} + \underset{\text{Char}}{0.21CH_{0.1}O_{0.15}} + \underset{\text{Gas}}{0.08CH_{0.44}O_{1.23}}$$

$$\underset{\text{Bio-oil}}{CH_{1.98}O_{0.76}} + 1.24H_2O \rightarrow CO_2 + 2.23H_2$$

The potential H_2 yield following this simple stoichiometric approach is, thus, about 13 mass% based on the biomass input. This is only slightly lower than gasification by use of oxygen. The bio-oil for hydrogen-rich gas combustion is in a subsequent process step gasified in, for example, an EF gasifier, either alone or together with finely dispersed char in it (Biomass to liquid 2 [BTL2] concept[99]). The concept is presented in Figure 6.17.

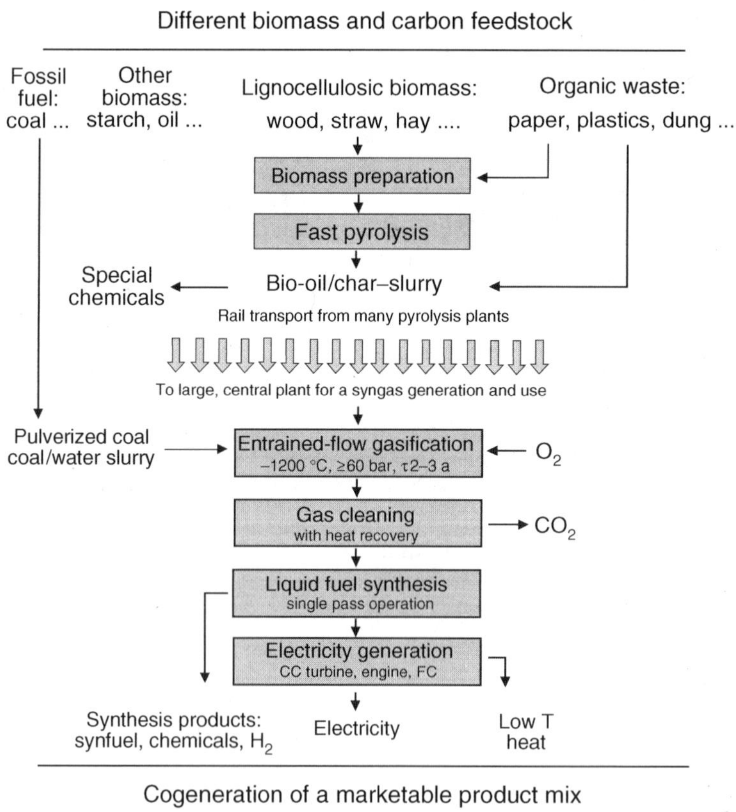

FIGURE 6.17
Syngas polygeneration based on a combination of decentralized pyrolysis of biomass and central gasification of pyrolysis oil/char slurry. (Reproduced from Henrich, E., Raffelt, K., Stahl, R., and Weirich, F., *Science in Thermal and Chemical Biomass Conversion*, CPL Press, Victoria, 2004. With permission.)

6.5 Hydrogen from Biomass-Derived Other Intermediates

Widely differing processes have been reported for hydrogen production based on biomass-derived intermediate compounds. Some of the main ones are reported in this section.

6.5.1 Pyrolysis to Hydrogen and Carbon

At Brookhaven National Laboratory in the 1980s, a two-step process was described for the coproduction of hydrogen and carbon, with methane being the intermediate; see Steinberg.[107,108] In the first step, carbonaceous material is hydrogasified to methane with a subsequent thermal decomposition of the methane to hydrogen and carbon. In the process, water is also formed from the oxygen present in the fuel. A typical overall conversion of biomass, as reported by Milne et al.,[5] would be

$$CH_{1.44}O_{0.66} \rightarrow C + 0.06H_2 + 0.66H_2O$$

The process was further elaborated in the 1990s by Dong and Steinberg[109] with the aim to produce methanol; see, for example, Dong and Borgwardt.[110]

6.5.2 Hydrogen Production from Biomass-Derived Methanol

Methanol derived from biomass gasification/pyrolysis through the syngas production route can in turn be converted to hydrogen. This option has the disadvantage of higher costs than direct hydrogen production; however, methanol is easily transportable to, for example, remote places for decentralized applications, whereas with hydrogen more techno-economic challenges are faced.

Steam reforming of methanol to produce hydrogen has already a long history and has been reviewed by Milne et al.;[5] a brief summary is presented here. A mechanistic description of this reaction has been given by Takahashi et al. (Japan)[111] describing Cu–Si catalysts. Studies on (Cu-containing) catalysts and systems for the methanol steam reforming with fuel cell applications as one of the main focusses[112] were continued in the 1990s of the past century until recently. Also, zeolites were studied in steam reforming studies; see, for example, Laniecki and Kazmierczak-Rosik.[113]

Catalyzed[114,115] and noncatalytic partial oxidation[116] have also been studied for the purpose of hydrogen production from methanol.

6.5.3 Hydrogen Production from Biomass-Derived Ethanol

Ethanol can be derived from biomass by means of acidic/enzymatic hydrolysis or also by thermochemical conversion and subsequent enzymatic ethanol formation. Likewise for methanol, hydrogen can be produced from ethanol with the ease of storage/transportation and an additional advantage of its nontoxicity. Apart from thermodynamic studies on hydrogen from ethanol steam reforming,[117–119] catalytic reaction studies were also performed on this reaction using Ni–Cu–Cr catalysts,[120] Ni–Cu–K alumina-supported catalysts,[121] Cu–Zn alumina-supported catalysts,[122,123] Ca–Zn alumina-supported catalysts,[122] and Ni–Cu silica-supported catalysts.[123]

Partial oxidation is also mentioned as a process to convert ethanol to hydrogen.[124] Another novel technology for ethanol to hydrogen has been described by Toci and Modica.[125] It is based on cracking ethanol vapors by "cold-plasma-chemical processing" in the presence of a Ni-based catalyst.

6.5.4 Hydrogen Production from Other Biomass-Derived Oxygenates

Recently, research is being done on other oxygenates from biomass to produce hydrogen subsequently.[126] The researchers report the catalytic steam reforming of acetic acid over Pt/ZrO_2 catalysts. The catalysts work bifunctionally. The Pt sites cause bond breaking in the acetic acid substrate molecule, thus producing H_2, CO, CH_4, and CO_2 in the gas phase. Carbonaceous residue formed herewith is catalytically broken down to syngas molecules on the ZrO_2 functionality by hydroxyl surface groups.

Although steam reforming of renewable alcohols has been studied widely in recent times, there are more processing steps before biomass is converted to these compounds, which make this process relatively energy intensive and costly. However, hydrogen can also be directly produced from sugars in an aqueous environment through the so-called aqueous phase process (APP). In this process, next to CO_2 and CO, alkanes are also produced ranging from C_1 to C_{15}.[127–133] The process temperatures are comparatively low, 200–260°C and the pressure is reported to be in the range 10–50 bar. Currently, Virent Energy Systems is working to commercialize this process. One of the advantages of this aqueous-phase reforming process is that it produces a product gas with low levels of CO (100–1000 ppm) in a single reactor, whereas conventional steam reforming requires multiple reactors to reduce the CO levels. According to Davda et al.,[131] the advantages of using the aqueous-phase reforming are

- The process eliminates the need to vaporize water and the oxygenated hydrocarbon, reducing the energy requirements for producing hydrogen.
- The feedstock compounds are nonflammable and nontoxic, allowing them to be stored and handled safely.
- The process is characterized by temperatures and pressures where the water–gas shift reaction is favorable for hydrogen.
- Production of hydrogen is at a pressure level where further purification can be effectuated with existing (PSA) and novel technologies (membranes).
- The process is conducted at low temperatures, minimizing undesirable decomposition reactions of sugar feedstock.
- A single reactor can be used versus multireactor steam reforming systems.

6.6 Hydrogen from Other Thermochemical Processes

6.6.1 Supercritical Water Gasification

One of the novel developments is the gasification of biomass under process conditions where water becomes *supercritical*, usually defined as *supercritical water gasification*.[13] Under severe conditions, by heating water containing the biomass to temperatures above 647 K and compressing the system to a pressure higher than 22, 1 MPa, biomass is rapidly decomposed leading to syngas components within a few minutes at a high conversion efficiency, using, for example, a FB configuration.[134] The comparatively very low amount of tar and char formed appear to be resulting products, which is highly interesting. Especially for wet biomass, this process seems to be attractive. The solubility of biomass components in hot compressed water had already first been studied by Mok et al.[135] They showed that already 40–60% of the biomass were converted under subcritical conditions. Minowa et al.[136–138] reported hydrogen production from cellulose gasification using a nickel catalyst, also under subcritical conditions. An important milestone was reached in the supercritical water gasification technology when the experimental results of Modell, Antal and co-workers were published.[139–141] Yu et al.[140] reported that the gasification of glucose under supercritical water process conditions (873 K, 34.5 MPa) was different from subcritical conditions. Gadhe and Gupta[142] performed experiments using methanol as feedstock for supercritical water reforming (973 K, 27.6 MPa) and produced hydrogen along with CO, CH_4, and CO_2. Main objective was to suppress methane formation. Experiments were targeted at this methane minimization by means of lowering the reactor residence time by addition of catalysts (KOH, K_2CO_3) and studying the effect of the reactorwall metal catalytic effect. All these approaches resulted in a significant reduction in the methane formation, and therefore, an increased hydrogen yield was achieved. Antal et al.[143] reported the complete gasification of biomass feeds—sugarcane bagasse, sewage sludge, corn- and potato starch gels, sawdust, potato waste—using a carbon catalyst at temperatures above 600°C.

Boukis et al.[144] report on a larger-scale *pilot plant* (100 kg/h capacity), where diluted biomass feed streams were gasified under supercritical water conditions. See Figure 6.18 for an overview of the plant. Using maize silage as model species, the researchers obtained a fuel gas with >50 vol% H_2, 40% CH_4, and up to 10% of higher hydrocarbon species with complete fuel utilization at temperatures higher than 923 K.[145] The catalytic role of potassium salts was identified; the authors indicate that the available 500 ppm of potassium was sufficient and that higher contents would not result in significant product changes anymore. Hydrogen and methane could be obtained in pure, pressurized form. Purification was easy because the heteroatom species remained in the aqueous phase.

Although in recent years, extensive research has been conducted toward this process, supercritical gasification of biomass is still in a stage of R&D.

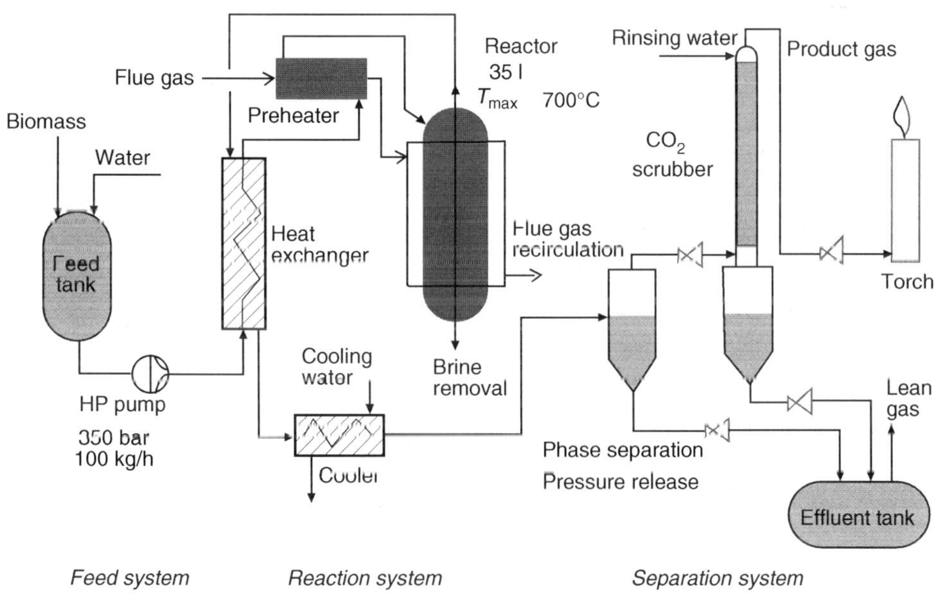

FIGURE 6.18
Process scheme of the supercritical water gasification pilot plant at FZ Karlsruhe. (Reproduced from Boukis, N., Galla, U., Diem, V., and Dinjus, E., *Science in Thermal and Chemical Biomass Conversion*, CPL Press, Victoria, 2004, 975–990. With permission.)

6.6.2 The Steam-Iron Process

Hydrogen can also be produced from biomass by making use of the steam-iron redox cycle that converts initially generated syngas into a hydrogen-rich gas stream. In this process, the syngas reacts with iron oxides (Fe_2O_3 and Fe_3O_4) to produce a reduced form of iron oxide (FeO) or Fe metal. The last-mentioned iron species (FeO and Fe) are reoxidized by steam to regenerate the original oxides and produce hydrogen. This process is one of the oldest "low-risk" methods for hydrogen production and it was developed in the late nineteenth/early twentieth century based on coal gasification.[5]

Hacker et al.[146] recently reviewed previous studies toward the process and performed a simulation of the process including a design for a 10 MW system. An advantage of the process was the demonstrated high purity of the produced hydrogen, pure enough for fuel cell application with CO concentrations being reported lower than 10 ppm. Sime et al.[147] and Biollaz et al.[148] explored the process based on biomass. The first step comprised the reducing of Fe_3O_4 by woodgas and the oxidation of the formed FeO was reported to be enhanced by the addition of other transition metal oxides. The authors expect that 90% of the low heating value gasification product gas can be transferred to hydrogen with suitable modified oxides. The cost of hydrogen was reported to be relatively high as competition with PSA is tough.

Currently, there is renewed interest in this process. At Twente University (The Netherlands), research is performed toward the use of bio-oil from (flash) pyrolysis as the reducing agent for the oxidized iron.[149] Therefore, a CO-containing gas stream and a relatively pure H_2 stream is generated separately from the biomass-derived bio-oil.

6.6.3 The Molten Bath Processes

There are processes under development that use molten iron–based baths, which are available in the steel industry. These processes are the HydroMax® (invented by Alchemix Corporation) and HyMelt® (invented by Marathon Ashland Petroleum LLC, licensed to EnviRes LLC) process. They comprise only one step, which consists of the dissociation of hydrocarbons in the molten bath at very high temperatures of 1200–1600°C. As feedstock are suggested petroleum coke, high sulfur coal, municipal waste, discarded tyres, and biomass (including sludges).[9] The process comprises two steps. In the first step, iron reacts with steam that has been generated by waste heat, here FeO and hydrogen are formed. In the second step, the iron oxide formed is reduced by carbon, which in itself consists of two steps by which iron oxide is first converted with CO to iron and CO_2 and then CO_2 and C react to CO by the well-known Boudouard reaction. It is claimed by Alchemix Corporation that the main advantage of their process is its capacity to create intimate reactant mixing and the efficiency of 82% for converting energy input to H_2 and steam.

6.7 Research Requirements

Considering the prospective hydrogen production cost, the main conclusion is that these still have to be decreased to compete with fossil fuel–based technology, and improved performance in this respect can be obtained by

- Application of new and especially for the short-term improved existing technologies
- Combining fuel and power production, possibly extended with (bulk) chemical production by advanced integrated biorefinery process concepts
- Using economy of scale

The technology development depends on the successful integration of the different process steps, and technical and economical analyses are required to match the optimum technology with available biomass feedstock to produce a product gas of the necessary quality for the target application. Although a lot is known about each of the different subsystems separately, this knowledge is often obtained during research aimed at very different applications. Because of this, experimental data on the main subsystems have to be obtained under conditions that are applicable to large-scale hydrogen production using thermochemical gasification of biomass.

Specifically, certain research and development aspects still need to be addressed for successful demonstrations, which are related to

- Feeding of all kinds of different biomasses
- The use, possibly with pretreatment, of high alkali and chlorine-containing agricultural, farming, and poultry residues, in view of agglomeration, slagging, and fouling problems they may cause, especially in the primary thermochemical conversion processes
- High-temperature gas cleaning (in particular, gas filtration using metal or ceramic filters and also the removal of especially sulfur species)
- Catalytic upgrading of the hydrogen-rich syngas (tar and hydrocarbon conversion, possibly in combination with filtration, also water gas shift catalyst use and development)
- The coprocessing of coal in gasification processes together with biomass (CO_2 separation and sequestration, sulfur species removal, and trace element emissions)

Special attention also has to be paid to scale effects to enable optimal sizing and a reliable assessment of the economies of scale. A great similarity exists between biomass-based processes producing hydrogen, methanol, and clean syngas for liquid fuel production, causing substantial synergy between the process development and the technology implementation in these areas. Currently, an increased number of large, government-funded R&D programs are ongoing in the United States, Japan, and European Union.

6.8 Conclusions

Biomass is a potentially sustainable and certainly renewable important source for hydrogen in a world moving toward the hydrogen based economy. It forms one of the substantial CO_2 neutral options for energy production. The world potential availability of biomass does not seem to be limiting large-scale bioenergy implementation, although local circumstances must be studied carefully to guarantee a really sustainable growth and use of the biomass. Thus, biodiversity, water usage, the fate of nutrients, agricultural practices, and their socioenvironmental impacts must be addressed.

With regard to the biomass conversion technology for large-scale hydrogen production, thermochemical conversion appears to be the most promising. Here, demonstration, the integration of the different process steps and the optimization of the whole process chain, should lead to lower production costs of hydrogen as these are still far from current fossil fuel (especially natural gas)–based figures.

Gasification is still the most mature technology for large-scale hydrogen production, but it needs integrated demonstration plants at sufficiently large scale including catalytic gas upgrading. This is now pursued in different demonstration projects outlined in this contribution. Process efficiencies for hydrogen production by gasification routes are higher than liquid transportation fuels. Pyrolysis processes with the aim to produce hydrogen gas are in the smaller scale demonstration phase. Supercritical water gasification is still in an early stage of development and research, as among other features, reactor material challenges are faced with the corrosive medium.

There is a large common area among systems producing hydrogen, methanol, and hydrogen-rich syngas from renewable biomass for the production of liquid transportation fuel. Exchange of information and integration of projects working on these processes is very important to make optimal use of available funding. The development of biorefineries, units that produce chemicals, materials, and (combined) heat and power in a flexible way, is to a large extent dependent on hydrogen production based on the biomass processed in a way that hydrogen also plays an important role in oil refineries.

List of Acronyms

AER	Absorption-enhanced reforming
APP	Aqueous phase process
BioHPR	Biomass heatpipe reformer
BFB	Bubbling fluidized bed
BTL	Biomass to liquid
BTU	British thermal unit
CFB	Circulating fluidized bed
DME	Dimethyl ether
ECN	Energy Research Centre of the Netherlands
EF	Entrained flow
EFR	Entrained flow reactor
FB	Fluidized bed
FC	fuel cell
FICFB	Fast internally circulating fluidized bed
HTW	High-temperature Winkler
IC	internal combustion
IGCC	Integrated gasification combined cycle
IEA	International Energy Agency
IPCC	Intergovernmental Panel on Climate Change
IPV	Integrierte Pyrolyse und Verbrennung
LHV	Lower heating value (= lower calorific value)
MCFC	Molten carbonate fuel cell
MHV	medium high calorific value
MPG	Multipurpose gasification
PAFC	Phosphoric acid fuel cell
PCFB	pressurized circulating fluidized bed
PEFC	Polymer electrolyte fuel cell
PEM	Polymer electrolyte membrane
PNL	Pacific Northwest Laboratory
PSA	Pressure swing adsorption
PTE	Process thermal efficiency
PV	photovoltaic

RTP	Rapid thermal processing
SRES	Special Report on Emission Scenarios
SOFC	Solid oxide fuel cell
SCGP	Shell coal gasification process
TU	Technical University

References

1. Knoef, H., *Handbook Biomass Gasification*. BTG (Biomass Technology Group), Enschede, 2005.
2. IPCC Climate Change 2007, The Physical Science Basis, Summary for Policymakers. http://ipcc-wg1.ucar.edu/wg1/docs/WG1AR4_SPM_Approved_05Feb.pdf (accessed on May 10, 2007).
3. Johansson, T. B.; Kelly, H.; Reddy, A. K. N.; Williams, R. H., *Renewable Energy*. Island Press, Washington, 1993.
4. Turner, J. A., A realizable renewable energy future. *Science* 1999, 285, 687.
5. Milne, T. A.; Elam, C. C.; Evans, R. J., *Hydrogen from Biomass—State of the Art and Research Challenges*. IEA/H2/TR-02/001, A Report for the International Energy Agency Agreement on the Production and Utilization of Hydrogen Task 16, Hydrogen from Carbon-Containing Materials, Golden, Co, National Renewable Energy Laboratory, 2001.
6. EC, *Hydrogen Energy and Fuel Cells—A Vision of Our Future*. European Commission—High level group on hydrogen and fuel cells, Brussels, 2003.
7. Momirlan, M.; Veziroglu, T. N., Current status of hydrogen energy. *Energy Education Science and Technology* 2002, 6, 141–179.
8. Czernik, S.; Elam, C. C.; Evans, R. J.; Milne, T. A., Thermochemical routes to hydrogen from biomass—a review. In *Science in Thermal and Chemical Biomass Conversion*, Bridgwater, A. V.; Boocock, D. G. B., Eds., CPL press, Victoria, 2004, pp. 1752–1761.
9. Collot, A.-G., *Prospects for Hydrogen from Coal*. CCC/78, IEA Clean Coal Centre, London, 2003, p. 76.
10. DOE, *Hydrogen, Fuel Cells and Infrastructure Technologies Program, Multi-year Research, Development and Demonstration Plan*. 2003 available at http://www1.eere.energy.gov/hydrogenandfuelcells/production/biomass_gasification.html (accessed March 2008).
11. Demirbas, M. F., Technological options for producing hydrogen from renewable resources. *Energy Sources, Part A: Recovery, Utilization, and Environmental Effects* 2006, 28(13), 1215–1223.
12. National Research Council, *The Hydrogen Economy: Opportunities, Costs, Barfriers, and R&D needs*. The National Academies Press, Washington, 2004.
13. Ni, M.; Leung, D. Y. C.; Leung, M. K. H.; Sumathy, K., An overview of hydrogen production from biomass. *Fuel Processing Technology* 2006, 87, 461–472.
14. Meadows, D. H.; Meadows, D. L.; Randers, J.; Behrens, W. W. (III), *The Limits to Growth—A report to the Club of Rome's project on the predicament of mankind*. Universe Books, New York. 1972.
15. UNFCC, *United Nations Framework Convention on Climate Change*, Kyoto Protocol by the 3rd Conference of the Parties, Kyoto 1997.
16. Huber, G. W.; Iborra, S.; Corma, A., Synthesis of transportation fuels from biomass: Chemistry, catalysts, and engineering. *Chemical Reviews* 2006, 106, 4044–4098.
17. Spath, P. L.; Lane, J. M.; Mann, M. K.; Amos, W. A., *Update of Hydrogen from Biomass—Determination of the Delivered Cost of Hydrogen*. NREL, Milestone Report for the U.S. Department of Energy's Hydrogen Program Process Task, Milestone Type C (Control), Golden, CO, 2001.
18. IEA, *Biofuels for Transport: An International Perspective*, IEA, Paris, 2004.
19. Berndes, G.; Hoogwijk, M.; Van den Broek, R., The contribution of biomass in the future global energy supply: A review of 17 studies. *Biomass and Bioenergy* 2003, 25, 1–28.
20. Hoogwijk, M.; Faaij, A. P. C.; Van den Broek, R.; Berndes, G.; Gielen, D.; Turkenburg, W. C., Exploration of the ranges of the global potential of biomass for energy. *Biomass and Bioenergy* 2003, 25(2), 119–133.

21. Demirbas, A., Effect of initial moisture content on the yields of oily products from pyrolysis of biomass. *Journal of Analytical and Applied Pyrolysis* 2004, 71(2), 803–815.
22. www.ecn.nl/phyllis (accessed May 10, 2007).
23. Werther, J.; Saenger, M.; Harge, E. U.; Ogada, T.; Siagi, Z., Combustion of agricultural residues. *Progress in Energy and Combustion Science* 2000, 26(1), 1–27.
24. Chen, G.; Fang, M.; Luo, Z.; Li, X.; Shi, Z.; Cen, K.; Ni, M.; Preto, I., Experimental research on rice husk combustion in CFB boiler and the design of a 35 t/h rice husk-fired CFB Boiler. *14th International Conference on Fluidized Bed Combustion*, ASME, Vancouver, 1997.
25. Balatinez, J. J., *The Potential of Densification of Biomass Utilization*. Plenum Press, London, 1983, pp. 181–189.
26. Van der Drift, A.; Van Ree, R.; Boerrigter, H.; Hemmes, K., Bio-syngas: Key intermediate for large scale production of green fuels and chemicals. In *2nd World Conference and Technology Exhibition on Biomass for Energy, Industry and Climate Protection*, Van Swaaij, W. P. M., Fjällström, T., Helm, P., Grassi, A., Eds., ETA-Florence, Rome, 2004, pp. 2155–2157.
27. Boerrigter, H.; Rauch, R., *Review of Applications of Gases from Biomass Gasification*. ECN-RX—06-066, ECN, Petten, 2006.
28. Boerrigter, H.; van der Drift, A., Bio Key Intermediate in Production of Renewable Transportation Fuel, Chemicals and Electricity, Optimum and Economic Properties of Fisher-Tropoch Plants, ETA, 2005.
29. Zaborsky, O. R., *BioHydrogen*. Plenum Press, New York, 1998.
30. Beenackers, A. A. C. M.; Van Swaaij, W. P. M., Gasification of biomass, a state of the art review. In *Thermochemical Processing of Biomass*, Bridgwater, A. V., Ed., Butterworth, London, 1984, pp. 91–136.
31. Klass, D. L., *Biomass for Renewable Energy; Fuels and Chemicals*. Academic Press, San Diego, CA, 1998.
32. Hofbauer, H.; Rauch, R.; Foscolo, P.; Matera, D., Hydrogen rich gas from biomass steam gasification. *1st World Conference on Biomass for Energy and Industry*, Sevilla, 2001, pp. 1997–2001.
33. Bridgwater, A. V., Renewable fuels and chemicals by thermal processing of biomass. *Chemical Engineering Journal* 2003, 19, 87–102.
34. Maniatis, K., State of the art on thermochemical conversion technologies. In *2nd World Conference on Biomass for Energy, Industry and Climate Protection*, Van Swaaij, W. P. M., Fjällström, T., Helm, P., Grassi, A., Eds., Vol. 1, ETA-Florence, Rome, 2004, pp. 16–21.
35. McKendry, P., Energy production from biomass (part 3): Gasification technologies. *Bioresource Technology* 2002, 83, 55–63.
36. Hamelinck, C.; Faaij, A. P. C., Future prospects for production of methanol and hydrogen from biomass. *Journal of Power Sources* 2002, 111(1), 1–22.
37. Arvelakis, S.; Gehrmann, H.; Beckmann, M.; Koukios, E. G., Effect of leaching on the ash behavior of olive residue during fluidized bed gasification. *Biomass and Bioenergy* 2002, 22, 55.
38. Demirbas, A., Hydrogen production from biomass by the gasification process. *Energy Sources* 2002, 24, 59.
39. Van der Drift, A.; Boerrigter, H.; Coda, B.; Cieplik, M. K.; Hemmes, K., *Entrained Flow Gasification of Biomass*. ECN-C—04-039, ECN, Petten 2004.
40. Prins, M. J., Thermodynamic Analysis of Biomass Gasification and Torrefaction. PhD Thesis, Technical University of Eindhoven, Eindhoven, 2005.
41. Zanzi, R.; Tito Ferro, D.; Torres, A.; Beaton Soler, P.; Björnbom, E., Biomass torrefaction. In *2nd World Conference on Biomass for Energy, Industry and Climate Protection*, Vol. 1, Van Swaaij, W. P. M., Fjällström, T., Helm, P., Grassi, A., Eds., ETA-Florence, Rome, 2004, pp. 859–862.
42. de Jong, W., *Nitrogen Compounds in Pressurised Fluidised Bed Gasification of Biomass and Fossil Fuels*. Delft University of Technology, Delft, 2005.
43. Olofsson, I.; Nordin, A.; Söderlind, U., *Initial Review and Evaluation of Process Technologies and Systems Suitable for Cost-Efficient Medium-Scale Gasification for Biomass to Liquid Fuels*, ETPC Report 05-02, University of Umeå/Mid Sweden University, Umeå, 2005, p. 90.

44. van der Nat, K.; Woudstra, N.; Spliethoff, H., Evaluation of several biomass gasification processes for the production of a hydrogen rich synthesis gas. *International Hydrogen Energy Congress and Exhibition IHEC*, Istanbul, 2005.
45. Hofbauer, H.; Rauch, R.; Loeffler, G.; Kaiser, S.; Fercher, E.; Tremmel, H., Six years experience with the FICFB-gasification process. *12th European Conference and Technology Exhibition on Biomass for Energy, Industry and Climate Protection*, Amsterdam, 2002.
46. Devi, L.; Ptasinski, K. J.; Janssen, F. J. J. G., Primary measures for tar elimination in biomass gasification processes. *Biomass and Bioenergy* 2003, 24, 125–140.
47. Paisley, M. A.; Welch, M. J., Biomass gasification combined cycle opportunities using the future energy Silvagas® gasifier coupled to Alstom's industrial gas turbines. *ASME Turbo Expo 2003*, ASME, Paper GT2003-38294, Atlanta, GA, 2003.
48. Paisley, M. A.; Overend, R. P., The SilvaGas® process from future energy resources—A commercialization success. *12th European Conference and Technology Exhibition on Biomass for Energy, Industry and Climate Protection*, Amsterdam, 2002.
49. Diho, R. J.; Patel, J. G., U-gas process for production of hydrogen from coal. *International Journal of Hydrogen Energy* 1983, 8(3), 175–182.
50. Rabah, M. A.; Eldighidy, S. M., Low-cost hydrogen production from waste. *6th World Hydrogen Energy Conference*, Vienna, 1986, pp. 1362–1370.
51. Larson, E. D.; Katofsky, R. E., *Production of Methanol and Hydrogen from Biomass*. Center for Energy and Environmental Studies, School of Engineering and Applied Science, Princeton University, Princeton, NJ, 1992.
52. Waldheim, L.; Morris, M.; Stahl, K.; Gardmark, L., CHRISGAS PROJECT—manufacture of a clean and hydrogen-rich gas through biomass gasification and hot gas upgrading. In *Science in Thermal and Chemical Conversion*, Bridgwater, A. V.; Boocomb, D. G. B., Eds., Victoria, 2006, pp. 930–941.
53. www.chrisgas.com (accessed May 9, 2007).
54. Marquard-Möllenstedt, T.; Sichler, P.; Specht, M.; Michel, M.; Berger, R.; Hein, K. R. G.; Höftberger, E.; Rauch, R.; Hofbauer, H., *New Approach for Biomass Gasification to Hydrogen*. In *2nd World Conference on Biomass for Energy, Industry and Climate Protection, Rome*, Van Swaaij, W. P. M., Fjällström, T., Helm, P., Grassi, A., Eds., ETA-Florence, Rome, 2004, pp. 758–762.
55. Dürrbeck, M.; Marquard-Möllenstedt, T.; Sichler, P.; Zuberbühler, U., Der AER-prozess: Neues thermochemisches biomasse-vergasungsverfahren mit flexibelen edukt- und produktströmen. *DGMK Conference Energetische Nutzung von Biomassen*, DGMK, Velen, 2006, pp. 59–66.
56. Lin, S. Y.; Harada, M.; Suzuki, Y.; Hatano, H., Hydrogen production from coal by separating carbon dioxide during gasification. *Fuel* 2002, 81, 2079–2085.
57. Hanaoka, T.; Yoshida, T.; Fujimoto, S.; Kamei, K.; Harada, M.; Suzuki, Y.; Hatano, H.; Yokoyama, S.-Y.; Minowa, T., Hydrogen production from woody biomass by steam gasification using a CO_2 sorbent. *Biomass and Bioenergy* 2005, 28, 63–68.
58. Anthony, E. J.; Wang, J., Capturing CO_2 in coal-fired combustors using CaO-based sorbents. *5th International Symposium on Coal Combustion*, Nanjing, 2003.
59. www.dm1-2.de/(accessed May 8, 2007).
60. Hamel, S.; Hasselbach, H.; Krumm, W.; Lotter, D.; Mertens, C., Betriebsergebnisse der IPV-Technikumsanlage. *Energetische Nutzung von Biomassen*, DGMK, Velen, 2004, pp. 65–72.
61. Metz, T.; Kuhn, S.; Karellas, S.; Stocker, R.; Karl, J.; Hein, D., Experimental results of the biomass heatpipe reformer. In *2nd World Conference on Biomass for Energy, Industry and Climate Protection*, Van Swaaij, W. P. M.; Helm, P.; Grassi, A., Eds., Vol. 1, ETA Florence, Rome, 2004, pp. 773–776.
62. Karellas, S.; Metz, T.; Kuhn, S.; Karl, J., Online analysis of the tar content of the product gas from biomass gasification. Application on the BIOHPR. *14th European Biomass Conference & Exhibition, Biomass for Energy, Industry and Climate Protection*, ETA-Renewable Energies, Paris, 2005, pp. 624–627.
63. Ratafia-Brown, J.; Manfredo, L.; Hoffmann, J.; Ramezan, M., *Major Environmental Aspects of Gasification-based Power Generation Technologies*. Science Applications International Corporation, Final Report for the DOE National Energy Technology Laboratory, La Jolla, CA, December 2002.

64. U.S. Department of Energy and Wabash River Coal Gasification Project Joint Venture, *The Wabash River Coal Gasification Repowering Project—A 262 MWe Commercial Scale Integrated Gasification Combined Cycle Power Plant (An Update)*; Topical Report Number 20, September 2000.
65. Hooper, M., Waste coal to ultra clean fuels applying the shell coal gasification process. In *Gasification Technologies Conference*, www.gasification.org/Presentations/2003.htm (accessed May 11, 2007), San Francisco, CA, 12–15 October, 2003.
66. Zuideveld, P., Overview of shell global solution's worldwide gasification developments. *Gasification Technologies Conference*, www.gasification.org/Presentations/2003.htm (accessed May 11, 2007), San Francisco, CA, October 12–15, 2003.
67. Van Dongen, A.; Kanaar, M., Co-gasification at the Buggenum IGCC power plant. *Energetische Nutzung von Biomassen*, DGMK, Velen, 2006, pp. 57–58.
68. www.choren.com (accessed May 10, 2007).
69. Hamelinck, C.; Faaij, A.; den Uil, H.; Boerrigter, H., Production of FT transportation fuels from biomass; technical options, process analysis and optimisation and development potential. *Energy, the International Journal* 2004, 29(11), 1743–1771.
70. Corella, J.; Aznar, M. P.; Gil, J.; Caballero, M. A., Biomass gasification in fluidized bed: Where to locate the dolomite to improve gasification? *Energy and Fuels* 1999, 13(6), 1122–1127.
71. Heidenreich, S.; Nacken, M., Katalytische filterelemente zum abbau von teeren in rohgasen aus der biomassevergasung. *Energetische Nutzung von Biomassen*, DGMK, Velen, 2004, pp. 187–194.
72. Boerrigter, H.; Den Uil, H.; Calis, H. P., Green diesel from biomass via Fischer-Tropsch synthesis: New insights in gas cleaning and process design. In *Pyrolysis and Gasification of Biomass and Waste, Strasbourg*, Bridgwater, A. V., Ed., CPL Press, Strasbourg, 2003, pp. 371–383.
73. Spiegel, R. J.; Preston, J. L., Test results for fuel cell operation on anaerobic digester gas. *Journal of Power Sources* 2000, 86, 283–288.
74. Bossart, S. J.; Cicero, D. C.; Zeh, C. M.; Bedick, R. C., *Gas Stream Cleanup*, Technical Status Report. Energy Technology Center, Morgantown, WV, 1990, p. 66.
75. *Fuel Cell Handbook 6th edition*, Report prepared by EG&G Services, Parsons, Inc. and Science Applications International Corporation under contract no. DE-AM26-99FT40575 for the U.S. Department of Energy, National Energy Technology Laboratory, Morgantown, WV, October 2002.
76. Wojcik, A.; Middleton, H.; Damopoulos, I.; Van Herle, J., Ammonia as a fuel in solid oxide fuel cells. *Journal of Power Sources* 2003, 118(1–2), 342–348.
77. Norheim, A.; Waernhus, I.; Broström, M.; Hustad, J. E.; Vik, A., Experimental studies on the influence of H_2S on solid oxide fuel cell performance at 800°C. *Energy and Fuels* 2007, 21(2), 1098–1101.
78. Aravind, P. V.; Woudstra, N.; Ouweltjes, J. P.; Andries, J.; De Jong, W.; Rietveld, G.; Spliethoff, H., *Integration of Solid Oxide Fuel Cells with Biomass Gasifiers*, In *2nd World Conference on Biomass for Energy, Industry and Climate Protection*, Van Swaaij, W. P. M., Fjällström, T., Helm, P., Grassi, A., Eds., ETA-Florence, Rome, 2004, pp. 1058–1061.
79. Aravind, P. V.; Ouweltjes, J. P.; de Heer, E.; Woudstra, N.; Rietveld, G., Impact of biosyngas and its components on SOFC anodes. In Vol. 7, 2005, pp. 1459–1467.
80. Dayton, D. C.; Ratcliff, M.; Bain, R., *Fuel Cell Integration—A Study of the Impacts of Gas Quality and Impurities*. NREL, Golden, CO, 2001.
81. Blackader, W.; Rensfelt, E., Synthesis gas from wood and peat: The mino process. In *Thermochemical Processing of Biomass*, Bridgwater, A. V., Ed., Butterworth, London, 1984, pp. 137–149.
82. Baker, E.; Mudge, L.; Wilcox, W. A., Catalysis of gas phase reactions in steam gasification of biomass. In *Fundamentals of Thermochemical Biomass Conversion*, Overend, R. P. et al., Ed., Elsevier Applied Science, London, 1985, pp. 1194–1208.
83. Baker, E.; Mudge, L.; Brown, M., Steam gasification of biomass with nickel secondary catalysts. *Industrial and Engineering Chemistry Research* 1987, 26, 1335–1339.
84. Mudge, L. K.; Baker, E. G.; Mitchell, D. H.; Brown, M. D., Catalytic steam gasification of biomass for methanol and methane production. *Journal of Solar Energy Engineering* 1985, 107, 88–92.
85. Mudge, L. K.; Baker, E. G.; Brown, M.; Wilcox, W., Catalytic destruction of tars in biomass-derived gases. In *Research in Thermochemical Biomass Conversion*, Bridgwater, A. V.; Kuester, J. L., Eds., Elsevier Applied Science, London, 1988, pp. 1141–1155.

86. Caballero, M. A.; Aznar, M. P.; Gil, J.; Martín, J. A.; Francés, E.; Corella, J., Commercial steam reforming catalysts to improve biomass gasification with steam-oxygen mixtures. 1. Hot gas upgrading by the catalytic reactor. *Industrial and Engineering Chemistry Research* 1997, 36(12), 5227–5239.
87. Aznar, M. P.; Caballero, M. A.; Gil, J.; Martín, J. A.; Corella, J., Commercial steam reforming catalysts to improve biomass gasification with steam-oxygen mixtures. 2. Catalytic tar removal. *Industrial and Engineering Chemistry Research* 1998, 37(7), 2668–2680.
88. Brown, R. C.; Sadaka, S. S.; Norton, G.; Xu, M.; Bown, N., Thermochemical production of hydrogen from biomass. In *2nd World Conference on Biomass for Energy, Industry and Climate Protection*, Van Swaaij, W. P. M., Fjällström, T., Helm, P., Grassi, A., Eds., ETA-Florence, Rome, 2004, pp. 1144–1147.
89. Narvaez, I.; Orio, A.; Aznar, M. P.; Corella, J., Biomass gasification with air in an atmospheric bubbling fluidized bed. *Industrial and Engineering Chemistry Research* 1996, 35(7), 2110–2120.
90. Zhang, R.; Brown, R. C.; Suby, A., Thermochemical generation of hydrogen from switchgrass. *Energy and Fuels* 2004, 18(1), 251–256.
91. Aznar, M. P.; Caballero, M. A.; Corella, J.; Molina, G.; Toledo, J. M., Hydrogen production by biomass gasification with steam-O_2 mixtures followed by a catalytic steam reformer and a CO-Shift system. *Energy and Fuels* 2006, 20, 1305–1309.
92. Caballero, M. A.; Aznar, M. P.; Gil, J.; Martín, J. A.; Corella, J., CO-shift catalytic beds after a biomass gasifier and a steam-reforming catalytic reactor to get new and interesting exit gas compositions. In *10th European Conference on Biomass for Energy and Industry*, Kopetz, H.; Weber, T.; Palz, W.; Chartier, P.; Ferrero, G. L., Eds., C.A.R.M.E.N., Würzburg, 1998, pp. 1789–1793.
93. Li, X.; Adris, A.-E. M.; Lim, C. J.; Grace, J. R.; Ellis, N., Enhanced hydrogen production from circulating fluidized bed biomass gasification by double equilibrium shift. In *Circulating Fluidized Bed Technology VIII*, Cen, K., Ed., Int. Academic Publishers, Hangzhou, 2005, pp. 499–506.
94. Spath, P. L.; Dayton, D. C., *Preliminary Screening—Technical and Economic Assessment of Synthesis Gas to Fuels and Chemicals with Emphasis on the Potential for Biomass-Derived Syngas*. NREL/TP-510-34929; NREL, Golden, CO, 2003.
95. Hamelinck, C. N.; Faaij, A. P. C., Future prospects of production of methanol and hydrogen from biomass. *Journal of Power Sources* 2002, 111(1), 1–22.
96. Meier, D., Flash-pyrolyse zur verflüssigung von biomasse—stand der technik. *Energetische Nutzung von Biomassen*, DGMK, Velen, 2002.
97. Bridgwater, A. V.; Peacocke, G. V. C., Fast pyrolysis processes for biomass. *Renewable and Sustainable Energy Reviews* 2000, 4, 1–73.
98. Mohan, D.; Pittman, C. U.; Steele, P. H., Pyrolysis of wood/biomass for bio-oil: A critical review, *Energy and Fuels* 2006, 20, 848.
99. Henrich, E.; Raffelt, K.; Stahl, R.; Weirich, F., Clean syngas from bio-oil/char-slurries. In *Science in Thermal and Chemical Biomass Conversion*, Bridgwater, A. V.; Boocock, D. G. B., Eds., CPL press, Victoria, 2004, pp. 1565–1579.
100. Bridgwater, A. V., Principles and practice of biomass fast pyrolysis processes for liquids. *Journal of Analytical and Applied Pyrolysis* 1999, 51, 3.
101. Meier, D.; Schöll, S.; Klaubert, H.; Markgraf, J., Betriebsergebnisse der ersten BTO-Anlage zur ablativen Flash-Pyrolyse von Holz mit Energiegewinnung in einem BHKW. In *Energetische Nutzung von Biomassen*, Velen, 2006, pp. 115 120.
102. Schöll, S.; Klaubert, H.; Meier, D., Bio-oil from a new ablative pyrolyser. In *Science in Thermal and Chemical Biomass Conversion*, Bridgwater, A. V.; Boocock, D. G. B. Eds., Vol. 2, pp. 1372–1379, CPL Press, Victoria, 2004.
103. Meier, D.; Faix, O., Fast pyrolysis: A route for energy and chemicals from wood—fluidized vs. ablative pyrolysis. In *Wood and Biomass Utilization for the Carbon Uptake*, Seoul, National University, Seoul, 2005, pp. 55–68.
104. Peacocke, G. V. C.; Bridgwater, A. V.; Brammer, J. G., Techno-economic assessment of power production from the Wellman and BTG fast pyrolysis processes. In *Science in Thermal and Chemical Conversion*, Vol. 2, Bridgwater, A. V.; Boocock, D. G. B., Eds., CPL Press, Victoria, 2004, pp. 1785–1802.
105. Rostrup-Nielsen, J. R., Syngas in perspective, *Catalysis Today* 2002, 71, 243.

106. Czernik, S.; French, R.; Feik, C.; Chornet, E., Hydrogen by catalytic steam reforming of liquid byproducts from biomass thermoconversion processes, *Industrial and Engineering Chemistry Research* 2002, 41, 4209.
107. Steinberg, M., *Clean Carbon and Hydrogen Fuels from Coal and Other Carbonaceous Raw Materials*, BNL-39630; Brookhaven National Laboratory, Upton, New York, 1987.
108. Steinberg, M., The conversion of carbonaceous materials to clean carbon and co-product gaseous fuel. *5th European Conference on Biomass for Energy and Industry*, Lisbon, 1989.
109. Dong, Y.; Steinberg, M., Hynol—an economical process for methanol production from biomass and natural gas with reduced CO_2 emission. In *10th World Hydrogen Energy Conference*, Block, D. L., Veziroglu, T. N. Eds., Beach, Florida, June 20–24, 1994, pp. 495–504.
110. Dong, Y.; Borgwardt, R. H., Biomass reactivity in gasification by the hynol process. *Energy and Fuels* 1998, 12(13), 479–484.
111. Takahashi, K.; Takezawa, N.; Kobayashi, H., The mechanism of steam reforming of vehicles. *Applied Catalysis* 1982, 2, 363–366.
112. De Wild, P. J.; Verhaak, M. J. F. M., Catalytic production of hydrogen from methanol. *Catalysis Today* 2000, 60, 3–10.
113. Laniecki, M.; Kazmierczak-Rosik, K., Steam-methanol reforming over zeolitic materials. In *12th World Hydrogen Energy Conference*, Bolcich, J. C.; Veziroglu, T. N. Eds., Buenos Aires, June 21–25, 1998, pp. 661–668.
114. Qi, A.; Hong, X.; Wang, S.; Fu, G.; Wu, D., Key factors in hydrogen generation system by methanol partial oxidation. In *13th World Hydrogen Energy Conference*, Mao, Z. Q.; Veziroglu, T. N., Eds., Beijing, 2000, pp. 246–252.
115. Newson, E.; Mizsey, P.; Truong, J.-B.; Hettinger, P.; Roth, F. von.; Schucan, T. H.; Geissler, K.; Schildhauen, T., Hydrogen from liquid energy carriers: The catalytic reforming of methanol and hydrocarbons. In *The International Hydrogen Energy Forum 2000, Policy-Business-Technology*, Munich, 2000, p. 351.
116. Leweler, S.; Stell, S.; George, R., Conversion of landfill gas, methanol, or natural gas to hydrogen for use in a hydrogen refueling system. In *10th Annual U.S. Hydrogen Meeting of the National Hydrogen Association*, Vienna, VA, 1999, pp. 293–298.
117. Garcia, E. Y.; Laborde, M. A., Hydrogen production by the steam reforming of ethanol: Thermodynamic analysis. *International Journal of Hydrogen Energy* 1991, 16(5), 307–312.
118. Vasudeva, K.; Mitra, N.; Umasankar, P.; Dhingra, S. C., Steam reforming of ethanol for hydrogen production. *International Journal of Hydrogen Energy* 1996, 21(1), 13–18.
119. Fishtik, I.; Alexander, A.; Datta, R.; Geana, D., A thermodynamic analysis of hydrogen production by steam reforming of ethanol via response reactions. *International Journal of Hydrogen Energy* 2000, 25, 31–45.
120. Luengo, C. A.; Ciampi, G.; Cencig, M. O.; Steckelberg, C.; Laborde, M. A., A novel catalyst system for ethanol gasification. *International Journal of Hydrogen Energy* 1992, 17(9), 677–681.
121. Marino, F. J.; Cerrella, E. G.; Duhalde, S.; Jobbagy, M.; Laborde, M. A., Hydrogen from steam reforming of ethanol. Characterization and performance of copper-nickel supported catalysts. *International Journal of Hydrogen Energy* 1998, 23(12), 1095–1101.
122. Cavallaro, S.; Freni, S., Ethanol steam reforming in a molten carbonate fuel cell. A preliminary kinetic investigation. *International Journal of Hydrogen Energy* 1996, 21, 465–469.
123. Cavallaro, S.; Freni, S., Syngas and electricity production by an integrated autothermal reforming molten carbonate fuel cell system. *Journal of Power Sources* 1998, 76, 190–196.
124. Verykios, X., Process for the Production of Hydrogen and Electrical Energy from Reforming of Bio-ethanol (Int. Pat. WO 99/61369). 1998.
125. Toci, F.; Modica, G., Hydrogen production by in-situ cracking of steam-bioethanol mixtures combining electronic stimulation with chemical catalysts. In *9th European Bioenergy Conference*, Chartier, Ph.; Ferrero, G. L.; Henius, U. M.; Hultberg, S.; Sachau, J.; Wiinblad, M. Eds., Copenhagen, 1996, p. 425.
126. Takanabe, K.; Aika, K.-I.; Inazu, K. T. B.; Seshan, K.; Lefferts, L., Steam reforming of acetic acid as a biomass derived oxygenate: Bifunctional pathway for hydrogen formation over Pt/ZrO_2 catalysts. *Journal of catalysis* 2006, 243(2), 263–269.

127. Huber, G. W.; Chheda, J. N.; Barrett, C. J.; Dumesic, J. A., Production of liquid alkanes by aqueous-phase processing of biomass-derived carbohydrates. *Science* 2005, 308(5727), 1446–1450.
128. Huber, G. W.; Dumesic, J. A., An overview of aqueous-phase catalytic processes for production of hydrogen and alkanes in a biorefinery. *Catalysis Today* 2006, 111, 119.
129. Cortright, R. D.; Davda, R. R.; Dumesic, J. A., Hydrogen from catalytic reforming of biomass-derived hydrocarbons in liquid water. *Nature* 2002, 418, 964.
130. Shabaker, J. W.; Davda, R. R.; Huber, G. W.; Cortright, R. D.; Dumesic, J. A., Aqueous-phase reforming of methanol and ethylene glycol over alumina-supported platinum catalysts. *Journal of Catalysis* 2003, 215, 344.
131. Davda, R. R.; Shabaker, J. W.; Huber, G. W.; Cortright, R. D.; Dumesic, J. A., A review of catalytic issues and process conditions for renewable hydrogen and alkanes by aqueous-phase reforming of oxygenated hydrocarbons over supported metal catalysts. *Applied Catalysis B* 2005, 56, 171.
132. Davda, R. R.; Shabaker, J. W.; Huber, G. W.; Cortright, R. D.; Dumesic, J. A., Aqueous-phase reforming of ethylene glycol on silica-supported metal catalysts. *Applied Catalysis B* 2003, 43, 13–26.
133. Shabaker, J. W.; Dumesic, J. A., Kinetics of aqueous-phase reforming of oxygenated hydrocarbons: Pt/Al_2O_3 and Sn-modified Ni catalysts. *Industrial and Engineering Chemistry Research* 2004, 43, 3105.
134. Matsumura, Y.; Minowa, T., Fundamental design of a continuous biomass gasification process using a supercritical water fluidized bed. *Journal of Hydrogen Energy* 2004, 29, 701.
135. Mok, W. S. L.; Jerry, M.; Antal, J., Uncatalyzed solvolysis of whole biomass hemicellulose by hot compressed liquid water. *Industrial and Engineering Chemistry Research* 1992, 31, 1157.
136. Minowa, T.; Zhen, F.; Ogi, T., Cellulose decomposition in hot-compressed water with alkali or nickel catalyst. *Journal of Supercritical Fluids* 1998, 13, 253.
137. Minowa, T.; Ogi, T., Hydrogen production from cellulose using a reduced nickel catalyst. *Catalysis Today* 1998, 45, 411.
138. Minowa, T.; Inoue, S., Hydrogen production from biomass by catalytic gasification in hot compressed water. *Renewable Energy* 1999, 16, 1114.
139. Modell, M., Gasification and liquefaction of forest products in supercritical water. In *Fundamentals of Thermochemical Biomass Conversion*, Overend, R. P.; Milne, T. A., Eds., Elsevier Applied Science Publishers, London, 1985, pp. 95–119.
140. Yu, D. H.; Aihara, M.; Antal, M. J., Hydrogen production by steam reforming glucose in supercritical water. *Energy and Fuels* 1993, 7, 574.
141. Xu, X.; Matsumura, Y.; Stenberg, J.; Antal, M. J., Carbon-catalyzed gasification of organic feedstocks in supercritical water. *Industrial and Engineering Chemistry Research* 1996, 35, 2522–2530.
142. Gadhe, J. B.; Gupta, R. B., Hydrogen production by methanol reforming in supercritical water: Suppression of methane formation. *Industrial and Engineering Chemistry Research* 2005, 44, 4577 4585.
143. Antal, M. J.; Allen, S. G.; Schulman, D.; Xu, X., Biomass gasification in supercritical water. *Industrial and Engineering Chemistry Research* 2000, 39, 4040–4053.
144. Boukis, N.; Galla, U.; Diem, V.; Dinjus, E., Biomass gasification in supercritical water: First results of the pilot plant, In *Science in Thermal and Chemical Biomass Conversion*, Bridgwater, A. V.; Boocock, D. G. B., Eds., CPL Press, Victoria, 2004, pp. 975–990.
145. Boukis, N.; Galla, U.; Jesus, P. D.; Müller, M.; Dinjus, E., Wasserstoff- und methanerzeugung aus maissilage in Überkritischem wasser. Ergebnisse von labor und pilotversuchen. In *DGMK Conference Energetische Nutzung von Biomassen*, DGMK, Velen, 2006, pp. 91–98.
146. Hacker, V., Fankhauser, R.; Faleschini, G.; Fuchs, H.; Friedrich, K.; Muhr, M.; Kordesch, K., Hydrogen production by steam-iron process. *Journal of Power Sources* 2000, 86, 531–535.
147. Sime, R.; Kuehni, J.; D'Souza, L.; Elizondo, E.; Biollaz, S., The redox process for producing hydrogen from woody biomass. *International Journal of Hydrogen Energy* 2003, 28, 491–498.
148. Biollaz, S.; Sturzenegger, M.; Stucki, S., Redox Process for the production of clean hydrogen from biomass. In *Progress in Thermochemical Biomass Conversion*, Seefeld, Tirol, 2000.
149. http://tccb.tnw.utwente.nl/(accessed May 10, 2007).

7

Use of Solar Energy to Produce Hydrogen

Neelkanth G. Dhere and Rajani S. Bennur

CONTENTS

7.1 Introduction

Hydrogen is an ideal, clean, carbon-free carrier of energy that produces only water vapor as a waste product and has potential applications in automobiles, airplanes, and also in home-heating. The total annual global consumption of energy in 1998 was estimated to be 112.5 terakilowatt hour (TkWh), that is, 384 quads (quadrillion British thermal unit [Btu]) equivalent to mean power generation during the year of 12.8 terawatt (TW), of which the United States consumed 29 TkWh, that is, 99 quads equivalent to mean power generation during the year of ~3.3 TW. The projected mean power generation for the years 2050 and 2100 has been estimated to be 28 and 46 TW, respectively. By far, sunlight provides the largest of all carbon-neutral energy sources. In fact, 123.05 TkWh, that is, 448 quads equivalent to mean power generation during the year of 14 TW of solar energy falls on the earth every hour. Thus, more energy from sunlight strikes the earth in 1 h than all the energy consumed on the planet in a year [1]. The most successful technologies taking advantage of this resource are photovoltaics (PVs; solar electricity), a $10 billion industry that is currently growing at a rate of 35–40% each year. Continued growth of the PV sector at a rate of ~25% would increase the production level from 1.7 GW in 2005 to 380 GW in 2030, and thus would satisfy a significant fraction of the world energy demand [2,3]. Moreover, among the renewable sources, PVs have the highest potential to reduce the costs compared to biomass, geothermal, wind, and solar thermal [1–3]. According to the World Petroleum Life-Cycle prediction by Duncan and Youngquist [4], world oil production may already have peaked in 2006, whereas the Organization of Petroleum Exporting Countries (OPEC) oil production will exceed non-OPEC production beginning in 2007. Of course, as recognized by them, the peak year may shift a little. Because the oil and natural gas prices are fluctuating in the range $55–110/bbl and $7.5/million Btu, respectively, electricity generation is shifting increasingly toward coal. It is therefore essential to accelerate the development of techniques for the production of hydrogen from an abundant source of energy, the sun, so as to slow down the threat of global warming [5,6]. Photoelectrolysis of water and especially the use of semiconductor–electrolyte interfaces illuminated with sunlight for the production of hydrogen from water and other suitable solvents have been reviewed in several excellent publications [7–18]. At present, only about 5% of the commercial hydrogen production is primarily via water electrolysis, whereas the other 95% is mainly derived from fossil fuels [16]. This does not represent in-house consumption of hydrogen such as oil refineries and ammonia plants where the bulk of hydrogen is consumed.

Fundamental studies of the semiconductor–electrolyte interface could only be initiated with the availability of well-defined single crystals of germanium and silicon around 1960, and of II-VI and III-V compounds, stable oxide semiconductors, and metal chalcogenides (layered compounds) afterward. Studies of microheterogeneous systems such as suspensions and colloidal solutions of semiconductor particles were initiated in 1976–1978. The semiconductors absorb light very effectively; the absorption coefficients of semiconductors range between 10^{-3} cm^{-1} for indirect bandgap semiconductors and 10^{-5}–10^{-6} cm^{-1} for direct bandgap semiconductors.

There are two principal methods for the electrolysis of water to generate hydrogen viz. active, that is, photoassisted and passive or dark and nonphotoassisted. The active method consists of utilization of photogenerated charge carriers in the electrolysis of water and other products. In the passive method, the electrolysis is carried out in the dark at low (80°C) temperatures using an alkaline electrolyte such as NaOH or proton exchange membrane cells involving polymeric sulfonic acids, at intermediate (200–500°C) temperatures for which suitable electrolytes are still being sought, and at high (>800°C) temperatures using oxide electrolytes such as yttria-stabilized zirconia. Photovoltaic technology including the electrochemical photovoltaic cells could be useful in this method as a source of electricity. This chapter deals specifically with the active method. The electrochemical photovoltaic cells are also discussed because of the compatible technique and the potential use as the source of electricity for the active as well as passive electrolyses.

7.2 Photoelectrochemical Devices

7.2.1 Semiconductor–Electrolyte Junction

A semiconductor–electrolyte junction is formed when a semiconductor is immersed in an appropriate electrolyte. The chemical potential of a semiconductor is given by the Fermi level, whereas the chemical potential of a liquid electrolyte is determined by its redox potential. It is instructive to compare the references for the energy scales for the two systems. In electrochemistry, the redox potentials are measured against the standard redox potential of the hydronium ion/hydrogen gas couple or normal hydrogen electrode (NHE) also termed *standard hydrogen electrode* (SHE). (Two other standards viz. saturated calomel [$HgCl_2$] electrode [SCE] or silver chloride [Ag/AgCl] [19] are also used.) However, the potential of vacuum is taken as the zero potential in solid-state physics. Conversions can be carried out easily by remembering that the standard H^+/H redox potential with respect to vacuum is +4.5 eV. (The potentials of standard calomel and Ag/AgCl electrodes with respect to the standard H^+/H redox potential are 0.234 and 0.197 V, respectively.) If the initial Fermi level of an n-type semiconductor is above (more negative with respect to) the initial electrolyte redox potential, electrons will flow from the n-type semiconductor to the electrolyte. The charge transfer leaves behind a depletion region in the semiconductor and a potential barrier. The charge in the depletion layer is compensated by a charge of the opposite sign that is induced in the electrolyte within a localized ~1 nm thick layer, known as the Helmholtz layer. It is composed of oriented water molecule dipoles and electrolyte ions adsorbed at the electrode surface. The potential drop, V_H, across the Helmholtz layer (also known as the Helmholtz barrier) is determined by the nature of the aqueous environment of the electrolyte and the properties of the photoelectrode surface. The performance characteristics of photoelectrochemical (PEC) cells depend, to a large extent, on the potential drop, V_H, across the Helmholtz layer. The junction has characteristics similar to those of a Schottky barrier formed between a semiconductor and a metal. The semiconductor has a net space charge of immobile, positively charged donor ions uncompensated by the mobile electron cloud. Figure 7.1 shows the formation of the depletion and Helmholtz layers, field, and potential across a semiconductor–electrolyte interface. The formation of the double layer leads to conduction and valence band bending and development of a field that opposes further charge transfer (Figure 7.1). Finally, equilibrium is reached when there is no net transfer of charge across the junction. The equilibrium potential barrier is formed with a built-in voltage V_B as a result of the band bending. The energy level diagram of a semiconductor–electrolyte junction and solid-state and electrochemical energy scales are depicted in Figure 7.2. The depletion layer width in a semiconductor diode of silicon having doping densities on p- and n-sides of 1.4×10^{15} and 1.4×10^{18} cm^{-3} is ~0.86 μm, the width being higher at lower doping densities. In contrast, in concentrated (>0.5 M) electrolytes, the charged Helmholtz layer is mostly localized. The flatband potential U_{fb} is an important parameter that is related to the properties of both the

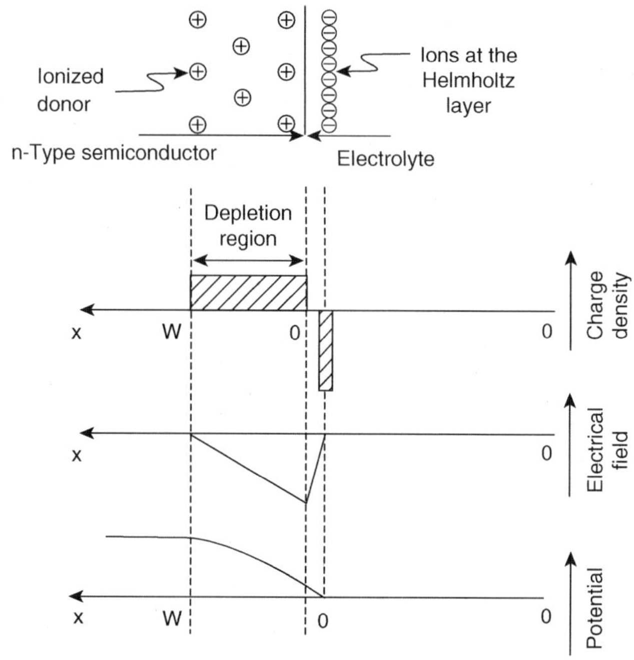

FIGURE 7.1
Depletion and Helmholtz layers, field and potential across a semiconductor–electrolyte interface.

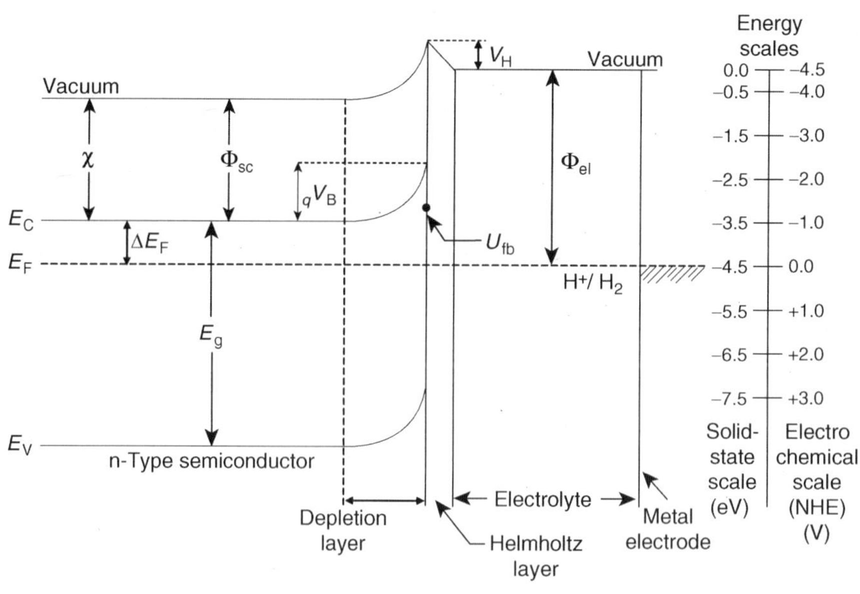

FIGURE 7.2
Energy level diagram of a semiconductor–electrolyte junction and solid-state and electrochemical energy scales.

semiconductor and the electrolyte. It is the potential, which if applied to the semiconductor would flatten the conduction and valence bands. It is given by

$$U_{fb} = (\chi + \Delta E_F + V_H) - 4.5 = (\varphi_{sc} + V_H) - 4.5 \quad (7.1)$$

where χ is the electron affinity; ΔE_F the difference between the Fermi level E_F and the majority carrier band edge of the semiconductor (conduction band in the present case); φ_{sc} and φ_{el} are work functions of the semiconductor and the electrolyte, respectively, and measure the Fermi level potential with respect to vacuum; and 4.5 is the scaling factor [7,14,19].

Because of the adsorption equilibrium for H^+ and OH^- ions between the surface of semiconductors and an aqueous (aq) solution, the semiconductor surface attains the point of zero charge (PZC). The flat-band potential U_{fb} of most semiconductors including all oxides and also other compounds such as n- and p-type GaAs, p-type GaP, and n- and p-type InP in an aqueous solution is determined solely by pH and shifts proportionately with pH with a slope of −59 mV/decade, that is, pH, for example,

$$S_s - OH + H^+_{aq} = S_s - OH_2^+ \quad (7.2)$$

where $S_s - OH$ refers to the OH group present at the semiconductor surface [13]. The flat band potential U_{fb} remains nearly constant for n- and p-type Si and metal chalcogenides such as n-CdSe and CdTe at pH below 6, and n-CdS for pH about 10, most probably because the semiconductor surface has no OH group in these pH ranges [13]. Moreover, for metal chalcogenides the flat-band potential U_{fb} becomes more negative in the presence of S^{2-} and Se^{2-} ions in the solution, whereas it becomes more positive in the presence of Cd^{2+} ions in the solution [13].

7.2.2 Photoinduced Charge Transfer

Light can be completely absorbed in a thin layer of a solid electrode. Only semiconductors are useful in photoelectrochemical cells because the lifetimes of excited states in metals are very short. Electron–hole pairs are generated when the semiconductor is illuminated with photons of energy $h\nu > E_g$, where E_g is the semiconductor bandgap. Electrons and holes excited to energy levels far away from the band edges are thermalized very quickly within 10^{-12}–10^{-13} s. Hence, all light quanta with larger energy than E_g contribute only the bandgap energy to the photoconversion. The sunlight that has a typical spectrum at the earth's surface is termed *air mass 1.5 global* (AM 1.5 global) as shown in Figure 7.3 [8,20]. Here air mass (number) defines the number of times sunlight traversed the thickness of the earth's air atmosphere; global defines the total radiation from direct and diffuse radiation. Therefore, semiconductors with bandgaps in the range 1–2 eV are more useful than those of higher or lower bandgaps for the conversion of solar energy. The minority charge carriers, viz. holes that are generated within a diffusion length of the depletion region and diffuse into it, together with those generated in the depletion region itself are separated by the built-in field and are swept to the semiconductor surface, and can produce anodic oxidation by recombining with the electrons provided by the negatively charged ions reaching the semiconductor–electrolyte interface. The majority charge carriers, viz. electrons that are carried to the semiconductor bulk, flow through the external circuit and can participate in a cathodic reduction reaction on being injected into the electrolyte at the counterelectrode. The semiconductor potential is driven toward the flat-band potential by illumination. As discussed in Section 7.2.3, the stability of a semiconductor electrode becomes an issue because of the possibility of strong oxidizing and reducing reactions due to holes and electrons.

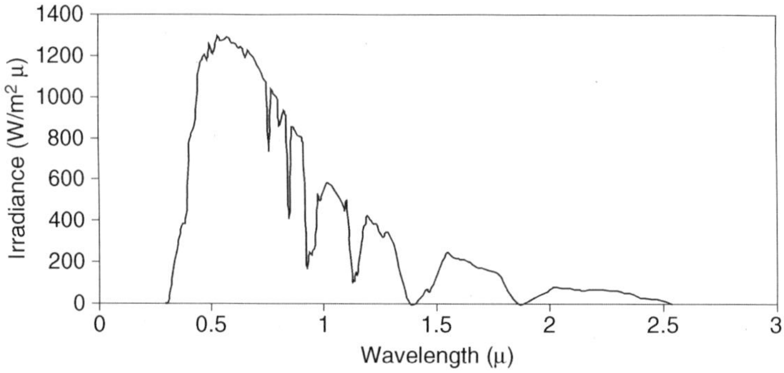

FIGURE 7.3
AM 1.5 global solar spectrum.

7.2.3 Liquid Junction Photovoltaic Cells

In the liquid junction PV cells, there is only one redox couple in the electrolyte. There is no chemical change; hence Gibb's free energy also does not change. The physics behind the operation of liquid junction solar cells is analogous to solid-state Schottky barrier cells. The carriers generated within the space-charge region and within a diffusion length of it are carried to the side where they become majority carriers. The carriers collected at the external contacts are available for work on an external load. The maximum possible open-circuit voltage V_{oc} is the difference between flat-band and redox potential of the predominant redox couple. The photocurrent depends on the bandgap of the semiconductor. The lower the bandgap the higher is the photocurrent. However, the open-circuit voltage decreases with decreasing bandgap. There is an optimum of the bandgap in the range 1.3–1.4 eV, where the photocurrent and photovoltage are optimum and the resultant PV conversion efficiency using sunlight at the earth's level attains the maximum value. The practical upper limit for the Schottky barrier–type PV solar cells is 10–12% because of the band bending, that is, the height of the potential barrier at the semiconductor/metal junctions are lower than the semiconductor bandgap [7]. In cells prepared with n-GaAs, surface recombination velocities have been reduced by a chemisorbed Ru^{3+} monolayer [7,21]. It can be exceeded in the liquid junction electrochemical solar cells because of the possibility of increasing the band bending by adjusting the redox potential of the electrolyte. Liquid junction cells are easy to fabricate. Polycrystalline materials usually present problems of reduced efficiency because of the imperfect contact between two solid phases. This problem can be overcome by an intimate contact between the electrolyte and the crystalline grains in a polycrystalline semiconductor.

It is essential to adjust the redox potentials carefully so as to avoid photocorrosion while providing enough band bending to collect the charge carriers efficiently. Hence, concentrations of redox components are kept high enough to scavenge carriers while preventing depletion near the electrode. It is necessary to carefully consider the constitution of the redox electrolyte. The electrolyte must be conductive and should absorb the incident solar radiation as little as possible. Thin cell structures are preferred for this purpose. Parkinson (table 1 in Ref. 15) has provided a compilation of some of the most efficient and stable liquid junction solar cells constructed by using single-crystal semiconductor electrodes in an aqueous electrolyte. It shows that several conventional semiconductors such as GaAs, Ga(AsP), InP, $CuInSe_2$, and CdSe has resulted in fairly high efficiencies.

Water is involved in most of the photodecomposition reactions. Hence, nonaqueous electrolytes such as methanol, ethanol, *N,N*-dimethylformamide, acetonitrile, propylene carbonate, ethylene glycol, tetrahydrofuran, nitromethane, benzonitrile, and molten salts such as $AlCl_3$-butyl pyridium chloride are chosen. The efficiency of early cells prepared with nonaqueous solvents such as methanol and acetonitrile were low because of the high resistivity of the electrolyte, limited solubility of the redox species, and poor bulk and surface properties of the semiconductor. Recently, reasonably efficient and fairly stable cells have been prepared with nonaqueous electrolytes with a proper design of the electrolyte redox couple and by careful control of the material and surface properties [7]. Results with single-crystal semiconductor electrodes can be obtained from table 2 in Ref. 15. Unfortunately, the efficiencies and stabilities achieved cannot justify the use of single-crystal materials. Table 2 in Ref. 15 summarizes the results of liquid junction solar cells prepared with polycrystalline and thin-film semiconductors [15]. As can be seen the efficiencies are fair. Thin films provide several advantages over bulk materials. Despite these possibilities, the actual efficiencies of solid-state polycrystalline thin-film PV solar cells exceed those obtained with electrochemical PV cells [22,23].

7.2.4 Photoelectrolysis Cells

In the photoelectrosynthetic cells, there are two effective redox couples in the electrolyte so that there is a net chemical change. When the net free energy change is positive, it is termed *photoelectrolysis*. However, when the net free energy change is negative, it is termed *photocatalysis*.

Splitting of water to generate hydrogen and oxygen is an endothermic reaction, and is accompanied by a large increase in Gibbs free energy $\Delta G°$ that can be provided by energetic photons and is expressed as

$$H_2O \rightarrow H_2 + \tfrac{1}{2}O_2 \quad \Delta G° = 237.1 \text{ kJ/mol equivalent to } 2.458 \text{ eV/molecule} \qquad (7.3)$$

Hydrogen generation from water is a two-electron process. However, for oxygen generation each atom releases two electrons, so that water splitting to generate a molecule of oxygen is a four-electron process. The average energy of the electron process is 118.5 kJ/mol, equivalent to $\Delta E_0^{H_2O} = 1.229$ eV/electron (lower hydrogen heating value reaction energy) based on liquid water conversion at 25°C.

In photoelectrolysis, the anodic reaction has a more positive redox potential than the cathodic reaction. An example is the photoassisted electrolysis of water to produce hydrogen and oxygen. In a typical photoelectrolysis cell, one electrode is a semiconductor, whereas the other is a metal. On illumination of the n-type semiconductor–electrolyte interface, the Fermi level in the semiconductor rises toward the flat-band potential. The maximum possible rise in the Fermi level with the electrodes shorted together is given by the flat-band potential U_{fb}. For direct water photoelectrolysis to occur on illumination, the flat-band potential U_{fb} must lie above (more negative compared to) the H^+/H_2 potential, whereas the O_2/H_2O potential should lie above the valence band of the n-type semiconductor. Usually, the valence band of the n-type semiconductor would lie below the O_2/H_2O potential. However, only a few semiconductors have U_{fb} above the H^+/H_2 potential, viz. n-$SrTiO_3$, Nb_2O_3, n-ZrO_2, n-Ta_2O_5, n-$BaTi0_3$, n-$KTaO_3$, n-InP, p-GaAs, n-CdS, n-SiC, and n-TiO_2 (in alkaline electrolyte) [7,17], hence direct hydrogen and oxygen could be generated in a Schottky barrier–type liquid junction cell. Table 7.1 provides values of flat-band potentials and positions of the valence band maximum E_{vs} and conduction band-minimum E_{cs} of oxide semiconductors, group IV and III/V semiconductors, and

TABLE 7.1

Flat-Band Potentials and Positions of the Valence Band Maximum E_{vs} and Conduction Band Minimum E_{cs} of Oxide Semiconductors, Group IV and III/V Semiconductors, and Mixed Oxide Semiconductors with Respect to the H^+/H_2 Scale, Where Minus Represents above Zero and Plus Represents below Zero

Material	Bandgap	U_{fb}	pH	E_{vs}	E_{cs}
n-Fe_2O_3	2.00	−0.1	14	+2.14	+0.14
n-ZnO	3.2	−0.85	14	+2.54	−0.66
n-$SrTiO_3$	3.2	−0.85	14	+2.54	−0.66
n-SnO_2	3.7	−0.45	14	+3.74	+0.04
n-TiO_2	3	−0.6	14	+2.64	−0.36
n-ZrO_2	5.0	−1.8	14	+2.44	−1.46
n-Ta_2O_5	~4.1	−1.2	14	+2.44	−0.66
p-SiC	2.9	+1.34	14	+1.34	−1.56
n-InP	1.25	−0.8	14	+0.45	−0.8
p-InP	1.25	−0.1	14	−0.1	−1.35
p-GaAs	1.43	0	14	0	−1.43
n-GaP	2.3	−1.9	14	+0.4	−1.9
p-GaP	2.3	+0.2	14	+0.2	−2.1
n-$BaTiO_3$	3.3	−0.46	14	+2.84	−0.46
n-$KTaO_3$	3.5	−1	14	+2.08	−1.42

Source: Part of the data has been extracted from Morrison, S.R., *Electrochemistry at Semiconductor and Oxidized Metal Electrodes*, Plenum Press, New York, 1984.

mixed oxide semiconductors with respect to the H^+/H_2 scale, where minus represents values above zero and plus represents values below zero. Some of the values have been extracted from Ref. 19. Neglecting the difference between the Fermi level and the majority carrier band edge of the semiconductor ΔE_F, the flat-band potential U_{fb} is the potential of the majority carrier band edge of the semiconductor with respect to the H^+/H_2 potential. Often the band-edge positions are provided at different pH values. Since it is known that the positions of band edges of oxides and mixed oxides become 59 meV more negative when the pH is increased by 1, majority carrier band edges of the oxide semiconductors can be calculated at any pH value. Figure 7.4 shows the positions of band edges of various semiconductors in aqueous electrolytes having a pH of 14. It shows the energy scales with respect to vacuum, the standard H^+/H redox potential, and the potentials of standard calomel and Ag/AgCl electrodes with respect to each other as well as the position of the O_2/H_2O potential of 1.229 V with respect to the H^+/H_2 potential. The flat-band potential of CdS is determined by the pH of the electrolyte as well as by the sulfide ion concentration [19].

An energy level diagram for an n-type semiconductor/metal photoelectrolysis cell in which the flat-band potential U_{fb} lies above the H^+/H_2 potential, whereas the O_2/H_2O potential lies above the valence band of the n-type semiconductor is shown in Figure 7.5. Here, the available photovoltage is in excess of the thermodynamic potential difference for water splitting (1.229 V at 25°C). It also overcomes the intrinsic energy level mismatches η_c and η_a, where η_c is the mismatch of the counterelectrode metal Fermi level E_F above the H^+/H_2 potential required to sustain the current flow and η_a the mismatch due to the difference between O_2/H_2O potential and the valence band edge at the interface [7]. The intrinsic energy level mismatch η_a is determined only by the bandgap and is independent of the current. Holes that are generated in the n-type semiconductor within a diffusion length of the depletion region and diffuse into it, together with those generated in the depletion region itself are separated and swept to the semiconductor surface by the built-in field. Because of the favorable band-edge positions and the available necessary overpotentials,

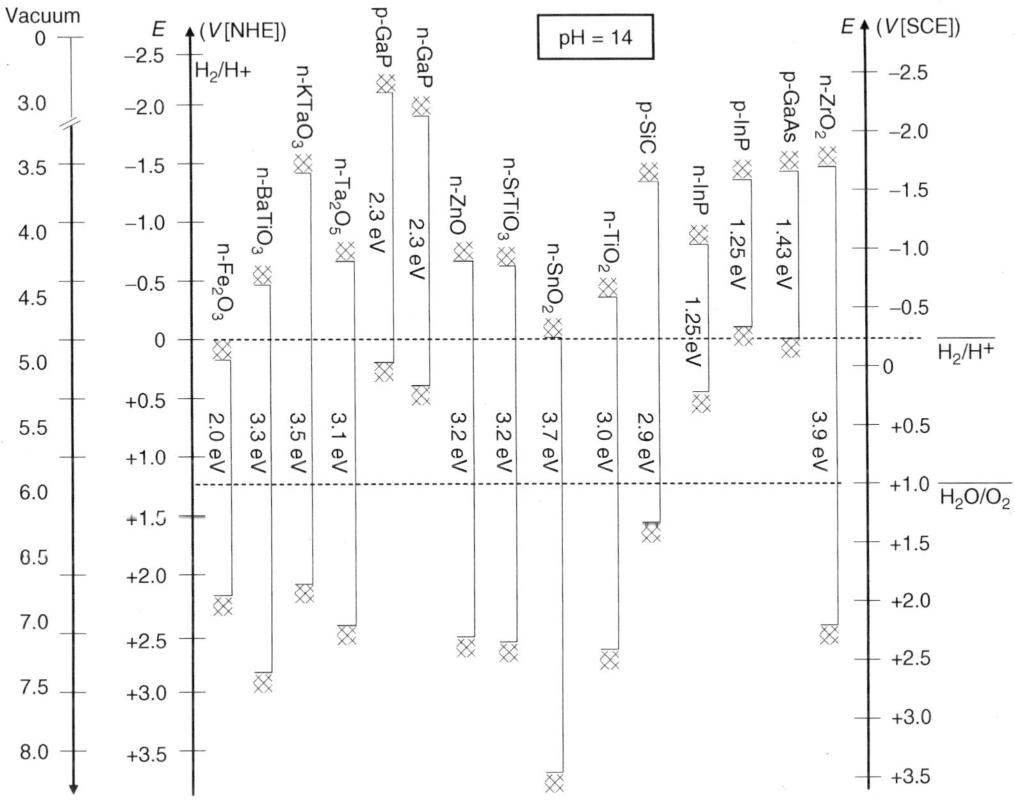

FIGURE 7.4
Positions of band edges of various semiconductors in aqueous electrolytes with pH = 14.

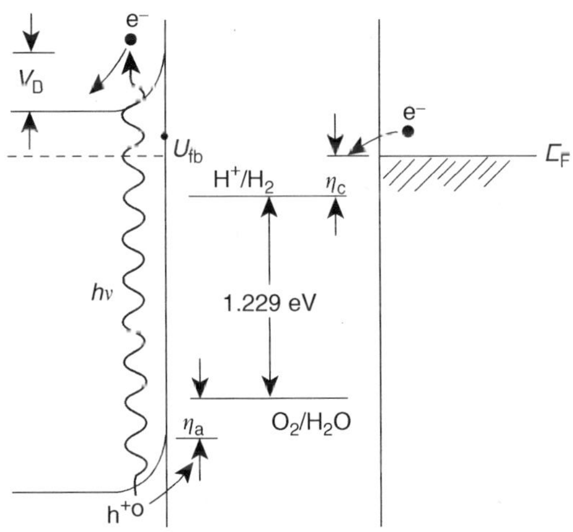

FIGURE 7.5
Energy level diagram for an n-type semiconductor–metal photoelectrolysis cell in which the flat-band potential U_{fb} lies above the H^+/H_2 potential, whereas the O_2/H_2O potential lies above the valence band of the n-type semiconductor.

they can produce anodic oxidation of water molecules and generate oxygen. However, the photogenerated electrons flow through the n-type semiconductor bulk to the back contact and from there to the metal counterelectrode where they are donated to reduce hydronium ion and to generate hydrogen. The excess photovoltage remaining after fulfilling the preceding minimum requirements is used to overcome the kinetic, diffusion, and ohmic overpotential losses which are collectively termed *polarization loss*. Often the intrinsic energy level mismatches η_c and η_a are considered as a part of the kinetic overpotential loss. The polarization loss determines the rate of generation of hydrogen and oxygen. In fact, collectively these losses could be significant when gas evolution occurs. As mentioned earlier, only a few semiconductors are able to provide such a photovoltage. Although $SrTiO_3$, Nb_2O_3, and ZrO_2 are suitable for direct water splitting to generate hydrogen and oxygen, their bandgaps are too high in the range 3.4–3.5 eV. Therefore, these wide bandgap semiconductors are illuminated by UV radiation that is hardly available in the solar spectrum and are not practical for use in solar hydrogen generation. CdS and SiC have favorable conduction and valence band positions. However, they have problems with photocorrosion.

As mentioned earlier, in the case of most n-type semiconductors, the flat-band potential lies below the H^+/H_2 potential at neutral pH, and therefore hydrogen and oxygen evolution would not occur. In these cases, it is essential to apply an external anodic bias to raise the Fermi level to generate H_2 and O_2. The bias can be applied with either an external voltage or by immersing the anode in a basic solution and the cathode in an acid, and the two compartments being separated by a membrane. Energy is saved in hydrogen generation because the photovoltage on the semiconductor supplements the external bias. Based on the importance of interface properties at the solid–solid interfaces (e.g., grain boundaries) and solid–liquid interfaces, Bak et al. [14] have emphasized the application of up-to-date materials processing technology for the development of economic PEC hydrogen generation using sunlight.

7.2.5 Electrode Stability and Photoelectrolysis Using Appropriate Bandgap Semiconductors

The charge carriers may reduce or oxidize the semiconductor itself leading to decomposition. This poses a serious problem for practical photoelectrochemical devices. Absolute thermodynamic stability can be achieved if the redox potential of oxidative decomposition reaction lies below the valence band and the redox potential of the reductive decomposition reaction lies above the conduction band. In most cases, usually one or both redox potentials lie within the bandgap. Then the stability depends on the competition between thermodynamically possible reactions. When the redox potentials of electrode decomposition reactions are thermodynamically more favored than electrolyte redox reactions, the result is electrode instability, for example, ZnO, Cu_2O, and CdS in an aqueous electrolyte.

Most semiconductors are prone to anodic photodecomposition because holes have a very high oxidizing power, especially in aqueous electrolytes. Therefore, it is easier to achieve semiconductor electrode stability using p-type semiconductors for which the minority carriers are electrons. In general, semiconductors are less vulnerable to cathodic corrosion than anodic. When the minority carriers (i.e., electrons) in p-type semiconductors are donated to the electrolyte for carrying out the reduction reaction, they also provide cathodic protection to the semiconductor surface. Heller and Vadimsky [24] successfully predicted that a monolayer of chemisorbed oxygen forming an indium oxide layer would passivate indium phosphide and reduce the surface recombination. When the surface of p-type InP is in contact with aqueous acids it is covered with a 0.6–1 nm pore-free, compact layer of hydrated indium oxide, through which electrons can tunnel freely.

The indium oxide layer together with the cathodic protection provides the needed stability [11,13,17,24,25]. This is especially true with other indium-containing semiconductors such as p-$CuIn_{1-x}Ga_xSe_{2-y}S_y$ that are also expected to be stable against electrocorrosion. It may be noted that p-type semiconductors are rare, and therefore the choice of materials for optimizing the energy position of band edges and other properties is limited. Moreover, at open-circuit conditions or under high photovoltages, the concentration of the majority carriers becomes relatively high at the surface and the anodic decomposition would proceed in the dark if the semiconductor is unstable against anodic decomposition. Bandgaps of the few p-type semiconductors available in nature are usually too small. Besides, they may also suffer from stability problems [8].

Typical decomposition reactions listed by Gerischer [8] are shown in the following where (aq) means a solvated species:

Anodic processes (p-type specimen or photodecomposition of n-type semiconductors) in aqueous electrolytes

$$\mathrm{Si} + 4\mathrm{h}^+ + 2\mathrm{H_2O} \rightarrow \mathrm{SiO_2(aq)} + 4\mathrm{H^+(aq)} \tag{7.4}$$

$$\mathrm{GaAs} + 6\mathrm{h}^+ + 3\mathrm{H_2O} \rightarrow \mathrm{Ga^{3+}(aq)} + \mathrm{AsO_3^{3-}(aq)} + 6\mathrm{H^+(aq)} \tag{7.5}$$

$$\mathrm{CdS} + 2\mathrm{h}^+ \rightarrow \mathrm{Cd^{2+}(aq)} + \mathrm{S} \tag{7.6}$$

$$\mathrm{ZnO} + 2\mathrm{h}^+ \rightarrow \mathrm{Zn^{2+}(aq)} + \tfrac{1}{2}\mathrm{O_2} \tag{7.7}$$

$$\mathrm{Cu_2O} + 2\mathrm{h}^+ + \mathrm{H_2O} \rightarrow 2\mathrm{CuO} + 2\mathrm{H^+(aq)} \tag{7.8}$$

$$\mathrm{MoS_2} + 14\mathrm{h}^+ + 9\mathrm{H_2O} \rightarrow \mathrm{MoO_3(aq)} + 2\mathrm{SO_3^{2-}(aq)} + 18\mathrm{H^+(aq)} \tag{7.9}$$

$$\mathrm{GaSe} + 7\mathrm{h}^+ + 3\mathrm{H_2O} \rightarrow \mathrm{Ga^{3+}(aq)} + 2\mathrm{SeO_3^{2-}(aq)} + 6\mathrm{H^+(aq)} \tag{7.10}$$

Cathodic processes (n-type specimen or photodecomposition of p-type semiconductors) in aqueous electrolytes

$$\mathrm{Si} + 4\mathrm{e}^- + 4\mathrm{H_2O} \rightarrow \mathrm{SiH_4} + 4\mathrm{OH}^- \tag{7.11}$$

$$\mathrm{CdS} + 2\mathrm{e}^- \rightarrow \mathrm{Cd} + \mathrm{S^{2-}(aq)} \tag{7.12}$$

$$\mathrm{ZnO} + 2\mathrm{e}^- + \mathrm{H_2O} \rightarrow \mathrm{Zn} + 2\mathrm{OH^-(aq)} \tag{7.13}$$

$$\mathrm{Cu_2O} + 2\mathrm{e}^- \rightarrow 2\mathrm{Cu} + 2\mathrm{OH^-(aq)} \tag{7.14}$$

$$\mathrm{GaSe} + 2\mathrm{e}^- \rightarrow \mathrm{Ga} + \mathrm{Se^{2-}(aq)} \tag{7.15}$$

The reactions that are more favored thermodynamically tend to be also favored kinetically. Semiconductor electrodes can be stabilized by using this effect. For this purpose, redox couples in the electrolyte are established with the redox potential more negative than the oxidative decomposition potential, or more positive than reductive decomposition potential in such a manner that the electrolyte redox reaction occurs preferentially compared to the electrode decomposition reaction.

Kinetic stability has been observed in n-type semiconducting compounds of transition metals having layered structures due to the d–d nature of the indirect electronic transition and also the inertness of van der Waals surface to chemical attack in these materials. Here, the valence band states are weakly bound (i.e., have little bonding character), whereas there is little weakening of bonds in the conduction band states. Overall, the electronic excitation does not affect the bond strength significantly. Thus, although the overall reaction affinity for corrosion is thermodynamically large in these systems, a high activation barrier for decomposition exists. Kinetic stability can also be achieved by a competing redox reaction that consumes the electronic charge carriers at the interface quickly enough to prevent the decomposition reactions. For example, photooxidation of CdS electrode in contact with the redox system is prevented by a competing reaction with polysulfide ion: $S_n^{2-} + S^{2-} + 2h^+ \rightarrow S_{n+1}^{2-}$ [8].

The lifetime of the electrode is a critical parameter for device applications. For practical applications, it is necessary that there should be no significant decline in the output, under load, during a time period that is at least 1000 times the time taken to totally decompose the moles of product in the electrode. As seen from table 1 in Ref. 15, although several of the conventional semiconductors present reasonable stability, they are far from meeting this rigorous criterion. As mentioned earlier, chalcogenides of transition metals of group VIa, such as WSe_2, $MoSe_2$, and $MoTe_2$, show very interesting results in terms of stability. Molybdenite exists as large natural crystals. The cells made with these have shown reasonable conversion efficiency. WSe_2 with a bandgap of 1.16 eV comes closest to fulfilling the criterion of stability with an equivalent lifetime of 3 years, whereas the desired lifetime is >20 years [15]. It has several limitations besides the lack of technological and industrial base. p-WSe_2 has an indirect bandgap and low minority carrier diffusion length. Moreover, the surface recombination velocities for this layered compound are excessive for all except the van der Waals planes that unfortunately act as mirrors for a part of the sunlight [24].

Besides the requirements of the formation of high energy barriers, low surface recombination velocity, high stability, being benign to the environment, and low cost, ideal semiconductor electrodes must possess high interfacial electron transfer rates or high catalytic activity of the semiconductor surface. However, a metal or metal complex catalyst deposited on the semiconductor surface can also act as a surface recombination center as well as a barrier. This can be overcome by depositing well-scattered nanodots of catalyst metal. For example, when nano-sized quantum dots are deposited on a silicon surface coated with a passivating layer of SiO_2, the photocurrent can be stable in the aqueous redox electrolyte. High open-circuit voltages and PV conversion efficiency were achieved with Si/SiO_2 samples coated with Pt quantum dots [13,25–28]. When the quantum dots are very small compared with the depletion region width of the semiconductor, that is, ~5 nm, there is almost no lowering of the energy barrier by the metal quantum dots. Here, the surface band modulation by quantum dots decays sharply toward the semiconductor bulk and is limited to the narrow surface region of approximately the quantum-dot size. Thus, the effective barrier height in the case of metal dot–coated semiconductor electrodes is almost equal to that on a bare semiconductor electrode. Moreover, the surface passivation layer that covers almost the entire surface effectively suppresses the surface carrier recombination. Open-circuit voltage of a photoelectrochemical cell prepared by using an n-Si(Pt) electrode, a platinum counterelectrode, and stirred 8.6 M HBr/0.1 M Br_2 electrolyte solution was only 0.28 V [26]. A (111) oriented n-type silicon wafer chip having a resistivity in the range 0.4–0.8 Ωcm chip was coated with vacuum-evaporated 3 nm-thick platinum to form the n-Si(Pt) electrode. High open-circuit voltage V_{oc} of 0.685 V was obtained after the n-Si(Pt) electrode was treated in 4 M NaOH solution at 90°C for ~3 min followed by treatment in 0.25 M NaOH solution containing 0.6 M 2-propanol at 90°C for ~3 min.

Current-density *J* versus voltage curves were stable during cyclic scans for at least 10 min. The platinum layer seemed to dissolve gradually in HBr/Br_2 solution. In contrast, bare n-Si electrodes degraded considerably in the first negative-to-positive scan in any redox solution. High efficiencies have also been reported with n-GaAs coated with Ru- or Os-metal dots [29]. Consequently, metal quantum-dot-coated semiconductor electrodes can be used for efficient hydrogen photoelectrolysis [13,24,30,31].

The following semiconductors that have more appropriate bandgaps in the range 1–2 eV for solar energy conversion and that have been used for photoelectrolysis cells to generate hydrogen and oxygen by water splitting have been listed by Parkinson [15] among the more interesting and efficient: p-InP(Ru) [24,32], n-Fe_2O_3 and p-Fe_2O_3 [33], and n-$Ru_xFe_{1-x}S_2$ [34,35]. Catalyzed p-type InP has been found to possess several interesting properties such as appropriate bandgap for efficient sunlight conversion, p-type conductivity, and chemical passivation from corrosion based on coverage by a hydrated and compact indium oxide layer, and nonleaking high barrier at the catalyst–semiconductor interface. The indium oxide layer reduces radiationless recombination of photogenerated charge carriers. Moreover, since the diffusion length of electrons on the InP surface is 100 nm, it becomes easy to cover the p-type InP surface by nano-sized dots spaced at distance smaller than the minority carrier diffusion length. Heller and Vadimsky [24] used the passivation by a monolayer of chemisorbed oxygen forming an indium oxide layer on indium phosphide to reduce the surface recombination and to improve photovoltage, photocurrent, and fill factor in photoassisted electrolysis. They coated 0.1–0.2 cm^2 sized (111)A, (111)B, and (100) faces of p-InP with <10 nm thick islands of ruthenium or rhodium by electroplating and then etched in dilute aqua regia to remove several hundred nanometers of the semiconductor to obtain Ru or Rh islands by undercutting. As thin oxide layer was formed by oxidation in aqua regia or in air followed by an extended immersion in 1 M HCl. Photoassisted electrolysis efficiency of 12% was obtained for a cell consisting of p-type InP(Ru) cathode and Pt(Rh) anode in aqueous HCl electrolyte cell under illumination with 87 mW/cm^2 sunlight to generate hydrogen. The required voltage was reduced from 1.3 to 0.65 V at the threshold. The efficiencies were calculated either by measuring the ratio of the electrical power conserved in electrolysis for solar-illuminated InP as compared with a smooth platinum electrode or by measuring the ratio of the electrical power generated in an ideal fuel cell (operating at the thermodynamic potential) from the photoelectrochemically derived hydrogen to the solar irradiance. The efficiency of p-InP(Ru)/1 M HCl cell was 12% when calculated by comparing with a Pt cathode and 8.6% when calculated by the power generation with an ideal fuel cell. The efficiency was estimated by reduction of the voltage. The quantum efficiency was estimated to be 82%. Indium phosphide was found to be stable in 1 M HCl in the absence of oxygen and chlorine. However, it was etched rapidly in air saturated acid. Reduction of indium oxide resulted in the loss of photocurrent. Performance declined after continuous hydrogen generation for 15–20 min. Photocurrent was recovered by reformation of oxide by exposure to air, by disconnecting the electrode, or by bringing to open-circuit conditions.

Solar-to-chemical conversion efficiency was increased by preparing a more uniform and thinner catalyst layer of hydrated indium oxide layer at pores in the metallic catalyst on 0.2 cm^2 sized (100) faces p-InP interfacing with highly pure 1 M $HClO_4$ electrolyte using spectroscopic-grade graphite counterelectrode. Care was taken to limit the impurities of other platinum group metals in rhenium to <0.05 ppm. The efficiencies of the p-InP(Rh–H alloy)/1 M $HClO_4$ cathode estimated by the first and second methods described earlier were 16.2% and 13.3%, respectively. In the case of p-InP(Re–H alloy)/1 M $HClO_4$ cathode, the respective efficiencies were 14.4% and 11.4%. It was essential to restore the passivating oxide layer by bringing the photocathodes to open-circuit conditions for 5–10 s after every ~5 min. Consequently, among the numerous photocathodes that have been explored, InP has been the only photocathode that converts solar energy to chemical energy stored in hydrogen at a Gibb's free energy efficiency of 13.3% calculated on the basis of the bias reduction [11,32].

Attempts have been made to deposit transparent, electrically conducting, robust, metallized, protective polymer films on the semiconductor surfaces by radio-frequency (RF) sputtering and reactive polymerization [36]. Initially, highly colored layers of unknown conductivity were obtained. Approaches at stabilizing semiconductor–electrolyte interface by metal-polymer RF cosputtering, plasma polymerization of reactive organosilane monomers with or without metals such as gold or platinum, and solution-phase coating of selected monomers and polymers resulted in occasional success including films that survived hundreds of hours in acidic aqueous electrolytes. However, reproducibility was a recurring problem with supposedly identical samples, prepared at the same time, failing (discoloring or peeling off) within 2–3 h, especially because of too many variables. Finally, it was conceded that because there was no material that would absorb solar radiation and generate highly reactive chemical species such as hydroxyl and peroxyl radicals at a competitive cost and because of the continued improvement in the efficiency, stability, and cost of solid-state PV cells, the more cost-effective approach would be the solar generation of electricity for general use and also for conventional electrolysis of water [36].

Because of the various difficulties in fulfilling the requirements for efficient photoelectrolysis using a single semiconductor photoelectrode, Gerischer reached a sobering conclusion that efficient water decomposition by direct solar photoelectrolysis using a single photoactive semiconductor electrode does not seem to be possible and that it will become necessary to provide assistance of an additional external voltage source from a PV cell or a second photovoltage-generating junction. This would need double the quanta and consequently a reduction in the overall efficiency by a factor of two. Gerischer also suggested the separation of the power generation in PV devices and the water decomposition in an electrolysis cell with electrodes of high catalytic activity and metallic character so as to optimize the requisite functions independently [8].

Several researchers have considered photosynthesis in nature as a model for electrochemical conversion of sunlight to chemical energy [8]. Photosynthesis is discussed in several textbooks [37–40]. The topic of photosynthesis and photoelectrochemistry is discussed in Section 7.5.

7.2.6 Two Photoelectrode Cells

Direct splitting of water can be accomplished by illuminating two interconnected photoelectrodes, a photoanode, and a photocathode as shown in Figure 7.6. Here, $E_g(\mathrm{n})$ and $E_g(\mathrm{p})$ are, respectively, the bandgaps of the n- and p-type semiconductors and $\Delta E_F(\mathrm{n})$ and $\Delta E_F(\mathrm{p})$ are, respectively, the differences between the Fermi energies and the conduction band-minimum of the n-type semiconductor bulk and valence band-maximum of the p-type semiconductor bulk. $U_{fb}(\mathrm{p})$ and $U_{fb}(\mathrm{n})$ are, respectively, the flat-band potentials of the p- and n-type semiconductors with the electrolyte. In this case, the sum of the potentials of the electron–hole pairs generated in the two photoelectrodes can be approximated by the following expression:

$$\begin{aligned} U_{fb} + \Delta E_F(\mathrm{p}) - \Delta E_F(\mathrm{n}) &= U_{fb}(\mathrm{p}) - U_{fb}(\mathrm{n}) + \Delta E_F(\mathrm{p}) - \Delta E_F(\mathrm{n}) \\ &= \frac{\Delta G}{nF} + \eta_a(\mathrm{n}) + \eta_c(\mathrm{p}) + iR \end{aligned} \tag{7.16}$$

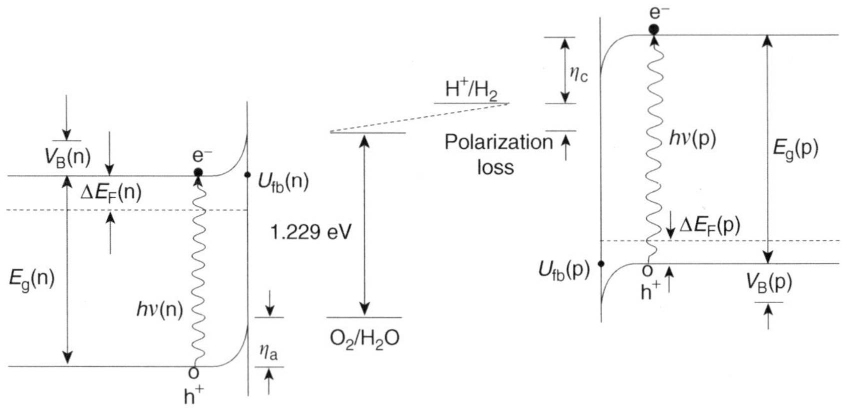

FIGURE 7.6
Direct splitting of water with illuminated and interconnected photoanode and photocathode.

All the terms in the preceding equation have the unit of energy per unit charge or electropotential. U_{fb} is the total flat-band potential which is the sum of band bending U_{fb}(p) in the p-type semiconductor and U_{fb}(n) in the n-type semiconductor. (The sign for U_{fb}(n) is negative because it is typically a negative number for the depletion layer junctions of interest.) The input energy of an absorbed photon must be greater than or equal to the semiconductor bandgap E_g. ΔG, n, and F are the Gibbs free energy in Joules per mole, the number equivalents of electrons per mole of material in equivalents per mole, and Faraday's constant in coulombs per equivalent, respectively. $\Delta G/nF$ in Joules per coulomb, that is, volts, is the portion of this potential energy per unit charge recovered as free energy of the net endoergic reaction in the electrolyte. In the case of the generation of hydrogen and oxygen by splitting water, the energy difference between the levels of H^+/H_2 and O_2/H_2O is 1.229 eV. The reduction of water to hydrogen occurs on the photocathode, which is the illuminated p-type semiconductor that provides an electron for this reduction. The oxidation of water to oxygen occurs on the photoanode, which is the illuminated n-type semiconductor that accepts an electron for this oxidation. η_a(n) and η_c(p) are the intrinsic, current-independent energy level mismatches of the n- and p-type semiconductor electrodes, respectively, and polarization loss is the total overpotential due to kinetic, diffusion, and ohmic losses consumed in the electrolyte for the flow of ions. The energy level mismatches, η_a(n), η_c(p), and polarization loss are the irreversible, entropy-producing terms that represent the lost potential energy per unit charge. Thus, it can be seen that the potential can be maximized by matching the band edges to the respective redox levels and by minimizing the polarization loss. Moreover, the efficiency can be increased by employing two different segments of the solar spectrum to illuminate the two photoelectrodes. In this case, an optimum efficiency is reached when one semiconductor has a bandgap of ~1.8 eV and the other has a bandgap of ~1 eV [41]. If such a tandem arrangement is employed, then the energetic photons having energy greater than 1.8 eV are absorbed by the higher bandgap semiconductor, whereas the lower energy photons pass through the higher bandgap semiconductor and the electrolyte and are incident on the lower bandgap semiconductor electrode, which absorbs the photons having energy greater than 1 eV. For efficient energy conversion, the photocurrents of the two electrodes must be equal. The maximum theoretical efficiency for dual photoelectrodes at AM 1.0 has been predicted to be 25% for the generation of hydrogen and oxygen by splitting water [7,11,42].

A p–n-type photoelectrochemical cell consisting of a p-type GaP crystal and an n-type TiO_2 crystal having ohmic contacts on one side, and connected back-to-back through the ohmic contacts has been developed. For this purpose, a TiO_2 crystal having the *c*-axis in the crystal plane was reduced in hydrogen to obtain a conductivity of 0.3/Ωcm. Ohmic contacts were formed on one face of each crystal by depositing a 300 nm thick layer of indium on n-type TiO_2 and a 300 nm thick layer of 1% Zn–99% Au alloy on p-GaP, followed by an overcoat of 1 μm thick Au and heat treatment in hydrogen at 600°C for 10 min. The two crystals were bonded using conductive silver epoxy. When the p–n photochemical diode was suspended in 0.2 N H_2SO_4 electrolyte and both p-GaP and n-TiO_2 faces were illuminated with simulated sunlight, H_2 and O_2 were found to evolve very slowly from p-GaP and n-TiO_2 faces, respectively [43]. This system mimics photosynthesis in which it also uses two photons for photoexcitation and then combines the carriers to carry out the reaction in two parts. Since then, the field of solid-state tandem solar cells has evolved considerably and >40% PV conversion efficiency has been achieved with a triple-junction tandem cell under high sunlight concentration [44]. Lattice-matched III-V multijunction cells have always been prepared using a wide range of subcell and device structure bandgaps that are grown with very high crystalline quality, thus with high minority carrier lifetimes. Recently, several groups are developing metamorphic multijunction cells in which the layers are not lattice-matched to the base layer on which they are grown. Instead a graded layer termed *metamorphic buffer* is deposited so as to enable transition from one lattice to the other. The metamorphic multijunction cells provide considerable flexibility in the selection of bandgaps because not all the subcell bandgaps of metamorphic multijunctions cells are required to be lattice-matched to the same growth substrate. Considerable increase in cell performance has resulted from the following developments: advances in the design of metamorphic subcells to reduce carrier recombination and to increase voltage, wide bandgap tunnel junctions (TJs) that can function at high concentrations, metamorphic buffers for transitioning from substrate lattice to active subcell lattice, antireflection (AR) coatings for concentrator cells, and integration into current-matched triple-junction cells for terrestrial spectrum. The PV parameters of the highest efficiency $Ga_{0.44}In_{0.56}P/Ga_{0.92}In_{0.08}As$ triple-junction cells of the two types, under AM 1.5 global solar irradiation at 25°C are as follows: lattice-matched—concentration 135x, 0.2547 cm^2 area, V_{oc} = 3.054 V, short-circuit current density J_{sc} measured at 135x concentrated intensity = 0.1492 A/W, voltage at the optimum power point V_{mp} = 2.755 V, fill factor = 0.881, efficiency = 40.1%; 1 sun, 4 cm^2 area, V_{oc} = 2.622 V, J_{sc} = 14.37 mA/cm^2, V_{mp} = 2.301 V, fill factor = 0.850, efficiency = 32%; metamorphic—concentration 240x, 0.267 cm^2 area, V_{oc} = 2.911 V, J_{sc} at 240x concentrated intensity = 0.1596 A/W, V_{mp} = 2.589 V, fill factor = 0.875, efficiency = 40.7%; 1 sun, 4 cm^2 area, V_{oc} = 2.392 V, J_{sc} = 15.99 mA/cm^2, V_{mp} = 2.055 V, fill factor = 0.819, efficiency = 31.3% [44].

7.2.7 Semiconductor Nanoparticles

Many researchers have prepared nano-sized semiconductors with very large reactive surface area so as to enhance surface adsorption and photocatalytic reduction and oxidation reactions for the generation of H_2 and O_2 by water splitting. There is neither a separate cathode for reduction of water to hydrogen nor a separate anode for oxidation of water to oxygen when particulate suspensions of suitable semiconductor materials such as TiO_2, $SrTiO_2$, CdS, and $MoSe_2$ in an electrolyte are employed to carry out photocatalytic activity for water splitting.

When a molecule in a solution absorbs light, an electron in the highest occupied molecular orbital (HOMO) is excited to the lowest unoccupied molecular orbital (LUMO) creating an electron vacancy, that is, a hole in the HOMO. The electron may be provided to a molecule in the solution to reduce it, whereas the hole in the HOMO may be provided to a molecule in the solution to oxidize it. This is similar to the reduction–oxidation process in the bulk semiconductor/electrolyte photoelectrolytic cell described earlier [13,17].

Optical properties of large (≥15 nm) semiconductor particles in suspension and colloidal solution are similar to those of extended (bulk) crystals. Quantization of energy levels occurs in small particles when the electrons and holes are limited by the potential walls of particles having size of the order of the De Broglie wavelength of the charge carriers [12,45]. Such particles are termed *quantized semiconductors* or *Q-semiconductors*. The De Broglie wavelength λ_B is given by

$$\lambda_B \sim \frac{h}{(2\pi k T m_e^*)} \tag{7.17}$$

where k is the Boltzmann constant, T the absolute temperature, and m_e^* the effective mass of electrons.

Although the De Broglie wavelength of free electrons is ~0.1 nm, the value of an electron in a small crystallite can be much larger because the effective mass of electrons in a small particle is considerably smaller. Energy levels evolve from HOMO and LUMO to those of clusters, Q-sized particles, and finally bulk semiconductor. Figure 7.7 shows the energy levels in bulk- and Q-sized particulate semiconductors.

Electrons and holes generated by energetic illumination in a small semiconductor particle can easily reach the surface and can carry out reduction and oxidation processes of adsorbed species, respectively, by providing and accepting electrons before they have a chance to recombine. Therefore, the quantum yield can be high.

However, both the reductive and oxidative reactions must occur simultaneously. Otherwise, the reaction would be stopped because of particle charging. Thus, a small semiconductor particle under bandgap excitation can act as a pair of short-circuited microelectrodes that always remain under open-circuit potential at which the anodic and cathodic currents are equal [12,46]. When the particle radius is much larger than the space-charge region width, the potential drop at the center of the particle φ_0 becomes equivalent to the potential drop within the space-charge region φ_{sc}. The reductive and oxidative currents are small because the built-in potential assists the transfer of minority carriers to the surface while opposing that of the majority carriers (Figure 7.7a). For small particles of radius ~10 nm, usually the whole particle is depleted, the very small potential drop at the center of the particle cannot aid in the separation of electron–hole pairs [12] (Figure 7.7b). A large band bending can be achieved when a metal particle of appropriate work function is deposited on a semiconductor (Figure 7.7c). Higher activity is achieved by establishing an ohmic contact either by deposition using a photocatalytic reaction or by heat treatment of the deposited metal on the semiconductor. A contour map of energy levels for an n-type semiconductor with deposited metal particulates obtained by numerical solution of the 2-D Poisson–Boltzmann equation shows that the electrons would be driven to the metal particles, whereas the holes will be driven to the semiconductor surface. The photocatalytic reactions with characteristic times at the semiconductor surface can be described by the following steps: (1) Absorption of a photon to generate an electron in the conduction band and a hole in the valence band typically occurring very fast in femtoseconds. (2) Transport of the electron and hole to the surface. (3) Stabilization of a hole at the surface to

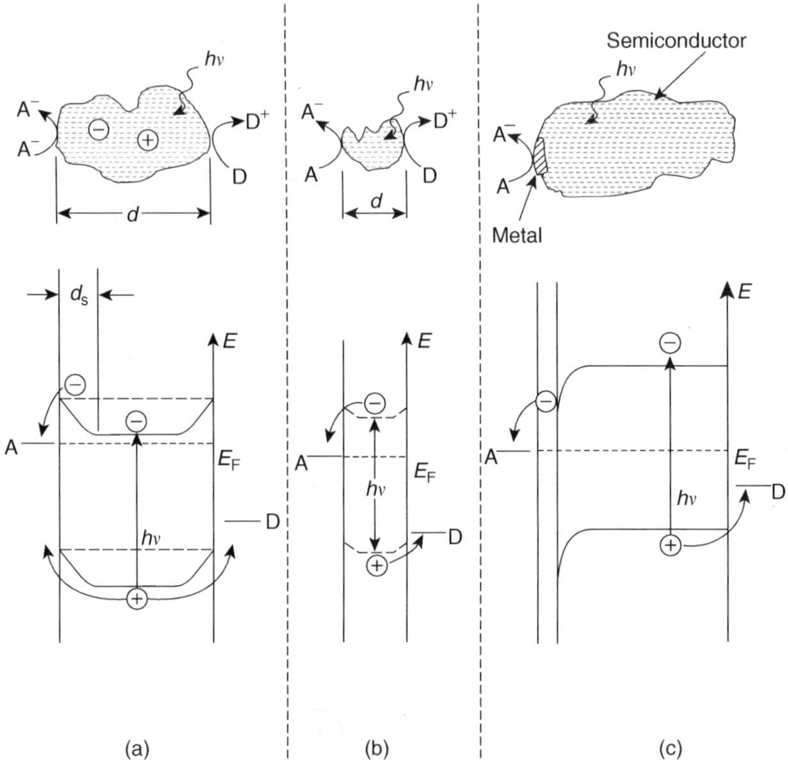

FIGURE 7.7
Energy levels in (a) bulk semiconductors, (b) Q-sized particulate semiconductors, and (c) semiconductor particle with metal islands.

form a trapped hole, occurring fast in 10 ns and dynamic equilibrium of an electron trapped at the surface with a shallow trap in 100 ps, or fast stabilization of the electron trapped at the surface with a deep trap in 10 ns. (4) Recombination of an electron with a trapped hole on the surface occurring slowly in 100 ns and recombination of a hole with a trapped electron occurring fast in 10 ns. (5) Reduction of molecules by the donation of an electron occurring slowly in 100 ns and oxidation of molecules at the surface by accepting an electron very slowly in milliseconds. (6) Exchange of a product at the surface with a reactant in the medium [47,48]. Of these the two key processes are (1) photogeneration of charge carriers and (2) reduction and oxidation of molecules. The photogeneration occurs very fast, the reduction occurs slowly, and the oxidation occurs very slowly.

7.2.8 Efficiency of Hydrogen and Oxygen Generation by Water Splitting

It is instructive to calculate the efficiency of generation of hydrogen and oxygen by splitting water using a simple PV cell and an electrolysis setup in the dark in which case the energy input is from sunlight incident on the PV cells and the output is the amount of hydrogen and oxygen generated by the setup. The energy return in the form of generated hydrogen is usually calculated by considering the thermodynamic potential difference of 1.229 V at 25°C for water splitting. As mentioned earlier, the losses are the material-dependant energy level mismatches η_c and η_a and the polarization loss that collectively represents the overpotentials due to kinetic, diffusion, and ohmic voltage losses for sustaining the current. Water splitting is an endothermic reaction. If it is carried out in a thermally isolated system in which the losses have been minimized, water will become progressively cooler and eventually freeze. The reaction can be carried out indefinitely by operating at the thermoneutral voltage for water splitting of 1.48 V, where the generated heat will sustain the endothermic reaction. In practice, if the system obtains the requisite heat energy from the ambient to which it is exposed and the cooling thus provided may even be welcome and therefore is not counted. The overall efficiency of the PV electrolysis system is then the product of efficiencies of the PV cells and the electrolysis setup at the operating voltage. Ideally, the operating voltage should be the optimum power point V_p of the PV cells at which their efficiency is maximum and hence is specified as the efficiency. For PV cells of total area A, incident AM 1.5 solar radiation of 100 mW/cm^2, optimum current I_p at the optimum power point V_p, the PV conversion efficiency η is given by

$$\text{PV efficiency,}\quad \eta\% = \frac{V_p \times I_p}{100\dfrac{\text{mW}}{\text{cm}^2} \times A} \times 100 \tag{7.18}$$

For the operating voltage at the optimum power point V_p and the Faradaic current for water splitting of I_p, the overall solar to hydrogen generation PV electrolysis efficiency is given by

$$\begin{aligned}\text{PV electrolysis efficiency}\,\% &= \frac{P_{out}}{P_{in}} \times 100\\ &= \frac{I_p}{P_{in}} E_0^{H_2O} \times 100\\ &= \frac{I_p\text{mA}}{100\dfrac{\text{mW}}{\text{cm}^2} \times A} \times 1.229 \times 100\\ &= \frac{1.229\eta}{V_p}\end{aligned} \tag{7.19}$$

Thus the PV electrolysis efficiency will be high if losses are minimized to operate the PV electrolysis setup at a voltage as low as possible compared to the thermodynamic potential difference of 1.229 V for water splitting at 25°C. In practice, the electrolyzer cost is reduced by increasing the rates of generation of hydrogen and oxygen, that is, by increasing the operating voltage to 2 V, thus resulting in higher polarization loss. Also in the case of some tandem solar cells, the optimum voltages are high. The operating voltages can be reduced to as low as 1.35 V by employing optimized electrodes and working with low currents obtained at low solar irradiances. Lastly, the operating voltages can be reduced to 1.28 V by providing additional excitation. For these three cases of operating voltages of 2, 1.35, and 1.28 V, the respective PEC efficiencies will be 61.5%, 91%, and 96% of the PV cell efficiency. It may often be noted that stirring is employed to improve PEC efficiency. A strict calculation should provide the efficiency including the energy input required for stirring.

7.3 Oxides

7.3.1 Titanium Oxide

Fujishima and Honda [17,49] demonstrated for the first time the possibility of oxygen and hydrogen evolution by photoelectrolysis of water using an n-type TiO_2 semiconductor anode in an electrolyte (anolyte) and a Pt counterelectrode in an electrolyte (catholyte). First, a ~1.5 mm thick, 1 cm^2 sized single-crystal wafer having an n-type TiO_2 rutile structure was treated at 700°C at 10^{-4}–10^{-5} Torr for ~5 h to increase its conductivity. A copper wire was connected to an indium layer evaporated on one side to ensure ohmic contact to the back surface that was later sealed with epoxy. When the n-type TiO_2 semiconductor is dipped in an electrolyte, a semiconductor–electrolyte junction is formed and conduction and valence bands are bent generating a built-in field. Current–voltage characteristics were measured with a static potentiometer in the dark and under illumination of the TiO_2 front surface. Anodic current proportional to the intensity of light began to flow for wavelengths shorter than 415 nm (equivalent to the bandgap of 3 eV of TiO_2). The anodic current reached saturation at positive potentials relative to SCE showing that photogenerated minority carriers, that is, holes were flowing into the electrolyte. Oxygen evolution was found to occur at −0.5 V (SCE) in an aqueous electrolyte of pH 4.7; this is more negative than the standard potential. This behavior was termed *photosensitized electrolyte oxidation*. Halogen ions introduced into the anolyte were also oxidized.

An electrochemical cell was constructed by connecting the copper wire attached at the back of the TiO_2 electrode to the platinum black cathode through a load. The two compartments were connected through an agar salt bridge that allows the exchange of ionic species.

Photogenerated electrons can flow to the counterelectrode, reduce the hydrogen ions, and generate hydrogen without an applied bias if the conduction band edge is at least −0.4 V (NHE) in an acid solution or at least −1.2 V (NHE) in an alkaline solution. As observed from the energy levels of TiO_2 in Figure 7.3, this condition is not met. Therefore, an external bias needs to be applied or a difference of pH between the compartments needs to be employed in such a way that the pH of the anolyte is higher than that of the catholyte. Thus, in the early work in the laboratory of Fujishima et al. [17,49,50], the TiO_2 anode and the platinum cathode were placed, respectively, in an anolyte having a higher pH value than that of the catholyte. Effectively, this decreased the required cell potential and provided the overpotential so as to drive the cell reactions to higher current densities and facilitating the hydrogen and oxygen evolutions.

Current flowed from the Pt electrode to the TiO_2 electrode, when the TiO_2 front surface was illuminated. The current direction revealed oxygen evolution at the TiO_2 anode and hydrogen evolution at the platinum black cathode. The starting potential of the oxidation reaction mostly corresponded to the flat-band potential that remained constant in an electrolyte solution of given pH. When the TiO_2 front surface was illuminated with light from a 500 W xenon lamp, a current of a few milliamperes flowed. The quantum efficiency was estimated to be ~10%. Thus, Fujishima and Honda accomplished the decomposition of water to generate oxygen and hydrogen by illumination of n-type TiO_2 by visible light without the application of an external bias, however, by maintaining a pH gradient that is equivalent to applying an external bias. This discovery could not have been made at a more opportune time. With the first oil crisis of 1973, there was worldwide awareness of the depleting petroleum resources, and major efforts to find alternative resources of energy were initiated.

7.3.2 Strontium Titanate

Watanabe et al. [51] were the first to observe decomposition of water into hydrogen and oxygen when strontium titanate single crystals were illuminated with energetic light of wavelength λ in the range 350–390 nm, that is, having optimum photon energy for intrinsic absorption by $SrTiO_3$ without the application of an additional bias. 1 mm thick, (001)-oriented $SrTiO_3$ perovskite single-crystal platelets having an area of 2.5 cm^2 were first heat-treated in high vacuum at 600°C for several hours to reduce their resistivity to ~10^3 Ωcm. The crystal turned blue probably due to oxygen vacancies. The face in contact with the electrolyte was illuminated with a light from a Xe lamp through a monochromator or filter. The opposite face was connected to a platinized platinum electrode having an area of 10 cm^2 in a separated compartment. For this purpose, a barrier-free contact was prepared by polishing followed by washing in nitric acid and in deionized water. The reduced crystal behaved like an n-type semiconductor. The n-type $SrTiO_3$ anode was placed in an anolyte and the platinum cathode was placed in a catholyte. The two compartments were connected through an agar salt bridge. The flat-band potential was found to increase with the pH at a rate of ~60 mV/pH as is expected with oxide electrodes because they adsorbed the same number of H^+ and OH^- ions, that is, they attain the PZC. The flat-band potential of $SrTiO_3$ electrode was found to be 0.3 V more cathodic than that of a TiO_2 electrode. Holes from the electron–hole pairs generated in the depletion region of the n-type $SrTiO_3$ crystal as well as those generated within a diffusion length from the depletion region and diffuse into it are propelled to the electrolyte interface, where they oxidize water to generate oxygen molecules, whereas the electrons flow toward the back contact and to the platinum cathode and reduce water to generate hydrogen molecules. The formal quantum efficiency was found to be 0.2 at a wavelength of 350 nm. The photoelectrochemical cell of n-type $SrTiO_3$ anode and platinum cathode worked efficiently even when the pH values of anodic and cathodic compartments were nearly the same. Moreover, the anodic dissolution of strontium titanate was negligible.

Wrighton et al. [52] were the first to show sustained conversion of water to hydrogen and oxygen using an n-type semiconductor $SrTiO_3$ anode illuminated with a light wavelength $\lambda < 330$ nm and a platinum cathode in an electrochemical cell without an applied external bias. Both the electrodes were placed in an aqueous electrolyte often without the SCE reference electrode. Squares of 5 mm × 5 mm size and 1 mm thick $SrTiO_3$ single crystals having cubic perovskite structure were heat-treated in hydrogen at 1050–1100°C for 4 h to obtain blue black n type semiconductivity. Illuminated $SrTiO_3$ anode was in contact with the electrolyte. A gallium–indium eutectic was rubbed onto the opposite surface of the $SrTiO_3$ crystal. A copper wire attached to it with conducting silver epoxy provided a connection to the platinum cathode consisting of a 1 mm diameter and 50 mm long platinum wire. The electrolyte was 9.1–9.5 M NaOH solution. The photoeffect was observed at photon wavelengths below 390 nm, the maximum of the photoeffect being at 330 nm. The formal quantum efficiency at wavelengths below 330 nm for electron flow was 1.0 ± 0.20 at applied potentials of 1.5–2 V. The quantum efficiency reduced to ~0.2 or lower on irradiation with wavelengths above 350 nm. This is because the less energetic photons are not absorbed in the depletion region or within a diffusion length and, therefore, the probability of their diffusing to the depletion region is low. Careful measurements in oxygen-18-labeled H_2O showed that the oxygen did not originate from the $SrTiO_3$ crystal. The photoelectrode stability was also confirmed by verifying that there was no weight loss.

Similar results of photoelectrolysis of water in cells with n-type $SrTiO_3$ single-crystal anode and platinized Pt cathode were reported around the same time in a preliminary communication by Mavroides et al. [53] who measured the maximum quantum efficiency at zero bias of 10% at a *hν* of 3.8 eV. The quantum efficiency is approximately an order of magnitude higher than that of TiO_2.

The preceding results of photoelectrochemical water splitting to generate H_2 and O_2 and photoelectrode stability are impressive if illumination with UV light is contemplated. However, the bandgap of $SrTiO_3$ is 3.2–3.4 eV and therefore excitation with a wavelength below 390 nm is required for photoelectrolysis of water. Since only ~3% of the energy is available from sunlight having wavelengths shorter than 400 nm, strontium titanate would not be an attractive choice for the generation of solar hydrogen [52].

Domen and colleagues [54–57] have achieved stoichiometric generation of hydrogen (24 μmol/h) and oxygen (11 μmol/h) by splitting of water using aqueous suspension of a few-micron-sized $SrTiO_3$ catalyst particles loaded with Ni/NiO_x cocatalyst having a size of a few tens of nanometers. Ni/NiO_x cocatalyst islands were loaded on $SrTiO_3$ particles by impregnation in $Ni(NO_3)_2$ solution. The dried powder was first reduced by heating in hydrogen at 500°C for 2 h to obtain metallic nickel islands on $SrTiO_3$ particles followed by a mild reoxidation in oxygen at 200°C for 1 h only to the external surface of Ni leaving behind a nickel core. It was found that H_2 reduction and O_2 reoxidation treatments were essential for activation [57]. Recently, hydrogen production by the photocatalytic overall water splitting on $NiO/Sr_3Ti_2O_7$ has been studied by Jeong et al. [58]. The catalyst $Sr_3Ti_2O_7$ prepared by the polymerized complex method (PCM) was found to have better activity than that prepared by the conventional solid-state reaction method (SSRM). Photocatalytic activity for water splitting of PCM $Sr_3Ti_2O_7$ increased substantially by loading with an optimum amount (3 wt%) of Ni treated by reduction–reoxidation treatments. The photocatalytic activity of PCM $Sr_3Ti_2O_7$ was maintained for a longer time (120 h), whereas that of the SSRM-based catalyst lasted only for 85 h. $SrTiO_3$ with rhodium cocatalyst has also been reported to show a reasonably high photocatalytic activity [59].

7.3.3 Oxides with NaOH and Carbonates

Following the pioneering research of Fujishima and Honda [17,49,50] on the generation of hydrogen and oxygen by the decomposition of water in a photoelectrochemical cell with TiO_2 anode and platinum cathode, considerable research has been carried out on titanium oxide and other oxide semiconductors for the direct conversion of solar to chemical energy, making TiO_2 the most studied photocatalyst. The advantages of titanium oxide are low cost, ease of preparation, stability, and high photocatalytic activity. Since it also has the major disadvantage of negligible photochemical activity with visible light, lowering its bandgap and thus improving its photocatalytic performance under visible light especially by doping with transition metals has been attempted by many groups [60]. Unfortunately, such efforts have only resulted in insignificant changes in the bandgap energy of TiO_2. Moreover, in most cases stoichiometric evolution of H_2 and O_2 has not been observed, the evolution of O_2 being very small. UV irradiation of Pt/TiO_2 suspensions in pure water generated only a small quantity of hydrogen. H_2 evolution rate decreased gradually with time [10,61]. Stoichiometric hydrogen and oxygen were simultaneously generated on NaOH-coated Pt/TiO_2, Rh/TiO_2, and Pd/TiO_2 photocatalysts from water vapor [62,63]. Two weight percent Pt, Rh, and Pd metals were photodeposited on anatase TiO_2 powder that was reduced at 600°C in hydrogen for 6 h to increase the electrical conductivity. An amount of 0.3 g grayish catalyst powder spread on a 16 cm^2 of quartz reaction cell was coated with NaOH by impregnation. It was outgassed before the introduction of water vapor and illuminated with filtered 500 W high-pressure Hg lamp. Rhodium was more active than platinum, whereas palladium was least active. The photolysis of Rh/TiO_2 was maximum at 15 wt% NaOH coating at ~2 Torr water vapor pressure. The liquid water could not be used in this system because the backward reaction was very fast. Sayama and Arakawa [61,64] discovered by chance that the addition of carbonate salts, especially sodium carbonate, to aqueous Pt/TiO_2 suspension is significantly effective in stoichiometric photodecomposition of liquid water. The photocatalytic reaction was performed in a closed, gas-circulating fused quartz system illuminated with an inner 400 W high-pressure Hg lamp covered with a water jacket (filter). A reaction mixture consisting of 0.3 g anatase TiO_2 powder, 0.3 wt% H_2PtCl_6, and additives of carbonates such as Na_2CO_3, K_2CO_3, $NaHCO_3$, and Li_2CO_3; or sodium compounds such as NaOH, NaCl, Na_2SO_4, Na_3PO_4, Na_2HPO_4, and Na_3PO_4 (0.003 M) with Na_2HPO_4 (0.15 M) in 350 mL distilled water were mixed well and deaerated. Argon was introduced into the circulation system. After the preparation of the Pt/TiO_2 photocatalyst by *in situ* photochemical deposition of Pt on TiO_2 irradiation, the evolved gases were pumped away. The evolution of hydrogen and oxygen was detected with an inline gas chromatograph. After continued irradiation, a small amount of hydrogen without any oxygen evolution was detected from pure distilled water when no additives were used. For the cases of sodium compound additions, the evolution rate of hydrogen increased slightly and decreased to zero with irradiation time, again without any oxygen evolution. For addition of carbonates, especially Na_2CO_3, the evolution rate of hydrogen increased significantly accompanied by oxygen evolution, the amount becoming stoichiometric (2:1) after 2 h. The evolution rates decreased gradually with irradiation time due to suppression by pressure buildup, and hence recovered after pumping away of the evolved gases. With each run the H_2 and O_2 evolution rates increased, saturating after the ninth run, that is, ~70 h. Thus the reaction had an induction period and the photocatalytic activity improved with irradiation time. The highest initial evolution rates for optimum amounts of Na_2CO_3 (0.76 M) at a pH of 11 and of K_2CO_3 (1.45 M) at a pH of 11.8 were 568 μmol/h of H_2 and 287 μmol/h of O_2 and 20 μmol/h of H_2 and 10 μmol/h of O_2, respectively. Overall, it was concluded that the only high concentration of carbonate ions and neither Na^+ cations nor pH value contributes significantly to the stoichiometric H_2 and O_2 evolution. The optimum loading of Pt in the presence of carbonate was 0.3 wt%. However, the optimum loading of Pt on TiO_2 in the presence of a sacrificial agent was 2 wt%. The improvement of reaction rate with irradiation time was found to be associated with the decrease in back-reaction ($H_2 + O_2 \rightarrow H_2O$) rate over the catalyst. The back-reaction rate over Pt/TiO_2 was maximum for pure water, intermediate for aqueous NaOH, and minimum for aqueous Na_2CO_3. Carbonate ion was found to aid H_2 and O_2 evolution as follows: in the presence of carbonate ions, Pt was covered with some titanium hydroxides, effectively reducing the back-reaction rate. Moreover, carbonate species covering the TiO_2 surface aided the irreversible desorption of O_2 through the dissociation of peroxycarbonates resulting in the desorption of CO_2 and O_2 [61,65]. The photocatalytic activity was found to be affected by the total pressure of oxygen or inert gases such as argon, helium, or neon in the reaction chamber, probably because the increase in retention time of H_2 and O_2 near the catalyst increases the back-reaction rate. Addition of Na_2CO_3 has been found to be effective in sustained stoichiometric H_2 and O_2 evolution for over 60 h in the case of the following kinds of aqueous metal- or metal-oxide-loaded semiconductor catalyst suspensions in 350 mL of water illuminated with a 400 W high-pressure Hg lamp: 1 g TiO_2 with Pt (0.3 wt%), RuO_2 (1 wt%), NiO_x (1 wt%); 1 g Ta_2O_5 with RuO_2 (1 wt%), NiO_x (1 wt%); 1 g ZrO_2 with Pt (0.3 wt%), RuO_2 (1 wt%), NiO_x (1 wt%); 1 g $SrTiO_3$ with Pt (1 wt%), Rh (0.1 wt%), NiO_x (1 wt%); $K_4Nb_6O_{17}$ with Pt (1 wt%); 1 g $Na_2Ti_6O_{13}$ with Pt (0.1 wt%), RuO_2 (1 wt%); 1 g $K_2Ti_6O_{13}$ with Pt (0.1 wt%), RuO_2 (1 wt%); and 1 g $BaTi_4O_9$ with RuO_2 (1 wt%). The photocatalytic activity was found to be enhanced in the case of Pt-loaded photocatalysts. Under the same conditions, the activity of $K_4Nb_6O_{17}$ with Rh (0.1 wt%) and NiO_x (1 wt%) was much better in pure water and was reduced after the addition of Na_2CO_3 [61,66]. This is because in some materials the carbonate anions adsorbed on the surface changing various parameters such as pH and ionic conductivity of solution that affect photocatalytic activity [61]. Some of the more significant results shown in Tables 7.2 and 7.3 have been extracted from the data presented by Arakawa and Sayama [61]. Experiments were also conducted for solar generation of hydrogen using the newly developed Na_2CO_3 addition method with three kinds of photocatalysts: 0.1 wt% Pt/TiO_2, 3 wt% NiO_x/TiO_2, and 3 wt% RuO_2/TiO_2. 50 mg of

TABLE 7.2

Rate of Photocatalytic H_2 and O_2 Evolution over Various Kinds of Metal-Loaded Simple Semiconductor Catalysts Suspended in Both Aqueous Solution Containing Optimum Quantity of Na_2CO_3 and Pure Water

		Rate of Gas Evolution/μmol/h			
		In Na_2CO_3 Solution		In Pure Water	
Semiconductor	**Loaded Material**	H_2	O_2	H_2	O_2
TiO_2	Pt (0.3 wt%)	78	38	2	0
	RuO_2 (1 wt%)	34	17	trace	0
	NiO_x (1 wt%)	64	32	1	0
Ta_2O_5	Pt (0.1 wt%)	1	0	1	0
	RuO_2 (1 wt%)	68	34	32	17
	NiO_x (1 wt%)	153	79	190	99
ZrO_2	Pt (0.1 wt%)	53	23	trace	0
	RuO_2 (1 wt%)	12	6	11	5
	NiO_x (1 wt%)	43	22	129	70

Note: Catalyst, 1 g (TiO_2: 0.3 g); water, 350 mL; Na_2CO_3, optimum quantity. An inner irradiation-type quartz cell, high-pressure Hg lamp (400 W) from Arakawa, H. and Sayama, K., *Photocatalysis, Science and Technology*, Kodansha/Springer, Berlin, 2002.

TABLE 7.3

Rate of Photocatalytic H_2 and O_2 Evolution over Various Mixed Oxide Semiconductor Catalysts Suspended in Both Aqueous Solution Containing 80 g of Na_2CO_3 and Pure Water

		Rate of Gas Evolution/μmol/h			
		In Na_2CO_3 Solution		In Pure Water	
Semiconductor	**Loaded Material**	H_2	O_2	H_2	O_2
$SrTiO_3$	Pt (1 wt%)	10	4	9	2
	Ru (0.1 wt%)	48	14	20	4
	NiO_x (1 wt%)	41	21	9	4
$K_4Nb_6O_{17}$	Pt (0.3 wt%)	451	217	trace	0
	NiO_x (1 wt%)	60	28	403	197
	RuO_2 (1 wt%)	41	20	211	100
$Na_2Ti_6O_{13}$	Pt (0.1 wt%)	15	5	trace	0
	RuO_2 (1 wt%)	55	25	5	2
$K_2Ti_6O_{13}$	Pt (0.1 wt%)	63	17	trace	0
	RuO_2 (1 wt%)	49	24	11	1
$BaTi_4O_9$	Pt (0.1 wt%)	2	trace	trace	0
	RuO_2 (1 wt%)	36	18	30	13

Note: Catalyst, 1 g; water, 350 mL; Na_2CO_3, 80 g. An inner irradiation-type quartz cell, high-pressure Hg lamp (400 W) from Arakawa, H. and Sayama, K., *Photocatalysis, Science and Technology*, Kodansha/Springer, Berlin, 2002.

photocatalyst, 40 mL of pure water, and a fixed amount of Na_2O_3 were introduced into a glass photoreactor having a 95 cm^2 quartz window. The photoreactor was evacuated up to 35 Torr, placed outdoor on a shaker, and exposed to sunlight for 6.5 h in summer at Tsukuba. For the total period of 6.5 h of outdoor exposure, 0.1 wt% Pt/TiO_2 photocatalyst produced only a small amount of hydrogen because of the inadequate suppression of the back reaction with sunlight. A 3 wt% RuO_2/TiO_2 photocatalyst produced 161 mL/m^2 of H_2 and 74 mL/m^2 of O_2 on a sunny day. A 3 wt% NiO_x/TiO_2 photocatalyst produced 420 mL/m^2 of H_2 and 184 mL/m^2 of O_2 on a sunny day and 170 mL/m^2 of H_2 and 68 mL/m^2 of O_2 on a cloudy day. The quantum efficiency was >1.79%. It has been estimated that a similar setup using 50 kg of NiO_x/TiO_2 photocatalyst in a large square-shaped 100 $m^2 \times 100$ m^2 pond would produce ~4200 L of H_2 and 1800 L of O_2 on a sunny day. These are the first instances of stoichiometric and steady-state water splitting using oxide semiconductor photocatalysts exposed to solar irradiance [61,66].

7.3.4 Oxynitrides and Doping with Other Anions

Recently, several researchers have experimented with anion doping to extend the photocatalytic activity to the visible portion of the solar spectrum. Doping of titanium oxide with transition metal cations has not been very successful because of the generation of deep localized *d* states that act as recombination centers or because of thermal instability [16]. Often, the photoactivity of the cation-doped TiO_2 decreased even in the UV region due to an increase in the recombination centers. Therefore, Asahi et al. [67] chose anionic species to dope TiO_2 so as to ensure that the conduction band-minimum and impurity states remain above the H_2/H_2O level and also that there is adequate overlap of bandgap states with the band states of TiO_2 so as to be able to transfer photoexcited carriers to reactive sites at the catalyst surface within their lifetime. Based on the calculations of densities of states for the substitutional doping with C, N, F, P, or S for O in anatase TiO_2 crystal, they found substitutional doping of N to be the most effective because its *p* states contribute to the bandgap narrowing by mixing with O 2*p* states. Doping with S also showed a similar bandgap narrowing but was deemed as not suitable because of its large ionic radius. The states introduced by C and P were found to be too deep in the gap to obtain adequate overlap of bandgap states with the band states. Substantial photocatalytic activity at wavelengths below 500 nm was observed in the case of yellowish and transparent $TiO_{2-x}N_x$ films prepared by reactive sputtering for decomposition of methylene blue but not for water splitting [67]. Nitrogen-doped titanium oxide films were deposited on aluminum substrates by reactive magnetron sputtering in a mixture of argon with O_2 and N_2. Crystallinity, surface morphology, and optical absorption were found to be influenced by the O_2/N_2 ratio, whereas the photocatalytic activity estimated by benzamide removal efficiency was found to be determined by the amount of doped N^{3-} in the film as measured by x-ray photoelectron spectroscopy (XPS). The benzamide removal efficiency of 89.9% was obtained for $TiO_{0.4}N_{0.6}$ film prepared with the optimum $N_2/(O_2 + N_2)$ ratio of 80% as compared to that of 61.8% obtained for sputtered TiO_2 film [68]. Nanocrystalline, porous, nitrogen-doped TiO_2 thin films prepared by reactive direct current (DC) magnetron sputtering in argon, oxygen, and nitrogen plasma, with varying nitrogen contents showed visible light absorption in the wavelength range from 400 to 535 nm [69]. XPS analysis was used to observe O–Ti–N bond formation in nitrogen-doped TiO_2 nanoparticles [70]. Dioxygen was generated from water at moderate anodic potentials by the illumination of the 850 nm thick, rutile $TiO_{2-x}N_x$ thin-film photoanodes prepared by reactive DC magnetron sputtering in Ar, O_2, and N_2. The new states were attributed to nitrogen doping and not to pure $TiO_{2-x}N_x$ [71]. Nitrogen-doped TiO_2 powders were prepared by heating titanium hydroxide with urea at a low temperature of ~350°C so as to maintain a relatively large specific surface area of 92 ± 2 m^2/g. The samples showed visible-light photocatalytic activity below the wavelength of ~550 nm [72]. Nanocolloidal $TiO_{2-x}N_x$ photocatalyst has been formed by the direct nitridation of anatase TiO_2 nanoparticles in colloidal solution and a partially agglomerated gel (solution) with alkylammonium salts. Depending on the degree of TiO_2 nanoparticle agglomeration, absorption onset of catalytically active $TiO_{2-x}N_x$ particles having anatase structure extended well into the visible region at ~550 nm. The introduction of a small quantity of palladium in the form of chloride or nitrate was found to facilitate further nitrogen uptake and produced a material absorbing well into the near-infrared (IR). The final $TiO_{2-x}N_x$ products that were free from organic incorporation were found to readily photodegrade methylene blue and lead to the catalytic oxidation of ethylene when they were placed as gels on surfaces [73].

Dark gray, chemically modified n-type TiO_2 films in which carbon substitutes for some of the lattice oxygen atoms, prepared by controlled combustion of Ti metal in a natural gas flame in 13 min, absorbed light at wavelengths below 535 nm and had a lower bandgap energy of 2.32 eV than that of rutile 3 eV. When illuminated with 40 mW/cm^2 radiation from a Xe lamp, they have been reported to have provided a total conversion efficiency of 11% and the maximum photoconversion efficiency of 8.35% for the generation of H_2 and O_2 by water splitting, at an applied potential of 0.3 V. This would have been considered a very impressive PEC photoconversion efficiency as compared to 1% for n-type TiO_2 biased at 0.6 V, except that the results have not been duplicated so far [74].

Recently, a photoelectrochemical cell has been designed using carbon-doped TiO_2 ($TiO_{2-x}C_x$) array as photoanode and platinum nanoparticles uniformly dispersed on titania nanotube array as cathode [75]. The photoanode consisting of uniform and conformable, self-organized, hexagonally ordered TiO_2 nanotube array having a wide range of nanotube structure was synthesized by sonoelectrochemical anodization technique using aqueous ethylene glycol and ammonium fluoride solution. The titania nanotubes were annealed under reducing hydrogen atmosphere to convert the amorphous nanotube arrays to photoactive anatase phase and also to dope TiO_2 with carbon from ethylene glycol to give a $TiO_{2-x}C_x$ photoanode. The cost of PEC cell was reduced by using an inexpensive cathode with only 0.4 wt% Pt on TiO_2 while maintaining the performance comparable to that of bulk platinum electrode. The cathode was prepared by synthesizing Pt nanoparticles (≤10 nm) obtained by the reduction of a Pt salt to Pt(0) into the titania nanotubular arrays by the incipient wet method. The PEC cell was found to provide reasonably good photocurrent density J_{ph} of 2.5–2.8 mA/cm^2 at a low external bias of –0.4 V with respect to Ag/AgCl electrode. The PEC cell was found to run continuously for 80 h without affecting the photocurrent for hydrogen generation by water splitting under the illumination of simulated 1 sun intensity. The efficiency of the PEC cell as measured by the reduction of electrically supplied energy under UV and visible light illumination was found to be 13.3% and 8.5%, respectively [75].

TiO_2-based polycrystalline powder doped with S were prepared by heating TiS_2 powder at 300°C or 600°C for 5 h in air. TiS_2 turned into anatase TiO_2 after annealing at 600°C. The residual S atoms occupied O-atom sites in TiO_2 to form Ti–S bonds that were found to cause a significant shift in the absorption edge to lower energy [76]. Chemically modified TiO_2 photocatalysts in which S (S^{4+}) substitutes for some of the lattice titanium atoms showed strong absorption for visible light and high activities for degradation of methylene blue, 2-propanol in aqueous solution, and partial oxidation of adamantine under illumination by light with wavelengths of >440 nm. The oxidation state of the S atoms incorporated into the TiO_2 particles was determined by x-ray photoelectron spectra to be mainly S^{4+} [77].

A novel multifunctional glazing composed of a simple bilayer coating of F-doped SnO_2 (SnO_2:F) as an inner layer and titanium dioxide (TiO_2) as an outer layer has been proposed to obtain a composite having both low-emissivity (low-e) of SnO_2:F and visible-light-responsive photocatalytic activity of TiO substitutionally doped with nitrogen. TiO layer having a thickness of 80–320 nm was deposited by reactive magnetron sputtering from a titanium target in an Ar/O_2 mixture on SnO_2:F-coated glass substrates at 370°C. Photocatalytic activity was modified by irradiation of TiO_2 films at 315°C with 0.2–4.0 keV ion beams of Ar, H, and N. After ion beam irradiation, an overcoat of very thin 20 nm TiO_2 film was deposited at 315°C. Coirradiation with 0.2 keV ions of H and N onto the TiO_2 was found to improve the photocatalytic activity under visible-light (≥370 and ≥410 nm) illumination by more than three times compared to that of nonirradiated samples. Although the bilayer coating was being developed to achieve both heat-reflecting and antibacterial properties in architectural, automotive, microwave oven, and freezing showcase-type applications, it is being included here to highlight the potential of doping by ion bombardment [78].

7.3.4.1 Tantalum Oxynitride

The TaON powder illuminated with visible irradiation having a wavelength in the range 420–500 nm has been found to function as a stable photocatalyst for the oxidation of water into O_2 in aqueous $AgNO_3$, which provides a sacrificial electron acceptor Ag^+ [79]. TaON was also used for the reduction of water to hydrogen in aqueous methanol, which is oxidized as a sacrificial agent to CO_2. Adding optimum amounts of cocatalyst Ru increased the hydrogen evolution rate significantly. However, loading with Pt, Rh, and Ir did not increase the rate [80]. β-TaON is yellow–yellowish-green in color and has the same structure as baddeleyite, monoclinic ZrO_2. TaON was prepared by heating Ta_2O_5 powder at 850°C for 10 or 15 h in flowing NH_3 with or without a small amount of nitrogen gas containing water vapor. The bandgap of TaON was estimated to be 2.5 eV.

The optical bandgap of $Y_2Ta_2O_5N_2$ powder synthesized by nitriding $YTaO_4$ under flowing ammonia was found to be 2.2 eV. $Y_2Ta_2O_5N_2$ photocatalyst illuminated by radiation having wavelengths of >420 nm from Xe lamp equipped with a cutoff filter was used to oxidize water to oxygen and reduce water to hydrogen in the presence of sacrificial agents viz. $AgNO_3$ and methanol. Incorporation of Pt and Ru increased the hydrogen evolution activity [81].

Although any specific method has not achieved a breakthrough, various techniques have been shown to be reasonably effective. Thus, a permutation and combination of some of them may produce the desired effect.

7.3.5 Niobates, Titanates, and Layered Compounds

Microporous crystals of compounds such as $K_4Nb_6O_{17}$ with a layered structure as in mica and $K_2La_2Ti_3O_{10}$ with a layered perovskite structure both have unique reaction sites (Figure 7.8) [57,82,83]. Therefore, they have shown remarkable photocatalytic activity on loading with Ni/NiO cocatalyst. $K_4Nb_6O_{17}$ is an ion-exchangeable layered compound consisting of macropolyanion sheets of niobates (NbO_{17}^{4-}) with K^+ ions located in the interlayer space. The niobate sheets have anisotropy along the stacking direction, that is, they have the front and back sides. The reactant water molecules are intercalated into the interlayer spaces when they are hydrated in an aqueous solution. Two different types of

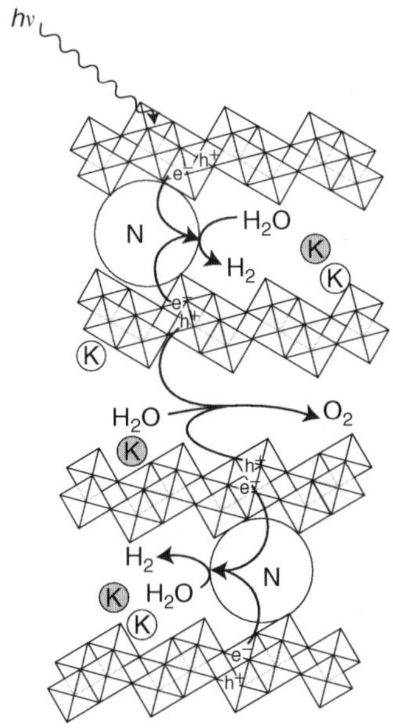

FIGURE 7.8
Water splitting in a layered structure.

interlayer spaces existing alternately exhibit different ion exchange properties. K^+ ions in type I interlayer space can be exchanged with Li^+ ions, Na^+ ions, and some polyvalent cations such as Ni^{2+}. However, those at the type II interlayer space can be exchanged only with monovalent cations. $K_4Nb_6O_{17}$ illuminated with energetic (>3.3 eV) photons showed low activity for stoichiometric hydrogen and oxygen evolution by water splitting in distilled water. Considerable enhancement with 5% quantum efficiency was obtained by loading with 0.1 wt% Ni/NiO cocatalyst. Even better results were obtained with $Rb_4Nb_6O_{17}$ that showed 10% quantum efficiency for 330 nm radiation [54,57,82,83].

Layered perovskite photocatalysts such as $A_2La_2Ti_3O_{10}$ (A = K, Rb) and $A_{2-x}La_2Ti_{3-x}Nb_xO_{10}$ (x = 0.5 and 1.0) having bandgaps in the range 3.4–3.5 eV, with Ni/NiO cocatalyst have also shown significant activity for stoichiometric H_2 and O_2 generation by water splitting [54,57,82]. All the perovskites were prepared by the conventional ceramic method and the Brunauer–Emmer–Teller (BET) surface area was ~1 m^2/g. It was found that hydration of interlayer space was essential for obtaining high activity.

Kudo and colleagues [54,84–88] surveyed new photocatalyst materials based on structural regularity. Photocatalytic activity of $K_3Ta_3Si_2O_{13}$ having a pillared structure for water splitting was found to increase substantially with the loading of a small amount of NiO cocatalyst. They found photocatalytic activity for H_2 and O_2 generation in alkali and alkaline earth tantalates, $LiTaO_3$ being more active without a cocatalyst. NiO-loaded $NaTaO_3$ was most active among all the tantalates. La doping of NiO–$NaTaO_3$ increased its photocatalytic activity further and the quantum yield for H_2 and O_2 generation by water splitting reached ~50%.

Reduction of particle size was found to increase the activity as long as the crystallinity was maintained or restored by calcination, etc. Surface features such as steps that act as active sites were also found to increase activity [54,57,84].

It is interesting to note that Domen et al. [54] also found mechano-catalytic activity in several catalysts where mechanical energy is provided instead of photon energy.

7.4 Cadmium Sulfide

The potential market for hydrogen from H_2S is quite large. Total sulfur recovered mostly from H_2S decomposition in the United States in 1991 was 6.6 million t equivalent to 4.7 teraliters (TL) of hydrogen [89–91]. Usual recourse to sacrificial agents to avoid photocorrosion of the semiconductor would be considered unattractive except when the reducing agents are waste products. Sulfide and sulfite are produced in large quantities as undesirable by-products from hydrodesulfurization and sour gas scrubbing processes in oil and natural gas refineries. Cadmium sulfide, which can be excited by a larger fraction of sunlight compared to titanium oxide because of its much lower bandgap of 2.42 eV than that of TiO_2, has been extensively studied for PEC hydrogen generation by water splitting, especially in the powder form. Williams [92], who studied the Becquerel PV effect at the semiconductor–electrolyte interface using CdS and other binary compound semiconductor crystals, showed that the photoelectromotive force (emf) generated at the illuminated CdS single crystal is accompanied by the chemical reaction of the crystal in which Cd^{++} ions go into the KCl solution and sulfur deposits on the CdS surface. Inoue et al. [93] have added reducing agents to avoid dissolution of CdS single crystal in the electrolyte. Harbour and colleagues [94,95] employed disodium salt of ethylenediaminetetraacetic acid (ETDA), as a sacrificial donor for photogeneration of hydrogen from illuminated CdS and platinized CdS particle dispersions while avoiding CdS photocorrosion. Aqueous dispersions of CdS powder have been shown to photosensitize the reduction of H_2O also by the sacrificial electron donor cysteine besides ETDA as discussed earlier. Platinized CdS was found to be 20 times efficient for hydrogen production [96,97]. A photocatalytic system for the production of hydrogen and sulfur via visible-light-induced decomposition of H_2S in aqueous solutions containing colloidal particles of platinized CdS is reported. Colloidal CdS was prepared before the experiments in the presence of polyvinyl alcohol. Both sodium sulfide and dissolved H_2S were used in the photocatalytic reaction. The maximum quantum yield of H_2 generation in the CdS/Pt/Na_2S photosystem was 12% [98]. Efficient hydrogen generation was observed from suspensions of platinized CdS particles in aqueous sodium sulfite solution by illumination with visible light. However, sulfite ions were oxidized to sulfate and dithionate ions [99]. Photocatalytic activity of illuminated Pt-loaded CdS powder in sulfite solutions was found to decline over long (200 h) periods probably due to the deposition of metal impurities on the platinum part of the photocatalyst. The deterioration was avoided by the addition of a small concentration of (5 mM) ETDA that probably formed stable complexes with the metal ions, preventing them from depositing on platinum [99,100]. Visible ($\lambda > 350$ nm) light-induced hydrogen production was achieved from *in situ*-generated colloidal Rh-coated CdS in dihexadecylphosphate (DHP) surfactant vesicles in the presence of thiophenol (PhSH) [101]. Photosensitized charge separation and hydrogen production was achieved from irradiated colloidal Pt–CdS prepared *in situ* in sodium bis-2-ethylhexylsulfosuccinate (Aerosol–OT) micelle-entrapped water pools in isooctane by thiophenol dissolved in organic phase [102]. Rafaeloff et al. [103] generated *in situ* ~4 nm diameter colloidal CdS and coated *in situ* with Rh in 80–100 nm diameter dioctadecyldimethylammonium chloride (DODAC) and dioctadecyldimethylammonium bromide (DODAB) surface vesicles and optimized hydrogen generation. Active photocatalysts have been prepared by platinum deposition on microcrystalline CdS powders. The efficiency of photochemical hydrogen production with SO_3^{2-} ion or a mixture of S^{2-} with SO_3^{2-} or $H_2PO_2^-$ ions was improved by photoetching the powders [103]. Illumination of aqueous alkaline CdS dispersion by visible light ($\lambda \geq 400$ nm) has been used to dissociate H_2S into hydrogen and sulfur. Hydrogen generation yield can be improved by coating CdS particles with small quantities of ruthenium oxide that acts as a catalyst for transferring electrons to hydrogen ions. Illuminated CdS dispersions containing both sulfite and sulfide ions can simultaneously generate hydrogen and thiosulfate. Hydrogen generation performance can be improved further in a system containing bare CdS in combination with RuO_2-loaded TiO_2 particles. The thiosulfate mechanism offers a potential route to a sustained, cyclic dissociation of H_2S that has been considered a potential source of hydrogen [104,105]. Here, sodium sulfide and sodium sulfite added to the electrolyte form thiosulfate as a result of photooxidation. Since thiosulfate is transparent, it does not block the light as does polysulfide [106,107]. H_2S formed by anaerobic bacterial digestion is also found in dilute form in underground aquifers [89]. Borrell et al. [108] developed various types of CdS-based photocatalysts for hydrogen generation from sulfide–sulfite solutions. Thermal treatment of CdS particles in air, argon, H_2 or H_2S, etching as well as deposition of small quantity of Pt or RuO_x catalyst were found to be beneficial. The best performance was obtained using ~1.4 μm sized CdS particles treated in air at 400°C and etched in concentrated HNO_3 and coated with 0.4 wt% Pt. Linkous et al. [89] have performed a series of experiments using platinized CdS powder suspensions, colloids, sedimentary dispersions, and immobilized particle beds as photocatalysts for the development of reactor design to produce hydrogen and sulfur using sunlight. They avoided the problem of absorption of sunlight from the build-up of polysulfide ion in an alkaline sulfide solution by the addition of sulfite ion SO_3^{2-} which results in the formation of a transparent thiosulfate solution. They found that on the photocatalyst weight basis, due to minimal scattering losses, photoactivity in immobilized and sedimentary systems could exceed that of colloidal and suspended systems, whereas on the basis of area, which is more important, photoactivity was greatest for colloidal systems followed by sedimentary, suspended, and immobilized systems. Immobilized systems were judged the best and colloidal systems the worst from the technological point of view. Sulfite addition was found to be inadvisable from practical considerations of sulfur recovery since the market for elemental sulfur greatly exceeds that of thiosulfate. Study of the kinetics of hydrogen production from illuminated $CdS/Pt/Na_2SO_3$ and $ZnS/Pt/Na_2SO_3$ dispersions showed that although the strong interaction permitted the use of low concentrations of sacrificial donor species, their concurrent interference adversely affected the reaction rate [109].

Subrahmanyam et al. [110] studied the effect of basic/acidic supports for CdS on the photocatalytic H_2S decomposition reaction. $Cd(NO_3)_2$ salt precursor for CdS preparation, 10 wt% CdS over basic support such as MgO, and 500 mg of photocatalyst such as Pt or RuO_2 gave a higher hydrogen yield. The efficiency of coprecipitation of CdS together with finely dispersed group VIII metals such as Pt, Pd, Ni, and Co on an ion exchange polymer was found to correlate with the overpotentials of cathodic hydrogen generation on the respective bulk metals. It has been shown using polymer-immobilized CdS that the activity could be further promoted by group I metals such as Cu [111]. Photocatalytic hydrogen evolution activity of CdS powder in aqueous Na_2SO_3 solution has been improved by the preparation of microcrystalline CdS powder by heat treatment with KCl powder [112].

7.5 Photosynthesis and Photoelectrochemistry, and Approaches Based on Photosynthesis

Photosynthesis is an energy-building process involving biosynthesis of simple carbohydrates such as glucose and starch from inorganic materials such as carbon dioxide and water, in the presence of light [8,37–40]. When a photosensitive molecule gets excited by absorbing a photon of appropriate energy an exciton is generated. An electron in its HOMO is excited to the LUMO creating an electron vacancy, that is, a hole in the HOMO. The electron–hole pair retains attraction and hence it is termed an *exciton*. The photoexcited molecule can donate the free electron in the HOMO to a neighboring molecule to which it is coupled. Moreover, the excited molecule can accept an electron from another molecule to which it is coupled. The electron thus accepted recombines with the hole in the LUMO. The electron transfer process continues through the chains of molecules until at one end the donated electron carries out a reduction reaction, whereas oxidative reaction is carried out at the other end through electron extraction. The overall result of the photosynthesis process is the conversion of light energy into chemical energy, releasing oxygen as a by-product, as shown in Equation 7.20.

$$CO_2 + H_2O \rightarrow (CH_2O) + O_2 \tag{7.20}$$

Studies using labeled oxygen have shown that oxygen evolved during the reaction is a product of water oxidation and is not derived from carbon dioxide. The overall reaction of synthesis of glucose ($C_6H_{12}O_6$) can be written as [8,37,113,114]

$$\text{Light} + 6CO_2 + 12H_2O^* \rightarrow 6O_2^* + 6H_2O + C_6H_{12}O_6 \tag{7.21}$$

A large number of glucose molecules are joined in a systematic manner to form the carbohydrate polymer starch having *n* number of glucose molecules—$(CH_2O)_n$.

The overall process takes place in two steps. In the first step, a photoexcited electron from an electron–hole pair generated by absorption of light is donated through intermediate steps for a reduction reaction. In the second step, the photogenerated reducing power is used to reduce CO_2 to carbohydrate. During the process, water is oxidized to O_2 and CO_2 is reduced to carbohydrate with the help of the hydrogen generated from water. Therefore, photosynthesis is described as a reduction–oxidation reaction or redox reaction [37,39].

In plants, the photosynthesis reaction takes place in specialized organelles termed *chloroplasts*. The chloroplasts are bounded in a two-membrane envelope with an additional third internal membrane called thylakoid membrane. This thylakoid membrane is a highly folded structure, which encloses a distinct compartment called thylakoid lumen. The chlorophyll found in chloroplasts is bound to the protein in the thylakoid membrane. The major photosensitive molecules in plants are the chlorophylls: chlorophyll *a* and chlorophyll *b*. They are coupled through electron transfer chains to other molecules that act as electron carriers. Structures of chlorophyll *a*, chlorophyll *b*, and pheophytin *a* are shown in Figure 7.9.

The initial conversion of light into chemical energy takes place in the thylakoid membrane. Besides the chlorophylls and series of electron carriers, the thylakoid membrane also contains the enzyme adenosine triphosphate (ATP) synthase. The enzymes that are responsible for the actual fixation of CO_2 and the synthesis of carbohydrate reside in the stroma that surround the thylakoid membrane. The stroma also contains deoxyribonucleic acid (DNA), ribonucleic acid (RNA), and ribosomes that are essential for protein synthesis [37].

7.5.1 Mechanism of Photosynthesis

The synthesis of carbohydrates takes place in two phases [8,37–39,114–121].

1. *Light reaction*. This is a photochemical phase and takes place in the thylakoid membrane of the chloroplast, where light energy is converted to chemical energy.

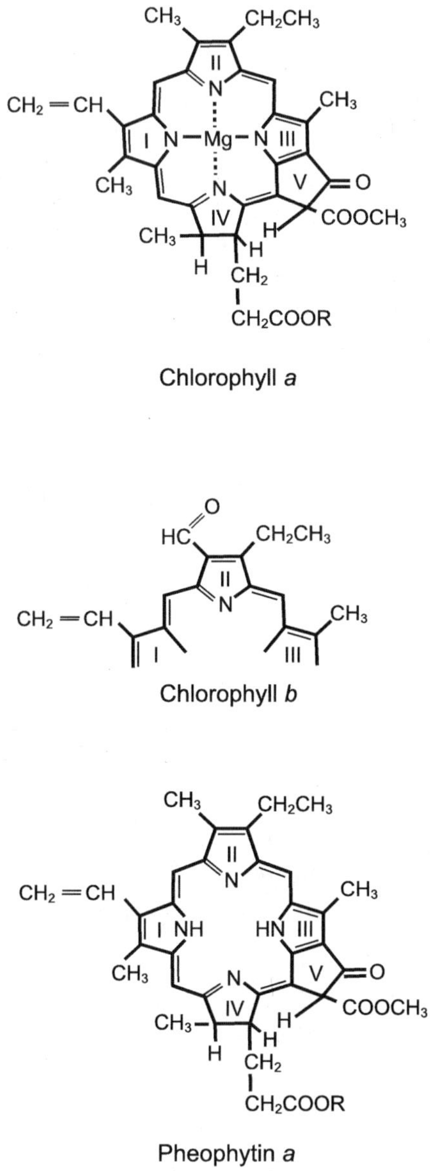

FIGURE 7.9
Structures of chlorophyll *a*, chlorophyll *b*, and pheophytin *a*.

Hence, the presence of light is essential for this step. In this reaction, chlorophyll, carotenoids, and several other pigments that are organized in clusters absorb light energy and store it in the form of ATP and nicotinamide adenine diphosphate (NADP) in its reduced state (NADPH).

2. *Dark reaction.* This is a biosynthetic phase that takes place in the stroma within the chloroplasts converting CO_2 to sugar. This process does not require light, but requires the products of the light reaction namely ATP and NADPH, which are

used to reduce CO_2 to form sugar and starch. This process is called the Calvin cycle [117].

During the process of photosynthesis, two chlorophyll molecules absorb light and get energized, that is, excited. The energetics of the redox reactions in the thylakoid membrane are performed by two excited chlorophyll molecules that are coupled by electron transfer chains through the membrane [8,37]. The circuit is closed by a proton transfer transport parallel to electron transfer through the membrane. This molecule can lose its energy by transferring it to nearby chlorophyll molecules. This energy is transferred from one chlorophyll to another till it reaches the reaction center which in turn gets excited and is ready to lose the electron. The flow of electrons sets up an electron gradient across the thylakoid membrane.

The trapping of light energy is the key reaction of photosynthesis. The light energy excites the photoreactive pigment from the ground level to an excited energy level by exciting an electron from HOMO to LUMO. This electron is then transferred to a nearby suitable electron acceptor. This process leaves behind a positively charged hole in the LUMO state of the electron donating molecule and a negatively charged electron in the acceptor molecule [37].

7.5.1.1 Types of Photosystems and Their Structures

In green plants, photosynthesis involving oxygen evolution depends on two photosystems, which differ slightly in the wavelength of light to which they respond—photosystem I (PS I) responds to light with a wavelength of 700 nm and photosystem II (PS II) responds to light with a wavelength of 680 nm. These two structures are membrane-bound light-sensitive components. During the process of photosynthesis, the electrons derived from water first flow through PS II, then through the cytochrome (cyt) b_6f that is also a membrane-bound system, and then through PS I [37,39].

PS II is a pheophytin–quinone type of system, which contains both chlorophyll *a* and chlorophyll *b* in equal quantity. The PS II has its excitation center at p. 680. The structure of PS II shows the presence of 10 polypeptide chains, 4 manganese ions, 1 chloride ion, and 1 tyrosine residue that forms a radical. X-ray crystallography studies have shown that manganese has been apparently selected for this role because of its ability to exist in multiple oxidation states Mn^{2+}, Mn^{3+}, Mn^{4+}, and Mn^{5+} and to form strong bonds with oxygen-containing species. The manganese center in the reduced form is at low energy level and so water donates its electrons to these manganese ions associated with p680.

The cyt b_6f complex (*f* stands for *feuille*, the French word for "leaf") consists of four subunits—one molecule of cyt *f*, one heme-containing cyt b_6, one iron–sulfur protein, and one bound plastoquinol. The cyt b_6f complex transports electrons from the outside to the inside of the thylakoid membrane.

PS I has a reaction center designated as p700 and has a high ratio of chlorophyll *a* to chlorophyll *b*. It consists of 13 polypeptide chains, 1 quinone molecule, and 4 Fe–4S clusters.

7.5.2 The Flow of Electrons from H_2O through PS I, Cyt b_6f, and PS II to NADP through the Z Scheme

The integration of PS I and PS II in chloroplasts occurs as shown in Figure 7.10 [37,39]. The overall reaction of the transfer of electrons starting from water takes place in three stages.

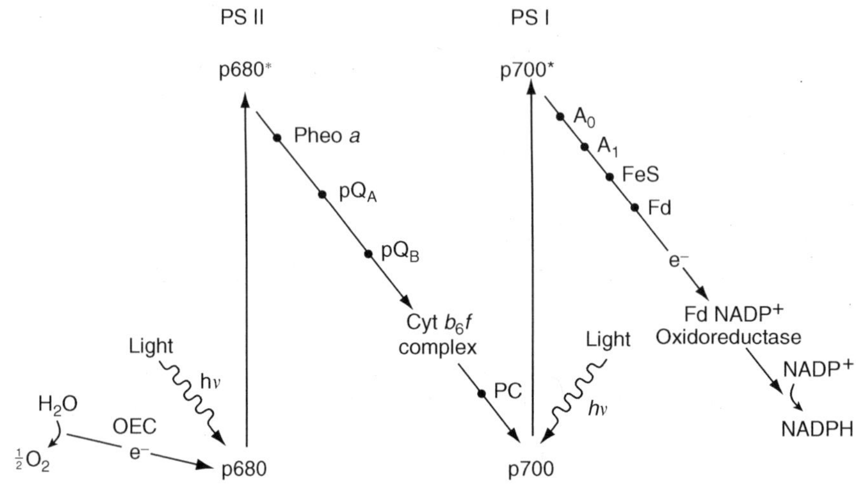

FIGURE 7.10
Integration of PS II and PS I via cyt b_6f in chloroplasts (Z scheme of photosynthesis). Pheo *a* represents pheophytin; pQ_A and pQ_B represent phytoquinone A and B, respectively; A_0 and A_1 represent electron acceptor chlorophylls, respectively; pc represents plastocyanin; and Fd represents ferridoxin.

During the first stage, the oxygen evolving complex (OEC) oxidizes two molecules of water (H_2O) to generate one molecule of oxygen (O_2) and four electrons [37,39]. This OEC is formed by binding together four protein-bound Mn ions associated with PS II together with two or three Ca^{2+} ions and four or five Cl^- ions. The OEC cycles through a series of oxidation states involving various combinations of Mn(III) and Mn(IV) while extracting protons and electrons from the H_2O molecules and releasing O_2 into the inner thylakoid space. A number of models have been proposed for the action of this OEC. These four electrons are taken up by the pigment p. 680 associated with PS II, in the presence of light and as a result the p680 gets excited shown in Figure 7.10 as p680*. The overall reaction catalyzed by PS II can be written as

$$2Q + 2H_2O \rightarrow O_2 + 2QH_2 \tag{7.22}$$

A single electron is transferred from p680* to a molecule of pheophytin *a* (Figure 7.9), represented as pheo *a* in Figure 7.10, chlorophyll *a* with its Mg^{2+} replaced by two protons, then to a plastoquinone (FeII) complex designated as Q_A. Subsequently, two electrons are transferred one at a time to a second plastoquinone molecule Q_B which takes up two protons at the stromal surface of the thylakoid membrane forming plastoquinol Q_BH_2. The equation showing the structural conversion of plastoquinone to plastoquinol is given in Figure 7.11. This particular reaction involves the transfer of two protons. All the remaining steps of the Z scheme, except the last step: NADP→NADPH, involve the transfer of only electrons (Figure 7.10).

1. From the plastoquinone pool, the electrons pass through the cyt b_6f complex, which generates much of the electrochemical proton gradient that drives the synthesis of ATP.

   The electrons subsequently pass to plastocyanin (PC), which is a copper-containing protein. The Cu-containing redox center of this 10.5 kD monomer cycles between Cu(I) and Cu(II) oxidation states. The structure of PC shows that

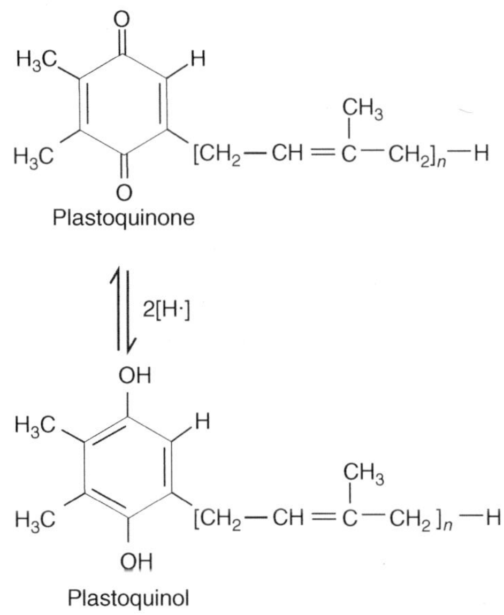

FIGURE 7.11
Coversion of plastoquinone to plastoquinol.

copper atom is coordinated with distorted tetrahedral geometry by four amino acids namely—a cysteine, a methionine, and two histidine residues. Plastoquinone brings about efficient electron transfer from cyt b_6f complex to PS I.

2. The final stage of the light reaction is catalyzed by PS I. A special pair of chlorophyll *a* molecules lies at the center of PS I and absorbs light at 700 nm and gets energized. The electrons are transferred down a pathway through the chlorophyll at site A_0, A_1, and a set of four Fe–4S clusters. From these the electrons are then transferred to ferridoxin (Fd), a soluble protein containing two Fe–2S clusters, coordinated to four cysteine residues. The positively charged hole in p700 is neutralized by the transfer of electrons from PC.

The overall reaction catalyzed by PS I is one electron oxidation–reduction reaction.

$$PC(Cu^+) + Fd_{ox} \xrightarrow{\text{light}} PC(Cu^{2+}) + Fd_{red} \tag{7.23}$$

The enzyme ferridoxin (Fd) NADP$^+$ oxidoreductase accepts the electron from Fd, one at a time, as it proceeds from its oxidized form through a semiquinone intermediate to its fully reduced form. The enzyme then transfers a hydride ion to NADP converting to its reduced state NADPH. Uptake of a proton in the reduction of NADP$^+$ further contributes to the proton gradient across the thylakoid membrane driving ATP synthesis.

$$2Fd_{red} + 2H^+ + 2NADP^+ \rightarrow 2Fd_{ox} + NADPH + H^+ \tag{7.24}$$

Thus the coordinate working of PS I and PS II creates the electron flow from H_2O to NADP.

The pathway of electron flow is called the *Z scheme of photosynthesis* because the diagram from p680 to p700 resembles the letter Z (Figure 7.10) [37,39].

7.5.3 Distribution of PS I, Cyt b_6f, and PS II and Its Significance

Freeze fracture electron microscopy has revealed the distribution of the preceding three components as

1. PS I occurs in the unstacked stroma lamellae in contact with the stroma, where it has access to $NADP^+$.
2. PS II is located almost exclusively between closely stacked grana, out of direct contact with the stroma.
3. Cyt b_6f is uniformly distributed throughout the molecule.

The high mobilities of plastoquinone (Q) and PC, the electron carriers that shuttle electrons between these particles permit the photosynthesis to proceed at a reasonable rate.

It has been observed that there are two valid reasons for the segregation of PS II and PS I by a distance of 10 nm. These are as follows:

1. The physical separation of PS II and PS I permits the chloroplasts to respond to changes in illumination. The relative amount of light absorbed by these two systems varies with the distribution of light harvesting complexes (LHCs) between the stacked and unstacked portion of the thylakoid membrane.

   Under the conditions of light illumination (under direct sunlight which has a high proportion of short-wavelength blue light), all other factors being equal, PS II absorbs more light than PS I. As such PS I cannot take electrons as fast as PS II can supply, leaving plastoquinone in its reduced state. This reduced plastoquinone activates a protein kinase that phosphorylates the threonine (Thr) residue of the LHCs that in turn migrate to the unstacked portion of the thylakoid membrane where it binds to PS I. As a result, a large portion of incident light is funneled to PS I.

   Under the conditions of low illumination (normally shady light, which has a high proportion of long-wavelength red light), PS I takes electrons faster than PS II can supply, leaving plastoquinone in its oxidized state. As a result, LHCs are dephosphorylated and migrate to the stacked portion of the thylakoid membrane where they drive to PS II.

   The chloroplast thus maintains a balance between the two photosystems by light-activated feedback mechanism.
2. If these two photosystems are located close to each other, the higher excitation energy of PS II (p680 versus p700) would pass a large fraction of its absorbed photons to PS I, via exciton transfer. These two particles are placed 10 nm away, thus eliminating this difficulty [37].

7.5.4 Photosynthesis in Algae and Bacteria

The process of photosynthesis also occurs in unicellular eucaryotes such as blue green algae and a number of bacteria such as cyanobacteria, purple bacteria, green sulfur bacteria, and *Rhodopseudomonas viridis*. The photochemical reactions that capture the energy of light are basically the same in bacteria, algae, and plants. Algae contain chloroplasts similar to higher plants. Photosynthetic bacteria contain either of the bacteriochlorophylls *a* or *b* or *d*, which have strong absorption bands in the region around 770 nm, depending on the bacterial species.

Reaction centers of bacteria contain polypeptides, bacteriochlorophylls, bacteriopheophytins, two quinines, and nonheme iron atom. In some bacterial species, both the quinones are ubiquinones, whereas in some others one of the quinones is menaquinone [37,39]. Depending on the bacterial species chloroplasts contain plastoquinone and phylloquinone. Structures of ubiquinone, menaquinone, and phylloquinone are provided in Figures 7.12 through 7.14, respectively.

The flow of electrons occurs in a similar manner from the excited pigment to cytochromes, quinones, pheophytins, ferridoxins, etc. The ATP synthase in the mitochondria of a bacterial system resembles that of the chloroplast—chloroplast proton translocating ATP synthase [37].

The circuit is closed by a proton transfer transport parallel to electron transfer through the thylakoid membrane. Thus a reductive and an oxidative redox occur on each side of the membrane with an optimal gain of 1.28 eV from two light quanta having the total energy input of a minimum of 3.6 eV [8]. Mother Nature carries out the energetically challenging task of oxidizing water to oxygen and reducing protons on to $NADP^+$ by splitting it into two more easily and consistently achievable photochemical reactions that can be performed

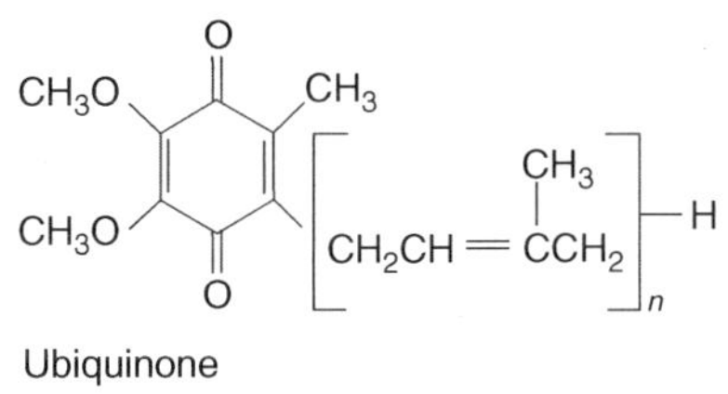

FIGURE 7.12
Structure of ubiquinone.

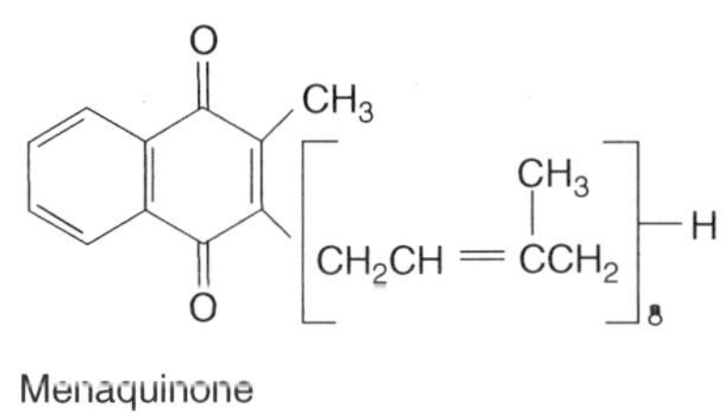

FIGURE 7.13
Structure of menaquinone.

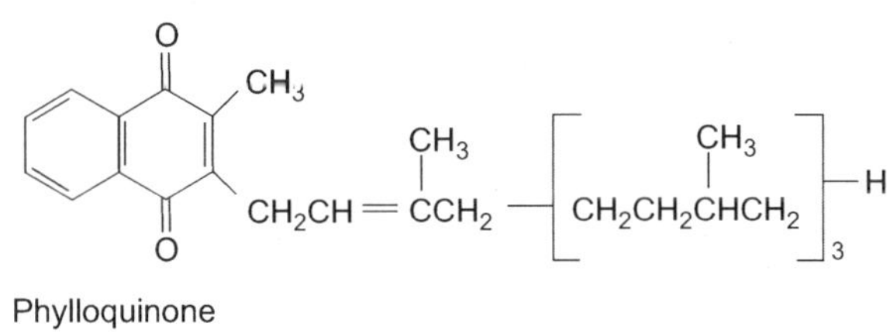

FIGURE 7.14
Structure of phylloquinone.

by using readily available, common materials and a more abundant portion of the solar spectrum [119]. Although such complex photosynthetic system optimized to perfection through evolution cannot be imitated in artificial photosynthesis, the following essential conditions for efficient photoelectrochemical solar energy conversion can be understood from it: efficient absorption of light with the minimum increase in entropy (entropy production), fast separation of charges after light absorption, effective separation of products so as to prevent the reverse reaction, adjustment of the redox potentials of the excited states to the redox reactions that store energy, and long-term stability through continuous reproduction.

The discussions in Sections 7.1 through 7.4 paint a picture of the difficult ordeal of generating hydrogen and oxygen by splitting water using available materials and the universally available sunlight in a stable and efficient manner.

7.5.5 Approaches Based on Photosynthesis

The different types of quinones active in photosynthesis are being used as electron acceptors in solar cells. The compounds such as Fd and NADP could also be used as electron/proton acceptors in the photoelectrochemical cells. Several researchers have attempted the same approach with a combination of two or more solid-state junctions or semiconductor–electrolyte junctions using bulk materials and powders. Here, the semiconductors can be chosen to carry out either oxygen- or hydrogen-evolving photocatalysis based on the semiconductor electronic band structure.

Tributsch [118] has pointed out the contradiction between the development of technical catalysts for water splitting based on transition metals and unusually stable compounds that resist a very positive oxidation potential and manganese complex of photosynthesis. The elemental manganese metal center is very labile and has not found application as a technical catalyst. An explanation based on a synergetic mechanism in which entropy is exported, that is, order is built, via autocatalytic step to generate a highly active, temporary activation state that is capable of catalyzing the multielectron transfer from water. A relatively simple model was used to simulate such a process and to deduce oscillating properties. The processes in oxygen evolution and other photosynthetic electron transfer chain indicate that photosynthetic membranes actually oscillate [120,121]. In a model experiment using a copper indium diselenide electrode in contact with a hydrogen peroxide–containing solution, it was found that photopotentials oscillated depending on the light intensity and that oscillations occurred synchronously with the liberation of hydrogen. Autocatalytic steps that lead, to oscillating, chaotic, or multistate mechanisms make the nonlinear photoelectrochemical system feasible. New phenomena exhibiting a very special entropy behavior occur far from equilibrium. A significant property of oscillating and structure formation mechanisms is the generation of order and export of entropy under certain conditions. In contrast, traditional hydrogen evolution can be interpreted as a prototype of a quasireversible reaction. Tributsch [118] has recognized the efficient entropy decrease or the improved degree of order reached in an open system during the process of electrocatalysis through a dynamic self-organization as an important quality factor. This would be possible if there exist autocatalytic processes in which the thermodynamic force such as photon energy or electrochemical potential induces a microscopic structural reorganization providing an optimized activation complex for charge separation or for multielectron transfer during water splitting. It is proposed to take advantage of such dynamic self-organization for achieving photoelectrochemical energy conversion, provided the reaction center and water-splitting complex optimize and perfect their structure while energy is flowing [118].

Linkous and colleagues [119,122–125] have obtained some success using alkaline iodate redox electrolyte and dispersed or immobilized powders of various semiconductor compound such as TiO_2 and InP. Although they demonstrated the proof of concept, they also found that it is essential to identify other photocatalysts due to various material problems [119]. Photocatalytic particles have been used in a dual-bed configuration for solar-driven generation of hydrogen by the decomposition of water. The system consisted of two shallow, flat, sealed chambers with immobilized micron-sized photocatalytic particles of titanium dioxide and indium phosphide in oxygen- and hydrogen-evolving beds. An aqueous solution containing redox mediator was circulated between the two chambers. Platinum catalysts were evaluated for promoting hydrogen evolution. A series of redox mediators with widely ranging electropotentials were tested. Although several electron donors facilitated hydrogen evolution, oxygen evolution was possible only with the most oxidizing catalysts. It was concluded that a single redox couple capable of charge exchange in both modules is desirable to minimize system complexity [122]. Photocatalytic O_2 evolution from oxidative decomposition of water was studied in alkaline media with TiO_2 as photocatalyst and iodate ion IO_3^- as a reversible electron acceptor. Significant rate enhancement was obtained by cocatalyst modification using Pt, Ir, and Ni/NiO_x. Photocatalyzed oxidation of iodide ion by O_2 was found to be the most important of the several possible recombination mechanisms and back reactions. The effect of carbonate ion, which has been shown to promote O_2 evolution in photocatalytic suspensions, was found to be reduced considerably due to competitive adsorption of IO_3^- [123]. Several organic pigments were also studied as possible water-splitting photocatalysts. Spectral properties of these compounds have been well studied. However, it was essential to determine their HOMO and LUMO positions on the redox potential scale using theory and experiment because their energy levels were unknown [124]. Selective gas evolution using visible wavelength light was observed for various organic pigments immobilized on acrylic flats along with noble metal cocatalysts. It was found that provision must be made for inhibiting the photocatalytic back reaction to ensure cyclical operation [125].

7.6 PEC Setups with Multijunction and Multiple PV Cells

7.6.1 PEC Setup with Tandem Cells

Several researchers have tried to emulate the PV electrolysis design by utilizing multijunction cells or multiple PV cells in combination with electrolysis systems. When the electrical field is generated in an illuminated solid-state device, it is not essential for the semiconductor to remain in direct contact with the electrolyte. Consequently, the various, often conflicting, requirements of formation of a semiconductor–electrolyte junction efficiently generating a photovoltage using the solar spectrum, stability of the semiconductor against corrosion by highly reactive electrolytic solution, and high catalytic activity at the semiconductor surface are eliminated. Khaselev et al. [126] integrated such a PV/electrolysis system using a highly efficient and more expensive $GaInP_2$/GaAs multijunction cell and a somewhat less efficient and low-cost, hydrogenated amorphous silicon, a-Si:H tandem PV solar cell, both of which generated voltages which exceeded that required for water splitting. The PV/electrolysis setup had two separate PV and electrolyzer components. Effects of current density, solar intensity variability, and light concentration on the overall efficiency for hydrogen production were studied. For this

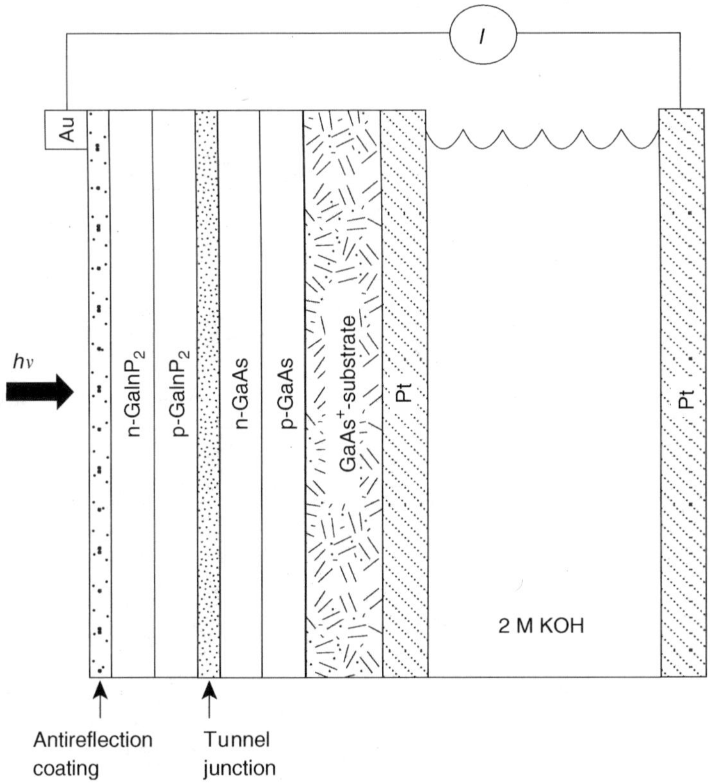

FIGURE 7.15
PV/electrolysis setup using the $GaInP_2$/GaAs dual-junction solar cell.

purpose, the areas of the electrolyzer electrodes were kept equal to those of the PV cells. The $GaInP_2$/GaAs cells were grown by atmospheric-pressure organometallic vapor-phase epitaxy (APOMVPE). The cell tandem structure was fabricated by the sequential growth of a p- and n-type GaAs layers on a p-type GaAs substrate, followed by heavily doped n^+-type GaAs layer and p^+-type $Ga_{0.52}In_{0.48}P$ ($GaInP_2$) layer further followed by p- and n-type $Ga_{0.52}In_{0.48}P$ layers. Here, the GaAs p–n junction forms the bottom PV cell, whereas the $GaInP_2$ p–n junction forms the top PV cell. Direct growth of a top $GaInP_2$ p–n junction on the bottom GaAs p–n junction will form a blocking n–p junction between the two p–n junctions. However, the grown-in tunnel junction (TJ) provides a low-resistivity, transparent connection between the top $GaInP_2$ p–n junction and the bottom GaAs junction and allows the passage of current by tunneling without a significant loss of power or light. The $GaInP_2$/GaAs cell area was 0.3 cm^2. The tandem cell was completed by the deposition of a gold or silver contact grid by electrodeposition and an AR coating in the front and the platinum layer ohmic contact at the back [126].

The amorphous silicon tandem solar cells consisted of three n–i–p a-Si:H cells grown by plasma-enhanced chemical vapor deposition (PECVD) [126]. The a-Si:H cell area was 0.5 cm^2.

The PV/electrolysis setup for the $GaInP_2$/GaAs dual-junction solar cell is shown in Figure 7.15. The top p–n $GaInP_2$ junction with a bandgap of 1.83 eV was designed to absorb the visible portion of the solar spectrum and the bottom p–n GaAs junction with a bandgap of 1.42 eV absorbed the near-IR portion of the spectrum that was transmitted through the top junction. The $GaInP_2/GaAs$ tandem junction solar cell was mounted with the platinum back-contact layer forming an anode in an electrolyte consisting of 2 M KOH prepared using analytical grade reagents dissolved in fresh deionized distilled water. The platinum foil counterelectrode was connected to the gold or silver contact grid with a copper wire through an external circuit. Illumination from a 75 W Xe arc lamp was made incident onto the AR coating and front grid of the $GaInP_2$ top cell. Electrons from electron–hole pairs photogenerated within a diffusion length from the p–n junction by high-energy photons with energy $h\nu$ greater than the $GaInP_2$ bandgap energy of 1.83 eV can diffuse into the p–n junction, and together with those generated in the depletion layer are transported to the n-type $GaInP_2$ layer by the built-in field of the top $GaInP_2$ junction. The electrons flow through the n-type $GaInP_2$ to the front contact grid and through the external circuit to the platinum cathode. However, the photogenerated holes flow through the p-type $GaInP_2$ toward the TJ. The photons with energy below the bandgap of $GaInP_2$ are transmitted to the bottom GaAs junction. Electrons from electron–hole pairs photogenerated by the photons with energy $h\nu$ greater than the GaAs bandgap energy of 1.42 eV are separated by the built-in field of the bottom GaAs junction, flow toward the TJ, and recombine with the holes that have flowed there from the top $GaInP_2$ junction. The holes from the photogenerated electron–hole pairs flow through the p-type GaAs toward the GaAs substrate. This low bandgap p-type GaAs layer makes an indirect contact with the electrolyte through the substrate and a thin platinum metal layer ohmic back contact and anode. The total voltage generated across the tandem $GaInP_2/GaAs$ cell V_{ph} is the sum of the voltages across the two junctions, V_{ph1} and V_{ph2}. If the total voltage produced by the $GaInP_2$/GaAs tandem cell is sufficient to drive the water redox reactions, and also to provide any additional voltage needed to overcome the overvoltage losses resulting from the hydrogen and oxygen evolution reactions, then the hydrogen and oxygen can be generated by splitting water. In such a case, the electrons generated in the top $GaInP_2$ cell flowing through the external circuit to the platinum cathode will reduce water to generate hydrogen. Two photons are, therefore, required to put one electron into the external circuit. A total of four electrons generated using four photons are needed to provide two electrons for generating a hydrogen molecule [126].

The electrons received by the bottom contact from the anode for oxidizing the water to oxygen recombine with the holes generated in the GaAs bottom cell at the interface with the metal contact. Again eight holes generated using eight photons will be needed to provide four holes to form an oxygen molecule at the anode. Thus a total of eight photons, four of higher energy $h\nu > 1.83$ eV and four of intermediate energy $1.83 > h\nu > 1.42$ eV, are needed to generate eight electron–hole pairs—four of the electrons flowing through the external circuit to complete the water splitting and reaching the back contact to recombine with the four holes reaching there. The remaining four electrons, after flowing through the bottom n-type GaAs layer and the n^+ GaAs layer of the TJ recombine with the four holes flowing through the p-type layer of the top GaInP cell and p^+ $GaInP_2$ layer of the TJ. Thus, eight photons are required to produce two molecules of H_2 and one molecule of O_2 by splitting two molecules of water. This is an example of a D4 scheme [127].

The total output photovoltage must exceed the thermodynamic potential difference for water splitting (1.229 V at 25°C), the energy level mismatches for the anodic and cathodic processes, and the polarization loss or overvoltages due to kinetic, diffusion, and IR potential losses in the bulk of electrolyte.

PV current–voltage and power–voltage characteristics of a standard $GaInP_2/GaAs$ tandem cell were measured under 100 mW/cm^2 illumination. The PV parameters under illumination were an open-circuit voltage $V_{oc} = 2.32$ V, a short-circuit current density $J_{sc} =$ 13.4 mA/cm^2, a fill factor FF = 0.88, and a PV conversion efficiency of 28.5%. Under the same illumination, the standard triple-junction a-Si:H solar cell exhibited an open-circuit voltage V_{oc} = 2.16 V, a short-circuit current density J_{sc} = 6.6 mA/cm^2, a fill factor FF = 0.62, and a PV conversion efficiency of 9%. Outdoor photocurrent measurements were monitored using an analog/digital data acquisition system, whereas the solar radiation was measured by the Solar Energy Research Facility Group of the National Renewable Energy Laboratory (NREL). Illumination for indoor experiments was provided by a 75 W Xe arc lamp. Photocurrent was measured with a potential-controlled potentiostat using a conventional two-electrode cell configuration.

The photocurrent was measured at various bias voltages for both the $GaInP_2/GaAs$ dual-junction solar cell and triple n–i–p junction a-Si:H cell setups, with the back-contact Pt layer anode, Pt foil cathode, and 2 M KOH electrolyte under 100 mW/cm^2 illumination. The hydrogen production was found to be proportional to the current density through Faradaic correlation. For the $GaInP_2/GaAs$ tandem junction cell setup, the light-limited current density was reached at the negative bias of approximately −0.57 V. Thus, the cell generated hydrogen, oxygen, and electrical power while operating at that potential. The efficiency for hydrogen production was calculated using the equation: efficiency = (power out)/(power in). The only input power was the incident light. Photocurrent electrolysis efficiency of 100% was assumed, that is, the entire flow of electrons in the external circuit and the electrolyte was assumed to be utilized for the production of hydrogen and oxygen. Thus, the output power was estimated by multiplying all the hydrogen production photocurrent by 1.23 V, the theoretical fuel cell limit at 25°C, to calculate the hydrogen production efficiency of 16.5% for the PV $GaInP_2/GaAs$ tandem cell electrolysis setup. Although this was the highest measured efficiency for group III-V tandem junction PV cell electrolysis setup, it was only 57.9% of the PV conversion efficiency of the 28.5% GaInP/GaAs tandem cell under AM 1.5 illumination. This was because electrical power generated when the cell is operating at −0.57 V was not considered. It was deemed impractical to configure such a system for producing hydrogen, oxygen, and electricity. Hence, it was suggested that the excess available power be allocated as a cushion against the loss of solar-to-hydrogen conversion efficiency under low solar irradiance or to compensate for the decrease in current density of the system due to reduction of dissolved oxygen rather than water at the cathode. While estimating the efficiency of a PEC setup that uses PV cells, it may be noted that the maximum efficiency of the PV cell is derived when it is operating at the optimum power point and that the PEC efficiency improves when the PEC cell is operated close to 1.23 V. First, PEC efficiency close to PV cell efficiency could be obtained by using a low-loss DC-to-DC voltage converter. Second, if PV electricity is supplied to a commercial electrolyzer with a typical efficiency of ~70%, the solar-to-hydrogen efficiency using an array of 28.5% efficient PV cells would be ~20%. Thus, the integrated system would actually give ~83% of the optimum.

The measured short-circuit current was found to follow the solar irradiance during the day. Moreover, copious amounts of gas were observed to evolve from both the electrodes during the entire period. The integrated ratio of input-to-output power efficiency of hydrogen production by solar irradiance during this experiment was calculated to be 16.4% at an average solar irradiance of 659 mWh/cm^2. This result was found to be in good agreement with indoor measurements under similar conditions that showed hydrogen generation throughout the experiment.

It would be possible to reduce the mismatch between the voltage at optimum power and the voltage needed for the efficient generation of hydrogen and oxygen by splitting water in PEC cells if the optimum power point voltage of 2.755 V from the recently developed 40.1% efficient, lattice-matched concentrator cell illuminated with 135x concentrated sunlight is used to provide the voltage to two PEC cells connected in series [44]. Although the efficiency of metamorphic triple-junction solar cells at 240 suns is higher (40.7%), it may not be possible to reduce the energy level mismatch and polarization losses of the two PEC cells low enough for operation with the present optimum power point voltage of 2.589 V. It would be preferable to use the triple-junction cells without concentration. For this purpose, it would be essential to develop these cells further so as to improve the present optimum power voltages for the lattice-matched and metamorphic solar cells of 2.3 and 2.06 V to ~2.7 V.

For the amorphous hydrogenated silicon, a-Si:H, three n-type, intrinsic i, and p-type region junction (n–i–p triple junction) PV cell electrolysis setup, the light-limited current was reached approximately at zero bias, that is, at short-circuit. Thus, the a-Si:H PV cell electrolysis setup produced only the chemicals hydrogen and oxygen without any electrical power. This represents the ideal condition of operation for the PV electrolysis system. The hydrogen production efficiency for the triple-junction a-Si:H PV cell electrolysis setup under incident irradiance of 100 mW/cm^2 was calculated to be 7.8%. Thus, the PV electrolysis efficiency was 86.7% of the PV conversion efficiency of 9% of the a-Si:H PV cell. This significantly high fraction of the total incident energy could be achieved because an effective electrolysis voltage of only 1.42 V was required to sustain the low-current density. Thus there was a better match between the maximum operating voltage of the a-Si:H PV cell and the voltage required for electrolysis [126]. Here again this is not the correct comparison and the comments made earlier must be applied also to this case while comparing the efficiencies.

The setups described earlier used a combination of PV and electrolysis. Earlier, Khaselev and Turner [128] developed an integrated, monolithic photoelectrochemical (PEC)/PV setup for a direct water electrolysis system. It also used a solid-state tandem cell consisting of a bottom GaAs p–n cell connected to a p-type $GaInP_2$ top layer through a transparent TJ interconnect. The solid-state tandem cell was grown at the NREL by APOMVPE. The surface of the p-type $GaInP_2$ layer was coated with a platinum catalyst. The complete PEC/PV setup consisted of the p-$GaInP_2$(Pt)/TJ/GaAs tandem immersed in a 3 M H_2SO_4 electrolyte with the back contact to the GaAs p–n junction connected to a platinum counterelectrode. The solid-state tandem cell was similar to that described earlier [126], except that instead of using the standard PV GaInP/GaAs tandem cell the PEC/PV setup used a modified configuration consisting of the bottom p–n GaAs junction/tunnel diode as in the standard PV cell, but, only with a p-type GaInP layer grown on top. The cell structure was grown as follows: initially a p–n GaAs junction was grown epitaxially on a p-type GaAs substrate. This was followed by the growth of a low-resistivity n^+ GaAs/p^+ p-$Ga_{0.52}In_{0.48}P$ (p-$GaInP_2$) TJ. Finally, a ~4 μm thick p-type $GaInP_2$ top layer was grown epitaxially. The p-type $GaInP_2$ layer formed a PEC Schottky-type semiconductor–electrolyte junction. Thus in this setup, the top p–n $GaInP_2$ junction was replaced by a p-type GaInP–electrolyte barrier junction. Thus, this configuration consists of a PEC device that is voltage-biased with an integrated p–n GaAs PV cell. When photons with energy greater than the $GaInP_2$ bandgap energy of 1.83 eV in the visible portion of the solar spectrum are absorbed in the p-type $GaInP_2$ absorber within a diffusion length of the PEC semiconductor–electrolyte junction, the electrons diffusing toward the p–n junction are propelled by the built-in field toward the electrolyte, whereas the holes flow through p $GaInP_2$ layer to the tunnel diode and then to the bottom GaAs p–n junction. In a similar manner, intermediate energy photons with energies $1.42 < h\nu < 1.83$ eV are transmitted through the top $GaInP_2$ region and absorbed in the p-type GaAs within a diffusion length of the p–n junction or in the depletion layer generating electron–hole pairs. The charge carriers are then separated by the built-in field of the p–n junction and transported to the respective GaAs regions where they are majority carriers. The electrons flow through the n GaAs regions to the tunnel diode where they recombine with the holes flowing there. The holes flow through the p-type GaAs toward the back ohmic contact [128].

The photovoltage generated across the illuminated bottom GaAs p–n junction biases the top $GaInP_2$–electrolyte Schottky-type barrier junction producing the sum of photovoltages generated by the two junctions. Although such a configuration can be designed for oxidation or reduction at the semiconductor–electrolyte interface for water-splitting reactions, hydrogen production at the semiconductor–electrolyte interface was preferred because the semiconductor surface is cathodically protected under illumination.

The PEC/PV configuration will work properly only if the GaAs bottom cell can provide sufficient voltage to boost the photovoltage across the semiconductor–electrolyte junction so as to fulfill the thermodynamics for water splitting (1.229 V) and the losses. When adequate voltage is provided, the electrons and holes propelled by the built-in fields will directly carry out water splitting to generate hydrogen and oxygen only with the energy input of incident light if all the required voltages are provided by the PEC/PV setup.

As explained earlier, of the two electrons, one generated in the GaAs bottom junction using the intermediate energy photon with energy >1.42 eV recombines at the TJ so that the other generated in the GaInP2 using higher-energy photon with energy >1.83 eV can flow into the external circuit with the added energy derived from the bottom cell. Therefore, the solid-state systems are optimized by matching photocurrents of top and bottom cells. However, when a photocurrent-matched configuration was placed in an aqueous environment, either no current was produced in the external circuit or it decomposed [129]. The problem was resolved by employing a thicker, p-type $GaInP_2$ top layer so that the bottom cell limited the electron supply. Consequently, the mismatch of electron–hole formation in the two-junction region appeared to be the key for proper operation.

The front p-$GaInP_2$(Pt) surface of the p-$GaInP_2$(Pt)/TJ/GaAs sample immersed in 3 M H_2SO_4 for photoelectrolysis was illuminated by concentrated light from a fiber-optic illuminator with a 150 W tungsten-halogen lamp having an irradiance of 1.196 W/cm^2 equivalent to 11x concentration of sunlight. Concentrated light was used because the III-V tandem cells are usually designed to operate efficiently at high (100x) solar concentrations. The illuminated p-$GaInP_2$ (Pt)/TJ/GaAs PEC cell showed an open-circuit voltage of 0.55 V, equivalent to the extra photovoltage generated by the GaAs cell. Copious amounts of hydrogen and oxygen in proportion of 2:1, H_2:O_2 evolved. Since large gas bubbles can act as miniature lenses causing pitting of the semiconductor electrode 0.01 M of a surfactant was added to promote the formation and rapid release of smaller bubbles. Initially, the photocurrent density decreased slightly and stabilized after ~5 min at ~120 mA/cm^2. It then mostly remained constant decreasing to 105 mA/cm^2 after 20 h. The sample appeared undamaged except for some localized damage due to gas bubble accumulation along the upper edge of the epoxy coating. The efficiency of H_2 production calculated by multiplying the H_2 production photocurrent of 120 mA/cm^2 by 1.23 V and dividing by the input power of incident light intensity of 1.196 W/cm^2 was 12.4% [128].

Rocheleau et al. [130] developed a reactor utilizing amorphous hydrogenated silicon a-Si:H, triple p–i–n junction solar cells in the glass/TCO superstrate structure provided by Solarex Thin Film Division. The PV conversion efficiency of the triple-junction a-Si:H cell under AM 1.5 conditions was 10.0–10.3%. It had an active area of 0.27 cm^2. A Ni/CoMo cathode was mounted against the metal back contact of the triple-junction PV cell. Good electrical contact was ensured using indium dots. A separate Ni/$FeNiO_x$ anode was connected to the front contact layer with external wires. The areas of the cathode and anode were 1 cm^2 each. The ratio of ~4:1 between the catalyst area to cell was found to be appropriate for operation under low solar concentration. The electrolyte was 1 N KOH solution. It may be noted that as in the preceding cases [126], the semiconductor is physically isolated from the electrolyte by a conductive and catalytic film and that there is no intimate contact of the semiconductor in the multijunction configuration with the electrolyte.

The overpotential required for oxygen evolution usually constitutes the rate-limiting step. The oxygen evolution overpotential at a current density of 100 mA/cm^2 was reduced by ~300 mV compared to undoped NiO_x anode. For this purpose, several iron-doped nickel oxide $NiFe_yO_x$ films were prepared by reactive sputter-deposition from elemental and alloy targets in an argon/oxygen 20% ambient and evaluated as oxygen evolution catalysts.

Solar-to-hydrogen efficiency of 7.8% was measured based on the lower heating value of hydrogen. Performance of the photoreactor was also tested on a clear day during the late afternoon in Hawaii. Solar-to-hydrogen conversion efficiency was 7.5% initially at 3:30 PM, decreased slowly to ~7% by 4:30 PM, and then dropped sharply to ~5.5% at 5:30 PM. It may be noted that in the case of multijunction cells, small changes in the incident spectra can have significant effects on the solid-state efficiency and consequently on the efficiency of the hydrogen production process. Extended life testing of catalysts of $NiFe_yO_x$ and CoMo films in 1 N KOH electrolyte carried out for 6500–7200 h did not show degradation of the catalyst activity even when low-purity KOH and drinking water were used to prepare the electrolyte.

Licht et al. [131] employed a solid-state bipolar multijunction AlGaAs/Si solar cell in a setup for water splitting. AlGaAs layers and GaAs buffer layer were grown sequentially on the bottom n–p$^+$ Si cell by metalorganic chemical vapor deposition (MOCVD). The top cell consisted of p$^+$–n junction between 300 nm thick $Al_{(0.3-0.15)}Ga_{(0.7-0.85)}As$ p$^+$ layers and 1 μm thick $Al_{0.15}Ga_{0.85}As$ n-type absorber layer. The highly doped top 50 nm thick $Al_{0.8}Ga_{0.2}As$ p$^+$ layer allowed the formation of an ohmic contact with the vacuum-evaporated Au–Zn/Au front contact fingers. The function of the 1.7 nm thick $Al_{0.15}Ga_{0.85}As$ n$^+$ layer and 20 nm thick GaAs n-type layer/10 nm GaAs buffer layer was to enable reasonably defect-free growth of the top AlGaAs cell as well as to provide a low impedance electrical contact with the bottom n–p$^+$ Si junction. These layers were also required to be highly transparent to avoid the loss of photons. Au-Sb/Au back contact was evaporated on highly n$^+$-doped Si so as to provide low resistivity contacts. An Ar–MgF_2/ZnS bilayer was finally deposited on the multijunction cell. The illuminated area of the AlGaAs/Si solar cell was 0.22 cm^2.

The front contact grid of the cell was connected to the RuO_2 anode, whereas the back contact was connected to the platinum cathode. The areas of anode and cathode were 10 cm^2 each. RuO_2 anodes were prepared by sintering ruthenium chloride paste applied on a titanium foil at 350°C in air [132]. Titanium (120 μm) foil was mechanically polished by sandblasting, then chemically etched with 20% HCl for 2 min and washed in water. Solvent was evaporated from 0.1 M $RuCl_3$ in 20% aqueous HCl and $RuCl_3$ power was redissolved in a minimum volume of 2-propanol. Titanium was then coated with $RuCl_3$/2 propanol paste using a brush and sintered for 15 min at 350°C. Such coating and annealing steps were repeated five times and then annealed at 350°C for 1 h to produce a RuO_2 layer having an optimum thickness of 2.3 μm. These optimized RuO_2 are termed in the following as 2.3 μm RuO_2. Thicker or thinner coatings, or other annealing temperatures, resulted in higher overpotential for oxygen evolution. The RuO_2 anodes annealed at ≥450°C or varying in thickness by 20% are termed *unoptimized RuO_2* in the following. Platinum black cathodes were prepared on a platinum mesh by using the following procedure: chloroplatinic acid was prepared with 3 g of H_2PtCl_6 and 0.25 g of $Pb(C_2H_3O_2)_2$ in 100 mL of water. A platinum mesh (30 × 30) having an open area of >60% was cleaned in dilute nitric acid, washed in water, and then immersed in the chloroplatinic acid solution applying a cathodic potential of 4 V for 5 min. This was followed by electrolysis in 1 N H_2SO_4 for 15 min to remove chloride impurities by washing in water.

The light was made incident on the top AlGaAs cell. Energetic photons with energy $h\nu > 1.6$ eV are absorbed by the top AlGaAs cell and generated electron–hole pairs, whereas the less energetic photons are transmitted. Most of the incident light is transmitted through the top 50 nm thick $Al_{0.8}Ga_{0.2}As$ p^+ layer having >2 eV indirect bandgap. The top and bottom p–n junctions transport photogenerated charge carriers to the respective doped regions where they are majority carriers. Holes from the bottom cell and electrons from the top cell thus transported recombine across the TJ and provide continuity for the current flow. The photogenerated holes in the top AlGaAs cell by energetic photons having energy $h\nu > 1.6$ eV flow toward the front through the highly doped $Al_{0.8}Ga_{0.2}As$ p^+ layer and recombine with the electrons provided to the RuO_2 anode in the process of oxidation of H_2O molecules and flowing to the front contact fingers. Four such holes are required to generate one oxygen molecule from two water molecules. Similarly, electrons photogenerated in the bottom cell by photons with energy in the range $1.6 > h\nu > 1.1$ eV flow to the back contact and the Pt black cathode. These electrons reduce the H_2O molecule and generate hydrogen atoms. As mentioned earlier, four such electrons will be required to generate two hydrogen molecules from the two water molecules. The photovoltages generated by the bottom and top cells are added to provide adequate potential difference for splitting water as well as the necessary overvoltage due to the polarization loss. However, for this purpose an equal number of electrons and holes must be consumed at the middle TJ.

Experiments were carried out to minimize the overpotentials for evolution of hydrogen and oxygen using platinum and platinum black cathodes and unoptimized RuO_2 and optimized 2.3 μm RuO_2 anodes in quiescent and stirred 1 M $HClO_4$ electrolyte. Since platinum is a poor electrocatalyst for oxygen evolution, an overpotential of ~500 mV can be saved by utilizing an optimized RuO_2 anode that provides more effective oxygen electrocatalysis by the intervening RuO_4/RuO_2 redox couple [132–134].

PV efficiency and open-circuit voltage of the multijunction AlGaAs/Si solar cell were 21% and 1.57 V, respectively. A cell with an area of 0.22 cm^2 in the horizontal position was connected to the vertically mounted 10 cm^2 Pt and RuO_2 electrodes. The photoelectrolysis current was measured for over 12 h and was found to remain constant. The cell illuminated with an irradiance of 135 mW/cm^2 that is equivalent to AM 0 irradiance obtained in space outside the atmosphere, provided photocurrents of 4.1–4.8 mA at photopotentials of 1.3–1.4 V to the 10 cm^2 Pt black cathode on the platinum mesh cathode and optimized 2.3 μm RuO_2 anode in stirred 1 M $HClO_4$ electrolyte. Measurements of electrolysis current and voltage for this optimized 2.3 μm RuO_2 anode and Pt black cathode showed that it required a voltage of 1.36 V for passing the Faradaic current of 0.5 mA/cm^2. Thus, the calculated Faradaic efficiency for this combination for the evolution of H_2 and O_2 by splitting water was greater than 18% [134]. These efficiency calculations ignore the energy needed for stirring.

Tributsch and colleagues [132] introduced the first photostable low-energy bandgap semiconductor, RuS_2, that is able to photooxidize water under IR illumination and studied oxygen catalysis mechanism of RuS_2 with RuO_2 as a reference compound. It was also shown that oxygen evolution current of $Ru_{0.99}Fe_{0.01}S_2$ semiconductor illuminated with 890 mW/cm^2 in 0.1 M H_2SO_4 increased significantly [135]. Ruthenium sulfide, RuS_2, is stable against anodic electrocorrosion because of the formation of one or two monolayers of ruthenium oxide, RuO_2, which is a metal-like conductor (semimetal). Formation of a semiconductor–metal Schottky barrier across RuS_2/RuO_2 tends to pin the quasi-Fermi level of holes to that of the RuO_2 layer [12].

Licht et al. [134] have considered the potential benefits of adding a third energetically matched RuS_2 and $Ru_{0.99}Fe_{0.01}S_2$ semiconductor to drive IR-assisted oxygen evolution. Their initial measurements utilizing plasma-deposited RuS_2 thin-film anode showed that a light-activated O_2 evolution current density of over 20 mA/cm^2 was generated at overpotential of 250 mV, or at 1250 mV versus the H_2 potential. Ruthenium sulfide has direct and indirect bandgaps of 1.85 and 1.3 eV, respectively. Thus, it cannot be illuminated by the IR portion of light transmitted through the multijunction AlGaAs/Si solar cell. However, when the compound $Ru_{0.99}Fe_{0.01}S_2$ is prepared by adding 1% Fe, the bandgap shifts to 1.1 eV. By comparing the quantum efficiencies of $Al_{0.37}Ga_{0.63}As$, GaAs, RuS_2, and $Ru_{0.99}Fe_{0.01}S_2$ Licht et al. [134] have shown that it would be feasible to activate the $Ru_{0.99}Fe_{0.01}S_2$ semiconductor.

7.6.2 PEC Setup Cell Combining Polycrystalline Thin-Film Solar Cells and Photoanodes

The operating voltage of highly efficient, individual single-junction PV cells cannot supply the total voltage required for water splitting. Dhere and colleagues [136–138] have carried out research on the development of multiple bandgap tandem photoelectrochemical cell. It uses a pair of polycrystalline thin-film solar cells that transmit IR photons unused in the PV conversion. The transmitted IR photons are made incident on a photoanode so as to reduce the overvoltage required for oxygen evolution (Figure 7.16).

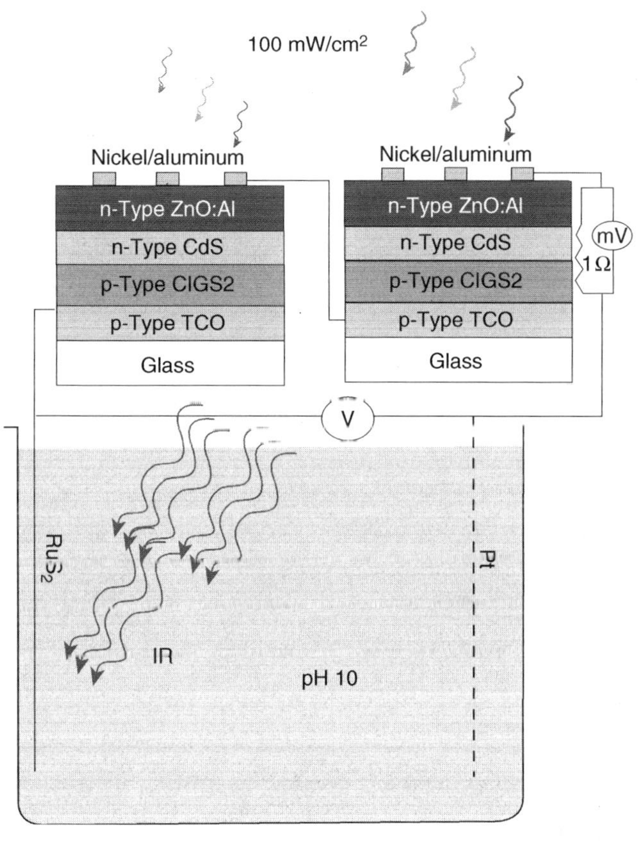

FIGURE 7.16
PEC cell that uses polycrystalline thin-film solar cells with a photoanode.

The chosen PV cells are amenable to economic mass production. More importantly, in this case, the photocatalyst need not be grown directly over the PV cell. As seen earlier, multiple-bandgap, solid-state tandem III-V compound semiconductor and a-Si:H PV cells have been used successfully for hydrogen generation by PV electrolysis. As discussed earlier, the best PV conversion efficiencies of 19.5% and 16.5% have been achieved with small-area, polycrystalline $CuIn_{1-x}Ga_xSe_2$ (CIGS) and CdTe thin-film solar cells [22,23]. Since tandem polycrystalline thin-film cells are more difficult to fabricate, it is essential to use a pair of cells to obtain the necessary voltage required for water splitting. Interconnected two or more cells can be compared to tandem junction solar cells because integrally interconnected a-Si:H, CdTe, and CIGS thin-film PV modules are already being manufactured in multimegawatt quantities annually. When the thin-film solar cells transmit the unused IR photons, the voltage needed for splitting water can be reduced significantly by illuminating an appropriate photoanode with the transmitted IR photons. Here, only the photoanode and cathode are in contact with the electrolyte. Moreover, this setup effectively integrates the PV cells and photoelectrolysis by using different parts of the solar spectrum for efficient water splitting [134,136,138].

$CuIn_{1-x}Ga_xS_2$ (CIGS2) thin-film cells were prepared as follows: transparent and conducting back-contact layers were prepared by the deposition of a thin (~10 nm) molybdenum layer by magnetron sputtering on a 500 nm thick layer of aluminum-doped zinc oxide ZnO:Al, or indium-tin oxide (ITO) deposited on soda lime glass substrates. Copper-gallium, CuGa (22 atomic%), and indium layers having the elemental ratio of Cu/(In+Ga) of 1.4 were sputter-deposited on soda lime glass substrates coated with the transparent and conducting layer. CIGS2 films having compactly packed, large grains were prepared by sulfurization of the Cu–In–Ga precursor layer in $Ar:H_2S$ (4%) gas flow at 475°C for up to 60 min. The process was optimized to form large, compact-grain CIGS2 films and at the same time sulfurize the thin Mo film on ZnO:Al or the ITO layer to form a transparent MoS_2 layer. The excess Cu_xS phase segregating on the surface was etched in 10% aqueous KCN solution. Solar cells were completed by the deposition of CdS heterojunction partner layer by chemical bath deposition, transparent-conducting ZnO/ZnO:Al window bilayer by RF sputtering, and vacuum deposition of Ni/Al contact fingers through a metal mask [139,140]. Conventional CIGS2 cells were also prepared on the usual opaque molybdenum back contacts for comparison.

RuS_2 photoanodes for oxygen generation were prepared as follows using a procedure similar to that used for the preparation of RuO_2 anodes [132]: a titanium (120 μm) foil was mechanically etched (sandblasted), then chemically etched with 20% HCl for 2 min, and washed in water. 0.1 M $RuCl_3$ solution was prepared in 20% aqueous HCl and evaporated. A paste of $RuCl_3$ was prepared by dissolving in a minimum volume of 2-propanol. Titanium foil was then coated with $RuCl_3$–2-propanol paste using a brush and was sintered in diluted H_2S ambient for 15 min at 350°C. This procedure of applying the paste and annealing was repeated five times and final annealing was carried out for 60 min at 350°C. Approximately, 2–2.5 μm thick RuS_2 layer was obtained on the titanium foil with this process. It would be possible to deposit a RuS_2 layer on the Ti foil using sputter-deposition of ruthenium in the diluted H_2S ambient for large-scale production.

The PEC setup consisted of two ~0.43 cm^2 CIGS2/CdS thin-film solar cells in series with the RuS_2 anode for oxygen generation and the platinum cathode for hydrogen generation (Figure 7.16). Here, the series-connected PV cells remained outside the electrolyte. IR rays not used in the PV conversion were made incident on the RuS_2 photoanode in contact with the electrolyte so as to boost the generation of hydrogen and oxygen by water splitting.

The PV efficiency of the CIGS2 cell prepared on glass with transparent conducting back contact, as measured at the NREL, was 5.95%. Calculated PEC efficiency of two such PV cells in series with the RuS_2 anode for oxygen generation and platinum cathode for hydrogen generation under AM 1.5 illumination was 2.99%.

The PV characteristics of the CIGS2 thin-film solar cell on opaque Mo back contact, as measured at the NREL under AM 1.5 conditions, were as follows: short-circuit current density J_{sc} of 20.88 mA/cm^2, open-circuit voltage V_{oc} of 830.5 mV, fill factor FF of 69.13%, and PV conversion efficiency η of 11.99%.

The PV electrolysis efficiency of a PEC setup in which two CIGS2 cells having an opaque molybdenum back contact were connected to the RuS_2 anode that remained in the dark and platinum cathode was 8.78%. This shows the potential for boosting the PEC efficiency using a combination of more efficient CIGS2 cells having transparent back electrodes with RuS_2 photoanodes illuminated with IR radiation transmitted through the PV cells [138].

7.7 Concluding Remarks

7.7.1 Comparison between Bulk and Particulate PEC and Solid-State Devices

It is instructive to compare bulk and particulate photoelectrochemical and solid-state devices. Considerable efforts are made to increase the surface area in particulate photoelectrochemical devices because of the beneficial effects of larger surface area on catalytic processes. However, it may be remembered that although a large surface area is beneficial in catalysis, it can be detrimental in photoelectrochemical cells because of the high recombination rates. In a solid-state or electrochemical PV cell as well as in a bulk photocatalytic cell, the photogenerated minority carriers are propelled efficiently by the built-in field to the side where they are majority carriers that together with the photogenerated majority carriers flow to the contacts or to a contact and the electrolyte. Consequently, the efficiency of collection of charge carriers generated in the depletion region and within a diffusion length of it is very high. Second, in bulk semiconductor–electrolyte devices and solid-state PV cells, reflection losses that can be substantial for the high refractive indices of most semiconductors can be minimized by using AR coatings. Therefore, the overall efficiency of 24% and >40% have been achieved for single and multiple-junction solid-state PV cells. Reflection losses cannot be minimized in particulate suspension-type photocatalytic cells. Moreover, in bulk photocatalytic cells, hydrogen and oxygen generation by splitting water and accumulation can take place at separate locations and thus the reverse reaction of generated hydrogen and oxygen to water can be mostly avoided. This again is not possible in particulate suspension-type photocatalytic cells that can suffer from high reverse reaction rates besides high recombination rates of photogenerated carriers.

7.7.2 Present Status and Promising Areas of Research and Development

Considerable progress has been made in the field of photoelectrolysis cells for hydrogen generation using sunlight. Particulate NiO_x/TiO_2 photocatalyst has already been used for stoichiometric production of hydrogen and oxygen on a sunny day and a cloudy day with the quantum efficiency of >1.79%. Layered compounds with Ni/NiO cocatalyst have also shown good conversion at 330 nm. Doping with other anions and ion implantation would assist in improving the PEC conversion efficiencies in combination with other processes.

Generation of hydrogen from H_2S using cadmium sulfide with waste products as sacrificial agents to avoid photocorrosion of the semiconductor is expected to grow as a niche application.

Group III-V compound and a-Si:H-based tandem cells especially when the cells are not in direct contact with the solar cells have already resulted in high conversion efficiencies. The achievement of >40% PV efficiency under concentration will benefit III-V PEC cells. Making the IR photons unused in the PV conversion incident on a photoanode so as to reduce the overvoltage required for oxygen evolution is an interesting concept and could result in considerable improvement of solar-to-hydrogen efficiency. Continued growth of PV cell and module market >30%/year would benefit PEC concepts based on c-Si, a-Si:H, III-V, CIGS, and CdTe cells all of which would allow the achievement of desirable quantum efficiencies and economic large-scale production.

Various types of quinones active in photosynthesis are already being used as electron acceptors in solar cells. The compounds such as Fd and NADP could also be used as electron/proton acceptors in the photoelectrochemical cells. The situation in terms of stability and solar-to-hydrogen conversion efficiency has not yet shown a significant improvement. The PEC cells based on stable semiconductors are not very efficient and those that can give efficient conversion do not have good stability in contact with the electrolyte. Facing the enormous challenge of economic and efficient solar-to-hydrogen conversion, it has become essential to learn from and to try to emulate Mother Nature. Now is the time to reconsider whether the basic premise of the choice of an absolute thermodynamic stability may have precluded high efficiency from the outset. In this case, attempts must be made toward developing an intelligent design that recovers order while generating hydrogen from sunlight.

Acknowledgments

The support by the NREL and the Thin-Film PV Partnership under contract no. XXL-5-44205-08 is acknowledged by Neelkanth G. Dhere. The authors are thankful to Dr Clovis A. Linkous and Dr Shankar Bennur for many useful discussions and for reviewing the entire manuscript and Section 7.5 of the manuscript, respectively; Dr John Turner and Dr H. S. Aparna for useful discussions; Dr Sucharita N. Dhere and Dr Ramesh G. Dhere for their help in manuscript preparation; Dr Rajeshwar Krishnan for providing the manuscript of his book chapter even before its publication; and Shirish Pethe, Sachin S. Kulkarni, Parag Vasekar, and Santosh Khatri for assistance.

References

1. Global Energy Perspective, http://nsl.caltech.edu.
2. Arvizu, D.E., Meeting the renewable energy challenge: What will it take to reach solar PV's ultimate potential, Presented at the *2006 IEEE 4th World Conf. on Photovoltaic Energy Conversion*, Waikoloa, Hawaii, NREL/PR-100-40011.
3. Dhere, N.G., *Towards GW/Year of CIGS Production Within the Next Decade, Solar Energ. Mater. Solar Cells,* 91, 1376, 2007.

4. Duncan, R.C. and Youngquist, W., The World Petroleum Life-Cycle, Based on the World Oil Forecasting Program, http://www.dieoff.com/page133.pdf, April 29, 2007.
5. Stern, N., Stern Review: The Economics of Climate Change, http://www.hm-treasury.gov.uk/independent_reviews/stern_review_economics_climate_change/sternreview_index.cfm, April 29, 2007.
6. U.N. Review of Global Warming, http://www.ipcc.ch/, April 29, 2007.
7. Nozik, A.J., Photoelectrochemical devices for solar energy conversion, in *Photovoltaic and Photoelectrochemical Solar Energy Conversion*, Cardon, F., Gomes, W.P., and Dekeyser, W. (Eds), Plenum, New York, 1981, 263.
8. Gerischer, H., Heterogeneous electrochemical systems for solar energy conversion, *Pure Appl. Chem.*, 52, 2649, 1980.
9. Kaneko, M. and Okura, I. (Eds), *Photocatalysis, Science and Technology*, Kodansha/Springer, Berlin, 2002.
10. Rajeshwar, K., Hydrogen generation from irradiated semiconductor–liquid interfaces, in *Solar Hydrogen Generation: Toward a Renewable Energy Future*, Rajeshwar, K., McConnell, R., and Licht, S., (Eds), Springer-Verlag, New York, 2008. 167.
11. Heller, A., Hydrogen evolving solar cells, *Science*, 223(4641), 1141, 1984.
12. Mattay, J., *Topics in Current Chemistry*, vol. 169, Springer, Berlin, 1994, 106.
13. Nakato, Y., Photoelectrochemistry at semiconductor/liquid interfaces, in *Photocatalysis, Science and Technology*, Kaneko, M. and Okura, I. (Eds), Kodansha/Springer, Berlin, 2002, Chap. 4.
14. Bak, T., Nowotny, J., Rekas, M., and Sorrell, C.C., Photo-electrochemical hydrogen generation from water using solar energy. Materials-related aspects, *Int. J. Hydrogen Energ.*, 27, 991, 2002.
15. Parkinson, B., On the efficiency and stability of photoelectrochemical devices, *Acc. Chem. Res.*, 17, 431, 1984.
16. Ni, M., Leung, M.K.H., Leung, D.Y.C., and Sumathy, K., A review and recent developments in photocatalytic water-splitting using TiO_2 for hydrogen production, *Renew. Sust. Energ. Rev.*, 11, 401, 2007.
17. Fujishima, A., Narasinga Rao, T., and Ohko, Y., Photoelectrochemical processes of semiconductors, in *Photocatalysis, Science and Technology*, Kaneko, M. and Okura, I. (Eds), Kodansha/Springer, Berlin, 2002, Chap. 2.
18. Domen, K. and Hara, M., Water photolysis by layered compounds, in *Photocatalysis, Science and Technology*, Kaneko, M. and Okura, I. (Eds), Kodansha/Springer, Berlin, 2002, Chap. 16.
19. Morrison, S.R., *Electrochemistry at Semiconductor and Oxidized Metal Electrodes*, 2nd printing, Plenum, New York, 1984.
20. Revised Air Mass 1.5 Spectral Distribution, *Terrestrial Photovoltaic Measurement Procedures*, ERDA/NASA/1022-77/16, NASA TM, 73702, NASA-Lewis Research Center, Cleveland, OH, 44134, 1977.
21. Heller, A., Parkinson, B., and Miller, B., Enhanced photoelectrochemical solar-energy conversion by gallium arsenide surface modification, *Appl. Phys. Lett.*, 33, 521, 1978.
22. Contreras, M.A., Ramanathan, K., AbuShama, J., Hasoon, F., Young, D.L., Egaas, B., and Noufi, R., Diode characteristics in state-of-the-art ZnO/CdS/Cu $(In_{1-x}Ga_x)Se_2$ solar cells, *Prog. Photovolt: Res. Appl.*, 13, 209, 2005.
23. Wu, X., Keane, J.C., Dhere, R.G., Dehart, D., Albin, D.S., Duda, A., Gessert, T.A., Asher, S., Levi, D., and Sheldon, P., *Proc. 17th European Photovoltaic Solar Energy Conf.*, 995, 2001.
24. Heller, A. and Vadimsky, R.G., Efficient solar to chemical conversion: 12% photoassisted electrolysis in the (p-type InP(Ru))/HCl KCl/Pt(Rh) cell, *Phys. Rev. Lett.*, 46, 1153, 1981.
25. Heller, A., Conversion of sunlight into electrical power and photoassisted electrolysis of water in photoelectrochemical cells, *Acc. Chem. Res.*, 14, 154, 1981.
26. Nakato, Y., Ueda, K., Yano, H., and Tsubomura, H., Effect of microscopic discontinuity of metal over layers on the photo voltages in metal-coated semiconductor–liquid junction photoelectrochemical cells for efficient solar energy conversion, *J. Phys. Chem.*, 92, 2316, 1988.
27. Nakato, Y. and Tsubomura, H., Silicon photoelectrodes modified with ultra fine metal islands, *Electrochim. Acta*, 37, 897, 1992.

28. Jia, J., Fujitani, M., Yae, S., and Nakato, Y., Hole diffusion length and temperature dependence of photovoltages for n-Si electrodes modified with LB layers of ultrafine platinum particles, *Electrochim. Acta*, 42, 431, 1997.
29. Parkinson, B.A., Heller, A., and Miller, B., Effects of cations on the performance of the photoanode in the n-gallium arsenide/potassium selenide (K_2Se)-potassium diselenide (K_2Se_2)-potassium hydroxide/carbon semiconductor liquid junction solar cell, *J. Electrochem. Soc.*, 126, 954, 1979.
30. Nakato, Y., Tonomura, S., and Tsubomura, H., The catalytic effect of electrodeposited metals on the photo-reduction of water at p-type semiconductors, *Ber. Bunsen., Phys. Chem.*, 80, 1289, 1976.
31. Nakato, Y., Yano, H., Nishiura, S., Ueda, T., and Tsubomura, H., Hydrogen photoevolution at p-type silicon electrodes coated with discontinuous metal layers, *J. Electroanal. Chem. Interfacial Electrochem.*, 228, 97, 1987.
32. Ahron-Shalom, E. and Heller, A., Efficient p-lnP (Rh–H alloy) and p-lnP (Re–H alloy) hydrogen evolving photocathodes, *J. Electrochem. Soc.*, 129, 2865, 1982.
33. Leygraf, C., Hendewerk, M., and Somorjai, G.A., Magnesium- and silicon-doped iron oxides for the photocatalyzed production of hydrogen from water by visible light (2.2 eV $\leq h\nu \leq$ 2.7 eV), *J. Catal.*, 78, 341, 1982.
34. Künhe, H.M. and Tributsch, H., Oxygen evolution from water mediated by infrared light on iron doped RuS_2 electrodes, *J. Electrochem. Soc.*, 130, 1448, 1983.
35. Künhe, H.M. and Tributsch, H., Oxygen and chlorine evolution on ruthenium iron disulfide mediated by low energy photons, *Ber. Bunsen. Phys. Chem.*, 88, 10, 1984.
36. Schwerzel, R.E. and Spahr, K.B., Hydrogen Production with Photoactive Semiconductor Catalysts Stabilized by Metallized Plasma Polymer Coatings, Final Report to U.S. Department of Energy, Brookhaven National Laboratory, Battelle, 1989.
37. Voet, D. and Voet, J.G., *Biochemistry*, 2nd ed., Wiley, New York, 1995.
38. Berg, J.M., Tymoczko, J.M., and Stryer, L., *Biochemistry*, 5th ed., Freeman and Company, New York, 2002.
39. Zubay, G.L., *Biochemistry*, 4th ed., WCB/Mc Graw-Hill, New York, 1998.
40. Blackenship, R.E., *Molecular Mechanism of Photosynthesis*, Blackwell, Boston, MA, 2002.
41. Fan, J.C.C., Tsau, B.Y., and Palm, B.J., Optimal design of high-efficiency tandem cells, *Proc. 16th IEEE Photovoltaic Specialists' Conf.*, 692, 1982.
42. Nozik, A.J., Energetics of photoelectrolysis, *Electrochem. Soc.*, 77–73, 272, 1977.
43. Nozik, A.J., Photochemical diodes, *Appl. Phys. Lett.*, 30, 567, 1977.
44. King, R.R., Law, D.C., Edmondson, K.M., Fetzer, C.M., Kinsey, G.S., Yoon, H., Krut, D.D., Ermer, J.H., Sherif, R.A., and Karam, R.H., Metamorphic concentrator solar cells with over 40% conversion efficiency, *Proc. of the 4th Int. Conf. on Solar Concentrators*, El Escorial, Spain, March 2007, p. 5.
45. Bahneman, D.W., Kormann, C., and Hoffman, M.R., Preparation and characterization of quantum size zinc oxide: A detailed spectroscopic study, *J. Phys. Chem.*, 91, 3789, 1987.
46. Kamat, P.V., Interfacial charge transfer processes in colloidal semiconductor systems, *Prog. React. Kinet.*, 19, 277, 1994.
47. Nosaka, Y., Photoelectrochemical reactions at semiconductor microparticles, in *Photocatalysis, Science and Technology*, Kaneko, M. and Okura, I. (Eds), Kodansha/Springer, Berlin, 2002, Chap. 5.
48. Hoffmann, M.R., Martin, T., Choi, W., and Bahnemann, D.W., Environmental applications of semiconductor photocatalysis, *Chem. Rev.*, 95, 69, 1995.
49. Fujishima, A. and Honda, K., Electrochemical photolysis of water at a semiconductor electrode, *Nature*, 238, 37, 1972.
50. Fujishima, A., Kobayakawa, K., and Honda, K., Hydrogen production under sunlight with an electrochemical photocell, *J. Electrochem. Soc.*, 122, 1487, 1975.
51. Watanabe, T., Fujishima, A., and Honda, K., Photoelectrochemical reaction at $SrTiO_3$ single crystal electrode, *Bull. Chem. Soc. Jpn.*, 49, 355, 1976.

52. Wrighton, M.S., Ellis, A.B., Wolczanski, P.T., Morse, D.L., Abrahamson, H.B., and Ginley, D.S., Strontium titanate photoelectrodes. Efficient photoassisted electrolysis of water at zero applied potential, *J. Am. Chem. Soc.*, 98, 2774, 1976.
53. Mavroides, G., Kafalas, J.A., and Kolesar, D.F., Photoelectrolysis of water in cells with $SrTiO_3$ anodes, *Appl. Phys. Lett.*, 28, 241, 1976.
54. Domen, K., Kondo, J.N., Hara, M., and Takata, T., Photo- and mechano-catalytic overall water splitting reactions to form hydrogen and oxygen on heterogeneous catalysts, *Bull. Chem. Soc. Jpn.*, 73, 1307, 2000.
55. Domen, K., Kudo, A., Onishi, T., Kosugi, N., and Kuroda, H., Photocatalytic decomposition of water into hydrogen and oxygen over nickel (II) oxide-strontium titanate ($SrTiO_3$) powder. 1. Structure of the catalysts, *J. Phys. Chem.*, 90, 292, 1986.
56. Domen, K., Kudo, A., and Onishi, T., Mechanism of photocatalytic decomposition of water into hydrogen and oxygen over nickel (II) oxide-strontium titanate ($SrTiO_3$), *J. Catal.*, 102, 92, 1986.
57. Kudo, A., Development of photocatalyst materials for water splitting with the aim at photon energy conversion, *J. Ceram. Soc. Jpn.*, 109, S81, 2001.
58. Jeong, H., Kim, T., Kim, D., and Kim, K., Hydrogen production by the photocatalytic overall water splitting on $NiO/Sr_3Ti_2O_7$: Effect of preparation method, *Int. J. Hydrogen Energ.*, 31, 1142, 2006.
59. Lehn, J.-M., Sauvage, J.-P., Ziessel, R., and Hilaire, L., Water photolysis by UV irradiation of rhodium loaded strontium titanate catalysts. Relation between catalytic activity and nature of the deposit from combined photolysis and ESCA studies, *Isreal J. Chem.*, 22, 168, 1982.
60. Sakthivel, S., Sankar, M.V., Palanichamy, M., Arabindoo, B., Bahnemann, D.W., and Murugesan, V., Enhancement of photocatalytic activity by metal deposition: Characterization and photonic efficiency of Pt, Au, and Pd deposited on TiO_2 catalyst, *Water Res.*, 38, 3001, 2004.
61. Arakawa, H. and Sayama, K., Water photolysis by TiO_2 particles—significant effect of Na_2CO_3 addition on water splitting, *Photocatalysis, Science and Technology*, Kaneko, M. and Okura, I. (Eds), Kodansha/Springer, Berlin, 2002, Chap. 14.
62. Yamaguchi, K. and Sato, S., Photolysis of water over metallized powdered titanium dioxide, *J. Chem. Soc., Faraday Trans. I*, 81, 1237, 1985.
63. Sato, S. and White, J.M., Photocatalytic water decomposition and water–gas shift reactions over NaOH-coated platinized TiO_2, *J. Catal.*, 69, 128, 1981.
64. Sayama, K. and Arakawa, H., Significant effect of carbonate addition on stoichiometric photodecomposition of liquid water into hydrogen and oxygen from platinum–titanium (IV) oxide suspension, *J. Chem. Soc., Chem. Commun.*, 150, 1992.
65. Sayama, K. and Arakawa, H., Effect of carbonate salt addition on the photocatalytic decomposition of liquid water over $PtTiO_2$ catalyst, *J. Chem. Soc., Faraday Trans.*, 93, 1647, 1997.
66. Arakawa, H. and Sayama, K., Oxide semiconductor materials for solar light utilization, *Res. Chem. Intermed.*, 26, 145, 2000.
67. Asahi, R., Morikawa, T., Ohwaki, T., Aoki, K., and Taga, Y., Visible-light photocatalysis in nitrogen-doped titanium oxides, *Science*, 293, 269, 2001.
68. Chen, S.Z., Zhang, P.Y., Zhuang, D.M., and Zhu, W.P., Investigation of nitrogen doped TiO_2 photocatalytic films prepared by reactive magnetron sputtering, *Catal. Commun.*, 5, 677, 2004.
69. Lindgren, T., Mwabora, J.M., Avendano, E., Jonsson, J., Hoel, A., Granqvist, C.G, and Lindquist, S.-E., Photoelectrochemical and optical properties of nitrogen doped titanium dioxide films prepared by reactive DC magnetron sputtering, *J. Phys. Chem. B*, 107, 5709, 2003.
70. Chen, X.B. and Burda, C., Photoelectron spectroscopic investigation of nitrogen-doped titania nanoparticles, *J. Phys. Chem.*, 108, 15446, 2004.
71. Torres, G.R., Lindgren, T., Lu, J., Granqvist, C.G, and Lindquist, S.E, Photoelectrochemical study of nitrogen-doped titanium dioxide for water oxidation, *J. Phys. Chem. B*, 108, 5995, 2004.
72. Kobayakawa, K., Murakami, K., and Sato, Y., Visible-light active N-doped TiO_2 prepared by heating of titanium hydroxide and urea, *Int. J. Photochem. Photobiol A: Chem.*, 170, 177, 2004.

73. Gole, J.L., Stout, J.D., Burda, C., Lou, Y., and Chen, X., Highly efficient formation of visible light tunable $TiO_{2-x}N_x$ photocatalysts and their transformation at the nanoscale, *J. Phys. Chem. B*, 108, 1230, 2004.
74. Khan, S.U.M., Al-Shahry, M., and Ingler W.B., Jr., Efficient photochemical water-splitting by a chemically modified n-TiO_2, *Science*, 297, 2243, 2002.
75. Mohapatra, S.K., Misra, M., Mahajan, V.K., and Raja K.S., Design of a highly efficient photoelectrolytic cell for hydrogen generation by water splitting: Application of TiO_2-xCx as a photoanode and Pt/TiO_2 as a cathode, to be published in, *J. Phys. Chem.* C_1 111, 8677, 2007.
76. Umebayashi, T., Yamaki, T., Itoh, H., and Asai, K., Band gap narrowing of titanium dioxide by sulfur doping, *Appl. Phys. Lett.*, 81, 454, 2002.
77. Ohno, T., Akiyoshi, M., Umebayashi, T., Asai, K., Mitsui, T., and Matsumura, M., Preparation of S-doped TiO_2 photocatalysts and their photocatalytic activities under visible light, *Appl. Catal. A: Gen.*, 265, 115, 2004.
78. Okada, M., Yamada, Y., Jin, P., Tazawa, M., and Yoshimura, K., Fabrication of multifunctional coating which combines low-e property and visible-light-responsive photocatalytic activity, *Thin Solid Films*, 442, 217, 2003.
79. Hitoki, G., Takata, T., Kondo, J.N., Hara, M., Kobayashi, H., and Domen, K., An oxynitride, TaON, as an efficient water oxidation photocatalyst under visible light irradiation (λ = <500), *Chem. Commun.*, 16, 1698, 2002.
80. Hara, M., Nunoshige, J., Takata, T., Kondo, J.N., and Domen, K., Unusual enhancement of H_2 evolution by Ru on TaON photocatalyst under visible light irradiation, *Chem. Commun.*, 24, 3000, 2003.
81. Liu, M., You, W., Lei, Z., Zhou, G., Yang, J., Wu, G., Ma, G., Luan, G., Takata, T., Hara, M., Domen, K., and Li, C., Water reduction and oxidation on Pt–Ru/$Y_2Ta_2O_5N_2$ catalyst under visible light irradiation, *Chem. Commun.*, 19, 2192, 2004.
82. Domen, K. and M. Hara, Water photolysis by layered compounds, *Photocatal. Sci. Technol.*, Kaneko, M. and Okura, I. (Eds), Kodansha/Springer, Berlin, 2002, Chap. 16.
83. Domen, K., Kudo, A., Tanaka, A., and Onishi, T., Overall photodecomposition of water on a layered niobate catalyst, *Catal. Today*, 8, 77, 1990.
84. Inoue, Y., Kubokawa, T., and Sato, K., Photocatalytic activity of alkali-metal titanates combined with Ru in decomposition of water, *J. Phys. Chem.*, 95, 4059, 1991.
85. Inoue, Y., Niiyama, T., and Sato, K., Photocatalysts using hexa- and octa-titanates with different tunnel space for water decomposition, *Top. Catal.*, 1, 137, 1994.
86. Ogura, S., Kohno, M., Sato, K., and Inoue, Y., Effects of RuO_2 on activity for water decomposition of a $RuO_2/Na_2Ti_3O_7$ photocatalyst with zigzag layer structure, *J. Mater. Chem.*, 8, 2335, 1998.
87. Kohno, M., Ogura, S., Sato, K., and Inoue, Y., Effect of tunnel structures of $BaTi_4O_9$ and $Na_2Ti_6O_{13}$ on photocatalytic activity and photoexcited charge separation, *Stud. Surf. Sci. Catal.*, 101, 143, 1996.
88. Domen, K., Hara, M., Kondo, J.N., Takata, T., Kudo, A., Kobayashi, H., and Inoue, Y., New aspects of heterogeneous photocatalysis for water decomposition, *Korean J. Chem. Eng.*, 18, 862, 2001.
89. Linkous, C.A., Muradov, N.Z., and Ramser, S.N., Considerations of reactor design for solar hydrogen production from hydrogen sulfide using semiconductor particulates, *Int. J. Hydrogen Energ.*, 20, 701, 1995.
90. Gary, L.H. and Handwerk, G.E., *Petroleum Refining, Technology and Economics*, Marcel Dekker, New York, 1975, 190.
91. Leffler, W.L., *Petroleum Refining for Non-Technical Persons*, Pennwell, Tulsa, OK, 1979, 115.
92. Williams, R., Becquerel photovoltaic effect in binary compounds, *J. Chem. Phys.*, 32, 1505, 1960.
93. Inoue, T., Watanabe, T., Fujishima, A., Honda, K., and Kohayakawa, K., Suppression of surface dissolution of CdS photoanode by reducing agent, *J. Electrochem. Soc.*, 124, 719, 1977.
94. Harbour, J.R. and Hair, M.L., Superoxide generation in the photolysis of aqueous cadmium sulphide dispersions, detection by spin trapping, *J. Phys. Chem.*, 81, 1791, 1977.
95. Harbour, J.R., Wolkow, R., and Hair, M.L, Effect of platinization on the photoproperties of CdS pigments in dispersion. Determination by H_2 evolution, O_2 uptake, and electron spin resonance spectroscopy, *J. Phys. Chem.*, 85, 4026, 1981.

96. Darwent, J.R. and Porter, G., Photochemical hydrogen production using cadmium sulphide suspensions in aerated water, *J. Chem. Soc., Chem. Commun.*, 4, 145, 1981.
97. Darwent, J.R., H_2 production photosensitized by aqueous semiconductor dispersions, *J. Chem. Soc., Faraday Trans. II*, 77, 1703, 1981.
98. Muradov, N.Z., Bazhutin, Yu., Bezuglaya, A., and Rustamov, M., Visible light induced decomposition of hydrogen sulfide in aqueous solutions using cadmium sulfide, *J. Phys. Chem.*, 54, 2563, 1982 (in Russian).
99. Matsumura, M., Saho, Y., and Tsubomura, H., Photocatalytic hydrogen production from solutions of sulphide using platinized cadmium sulphide powder, *J. Phys. Chem.*, 87, 3807, 1983.
100. Matsumura, M., Ohnishi, H., Hanafusa, K., and Tsubomura, H., Stabilization of platinum-loaded CdS photocatalyst by addition of EDTA, *Bull. Chem. Soc. Jpn.*, 60, 2001, 1887.
101. Tricot, Y.-M. and Fendler, J.H., Colloidal catalyst-coated semiconductors in surfactant vesicles: *In situ* generation of Rh-coated CdS particles in dihexadecylphosphate vesicles and their utilization for photosensitized charge separation and hydrogen generation, *J. Am. Chem. Soc.*, 106, 7359, 1984.
102. Meyer, M., Wallberg, C., Kurihara, K., and Fendler, J.H., Photosensitized charge separation and hydrogen production in reversed micelle entrapped platinized colloidal cadmium sulphide, *J. Chem. Soc., Chem. Commun.*, 2, 90, 1984.
103. Rafaeloff, R., Tricot, Y.-M., Nome, F., and Fendler, J.H., Colloidal catalyst coated semiconductors in surfactant vesicles: *In situ* generation of rhodium-coated cadmium sulfide particles in dioctadecyldimethylammonium halide surfactant vesicles and their utilization for photosensitized charge separation and hydrogen generation, *J. Phys. Chem.*, 89, 533, 1985.
104. Bühler, N., Meier, K., and Reber, J.-F., Photochemical hydrogen production with cadmium sulphide suspensions, *J. Phys. Chem.*, 88, 3261, 1984.
105. Borgarello, E., Serpone, N., Grätzel, M., and Pelizzetti, E., Hydrogen production through microheterogeneous photocatalysis of hydrogen sulphide cleavage. The thiosulfate cycle, *Int. J. Hydrogen Energ.*, 10, 737, 1985.
106. Borgarello, E., Desilvestro, J., Grätzel, M., and Pelizzetti, E., Photoreduction of thiosulfate in semiconductor dispersions, *Helv. Chim.*, 66, 1827, 1983.
107. Borgarello, E. and Grätzel, M., Hydrogen energy progress, IV, *Proc. 4th World Hydrogen Energy Conf.*, vol. 2, Veziroglu, T.N., von Vorst, W.D. and Kelley, J.H. (Eds), p. 739, Pergamon Press, New York, 1982.
108. Borrell, L., Cervera-March, S., Gimenez, J., Simarro, R., and Andújar, J.M., A comparative study of CdS-based semiconductor photocatalysts for solar hydrogen production from sulphide + sulphite substrates, *Solar Energ. Mater. Solar Cells*, 25, 25, 1992.
109. Hayes, D., Grieser, F., and Furlong, D.N., Kinetics of hydrogen production form illuminated $CdS/Pt/Na_2SO_3$ and $ZnS/Pt/Na_2SO_3$ dispersions, *J. Chem. Soc. Faraday Trans.*, 86, 3637, 1990.
110. Subrahmanyam, M., Supriya, V.T., and Reddy, P.R., Photocatalytic H_2 production with CdS-based catalysts from a sulphide/sulphite: An effort to develop MgO-supported catalysts, *Int. J. Hydrogen Energ.*, 21, 99, 1996.
111. Gruzdkov, Yu.A., Savinov, E.N., and Parmon, V.N., Photocatalytic decomposition of hydrogen sulfide in the presence of polymer immobilized cadmium sulfide. Promotion by I and VIII group metals, *Int. J. Hydrogen Energ.*, 12, 393, 1987.
112. Kobayakawa, K., Miura, T., Suzuki, A., Sato, Y., and Fujishima, A., Enhancement of the photocatalytic activity of CdS powder for hydrogen evolution from aqueous sulfide solution by heat treatment with KCl, *Solar Energ. Mater. Solar Cells*, 30, 201, 1993.
113. Govindjee and Coleman, W.J., How plants make oxygen, *Sci. Am.*, 262, 50, 1990.
114. Govindjee, Beatty, J.T., Gest, H., and Allen, J.F. (Eds), *Discoveries in Photosynthesis. Advances in Photosynthesis and Respiration*, vol. 20, Springer, Berlin, 2005.
115. Witt, H.T., Coupling of quanta, electrons, fields, ions, and phosphorylation in the functional membrane of photosynthesis. Results by pulse spectroscopic methods, *Quart. Rev. Biophys.*, 4, 365, 1971.
116. Barber, J., *Topics in Photosynthesis, Vol. 2: Primary Processes of Photosynthesis*, Elsevier, Amsterdam, 1977.

117. Miziorko, H.M. and Lorimer, G.H., Ribulose-1, 5-biphosphate carboxylase-oxygenase, *Ann. Rev. Biochem.*, 52, 507, 1983.
118. Tributsch, H., Challenges for (photo) electrocatalysis research, *Catal. Today*, 39, 177, 1997.
119. Linkous, C.A. and Slattery, D.K., Solar hydrogen via photosynthetic Z-scheme analogue based on semiconducting powders, *Am. Chem. Soc. Fuel Chem.*, 43, 378, 1998.
120. Walker, D.A., Concerning oscillations, *Photosynth. Res.*, 34, 384, 1992.
121. Walker, D.A., Excited leaves, *New Phytol.*, 121, 325, 1992.
122. Linkous, C.A., Slattery, D.K., Ouellette, A.J.A., McKaige, G.T., and Austin, B.N.C., Solar photocatalytic H_2 from water using a dual bed photosystem, *Proc. 11th World Energy Conf.*, Stuttgart, Germany, 1996.
123. Linkous, C.A., Slattery, D.K., and Mraz, M.E., Hydrogen production via photocatalysis. Catalysis of O_2 evolution, *Proc. 15th World Hydrogen Energy Conf.*, Yokohama, Japan, 2004.
124. Slattery, D.K., Linkous, C.A., and Gruhn, N.E., Organic pigments as photocatalytic agents in a solar water-splitting scheme, *Proc. 13th World Hydrogen Energy Conf.*, Beijing, China, 1, 390, 2000.
125. Linkous, C.A., Slattery, D.K., and Fowler, J.R., Coupling of organic pigment photocatalytic systems to produce hydrogen from water, *Proc. 14th World Hydrogen Energy Conf.*, Montreal, Canada, 2002.
126. Khaselev, O., Bansal, A., and Turner, J.A., High-efficiency integrated multijunction photovoltaic/electrolysis systems for hydrogen production, *Int. J. Hydrogen Energ.*, 26, 127, 2001.
127. Bolton, J.R., Strickler, S.J., and Connolly, J.S., Limiting and realizable efficiencies of solar photolysis of water, *Nature*, 316, 495, 1985.
128. Khaselev, O. and Turner, J.A., A monolithic photovoltaic-photoelectrochemical device for hydrogen production via water splitting, *Science*, 280, 425, 1998.
129. Kocha, S.S., Montgomery, D., Peterson, M.W., and Turner, J.A., Photoelectrochemical decomposition of water utilizing monolithic tandem cells, *Solar Energ. Mater. Solar Cells*, 52, 389, 1998.
130. Rocheleau, R.E., Miller, E.L., and Misra, A., High-efficiency photoelectrochemical hydrogen production using multijunction amorphous silicon photoelectrodes, *Energy Fuels*, 12, 3, 1998.
131. Licht, S., Wang, B., Mukerji, S., Soga, T., Umeno, M., and Tributsch, H., Over 18% solar energy conversion to generation of hydrogen fuel; theory and experiment for efficient solar water-splitting, *Int. J. Hydrogen Energ.*, 26, 653, 2001.
132. Alonso-Vante, N., Colell, H., Stimming, U., and Tributsch, H., Anomalous low-temperature kinetic effects for oxygen evolution on RuO_2 and Pt electrodes, *J. Phys. Chem.*, 97, 7381, 1993.
133. Lodi, G., Sivieri, E., Debattisti, A., and Trasatti, S., Ruthenium dioxide-based film electrodes. III. Effect of chemical composition and surface morphology on oxygen evolution in acid solutions, *J. Appl. Electrochem.*, 8, 135, 1978.
134. Licht, S., Ghosh, S., Tributsch, H., and Fiechter, S., High efficiency solar energy water-splitting to generate hydrogen fuel: Probing RuS_2 enhancement of multiple band electrolysis, *Solar Energ. Mater. Solar Cells*, 70, 471, 2002.
135. Kuhne, H.M., Jaegermann, W., and Tributsch, H., The electronic band character of ruthenium dichalcogenides and its significance for the photoelectrolysis of water, *Chem. Phys. Lett.*, 112, 160, 1984.
136. Dhere, N.G. and Jahagirdar, A.H., Photoelectrochemical water splitting for hydrogen production using multiple bandgap combination of photovoltaic cell and thin-film-photocatalyst, *Thin Solid Films*, 480–481, 462, 2005.
137. Avachat, U.S., Jahagirdar, A.H., and Dhere, N.G., Multiple bandgap combination of thin film photovoltaic cell and photoanode for efficient hydrogen and oxygen generation by water splitting, *Solar Cells Solar Energ. Mater.*, 90, 2464, 2006.
138. Jahagirdar, A.H. and Dhere, N.G., Photoelectrochemical water splitting using $CuIn_{1-x}Ga_xS_2$/CdS thin film solar cells for hydrogen generation, *Solar Energ. Mater. Solar Cells*, 91, 1488, 2007.
139. Dhere, N.G., Kadam, A.A., Kulkarni, S.S., Bet, S.M., and Jahagirdar, A.H., Large area CIGS2 thin film solar cells on foils: Nucleus of a pilot plant, *Solar Energ. J.*, 77, 697, 2004.
140. Dhere, N.G., Ghongadi, S.R., Pandit, M.B., and Jahagirdar, A.H., CIGS2 thin-film solar cells on flexible foils for space power, *Prog. Photovolt: Res. Appl.*, 10, 407, 2002.

8

Hydrogen Separation and Purification

Ashok Damle

CONTENTS

8.1 Introduction

Realization of the vision of "hydrogen economy" and utilization of hydrogen as an energy carrier will require increasing the hydrogen production by more than an order of magnitude over the current production levels. About 42 million t of hydrogen is currently produced per year worldwide. Most of the hydrogen is used captively on-site in "nonenergy" uses for the production of ammonia, methanol, and other chemicals and for hydroprocessing in petroleum refineries primarily to remove sulfur and upgrade the heavier fractions into more valuable products (ORNL, 2003). The hydrogen production in the United States is estimated to be about 9 million t/year, 7.5 million of which are consumed on-site. The remaining 1.5 million t/year of hydrogen production may be considered as transportable and available for other applications (National Academies Press, 2004). For utilization in hydrogen-powered cars, almost 40 million t/year of hydrogen will have to be produced to be able to support 100 million cars in the United States (U.S. Department of Energy, 2002). The demand for hydrogen will escalate even more if hydrogen is used as an energy carrier for power generation applications, for example, portable or distributed power. Central as well as distributed hydrogen generation plants will probably be necessary to meet the projected hydrogen demand. For successful transition to hydrogen economy, hydrogen must also be affordable.

8.2 Hydrogen Production

Hydrogen can be produced from a variety of sources. Fossil fuels such as liquid hydrocarbons, coal, and natural gas may be used to produce hydrogen by first converting their fuel value to gas phase by reaction with steam, oxygen, or air (gasification/reforming) followed by hydrogen enrichment and separation. Renewable carbonaceous sources such as biomass may be used to produce hydrogen in a similar way as fossil fuels through their gasification. Renewable sources such as wind and solar energy may be used to produce hydrogen by first converting the energy form to electricity followed by electrolysis of water. Nuclear or solar heat may be used to drive thermochemical reaction cycles that produce hydrogen as a product. Solar energy can also be converted directly to hydrogen using photoelectrochemical or photobiological methods. A number of these processes are discussed in detail in this sourcebook. Feasibility of individual methods to produce hydrogen will be driven by the cost of hydrogen production and its purification for the intended use, availability of resources, and demand for hydrogen. Currently over 80% of hydrogen is produced by steam reforming of natural gas followed by hydrogen separation as the most economical option depending on the availability of natural gas (Ritter and Ebner, 2005). Hydrogen will ultimately need to be produced totally from renewable energy sources for sustainable hydrogen economy; however, for the near-term transition period, hydrogen will need to be produced from fossil fuels, primarily, natural gas and coal.

A portfolio of feedstocks and technologies for hydrogen production will be necessary to meet the hydrogen demand envisioned for hydrogen economy. The near-term focus will be on distributed reforming of natural gas and renewable liquid fuels and on electrolysis of water to meet initial lower volume hydrogen needs with the least capital equipment costs (U.S. Department of Energy, 2007). For the longer term, hydrogen will probably come from renewable feedstocks and energy sources, with emphasis on centralized options to take advantage of economies of scale when an adequate hydrogen delivery infrastructure is in place.

Currently, a large number of hydrogen generation plants are located at captive utilization sites, for example, in ammonia and methanol plants to produce an appropriate synthesis gas mixture, and in refineries to generate hydrogen needed for hydrotreating or hydrocracking operations. A large fraction of these hydrogen plants are designed to produce synthesis gas rather than pure hydrogen with a H_2:CO ratio in the synthesis gas suitable for the intended product, for example, for methanol production, a H_2:CO ratio of 2:1 is desirable, whereas for ammonia production, partial oxidation (POX) of hydrocarbon fuels with air is used to produce the N_2:H_2 ratio of 3:1 with no CO. Hydrogen separation and purification requirements in the existing plants depend on the intended use of hydrogen, for example, for ammonia production it is customary to convert almost all of the CO to hydrogen using low-temperature water–gas shift (WGS) reaction followed by CO_2 scrubbing and final methanation step to convert residual CO and CO_2 to methane, whereas CO_2 removal may not even be necessary for the utilization of synthesis gas for producing methanol or liquid fuels.

Increasing the utilization of hydrogen as an energy carrier in transportation and distributed power generation applications based on proton exchange membrane (PEM) fuel cells will create a demand for plants that produce high-purity hydrogen as the primary product.

Steam reforming of natural gas is most commonly used for producing pure hydrogen as a product accounting for almost 95% of hydrogen produced in the United States and about 80% of hydrogen produced worldwide (ORNL, 2003). Steam methane reforming (SMR) is the most commonly used process for industrial hydrogen production from methane (natural gas) when compared with the POX of methane or autothermal reforming (ATR) of methane processes except for very large installations when ATR becomes more cost effective (Ritter and Ebner, 2005). Petroleum feedstocks, for example, naphtha and fuel oils, are also widely used for the generation of hydrogen instead of natural gas depending on availability and cost of those feedstocks especially in petroleum refinery operations.

In hydrogen generation processes based on natural gas and liquid hydrocarbon fuels, the feedstock is first desulfurized by hydrodesulfurization and H_2S removal to reduce the sulfur levels to below 2 ppm to protect catalysts used in the downstream reforming process. After desulfurization, SMR is carried out at 800–900°C temperature and 25–35 atm pressure using an alkali promoted Ni-based catalyst supported on alumina. SMR is a highly endothermic process and heat is supplied by the combustion of a part of the fuel. In some cases, especially when using heavier hydrocarbon feedstocks, a prereformer is used upstream of the primary reformer to first convert the heavy hydrocarbons to methane. A prereformer typically operates at a lower temperature of 400–500°C and may also convert some of the methane to CO and H_2. SMR produces a synthesis gas constituting primarily hydrogen, carbon monoxide, carbon dioxide, and unreacted methane. Alternatively, oxygen or air may be used in addition to steam in POX or ATR processes, which combine heat generation by fuel combustion with the steam reforming process. Typical hydrogen production plants use SMR process, whereas syngas production plants use either ATR or POX, for example, methanol and ammonia plants tend to use ATR (Ritter and Ebner, 2005).

Solid fuels such as coal and biomass are gasified again using steam, oxygen, or air. A variety of commercial designs are available for the gasification process. The synthesis gas generated by coal or biomass contains much higher $CO{:}H_2$ ratio due to greater carbon content in the feedstock. The synthesis gas produced by gasification is then cleaned to remove particulates, H_2S, ammonia, HCl, and any other contaminants that may be present depending on the raw material used (e.g., tar when using certain biomass materials) in the gasification process.

The hydrogen content of the synthesis gas produced by using any of the feedstock is further increased by WGS reaction, in which steam reacts with carbon monoxide to produce additional hydrogen and carbon dioxide. Hydrogen is then separated from other gas-phase constituents, primarily carbon dioxide. The separated hydrogen may need to be purified further to meet the requirements of the specific intended application. The level of hydrogen purification depends strongly on the mode of hydrogen utilization, for example, to use in a PEM fuel cell much greater purity hydrogen will be required than for its utilization in a solid oxide fuel cell (SOFC) or in a gas turbine for power generation. Hydrogen generation processes based on water electrolysis using electricity generated from renewable solar or wind energy will produce relatively much purer hydrogen with only water vapor as a diluent. Very little hydrogen purification will be needed for such hydrogen generation approaches. This chapter will, therefore, focus on synthesis gas-based hydrogen generation processes and the pertinent hydrogen separation and purification approaches.

All processes using carbonaceous feedstocks, including hydrocarbon fuels and biomass to generate pure hydrogen will involve a suitable combination of processes to produce synthesis gas followed by WGS reaction. Subsequent hydrogen separation and purification options will, thus, be similar to all carbonaceous feedstock-based hydrogen production. Steam reforming of methane is, therefore, used as an example in the following section to describe reforming of hydrocarbon fuels in general.

8.2.1 Steam Reforming of Methane and Hydrocarbon Fuels

Steam reforming of methane and petroleum feedstocks, for example, naphtha, is an industry standard process for generating hydrogen commercially, and extensive information is available on thermodynamics, kinetics, and catalysis of the reforming reactions (Bridger, 1980; Ridler and Twigg, 1989; Rostrup-Nielsen, 1984; Van Hook, 1980). Steam reforming is a relatively simple, efficient, well-proven technology; it provides a cost-effective product and offers significant flexibility in operating with different hydrocarbon feedstocks (Steinberg and Cheng, 1989). Steam reforming of hydrocarbons, generally, is a metal-catalyzed reaction described by three separate stoichiometric reactions. First, the hydrocarbon dissociates on the metal surface, and the hydrocarbon fragments react with adsorbed steam to produce CO and H_2:

$$\text{Reforming}\quad C_nH_m + nH_2O \rightarrow nCO + (n + m/2)\,H_2 \quad (\text{for } CH_4{:}\Delta H^{\circ}_{298} = +206\ \text{kJ/mol}) \tag{8.1}$$

Following reaction 8.1, equilibrium concentrations of CO, H_2, CH_4, CO_2, and H_2O are established according to the reaction equilibria for the methanation and WGS reactions:

$$\text{Methanation}\quad CO + 3H_2O \Leftrightarrow CH_4 + H_2O \quad (\Delta H^{\circ}_{298} = +206\ \text{kJ/mol}) \tag{8.2}$$

$$\text{WGS}\quad CO + H_2O \Leftrightarrow CO_2 + H_2 \quad (\Delta H^{\circ}_{298} = -41.2\ \text{kJ/mol}) \tag{8.3}$$

Although reaction 8.1 is strongly endothermic, the overall heat of reaction may be positive or negative, depending on the process conditions; most typically, it is strongly endothermic, and heat must be supplied to the reformer, usually by burning part of the fuel. In a POX (also termed as autothermal reformer process), fuel combustion and subsequent steam reforming of the remaining fuel is conducted in a single reactor. Because air is often used in an autothermal reformer reactor, the resulting reformate gas contains much lower concentration of hydrogen than that obtainable by steam reforming of the hydrocarbon fuel. In the SMR process, depending on the pressure, temperature, and steam to methane ratio used, typically, 75–80% of methane is converted to CO and H_2 in the reforming reaction resulting in a dry basis composition of the synthesis gas of about 75% hydrogen, 7% methane, 12% CO, and 6% CO_2 for SMR conducted at 800°C, 300 psig, and steam to methane ratio of 3. Although the preceding composition is for synthesis gas produced from methane, the composition depends on the hydrocarbon feedstock used, for example, for synthesis gas generated with coal gasification, the H_2:CO ratio in the synthesis gas will be closer to 1:1 and that for naphtha or higher hydrocarbon feedstocks, it will be closer to 2:1.

The amount of hydrogen in the synthesis gas may further be increased by reacting CO with steam in the WGS reaction (reaction 8.3) to produce additional hydrogen and CO_2. Because the WGS reaction is an equilibrium-limited reaction favored at lower temperatures, the WGS reaction is typically conducted in two steps: high-temperature shift (HTS) reaction for faster kinetics is typically conducted on a chromium iron oxide catalyst at 340–455°C in a fixed-bed adiabatic reactor. Approximately, 94% of the CO in the raw gas is reacted (Kirk-Othmer, 1999). Converted gases are cooled outside of the reactor by producing steam and sent to a second low-temperature shift (LTS) converter at about 200–215°C to complete the shift reaction over a copper–zinc catalyst supported on alumina. The exit gas typically contains about 87.3% H_2, 10.7% CO_2, 0.70% CO, and 1.3% CH_4 (dry basis) and is further purified by removing CO_2. The resulting product gas mixture may contain other contaminants depending on the original feedstock used for synthesis gas generation.

8.3 Hydrogen Separation

Four technologies have been most commonly used for separating hydrogen, carbon dioxide, and carbon monoxide from synthesis gas:

1. Adsorption of gas species other than hydrogen, which produces pure hydrogen as a product
2. Absorption (physical or chemical) of CO_2, which produces pure carbon dioxide as a product
3. Polymeric membranes, which offer bulk separation of hydrogen
4. Cryogenic separation, which can provide multiple pure products and is especially used for separation of pure carbon monoxide

Absorption of CO_2 in suitable solvents followed by methanation of residual CO and CO_2 was the technology of choice before 1980 for producing pure hydrogen as a product. For plants built after 1980, pressure swing adsorption (PSA) has been the most commonly used technology for producing high-purity hydrogen as a product by SMR (Ritter and Ebner, 2005). Retrofit PSA separation has also replaced CO_2 scrubbing systems in a number of existing installations for hydrogen production. Scrubbing of CO_2 is still primarily used in producing N_2–H_2 synthesis gas for ammonia production and a number of solvents are available for CO_2 absorption, both physical and chemical. Amine absorption processes also remove sulfur compounds that may be present in the synthesis gas. Membrane separation is a relatively new technology for separating hydrogen. The prism membrane was introduced for hydrogen recovery in 1979 by Monsanto (Gunardson, 1998) (now available from Air Products and Chemicals, Inc.). Hydrogen separation membranes have also been introduced by several other companies, for example, PolySep by UOP, Inc., and are commonly used for recovering hydrogen from a variety of waste streams containing a high content of hydrogen in refinery and chemical processing operations. Polymeric membrane processes do not produce high-purity hydrogen in a single stage, and multiple stages with a low overall hydrogen recovery are necessary to produce a high-purity hydrogen product. Polymeric membranes are, however, used for bulk separation of hydrogen and may be combined with another process, for example PSA, in a hybrid configuration for high-purity hydrogen production. Cryogenic separation processes are now largely replaced by PSA systems (Gunardson, 1998) and are only used for specialized applications to produce multiple pure products, for example, CO.

The overall processes utilizing SMR with PSA, CO_2 absorption, and membrane-based hydrogen separation systems are shown schematically in Figures 8.1 through 8.3, respectively. The choice of applicable hydrogen separation technology often depends on the hydrocarbon feedstock used and the resulting synthesis gas composition. PSA has been dominant for hydrogen production plants utilizing natural gas as the feedstock, which is currently often the best option based on the least cost and availability. The synthesis gas produced by SMR contains a high concentration of hydrogen making the economics favorable to a PSA system. For feedstocks containing high carbon to hydrogen ratio, for example, heavy hydrocarbon fuels and coal, other hydrogen separation systems such as CO_2 absorption–based systems or bulk membrane separation systems may become attractive either as stand alone or in hybrid configuration combining two or more approaches. The current emphasis on global climate control will also have a strong impact on the selection of hydrogen separation approaches; hydrogen separation processes that can also capture

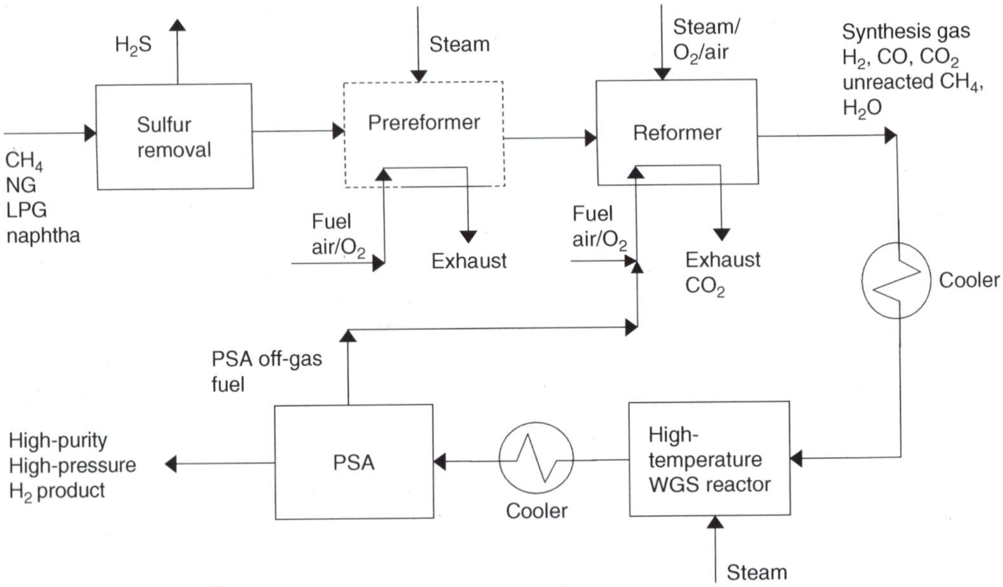

FIGURE 8.1
Steam methane reforming with PSA for hydrogen production.

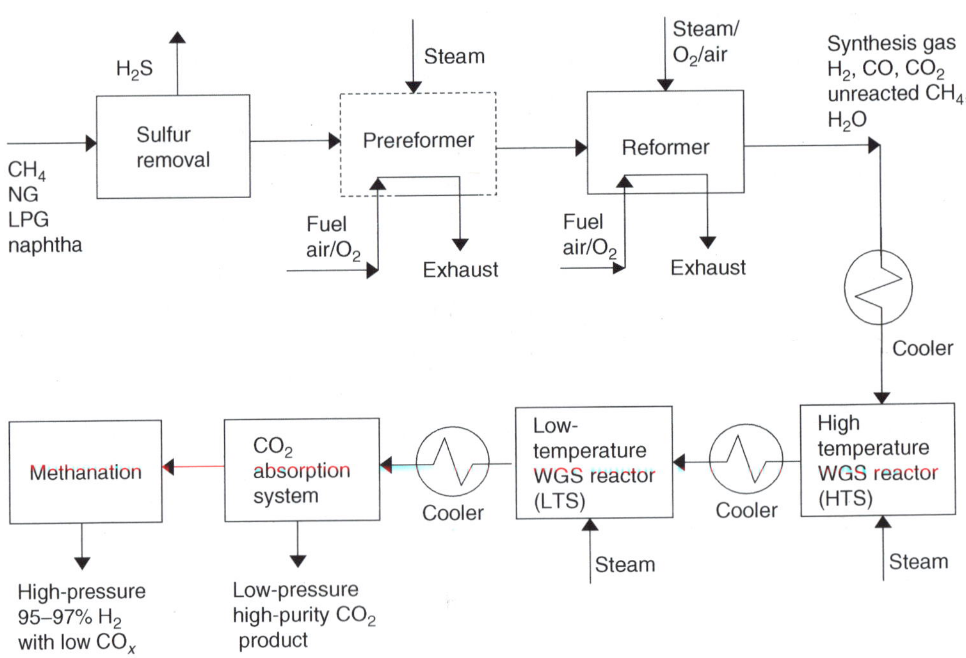

FIGURE 8.2
Steam methane reforming with CO_2 absorption for hydrogen production.

the CO_2 in a pure form suitable for sequestration are, therefore, likely to become attractive in the new generation of pure hydrogen plants.

Although, in large-scale hydrogen plants, the cost of producing hydrogen is dominated by the cost of feedstock (most often natural gas) used and the cost of fuel used to supply heat

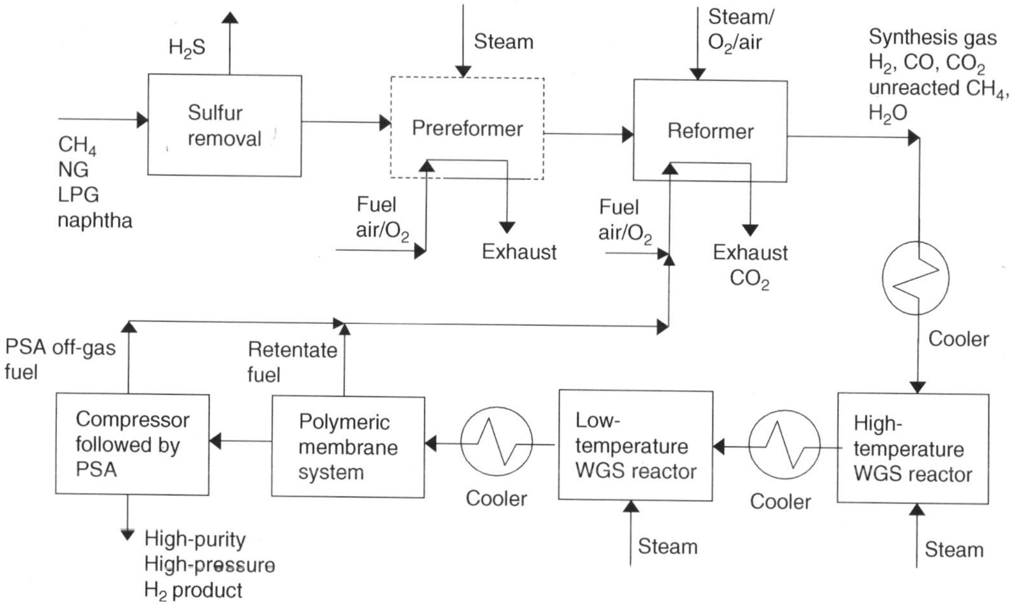

FIGURE 8.3
Steam methane reforming with polymeric membrane for hydrogen production.

for the reforming reaction, the hydrogen separation and purification equipment contributes at least 50%, and sometimes up to 80%, of the capital investment costs. Thus, hydrogen separation and purification component will contribute significantly to the depreciation and investment costs in plant operation. The total in-plant energy consumed in the SMR process for hydrogen production is estimated at about 56,200 Btu/lb. of H_2 produced with energy used for separation process accounting for about 12,400 Btu/lb. of H_2 produced (ORNL, 2005). The hydrogen separation and purification process, thus, contributes significantly to the operating costs of hydrogen production.

Reducing the energy requirement for the separation and purification of hydrogen and the separation equipment cost will significantly reduce the cost of producing hydrogen. Substantial research efforts are, therefore, underway to investigate new concepts for hydrogen separation as well as to improve existing proven processes especially PSA (U.S. Department of Energy, 2007). The following sections describe the existing processes that are commercially practiced for hydrogen separation and purification as well as promising new concepts that are being investigated.

8.3.1 Pressure Swing Adsorption

Since 1980, PSA systems have largely replaced the CO_2 removal systems based on amine absorption followed by methanation as well as cryogenic separation systems as the technology of choice in new plants for producing hydrogen. PSA systems have also replaced CO_2 absorption systems in existing plants producing hydrogen. Advantages of PSA systems over the amine absorption system include lower equipment cost, simpler operation, less operational maintenance, and ability to deliver a high-purity hydrogen product. As seen in Figure 8.1, in a PSA-based process, the LTS reactor is generally eliminated along with the final methanation step for hydrogen purification. The PSA system is installed following the HTS reactor and gas cooler in the conventional SMR hydrogen production process. Although PSA systems produce a high-purity hydrogen product, the hydrogen recovery is correspondingly low. The waste residual PSA off-gas, however, has significant fuel value since it contains unreacted methane, unrecovered H_2, as well as unreacted CO remaining after the HTS reactor. The tail gas is, therefore, burned in the reformer to meet almost all heat requirements for the hydrocarbon fuel reforming process.

The PSA units are extensively used not only for hydrogen production by SMR but for hydrogen recovery from a variety of hydrogen-rich off-gas streams containing a wide range of impurities. PSA is well suited for hydrogen separation from hydrogen-rich gas streams since hydrogen is hardly adsorbed on most adsorbents. Thus, high-purity hydrogen may be recovered using almost any adsorbent. For various applications, adsorbents are selected based on the species other than hydrogen that are present in the gas stream for their efficient removal from hydrogen. The PSA process for hydrogen purification was first commercialized by UOP in 1966. Since then, over 530 units have been installed worldwide, 40% of which were installed for hydrogen production by SMR (Stocker et al., 1998). Because a PSA process for hydrogen separation adsorbs all gas species except hydrogen, PSA hydrogen separation is attractive when the feed gas contains a high concentration of hydrogen. With a dilute hydrogen stream, PSA may still be effectively used in a hybrid process configuration by employing another process for hydrogen enrichment, for example, CO_2 absorption, to produce a hydrogen-rich high-pressure feed gas for the PSA system.

Solid adsorbents, for example, molecular sieves, 5A zeolite, activated carbon, alumina, and silica gel, adsorb more adsorbates at a higher gas partial pressure than at a lower partial pressure. In addition, some of the compounds are adsorbed more strongly than others allowing the separation of strongly adsorbed compounds from the weakly adsorbed compounds when their mixture is passed over a fixed adsorbent bed. Hydrogen essentially is not adsorbed by the adsorbents used in the PSA process for hydrogen purification. In principle, for purification of hydrogen from SMR product gas, for example, all other gas species besides hydrogen (i.e., CO, CO_2, CH_4, and H_2O) are adsorbed at a high pressure producing high-purity hydrogen product. Using sufficient quantity of the adsorbent material and by regulating gas flow rate and adsorption cycle time, very high purity of hydrogen can be achieved. After saturation of the fixed bed by the contaminants, the adsorbent bed is regenerated by removing the contaminants by lowering the pressure (pressure swing) in the bed. The contaminants may also be removed by increasing the temperature of the bed at constant pressure in a thermal swing adsorption (TSA) process; however, the cycle times for TSA are typically much longer due to slow heat transfer to/from a solid bed. Thus, PSA systems are preferred over TSA for hydrogen separation applications. The principles and theory of the PSA process operation is well established and is described in literature (e.g., Thomas and Crittenden, 1998; Ruthven et al., 1994).

Although PSA is a batchwise process, by using multiple beds in a sequential manner the overall process is operated in a continuous fashion. Each bed may contain layers of different adsorbent materials selective for specific contaminants in the hydrogen gas stream to be purified. Each bed undergoes a sequence of four basic steps in a PSA cycle: adsorption, depressurization, purge at low pressure, and repressurization. This sequence of cyclic operations for each bed is shown schematically for a four-bed PSA process in Figure 8.4 (Yang, 1987; Cassidy, 1980; Miller and Stocker, 1999).

The feed gas is introduced at the high adsorption pressure in the adsorption step, impurities are adsorbed on the internal surfaces of the adsorbent leaving high-purity hydrogen product gas in the void spaces of the vessel, and high-purity hydrogen is withdrawn continuously as product from the top of the vessel under pressure. The gas flow is normally

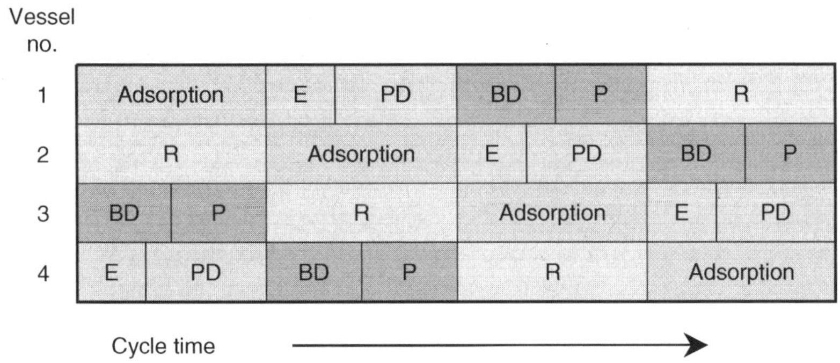

E = Equalization and cocurrent depressurization
PD = Purge and cocurrent depressurization
BD = Blowdown and countercurrent depressurization
P = Purge
R = Repressurization

FIGURE 8.4
Pressure swing adsorption cycle sequence for a four-bed system. (Adapted from Miller, G.Q. and Stocker, J., Selection of a Hydrogen Separation Process, UOP Report, January 1999, available at http://www.uop.com/objects/SelOfHydroSepProc.pdf; Cassidy, R.T., Adsorption and ion exchange with synthetic zeolites, in *ACS Symposium Series*, ed. W.H. Flank, American Chemical Society, Washington, Vol. 135, p. 275, 1980.)

in the upward direction. When the adsorber bed reaches its adsorption capacity, it is taken off-line and the feed is automatically switched to another fresh adsorber bed.

The pressure in the saturated adsorber is then reduced and the product gas remaining in the void spaces of the vessel is first recovered in a cocurrent depressurization, that is, gas is withdrawn in the same direction as that of feed gas flow in the adsorption cycle. The recovered intermediate product gas stream is used internally for purging and repressurization. The adsorbed impurities are then removed in a countercurrent depressurization step (i.e., the gas is withdrawn in the opposite direction as that of the feed gas flow in the adsorption cycle) as a PSA off-gas. The adsorber vessel is then purged with a small amount of high-purity product gas withdrawn during the cocurrent depressurization step to complete the regeneration of the adsorbent bed. Finally, the vessel is repressurized with a part of the product gas and the high-purity hydrogen stream from the cocurrent depressurization (Babicki and Hall, 2003).

Because each of the cyclic process step is a batch process, at least four beds are needed undergoing the cyclic steps described earlier in a sequential manner to provide a nearly continuous product gas flow. In addition, large surge tanks are used to dampen variations in flows of feed, product, and exhaust streams. To improve product gas yields, more complex PSA cycles can be used with more number of adsorbent beds. A polybed PSA process was commercialized by UOP in 1976 with the utilization of up to 12 beds (Stocker et al., 1998). Increased number of beds leads to shorter cycle time, increased hydrogen recovery and greater hydrogen capacity per unit adsorbent weight, and more operational flexibility. The optimal number of beds is determined by a trade-off with greater product gas yield and recovery efficiency and higher capital cost for additional beds with additional flow switching equipment (Miller and Stocker, 1999). To switch gas flows between beds during a typical cycle, a complex network of valves is used along with associated instrumentation, control systems, and process piping. The valves are actuated by solenoids using a logic controller. As the number of beds increase, the number of mechanically operated valves increases significantly. For example, for an eight-bed hydrogen PSA process, 42 valves are needed. The network of valves adds complexity and cost to the PSA system. In addition, the valves and piping also add dead volume to the overall PSA system reducing product yield and process efficiency. The optimal selection of the number of beds is determined by the required hydrogen capacity and percentage recovery. For small capacity systems of <5,000 Nm^3/h, four beds are normally used, whereas more than eight beds may be used for capacities exceeding 50,000 Nm^3/h (Stocker et al., 1998).

Hydrogen PSA units typically operate with a feed gas pressure in the 200–700 psig range and are operated at a pressure slightly greater than the desired hydrogen delivery pressure. The normal operating temperature range is 100–150°F (Gunardson, 1998). Typical hydrogen purity ranges from 99.5% to 99.9999%, depending on the hydrogen recovery. The required hydrogen product purity also has a minor impact on the hydrogen recovery achievable and the amount of adsorbent needed for a given capacity. Hydrogen recovery defined as the ratio of the amount of hydrogen in the product to hydrogen in the feed is not very sensitive to feed gas pressure but is very sensitive to the purge gas pressure. The purge gas pressure is set to be sufficient for feeding the off-gas to the reformer burner directly. With a feed gas pressure of 400–500 psig and purge gas pressure of 5 psig, the hydrogen recovery will be in the 80–90% range. Increasing the purge gas pressure to 50 psig, for example, will reduce the recovery to 65%. The optimum feed gas pressure for maximum hydrogen recovery is typically 300 psig.

8.3.1.1 Improvements in Pressure Swing Adsorption Process

Despite PSA's widespread use for hydrogen as well as other separation applications, utilization of beaded sorbent materials and mechanical switching valves do introduce certain limitations. The maximum gas flow through the adsorbent bed is limited by the fluidization velocity for the adsorbent beads used. Increasing the bead size will allow greater flow velocity but will also result in a greater mass transfer resistance for the adsorption process. An optimal adsorbent bead size must be selected to match the throughput rate with the rate of removal of impurities from the hydrogen gas stream. Although the PSA process is simple in operation, the numerous switching valves needed for its operation are the most complicated part of the system and is also the most vulnerable to mechanical failure (Gunardson, 1998).

Although the basic principle of PSA operation is simple, there are a number of operational variables due to large number of beds and sequencing steps, and a number of operational improvements have been reported to reduce the cycle time and increase hydrogen throughput capacity per unit of adsorbent weight. For example, by extending the duration of the purge step, the production capacity of a PSA unit was increased exceeding 110 Nm^3/h with 16 beds (Whysall and Wagermans, 2001). Using storage tanks to collect high-purity hydrogen and reusing it as purge gas during cycle steps was found to increase hydrogen capacity and decrease the number of PSA beds required (Baksh and Terbot, 2003). Cycle time was also found to be reduced by manipulating pressure equalization steps (Xu et al., 2003). Performance of a PSA process and hydrogen throughput may be improved by reducing the cycle time, made possible by process innovation for handling gas flows (Keefer and Doman, 2004). Improvements in a PSA system are also sought by modeling and simulation for the optimization of various process parameters (Jiang et al., 2004; Warmuzinski and Tanczyk, 1997; Waldron and Sircar, 2000).

Using three-dimensionally structured adsorbent with a large number of uniform gas channels, the limitations of mass transfer kinetics and fluidization of the conventional beaded sorbents may be removed. Such structured adsorbents have been described in recent patents (Golden et al., 2003; Keefer et al., 2004). A novel rotary valve design with rotating beds instead of switching valves has been shown to increase the cycle speeds by an order of magnitude to 100 cycles/min. Both of these improvements have been incorporated in a "rapid cycle PSA" claimed to reduce the size of conventional PSA by a factor of 50 and the cost by a factor of 2 (Connor et al., 2002; Babicki and Hall, 2003).

8.3.2 CO_2 Absorption Systems

Although PSA systems have largely replaced hydrogen separation based on CO_2 absorption in the SMR process, PSA systems are not well suited for capturing the CO_2 as may be mandated by the need to control greenhouse gas emissions (primarily CO_2) for stemming global warming. As seen in Figure 8.1 process schematic for PSA-based hydrogen production, all of the carbon initially present in the feedstock used for hydrogen generation is accumulated in the PSA off-gas, which is burned for its fuel value to provide heat in the reformer. If carbon capture is desired in processes for hydrogen generation using carbonaceous feedstocks, CO_2 absorption–based systems would be attractive since they allow the production of pure hydrogen and pure CO_2 product streams. For CO_2 absorption–based bulk hydrogen separation systems, a final PSA-based hydrogen purification may be used for utilizing the product hydrogen stream in applications requiring stringent hydrogen product specification (e.g., power generation by PEM fuel cells). The size of such polishing units will, however, be much smaller than the conventional PSA units for bulk contaminant removal. Methanation process is commonly used to convert any CO or CO_2 present in the product hydrogen leaving CO_2 absorption process to methane, which is often a benign diluent, for example, in ammonia synthesis or for a PEM fuel cell. A PSA polishing bed may be preferred if any other trace contaminants are present in the product gas, for example, sulfur compounds. CO_2 absorption–based systems are currently used for hydrogen and N_2/H_2 synthesis gas generation in ammonia plants, and the technology is considered mature due to its widespread use in pre-1980 hydrogen plants.

An excellent account of the CO_2 absorption technology is provided in the earlier version of the hydrogen sourcebook (Kermode, 1977). Literature regarding CO_2 capture has also been reviewed in recent reports (Smith, 1999; Collot, 2003). Because sulfur present in the carbonaceous feedstocks, including natural gas, is usually removed before their reforming, the hydrogen separation from the product gas primarily needs to focus on CO_2 removal rather than CO_2 and H_2S removal. Liquid sorbents that have been commercially used for CO_2 removal include ethanol amine solutions, ammonia solutions, potassium carbonate solutions, methanol, and water (Kermode, 1977). As seen in Figure 8.2 process schematic, CO_2 absorption is preceded by both HTS and LTS to convert almost all of the CO in the synthesis gas to hydrogen and CO_2. A liquid sorbent chosen to remove CO_2 from the product gas after LTS should

- Be highly selective for absorbing CO_2 relative to H_2, CH_4, and CO
- Have a low vapor pressure, which will minimize absorbent losses in the regeneration step
- Not be irreversibly poisoned by other contaminants in the gas stream, for example, COS and HCN
- Be readily regenerated by steam stripping
- Be noncorrosive to mild steel
- Be inexpensive

TABLE 8.1

Commercially Available CO_2 Absorption Processes

Process—Trade name	Provider—Developer	Absorbent
ADIP-X	Shell Global Solutions	MDEA + accelerator
Sulfinol-X	Shell Global Solutions	Diisopropanolamine/MDEA + sulfolane + accelerator
Econamine FG Plus	Fluor Corporation	MEA, DGA
Fluor solvent	Fluor Corporation	Propylene carbonate
Giammarco-Vetrocoke	Giammarco-Vetrocoke	Hot potassium carbonate + promoter (arsenic trioxide)
Benfield Process	UOP	Hot potassium carbonate + proprietary promoter
Catacarb	Eickmeyer and Associates	Hot potassium carbonate + proprietary promoter
aMDEA	BASF Corp.	MDEA + activator
UCARSOL Series solvents	Dow Chemical Company	Proprietary amine formulations
GAS/SPEC solvents	INEOS Oxide	MDEA formulations + additives, for example, piperizine
Selexol	UOP	Polyethylene glycol
Purisol	Lurgi/Air Liquide	*N*-methyl-2-pyrrolidone (NMP)
Rectisol	Lurgi/Air Liquide	Chilled methanol

Some of the commercial processes offered for removal of CO_2 are presented in Table 8.1. These processes use either the chemical interaction between CO_2 and the absorbent (e.g., amine, alkali metal hydroxide, or carbonate) or the physical solubility of CO_2 in the absorbent (e.g., methanol and polyethylene glycol). In either case, CO_2 is removed from the gas stream into the sorbent at a lower temperature or higher pressure and is released later at a higher temperature or lower pressure regenerating the sorbent. Although absorption by chemical interaction is more efficient, the process typically requires greater energy for regeneration compared to physical solvents. Thermal regeneration of chemical absorbents may also lead to degradation over a number of cycles due to irreversible reactions with contaminants, for example, sulfur compounds. Physical solvents can also degrade over time and usually are more volatile leading to losses due to evaporation. Therefore, physical solvents, for example, methanol, are typically used at lower temperatures than the amine-based systems and are regenerated by depressurizing and heating.

Absorbents based on a variety of amines are by far the most common. Amines that have been used include monoethanolamine (MEA), diethanolamine (DEA), triethanolamine (TEA), diisopropanolamine (DIPA), methyl diethanolamine (MDEA), and diglycolamine (DGA). CO_2 is typically absorbed at 80–140°F temperature and up to 1000 psig pressure by chemically reacting with the basic amine functional group in the absorbent, for example, primary amine, MEA, reacts with CO_2 by the following reaction forming a carbamate salt:

$$2HOC_2H_4NH_2 + CO_2 \Leftrightarrow HOC_2H_4NHCOO^{-}\,{}^{+}NH_3C_2H_4OH \tag{8.4}$$

All amines are relatively weak bases, which react with CO_2, H_2S, and COS forming bonds that can easily be broken at elevated temperatures. Primary amines are chemically stronger than secondary, which in turn are stronger than tertiary. Weaker amines have greater selectivity for H_2S, which is especially high for tertiary amines where the nitrogen atom does not have a free hydrogen association to form the carbamate ion by direct reaction with CO_2. CO_2 reacts with secondary amine by forming carbamate ion similar to the reaction with primary amine:

$$2R_2NH + CO_2 \Leftrightarrow R_2NCOO^- \, ^+H_2NR_2 \tag{8.5}$$

However, reaction of CO_2 with a tertiary amine results in the formation of bicarbonates:

$$R_3N + CO_2 + H_2O \Leftrightarrow HCO_3^- \, ^+HNR_3 \tag{8.6}$$

The product salts formed by reactions 8.5 and 8.6 are thermally unstable and decompose at higher temperatures releasing CO_2 and regenerating the MEA absorbent. The amine regeneration is typically conducted at about 240°F and at low pressure of 10–15 psig. A number of variations of this basic process exist based on specific absorbent and its concentration, process conditions, solvent, and additives used as seen in the variety of amine-based commercial processes in Table 8.1. Compared to the PSA process, amine-based hydrogen separation encounters more operational issues and are generally more expensive to operate. Some of the disadvantages of the amine-based processes include

- Severe corrosion problems especially with high CO_2 concentration in the feed gas
- Ethanolamines are relatively expensive
- Significant loss of absorbent due to vaporization and irreversible reactions

Reactions of amine with stronger acids such as thiocyanate (SCN^-) that may be present in the gas being treated produce irreversible degradation products, which are thermally stable. Degradation products are also formed by the reaction among amines themselves. MEA is currently the most widely used solvent for CO_2 capture and can achieve CO_2 recovery rates of 98% with product purity in excess of 99% (Smith, 1999). However, absorption processes based on activated MDEA are increasingly used due to higher CO_2 absorption capacity and biodegradability of solvent. BASF (aMDEA), Dow chemical company (UCARSOL series), INEOS (Gas/Spec CS 2000), and Shell (ADIP-X), for example, offer commercial accelerated MDEA processes for CO_2 separation from synthesis gas in ammonia plants.

Another successful CO_2 absorbent utilizing chemical interaction has been potassium carbonate used in the "hot carbonate process." The CO_2 absorption and absorbent regeneration occur at about the same temperature of 225°F in this process. CO_2 absorption occurs at high pressure and the sorbent regeneration occurs due to lowering of pressure with steam stripping. The absorption/regeneration reactions are represented as

$$K_2CO_3 + CO_2 + H_2O \Leftrightarrow 2KHCO_3 \tag{8.7}$$

Different variations of the hot carbonate process exist (e.g., Catacarb and Benfield) based on proprietary catalyst used to enhance the rate of reactions, for example, formaldehyde, arsenite, and hypochlorite. The activation by catalysts is apparently necessary to make the process economical compared with the amine-based processes.

Recent developments for reactive CO_2 sorbents include sterically hindered amines such as 2-amino-2-methyl-1-propanol (AMP) and 1,8-*p*-methanediamine (MDA) and 2-piperidine ethanol (PE), which are claimed to have good reversible CO_2 capacity (Veawab et al., 1998) and low-temperature molten salts called ionic liquids (Bates et al., 2002). Ionic liquids are attractive due to their negligible vapor pressure up to their decomposition at

>300°C, however, the materials currently used, for example, $(CH_3)_4$ NF-, $4H_2O$-, and TSL1-containing imidazolium anion need to be investigated further.

Three prominent commercial processes that use physical sorbents include Selexol by UOP utilizing dimethylether polyethylene glycol (Holbrook, 2006), Rectisol process using cold methanol developed by Lurgi (Mak et al., 2004), and Purisol using NMP also developed by Lurgi (Smith, 1999). These processes are based on a higher solubility of CO_2 in these solvents for CO_2 when compared with H_2, CO, and CH_4. Physical absorption processes require much less energy for solvent regeneration and often the solvent can be regenerated just by decreasing the pressure. Physical absorption systems are typically used when the partial pressure of CO_2 is high. Rectisol process is especially suited for removing CO_2 from the synthesis gas mixture used in methanol plants since the solvent is also methanol, and the vaporized solvent present in the synthesis gas is not a problem. One advantage of the Selexol process is the low vapor pressure of its solvent permitting its operation at near ambient temperature instead of cryogenic or chilled temperature as used in Rectisol and Purisol systems. The selexol solvent is a mixture of polyethylene glycol/dimethyl ethers ($CH_3O(C_2H_4O)_nCH_3$) with n from 3 to 9 (Kubek et al., 1996). A combination of Selexol with PSA was demonstrated to be lower in capital and operating cost compared to a conventional Rectisol/liquid N_2 cryogenic system at the Coffeyville Resources Gasification Ammonia Complex in Coffeyville, Kansas, for the production of ammonia and urea granules (Mak et al., 2004). Recently, CO_2 absorption by chilled ammonia is being investigated for CO_2 removal from power plant flue gases as a lower cost option compared to the amine absorption process (Rhudy and Black, 2007).

Although the preceding processes described for CO_2 separation are based on liquid absorbent solutions operating at low (cold methanol) to moderate temperatures (hot carbonate), CO_2 may also be separated by solid sorbents, typically operating at high temperatures, for example, base metal oxide such as CaO (Fan et al., 2005; Harrison and Peng, 2003) can absorb CO_2 at 600°C temperatures by reversible carbonation/decarbonation reaction:

$$CaO + CO_2 \Leftrightarrow CaCO_3 \tag{8.8}$$

Sorbents such as lithium silicate and lithium zirconate can absorb CO_2 at somewhat lower temperature of 500°C (Kato et al., 2002; Li et al., 2006),

$$Li_4SiO_4 + CO_2 \Leftrightarrow Li_2CO_3 + Li_2SiO_3 \tag{8.9}$$

whereas, highly active sorbents such as sodium carbonate may be used for removing CO_2 from flue gases at moderate temperatures of 100–150°C (Nelson et al., 2007).

$$Na_2CO_3 + CO_2 + H_2O \Leftrightarrow 2NaHCO_3 \tag{8.10}$$

CO_2 is removed and the solid sorbent is regenerated by heating it at a temperature greater than the carbonation temperature reversing the equilibrium.

Pure solid sorbents have lower percentage of utilization, typically about 30%, due to the accumulation of the product on the active surface. The degree of utilization of the active sorbent may be increased by using smaller porous particles or loading the active sorbent onto the surface of suitable nonactive supports such as alumina. The solid sorbents also present operational challenges for moving the solids (e.g., in a transport reactor) or otherwise providing good gas/solid contact for effective mass transfer in fixed/fluidized beds.

The solid sorbent material must also be heated and cooled for regeneration and absorption cycles. Because the solid CO_2 sorbents operate at higher temperatures than the liquid absorption systems, they provide opportunities for simultaneous WGS reaction with CO_2 removal improving the CO conversion in sorbent-enhanced reaction process (SERP). A number of studies have been reported on the SERP concept and are currently underway as discussed in later sections.

8.3.3 Membrane Separations

Some of the polymeric membranes are suitable for bulk separation of hydrogen from impurities to enrich a dilute hydrogen stream. Dense polymers permeate gases by solution diffusion mechanism. The permeation rate of a gas species through a polymer membrane is, therefore, proportional to the solubility of the gas species in the polymer as well as the diffusivity of the dissolved species in the polymer. Hydrogen, being a small molecule usually is much more permeable through a polymer compared to other species present in the reformer gas streams: CO_2, CH_4, and H_2O, however, its solubility is often lower than other condensable and polar species such as CO_2 and water. Glassy polymers are dominated by diffusivity selectivity, and therefore, exhibit greater separation factors for hydrogen, for example, polyimide membranes. Rubbery polymers, however, tend to have lower separation factors for hydrogen. The polymeric membranes are temperature sensitive and are generally operated at ambient or only a slightly elevated temperature. Membrane permeability generally increases with temperature for all species, whereas membrane selectivity often decreases with temperature due to a decrease in solubility selectivity with temperature.

Because polymeric membranes separate hydrogen from other species based on relative permeability, they are not capable of providing high-purity hydrogen or absolute separation of species. Purity of the hydrogen product strongly depends on the hydrogen recovery by the membrane. Greater the hydrogen recovery by a membrane, lesser would be the purity of the hydrogen product. The performance of a membrane system, that is, its relationship between product purity and product recovery for a given feed, primarily depends on the feed to permeate pressure ratio and is only slightly dependent on the absolute pressure level. The membrane area required for a desired separation, however, is inversely proportional to the partial pressure differential, which depends on the absolute pressure level. Greater the feed side pressure, lesser membrane area will be required for the same feed to permeate pressure ratio used for a desired membrane performance. The hydrogen permeate is produced at a lower pressure, often slightly above the ambient pressure. The product hydrogen may, therefore, need to be recompressed depending on the intended hydrogen application.

Hydrogen permeability of the membrane is inversely proportional to the membrane thickness. Therefore, it is desirable to reduce the thickness of the membrane layer providing the gas separation. Yet, the membrane must have sufficient mechanical strength and chemical/thermal stability. Two types of membrane synthesis are, therefore, used to provide the desired membrane properties: asymmetric and composite membranes (Zolandz and Fleming, 1992). An asymmetric membrane consists of a thin dense layer of the polymer providing gas separation supported on a thicker but microporous layer of the same polymer material that provides the mechanical strength. Such membranes are prepared by a phase inversion technique used in polymer film synthesis. In a composite membrane, two distinct layers, possibly of two different materials, are combined consisting of a thin dense polymer layer providing the gas separation and a thicker microporous substrate polymer layer. The thinner layer may be coated on to the substrate layer or laminated with it. The composite membrane concept allows using expensive polymers with desired gas separation properties since only a thin layer of this material needs to be fabricated and the support layer can be made of an inexpensive material. Multiple layers of materials are often used in a composite membrane with graded density to obtain high permeability without loss of selectivity and provide the necessary mechanical structural strength.

Polymeric membranes are fabricated as flat sheets or as hollow fibers for the composite- and the asymmetric membrane configuration (Zolandz and Fleming, 1992). Flat sheet membranes are typically packaged as spiral wound modules to provide a compact high surface area packaging. Flat membranes are sandwiched between a porous feed spacer and a porous permeate spacer. The membrane assembly sheets are wound around a central permeate tube. All permeate channels are connected to the permeate tube. Feed gas flows axially through the feed spacers, whereas the permeate flow passing through the membrane flows perpendicular to the feed flow toward the central collection tube. Hollow fiber membranes inherently provide high surface area to volume ratio. Hollow fiber modules are similar to a shell and tube heat exchanger. Hollow fibers are arranged in fiber bundles with ends potted in tube sheet at each end. The feed generally flows through the shell side outside the fibers and the permeate is collected in the tubesheet manifolds. The fibers may be wound on porous support tubes for structural strength. Both types of modules can provide a large membrane surface area in a compact volume. Although polymeric membranes are generally less expensive, they are modular in nature and the scale of economy is not available for larger capacities. Large capacity must be provided by simply adding more membrane modules.

The first commercial "PRISM" hydrogen separation membrane based on polysulfone membrane material was introduced in 1979 by Monsanto and is now available from Air Products and Chemicals, Inc. (Bollinger et al., 1982). Other commercially available hydrogen separation membranes include a polyimide-based membrane by Ube (Makino et al., 1983), polyaramide and polyimide-based membrane by MEDAL (Air Liquide, 2007), and PolySep/Separex family of membranes by UOP (Miller and Stocker, 1989). These membranes are extensively used to recover hydrogen from high-pressure waste or recycle streams. For the hydrogen generation application, polymeric membranes may be used to provide bulk hydrogen separation (Figure 8.3) and in hybrid configuration with PSA or CO_2 absorption system, for example, to recover hydrogen from PSA off-gas stream.

8.3.4 Cryogenic Separation

Cryogenic separation uses the difference in boiling points of the components in a feed gas to separate condensable components by partial condensation of the gas stream. This technique is often used for separating hydrogen from C_2+ hydrocarbons. Cryogenic separation has also been used to separate high-purity CO as a product from the synthesis gas stream. Although cryogenic separation can be used for producing pure hydrogen from synthesis gas along with other fractionated products, primarily CO, earlier cryogenic systems have largely been replaced by PSA systems when pure hydrogen is desired as the primary product (Gunardson, 1998).

8.3.5 Developmental Hydrogen Separation Processes

A realistic possible pathway for the evolution of hydrogen economy starts with the current industry standard process of SMR. Hydrogen produced by SMR at a distributed scale is projected to be cost-effective compared to the cost of hydrogen based on central large-scale production with delivery by compressed gas tube trailers or cryogenic trucks. The cost of distributed hydrogen production by SMR is estimated to be about $5/gasoline gallon equivalent (gge) (U.S.$2003) based on SMR technology available commercially, which utilizes PSA as the hydrogen purification process. Based on production and installation of 500 distributed-scale (forecourt) SMR units/year, the cost of hydrogen as predicted by U.S. DOE's hydrogen analysis model (H2A) (U.S. Department of Energy, 2007) is \$3/gge, which is significantly greater than the U.S. DOE Year 2010 target of \$2.5/gge and the U.S. DOE 2015 target of \$2/gge (U.S. Department of Energy, 2006). (One gge is a considered equivalent to 1 kg of hydrogen based on the current performance of a 100 kW PEM fuel cell stack for automotive power.) This analysis assumes the utilization of the SMR and PSA purification technology that is commercially available today. The cost targets are based on producing hydrogen cost competitively compared to the currently available transportation fuel, that is, gasoline. The cost of the currently available SMR technology for hydrogen production must, therefore, be reduced to make the hydrogen cost competitive for the transportation fuel market as envisioned in the transition to the hydrogen economy.

To reduce the cost of hydrogen produced by SMR, approaches are being pursued for improving the reforming process itself as well as to reduce the cost of hydrogen separation. All of the commercially available hydrogen separation approaches described in earlier sections require cooling the product gas to a lower separation process temperature (e.g., ambient for PSA and amine absorption). If hydrogen separation could be conducted at high temperatures suitable for WGS reaction (e.g., 250–450°C) or even at a higher temperature suitable for methane reforming (e.g., ~ 600°C or greater) reaction, hydrogen separation process could be integrated with either the WGS reaction or with methane reforming and WGS reactions together. The WGS reaction as well as methane reforming/methanation reactions are equilibrium limited and hydrogen separation simultaneous with the reaction can provide an additional benefit of increasing reaction conversions by removing one of the products from the reaction environment. Separation of one of the reaction products (H_2) will also allow conducting reaction at lower temperatures and with lower, near stoichiometric, steam to carbon ratios leading to less steam consumption. Separation of CO_2 from the reaction environment also offers similar advantage of increasing reaction conversion. Integrating hydrogen separation (or CO_2 separation) either with the WGS reaction or both the methane reforming and WGS reactions will reduce the number of unit operations involved in hydrogen production and is likely to reduce the capital cost of the hydrogen plant equipment. Therefore, approaches are being investigated for high-temperature separation of hydrogen (or CO_2) and integrating hydrogen separation with the hydrogen generation process, by combining either WGS shift reaction or the methane reforming and WGS reactions together with hydrogen separation. Membrane reactor processes for conducting WGS reaction and methane reforming reaction with simultaneous hydrogen separation are shown schematically in Figures 8.5 and 8.6.

Reducing the energy requirement for separation and purification of hydrogen and reducing the separation equipment cost will significantly decrease the overall cost of producing hydrogen. Substantial research effort is, therefore, underway to investigate new concepts for hydrogen separation as well as to improve existing proven processes especially the PSA. Hydrogen separation membranes have been shown to have potential to reduce separation costs by over 30% compared to PSA hydrogen purification (Lasher et al., 2002). Although SMR is expected to be the near-term technology for hydrogen production, gasification of coal and biomass as well as reforming of biomass-derived liquid fuels such as ethanol are the first candidates likely to replace SMR in the future. All of these hydrogen production technologies first produce synthesis gas followed by hydrogen enrichment by WGS reaction and hydrogen separation. The developmental concepts based on improving hydrogen separation process or integration of hydrogen separation with hydrogen generation will, thus, be applicable to all of the various options that are likely to be used for hydrogen production in the foreseeable future.

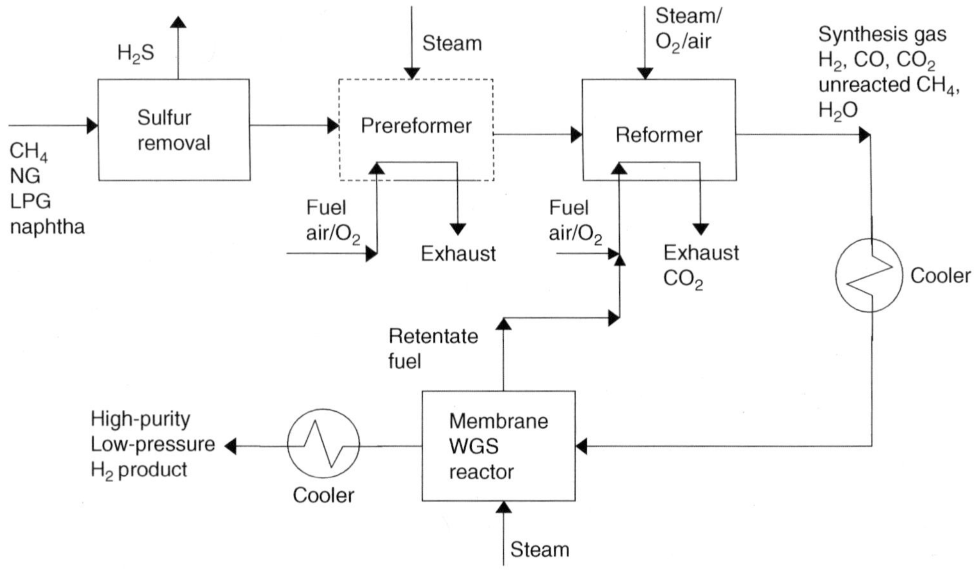

FIGURE 8.5
Steam methane reforming with membrane WGS reactor for hydrogen production.

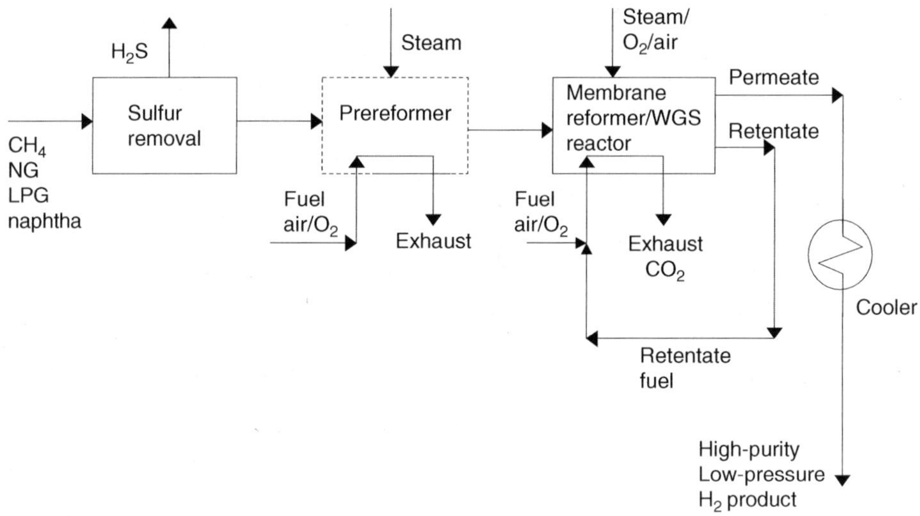

FIGURE 8.6
Steam methane reforming with membrane reformer/WGS reactor for hydrogen production.

Several inorganic materials are being pursued as membrane candidates for high-temperature hydrogen separation: (1) dense metallic membranes, (2) dense ceramic ion transport membranes, and (3) porous inorganic membranes including molecular sieve-based carbon and zeolite membranes. Two types of polymer membranes are being investigated: high-temperature polybenzimidazole (PBI) and similar polymer membranes and reverse-selective CO_2 separation membranes. In addition, high-temperature CO_2 sorbents are being investigated for simultaneous reforming/WGS reaction and CO_2 separation in SERP. Among these developmental concepts, palladium-based metallic membranes have received the most attention and may be considered as the most promising candidate for high-temperature hydrogen separation membrane (Ritter and Ebner, 2005). Most of these concepts are being demonstrated on a small laboratory or bench-scale systems. Long-term stability and durability of the materials at an industrial scale have not yet been demonstrated for most of these developmental processes including the palladium-based membranes. One disadvantage of all membrane materials preferentially permeating hydrogen is that the hydrogen product stream is produced at a much lower pressure compared to the feed gas pressure. If a high product hydrogen pressure is desired, for example, for transportation and storage, the cost of an additional compressor must be taken into account in determining the cost of hydrogen. Membrane reactor concept using hydrogen separation membranes is, thus, more attractive for applications utilizing hydrogen at a lower pressure, for example, in a distributed power generation using PEM fuel cells with on-site hydrogen generation.

The developmental hydrogen separation concepts are discussed in the following sections regarding their mechanisms for hydrogen separation and the current status of development.

8.3.5.1 Dense Metallic Membranes

Hydrogen permeates through solid metals in atomic form by solution—diffusion mechanism. Hydrogen molecules first adsorb on the metal surface and are dissociated into atoms, which then diffuse through the bulk metal under a concentration gradient. After passing through the bulk metal, the hydrogen atoms recombine into hydrogen molecules at the opposite metal membrane surface and the molecular hydrogen gas exits from the surface to the gas phase. The diffusive flux of hydrogen through a bulk metal membrane of thickness L is generally described by Sievert's law (Ward and Dao, 1999):

$$J = \frac{-k(\sqrt{P_f} - \sqrt{P_p})}{L} \tag{8.11}$$

where J is diffusive flux, mol/m^2-s, k is the hydrogen permeability of the metal, mol/m-s-Pa$^{0.5}$, L is the membrane thickness, m, P_f and P_p are the hydrogen partial pressures on the feed side and permeate side of the membrane surface, Pa, and "*–ve*" sign indicates hydrogen flux toward lower concentration.

The square root proportionality of hydrogen flux with respect to the partial pressure driving force is indicative of the bulk-phase diffusion as the rate-limiting step. In practice, for thin membranes, the proportionality can be different with the power law exponent between 0.5 and 1 as other factors such as gas-phase mass transfer and surface dissociation may contribute to the overall permeation. Because hydrogen is transported through a unique dissociated atomic form through the metal, the metal membranes are 100% selective to hydrogen and allow the permeation of other gas species only through pinholes and defects in the metal membrane. Palladium is the most widely researched metal membrane since it is highly permeable to hydrogen, resistant to ambient conditions of air, oxygen, and moisture, and is able to rapidly dissociate hydrogen molecules into atoms on the surface.

8.3.5.1.1 *Palladium-Based Membranes*

Palladium-based dense metallic membranes have been known to be completely selective for hydrogen permeation and are used in commercially available small-scale hydrogen purification units (e.g., Johnson Matthey, 2007; REB Research, 2007; Power + Energy, 2007; ATI Wah Chang, 2007). These hydrogen purification units typically use palladium-alloy foils 25–30 μm thick or tubes (60–100 μm thick) to produce high-purity hydrogen (e.g., 99.99999%) for specialized applications using industrial-grade hydrogen. The small-scale hydrogen purifiers are expensive on unit hydrogen flow basis. Because membrane systems are modular in nature, that is, larger capacities are provided by adding more modules or increasing membrane surface area linearly, palladium foil or tube-based hydrogen separation is likely to be cost prohibitive for large-scale hydrogen separation applications. Palladium-alloy foils or tubes have also been incorporated in small hydrocarbon fuel reformers that are being demonstrated (e.g., IdaTech, 2007; InnovaTek, 2007; Intelligent Energy, 2007). Palladium-alloy membranes are used either to purify the product hydrogen after a fuel reformer (e.g., IdaTech and InnovaTek) or integrated within the fuel reforming process (e.g., MesoGen unit by Intelligent Energy). The hydrogen capacity of such available units is typically designed to support a PEM fuel cell power rating up to a few kilowatts or about 100 standard liters of H_2 per minute or less.

The cost of Pd-alloy membranes used for hydrogen separation may be reduced by depositing a thin Pd-alloy film on a suitable porous substrate to form a composite membrane. Almost all of the Pd-alloy membrane development efforts are, thus, focused on preparing thin yet defect-free Pd-alloy composite membranes (e.g., Hopkins, 2007; Coulter, 2007; Delft et al., 2005; Damle et al., 2005; Mardilovich et al., 2002). A detailed review of the Pd-alloy membrane research has been prepared by Paglieri and Way (2002) with an extensive bibliography of the palladium membrane research to date. An updated review has been recently prepared by Collot (2003) and Paglieri (2006).

Pd-alloy composite membranes may be prepared by depositing a thin Pd-alloy film on a suitable porous support. A porous support provides the mechanical and structural strength to the composite membrane allowing the selective gas separation metal film to be thin. Porous ceramic (e.g., alumina and zirconia) and metallic (e.g., stainless steel) supports have been used for the preparation of composite membranes. Porous supports generally have graded pore size structure with a coarse pore size bulk support and finer pore size layers near the surface and are typically used in liquid filtration applications. The thickness of the metal film deposition needed to assure a defect-free composite membrane strongly depends on the porous support characteristics such as the pore size in the top most surface layer (Mardilovich et al., 2002). Ceramic supports are generally prepared by depositing a layer of fine ceramic particles prepared by sol–gel technique, producing pores that are small and have a narrow distribution. The disadvantage of ceramic supports is their fragility and difficulty in providing connections and seals especially in the high-temperature operation. Porous metallic supports are sturdier but are generally coarser in pore size, have a low overall porosity, and generally require much thicker Pd-alloy film deposition to produce high hydrogen selectivity. Another disadvantage of metallic supports is the possibility of intermetallic diffusion, especially at high operating temperatures, between Pd-alloy and the porous metal itself. A diffusion barrier is, thus, needed between the porous metal and the Pd-alloy film for long-term stability. Approaches for introducing a suitable barrier include oxidizing the surface of the porous metal or deposition of a fine pore size ceramic layer on the porous metal surface. The latter approach can combine the advantages of both ceramic (smaller pore size) and metallic (structural strength) supports and is recently being investigated using zirconia/porous stainless-steel composite tubular substrates (Hopkins, 2007; Damle et al., 2007).

Pure palladium becomes brittle in the presence of hydrogen during thermal cycling due to dimensional changes caused by a transformation between two phases (α and β) of palladium hydride around 300°C. To avoid metal embrittlement and resulting membrane cracking or distortion, pure palladium membrane should not be exposed to hydrogen at temperatures below 300°C. To increase resistance to embrittlement, Pd is alloyed with other metals such as Ag and Cu. The most commonly used alloy in the commercially available membrane modules contains 23% Ag and 77% Pd by weight; this composition is also claimed to maximize the hydrogen permeability of the Pd–Ag alloy (Knapton, 1977; Shu et al., 1991). In addition to providing thermal stability, alloying Pd with other metals also increases its resistance to chemical contaminants. Among the contaminant species present in hydrocarbon fuels used for hydrogen production, Pd is especially vulnerable to sulfur (H_2S) present in fuel gases. Alloying of Pd with Cu has been reported to offer significant tolerance to the presence of H_2S in the hydrogen gas stream (Edlund, 1996; McKinley, 1967; Roa et al., 2003; Kamakoti et al., 2005). Various compositions of Pd–Cu alloys were evaluated in a systematic study at U.S. DOE NETL in which 80% Pd–20% Cu alloy was indicated to be an optimized composition with regard to contaminant tolerance including no significant effect of 1000 ppm H_2S on the hydrogen permeation (Killmeyer et al., 2003, 2005; Morreale et al., 2004). Minimal effect of CO or H_2O was observed on the hydrogen permeation by the membrane of the same composition except for CO exposure at about 635°C, which was attributed to carbon deposition on the surface due to CO reduction. The membrane surfaces, however, were observed to experience pitting and surface roughening during the exposure tests. The sulfur tolerance of 60% Pd and 40% Cu alloy was found to depend on temperature of operation with marked permeability reduction at temperatures 500°C or lower in the presence of H_2S. This temperature also corresponds to the transition between face centered cubic (fcc) lattice structure above 500°C to body centered cubic (bcc) lattice structure below 500°C. Among all Pd–Cu alloy compositions, the hydrogen permeability is thought to be the highest for the 60% Pd and 40% Cu composition and for this composition, it passes through a maximum at about 450°C (Killmeyer et al., 2003; Shu et al., 1991; Knapton, 1977). Several other alloying elements, for example, Au, Y, and Ce have been identified with significant increase in hydrogen permeability with 8% Y and 8% Ce alloys (Knapton, 1977). Ternary alloy compositions (Pd–Cu-transition metal) are also being investigated (Vanderspurt et al., 2007) for maximizing membrane permeability and stability.

Several approaches have been used to produce thin-film Pd-alloy composite membranes: physical vapor deposition (PVD), chemical vapor deposition (CVD), electroplating, compression cladding, sputtering, spray pyrolysis, and electroless plating. Of these methods, CVD and electroless deposition methods are considered as the promising routes for thin composite membrane fabrication. The electroless plating method is generally preferred over CVD for several reasons: electroless deposition provides thin metal and metal-alloy films that have excellent adhesion properties; the deposition rate is high, and it can be easily controlled. Further, this is an autocatalytic process that does not require any expensive equipment. The CVD process requires appropriate stable organometallic precursors, which are expensive. It is also difficult to deposit two metal alloying species simultaneously by CVD for the fabrication of palladium-alloy membranes. Techniques for forming thin free-standing Pd-alloy films by sputtering on smooth, high surface finish substrates, for example, polymers and silicon wafers, are also being investigated (Arps, 2006; Klette et al., 2006). The film formed on a smooth surface is later separated from the substrate to yield a free-standing dense Pd-alloy film, which is then supported on a suitable porous support (e.g., porous stainless steel) to prepare a functional composite membrane. The porous material provides only the mechanical support for the film. The advantage of this technique is that the Pd-alloy film is formed without any influence of the porous support characteristics; however, the thin free-standing film is highly flexible and must be handled appropriately to prevent warping, creases, and stresses.

Because Pd-alloy membranes operate at high temperatures in the range of WGS reaction and on the lower end of methane reforming reaction, they can be used in a membrane reactor configuration for the simultaneous separation of hydrogen. As discussed earlier, a membrane reactor configuration is also beneficial for shifting reaction equilibrium to increased conversion as well as for allowing milder reaction conditions and utilization of stoichiometric amount of steam in the reaction. The potential of palladium membrane reactors for conducting reactions involving the generation or consumption of hydrogen has long been investigated (e.g., Gryaznov et al., 1983; Gryaznov, 1986; Itoh, 1987; Itoh and Govind, 1989; Uemiya et al., 1988, 1991a, 1991b; Edlund and Pledger, 1993; Collins and Way, 1994). A review of membrane reactors for hydrogenation and dehydrogenation processes was prepared by Dittmeyer et al. (2001). Utilization of palladium membrane reactors specifically for conducting methane reforming and WGS reaction has clearly indicated feasibility of using hydrogen separation simultaneously with the reaction to increase reaction conversion and conducting methane reforming under milder conditions of 500°C, while still achieving up to 80% methane conversion to hydrogen (Uemiya et al., 1991a, 1991c; Shu et al., 1994). An economic feasibility analysis of using the membrane reactor concept for the WGS reaction was conducted by Criscuoli et al. (2001) along with experimental evaluation of the concept, for example, by Basile et al. (2001), Lin and Rei (2001a), and Damle (2002). Feasibility of conducting the WGS reaction in a membrane reactor at high temperatures was investigated by Killmeyer et al. (2003). Feasibility of utilizing a palladium membrane reactor has also been investigated for methanol reforming with simultaneous hydrogen separation (e.g., Lin and Rei, 2001b; Flytzani-Stephanopoulos et al., 2004).

Although Pd-alloy composite membranes have been studied extensively over the past 20 years, the scale of the membrane used has been very small and the durability and stability of such membranes has not yet been demonstrated on an industrial scale. Ma et al. (1998) reported a composite palladium membrane, prepared on a porous stainless-steel support with an oxide barrier layer, which was stable in the temperature range 350–450°C for over 6000 h. Pd–Ag alloy tubular composite membranes up to 80 cm length were recently prepared using alumina support tubes and were tested for hydrogen permeation from the reformate gas stream and also for direct methane reforming with hydrogen separation for up to 100 days (Delft et al., 2005). The temperature during these tests ranged from 400 to 550°C. The membrane reformer tests indicated methane conversion greater than that based on equilibrium at feed conditions (650°C, methane:steam ratio of 1:3) especially at low methane feed flow rate and low feed pressure of 10 bar.

In a national project, Japan Hydrogen and Fuel Cell Demonstration project, sponsored by New Energy and Industrial Technology Development Organization (NEDO), Tokyo Gas Company developed an industrial-scale membrane reformer with a rated hydrogen production capacity of 40 Nm^3/h (Yasuda et al., 2005). This unit housed 224 membrane modules of 4 cm × 46 cm Pd-alloy membranes with <20 μm thickness (total Pd-alloy membrane area ~4.1 m^2). The reformer system was operated at an average membrane temperature of 550°C, steam to methane ratio of ~3:1, feed side pressure of 0.8 MPa (~120 psig), permeate side pressure of 10 psig, and methane feed rate up to 12 Nm^3/h. With natural gas feed rate of 11.6 Nm^3/h, the system achieved the design hydrogen production rate at 40 Nm^3/h with 76.2% system efficiency. The membrane residual gas was burned in the reformer to provide all of the heat requirement. The system was operated continuously at a natural gas feed rate of 8 Nm^3/h for a total cumulative operation for 3310 h with 61 start-up and shutdown cycles until development of leaks in the reactor modules.

An integrated proof-of-concept (POC) size fluidized-bed methane reformer with embedded palladium membrane modules for simultaneous hydrogen separation is being developed for demonstration (Tamhankar et al., 2007). The membrane modules will use two 6 in. × 11 in. Pd-alloy membrane foils, 25-μm thick, supported on a porous support. The developmental fluidized-bed reactor will house a total of five (5) membrane modules with a total membrane area of about 0.43 m^2 and is scheduled for demonstration by September 2007.

Although considerable research has been conducted with Pd-alloy foils, tubes, and thinner composite membranes, long-term durability and stability need to be further demonstrated, especially in the fuel reforming or WGS operating conditions, for acceptance of this technology in a commercial sector. Furthermore, mass-scale and cost-effective production of industrial-scale Pd-alloy thin-film composite membranes need to be demonstrated to be competitive in the hydrogen production and purification market.

8.3.5.1.2 Group V Elements

Group V elements of niobium, tantalum, and vanadium have over an order of magnitude more permeability for hydrogen than palladium as seen in Figure 8.7. These membranes, termed "superpermeable" in the nuclear industry, have been used successfully to separate isotopes of hydrogen from helium in plasmas. A background review of the properties of these metals with regard to hydrogen permeation was recently prepared by Mundschau et al. (2006). Because hydrogen is transported in its dissociated atomic form through the dense metals, it is necessary to dissociate hydrogen on the metal surface rapidly. In addition, unlike noble metals such as Pd, the reactive metals including Nb, Ta, Ti, V, and Zr readily get oxidized effectively destroying the membrane's hydrogen transport property. It is, therefore, necessary to cover the surfaces of these metals by a protective as well as a catalytic layer of suitable materials capable of dissociating hydrogen. The catalytic protective layer must be sufficiently thin for not introducing significant additional resistance to hydrogen permeation and must adequately adhere to the base metal with intimate contact to allow the transfer of dissociated hydrogen. In addition, there needs to be a good match in the thermal expansion property between the catalyst layer material and the base metal to avoid cracking and delamination during thermal cycling. Among the suitable

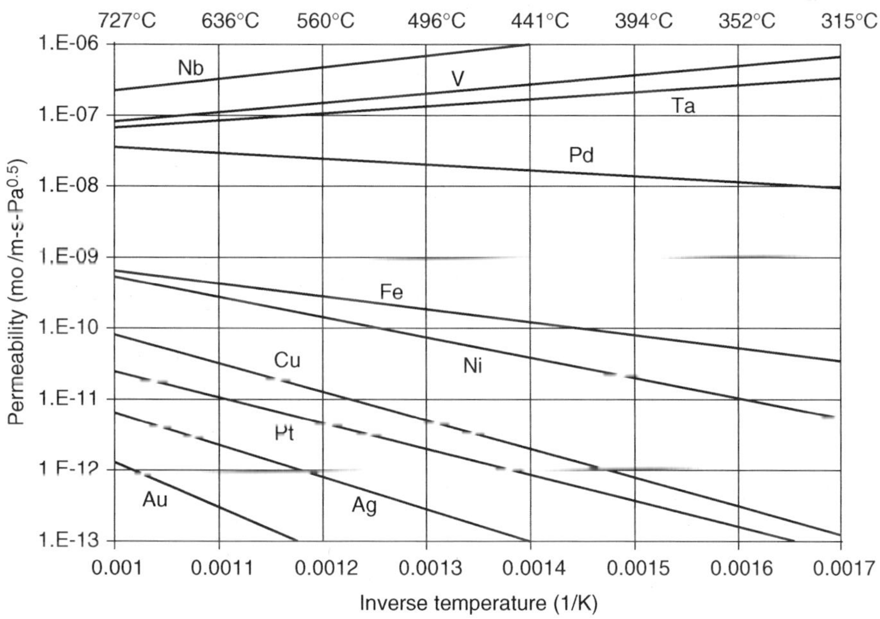

FIGURE 8.7
Hydrogen permeability of various metals. (Based on data from Steward, S.A., Review of Hydrogen Isotope Permeability through Materials, Lawrence Livermore National Laboratory Report, UCRL-53441, 1983.)

candidates of noble metals such as palladium, platinum, iridium, and rhodium, a thin layer of palladium alloy is most commonly applied, for example, by electroless plating technique, to protect the metal membrane surface as well as to catalyze hydrogen dissociation on the surfaces.

Hydrogen permeabilities of Pd-coated Nb, V, and Ta disks and tubes were reported by Buxbaum and Marker (1993) to be much greater than comparable Pd-alloy foils or tubes or other hydrogen transport materials. Pd was coated by electroless plating in these studies. Edlund and Pledger (1993) used 25-μm thick palladium or platinum foils to laminate the base metal membrane on both sides by compression cladding to prepare hydrogen separation membranes. These membranes were used for hydrogen generation by H_2S decomposition (with Pt foil facing H_2S) as well as hydrogen purification membranes. Hydrogen permeation through palladium-coated binary alloys was studied by Hara et al. (2002). A common problem with the Pd-coated metallic membranes is the intermetallic diffusion of the coated Pd with the base metal reducing the catalytic and protective ability of the coated layer especially at a high-temperature operation above 500°C. A diffusion barrier is, therefore, needed to prevent intermetallic diffusion. Evaluation of the Pd-coated vanadium membrane at high temperatures over 600°C indicated a nearly complete loss of distinct Pd layers over time (Rothenberger et al., 2003).

The group V elements, Nb, Ta, and V have a very high solubility for hydrogen resulting in a high H/metal ratio during hydrogen permeation, which can lead to embrittlement of the metal that may lead to subsequent cracking of the membrane. Long-term durability of a proprietary membrane/catalyst formulation of this type was recently determined over 300 days in a reformate gas environment at 340°C containing 41% H_2, 37% H_2O, 18% CO_2, and 3% CO in which the hydrogen permeability was found to decrease by a factor of three over the test duration (Grimmer, 2006). Further improvement in membrane/catalyst formulations for durability and stability are ongoing (Jack, 2007). Small-scale brazable vanadium-alloy membranes are being evaluated for WGS membrane reactor applications (Barton and Argyle, 2007). Systematic evaluation of a variety of binary transition metal alloys with protective catalytic coatings is being conducted to develop a dense metallic membrane with superior hydrogen separation properties and long-term durability and resistance to embrittlement (Buxbaum, 2007).

8.3.5.2 Dense Ceramic Ion Transport Membranes

In dense ceramic membranes, hydrogen permeation occurs in its ionic (protonic) form. Molecular hydrogen first dissociates into protons and electrons on the surface of the membrane. The protons move through the membrane by an ionic transport mechanism under a proton concentration gradient across the membrane to the opposite face. The protons must then recombine with electrons to form molecular hydrogen, which then dissociates from the membrane surface. As a result, the ceramic membrane must also be able to transport electrons through the membrane, that is, should have the protonic and electronic conductivity to be able to permeate hydrogen. Because the membranes are dense, they are 100% selective for hydrogen and allow the permeation of other gas species only through pinholes and defects in the ceramic membrane. The ceramic materials that exhibit such properties are perovskite-based oxides with a general formula ABO_3. For improved permeation properties, dopant elements are often used to replace a small part of either A or B or both elements and create oxygen deficiency represented by $A_{1-x}A'_xB_{1-y}B'_yO_{3-\delta}$, where A′ and B′ are dopant elements, x and y are fractions of dopants in A and B, respectively, and δ is the number of oxygen vacancies in the material formula. Barium and strontium cerates and zirconates are the most commonly studied perovskite materials with trivalent dopants such as Y, Yb, and Gd (Roark et al., 2002), since these materials have the necessary mixed proton/electron conductivity.

The hydrogen flux of the cerate and zirconate perovskite materials increases with temperature and becomes appreciable at high temperatures of 800–900°C. Hydrogen flux of such materials, however, is limited by low electronic conductivity (Guan et al., 1998). A number of studies are ongoing to improve the hydrogen permeability of these materials with different approaches (Dorris et al., 2002; Elangovan et al., 2003; Lin, 2002; Siriwardane et al., 2003). One promising approach is to mix the ceramic phase with a metallic phase to form "cermet" materials. The metal component in cermet provides a high electronic conductivity, whereby ceramic materials with high protonic conductivity could be used regardless of their electronic conductivity (Balachandran et al., 2002). Alternatively, a hydrogen permeable metal such as palladium may be incorporated in the ceramic matrix to form metal–ceramic composite materials, whereby the hydrogen transport occurs through the metal phase alone with the ceramic material acting only as a support matrix (Balachandran, 2005). The hydrogen permeation performance in such metal–ceramic composite membranes essentially depends on the metal used and will be similar to the dense metallic membranes discussed earlier, taking into account the fraction of metal in the composite.

One major concern of perovskite membranes and cermets where the perovskite materials provide the protonic conductivity is their chemical stability in the presence of major syngas components of CO_2 and H_2O as they can form carbonates and oxides (Schwartz, 2002). In metal-ceramic composite materials, where the ceramic provides only a structural support, materials with most resistance to the syngas components including H_2S may be chosen. Metal ceramic composites using alumina and yttria-stabilized zirconia were reported to be stable in syngas environments with H_2S (Balachandran, 2005). Developments are continuing in dense ceramic membranes to decrease the thickness of metal–ceramic composite as well as cermet membranes and support the thinner membranes on another porous ceramic material to increase overall hydrogen flux rates. There is also a need to reduce the operating temperature below 700°C while maintaining high hydrogen flux rates for practical hydrogen separation applications. The scale of the development of these types of membranes may be considered as laboratory scale.

8.3.5.3 *Porous Inorganic Membranes*

Porous ceramic membranes can separate gas components by various mechanisms that depend on the pore size and gas species properties. When the pore size is much smaller than the mean free path of gas molecules, the gas species permeate by Knudsen diffusion where the gas species permeability is inversely proportional to the square root of its molecular weight. Depending on the temperature of operation (mean free path of gas molecules increases with temperature), pore sizes in the range 2–10 [nm pore diameter] can separate gas species by Knudsen diffusion. Hydrogen, being a low-molecular-weight species compared to the other syngas components, permeates faster through the small pores than other gas species in the Knudsen diffusion regime. However, the separation selectivity offered by Knudsen diffusion is quite low, for example, H_2 to CO_2 selectivity can only reach 4.69, which is not sufficient to achieve a high hydrogen purity in the permeate with high hydrogen recovery in a single stage. Multiple stages of separation would be necessary to increase hydrogen purity in the permeate of each successive stage, however, with considerable cost and low hydrogen recovery (Bischoff and Judkins, 2006). Advantages of porous ceramic membranes, for example, alumina, include chemical stability against contaminants and high permeation rates. Ceramic porous membranes, however, do not provide high selectivity and, therefore, high purity hydrogen product. Control of pore size distribution or its narrowness is necessary for assuring Knudsen diffusion-based selectivity. Ceramic membranes, especially silica-based membranes and γ-alumina membranes (with an initial nominal pore size of 4 nm), exhibit pore coarsening in the presence of steam, which decrease their selectivity over time.

Porous ceramic membranes with a pore size in the 1–10 nm range are generally prepared by sol–gel technique (Hsieh, 1988) and have commonly been prepared with alumina, silica, zirconia, and titania materials. To obtain greater gas separation selectivity than that offered by Knudsen diffusion mechanism, the pore size generally needs to be reduced below 3 nm. A number of studies have been undertaken to evaluate permeability and selectivity of microporous (International Union of Pure and Applied Chemistry [IUPAC] defines pores <2 nm as micropores) membranes (Prabhu and Oyama, 2000; Benson, 2000; Collot, 2003). These membranes have also been studied for gas/vapor separation and pervaporation applications. High hydrogen selectivity has been reported with silica membranes (e.g., Wu et al., 1994), however, these membranes have been known to be affected by steam. Porous inorganic membranes, for example, 4 nm pore size γ-alumina, generally undergo pore coarsening in the presence of steam with decrease in selectivity over time (Gallaher and Liu, 1994; Lin et al., 1994). Steam reforming of methane using porous membranes was evaluated theoretically and experimentally with microporous silica membranes (Tsuru et al., 2004). Silica membranes were prepared by the sol–gel technique with a pore size in the 1–2 nm range with H_2/N_2 selectivity in the range 42–87 and H_2/H_2O selectivity in the range 8–10. Methane conversions greater than those predicted for equilibrium at feed conditions were obtained by selectively permeating hydrogen. The selectivity of the silica membrane was, however, observed to decrease over time in the reaction environment due to the presence of steam (Tsuru et al., 2004).

As the pore size is reduced to 1 nm or less, gas permeation may exhibit a thermally activated diffusion phenomena. For example, in studies at Oak Ridge National Laboratory, for a certain proprietary membrane material and configuration, permeation of helium appeared to increase much faster than other gases resulting in an increase in Helium to CO_2 selectivity from ~5 at 25°C to about 48.3 at 250°C (Bischoff and Judkins, 2006). Hydrothermal stability of this membrane in the presence of steam, however, was not reported.

At sufficiently low temperatures (e.g., near ambient), certain gas species may adsorb on the pore surfaces or condense in small enough pores by capillary condensation. Strongly adsorbed species or species that are able to condense in the pores can effectively block permeation of other gas species. Membrane surfaces having an affinity for one of the gas species can separate that species by selective surface flow (SSF). SSF membranes utilizing high surface area carbonaceous material with uniform pores in the molecular sieving range allow the adsorption of a heavier adsorbable gas such as CO_2, which decreases the pore size excluding permeation of gas species that would have otherwise permeated. The heavy component then simply diffuses along the surface of the walls (Rao and Sircar, 1993a). The selectivity of such membranes, however, is not very high (e.g., CO_2/H_2 selectivity ~5 at 20°C; McCarley and Way, 2001) and decreases with increasing temperature.

When the membrane pore size is reduced further to molecular dimensions, gas species separation can occur by molecular sieving. To separate hydrogen selectively from the other syngas components (CO, CO_2, CH_4, and H_2O), porous membranes need to be able to discriminate molecules in the 0.3–0.4 nm size with 0.1 nm or less in size difference.

With appropriate membrane pore size and a narrow distribution, membrane selectivity for smaller gas molecules can be high but the overall permeability is generally low due to a high flow resistance in fine pores. Several studies are being conducted to develop molecular sieve-type membranes using different inorganic materials, for example, those based on carbon (Liu, 2007), silica (Pex and van Delft, 2005), and zeolites (Lin, 2007).

8.3.5.3.1 Carbon Membranes

Carbon molecular sieve (CMS) membranes are generally prepared by controlled pyrolysis of organic precursors in an inert atmosphere (Saufi and Ismail, 2004). CMS membranes with pore diameters of 3–5 Å (0.3–0.5 nm) have been prepared for various gas separation applications. Polymeric precursors used include thermosetting polymers that harden during heating (instead of melting), such as polyacrylonitrile, phenolic resins, polyfurfural alcohol, and polyimides. The permeation characteristics of the molecular sieve membranes strongly depend on the pyrolysis conditions of temperature and duration as well as posttreatment (Ismail and David, 2001). The pore system of CMS membranes consists of relatively wide openings with narrow constrictions. The openings contribute the major part of pore volume and are, thus, responsible for the adsorption capacity, whereas the constrictions of molecular dimensions are responsible for the species selectivity. Accurate control of the pore structure is needed to develop CMS membranes with desired properties. As a pore structure usually has a certain distribution associated with a mean pore size, the separation is not absolute and the selectivity depends on the narrowness of the pore size distribution. Because the membranes are generally several microns thick, the permeability of CMS is low due to the fine pore size and low porosity. Polymer layers may be deposited on porous substrates such as porous carbon before pyrolysis for synthesizing supported CMS membranes. Polymer precursors may be used in their hollow fiber form to produce carbon hollow fiber membranes.

Because carbon has a natural affinity for adsorption of heavy hydrocarbon species and polar molecules, CMS membranes need to be used at a sufficiently high temperature to eliminate contribution/interference of the adsorption. In contrast, strong adsorption of heavier molecules may be used to separate those species by adsorption as discussed earlier by the SSF mechanism (Rao and Sircar, 1993b). The SSF carbon membranes typically have pore dimensions much greater than those needed for CMS membranes since the separation is based on the adsorbed species effectively blocking permeation of other components (Fuertes, 2000). Carbon membranes are resistant to contaminants such as H_2S and are thermally stable and can be used at higher temperatures compared to the polymeric membranes. For the synthesis gas environment, the hydrothermal stability of carbon in the presence of steam will be a concern limiting its operation temperature.

Carbon membranes have been studied since the early 1980s (Koresh and Sofer, 1980, 1983) and have been evaluated for hydrocarbon separations based on their adsorption properties (Suda and Haraya, 1997; Okamoto et al., 1999) and air separation (Singh-Ghosal and Koros, 2000). CMS membranes prepared from polyimide precursor were recently evaluated for their hydrogen separation potential (Hatori et al., 2004). When tested at 50°C, a very high H_2/CO selectivity, over 1000, was observed. The H_2/CO_2 selectivity was relatively lower, which is thought to be due to the greater adsorption of CO_2 compared to CO. Application of CMS membranes for conducting low-temperature WGS reaction at ~200°C with simultaneous hydrogen separation is being investigated at pilot scale (Liu, 2006) with membranes exhibiting H_2/N_2 selectivity of ~50–100 and hydrogen permeance of 1 m^3/m^2-h-bar at 220°C. The CMS membrane was demonstrated to be stable in the presence of H_2S, NH_3, and higher hydrocarbon contaminants. Although the hydrogen selectivity with respect to other gases is not very high, it is considered to be adequate for high hydrogen recovery with sufficient hydrogen product purity (Liu, 2007). For example, with feed gas containing 44% H_2, typical of reformate gas streams, permeate hydrogen purity of ~90% was obtained with ~80% hydrogen recovery. These membranes also have a high permeability for water, comparable or greater than that of hydrogen, which must be taken into account in designing a membrane reactor system.

8.3.5.3.2 Zeolite Membranes

Zeolites are crystalline microporous aluminosilicates with well-defined cagelike pore structures of molecular dimensions (<1 nm). Zeolite membranes separate gas molecules by molecular sieving depending on the pore size of their structure. Among various types of zeolite membranes synthesized to date, the 10-member ring MFI structure (silicalite-1 and ZSM-5) is the most extensively studied for gas separation applications. MFI-type membranes have straight, circular channels of 0.54 nm × 0.56 nm running perpendicular to simusoidal and elliptical channels of 0.51 nm × 0.54 nm, and are used for several industrial separations (Hong et al., 2005). The kinetic diameters of the synthesis gas components are, however, in the 0.28–0.38 nm range: 0.28 nm for H_2, 0.33 nm for CO_2, 0.37 nm for CO, and 0.38 nm for CH_4. Thus, MFI-type zeolite membranes have too large pores to provide a good separation. To separate hydrogen from the syngas components, the zeolite membranes need to be 0.3–0.4 nm in size with a narrow pore size distribution within 0.1 nm. At the same time, nonzeolite large pores or defects need to be eliminated to improve membrane selectivity. Small pore size zeolite membranes include zeolite A, SAPO-34, ETS-4, SSZ-13, and sodalite membranes. SAPO-34 membranes with an eight-member ring chabazite structure and pore size of 0.38 nm have been synthesized; however, they appear to contain a significant number of bigger nonzeolite defects (Hong et al., 2005) reducing their gas species selectivity.

Zeolite materials are commonly prepared by precipitation (crystallization) under hydrothermal conditions in the presence of a template that dictates the zeolite structure during crystal growth. A common template is tetrapropyl ammonium hydroxide (TPAOH), which is burned off after zeolite crystal growth resulting in significant lattice contraction and mesoporous defects. In the secondary growth, template-free method, a zeolite seed layer is dip-coated on the support followed by the growth of seeds into a continuous zeolite layer (Lin, 2007; Verweij et al., 2006). To make a good quality–supported zeolite membrane, it is necessary to select the synthesis conditions so that the membrane structure nucleates and grows on the surface of the porous support, for example, porous alumina or zirconia, forming an integral zeolite/support composite membrane. Homogeneous nucleation and zeolite crystal growth in the bulk solution needs to be avoided as it forms detached particles that deposit on the support leading to defects.

Zeolite membranes can be synthesized as a dense layer on a porous support surface (α-alumina, γ-alumina, or stainless steel) or within the pores of a support. The porous support can be thick but with large pores (0.2–0.5 μm), so that mechanical stability is excellent without introducing additional mass transfer resistance. Because of the inorganic nature of zeolite materials, these membranes are robust and usable in high pressure and high temperature (at least up to 750 K) environments and are resistant to chemical contaminants including H_2S. The selectivity depends on the particular zeolite used for the membrane, its chemical composition (e.g., Si/Al ratio in ZSM-5), the crystal orientation, the identity of the charge neutralization ion, and the quality of the membrane, and these properties can be changed by preparation conditions or by posttreatment (Nenoff et al., 2003).

At low temperatures, adsorptive separation becomes important for zeolite membranes as sorption of one species can effectively hinder permeation of other species.

For example, for a ZSM-5 membrane at room temperature the CO_2/H_2 permeation ratio was found to be ~12. At a higher temperature, as CO_2 desorbs the membrane becomes hydrogen selective as hydrogen begins to permeate faster due to its smaller size (Kapteijn et al., 1995). Because the most common MFI-type zeolite membranes have larger pore size than needed for effective hydrogen separation, recent research has focused on reducing the crystal pore size by the deposition of precursor materials in the pores by absorption followed by cracking/sintering of the absorbed precursors or by CVD of a cracked/sintered precursor directly in the pores. For example, methyldiethoxysilane (MDES) was chemisorbed in the MFI membrane at 350°C followed by its cracking at 550°C resulting in an increase in H_2/CO_2 selectivity (determined with mixed gas tests at 300°C) from ~2 before treatment to ~48 after treatment. However, hydrogen permeability decreased by more than an order of magnitude (Hong et al., 2005). Alternatively, one-step cracking of MDES over the MFI membrane at 450°C was found to increase H_2/CO_2 selectivity at 450°C from ~2 before treatment to ~9 after treatment with a decrease in hydrogen permeability by half. Continued annealing at 450°C increased selectivity slightly to ~12 with a stable hydrogen flux (Lin, 2007). The zeolite membrane development is at a laboratory scale. For commercial applications, greater hydrogen flux and selectivity needs to be demonstrated.

8.3.5.3.3 Silica Membranes

Amorphous silica membranes with pore sizes <1 nm may be formed by sol-gel or CVD techniques (Verweij et al., 2006). Silica polymeric sols are formed by acid-catalyzed hydrolysis and the polymerization of tetraethylorthosilicate (TEOS) or methyltriethoxysilane under controlled reaction conditions of temperature, pH, and duration (Leenars et al., 1984; Leenars and Burggraaf, 1985). In a CVD technique, silica particles may be formed by chemical reactions, for example, gas-phase silane hydrolysis or thermal decomposition of precursors (Tsapatsis and Gavalas, 1994). Although silica membranes have been known to be selective for hydrogen at high temperatures with thermally activated diffusion (Wu et al., 1994), it has also been known to be reactive with water with degradation of separation performance over time in the presence of steam at high temperatures. Because silica membranes are porous, the hydrogen selectivity is not very high. Owing to low selectivity and concerns about hydrothermal stability in the synthesis gas environment containing steam, a recent study concluded that porous silica membranes are not suitable for hydrogen separation from the synthesis gas (Pex and van Delft, 2005).

8.3.5.4 Polymeric Membranes

As discussed earlier, polymeric membranes are commercially used for hydrogen recovery from waste gas streams, for example, an ammonia recycle stream. There are two primary shortcomings of the polymeric membranes when used for hydrogen separation. (1) Polymeric membrane materials are highly temperature sensitive and are typically used at ambient temperature, which requires cooling of the product hydrogen stream from upstream operations before hydrogen can be separated. Low-temperature polymeric membranes cannot, therefore, be used as membrane reactors even for the low-temperature WGS reaction. (2) Because hydrogen preferentially permeates through these membranes, the product hydrogen is produced at a lower pressure, whereas a high pressure may be desirable depending on the intended use and downstream operation. Although there is still some development of new hydrogen selective membrane materials, the primary research effort in polymeric membranes relevant to hydrogen production application is focused on alleviating the preceding two shortcomings: high-temperature polymer membrane with selectivity for hydrogen compared to other synthesis gas components and reverse-selective membranes that selectively permeate CO_2 leaving hydrogen in the residue gas at a high feed gas pressure.

Polymer PBI is a thermally stable polymer, reportedly stable up to 450°C temperature. PBI has been shown to be selective for hydrogen compared to other syngas components at several temperatures up to 400°C (Young, 2003). PBI is also being actively pursued as a material for high-temperature PEM fuel cells by Celanese Ventures GMBH and Plug Power (Staudt and Boyer, 2004). PBI is a glassy polymer with a glass transition temperature of 450°C and is claimed to be similar to polyimides with good chemical stability. A recent evaluation of permeation properties of a PBI composite membrane prepared with stainless-steel microporous support indicated a H_2/CO_2 selectivity of about 20 and H_2/CO selectivity of about 35°C at 300°C (Freeman, 2006). The hydrogen permeability of the PBI membrane, however, is low and an economic analysis of the composite membrane indicated that the PBI composite membrane prepared with a stainless-steel support will not be able to meet the hydrogen cost targets, and the study recommended an all-polymer asymmetric PBI composite membrane development (Freeman, 2006) with thin dense PBI film supported on a porous PBI support. PBI is much more permeable to water vapor compared to hydrogen with water to hydrogen selectivity of ~7. The water vapor permeability must, therefore, be taken into account for utilizing PBI-based membranes for WGS membrane reactor applications where water is one of the active reactant in the feed gas. Other thermally stable polymers that may potentially be useful as high-temperature hydrogen separation membranes include polybenzoxazole and polyimide with glass transition temperatures of ~330°C.

Approaches to make a polymeric membrane selective to CO_2 attempt to enhance the solubility selectivity of the polymer material for CO_2 and reduce the diffusivity selectivity of the polymer that favors smaller hydrogen molecule. The permeability of a polymer membrane for species A, P_A, is often expressed as (Ghosal and Freeman, 1994)

$$P_A = D_A \cdot S_A \tag{8.12}$$

where D_A is the diffusion coefficient and S_A the solubility coefficient of gas A in the membrane. The ability of a membrane material to separate two components, A and B, is often characterized in terms of the ideal selectivity, $\alpha_{A/B}$, which is defined as the ratio of their permeabilities (Ghosal and Freeman, 1994)

$$\alpha_{A/B} = \frac{P_A}{P_B} = \left(\frac{D_A}{D_B}\right)\left(\frac{S_A}{S_B}\right) \tag{8.13}$$

It follows from Equation 8.13 that $\alpha_{A/B}$ can be expressed as the product of the diffusivity selectivity, D_A/D_B, and the solubility selectivity, S_A/S_B. Diffusion (or mobility) selectivity is governed primarily by the size difference between gas molecules and always favors smaller gas molecules. Solubility selectivity is controlled by the relative condensability of the gases in the polymer and their relative affinity for the polymer. Solubility selectivity typically favors larger, more condensable molecules. From Equation 8.13, it is seen that the product of gas mobility and solubility selectivity determines the overall membrane selectivity. It is clear that for a membrane to be CO_2 selective, it must have high diffusivity selectivity based on the affinity for CO_2 but it should be flexible enough to permeate larger molecules such as CO_2 as easily as the H_2 molecule, that is, it would be necessary to reduce the diffusivity selectivity for hydrogen to nearly one. Among various organic functional groups, ethylene oxide (EO) has been identified as the best functional group to interact favorably with CO_2 resulting in a high solubility selectivity for CO_2 (Lin and Freeman, 2005). If polymers containing EO can be made highly flexible leading to weak size-sieving ability, the diffusivity selectivity of the polymer for hydrogen can be reduced considerably. Although pure poly ethylene oxide (PEO) has high CO_2 selectivity as expected, it has a very low permeability for CO_2 (~12 Barrer at 35°C) as a result of high crystallinity levels. To inhibit crystallinity, short non-PEO segments and chain branches of short noncrystallizable EO segments were introduced in the PEO polymer backbone resulting in increased permeability while still preserving CO_2 selectivity in a family of amorphous, high-molecular-weight, cross-linked network copolymers (Lin et al., 2006). The family of copolymers contained different composition ratios of poly(ethyleneglycol) diacrylate (PEGDA) and poly(ethyleneglycol) methyl ether acrylate (PEGMEA) with 70 wt% PEGMEA/30 wt% PEGDA copolymer, providing CO_2 permeability of ~400 Barrer and CO_2/H_2 selectivity of 10 at 35°C. Contrary to commonly observed behavior, the CO_2/H_2 selectivity increased to ~30 as the temperature decreased to −20°C while maintaining nearly the same CO_2 permeability (Lin et al., 2006).

Another way of reducing diffusivity selectivity for small molecules is to introduce nanoparticles in the matrix of a polymer that has a high solubility selectivity for CO_2 (Merkel et al., 2002). Intuitively, incorporation of nanoparticles as fillers is expected to reduce gas species permeability due to a reduction in the proportion of polymer through which the gas species diffuses. However, incorporation of 13 nm size fumed silica particles containing trimethylsilyl groups in a high free-volume glassy polymer poly-4-methyl-2-pentyne (PMP) was found to increase the selectivity of large organic molecules such as *n*-butane over smaller molecules such as methane. By adding 45 wt% silica in PMP increased *n*-butane selectivity from 13 to 22 while also increasing *n*-butane permeability from 7,000 to 25,000 Barrers (Merkel et al., 2002). This concept is being explored for polymer materials with high CO_2 solubility selectivity by decreasing the diffusivity selectivity of the polymer for hydrogen and, thus, increasing overall selectivity. Reverse selective (i.e., polymers selective to larger molecules compared to smaller molecules) rubbery polymer membranes are commercially used for separation of heavier hydrocarbons from natural gas (e.g., MTR, 2007). In principle, such membranes could be used to selectively remove CO_2 from synthesis gas components if the diffusivity selectivity for hydrogen could be reduced either by adding nanoparticles as indicated earlier or simply by adsorptive blockage.

Both the high-temperature and reverse selective polymeric nanocomposite membranes are being investigated on a laboratory scale.

8.3.5.5 High Temperature CO_2 Sorbents

Similar to high-temperature hydrogen separation-based membrane reactors, separation of CO_2 simultaneously with the WGS or methane reforming reaction may also be used to shift the reaction equilibrium toward greater hydrogen production. Separation of CO_2 has an added benefit of keeping the product hydrogen at a high pressure. Separation of CO_2 in a high purity stream will also be helpful for its sequestration if so desired. Sorbent materials that are being developed and evaluated for separating CO_2 at high temperatures may be grouped into two categories based on the mode of sorbent regeneration. Sorbent materials adsorbing CO_2 at high pressures by chemisorption may be regenerated by reducing the CO_2 pressure over the sorbent, that is, by pressure swing. Sorbent materials chemically reacting with CO_2 at high temperatures such as CaO and lithium silicate/zirconate must be regenerated by heating them at a higher temperature, that is, by temperature swing.

Hydrotalcites and double salts containing MgO have been shown to exhibit pressure reversible CO_2 capacity at temperatures of 300–500°C suitable for WGS or methane reforming reaction (Hufton et al., 1999; Waldron et al., 2001; Ding and Alpay, 2000a,b; Mayorga et al., 2001). Hydrotalcites are of the general formula $M^{2+}_{1-x}M^{3+}_{x}(OH)_2(A^{n-})_{x/n} \cdot mH_2O$, where M^{2+} and M^{3+} are divalent and trivalent metal ions, and A^{n-} is an anion. The value of x is in the range 0.2–0.33. The metal ions and the anions appear in a layered structure. A naturally occurring mineral and the most commonly studied "hydrotalcite" is of the formula $Mg_{0.75}Al_{0.25}(OH)_2(CO_3)_{0.125}\ 0.5H_2O$. The reversible CO_2 adsorption capacities of potassium carbonate–promoted hydrotalcites range from 0.4 to 0.7 gmol/kg (<3% by weight) at 300–400°C, even in the presence of steam. Adsorbent materials containing an alkali metal salt and a magnesium salt are referred to as double salts with a general formula: $\{(M_2CO_3)_m(2MHCO_3)_{1-m}\}_n(MgCO_3)_p(MgO)_{1-p}xH_2O$, where M is an alkali metal such as Li, Na, K, Rb, and Cs. Double salts have been shown to have even greater CO_2 adsorption capacities in the range 1–10 gmol/kg at 350–400°C (Mayorga et al., 2001).

Although double salts appear to have a greater hydrogen capacity, their stability has not been studied in detail and they are also difficult to pelletize for a fixed-bed reactor application (Reijers et al., 2005). Potassium carbonate–promoted hydrotalcite in a pelletized form has been evaluated at the bench- and pilot-scale system for SERP for conducting methane steam reforming with simultaneous adsorption of CO_2 (Hufton et al., 1999; Waldron et al., 2001; Reijers et al., 2005). Commercial SMR catalyst was mixed with the CO_2 sorbent and SMR was conducted at a low temperature of 450°C and 55 psig pressure with steam to methane ratio of six resulting in 82% conversion of methane and a product stream containing 95% hydrogen, 5% unreacted methane, and <50 ppm of carbon oxides (Hufton et al., 1999). Utilization of hydrotalcite adsorbent in the form of a thin film membrane deposited on a porous support for a continuous membrane reactor process is also being investigated (Liu, 2005; Feuillade and Haije, 2006), however, such membranes appeared to give low overall methane conversion (Jansen et al., 2006).

Alkali metal oxides such as CaO and MgO have been known to react with CO_2 at high temperatures. The oxides can be regenerated by calcining respective carbonates at a greater temperature with carbonation/calcination equilibrium depending on the reaction temperature and CO_2 partial pressure (Han and Harrison, 1994; Gupta et al., 2002). Because CO_2 is absorbed by chemical reaction, the theoretical material weight–based CO_2 absorption capacity as determined by reaction stoichiometry is high. Formation of the carbonate product layer on the active sorbent surface, however, limits the CO_2 absorption capacity typically to 30–40 wt%. The active sorbent weight–based CO_2 capacity may be increased substantially to up to 80% by improving sorbent particle morphology (Fan et al., 2005) or by loading the active sorbent on a porous support. Because CaO reacts with CO_2 at temperatures suitable for methane reforming, for example, 650°C, CaO has been investigated for sorbent-enhanced methane reforming with simultaneous CO_2 removal (Ortiz and Harrison, 2001). CaO has also been investigated for sorbent-enhanced WGS reaction at 600°C (Fan et al., 2005). Although greater than equilibrium CO conversion was observed, the effectiveness of the regenerated sorbent in the second cycle was significantly lower than the virgin sorbent. Also the CO_2 absorption capacity was found to decrease with each regeneration cycle (Fan et al., 2005). As the metal oxide-based sorbents are regenerated by thermal decomposition at temperatures of >700°C, efficient solids heating system must be designed for fixed-bed reactors. Alternatively, fluidized-bed absorption and regeneration reactors may be used with circulation of solids between the two reactors at high temperatures and interstage solids heating and cooling making the overall system more complex. The solid particles must be made attrition-resistant, and heat-resistant alloys must be used for the reactor vessels due to high temperatures involved (Reijers et al., 2005).

Lithium metal-based reagents such as lithium zirconate and lithium silicate react with CO_2 at lower temperatures than CaO, for example, at 400–500°C with nearly 30 wt% CO_2 absorption capacity. Alkali carbonates are typically added as promoters (Kato et al., 2002, 2004; Xiong et al., 2003; Ida et al., 2004). Because CO_2 absorption and sorbent regeneration occur at lower temperatures than those involved with CaO sorbent, lithium-based sorbents would be more attractive for SERPs especially for WGS reaction. The capacity of lithium silicate–based materials was also shown to be reproducible over a number of carbonation/regeneration cycles and the sorbent may be regenerated by a temperature swing or by inert purge at the absorption temperature (Li et al., 2005). In further developments, attrition resistant and fluidizable sorbent formulations were being prepared to allow the sorbent carbonation and regeneration processes to be carried out in more efficient fluidized beds (Li et al., 2006).

Regenerable CO_2 sorbents capable of removing CO_2 at 300–400°C have recently been developed (Siriwardane et al., 2007). This sodium-based sorbent shows a very high carbon dioxide capture capacity and greater than 99% carbon dioxide removal efficiency at 315°C. The sorbent can be regenerated at 700°C. A 10-cycle test conducted at U.S. DOE NETL showed that the capture capacity increased during the cyclic testing. Such sorbents will especially be attractive for conducting LTS reaction with simultaneous CO_2 removal in a sorbent-enhanced reactor process similar to the earlier described sorbent materials suitable for higher temperatures.

Developments of sorbent materials for high-temperature CO_2 removal from synthesis gas and their utilization as SERPs appears attractive, however, the development scale may be considered as laboratory/bench scale.

8.4 Conclusion

The preceding sections discussed the developmental hydrogen separation processes being investigated. Although a large amount of research effort is ongoing, it is clear that substantial progress still needs to be made for developing commercially viable alternatives to the established hydrogen separation processes of PSA, CO_2 absorption, and currently available polymeric membranes. Among the developmental activities, most effort is being devoted to palladium-alloy composite membrane as being the most promising. The merits and shortcomings of the different approaches and their current development status is summarized in Table 8.2.

Among commercially available processes, PSA is often the most cost-effective option for hydrogen separation and purification, especially for hydrogen production by steam reforming of natural gas. For coal and biomass-based feedstocks for hydrogen generation, CO_2 absorption either alone or in combination with PSA may become attractive due to the greater content of CO_2 in the product hydrogen stream. With increasing emphasis on carbon sequestration, producing pure CO_2 as well as hydrogen will become important, increasing the attractiveness for CO_2 absorption and hybrid CO_2 removal/PSA configurations.

TABLE 8.2

Developmental Hydrogen Separation Approaches

Technology	Advantages	Drawbacks	Status
Pd-alloy composite membranes	High flux and high selectivity for hydrogen	Vulnerable to sulfur species	Pilot-scale testing in WGS and methane reforming membrane reactor configuration
		Phase transformations during thermal cycling	Need demonstration of long-term stability and durability in practical applications
Group V metal composite membranes	High flux and high selectivity for hydrogen	Need catalyst for hydrogen dissociation and base metal protection	Pilot-scale testing for hydrogen separation
		Pd coating vulnerable to sulfur and alloys with base metal	Need demonstration of long-term stability and durability in practical applications
Dense ceramic ion transport membranes	High selectivity for hydrogen	Low permeability for hydrogen	Laboratory scale
	Resistant to contaminants	Vulnerability to CO_2 and steam	Need to increase hydrogen flux, for example, by incorporating hydrogen permeable metals—cermets
Porous inorganic membranes	High flux for hydrogen	Low selectivity for hydrogen	Pilot-scale testing for hydrogen permeation
	Low cost materials	Low hydrogen recovery for high purity	
	Resistant to contaminants	Pore coarsening by steam	Need demonstration of long-term stability and durability with practical hydrogen recovery
Carbon molecular sieve membranes	Resistant to contaminants	Intermediate hydrogen flux and selectivity	Pilot-scale testing in low temperature WGS membrane reactor application
	Intermediate hydrogen flux and selectivity	High water permeability	Need demonstration of long-term stability and durability in practical applications
Zeolite membranes	Resistant to contaminants	Intermediate hydrogen flux and selectivity	Laboratory scale
	Intermediate hydrogen flux and selectivity		Need to increase hydrogen flux as well as selectivity, for example, by thinner yet defect-free zeolite membrane layers
High-temperature or reverse CO_2 selective polymeric membranes	Potential for lower cost and high surface area modules, for example, hollow fiber or spiral wound	Low permeability and selectivity for hydrogen or CO_2 separation	Laboratory scale
			Need to develop practical composite membrane modules with promising materials
High-temperature CO_2 sorbents (SERP)	Hydrogen produced at high pressure	High-temperature regeneration with solids handling, attrition, and materials issues	Laboratory/bench scale
		Decreasing capacity with number of cycles	Need to demonstrate long-term stable CO_2 capacity and sorbent durability

References

Air Liquide, MEDAL Hydrogen Membrane, available at http://www.medal.airliquide.com/en/membranes/hydrogen/index.asp (accessed July 2007).

Arps, J., Cost-Effective Method for Producing Self-Supporting Pd Alloy Membrane for Use in the Efficient Production of Coal-derived Hydrogen, *Proceedings of 2006 U.S. DOE Hydrogen Annual Merit Review Meeting*, Arlington, VA, May 2006.

ATI Wah Chang, Palladium-Based Metal Membrane, available at http://www.wahchang.com/pages/products/data/pdf/Hydrogen%20Metal%20Membrane.pdf (accessed July 2007).

Babicki, M. and A. Hall, PSA Technology Hits the Fast Lane, Chemical Processing Digital Edition, 9/10/2003, available at http://www.chemicalprocessing.com/articles/2003/322.html.

Baksh, M.S.A. and C.E. Terbot, Pressure Swing Adsorption Process for the Production of Hydrogen, U.S. Patent 6,503299 B2, 2003.

Balachandran, U., Development of Dense Ceramic Membranes for Hydrogen Separation, *Proceedings of 2005 U.S. DOE Hydrogen Annual Merit Review Meeting*, Arlington, VA, May 2005.

Balachandran, U., T.H. Lee, S. Wang, G. Zhang, and S.E. Dorris, Current Status of Dense Ceramic Membranes for hydrogen Separation, *27th International Technical Conference on Coal Utilization and Fuel Systems*, Clearwater, FL, March 2002.

Barton, T. and M. Argyle, Integration of a Structural Water Gas Shift Catalyst with a Vanadium Alloy Hydrogen Transport Device, *Proceedings of 2007 U.S. DOE Hydrogen Annual Merit Review Meeting*, Arlington, VA, May 2007.

Basile, A.G. Chiappetta, S. Tosti, and V. Violante, Experimental and simulation of both Pd and Pd/Ag for a water gas shift membrane reactor, *Sep. Purif. Technol.*, 25, 549–571, 2001.

Bates, E.D., R.D. Mayton, I. Ntai, and J.H. Davis Jr., CO_2 capture by a task-specific ionic liquid, *J. Am. Chem. Soc.*, 124(6), 926–927, 2002.

Benson, S., Ceramics for Advanced Power Generation, IEA Clean Coal Center Report CCC/37, IEA Coal Research, London, August 2000.

Bischoff, B.L. and R.R. Judkins, Scale-Up of Microporous Inorganic Hydrogen-Separation Membranes, *Proceedings of 2006 U.S. DOE Hydrogen Annual Merit Review Meeting*, Arlington, VA, May 2006.

Bollinger, W.A., D.L. Maclean, and R.S. Narayan, Applications of prism separators in oil refining and production, *Chem. Eng. Progress*, 78(10), 27, 1982.

Bridger, G.W., Steam reforming of hydrocarbons, in *Catalysis*, eds. C. Kemball and D.A. Dowden, The Chemical Society, London, Vol. 3, p. 39, 1980.

Buxbaum, R.E., High Flux Metallic Membranes for Hydrogen Recovery and Membrane Reactors, *Proceedings of 2007 U.S. DOE Hydrogen Annual Merit Review Meeting*, Arlington, VA, May 2007.

Buxbaum, R.E. and T.L. Marker, Hydrogen transport through non-porous membranes of palladium-coated niobium, tantalum and vanadium. *J. Membr. Sci.*, 85, 29–38, 1993.

Cassidy, R.T., Adsorption and ion exchange with synthetic zeolites, in *ACS Symposium Series*, ed. W.H. Flank, American Chemical Society, Washington, Vol. 135, p. 275, 1980.

Collins, J.P. and J.D. Way, Catalytic decomposition of ammonia in a membrane reactor, *J. Membr. Sci.*, 96, 259–274, 1994.

Collot, A.G., Prospects for Hydrogen from Coal, IEA Clean Coal Center Report, IEA Coal Research, London, September 2003.

Connor, D.J., D.G. Doman, L. Jeziorowski, B.G. Keefer, B. Larisch, C.R. McLean, and I. Shaw, Rotary Pressure Swing Adsorption Apparatus, U.S. Patent 6,406,523 B1, 2002.

Coulter, K., Cost-Effective Method for Producing Self-Supporting Pd Alloy Membrane for Use in the Efficient Production of Coal-Derived Hydrogen, *Proceedings of 2007 U.S. DOE Hydrogen Annual Merit Review Meeting*, Arlington, VA, May 2007.

Criscuoli, A., A. Basile, E. Drioli, and O. Loiacono, An economic feasibility study for water gas shift membrane reactor, *J. Membr. Sci.*, 181, 21–27, 2001.

Damle, A.S., Separation of Hydrogen and Carbon Dioxide in Advanced Fossil Energy Conversion Processes using a Membrane Reactor, *2002 Pittsburgh Coal Conference*, Pittsburgh, PA, September 2002.

Damle, A.S., J. Schwartz, and P. Apte, Palladium-alloy based Membrane Reactor Process for Hydrogen Generation, *Proceedings of 2005 Fuel Cell Seminar,* Palm Springs, CA, November 2005.

Damle, A.S., C. Richardson, C. Love, T. Powers, and J. Aquaviva, High performance palladium-alloy based composite membranes for hydrogen production, *2007 NHA Annual Meeting,* San Antonio, TX, March 2007.

Ding, Y. and E. Alpay, Equilibria and kinetics of CO_2 adsorption on hydrotalcite adsorbent, *Chem. Eng. Sci.*, 55, 3461–3474, 2000a.

Ding, Y. and E. Alpay, Adsorption-enhanced steam methane reforming, *Chem. Eng. Sci.*, 55, 3929–3940, 2000b.

Dittmeyer, R., V. Hollein, and K. Daub, Membrane reactors for hydrogenation and dehydrogenation processes based on supported palladium, *J. Mol. Catal. A: Chem.*, 173, 135–184, 2001.

Dorris, S.E., T.H. Lee, S. Wang, J.J. Picciolo, J.T. Dusek, and U. Balachandran, Dense Cermet Membranes for Hydrogen Separation, *19th Annual International Pittsburgh Coal Conference,* Pittsburgh, PA, September 2002.

Edlund, D., A Membrane Reactor for H_2S Decomposition, *Proceedings of the U.S. DOE Advanced Coal Fired Power Systems Review Meeting,* Morgantown, WV, July 1996.

Edlund, D.J. and W. Pledger, Thermolysis of hydrogen sulfide in a metal membrane reactor, *J. Membr. Sci.*, 77, 255–264, 1993.

Elangovan, S., B. Nair, J. Hartvigsen, and T. Small, Mixed Conducting Membranes for Pressure Driven Hydrogen Separation from Syngas, *225th American Chemical Society National Meeting, Fuels Division,* New Orleans, LA, March 2003.

Fan, L.S., H. Gupta, M. Iyer, and B. Sakadjian, Enhanced Hydrogen Production Integrated with CO_2 Separation in a Single-Stage Reactor, *Proceedings of 2005 U.S. DOE Hydrogen Annual Merit Review Meeting,* Arlington, VA, May 2005.

Feuillade, V.C. and W.G. Haije, Characterization of Hydrotalcite Materials for CO_2 Selective Membranes, *Proceedings of the 9th International Conference on Inorganic Membranes,* Lillehammer, 25–29, June 2006.

Flytzani-Stephanopoulos, M., X. Qi, and S. Kronewitter, Water-Gas Shift with Integrated Hydrogen Separation Process, Final Report, U.S. DOE Contract # DE-FG2600-NT-40819, February 2004.

Freeman, B.D., High Temperature Nano-Composite Membranes for Hydrogen Production for Fuel Cells, Final Report, TCEQ/NTRD Grant Contract # 582-5-65591-0002, August 2006.

Fuertes, A.B., Adsorption-selective carbon membrane for gas separation, *J. Membr. Sci.*, 177, 9, 2000.

Gallaher, G.R. and P.K.T. Liu, Characterization of ceramic membranes I. thermal and hydrothermal stabilities of commercial 40 Å membranes, *J. Membr. Sci.*, 92(1), 29–44, 1994.

Ghosal, K. and B.D. Freeman, Gas separation using polymer membranes: an overview, *Polym. Advan. Technol.*, 5, 673–697, 1994.

Golden, T.C., C.M.A. Golden, and D.O. Zwilling, Self-Supported Structured Adsorbent for Gas separation, U.S. Patent 6,656,627 B1, 2003.

Grimmer, P.J., Scale-up of Hydrogen Transport Membranes for IGCC and FutureGen Plants, *Proceedings of 2007 U.S. DOE Hydrogen Annual Merit Review Meeting,* Arlington, VA, May 2006.

Gryaznov, V.M., Hydrogen permeable palladium membrane catalysts, an aid to the efficient production of ultra purechemicals and pharmaceuticals, *Plat. Met. Rev.*, 30, 68–72, 1986.

Gryaznov, V.M., M.M. Ermilova, L.S. Morozova, and N.V. Orekhova, Palladium alloys as hydrogen permeable catalysts in hydrogenation and dehydrogenation reactions, *J. Less-Common Metals,* 89, 529–535, 1983.

Guan, J., S.E. Dorris, U. Balachandran, and M. Liu, The effects of dopants and A:B site nonstoichiometry on properties of perovskite-type proton conductors, *J. Electrochem. Soc.*, 145, 1780, 1998.

Gunardson, H., *Industrial Gases in Petrochemical Processing,* Chapter 3, Marcel Dekker, New York, 1998.

Gupta, H., M. Iyer, B. Sakadjian, and L.S. Fan, Separation of CO_2 from flue gas by high reactivity calcium-based sorbents, *Pittsburgh Coal Conference,* Pittsburgh, PA, September 2002.

Han, C. and D.P. Harrison, Simultaneous shift reaction and carbon dioxide separation for the direct production of hydrogen, *Chem. Eng. Sci.*, 49, 5975–5883, 1994.

Hara, S., N. Hatakeyama, N. Itoh, H.-M. Kimura, and A. Inoue, Hydrogen permeation through palladium-coated amorphous Zr–M–Ni (M=Ti, Hf) alloy membranes, *Desalination*, 144, 115–120, 2002.

Harrison, D.P. and Z. Peng, Low-carbon monoxide hydrogen by sorption-enhanced reaction, *Int. J. Chem. Reactor Eng.*, 1, 1–9, 2003.

Hatori, H., H. Takagi, and Y. Yamada, Gas separation properties of molecular sieving carbon membranes with nanopore channels, *Carbon*, 42, 1169–1173, 2004.

Holbrook, D., Technology Integration to Maximize Value of Oil Sands Product, *Gasification Technology Conference*, Washington, October 2006.

Hong, M., R.D. Noble, and J.L. Falconer, Highly selective H_2 Separation Zeolite Membranes for Coal Gasification Membrane Reactor Applications, Annual Technical Progress Report, U.S. DOE Contract # DE-FG26-02NT41536, December 2005.

Hopkins, S., High-Performance, Durable, Palladium-Alloy Membrane for Hydrogen Separation and Purification, *Proceedings of 2007 U.S. DOE Hydrogen Annual Merit Review Meeting*, Arlington, VA, May 2007.

Hsieh, H.P., Inorganic membranes, *AIChE Symp. Ser.*, 261, 1–18, 1988.

Hufton, J.R., S. Mayorga, and S. Sircar, Sorption-enhanced reaction process for hydrogen production, *AIChE J.*, 45(2), 248–255, 1999.

Ida, J.-I., R. Xiong, and Y.S. Lin, Synthesis and CO_2 sorption properties of pure and modified lithium zirconate, *Sep. Purif. Technol.*, 36, 41–51, 2004.

IdaTech, Fuel Processing Technology and ElectrGen, available at http://www.idatech.com/

InnovaTek, InnovaGen Fuel Processor, available at http://www.tekkie.com/docs/InnovaGen.pdf (accessed July 2007).

Intelligent Energy, Fuel Processor and Hydrogen Generator, available at http://www.intelligent-energy.com/images/uploads/compact_hydrogen_generator_a4.pdf (accessed July 2007).

Ismail, A.F. and L.I.B. David, A review on the latest development of carbon membranes for gas separation, *J. Membr. Sci.*, 193, 1–18, 2001.

Itoh, N., A membrane reactor using palladium, *AIChE J.*, 33, 1576–1578, 1987.

Itoh, N. and R. Govind, Combined oxidation and dehydrogenation in a palladium membrane, *Ind. Eng. Chem. Res.*, 28, 1554–1557, 1989.

Jack, D.S., Scale-up of Hydrogen Transport Membranes for IGCC and FutureGen Plants, *Proceedings of 2007 U.S. DOE Hydrogen Annual Merit Review Meeting*, Arlington, VA, May 2007.

Jansen, D., W. Haije, M. Carbo, V. Feuillade, J.W. Dijkstra, and R. Brink, Advanced membrane reactors for carbon-free fossil fuel conversion, *ECN GCEP Project Presentation*, GCEP Energy Research Symposium, Stanford, September 2006.

Jiang, L., V.G. Fox, and L.T. Biegler, Simulation and optimal design of multi-Bed pressure swing adsorption systems, *AIChE J.*, 50, 2904–2917, 2004.

Johnson Matthey, *Gas Purification Technology*, available at http://www.jmgpt.com/ (accessed July 2007).

Kamakoti, P., B.D. Morreale, M.V. Ciocco, B.H. Howard, R.P. Killmeyer, A.V. Cugini, and D.S. Sholl, Prediction of hydrogen flux through sulfur tolerant binary alloy membranes, *Science*, 307, 569–573, 2005.

Kapteijn, F., W.J.W. Bakker, J.v.d. Graaf, G. Zheng, J. Poppe, and J.A. Moulijn, Permeation and separation behavior of a silicalite-1 membrane, *Catal. Today*, 25, 213, 1995.

Kato, M., K. Nakagawa, T. Ohashi, S. Yoshikawa, and K. Essaki, Carbon Dioxide Gas Absorbent Containing Lithium Silicate, U.S. Patent No. 6,387,845 B1, May 2002.

Kato, M., S. Yoshikawa, K. Essaki, and K. Nakagawa, Carbon Dioxide Gas Absorbent and Carbon Dioxide Gas Separating Apparatus, U.S. Patent No. 6, 712.879 B2, March 2004.

Keefer, B.G., A. Carel, B. Sellers, I. Shaw, and B. Larisch, Adsorbent Laminate Structures, U.S. Patent No. 6,692,626, B2, 2004.

Keefer, B.G. and D.G. Doman, Flow Regulated Pressure Swing Adsorption System, U.S. Patent No. RE38,493E, 2004.

Kermode, R.I., Hydrogen from fossil fuels, in *Hydrogen: Its Technology and Implications*, Section 3.2.2, Chapter 3, Vol. 1—Hydrogen Production Technology, eds., K.E. Cox, and K.D. Williamson, CRC Press, Cleveland, OH, 1977.

Killmeyer, R., B. Howard, K. Rothenberger, M. Ciocco, B. Morreale, R. Enick, and O. Iyoha, Hydrogen Separation, *Proceedings of 2005 U.S. DOE Hydrogen Annual Merit Review Meeting*, Arlington, VA, May 2005.

Killmeyer, R., K. Rothenberger, B. Howard, M. Ciocco, B. Morreale, R. Enick, and F. Bustamante, Water-Gas Shift Membrane Reactor Studies, *Proceedings of 2003 U.S. DOE Hydrogen Annual Merit Review Meeting*, Berkeley, CA, May 2003.

Kirk-Othmer, Concise encyclopedia of chemical technology, Fourth Edition, Wiley, New York, Vol. 12, 950–982, 1999.

Klette, H., T. Peters, A. Mejdell, and R. Bredesen, Development of Palladium-Based Hydrogen Membranes for Water Gas Shift Conditions, *Proceedings of Eight International Conference on Greenhouse Gas Control Technologies (GHGT-8)*, Trondheim, June 2006.

Knapton, A.G., Palladium alloys for hydrogen diffusion membranes—A review of high permeability materials, *Plat. Met. Rev.*, 21, 44–50, 1977.

Koresh, J.E. and A. Sofer, Study of molecular sieve carbon membranes, Part 1. Pore structure, gradual pore opening, and mechanism of molecular sieving, *J. Chem. Soc., Faraday Trans. I*, 76, 2457, 1980.

Koresh, J.E. and A. Sofer, Molecular sieve carbon permselective membrane, Part I. Presentation of a new device for gas mixture separation, *Sep. Sci. Technol.*, 18, 723, 1983.

Kubek, D.J., E. Polla, and F.P. Wilcher, Purification and Recovery Options for Gasification, *Gasification Technology Conference*, San Francisco, CA, October 1996.

Lasher, S., M. Stratonova, and J. Thijssen, Hydrogen Technical Analysis, *Proceedings of 2002 U.S. DOE Hydrogen Annual Merit Review Meeting*, Golden, CO, May 2002.

Leenars, A.F.M. and A.J. Burggraaf, The preparation and characterization of alumina membranes with ultra-fine pores, Part 2. The formation of supported membranes, *J. Coll. Interface Sci.*, 105(1), 27–40, 1985.

Leenars, A.F.M., K. Keizer, and A.J. Burggraaf, The preparation and characterization of alumina membranes with ultra-fine pores, Part 1. Microstructural investigations on non-supported membranes, *J. Mater Sci.*, 19, 1077–1088, 1984.

Li, W., S.K. Gangwal, R.P. Gupta, and B.S. Turk, Development of Fluidizable Lithium Silicate-Based Sorbents for High Temperature Carbon Dioxide Removal, *2006 Pittsburgh Coal Conference*, Pittsburgh, PA, September 2006.

Li, W., B.S. Turk, S.K. Gangwal, T.O. Nelson, R.P. Gupta, S. Park, and S. Tamhankar, High Temperature Carbon Dioxide Removal from Syngas Using Lithium Silicate-Based Sorbents, *2005 Pittsburgh Coal Conference*, Pittsburgh, PA, September 2005.

Lin, C.L., D.L. Flowers, and P.K.T. Liu, Characterization of ceramic membranes II. Modified commercial membranes with pore size under 40 Å, *J. Membr. Sci.*, 92(1), 45–58, 1994.

Lin, H. and B.D. Freeman, Materials selection guidelines for membranes that remove CO_2 from gas mixtures, *J. Mol. Struct.*, 739, 57, 2005.

Lin, H., E.V. Wagner, B.D. Freeman, L.G. Toy, and R.P. Gupta, Plasticization-enhanced hydrogen purification using polymeric membranes, *Science*, 311, 639–642, 2006.

Lin, J.Y.S., Proton conducting dense ceramic membranes for hydrogen separation, Annual Progress Report, U.S. DOE Contract # DE-FG26-00NT40818, December 2002.

Lin, J.Y.S., Zeolite Membrane Reactor for Water-Gas-Shift Reaction for Hydrogen Production, *Proceedings of 2007 U.S. DOE Hydrogen Annual Merit Review Meeting*, Arlington, VA, May 2007.

Lin, Y.M. and M.H. Rei, Separation of hydrogen from the gas mixture out of a catalytic reformer by using supported palladium membrane, *Sep. Purif. Technol.*, 25, 87–95, 2001a.

Lin, Y.M. and M.H. Rei, Study on the hydrogen production from methanol steam reforming in supported palladium membrane reactor, *Catal. Today*, 67, 77–84, 2001b.

Liu, P.K.T., CO_2 Selective ceramic membrane for water-gas-shift reaction with concomitant recovery of CO_2, Final Report, U.S. DOE Contract # DE-FC26-00NT40922, July 2005.

Liu, P.K.T., Carbon Molecular Sieve Membrane as Reactor for Water Gas Shift Reaction, *Proceedings of 2006 U.S. DOE Hydrogen Annual Merit Review Meeting*, Arlington, VA, May 2006.

Liu, P.K.T., Carbon Molecular Sieve Membrane as Reactor/Separator for Water-Gas-Shift Reaction, *Proceedings of 2007 U.S. DOE Hydrogen Annual Merit Review Meeting*, Arlington, VA, May 2007.

Ma, Y.H., P.P. Mardilovich, and Y. She, Stability of hydrogen flux through Pd/Porous stainless steel composite membranes, *Proceedings of International Conference on Inorganic Membranes, ICIM5*, Nagoya, 1998.

Mak, J.Y., D. Heaven, D. Kubek, C. Sharp, and M. Clark, Synthesis Gas Purification in Gasification to Ammonia/Urea Complex, *Gasification Technology Conference*, October 2004.

Makino, H., Y. Kusuki, H. Yoshida, and A. Nakamura, Process for Preparing Aromatic Polyimide Semipermeable Membranes, U.S. Patent No. 4,378,324, March 1983.

Mardilovich, I.P., E. Engwall, and Y.H. Ma, Dependence of hydrogen flux on the pore size and plating surface topology of asymmetric Pd-porous stainless steel membranes, *Desalination*, 144, 85–89, 2002.

Mayorga, S.G., S.J. Weigel, T.R. Gaffney, and J.R. Brzozowski, Carbon Dioxide Adsorbents Containing Magnesium Oxide Suitable for Use at High Temperature, U.S. Patent No. 6,280,503 B1, August 2001.

McCarley, K.C. and J.D. Way, Development of a model surface flow membrane by modification of porous γ-alumina with octadecyltrichlorosilane, *Sep. Purif. Technol.*, 25, 195–210, 2001.

McKinley, D.L., Metal Alloy for Hydrogen Separation and Purification, U.S. Patent No. 3,350,845, 1967.

Merkel, T.C., B.D. Freeman, R.J. Spontak, Z. He, I. Pinnau, P. Meakin, and A.J. Hill, Ultrapermeable, reverse-selective nanocomposite membranes, *Science*, 296, 519–522, 2002.

Miller, G.Q. and J. Stocker, Selection of a Hydrogen Separation Process, Presented at NPRA Annual meeting, San Francisco, March 19–21, 1989, available at http://www.uop.com/objects/SelOfHydroSepProc.pdf. (accessed January 1999)

Morreale, B.D., M.V. Ciocco, B.H. Howard, R. Killmeyer, A.V. Cugini, and R.M. Enick, Effect of hydrogen sulfide on the hydrogen permeance of palladium-copper alloys at elevated temperatures, *J. Membr. Sci.*, 241, 219–224, 2004.

MTR, Membrane Technology Research, Inc. VaporSep, available at http://www.mtrinc.com/ (accessed July 2007).

Mundschau, M.V., X. Xie, and C. Evenson, Superpermeable hydrogen transport membranes, Nonporous Inorganic Membranes, Chapter 4, eds. A.F. Sammells and M.V. Mundschau, Wiley VCH Verlag GmBH, Weinheim, 2006.

National Academies Press, The Hydrogen Economy: Opportunities, Costs, Barriers, and R&D Needs, Chapter 2, 17, Washington, 2004.

Nelson, T.O., P.D. Box, D.A. Green, and R.P. Gupta, Carbon Dioxide Recovery from Power Plant Flue Gas using Supported Carbonate Sorbents in a Thermal Swing Process, *Sixth Annual Conference on Carbon Capture and Sequestration*, Pittsburgh, PA, May 2007.

Nenoff, T.M., M.E. Welk, and F. Bonhomme, Defect-free thin film membranes for H_2 separation and isolation, *2003 Annual National Hydrogen Association Meeting*, Washington, March 2003.

Okamoto, K.,S. Kawamura, M. Yoshino, H. Kita, Y. Hirayama, N. Tanihara, and Y. Kusuki, Olefin/paraffin separation through carbonized membranes derived from an asymmetric polyimide hollow fiber membrane, *Ind. Eng. Chem. Res.*, 38, 4424, 1999.

ORNL (Oak Ridge National Laboratory), *Transportation Energy Data Book: Edition 23*. Center for Transportation Analysis, Oak Ridge National Laboratory. ORNL-6970. UT-Battelle, LLC, Oak Ridge, TN, 2003.

ORNL (Oak Ridge National Laboratory) and BCS. Inc., Materials research for separations technologies: Energy and emission reduction opportunities, U.S. DOE Office of Industrial Technologies Program Report, May 2005.

Ortiz, A.L. and D.P. Harrison, Hydrogen production using sorption-enhanced reaction, *Ind. Eng. Chem. Res.*, 40, 5102–5109, 2001.

Paglieri, S.N., Palladium membranes, Nonporous Inorganic Membranes, Chapter 3, eds. A.F. Sammells and M.V. Mundschau, Wiley-VCH Verlag GmBH, Weinheim, 2006.

Paglieri, S.N. and J.D. Way, Innovations in palladium membrane research, *Sep. Purif. Methods*, 31(1), 1–169, 2002.

Pex, P.P.A.C. and Y.C. van Delft, Silica membranes for hydrogen fuel production by membrane water gas shift reaction and development of a mathematical model for a membrane reactor, in *Carbon Dioxide Capture for Storage in Deep Geologic Formations—Results from the CO_2 Capture Project; Capture and Separation of Carbon Dioxide from Combustion Sources*, eds., D. Thomas, and B. Sally, Vol. 1, Chapter 17, 2005.

Power + Energy, Hydrogen Separator, available at http://www.powerandenergy.com/products.shtml (accessed July 2007).

Prabhu, A.K. and S.T. Oyama, Highly hydrogen selective ceramic membranes: Application to the transformation of greenhouse gases, *J. Membr. Sci.*, 176, 233–248, 2000.

Rao, M.B. and S. Sircar, Nanoporous carbon membranes for separation of gas mixtures by selective surface flow, *J. Membr. Sci.*, 85(3), 253–264, 1993a.

Rao, M.B. and S. Sircar, Nanoporous carbon membranes for gas separation, *Gas Sep. Purif.*, 7, 279, 1993b.

REB Research, *Hydrogen Purifier*, available at http://www.rebresearch.com/catalog.html (accessed July 2007).

Reijers, H.Th.J., S.E.A. Valster-Schiermeier, P.D. Cobden, and R.W. van den Brink, Hydrotalcite as CO_2 Sorbent for Sorption-Enhanced Steam Reforming of Methane, ECN Report RX-05-122, July 2005.

Rhudy, R. and S. Black, Chilled ammonia process update, *CO_2 Capture Network*, Lyon, May 2007.

Ridler, D.E. and Twigg, M.V., in Steam reforming, *Catalyst Handbook*, ed. M.V. Twigg, Chapter 5, Wolfe Publishing Co., London, 1989, pp. 225–280.

Ritter, J.A. and A.D. Ebner, Separation technology R&D needs for hydrogen production in the chemical and petrochemical industries, U.S. DOE Office of Industrial Technologies Program Report, December 2005.

Roa, F. and J.D. Way, Influence of alloy composition and membrane fabrication on the pressure dependence of the hydrogen flux of palladium-copper membranes, *Ind. Eng. Chem. Res.*, 42(23), 5827–5835, 2003.

Roa, F., J.D. Way, R.L. McCormick, and S.N. Paglieri, Preparation and characterization of Pd-Cu composite membranes for hydrogen separation, *Chem. Eng. J.*, 93(1), 11–22, 2003.

Roark, S.E., R. Mackay, and A.F. Sammells, Catalytic Membrane Recators for Hydrogen Separation in Hydrocarbon Feedstreams, *2002 Pittsburgh Coal Conference*, Pittsburgh, PA, September 2002.

Rostrup-Nielsen, J.R., Catalytic steam reforming, in *Catalysis, Science and Technology*, eds. J.R. Anderson and M. Boudart, Springer, New York, 1984.

Rothenberger, K.S., B.H. Howard, R.P. Killmeyer, A.V. Cugini, R.M. Enick, F. Bustamente, M.V. Ciocco, B.D. Morreale, and R.E. Buxbaum, Evaluation of tantalum-based materials for hydrogen separation at elevated temperatures and pressures, *J. Membr. Sci.*, 218, 19–37, 2003.

Ruthven, D.M., S. Farooq, and K.S. Knaebel, *Pressure Swing Adsorption*, VCH Publishers, Inc., New York, 1994.

Saufi, S.M. and A.F. Ismail, Fabrication of carbon membranes for gas separation—a review, *Carbon*, 42, 241–259, 2004.

Schwartz, M., S. Gade, R. Schaller, and B. Berland, Novel Composite Membranes for Hydrogen Separation in Gasification Processes in Vision 21 Plants, *19th Annual International Pittsburgh Coal Conference*, Pittsburgh, PA, September 2002.

Siriwardane, R.V., C. Robinson, M. Shen, and T. Simonyi, Novel regenerable sodium based sorbents for CO_2 capture at warm gas temperatures, *Energy Fuels J.*, 21(4), 2088–2097, 2007.

Siriwardane, R.V., J.A. Poston, E.P. Fisher, T.H. Lee, S.E. Dorris, and U. Balachandran, Characterization of ceramic-metal composite hydrogen separation membranes consisting of barium oxide, cerium oxide, yttrium oxide, and palladium, *Appl. Surf. Sci.*, 217, 43–49, 2003.

Shu, J., B.P.A. Grandjean, and S. Kaliaguine, Methane steam reforming in asymmetric Pd- and Pd-Ag/porous SS membrane reactors, *Appl. Catal. A: General*, 119, 305–325, 1994.

Shu, J., B.P.A. Grandjean, A. Van Neste, and S. Kaliaguine, Catalytic palladium-based membrane reactors: A review, *The Canadian J. Chem. Eng.*, 69, 1036–1060, 1991.

Singh-Ghosal, A. and W.J. Koros, Air separation properties of flat sheet homogeneous pyrolytic carbon membranes, *J. Membr. Sci.*, 174, 177, 2000.

Smith, I., CO_2 Reduction Prospects for Coal, IEA Coal research Report, IEA Coal Research, London, ccc/26, London, 1999.

Staudt, R. and J. Boyer, Development of polybenzimidazole-based, high temperature membrane and electrode assemblies for stationary and automotive applications, *U.S. DOE 2004 Annual Program Review Proceedings*, Philadelphia, PA, May 2004.

Steinberg, M. and H.C. Cheng, Modern and prospective technologies for hydrogen production from fossil fuels, *Int. J. Hydrogen Energ.*, 14(11), 797–820, 1989.

Steward, S.A., Review of Hydrogen Isotope Permeability through Materials, Lawrence Livermore National Laboratory Report, UCRL-53441, 1983.

Stocker, J., M. Whysall, and G.Q. Miller, 30 Years of PSA Technology for Hydrogen Production, UOP Report, July 1998, available at http://www.uop.com/objects/30YrsPSATechHydPurif.pdf.

Suda, H. and Haraya, Alkene/Alkane permselectivities of a carbon molecular sieve membrane, *J. Chem. Soc. Chem. Commun.*, 93, 1997.

Tamhankar, S., R. Sethna, T. Boyd, A. Gulamhusein, M. Golben, and D. DaCosta, Integrated Hydrogen Production, Purification and Compression System, *Proceedings of 2007 U.S. DOE Hydrogen Annual Merit Review Meeting*, Arlington, VA, May 2007.

Thomas, W.J. and B. Crittenden, *Adsorption Technology and Design*, Butterworth and Heinemann Publishers, Oxford, 1998.

Tsapatsis, M. and G. Gavalas, Structure and aging characteristics of H_2-permselective SiO_2-Vycor membranes, *J. Membr. Sci.*, 87, 281, 1994.

Tsuru, T., K. Yamaguchi, T. Yoshioka, and M. Asaeda, Methane steam reforming by microporous catalytic membrane reactors, *AIChE J.*, 50(11), 2794–2805, 2004.

Uemiya, S., Y. Kude, K. Sugino, N. Sato, T. Matsuda, and E. Kikuchi, A palladium/porous glass composite membrane for hydrogen separation, *Chem. Lett.*, 10, 1687–1690, 1988.

Uemiya, S. et al., Separation of hydrogen through palladium thin film supported on a porous glass tube, *J. Membr. Sci.*, 56, 303, 1991a.

Uemiya, S., T. Matsuda, and E. Kikuchi, Hydrogen permeable palladium-silver alloy membrane supported on porous ceramics, *J. Membr. Sci.*, 56, 315, 1991b.

Uemiya, S., N. Sato, H. Ando, T. Matsuda, and E. Kikuchi, Steam reforming of methane in a hydrogen permeable membrane reactor, *Appl. Catal.*, 67, 223–230, 1991c.

U.S. Department of Energy, National Hydrogen Energy Roadmap, U.S. Department of Energy, Washington, November 2002.

U.S. Department of Energy, Hydrogen Posture Plan, U.S. Department of Energy, Washington, December 2006.

U.S. Department of Energy, Washington, U.S. DOE Hydrogen Analysis Model (H2A), available at http://www.hydrogen.energy.gov/h2a_analysis.html (accessed July 2007).

U.S. Department of Energy, Washington, Multi-Year Research, Development and Demonstration Plan: Planned Program Activities for 2004–2015, updated 4/27/2007.

van Delft, Y.C., L.A. Correia, J.P. Overbeek, D.F. Meyer, P.P.A.C. Pex, J.W. Dijkstra, and D. Jansen, Hydrogen Membrane Reactor for Industrial Hydrogen Production and Power Generation, ECN-RX-05-154, *7th International Conference on Catalysis in Membrane Reactors*, Cetraro, 11–14, September 2005.

Vanderspurt, T.H., S.C. Emerson, Z. Dardas, S. Opalka, R. Radhakrishnan, S. Seiser, S. Tulyani, H. Wen, R. Willigan, W. Huang, T.B. Flanagan, and D. Wang, Advanced Water Gas Shift Membrane Reactor, *Proceedings of 2007 U.S. DOE Hydrogen Annual Merit Review Meeting*, Arlington, VA, May 2007.

Van Hook, J.P., Methane-steam reforming, *Catal. Rev. Sci. Eng.*, 21, 1–5, 1980.

Veawab, A., P. Tontiwachwuthikul, and A. Chakma, Corrosion Behavior in Sterically-Hindered Amine for CO_2 Separation, *Proceedings of Fourth International Conference on Greenhouse Gas Control technologies*, Interlaken, September 1998.

Verweij, H., Y.S. Lin, and J. Dong, Microporous silica and zeolite membranes for hydrogen purification, *MRS Bull.*, 31(10), 756–769, 2006.

Waldron, W.E. and S. Sircar, Parameteric study of a pressure swing adsorption process, *Adsorption*, 6, 179–188, 2000.

Waldron, W.E., J.R. Hufton, and S. Sircar, Production of hydrogen by cyclic sorption enhanced reaction process, *AIChE J.*, 47(6), 1477–1479, 2001.

Ward, T.L. and T. Dao, Model of hydrogen permeation behavior in palladium membranes, *J. Membr. Sci.*, 153, 211, 1999.

Warmuzinski, K. and M. Tanczyk, Multicomponent pressure swing adsorption. Part I. Modeling of large-scale PSA installations, *Chem. Eng. Process.*, 36, 89–99, 1997.

Whysall, M. and L.J.M. Wagermans, Very Large Scale Pressure Swing Processes, U.S. Patent No. 6,210,466 B1, 2001.

Wu, J.C.S., H. Sabol, G.W. Smith, D.L. Flowers, and P.K.T. Liu, Characterization of hydrogen-permselective microporous ceramic membranes, *J. Membr. Sci.*, 96(3), 275–287, 1994.

Xiong, R., J. Ida, and Y.S. Lin, Kinetics of carbon dioxide sorption on potassium-doped lithium zirconate, *Chem. Eng. Sci.*, 58, 4377–4385, 2003.

Xu, J., D.L. Rarig, T.A. Cook, K.K. Hsu, M. Schoonover, and R. Agrawal, Pressure Swing Adsorption Process with Reduced Pressure Equalization Time, U.S. Patent 2003/0015091 A1, 2003.

Yang, R.T., *Gas Separation by Adsorption*, Butterworth Publishers, Stoneham, MA, 1987.

Yasuda, I., T. Tsuneki, and S. Shiraski, Development of Membrane Reformer System for Highly-Efficient Hydrogen Production from Natural Gas, *World Conference on Wind Energy, Renewable Energy, Fuel Cell* (WCWRF 2005), Hamamatsu, Japan, June 2005.

Young, J.S., CO_2 Separations using High-Temperature Polymeric-Metallic Composite Membranes, *2nd Annual Conference on Carbon Sequestration*, Alexandria, VA, May 2003.

Zolandz, R.R. and G.K. Fleming, Gas permeation, in *Membrane Handbook*, eds. W.S.W. Ho and K.K. Sirkar, Van Nostrand Reinhold, New York, 17, 1992.

Section II

Transportation and Storage of Hydrogen

9

Targets for Onboard Hydrogen Storage Systems: An Aid for the Development of Viable Onboard Hydrogen Storage Technologies

Sunita Satyapal and George J. Thomas

CONTENTS

9.1 Introduction

The transportation segment of the U.S. economy relies heavily on petroleum, accounting for more than two-thirds of all the oil used in the nation [1]. More than 50% of the total petroleum consumption in the United States is imported and this percentage is projected to increase to about 60% by 2025 [2]. It is clearly essential that technology options, complemented by policy approaches, to the petroleum-based transportation system be developed and successfully commercialized to transform the transportation infrastructure in the United States.

The challenges in this process are significant. Gasoline-fueled (and diesel-fueled) vehicles powered by internal combustion engines (ICEs) won the market competition over other motive systems and their performance has since been improved and refined for more than 100 years. Furthermore, a vast infrastructure has been established that manufactures, maintains, and fuels the current vehicular fleet in the United States.

How then does an alternative approach gain a foothold in such an established and competitive market? First, it should offer, at a minimum, performance comparable to current vehicles and, ideally, should result in improved performance over current vehicles. The term "performance" encompasses a number of properties and attributes that will be discussed in detail in this chapter. Hydrogen fuel-cell powered electric vehicles do have the potential to offer a number of performance improvements over ICEs, such as better fuel efficiency and quiet operation, but one of the key technical obstacles is onboard hydrogen storage.

In the United States, President Bush launched the Hydrogen Fuel Initiative in 2003, pledging $1.2 billion over 5 years (from the fiscal year 2004 to 2008) to accelerate the research, development, and demonstration of hydrogen and fuel-cell technologies. The increased funds, administered primarily by the U.S. Department of Energy (DOE) complemented substantial private investment by numerous companies in the development of a fuel-cell industry. Starting in about 2004, significant effort at an unprecedented level has been initiated in hydrogen storage, among other hydrogen and fuel-cell-related technologies. The rapid progress in novel hydrogen storage research and development (R&D) is particularly noteworthy because funding before the initiative was primarily focused on onboard reforming of gasoline. Both industry and government funding was based on the assumption that early market penetration would be through hydrogen fuel-cell vehicles that could directly utilize gasoline at the present gasoline refueling stations. This would allow more time for hydrogen production, delivery and storage technologies to be developed, and gradually replace the existing gasoline infrastructure. However, onboard fuel processing of a complex hydrocarbon such as gasoline within the constraints of onboard vehicular operation—performance, weight, volume, start-up time, cost, durability, etc.—without producing contaminants that could poison proton exchange membrane (PEM) fuel cells proved to be too challenging. Furthermore, the overall energy efficiency and greenhouse gas emissions are not as favorable compared to direct hydrogen fuel-cell vehicles. With increased urgency in reducing imported oil and resolving environmental concerns, the development efforts were shifted to direct hydrogen fuel cells by about 2004. The DOE issued a formal "no-go" decision based on independent technical reviews and terminated funding for onboard fuel processing of gasoline. From this point onward, one of the most pressing technical challenges for fuel-cell vehicles was the onboard storage of hydrogen.

In an effort to guide R&D activities in hydrogen storage, the DOE established a comprehensive set of targets describing the performance and cost requirements to help enable the commercial viability of hydrogen-fueled vehicles. These targets were developed through FreedomCAR, a collaborative effort between the DOE and the U.S. Council for Automotive Research (USCAR) partners—DaimlerChrysler Corporation, Ford Motor Company, and General Motors Corporation [3]. The "CAR" in FreedomCAR stands for Cooperative Automotive Research and one of the goals of the partnership is to examine and advance collaborative R&D of technologies to enable high-volume production of affordable hydrogen fuel-cell vehicles and the national hydrogen infrastructure to support them. FreedomCAR was expanded in 2003 to include the energy companies—BP America, Chevron Corporation, ConocoPhillips, Exxon Mobil Corporation, and Shell Hydrogen (United States)—and has since been called the FreedomCAR and Fuel Partnership [4]. The targets were originally developed in the 2002–2003 time frame by technical representatives of DaimlerChrysler, Ford Motor Company, and General Motors after extensive consultation with automotive engineers, DOE and DOE National Laboratory scientists and engineers as well as others with knowledge or expertise in this area.

The targets were derived from a consideration of design approaches, market requirements, and the "typical" performance of vehicles at that time. They are market driven and not a technology "push;" in fact, the targets represent a challenging goal for scientists and

TABLE 9.1

Summary of Market-Driven Performance Expectations and Their Relationship to Specific Hydrogen Storage System Properties

Vehicle Parameter	Expected Performance	Hydrogen Storage Property
Driving range	Greater than 300 mi.	High-energy storage capacity
Acceleration/handling	Similar to or better than current ICE vehicles	Minimal weight (high gravimetric energy density) Rapid hydrogen release kinetic properties
Safety	Meets all the current codes and standards	Codes and standards to be developed for hydrogen fuel usage No environmental, health, or safety issues
Passenger/trunk space	No intrusion into current passenger and trunk volumes	Minimal volume (high volumetric energy density) Conformability
Initial cost	Comparable to current ICE vehicles	Storage system cost similar to gas tank Minimal weight increment to overall vehicle
Reliability/longevity	Consistent performance over life of vehicle	Durability, cycle life, and minimal performance loss due to contaminants and a range of operating conditions over time
Refueling time	About 3 min	Rapid kinetic properties
Fuel operating cost	Comparable to current ICE vehicles	Energy-efficient hydrogen storage
Operating temperature range	Ambient temperature can range from −40°C to 60°C	Operational over ambient temperature range, including high solar loads while parked, cold starts, or high-altitude conditions

engineers to achieve. Table 9.1 lists some of the performance parameters of current ICE vehicles and shows how vehicle performance relates, qualitatively, to specific hydrogen storage system requirements in a hydrogen fuel-cell powered vehicle.

9.2 Basis for Target Development

System-level targets were developed in six broad categories of performance: (1) hydrogen capacity; (2) cost; (3) durability/operability; (4) hydrogen charging/discharging rates; (5) fuel quality; and (6) environmental, safety, and health. Table 9.1 shows the DOE's specific technical targets, the current focus being for 2010. The targets are subject to change as more is learned about system-level requirements and as progress in fuel-cell, power electronics, and materials technology is made. In addition, limited trade-offs between targets are being explored. As targets are updated, if and when necessary, information will be available at www.eere.energy.gov/hydrogenandfuelcells/.

Table 9.2 shows examples of the DOE targets currently in use to guide the R&D community toward the development of viable hydrogen storage technologies. Although longer-term targets have also been developed (e.g., for the 2015 time frame) to enable mass market

TABLE 9.2

Technical Targets for Onboard Hydrogen Storage Systems

Storage Parameter	Units	2010	2015
System gravimetric capacity. Usable, specific energy from H_2 (net useful energy/max system mass)[a]	kWh/kg (kg H_2/kg system)	2 (0.06)	3 (0.09)
System volumetric capacity. Usable energy density from H_2 (net useful energy/max system volume)	kWh/L (kg H_2/L system)	1.5 (0.045)	2.7 (0.081)
Storage system cost[b] (and fuel cost)[c]	\$/kWh net (\$/kg H_2) \$/gge at pump	4 (133) 2–3	2 (67) 2–3
Durability/operability			
Operating ambient temperature[d]	°C	−30/50 (sun)	−40/60 (sun)
Minimum/maximum delivery temperature	°C	−40/85	−40/85
Cycle life (one-fourth tank to full)[e]	Cycles	1000	1500
Cycle life variation[f]	% of mean (min) at % confidence	90/90	99/90
Minimum delivery pressure from storage system; FC = fuel cell, ICE = internal combustion engine	atm (absolute)	4FC/35 ICE	3FC/35 ICE
Maximum delivery pressure from storage system[g]	atm (absolute)	100	100
Charging/discharging rates			
System fill time (for 5 kg)	min	3	2.5
Minimum full-flow rate	(g/s)/kW	0.02	0.02
Start time to full flow (20°C)[h]	s	5	5
Start time to full flow (−20°C)[h]	s	15	15
Transient response 10%–90% and 90%–0%[i]	s	0.75	0.75
Fuel purity (H_2 from storage)[j]	% H_2	99.99 (Dry basis)	
Environmental, health, and safety		Meets or exceeds applicable standards	
Permeation and leakage[k]	Scc/h		
Toxicity	—		
Safety	—		
Loss of usable H_2[l]	(g/h)/kg H_2 stored	0.1	0.05

Note: 0.2778 kWh/MJ, ~33.3 kWh/gge.

[a] Generally, the "full" mass (including hydrogen) is used, for systems that gain weight the highest mass during discharge is used.

[b] 2003 U.S.\$; total cost includes any component replacement if needed over 15 years or 150,000 mi. life.

[c] 2005 U.S.\$; includes offboard costs such as liquefaction, compression, and regeneration; 2015 target based on H_2 production cost of \$2–3/gge untaxed, independent of the production pathway.

[d] Stated ambient temperature plus full solar load; no allowable performance degradation from −20°C to 40°C. Allowable degradation outside these limits is to be determined.

[e] Equivalent to 100,000; 200,000; and 300,000 mi., respectively (the current gasoline tank specification).

[f] All targets must be achieved at the end of life.

[g] For delivery to the storage system, in the near term, the forecourt should be capable of delivering 10,000 psi (700 bars) compressed hydrogen, liquid hydrogen, or chilled hydrogen (77 K) at 5,000 psi (350 bars). In the long term, it is anticipated that delivery pressures will be reduced to between 50 and 150 atm for solid-state storage systems based on the current knowledge of sodium alanates.

[h] Flow must initiate within 25% of target time.

[i] At operating temperature.

[j] The storage system will not provide any purification, but will receive incoming hydrogen at the purity levels required for the fuel cell. For fuel-cell systems, purity meets SAE J2719, Information Report on the Development of a Hydrogen Quality Guideline in Fuel Cell Vehicles. Examples include total nonparticulates, 100 ppm; H_2O, 5 ppm; total hydrocarbons (C_1 basis), 2 ppm; O_2, 5 ppm; He, N_2, Ar combined, 100 ppm; CO_2, 1 ppm; CO, 0.2 ppm; total S, 0.004 ppm; formaldehyde (HCHO), 0.01 ppm; formic acid (HCOOH), 0.2 ppm; NH_3, 0.1 ppm; total halogenates, 0.05 ppm; maximum particle size, <10 μm; particulate concentration, <1 μg/L H_2. These are subject to change. See Appendix on Hydrogen Quality of DOE Multiyear Research, Development and Demonstration Plan (www.eere.energy.gov/hydrogenandfuelcells/mypp/) to be updated as fuel purity analyses progress. Note that some storage technologies may produce contaminants for which effects are unknown; these will be addressed as more information becomes available.

[k] Total hydrogen lost into the environment as H_2; relates to hydrogen accumulation in enclosed spaces. Storage system must comply with CSA America's Hydrogen Gas Vehicle (HGV) fuel containers draft (CSA/HGV2) standards for vehicular tanks. This includes any coating or enclosure that incorporates the envelope of the storage system.

[l] Total hydrogen lost from the storage system, including leaked or vented hydrogen; relates to loss of range.

penetration of hydrogen fuel-cell vehicles across all vehicle makes and models, the focus of this chapter will be on the targets set for the 2010 time frame that represent what is required for at least some level of commercially viable market penetration for certain vehicle platforms.

9.2.1 Hydrogen Storage System Capacity: Weight and Volume

One of the most widely quoted DOE targets is hydrogen storage system capacity, primarily in weight percent (wt%). This target defines the net usable specific energy from the standpoint of the total onboard storage system, not just the storage medium. The storage system includes the media; any catalyst; the storage tank; interfaces with the refueling infrastructure; safety features; any required insulation or shielding; all necessary temperature/humidity management equipment; any regulators, electronic controllers, and sensors; all onboard conditioning equipment necessary to store the hydrogen (compressors, pumps, filters, etc.) as well as mounting hardware and delivery piping.

The term "net usable energy" is used to account both for unusable fuel (e.g., hydrogen left in a tank below the minimum power train system pressure requirement) and for hydrogen-derived energy used to extract the hydrogen from the storage medium (e.g., fuel used to heat a hydride to initiate or sustain hydrogen release). The term "maximum system mass" in the denominator implies that all of the equipment enumerated earlier plus the maximum charge of hydrogen are included in the calculation. Furthermore, reactive systems can increase in mass as they discharge hydrogen because of reaction products. For example, the metal oxides or hydroxides formed in hydrolysis-based systems are heavier than the corresponding metal hydrides and they also may be larger in volume. In such systems, the discharged mass should be used in calculating energy densities.

The targets are based on a reference defined as the U.S.-weighted average corporate vehicle (WACV) that includes minivans, light trucks, economy cars, and sports utility vehicles (SUVs)/crossover vehicles in proportion to their sales. The wt% target is derived directly from the WACV that weighs 1740 kg and has a fuel capacity of 19.8 gal. This "average" vehicle has a range of 370 mi. (fuel economy of 18.7 mpg). The "average" fuel system has a mass of about 74 kg including 55.4 kg of gasoline. Clearly, most of the fuel system mass is of the gasoline. The energy in this fuel totals 659 kWh (33.3 kWh/gal. × 19.8 gal.), and the resultant specific energy is (659/74) 8.9 kWh/kg. (Note: The lower heating values of all fuels are used throughout this chapter.) If one assumes fuel economy gains of 2.5 and 3.0X, using fuel-cell vehicles, the corresponding specific energies for the average vehicle described earlier are 3.5 and 3 kWh/kg, respectively. The less stringent storage target of 3 kWh/kg (or 9 wt%) was selected as the long-term target. For 2010, some vehicles with a driving range of more than 300 mi. would require 2 kWh/kg net specific energy (or 6 wt%). Obviously, these targets include an expectation that vehicle and power train efficiency improvements will be forthcoming. Light-duty vehicles currently achieve fuel-cell efficiencies up to about 50%.

Although system wt% is the most widely quoted number in the literature, volumetric capacity (or system energy density) is, perhaps, a more important capacity target. As mentioned earlier, the onboard hydrogen storage system includes every component required to safely accept hydrogen from the delivery infrastructure, store it onboard, and release conditioned hydrogen to the power plant. Given vehicle constraints and customer requirements (i.e., driver/rider comfort, aerodynamics for fuel economy, and luggage capacity), the storage system cannot take up too much volume. The "shape factor" of the storage system also impacts the volume required to accommodate it and, as mentioned earlier, any unusable fuel must be taken into account. The targets assume both increased vehicle efficiency and the ability to store hydrogen in volumes that are currently dedicated to systems that may not be required in a hydrogen-fueled vehicle (i.e., a catalytic converter or muffler). For reference, the WACV described earlier (with a roughly 20 gal. tank) has a total fuel system volume of about 107 L (including tank with vapor space, filler tube, pump, filter, fuel lines, vapor canister, valves, and mounting straps), stored energy of about 659 kWh, and a resultant usable energy density of about (659/107) 6.15 kWh/L net. For fuel economy gains of 2.5 and 3.0X, the energy density is approximately 2.5 and 2 kWh/L, respectively.

One other factor is considered in calculating the volumetric capacity target based on the fact that the present gasoline tanks are considered conformable (or conformal). Conformability requires a tank to take irregular shapes, and to fill the space available in the vehicle, but right-angle bends and inchwide protuberances are not required. For conformable fuel tanks, the required volumetric energy density may be reduced up to 20% because space not allocated for fuel storage may be used without a penalty. Because early hydrogen storage systems may not be completely conformable, an additional 20% is added to these targets to give approximately 3 and 2.5 kWh/L for fuel economy gains of 2.5 and 3.0X, respectively. The long-term target was set at 2.7 kWh/L (81 g/L) as an average. By contrast, liquid hydrogen by itself has a density of 2.35 kWh/L (~70 g/L). Given the challenge of this target and an understanding of the current status, the 2010 target was set at 1.5 kWh/L or 45 g/L and represents what is needed to at least achieve partial market penetration of hydrogen fuel-cell vehicles.

9.2.2 Cost: Storage System Cost and Fuel Cost

Although R&D is primarily focused on performance targets, cost is an inherent requirement and cannot be ignored even in the early development phase of hydrogen storage materials, catalysts, and other components that may be required. The storage system cost target refers to the total projected cost of the entire onboard hydrogen storage system when mass-produced in quantities of 500,000 units/year—typical of current automobile manufacturing. It includes all hardware and storage media, plus an amortized estimate for any components or media that would have to be replaced for the system to demonstrate a useful life of 150,000 mi. in a vehicle. The target is in units of 2003 U.S. dollars per kilowatt hour of usable energy capacity ("net usable energy" has been previously defined). The use of constant dollars is to facilitate direct comparisons. For reference, the example WACV would have a system cost of $269 and a usable capacity of 659 kWh for a resultant-specific storage system cost of $0.41/kWh. Accounting for 2.5 and 3.0X fuel economy gains, the cost becomes $1.03–1.23/kWh. Note that the cost of the *first* charge (and any additional costs associated with the first charge such as preconditioning cost) is included in the specific storage system cost, regardless of the storage method (e.g., high-pressure tanks, chemical storage, and metal hydrides). For example, if the first charge is 8 kg of H_2 at a cost of $2.5/kg (within the cost target of $2–3/kg H_2) the specific storage system cost is approximately $1.3/kWh, assuming a 3.0X fuel economy gain. Targeting for cost competitiveness in 2015, the cost target has been set at $2/kWh. For greater flexibility in the near term (2010 target), when subsidies are most likely to reduce cost to the consumer, the storage system cost target is $4/kWh. At 33.3 kWh/kg for the lower heating value of hydrogen, these system cost targets translate to $133 and $67/kg H_2, respectively, for 2010 and 2015, but note that this is for the complete storage system rather than the cost of just the fuel; hence, the terminology of dollars per kilowatt hour is preferred.

Although fuel cost is a small fraction of the total storage system cost, it is certainly an important overarching challenge. Because a kilogram of hydrogen has approximately the same energy content as a gallon of gasoline, the unit of dollars per gallon gasoline equivalent (gge) is used interchangeable with dollars per kilogram of hydrogen. The fuel cost target is defined through the hydrogen production goals that take into account the fuel economy improvements of hydrogen fuel-cell vehicles [5]. Fuel-cell vehicles were assumed to be 1.66 times more efficient than hybrid ICE vehicles and 2.4 times more efficient than conventional ICEs [6]. Based on the projected cost of gasoline in 2015 and the two values of fuel economy improvements, the cost target for hydrogen production (untaxed, delivered at the pump) was set at a range of \$2–3/kg (or per gge) H_2 [5]. It includes the costs for compression, liquefaction, delivery, and chemical recovery, as required. This target is meant to provide guidance for chemical storage systems that are regenerated offboard, where the cost of regenerating by-products must be included in the total fuel cost. The cost target is essentially independent of the production pathway (e.g., natural gas reforming, biomass, electrolysis, and photobiological). Before achieving the long-term cost target of \$2–3/kg, an approximate hydrogen cost of \$4–5/kg can be used for estimating storage system cost in the near term.

Energy efficiency is inherently included in the fuel cost. For example, if a chemical hydride system requires significant energy input for regeneration, this will be reflected in a higher cost of the fuel. Additionally, a guideline is provided specifically for energy efficiency to emphasize that systems must be energy efficient. For reversible systems, >90% energy efficiency for the energy delivered to the power plant from the onboard storage system is the goal. This means that only 10% or less (e.g., 24 kJ/mol) of the lower heating value of hydrogen may be used to discharge the hydrogen from the onboard storage system. Setting such an objective early on in the R&D phase can help eliminate a number of potentially energy-inefficient materials. If waste heat from the PEM fuel cell can be used to release hydrogen from the storage medium, more options become available. Of course, this requires operating the storage system at <80°C, the temperature at which automotive fuel cells currently operate. This is a major challenge since many high-capacity materials require higher temperatures to release the hydrogen. If a subsystem to provide higher temperatures is used (e.g., a burner), its weight and volume must be included as a part of the overall storage system. For systems regenerated offboard, the energy content of the hydrogen delivered to the automotive power plant should be >60% of the total energy input to the process, including the input energy of hydrogen and any other fuel streams for generating process heat and electrical energy. This is comparable to other approaches that include the efficiency losses for hydrogen production and delivery.

9.2.3 Durability and Operability Requirements

9.2.3.1 Temperature and Pressure

Although the R&D community tends to focus on hydrogen storage capacity, performance requirements such as durability and operability must clearly be met as well. The storage system, and hence any material (such as metal hydride) that stores the hydrogen, must dependably store and deliver hydrogen at all conditions experienced by the vehicle. These include temperatures ranging from freezing conditions (down to −40°C) to a "hot soak" condition in direct sun (up to 60°C). Less severe ambient conditions will be selected for early demonstration fleets and therefore interim targets are less stringent, but ultimately consumers will expect vehicles to operate in any weather and not be a limiting factor as they drive, for instance, from Canada to Florida.

Besides operating requirements under external conditions, there are also requirements internal to the vehicle in terms of temperature and pressure. PEM fuel cells currently operate at approximately 80°C and to maintain performance, the hydrogen entering the fuel cell should not be much too hot or too cold. Based on industry input from both the fuel-cell and hydrogen storage technical teams of the FreedomCAR and Fuel Partnership, the lowest temperature acceptable for the hydrogen exiting the storage system is −40°C, comparable with the lowest ambient temperature specification as described earlier (see Table 9.2). This means that for cryogenic hydrogen storage systems (e.g., 20 or 77 K), the hydrogen must be heated before the fuel-cell inlet. The maximum acceptable delivery temperature from the storage system is 85°C. If the temperature is much higher, significant performance and degradation losses can result because the water required for proton transport within the fuel-cell membrane would not be sufficient. High temperatures also add to the already significant heat rejection problem. Over time, a higher value such as 100°C may be substituted because fuel-cell R&D is focusing on developing membranes that can operate at a higher temperature (e.g., up to 120°C according to the DOE targets for PEM fuel cells) [7].

Targets for pressure provide critical information for material developers to tailor their materials, hydrogen-binding energies, and thermochemistry as well as to system designers to ensure adequate pressure regardless of the system state of charge. The two pressure targets related to onboard storage are the minimum and maximum pressure limits for hydrogen delivered from the storage system to the power plant. The storage system must provide a minimum pressure to the power plant simply to move the hydrogen from the onboard system to the power plant. This target is different for fuel cells and for ICEs. If only low-pressure hydrogen were available at the entrance to the power plant, any pumps necessary to push or draw that fuel through would be considered part of the fuel storage system and add to the weight and volume penalties. Based on the input from current fuel-cell developers and OEMs, the present fuel-cell systems require inlet pressures of 8 bars (absolute) to ensure that adequate performance is achieved. This is relatively high and it is expected that fuel-cell performance will improve, leading to the 2010 target of 4 and 3 bars as the longer-term target. For ICEs, an even higher pressure is needed (35 bars) because ICE technology relies on pressurized fuel injection, and is envisioned to advance from low-pressure central or port injection to high-pressure in-cylinder direct injection by 2010. The pressure targets place a limitation on the equilibrium overpressure within the operating temperature range for reversible hydrides. It may also impact the quantity of available hydrogen for a given material system.

In addition, the storage system's charging needs dictate the pressures required from the hydrogen dispensing system at a hydrogen refueling station. Since physical storage (high pressure or cryogenic hydrogen) is likely in the near term, the refueling station should be capable of delivering 10,000 psi (700 bars) compressed hydrogen or liquid hydrogen to the vehicle. If cold gas or adsorbents that require low temperatures are used, then the station must also be able to supply chilled hydrogen (77 K) at 5000 psi (350 bars). In the long term, it is anticipated that delivery pressures will be reduced to between 50 and 150 bars for solid-state storage systems, based on the current knowledge of sodium alanates.

There is also a maximum acceptable pressure, 100 bars (absolute), for the gas delivered from the onboard hydrogen storage system to the automotive power plant. This target ensures that excessive pressures are not applied to the fuel cell or connecting hardware that could result in the failure of these components. Typically, a pressure regulator is used to limit the maximum delivery pressure from the storage system and must be included in the total storage system weight and volume.

9.2.3.2 Cycle Life and Durability

Cycle life, or the ability to maintain consistent performance over many refueling cycles, is clearly another requirement for the storage system. Customers expect the fuel system to be comparable with a gasoline tank (which requires little if any maintenance in current automobiles) and to last the life of the vehicle, typically 150,000 mi, assuming a 300 mi. driving range between fill-ups, which corresponds to 500 full fill cycles. Of course, drivers generally refuel when the tank is partially rather than completely empty, which requires a higher target for the number of fill cycles. Assuming customers refuel at about one-fourth tank capacity and with an engineering factor included to ensure product reliability, the long-term target of 1500 cycles was established. Early demonstration fleets may not require this level of durability; therefore, 1000 cycles is the 2010 target. Storage material developers, however, should target material stability consistent with at least 1500 hydrogen charge and discharge cycles to meet long-term goals.

Another target, used by industry manufacturers, although not as relevant for the basic materials research community, is cycle life variation. This relates to the fact that manufactured items have item-to-item variation. It is expected that only one or two systems will be fabricated to test cycle life of early concepts. Therefore, the data generated from such a limited sample size has significant uncertainty associated with it and hence a factor is required to account for this uncertainty. The effect is to increase the required cycle life based on normal statistics using the number of samples tested. The value is given in the form XX/YY, where XX is the acceptable percentage of the target life (90 means 90%) and YY is the percent confidence that the true mean will be inside the XX% of the target life (95 indicates 95% confidence or an alpha value of 0.05). Systems beyond the R&D phase (such as high-pressure tanks) are covered in this target. It is not expected that many thousands of cycles are tested during the early materials R&D phase. However, it is expected that this target will guide technology developers away from designs that are very complex with very fine tolerances and many active parts. Variation testing needs to be included for general sales and this entire sequence is standard practice in the mass production of automobiles and their components.

9.2.4 Charging and Discharging Rates

Two essential requirements for storage systems are (1) to supply hydrogen to the fuel cell at rates sufficient to meet the power demands of the driver and (2) to absorb hydrogen fast enough to limit the time needed to refuel the vehicle. These provide important guidance to researchers related to material kinetic properties. The fill time target was designed to parallel the current consumer experience with gasoline vehicles that are filled in about 2–5 min (large vehicles with larger gas tanks clearly take longer to refuel). The target for fill rate was based on systems with 5 kg H_2 or less. Since current estimates indicate that 5–13 kg of H_2 will be needed over the entire light-duty vehicle fleet, it is acceptable for larger systems to need proportionately longer fill times. Also, early demonstration fleets could operate with longer fill times so this target is not crucial in the near term. A rapid fill also poses a significant heat rejection challenge for systems containing materials that have high hydrogen binding energies. For instance, a 5 kg system containing a hydride with a formation enthalpy ΔH of 40 kJ/mol of H_2 would have to reject a total of 100 MJ of energy to completely fill the system. If the fill time were 3 min, then the average rate of heat generated would be 550 kW, or just over half a megawatt! The added complexity in rejecting so much heat in such a short time can be an issue, although there is room for ingenuity (e.g., utilizing the waste heat at the refueling station).

One target that does not change with regard to time frame is called the "minimum full-flow rate" of hydrogen. It is critical to early vehicles as well as to the mass-market vehicles envisioned beyond 2015. The full-flow rate is related to the kinetics of hydrogen release from the storage system and while the maximum rate of hydrogen flow can be calculated from the fuel-cell power and efficiency, the target goes beyond this simple calculation.

It should not be considered only as a transient phenomenon, but rather is based on the actual measured maximum gasoline fuel flow required to achieve a desired level of performance in an average 3000 lb. current production vehicle with a power plant of about 150 kW. It corresponds to the flow that may be required for a vehicle to accelerate or tow a heavy trailer up a long grade (e.g., the 18 mi. grade, on Interstate 5 near Baker, California). The hydrogen storage target is 0.02 (g/s)/kW and is scaled to the power plant size. Thus, for a fuel-cell power plant of 80 kW, the minimum required full-flow rate of hydrogen is 1.6 g/s (about 1075 standard liters of hydrogen per minute). Because fuel-cell efficiency is poorest at full load, this is an important target. The target ensures that, whatever the motive technology, storage systems will be capable of meeting maximum power demands.

9.2.5 Start-Up Time and Transient Response for Storage Systems

During start-up of a vehicle, hydrogen must be available to the power plant at essentially full flow almost instantly. Automobile drivers are accustomed to start times of less than a second and full power is available on demand at any time immediately after start-up. Early demo fleets may not require starting times that rival current ICE technology, but once larger-scale production is started, performance similar to an ICE would be assumed. The start-up time target is based on the fuel-cell target of 5 s (at ambient temperature) as the start time needed to achieve 50% rated power for the complete fuel-cell system (2010 and 2015). The hydrogen storage target for start-up time to full flow is, thus, also 5 s (at ambient temperature). However, this need not mean that the entire storage system must start in 5 s—only that it is capable of delivering fuel at the maximum flow if requested. A small, moderate pressure buffer could serve to lengthen the true start-up time (but would add to the overall system volume and weight).

At low temperature (−20°C), the start time to full flow reflects the current customer acceptance of the fact that in cold weather, starting is more difficult and can take longer. Ideally, cold start should not require a significant energy input penalty. Although a battery assist may be an option, it is important to note that batteries are at their worst power capabilities at very low temperature. The target start time to achieve 50% rated power for the complete fuel-cell system at −20°C is 30 s (for 2010 and 2015) [7]. Hence, the storage system target for start time to full flow (at −20°C) is 15 s to ensure that sufficient hydrogen is available to the fuel cell during the difficult cold start conditions. Consistent with the preceding target, some flow will be required to the power plant within 25% of the full-flow target time. Given the possibility that some hydrogen may be used to assist with cold start of the power plant, the storage system is set to achieve full flow within 50% of the start time for the power plant.

Transient response is one of the most significant challenges a vehicle's power train faces. The storage system must track the needs of the fuel cell closely to provide adequate power and a suitable driving experience. The fuel-cell system transient response requirement is 1 s (2010 and 2015 targets) [7] and the storage system must be at least slightly more responsive to accommodate the fuel cell's needs, hence the target is 0.75 s (2010 and 2015 targets). It should be noted that the transient response is not symmetric. The 10–90% transient target is to meet the demand of the fuel cell or ICE during acceleration and means that in 0.75 s, the storage system must be able to go from 10% to 90% of full flow. The 90–0% transient reflects the fact the fuel cell can stop using hydrogen almost instantly and the fuel supply must avoid overpressurizing any part of the system. In other words, within 0.75 s, the hydrogen storage system must be able to go from 90% to 0% of hydrogen output—that is, all hydrogen flow must stop if needed within 0.75 s. This parameter impacts performance, fuel-cell durability, and vehicle control.

9.2.6 Hydrogen Quality

Polymer fuel cells currently require extremely pure hydrogen or else system performance and durability degrade. It should also be noted that, although ICEs are more tolerant to fuel impurities an exhaust after-treatment system may not be. Even inert impurities can degrade fuel-cell performance by progressively diluting the hydrogen at the anode and necessitating venting. If it is assumed that impurity levels in the hydrogen refueling source meet the requirements, then the hydrogen output from the storage system should be able to meet fuel-cell purity targets by not adding contaminants to the hydrogen.

Fuel purity requirements are being discussed by a number of groups around the world such as the Society of Automotive Engineers (SAE). The SAE J2719 Information Report on the Development of a Hydrogen Quality Guideline for Fuel Cell Vehicles provides specific examples of contaminants and acceptable concentrations [8]. It should be emphasized that the SAE report is an interim document and, as additional operational experience with fuel-cell vehicles and hydrogen production and delivery systems is gained, it will likely be modified. Currently, the target (for 2010 and 2015) is 99.99% (volume percent on a dry basis) of pure hydrogen. Examples of contaminants and concentration limits are total nonparticulates, 100 ppm; H_2O, 5 ppm; total hydrocarbons (C_1 basis), 2 ppm; O_2, 5 ppm; He, N_2, Ar combined, 100 ppm; CO_2, 1 ppm, CO, 0.2 ppm; total S, 0.004 ppm; formaldehyde (HCHO), 0.01 ppm; formic acid (HCOOH), 0.2 ppm; NH_3, 0.1 ppm; total halogenates, 0.05 ppm; maximum particle size, <10 μm; particulate concentration, <1 μg/L H_2. These are subject to change and are only provided as examples as of date. Details are also available in the DOE's Multiyear Research, Development and Demonstration Plan (www.eere.energy.gov/hydrogenandfuelcells/mypp/) in the Appendix on Hydrogen Quality [9]. In addition, the DOE has established a Hydrogen Quality Working Group with key experts from various organizations to investigate trade-offs between fuel-cell performance and durability versus the cost and efficiency of producing and delivering various grades of hydrogen.

During the current early stages of R&D, some storage technologies may produce contaminants for which effects are unknown, and these will be addressed as more information becomes available. The most obvious implication experienced by hydrogen storage material developers in the past few years is the issue of ammonia production from materials such as amides. Currently, any storage material that releases ammonia at concentrations above 0.1 ppm in the hydrogen stream would be detrimental to PEM fuel-cell systems and would not likely to be considered for use. As other materials are being developed, such as borohydride-based compounds, the impact of additional contaminants (e.g., boron based) will need to be determined on fuel-cell performance and durability. Presently, it is assumed that 99.99% pure hydrogen will enter the storage system and 99.99% pure hydrogen will exit the storage system to be delivered to the fuel-cell power plant.

9.2.7 Environmental, Health, and Safety

Environmental, health, and safety-related targets are implicit in all storage options. A key target in this category is the permissible permeation or leakage of hydrogen from the storage system. Clearly, the ideal situation is zero leakage or permeation and in all cases storage systems must meet or exceed applicable standards. As an example of the HGV2 standard for type-4 hydrogen storage tanks, 2 Scc/h (standard cubic centimeters per hour) and 2.8 Scc/h are the upper limits for 350 and 700 bars tanks, respectively [10]. Permeation and leakage do not include hydrogen that leaves the storage system and is transformed into another species (e.g., water, via catalytic oxidation in a vent line) because it pertains to the possibility of generating a combustible hydrogen–air mixture outside the storage tank. However, if lost hydrogen is converted to noncombustible species, the loss is still of relevance and is defined by another target—hydrogen loss. This target protects against the loss of range after extended periods of rest, for example, parking over a vacation period. Consumers expect to have a minimal perceptible loss of range after a week or two of parking, similar to current gasoline vehicles. Because the targets are normalized to mass of hydrogen stored (in units of grams of hydrogen loss per hour, per kilogram of hydrogen stored) this target can be used for all tank sizes. Demonstration fleets are not expected to operate extensively in the normal consumer cycle, and the drivers are better prepared to deal with low-fuel situations; thus, a lower standard is required in the early years. The 2010 target is 0.1 (g/h)/kg H_2 stored. However, to truly meet consumer expectations, the longer-term target is 0.05 (g/h)/kg H_2 stored. It should be noted that hydrogen lost via all routes, whether it is leakage and permeation of pure hydrogen or lost hydrogen that is converted to a noncombustible species, is included in this target.

Toxicity covers the possibility of consumer exposure to the storage material in normal, or abnormal conditions as well as worker exposure during manufacture and assembly. In addition to consideration of the toxicity of storage materials alone, any by-products that may be produced under certain conditions (such as diborane, from boron-based species) must be considered. Material toxicity will clearly be a factor in the downselection of potential materials that may be used in hydrogen storage systems. Just as one example, although beryllium is quoted in theoretical studies as an attractive light element to enable high-capacity storage compounds; its toxicity makes it unlikely to be considered as a commercially viable option. Safety covers all the typical safety statutes including certification and operation of vehicles, manufacture, transport, dispensing of fuel, and end-of-life issues. In each of these categories, compliance with federal standards and potentially state and local standards will be required.

9.3 Current Status Compared to Targets

Figure 9.1 shows the current status of hydrogen storage systems in comparison to the capacity targets and includes R&D data and modeled estimates provided by developers as well as cost estimates [10]. The plot is available on the DOE Web site and is periodically updated by the DOE as more system data become available. Figure 9.1 also includes tank data validated through the DOE's "learning demonstration" effort that includes more than 70 hydrogen-fueled vehicles and 14 fueling stations operating to date [11]. Based on the operation of these vehicles under real-world conditions, a driving range of 103–190 mi. has been achieved thus far (assuming an Environmental Protection Agency (EPA) drive cycle). It is evident that none of the current vehicular hydrogen storage systems meet the combined gravimetric, volumetric, and cost targets for either 2010 or 2015. Recent system developments suggest that, depending on the material and on-the-system design, material capacities may need to be a factor of 1.2–2 times higher than system capacity targets.

9.4 Conclusion

The DOE targets for hydrogen storage, as developed through the FreedomCAR and Fuel Partnership, represent the first truly comprehensive targets for onboard vehicular hydrogen storage systems. Rather than being technology driven and based on the current state of the

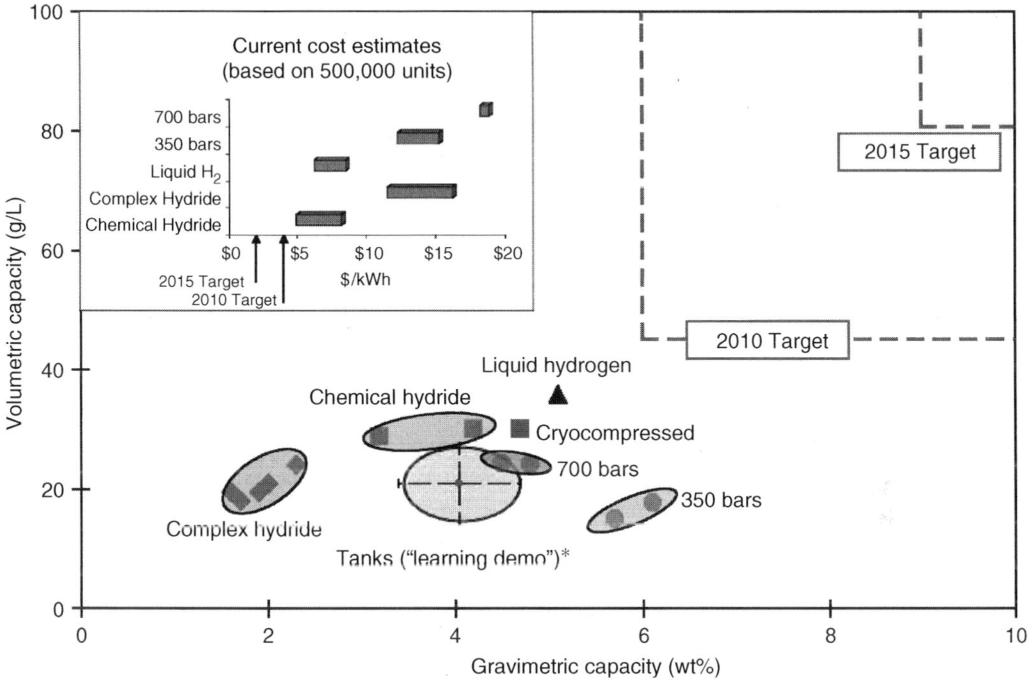

FIGURE 9.1
Current status of hydrogen storage systems and future targets. (Cost exclude regeneration/processing. Data based on R&D projections and independent analysis (FY05–FY06). To be periodically updated by the DOE.)
*Learning demo refers to DOE's technology validation activity with tank delta from >70 vehicles to date.

art, the targets are based on consumer expectations for commercially viable passenger vehicles. Through significant effort and the cooperation of technical experts in hydrogen storage from DaimlerChrysler, Ford Motor Company, and General Motors working with national laboratory experts, the operating requirements for hydrogen-fueled vehicles were translated into specific targets for the performance parameters of hydrogen storage systems.

Clearly, some of these targets are extremely challenging. However, a revolutionary shift from the established gasoline-based vehicular infrastructure to an environmentally beneficial hydrogen-fueled economy requires more than just the expected environmental benefits to gain acceptance. The performance and cost of fuel-cell vehicles must also meet consumer expectations to encourage industrial investment in the fuel and vehicular infrastructure. In the past couple of years, significant progress has been made and a number of new materials and approaches have been developed [12]. However, there is yet no technology that meets all the targets. The global R&D community is encouraged to continue to reach for innovative, high-risk–high pay-off concepts for developing energy-efficient, safe, and cost effective onboard hydrogen storage systems.

References

1. Oak Ridge National Laboratory, Transportation Energy Data Book: Edition 26, Table 1.13 Consumption of Petroleum by End-Use Sector, 1973–2006, pp. 1–17, http://cta.ornl.gov/data/tedb26/Edition26_Chapter01.pdf (2007).

2. U.S. Department of Energy, *Energy Information Administration*, Annual Energy Outlook 2006: With Projections to 2030, DOE/EIA-0383 (2006), p. 8, http://www.eia.doe.gov/oiaf/aeo/pdf/0383(2006).pdf (February 2006).
3. United States Council for Automotive Research (USCAR), http://www.uscar.org/guest/index.php (2003).
4. FreedomCAR and Fuel Partnership, http://www1.eere.energy.gov/vehiclesandfuels/about/partnerships/freedomcar/index.html (2004).
5. DOE, *Hydrogen Program Record 5013, Hydrogen Cost Goal*, http://hydrogen.energy.gov/pdfs/5013_h2_cost_goal.pdf (December 2005).
6. The Hydrogen Economy: Opportunities, Costs, Barriers and R&D Needs, Committee on Alternatives and Strategies for Future Hydrogen Production and Use, National Research Council and National Academy of Engineering (2004), pp. 26, 66.
7. DOE, *Hydrogen, Fuel Cells and Infrastructure Technologies Multiyear Research, Development and Demonstration Plan*, Chapter 3.4, Fuel Cells, http://www1.eere.energy.gov/hydrogenandfuelcells/mypp/pdfs/fuel_cells.pdf (revised April, 2007).
8. The SAE J2719, Information Report on the Development of a Hydrogen Quality Guideline for Fuel Cell Vehicles, http://www.sae.org/technical/standards/J2719_200511 (November, 2005).
9. DOE, *Hydrogen, Fuel Cells and Infrastructure Technologies Multiyear Research, Development and Demonstration Plan*, Appendix C—Hydrogen Quality, http://www1.eere.energy.gov/hydrogenandfuelcells/mypp/pdfs/appendix_c.pdf (revised April, 2007).
10. Status of Hydrogen Storage Technologies, U.S. Department of Energy, *Hydrogen, Fuel Cells and Infrastructure Technologies*, website at http://www1.eere.energy.gov/hydrogenandfuelcells/storage/tech_status.html (2007).
11. DOE, National Hydrogen Learning Demonstration, U.S. Department of Energy, *Hydrogen, Fuel Cells and Infrastructure Technologies*, website at http://www1.eere.energy.gov/hydrogenand fuelcells/tech_validation/fleet_demonstration.html (2007).
12. S. Satyapal, J. Petrovic, C. Read, G. Thomas, and G. Ordaz. The U.S. Department of Energy's National Hydrogen Storage Project: Progress towards meeting hydrogen-powered vehicle requirements, *Catalysis Today*, 120, 246 (2007).

10

Hydrogen Transmission in Pipelines and Storage in Pressurized and Cryogenic Tanks

Ming Gao and Ravi Krishnamurthy

CONTENTS

10.1 Introduction

There are three primary options [1] for hydrogen transport: (1) pipelines and tube trailers for gaseous hydrogen or a mixture of hydrogen and natural gas transportation; (2) trucks, rails, barges, and ships with cryogenic tanks for liquefied hydrogen delivery; and (3) high-energy-density carriers such as ethanol, methanol, and other liquids derived from renewable biomass that can be transported and reformed to hydrogen at the end point of use.

Studies have shown that a pipeline network infrastructure is the most cost-effective and energy-efficient method to transport hydrogen in large quantities over long distance—as

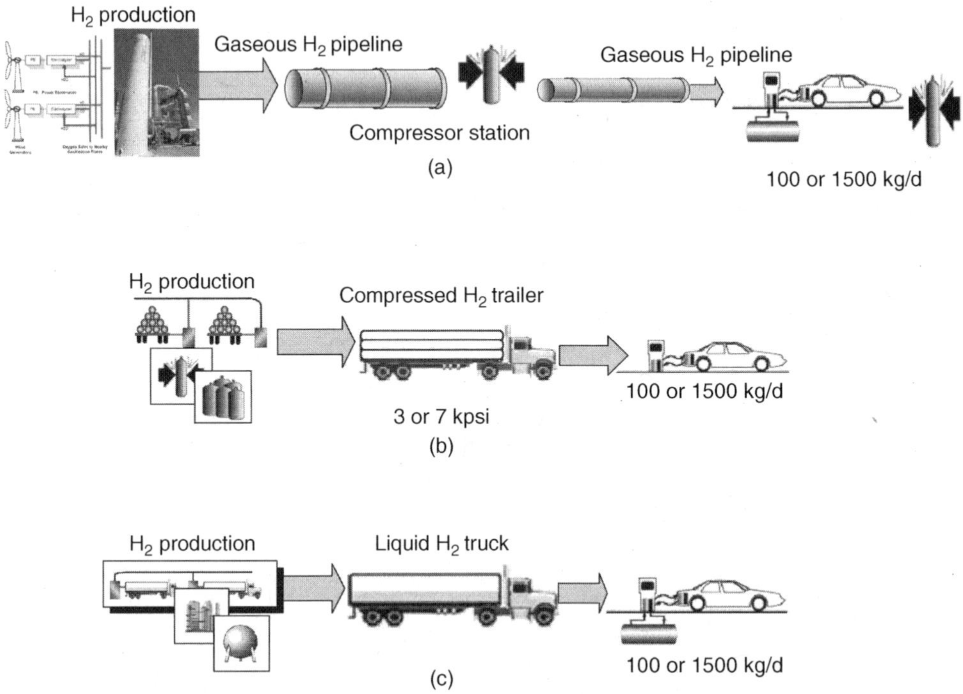

FIGURE 10.1
Illustration of hydrogen delivery methods by (a) pipeline, (b) gas tube trailer, and (c) cryogenic liquid tank truck. (Adapted from Ringer, M., Mintz, M., Gillette, J., Elgowairy, A., Brown, D., and Paster, M., Cost installing and operating hydrogen pipelines, *Second Panel Forum—Hydrogen Pipeline Transmission*, IPC2006, Calgary, September 2006.)

is done with natural gas currently [1–6]. However, other methods, such as the use of gas tube trailers, cryogenic liquid tank delivery, or carriers, will be needed for the initial transportation market, and may also play a long-term role for hydrogen delivery. Figure 10.1 illustrates hydrogen delivery by pipelines, gas tube trailers, and cryogenic liquid tanks [7]. Different delivery approaches may be used to best handle the inherent difference in delivery distances (short and long) and difference between high-population-density urban areas and low-density rural areas, and interstate refueling stations. In any event, safe, cost-effective, and energy-efficient delivery technologies are needed.

10.2 Pipelines for Hydrogen Transport

Currently, there is a small hydrogen pipeline transport infrastructure, approximately 700 mi., in the United States. These pipelines predominantly serve petroleum refineries and chemical plants in the Gulf Coast region, with some additional pipelines in southern California and Chicago [2]. The success and safety in operating these pipelines is encouraging, but they are relatively short, small diameter, and operating at low and constant pressures. They are not subject to the technical demands of centralized and large-volume hydrogen production, and the economic challenge of delivering low-volumetric-energy-density gaseous hydrogen over hundreds or thousands of miles to compete with other hydrogen sources at the destination. Figure 10.2 illustrates hydrogen transmission via

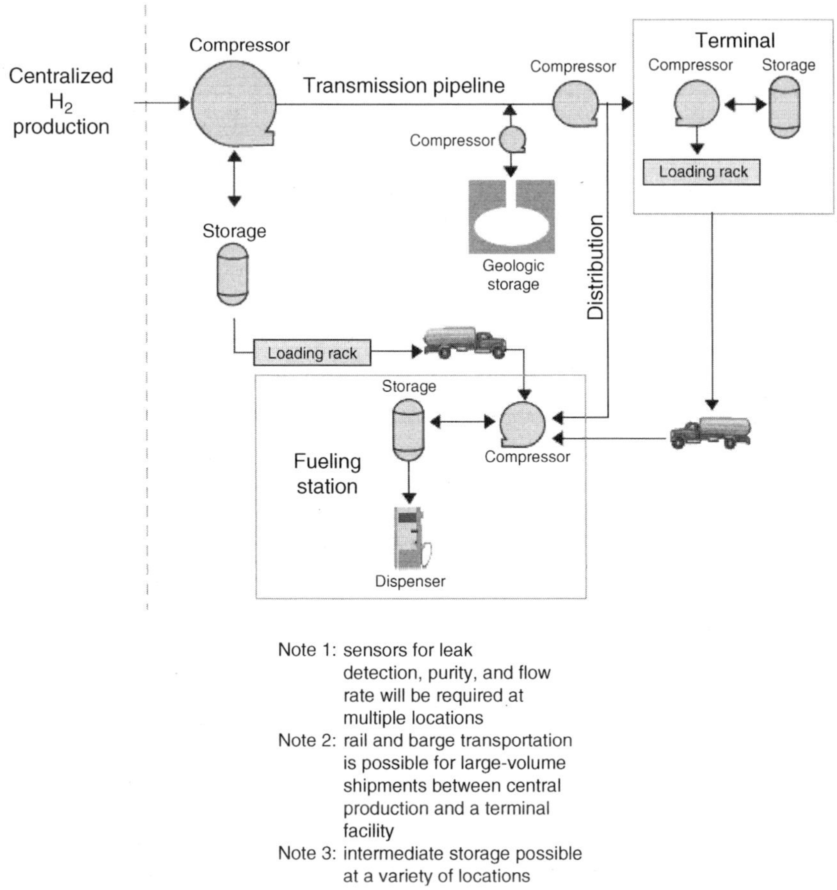

FIGURE 10.2
Gaseous hydrogen delivery pathway via pipelines and tube trailers. (After U.S. Department of Energy: Hydrogen, fuel cells and infrastructure technologies program multi-year research, development and demonstration plan, Section 3.2, Hydrogen Delivery, January 21, 2005.)

pipelines from the centralized hydrogen production to the terminals or storage tanks, then distributing hydrogen to the fueling stations for dispensing with the aid of tube trailers [1]. Various compressors and sensors are used for pressurizing hydrogen along the pipelines between the centralized production plant and terminals/storage tanks. There are four major issues for gaseous hydrogen delivery via pipelines [8]:

1. *High initial capital investment costs.* Pipeline transmission of hydrogen offers important technical and economic advantage and synergies as compared to other methods; however, new pipeline design and construction imply high initial capital costs. This is a major barrier for building new pipelines, requiring extensive research efforts for cost reduction and technical justifications. This has resulted in a large research and development (R&D) program under the U.S. Department of Energy's (DOE's) Multi-Year Research, Development and Demonstration Plan and other private foundations supporting programs [3].

   Transporting gaseous hydrogen via the existing pipelines is the low-cost option for delivering large volumes of hydrogen. Converting natural gas pipelines to carry a blend of natural gas and hydrogen (up to about 20% hydrogen) may require only modest modifications to the pipeline; however, more substantial modifications may be required for delivering pure hydrogen. Analyzing, investigating, and developing technologies for using the existing natural gas pipelines for transporting hydrogen–natural gas mixture and for upgrading natural gas pipelines for pure hydrogen are also part of the DOE program.

2. *Material challenge—hydrogen embrittlement of steel pipelines and alternatives.* Hydrogen delivery via pipelines would require the gaseous hydrogen to be transmitted under very high pressure levels (up to 3000 psi) and the use of economically viable materials such as pipeline steels. Hydrogen embrittlement (HE) of steels at such high pressures is not completely understood. Current welding technology for pipeline impacts the steel microstructure that can exacerbate HE issues. Strategies for elimination or mitigation of HE is the key to ensure the safe, cost-effective, and long-term reliability of the hydrogen delivery pipeline infrastructure [9]. The use of composites pipelines, as recently introduced for natural gas for gathering at wellheads, may be an alternative to resolve these issues and is being studied in parallel to pipeline steels.
3. *Hydrogen leakage and integrity monitoring sensors.* The hydrogen molecule is very small and diffuses rapidly compared with other gases such as natural gas. This makes it more challenging to materials, seals, valves and fittings, and equipment design to avoid leakage. The delivery pipelines for hydrogen use will not only rely on materials and robust designs and engineering but also heavily on sensors. Low-cost leak detection sensors are needed. Suitable odorant technology for hydrogen leak detection may be needed for hydrogen distribution pipelines. Use and further development of mechanical integrity sensors that can be built into pipelines and vessels are essential for protection against third-party damage-related and other mechanical failures [1].
4. *Hydrogen compression.* Compression of natural gas is a well-established technology. Utilizing natural gas compression technologies for hydrogen is unreliable because (a) the hydrogen molecule is much smaller and lighter than natural gas and (b) gaseous hydrogen contains only one-third the energy of natural gas by volume (13 MJ versus 40 MJ/Nm^3). These create significant challenges for hydrogen compression. For example, centrifugal compression is the lowest-cost approach for natural gas compression, but it becomes very difficult to compress hydrogen with its low molecular weight. This requires up to 60 stages of centrifugal compression of hydrogen as compared with four to five stages for natural gas. As a consequence, hydrogen transmission requires significantly greater power and energy for compressors. Also, lubricants used in normal compression application can result in unacceptable contamination problem for fuel cell use. More reliable, low-cost, and more efficient compression technologies for pipelines are needed. Alternatively, compressorless hydrogen transmission pipelines are being explored based on the current technology (up to 100 bars output) of high-pressure-output electrolyzers. For pressure output higher than 100 bars, further R&D effort is needed.

Detailed discussions on each of the four issues are given in Sections 10.2.1 through 10.2.5.

10.2.1 Economical Analyses

A complete hydrogen pipeline infrastructure would include both transmission and distribution to minimize the overall hydrogen transport cost [6]. When one considers the cost for pipeline infrastructure, there will be some semioptimized operating pressure range that determines the overall costs, including the cost of compression based on the distance and volume that hydrogen will need to be transported, the cost for compression (based on the pressure needs) at refueling sites, and the cost for safety measures. Cost models, such as H2A Delivery Components and Scenarios Model [10], have been developed for the analyses. The initial analysis efforts based on H2A Models [10] suggested that line sizes nominally similar to natural transmission and distribution line sizes with line pressures on the order of 31–150 bars (500–2200 psi) may be advantageous for minimizing cost. Safety considerations, codes, and standards might reduce the pressure allowed in urban distribution line. Current urban regulations restrict natural gas distribution line pressure to 8.5 bars (125 psi) for nonindustrial users. It is desirable to develop sufficient information and technology to permit distribution lines delivering gaseous hydrogen at higher pressures. More analysis is required to better understand the cost and other trade-offs for a hydrogen pipeline infrastructure before a semioptimized pressure range can be identified.

Examples of using H2A Delivery Components and Scenarios Models are illustrated in Figures 10.3 and 10.4 [11]. Figures 10.3 and 10.4 show representative results for hydrogen delivery to urban markets of two different population sizes: 250,000 and 1,000,000 people. Since urban markets will require sufficient hydrogen, that the central large production plants will be located within a few 100 km from the city. Therefore, Figures 10.3 and 10.4 assumed the central plants to be located 100 km (62 mi.) from the edge of the city. The analyses were based on the costs of 2005 for delivery technology and provide a cost comparison of delivery technology between pipeline, gas tube trailer, and cryogenic liquid truck. Implications of Figures 10.3 and 10.4 are

- At <10% market penetration, delivery costs from a central plant located some distance from the city are very high. During this period, local production will likely play an important role. Another alternative would be to utilize gaseous tube trailer delivery from nearby central hydrogen production facilities if such facilities exist near or in that urban area. This could be cost-effective if the ultimate tube trailer carrying capacity target of 1100 kg could be achieved.

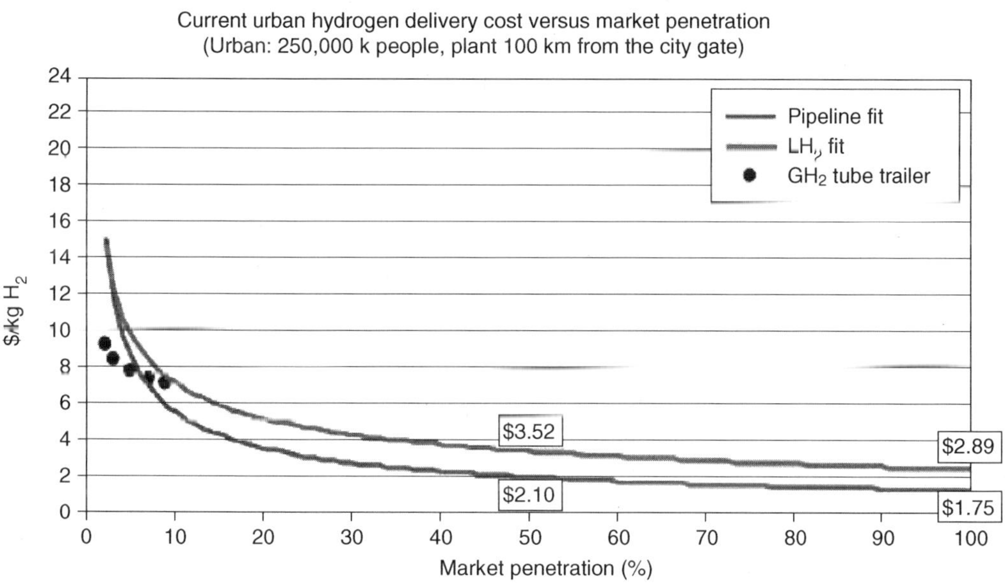

FIGURE 10.3
H2A delivery scenario model for H_2 delivery to urban market (250,000 people).

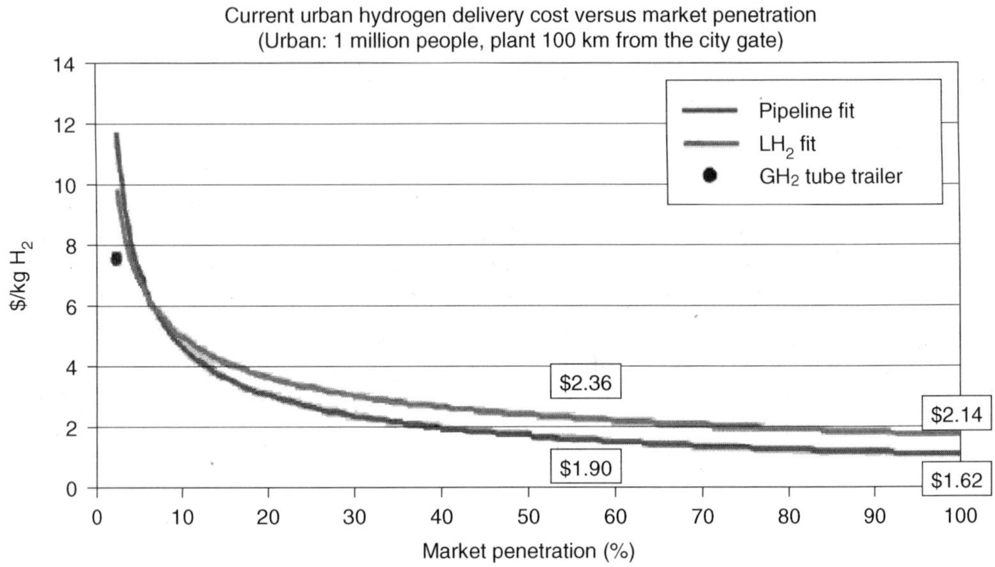

FIGURE 10.4
H2A delivery scenario model for H_2 delivery to urban market (1 million people).

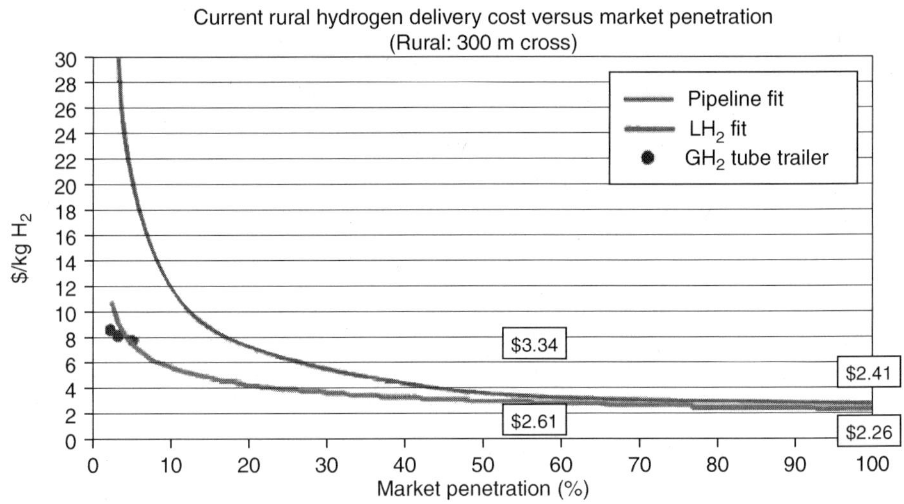

FIGURE 10.5
H2A delivery scenario model for rural/interstate case.

- The cost of pipeline delivery drops rapidly as market penetration increases and tends to flatten out above 25%. The difference in delivery cost for a city of 1,000,000 and 250,000 people is not large. Although not shown, the costs for delivery for cities <250,000 people do start to increase significantly.
- Pipelines are the low-cost delivery pathway with current technology when the market penetration is >10%.

For intercity/rural markets, again, the costs are a function of hydrogen demand. Note that delivery cost tends to be higher and pipelines tend to be less competitive in this market (Figure 10.5).

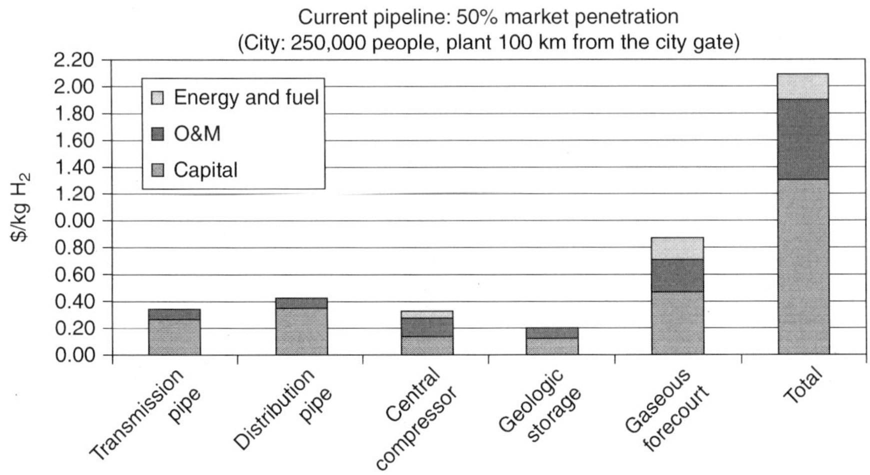

FIGURE 10.6
Current delivery costs for pipeline pathway.

A breakdown of the current delivery costs for pipelines for an urban market of 250,000 people is shown in Figure 10.6. The forecourt/refueling site cost represents a large fraction of the total delivery cost (40%). This is predominantly the cost of compression and storage at the refueling site and highlights the importance of these areas for cost reduction. The reminder of the cost is split relatively evenly over the cost of the other parts of the pipeline infrastructure. This means that some cost reduction is needed in all of these areas.

The effect of compression and storage cost at the refueling site on the total cost of pipeline infrastructure may also impact the cost advantage of converting an existing oil and natural gas pipeline to transport hydrogen. It has been shown that the cost saving from such conversion is very marginal because the saving from utilizing the existing transmission pipeline is only a small portion of the total cost of hydrogen pipeline infrastructure [12] (Figure 10.7).

To reduce the cost, the economical advantage of using compressorless transmission pipelines deliver large-scale stranded renewable energy, mainly wind power, is being explored [5,13]. The economical analysis is based on a 1000 MW (1 GW) wind plant in the large wind resource of the Great Plains of North America, exclusively delivering hydrogen fuel via a *new* gaseous hydrogen pipeline to an urban market at least 300 km distant and all electric energy produced by wind power to be converted at the wind plant to hydrogen via high-pressure (100 bars) electrolyzers, and directly feeding the gaseous hydrogen transmission pipeline without compressor stations at inlet or at midline.

The analysis showed, with various "value adders" (e.g., oxygen sales and carbon-emission-offset credits), the cost of *wind-source gaseous hydrogen* delivered by pipelines from production point to distant markets (about 200–1000 mi.) at an untaxed wholesale energy unit cost will be competitive with market prices (in 2005) of gasoline and hydrogen fuel made from natural gas by steam methane reforms (SMRs).

Overall, the DOE has set a technical and cost target for hydrogen delivery via pipelines. With the current total capital investment for transmission pipelines being $700,000/mi. for a 16 in. pipe, the target cost is $600,000/mi. for 2012, and $490,000/mi. for 2017. For distribution lines it is $320/mi. for 2005, $270 for 2012, and $190/mi. in 2017, on the completion of milestones of the DOE's R&D programs [1].

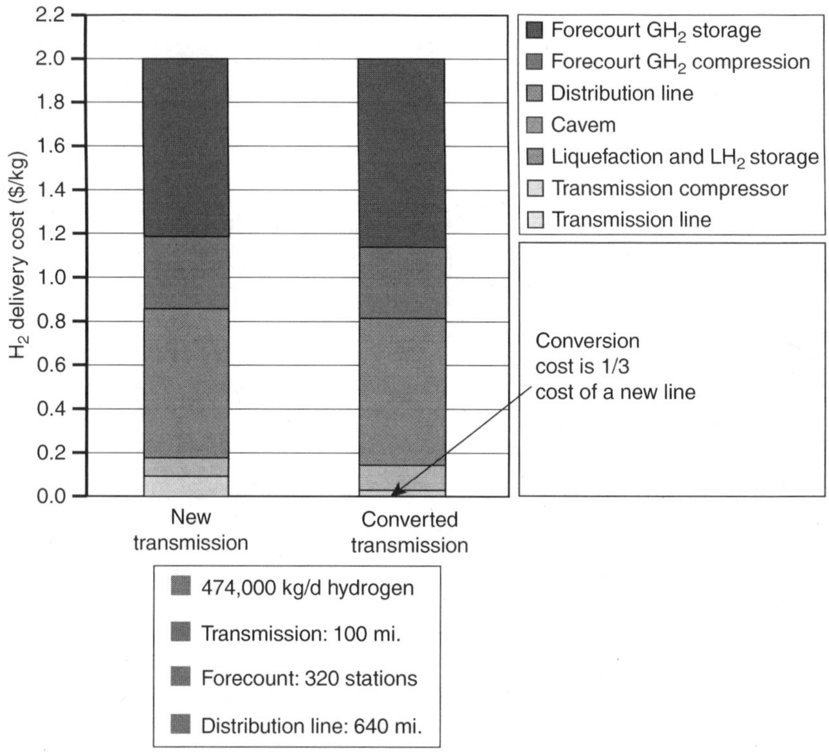

FIGURE 10.7
Economics of converting pipelines.

10.2.2 Material Challenges: Pipeline Steels and Hydrogen Embrittlement

Although the industry has been safely operating gaseous hydrogen pipelines for decades [2], and the early studies of hydrogen effect on pipeline steels with pipeline test loop showed no degradation in the mechanical and physical properties [14–17], the oil and gas industry has always been troubled by internal and external hydrogen attack in the form of hydrogen-induced cracking (HIC), HE, sulfide stress cracking (SSC), and stress corrosion cracking (SCC) on steel pipelines [4]. The predominant modes of threats on gas transmission pipelines are external; however, in hydrogen transmission pipelines the internal threat may be the more challenging problem. For the delivery of large-volume and high-pressure (up to 3000 psi) hydrogen, potential HE and HIC of the pipeline material, particularly in the weld region, is one of the major concerns, and needs to be thoroughly addressed. Research programs are currently in progress to develop fundamental understanding and practical solution of this issue [1,4,9,18–20]. Considering the specifics of materials' embrittlement in high-pressure gaseous hydrogen, the following four areas are discussed in detail: hydrogen entry and permeation, embrittlement and degradation, damage tolerance, and steel grades and metallurgy.

10.2.2.1 Hydrogen Entry and Permeation

Hydrogen entry into and permeation through the steel is the precursor of HE. Presently, limited knowledge is available about the entry, rate permeation, and amount of hydrogen in the steel under the high-pressure gaseous environment in pipeline steels. The issues of hydrogen entry, permeation, and the amount of hydrogen in the steel are interlinked.

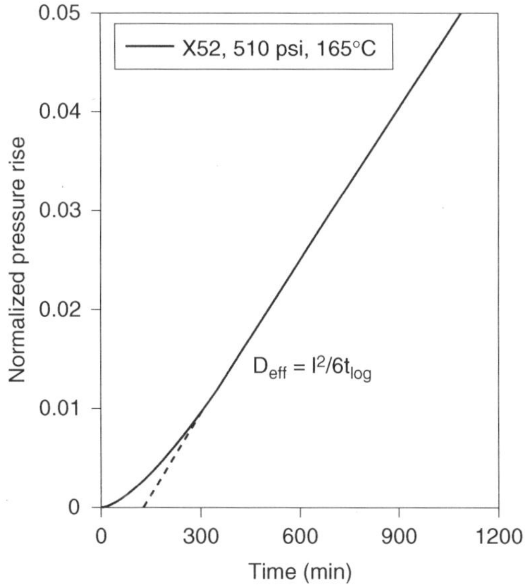

FIGURE 10.8
Pressure rise curve for X52 steel during high-pressure hydrogen permeation measurement. The measured pressure is normalized to the charging pressure on the upstream side. Here, the hydrogen charging pressure is 510 psi and the test temperature is 165°C.

For example, hydrogen entry may be the limiting process when the rate of entry is slower than that of permeation (diffusion), and the amount of hydrogen accumulated in the steel is determined by a combination of the amounts of hydrogen entry, permeation, and hydrogen finally leaving the steel from the pipeline's external surface. The study of hydrogen permeation usually gauges the mechanism and kinetics of hydrogen entry, the amount of hydrogen coverage on the surface (for pipelines, internal surface), the hydrogen diffusivity through the material, and hydrogen trapping (reversible and irreversible).

Studies of high-pressure hydrogen permeation in pipeline steels were recently performed under the programs supported by the DOE [9]. Two steel grades, namely, API X52 that is a typical 1950s pipeline steel grade and API X65 that is a typical 1990s pipeline steel grade were evaluated. The steel samples used were removed from natural gas transmission pipelines. To quantify the effect of surface coating and conditions for hydrogen entry, the specimens were not electroplated with Pd as usually done for permeation studies to eliminate the flux-limiting surface coating.

Figure 10.8 is the rise curve for X52 steel at a pressure of 510 psi (35 bars) and a temperature of 165°C. In Figure 10.8, the pressure transient measured on the downstream side of the specimen was normalized to the hydrogen charging pressure (i.e., pressure at the entry side of the specimen). As shown in Figure 10.8, it only took about 30 min before hydrogen broke through the 0.5 mm thick sample and 600 min (100 h) to reach the steady state of permeation flux where the hydrogen concentration in the sample is saturated. The transient pressure rise shown in Figure 10.8 is used to determine the effective diffusivity by the time lag technique [21]. The solubility of hydrogen in the steel was calculated from the steady state and the effective diffusivity.

From the pressure rising curve, the diffusivities of X52 and X65 can be calculated. Figure 10.9 shows the calculated diffusivities as compared to the diffusivity data compiled by Alefeld and Volkl [22] from open literature. It is seen that the measured X52 and X65 diffusivities [9] under both low- and high-pressure conditions are significantly lower than those collected by Alefeld and Volkl. The causes of this difference may be attributed to the presence of a passive

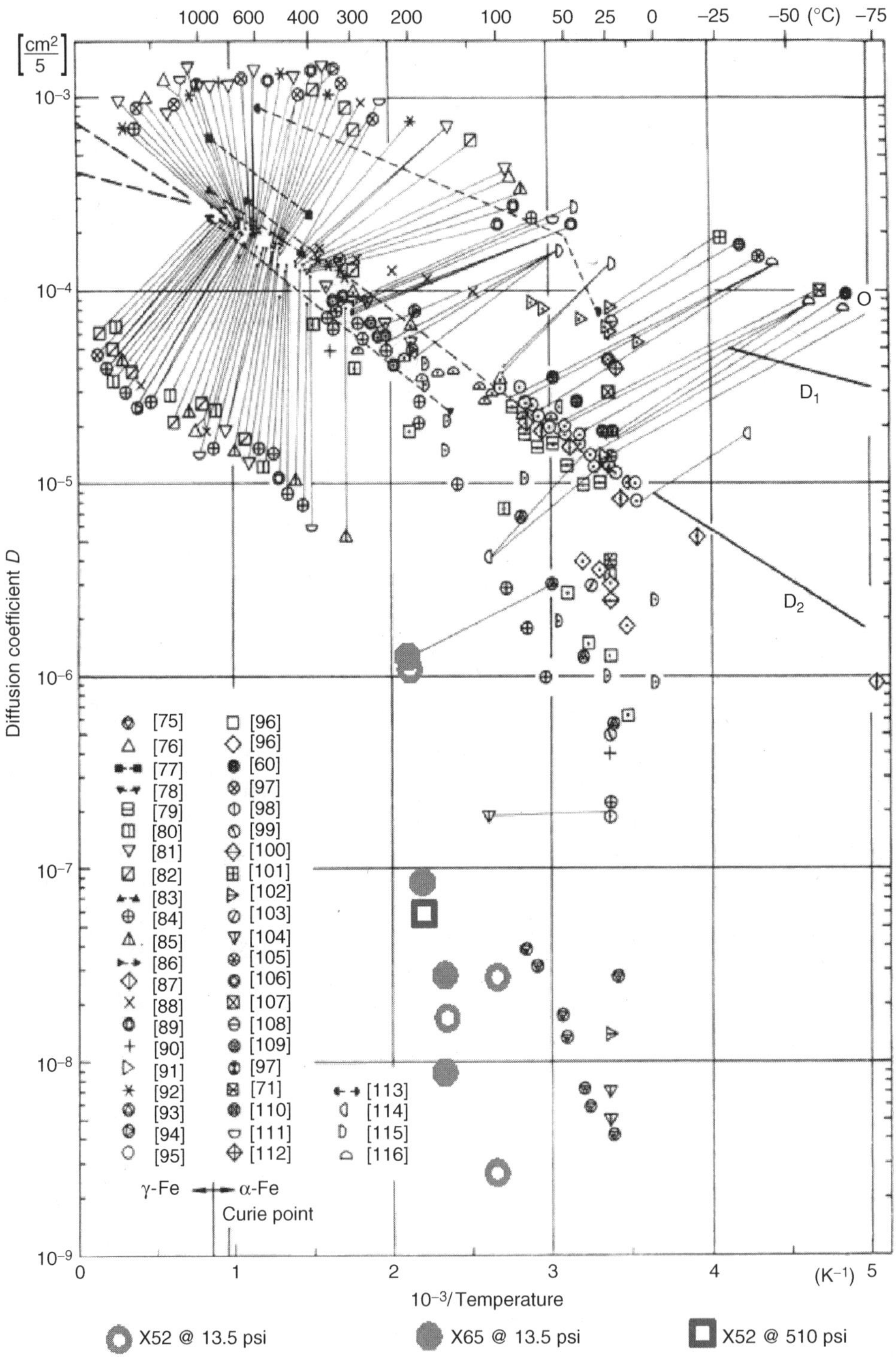

FIGURE 10.9
Hydrogen diffusivity as a function of temperature. The effective diffusivity data of X52 and X65 pipeline steels obtained under gaseous hydrogen charging conditions are compared with the compiled results for the literature. (After Alefeld, G. and Volkl, J. *Hydrogen in Metals I—Basic Properties*, Springer, New York, 1978.)

surface oxide layer that generally affects hydrogen absorption and dissociation and retards hydrogen permeation and are being investigated [9]. In addition, contribution of increased trap density in commercial pipeline steel could cause a reduction in diffusivity.

Comparing the permeation data between high- and low-pressure tests suggests that the effect of hydrogen-charging pressure on the diffusivity are relatively insignificant, at least from these preliminary data. However, the amount of hydrogen in the steel (i.e., hydrogen concentration) is highly pressure dependent, as expected by Sievert's law. For example, the saturated hydrogen concentration at 510 psi and 165°C is about 200 ppm, whereas it is about 70 ppm at 13.5 psi at the same temperature. It is expected that higher hydrogen concentration in the steel is more detrimental to the structural integrity of pipeline steels.

The data reported so far are preliminary. The pressures tested are significantly lower than the pressures used for future pipeline transmission. Using the test apparatus developed by Oak Ridge National Laboratory (ORNL), it is possible to conduct hydrogen permeation tests at very high pressures (up to 12,000 psi, i.e., 827 bars) with temperatures up to 1000°C. More data would be available in the future following the progress of DOE hydrogen projects.

The effect of surface condition (such as oxide layer) and coating (such as glasses) on the hydrogen entry can be assessed by hydrogen permeation tests, which will assist in developing appropriate barriers (coatings or surface treatment) to minimize hydrogen entry. Understanding the increased trapping of hydrogen in pipeline steels is essential to mitigating and managing the possibility of hydrogen damage from a metallurgical aspect.

10.2.2.2 Embrittlement and Degradation

It is generally agreed that hydrogen almost always has a deleterious effect on materials' mechanical properties. Extensive work has been done in the past on steels in gaseous hydrogen at low pressures [23–25]; however, very limited research work has been done on pipeline steels in high-pressure gaseous hydrogen environments. Because of the lack of understanding of the interactions between material and gaseous hydrogen at high pressures, systematic research work has been planned and being conducted to characterize mechanical properties of API pipeline steel grades, including X52, X60, X65, X70, and X80 [1,18,19]. The preliminary results from X70 and X80 tests show insignificant effect of high-pressure hydrogen charging on tensile properties (YS and UTS) and failure strain for steels charged with hydrogen at 20,000 psi (1379 bars) at 100°C for 8 days. Figure 10.10 and Table 10.1 are the stress–strain curves and detailed data obtained from the tests. Since the work is preliminary and additional testing is in progress, no conclusions can be made at this time.

Yet, little information is available for the degradation of fracture toughness of pipeline steels in high-pressure gaseous hydrogen. Work on the low-alloy pressure vessel steels back to the 1970s [4,5,26–28] may have provided some insights into the influence of high-pressure gaseous hydrogen on pipeline steels. Figure 10.11a through 10.11c illustrates the effects of hydrogen pressure (up to about 950 bars, i.e., 13,700 psi), temperature and materials' strength on K_{IH} for three vessel steels, namely, AISI 4130, AISI 4145, and AISI 4147, where K_{IH} is the fracture toughness, that is, the critical stress intensity factor for failure, the subscripts "I" denotes Mode I crack and "H" denotes in gaseous hydrogen. From Figure 10.11, it is seen that

- Fracture toughness K_{IH} decreases with increasing hydrogen pressure, about three times lower at 950 bars than that at 1 bar ($K_{IH} \approx 40$ ksi in.$^{1/2}$ at 950 bars) [26].
- Fracture toughness K_{IH} decreases with increasing steel yield strength, about four times lower for YS = 1200 MPa than that for YS = 620 MPa [26]. It is therefore expected that the degree of degradation by high-pressure hydrogen to the fracture toughness of pipeline steels (X52–X65) will be less or significantly less because

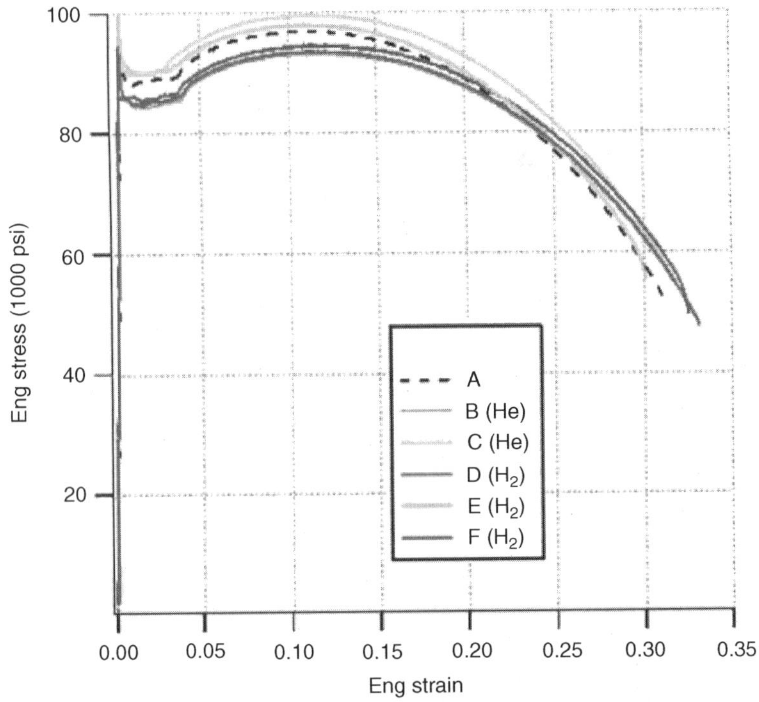

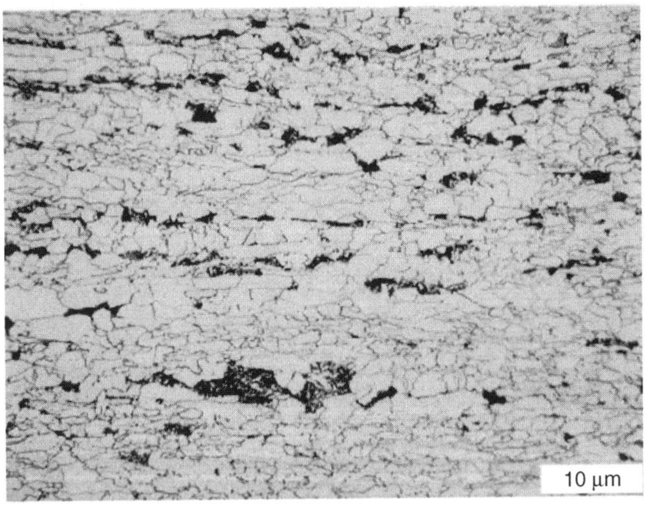

Microstructure: ferrite + pearlite

Yield stress: Intermediate

Note: stress–strain curves variability

(a)

FIGURE 10.10
Stress–strain curves and microstructure from specimens of pipeline steels precharged with high-pressure gaseous hydrogen at 20,000 psi and 100°C for 8 days: (a) API X70, (b) API X70/X80 type, and (c) API X70/X80 type.

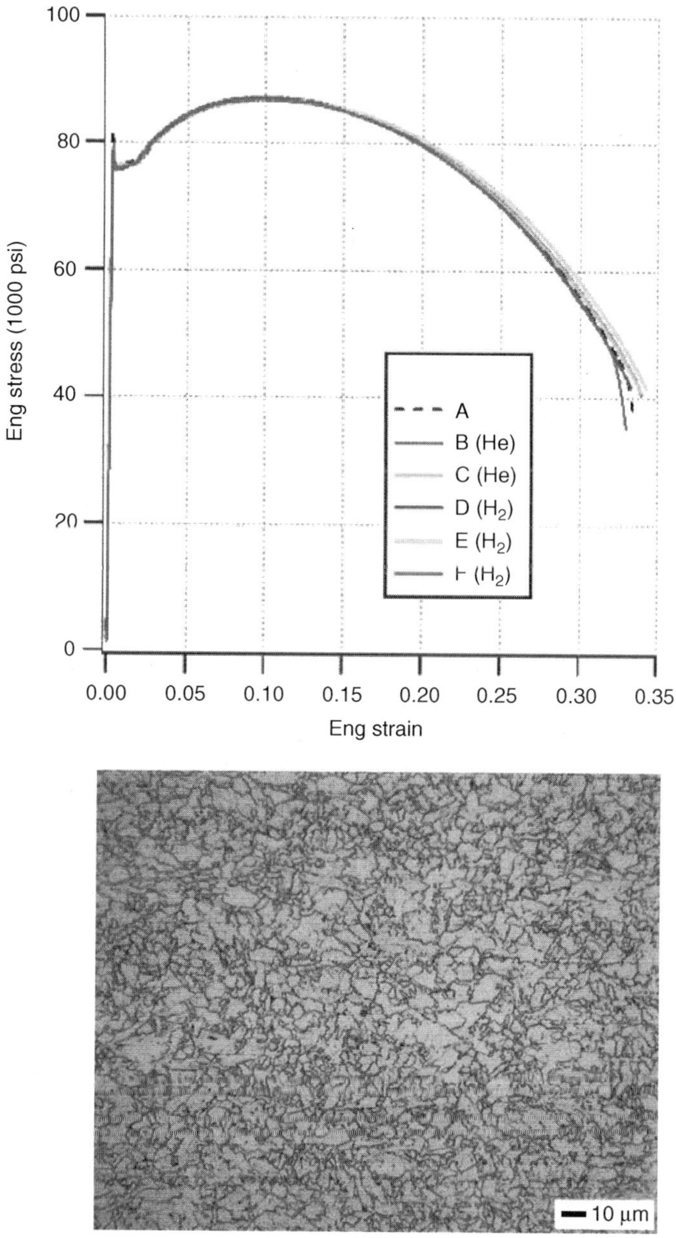

Microstructure: ferrite + acicular ferrite

Yield stress: low

Note: consistent stress–strain curves

(b)

FIGURE 10.10 (Continued)

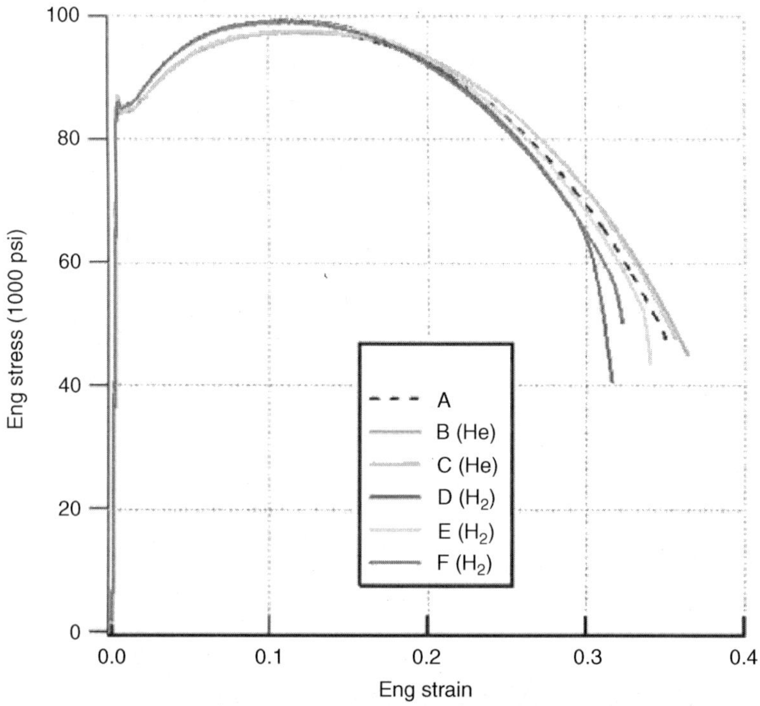

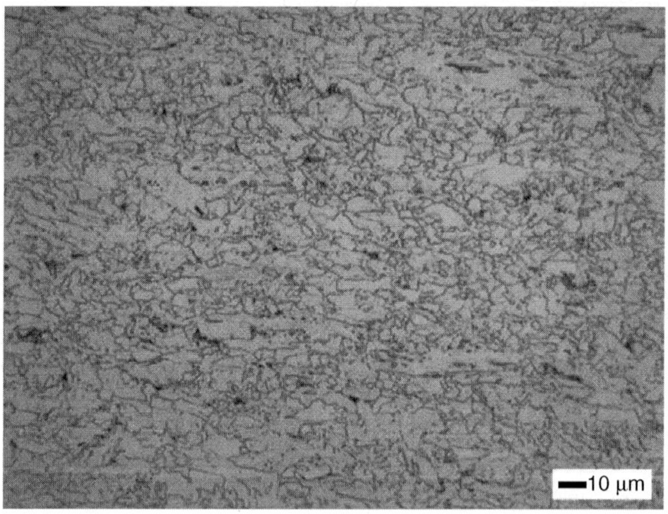

Microstructure: ferrite/acicular ferrite + small quantity pearlite
Yield stress: high
Note: stress–strain curve with slight variability

(c)

FIGURE 10.10 (Continued)

TABLE 10.1

Summary of Mechanical Testing Results of Precharged Steels with High-Pressure Gaseous Hydrogen at 20,000 psi and 100°C for 8 Days

Material	ID	Note	Min Diameter (in.)	Failure Strain	Tensile Modulus (10^6 psi)	
X70 A	A	As is	0.1605	0.31	26.6	
	B	He	0.1605	0.33	29.2	
	C	He	0.1612	0.32	31.9	Medium YS
	D	H_2	0.1608	0.32	31.4	
	E	H_2	0.1607	0.30	29.5	
	F	H_2	0.1605	0.33	28.7	
X80 B	A	As is	0.1601	0.33	30.5	
	B	He	0.1605	0.34	33.8	
	C	He	0.1603	0.34	32.7	Low YS
	D	H_2	0.1603	0.33	29.8	
	E	H_2	0.1606	0.34	29.9	
	F	H_2	0.1605	0.33	29.6	
X80 C	A	As is	0.1609	0.35	28.9	
	B	He	0.1607	0.36	31.3	
	C	He	0.1602	0.36	30.4	High YS
	D	H_2	0.1601	0.32	28.0	
	E	H_2	0.1602	0.34	29.5	
	F	H_2	0.1603	0.32	29.8	

their yield strengths are lower than those of low-alloy high-strength pressure vessel steels. The K_{IH} for pipeline steel grades might be expected to be higher than 80 ksi in.$^{1/2}$ at 6000 psi of hydrogen.

- K_{IH} essentially remains constant in the range between −40°C and 20°C and then increases with increasing temperature [28].

To develop fracture criteria for safe operation under high hydrogen pressures, Sofronis [19] adopted an integrated approach that combines mechanical property testing at macroscale, microstructural and deformation process analyses at micro- and nano-scale with the aid of transmission microscopy; the first-principle calculation of interfacial cohesion at the atomic scale; and finite element modeling and simulation at the micro- and macro-level. The mechanical testing characterizes the elastic deformation and flow characteristics of pipeline materials in the presence of hydrogen. The microstructural and deformation analyses assists in understanding the interaction of hydrogen transient transport kinetics with materials elastoplastic deformation ahead of the crack tip. The first-principles calculation and a thermodynamics methodology calculate the reversible work of separation at interfaces between precipitate/second phase and matrix as a function of the hydrogen concentration and establish hydrogen-dependent traction-separation laws. The results from the first principles calculation in conjunction with the finite element simulation of the hydrogen concentration profiles ahead of a crack tip allow for the development of engineering fracture criteria in terms of macroscopic parameters. When combined with information of fracture toughness, fatigue, and subcritical crack growth rates, the hydrogen profiles further helps to establish the regime of critical hydrogen concentrations and critical elapse time for a crack to remain stable under high hydrogen pressure. Figures 10.12 and 10.13a are the examples of the finite element calculations of transient hydrogen transport under plane strain conditions ahead of the crack tip and the uptake of hydrogen in the interior surface from the crack tip of the pipe in a hydrogen environment of 7 MPa (70 bars) pressure.

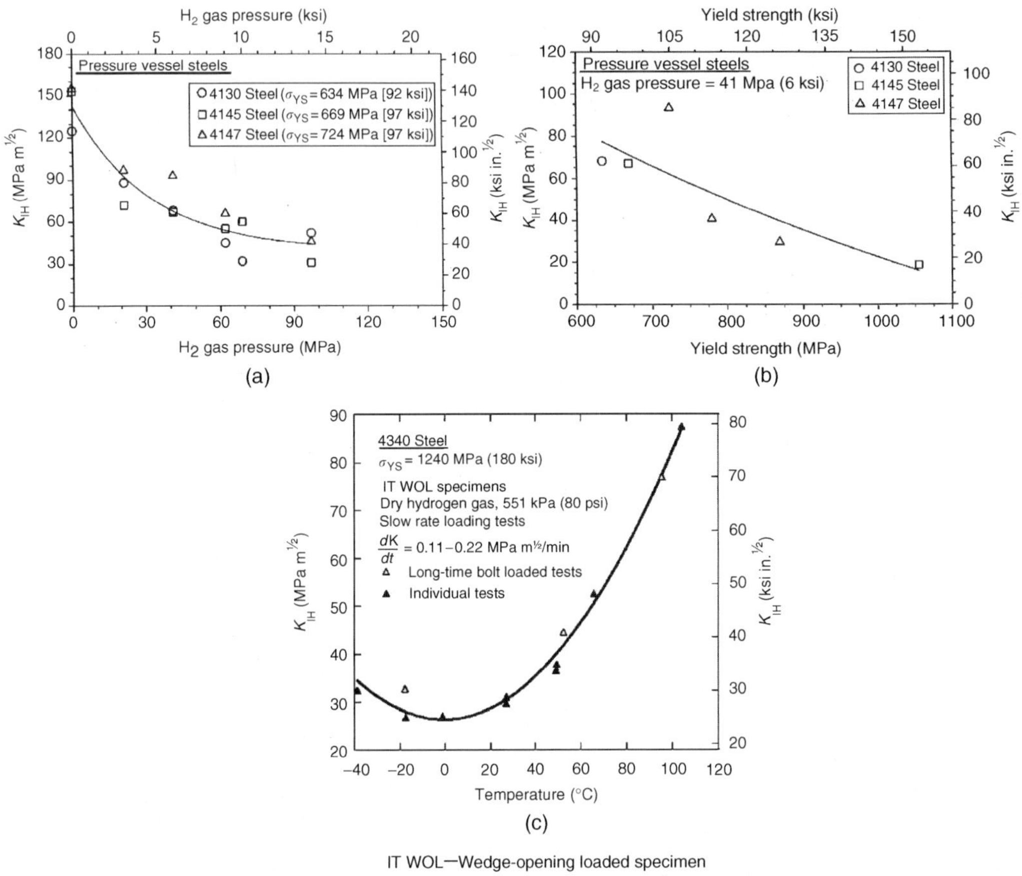

FIGURE 10.11
Effects of hydrogen pressure, materials yield strength, and temperature on the fracture toughness (K_{IH}) of three low-alloy vessel steels: (a) pressure, (b) yield strength, and (c) temperature.

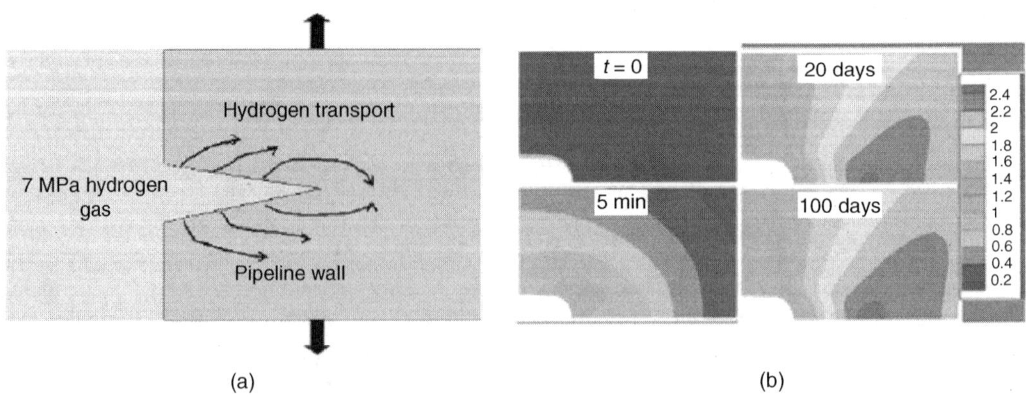

FIGURE 10.12
(a) Schematic of hydrogen uptake through the faces of a crack in a pipeline wall and (b) contour plots of the normalized hydrogen concentration in normal NILS near the crack tip.

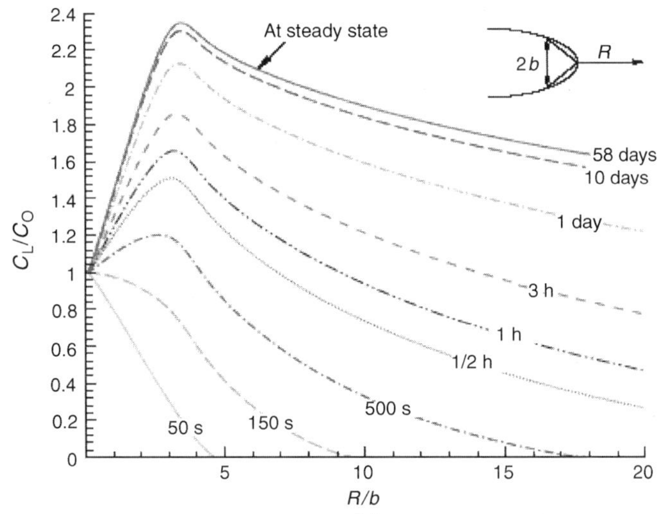

(a)

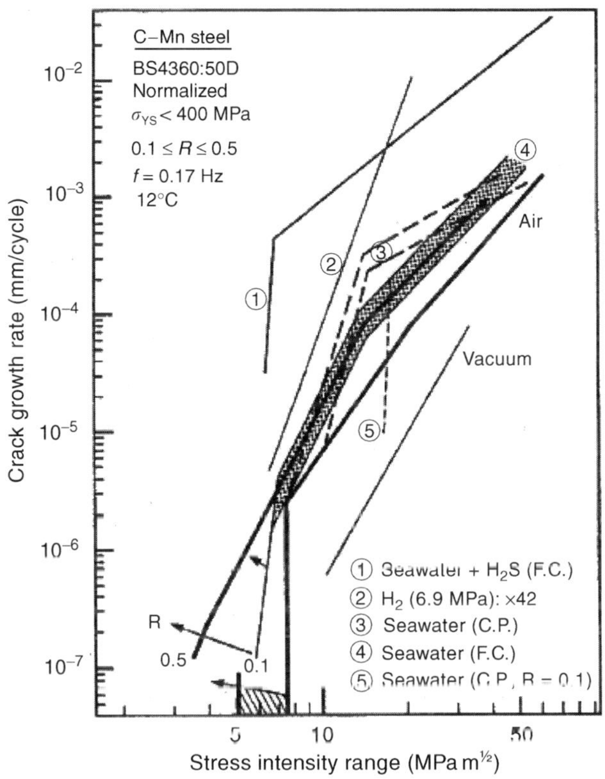

(b)
C.P—cathodic protection; F.C—free corrosion

FIGURE 10.13
(a) Plots of the normalized hydrogen concentration at NILS head of the crack tip as a function of the normalized distance from the tip at various time after hydrogen is absorbed on the crack faces and beginning to diffuse in the material. The insert shows the crack parameters used for Finite Element Analysis: $b = 18.13\ \mu$, the crack opening displacement associated with the applied load of 55 MPa $m^{1/2}$ and R is the distance from the crack tip. (b) Fatigue crack growth in various environments including vacuum, moist air, aqueous solutions, gaseous hydrogen, and H_2S. (c) Fatigue crack growth rates as a function of various steels and hydrogen pressures.

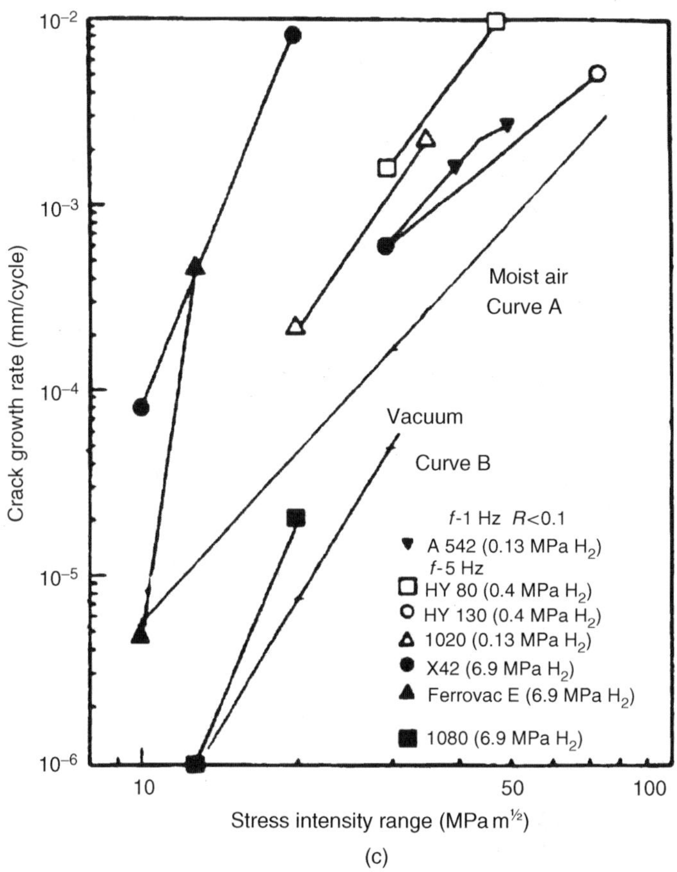

FIGURE 10.13 (Continued)

Figure 10.13a shows that a steady state is reached after about 58 days from the initiation of hydrogen diffusion in the system. The maximum of the normal interstitial lattice sites (NILS) profile coincides with the maximum hydrostatic stress peak located at the crack tip. From Figure 10.13a, it is seen that, within a good approximation, the maximum concentration at the maximum hydrostatic stress peak location reaches its steady-state value in 20 days. The time to reach the steady state is of engineering significance as it sets an upper bound on the time period during which HE phenomena may take place after hydrogen absorption in the highly strained volume ahead of the crack tip.

It is noted that the time required for reaching the steady state of hydrogen diffusion is stress, specimen geometry, and dimension dependent, and cannot be used to compare with the permeation data as discussed earlier, which was measured from stress- and crack-free thin membrane specimens.

From the preceding discussion, it is seen that characterization of hydrogen uptake and embrittlement of pipeline steels in high hydrogen pressure environment is at its initial stage. In particular, there is little data on weld and heat-affected zone. Much more work needs to be done in the forthcoming years.

Another damage mechanism that should be addressed is fatigue of line pipe steel in the presence of high-pressure hydrogen. Enhancement of crack growth rates of mild-pressure vessel steels in the presence of high-pressure hydrogen has been documented in the literature. Gas transmission pipelines experience low-frequency fatigue; however, the R ratios are <0.5. The fatigue mechanism could cause failure of the surface oxide layer and enhance hydrogen entry. In addition, crack-tip cycling would cause increased trapping of hydrogen and hence enhance the growth rate. As shown in Figure 10.13b and 10.13c, the presence of hydrogen gas substantially enhances crack growth rate. In the presence of vacuum, fatigue damage is primarily crack-tip strain accumulation. However, as it gets exposed to moist air there is an order of magnitude increase in rates. In the presence of hydrogen gas, the crack growth rate is higher as compared to aqueous solutions; however, lower than H_2S (Figure 10.13b). The presence of H_2S inhibits the recombination of hydrogen atoms, and enhances H atom diffusion into steel. However, the hydrogen gas environment is not as deleterious as H_2S; however, substantially higher than air fatigue (Figure 10.13c). The enhancement of fatigue has to be studied as a function of the *R* ratio and frequency, but has to be managed in the hydrogen transmission pipeline. The degree of this problem and its impact on operability of hydrogen pipelines should be addressed in future research.

10.2.2.3 Damage Tolerance

Cracklike flaws are often present in the pipeline. Flaws in the pipeline can result from manufacturing, handling, welding, and servicing. Consideration of pipeline damage tolerance in high-pressure hydrogen environments is one of the essential components in the steel and welding technology development, pipeline design, and maintenance. Pipeline damage tolerance is a function of mechanical loading, material, and hydrogen pressure, and can be assessed using fracture mechanics–based industry standard procedures such as API 579 and BS 7910. For example, the maximum tolerable (allowable) cracklike flaw can be calculated for the maximum allowable operating pressure (MAOP), pipe dimensions (OD and wall thickness), and orientation of the flaw using these procedures and the measured fracture toughness K_{IH}, YS, UTS, and stress–strain curves [29,30]. Exercise has been performed and illustrated in Figure 10.14 using a simple $K_I \leq K_{IH}$ failure criterion [4].

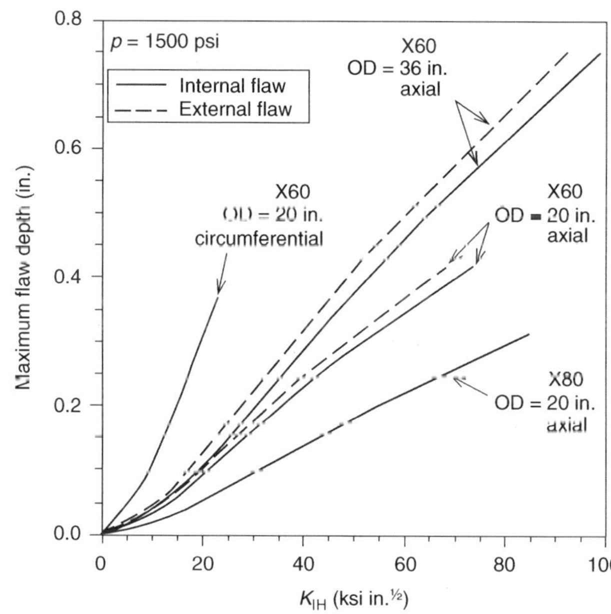

FIGURE 10.14
Plots of maximum tolerable flaw depth as a function of steel's fracture toughness K_{IH} for different pipe dimension, steel grades, flaw orientation, and flaw locations.

TABLE 10.2

Pipeline Parameters Used in the Fracture Mechanics Calculations

Material	Yield Strength (S_y [psi])	Pressure (p [psi])	Design Factor (F)	OD (in.)	Wall Thickness (t [in.])[a]
X60	60,000	1500	0.72	20	0.46
X60	60,000	1500	0.72	36	0.83
X80	80,000	1500	0.72	20	0.35

[a]*Note:* Wall thickness (t) was determined from the p, OD, S_y, and F, i.e. $t = pd/(2S_yF)$. Calculation employed the maximum design factor of 0.72, which yielded the lowest value of wall thickness.

In contrast, the burst pressure or the required fracture toughness of steel or weld can be estimated when the maximum flaw size or the desired MAOP are given, respectively. The pipeline parameters used for the calculation are shown in Table 10.2. The flaws for the calculation are assumed to be semiellipse with the depth-to-length ratio of 1:10.

Figure 10.14 shows

- Pipeline with the higher fracture toughness K_{IH} can tolerate larger flaw size (depth).
- Pipeline can tolerate larger circumferential flaws as compared to axial flaws (see the curves for X60 steel, OD 20 in.).
- Larger flaws can be tolerated in X60 as compared to X80 for the same diameter of the pipe. This is because the wall of the X80 pipe is thinner than X60 at the same MAOP = 1500 psi for the same design factor DF = 0.72 or SF = 1.39.
- For the same steel (X60), larger flaws can be tolerated in the 36 in. diameter pipe as compared to the 20 in. diameter pipe. This is also due to the wall of the 36 in. pipe being thicker than that of the 20 in. pipe when both the pipes are operated at the same 0.72 design pressure of 1500 psi.

From the preceding discussion, it is seen that for a given operating pressure and pipe dimensions, pipeline's damage (flaw) tolerance depends on the material's fracture toughness K_{IH}. However, the information on the values of K_{IH} for line pipe steels for high-pressure hydrogen transmission is extremely limited. Therefore, the establishment of a material database of pipeline steel grades and new materials is an urgent task and precursor prior to the design.

Another aspect of high-pressure hydrogen effect that needs to be addressed is the subcritical crack growth under sustained and fatigue load conditions. To prevent subcritical crack growth of a given flaw size (e.g., the maximum flaw that may have been missed by the mill inspection), the threshold stress intensity factor K_{IHAC} and fatigue threshold ΔK_{th} in high-pressure hydrogen should not be exceeded. Moreover, for life cycle predictions, the crack growth rates da/dt under sustained load conditions and da/dN for fatigue loading are essential. Again, currently there are no data available for high-pressure hydrogen. Therefore, it may be important to develop a material testing program for characterizing threshold stress intensity factors and crack growth rates under both sustained and fatigue loading conditions.

Considering the technical difficulties in establishing the material database, literature data for H_2S cracking of pipeline steels may be used for damage tolerance assessment in the interim. Utilizing data measured from H_2S environments most likely produces conservative results because hydrogen damage resulted from H_2S is more severe than that from dry gaseous hydrogen.

10.2.2.4 Steel Grades and Metallurgy

As mentioned earlier, the DOE has research programs to fully characterize the mechanical behavior of API steel grades for pipelines, ranging from X52 to X70/X80 [18]. The scope of the program includes (1) the development of barrier thin-film coatings that would minimize hydrogen permeation in the current natural gas pipelines and (2) the identification of steel compositions and associated welding filler wires and process that would be suitable for new pipeline infrastructures for the transport of hydrogen at high pressures. The programs also include the characterization of pipeline steel grades that have experienced extensive field services for natural gas transmission [31]. The on-going programs aim at investigating the ability of the existing distribution pipeline steels to transport pure gaseous hydrogen or mixed hydrogen–natural gas (i.e., 20% hythane contains 20% H_2 and 80% natural gas) using advanced fracture mechanics measurement techniques and develop environment controlled/modified failure assessment diagram [29,30] that can be used to predict flaw tolerance and stability in the pipelines. In addition, fatigue behavior of pipes in distribution service environments and FEA analysis will be conducted for life and burst failure predictions. On completion of the programs, steel grades and metallurgy for high-pressure hydrogen transport, both main and distribution pipelines, should be identified. The identification of a maximum operating pressure in a hydrogen transmission pipeline will have a great impact on the economics and hydrogen damage.

It is noted that there is a survey on the existing gaseous hydrogen pipelines [4]. The purpose of the survey was to identify candidate steel grades and issues associated with the large size of hydrogen pipelines for wind power renewable energy sources. The survey showed that a variety of steels, but primarily mild steels, are in use [32–34]. The survey shows that, in the absence of a gaseous specification, the Canadian regulatory board requires a gaseous hydrogen pipeline built in the 1980s to meet [33]

- Maximum material grade 290, design factor <0.6
- Toughness 30% higher than line pipe steels for comparable natural gas service
- Operating temperature $<40°C$
- Limit pressure reversal >3 MPa to fewer than 100 cycles per year

Furthermore, the survey also suggests if a steel line pipe is to be used for 100% gaseous hydrogen, it should generally meet

- Lower strength, probably X65 maximum grade
- Sour-service standards, which specifies alloying and processing, homogeneous microstructure, and segregation control
- Very high quality: very few, very small inclusions
- Low Mn, very low S (<10 ppm), relatively low C and low hardenability
- Thicker wall than for natural gas pipeline to allow mild HE to the inside surface region while retaining the overall pipe strength and toughness

Other options for steel pipe gaseous hydrogen service [4] may be

- Al–Fe (aluminum–iron) alloy, whereby the aluminum provides an inner hydrogen barrier, preventing hydrogen from permeating the bulk of the steel
- Variable-hardness pipe, with the harder material in the interior, softer toward the exterior, so that any hydrogen that diffused into the interior steel rapidly diffuses outward and escapes

10.2.3 Material Challenges for Composites

Because of high capital costs and potential HE of steel pipelines for high-pressure hydrogen transport, the investigation of fiber reinforcement or other plastic composites as substitutes for steels is one of the major R&D tasks within the DOE hydrogen delivery program. The purpose of the work is to achieve reduction in installation costs, better reliability, and safer operation of hydrogen pipelines.

The initial evaluation showed that utilizing fiber-reinforced polymer (FRP) for pipelines is a feasible alternative to steel pipelines with regard to performance and cost [35]. From the cost analysis, an FRP pipe is quite attractive, especially in the regional or distributed service. Currently, spoolable piping manufacturers could install a composite pipeline for serving a 100,000 population for a cost of $250,000–500,000/mi. (does not include the cost for right-of-way), which is well below the DOE's capital cost target in 2017 of $800,000/mi. [35]. From this estimate and cost analyses, it is seen that FRP pipe economics is very attractive, especially for the distribution service.

An FRP pipeline typically consists of (1) an inner nonpermeable barrier tube that transports the pressurized gas, (2) a protective layer over the barrier tube, (3) an interface layer over the protective layer, (4) multiple glass or carbon-fiber composite layers, (5) an outer pressure barrier layer, and (6) an outer protective layer. Each of the layers provides a distinct function and the interaction between the layers produces a pipe with exceptional performance.

The challenge for adapting FRP pipeline technology to hydrogen service consists of the following [35]:

- Evaluating the pipeline materials for hydrogen compatibility
- Developing a method for manufacturing large-diameter pipelines
- Developing a plastic liner with acceptably low hydrogen permeability
- Determining the necessary modifications to the existing codes and standards to validate the safe and reliable implementation of the pipeline

For developing a nonpermeable barrier or an acceptably low hydrogen permeability layer, polymer-layered silicate (PLS) nanocomposites are being formulated. PLS nanocomposites are formed by mixing organomodifier clay (montmorillonite) with a polymer such as poly(ethylene) terephthalate (PET) in melt or solution. If the polymer and modified clay are well matched for ion exchange, exfoliation of the layered clay is achieved. The exfoliated morphology is critical for achieving a dramatic decrease in hydrogen permeability in the polymer. Figure 10.15 shows the small angle x-ray scattering (SAXS) analysis results and transmission electron microscopy (TEM) micrographs of the nanocomposites (PET/5% clay and PET/10% clay) as compared with pure PET. The SAXS showed some intercalation of silicates present. The TEM showed the presence of a partial exfoliation. Figure 10.16 is a comparison of the diffusion rate of hydrogen in thin films of pure PET and PET/10% clay measured using the permeation method. The results showed that the diffusion rate in the PET/10% clay film exhibited a 60% reduction when compared to that in the pure PET film. This is a significant improvement in the already good hydrogen barrier property of the pure PET. To achieve a higher degree of exfoliation, further improvement is being developed using the partial sulfonation of polytechnique introduced by Chisholm et al. [36]. Results showed that a better exfoliation has been achieved [35].

In addition to the FRP pipeline technology, TransCanada Pipelines Limited (TCPL) is developing the composite reinforced line pipe (CRLP^{TM}) under the license from NCF industries [4]. The CRLP consists of mainly a high-performance composite material and a

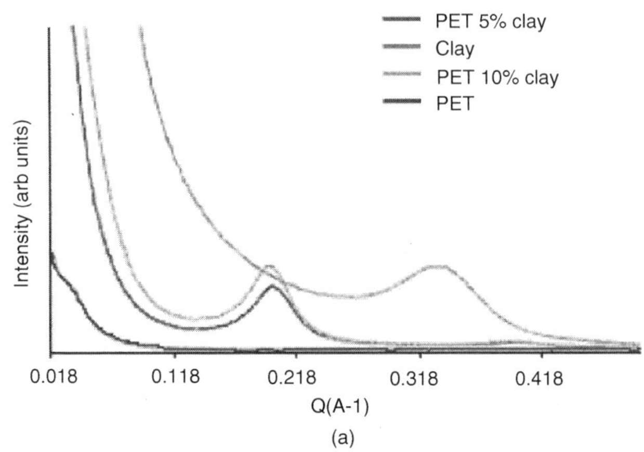

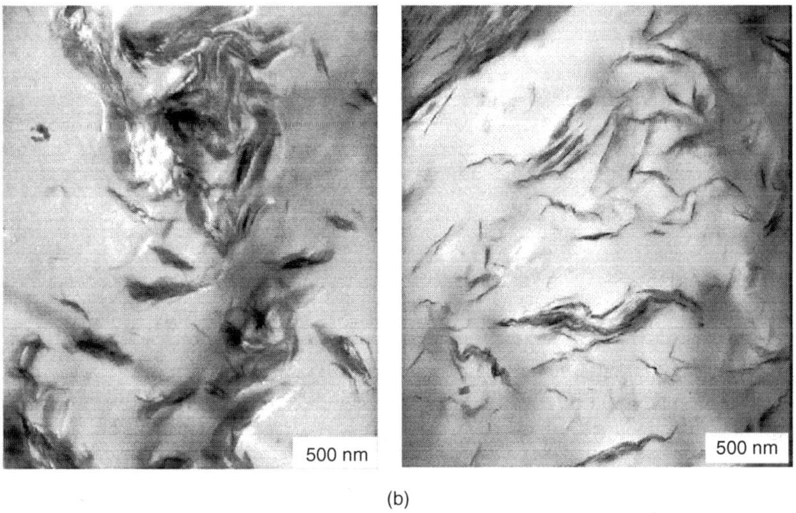

FIGURE 10.15
(a) SAXS profiles of PET/clay nanocomposites, pure clay, and pure PET—the peaks of PET/clay indicating the presence of intercalation, and (b) TEM micrographs showing some degree of exfoliation (black ridges of clay and light shadowed PET) formed.

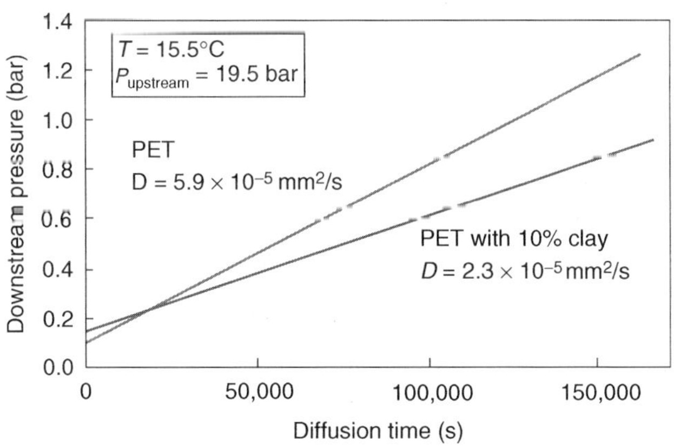

FIGURE 10.16
Permeation measurement of PET and PET/10% clay nanocomposite.

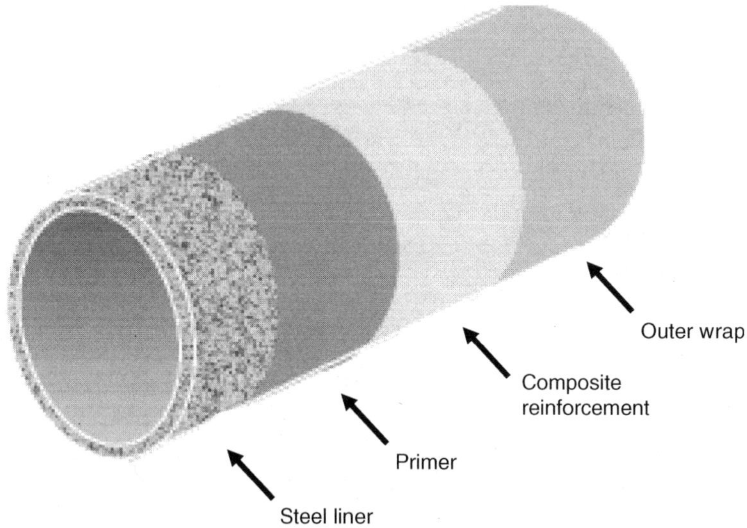

FIGURE 10.17
Composite reinforced line pipe (CRLP™). (After Zimmerman, T., Stephen, G., and Glover, A., Composite reinforced line pipe (CRLP™) for on shore gas pipelines, IPC2002, Paper No. 27125, *Proceedings of IPC02*, Calgary, October 2002.)

thin-wall, high-strength low-alloy (HSLA) (X42–X80) steel pipe. The steel and composites together creates a hybrid that provides an economical alternative to a higher-strength all-steel pipe. TPCL estimates that the total installation cost of the pipeline needed for large-scale hydrogen transmission, OD >1.5 m (59 in.), pressure >14 MPa (140 bars), would be 3–8% cost saving for a solid steel pipeline. Figure 10.17 is a sketch of the CRLP [37].

Issues related to HE of steel liner may still exist; however, due to the thin steel liner carrying lower stresses, potential for hydrogen cracking could be reduced.

From the preceding discussion, it is seen that the issues related to materials for high-pressure hydrogen pipelines present major concerns, which require extensive investigation. With the experiences gained from the existing hydrogen and natural gases pipelines and fundamental studies of HE, an engineering and economic solution will be achieved timely for constructing both large and small sizes of transmission and distribution pipelines.

10.2.4 Sensors: Leak Detection and Pipeline Integrity Monitoring

Hydrogen delivery infrastructure requires sensors for detecting hydrogen leaks and monitoring pipeline integrity. A brief discussion for each type of sensor technology is given in Sections 10.2.4.1 and 10.2.4.2.

10.2.4.1 Leak Detection

Hydrogen is odorless, colorless, and tasteless and therefore undetectable by human senses. Adding odorants to hydrogen is one of the options for leak detection because people can smell it. The most common odorants for natural gas that is also odorless, colorless, and tasteless are sulfur containing. These odorants, however, are not appropriate for hydrogen

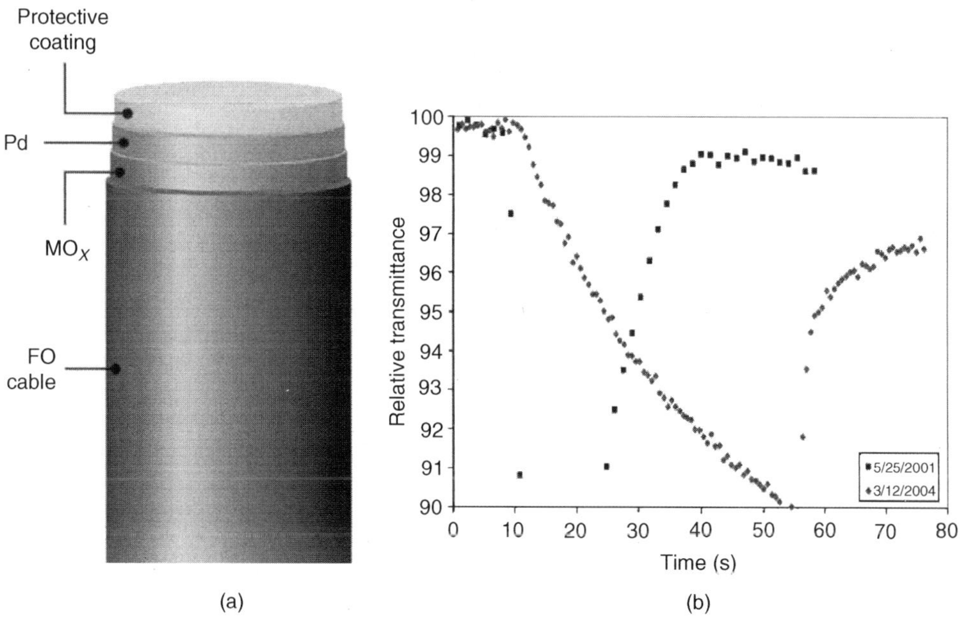

FIGURE 10.18
(a) Thin-film/optic-fiber hydrogen sensor schematic and (b) response of optical hydrogen sensor to the lower detection limit of hydrogen (0.1%) after 3 years, compared to its original response after fabrication. (After Pitts, R.R., Smith, D., Lee, S., and Tracy, E., Interfacial Stability of Thin Film Hydrogen Sensors, FY 2006 Annual Progress Report, VI.3, National Renewable Energy Laboratory, 2004.)

because they are not light enough to "travel with" hydrogen and at the same dispersion rate. Moreover, odorants also contaminate fuel cells—a popular hydrogen application. Therefore, odorants may be problematic for hydrogen and odorizing hydrogen gas is a particular challenge [6,8].

As part of the hydrogen program, however, inexpensive and reliable hydrogen sensors, such as optical fiber leak detection technologies, are being developed by the National Renewable Energy Laboratory [38]. One example is the thin-film optic-fiber sensor technology. Figure 10.18 is a schematic illustration of the sensor configuration and plots of its detection performance.

The sensor uses chromogenic materials as indicators of the presence of hydrogen. At concentrations above 0.02% hydrogen in air, these materials undergo optical changes either by changing color or by changing the transmittance through the film as atomic hydrogen is incorporated. A number of materials exhibit such optical changes, including WO_3, NiO_x, V_2O_5, and an assortment of metal hydrides. Dissociation of hydrogen gas is accomplished on the surface of a thin catalyst top layer (e.g., Pd). Some of the atomic hydrogen diffuses rapidly through the catalyst and into the chromogenic layer, changing its optical properties. The optical changes can be read easily by a light beam, either by measuring transmission through the stack of thin films, or by measuring the reflectance off the catalyst layer. The optical thin films may be deposited on the end of a fiber-optic (FO) cable, as depicted in Figure 10.18. A beam of light is propagated down the cable, and the intensity of either the reflected beam or the transmitted beam is monitored to provide the detection of hydrogen gas.

Further, the data showed that the optical sensors exhibit satisfactory lifetime. Figure 10.18b shows that the sensors fabricated in May 2001 continued to be functional for three years (at the time of report) in sensing hydrogen at the lower detection limit of 0.1% H_2 in air. Although the response time has increased and the dynamic range of coloration has decreased as compared to its original performance, but the sensor itself is still functional.

For leak detection, the sensor must be sensitive and fast enough to provide early leak detection so that action can be taken before the explosive limit in air is reached. Utilizing a fiber-optic sensor configuration could provide the best chance of meeting fast response and inexpensive and reliable goals.

It should be noted, however, that applicability of the thin-film optic-fiber sensor technology to pipelines for leak detection is still a challenge because pipelines are hundreds (even thousands) mile long. Additional research is needed to identify and develop sensors technologies for hydrogen leak detection along the pipeline.

10.2.4.2 Pipeline Integrity Monitoring Sensor Technologies

Generally, there have been nine major potential integrity threats present in the natural gas pipelines as indicated by American Society of Mechanical Engineers in the industry code ASME B31G.8S, that is, internal and external corrosion, SCC, third-party damage, soil movements, manufacture and construction-related threats, incorrect operation, weather, and outside force threats. Among them, the highest integrity risk is from third-party damage. The third-party damage is mainly due to contractors inadvertently damaging the pipeline during construction and excavation. This can result in very serious consequences. These traditional natural gas pipeline integrity threats will also be present in hydrogen transmission pipelines.

There are several sensor technologies currently available for monitoring mechanical integrity of pipelines (and pressure vessels as well). Fiber-optic sensors and other sensors have been developed which can monitor time-dependent defects including internal corrosion, external corrosion, SCC, pipe movement, pipe stress, and buckling strains due to pipeline slope instability, ground settling, and currents acting on exposed pipeline. These types of sensing technologies are particularly well adapted to composite structures, but can also be applied to steel pipelines or vessels. Such technologies might prove very valuable for the hydrogen delivery infrastructure and could compliment leak detection. It might prove valuable as an early detection approach that could avoid mechanical failures and significant hydrogen leakage [39,40].

Figure 10.19 is an example that illustrates the Brillouin fiber-optic sensor system that could be applied to pipelines for integrity monitoring [39].

The system used to measure the optical fiber signals employs two separate frequency tunable laser light sources operating at about 1320 nm wavelength. One laser acts as a pump laser, whereas the other serves as the probe laser that sends optical pulses down the fiber to interact with the counterpropagating laser light wave pumped into the fiber from its opposite end.

Preliminary research has shown that Brillouin fiber-optic sensing systems provide a possible method to detect leaks and third-party intrusion on a pipeline over distances of 25 km or more. Their intrinsic response to both temperature and mechanical strain allows for the separation of these parameters and the detection of anomalies in the scan profiles. In addition, the same sensor could be integrated into the pipeline system to detect possible ground movement relative to fixed reference points. Limited test results on surface loads associated with the intrusion of vehicles and people on a pipeline have demonstrated the sensitivity of the system and its ability to discriminate loads at different soil depths.

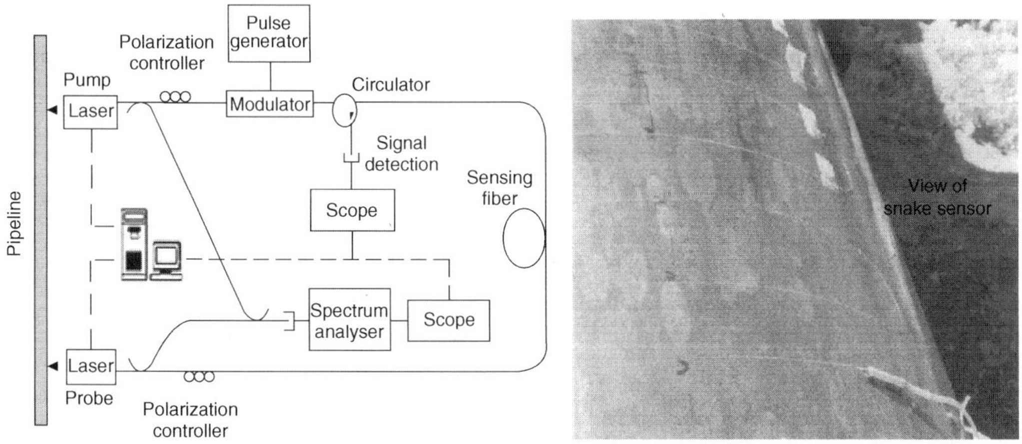

FIGURE 10.19
Typical Brillouin instrument system.

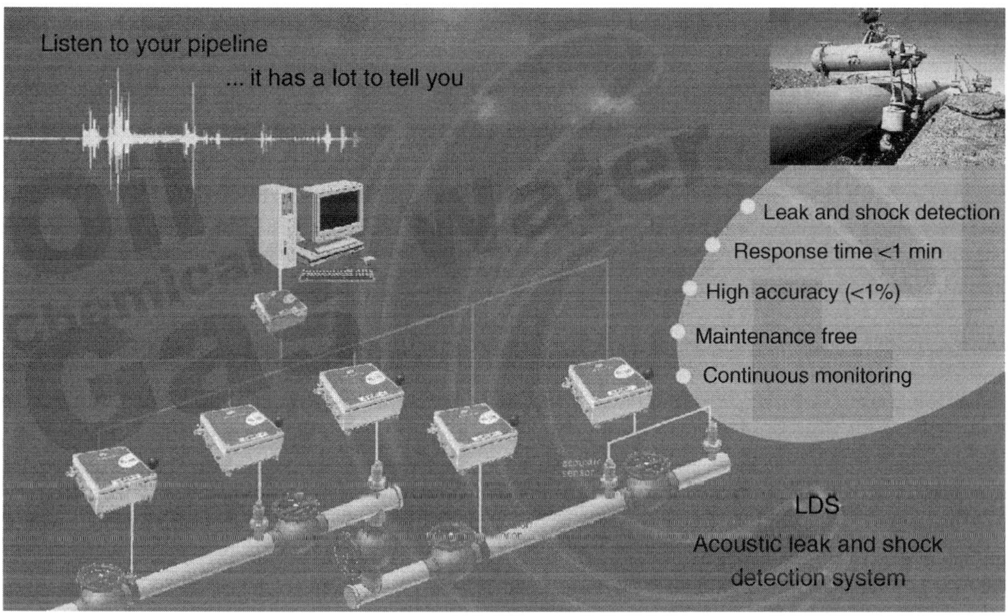

FIGURE 10.20
Illustration of acoustic/hydrophone system for leak and shock detection.

Installation of a Brillouin system is most cost-effective when applied to a new pipeline construction when the trenching is available for inserting the optical fiber cables. If the existing sections of buried pipeline require replacement or repair, then the Brillouin system can also be used selectively in high consequence areas. In general, total system costs will vary for each application.

For third-party damage, other sensors such as acoustic (hydrophone)-based technology is also being developed (Figure 10.20).

Although extensive research and industry efforts have been made to develop sensor technologies for third-party mechanical damage, more work is needed to resolve unique issues related to hydrogen pipelines, such as the accuracy of alert calls (fault call), special distance resolution along the pipeline, response time, and characteristics of leak signal from light atomic weight of hydrogen gas.

10.2.5 Hydrogen Compression

Compression is an integral aspect of gaseous hydrogen delivery via pipelines. Figure 10.21 illustrates various sizes of compressors to be used in the pipeline [6].

For natural gas service, five types of compressors are typically used, that is, diaphragm, piston, rotary, screw, and turbo compressors. The characteristics of each type of compressors are a function of throughput and pressure, as shown in Figure 10.22 [6]. Currently, piston (or reciprocating) and rotary compressors are used to compress hydrogen that are similar to those used for natural gas, with modifications of design and materials.

For large-volume and large-diameter cross-country pipelines for gaseous hydrogen transmission, candidate compression technologies have been suggested. Table 10.3 is a list of these technologies [5] that summarizes the benefits and problems for each of the candidate technologies.

It is noted that all the candidate compressors mentioned earlier are costly in capital, energy, and operations and maintenance (O&M). Advances have centered on the optimization of subsystems for cost reduction and improvement of the energy efficiency of gaseous hydrogen compression. Centrifugal compression and electrochemical hydrogen

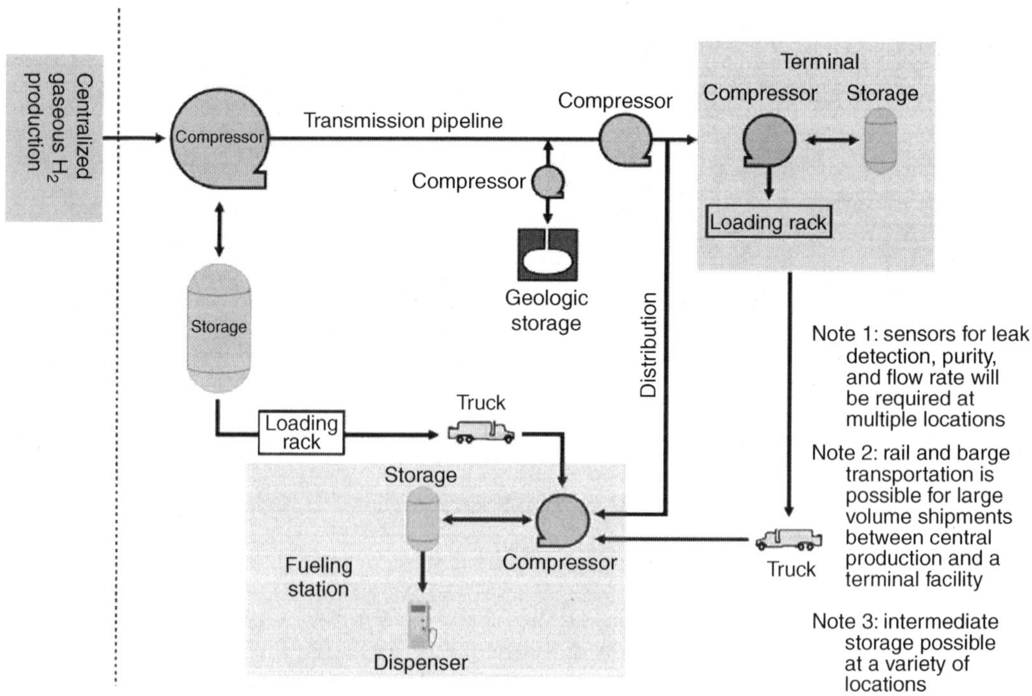

FIGURE 10.21
Gaseous hydrogen delivery pathway.

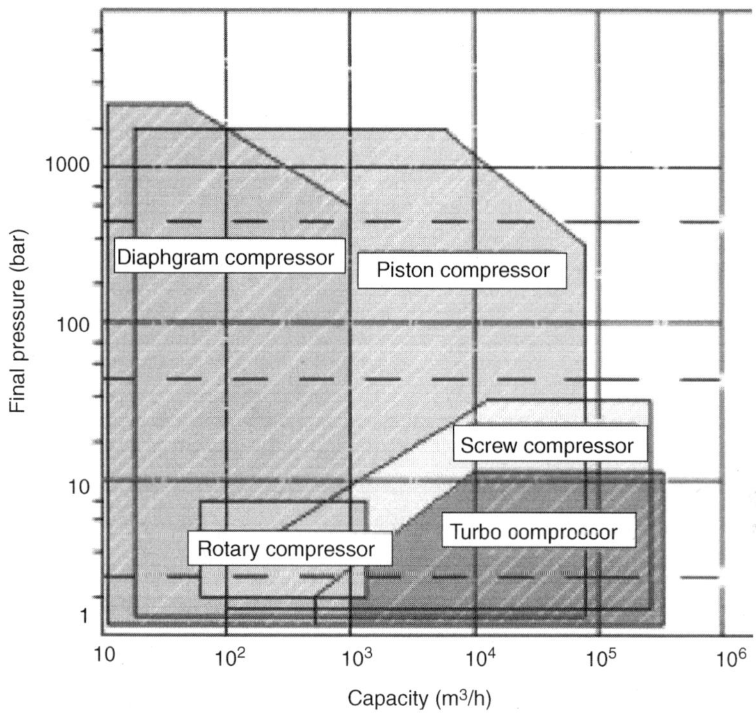

FIGURE 10.22
Operating characteristics of various compressors. (After FreedomCar FuelPartnership. Hydrogen Delivery Technology Roadmap, Final Report, 2007.)

compression technologies are the major focus for the development. The total target costs for fiscal year (FY) 2017 are $9 million for capital investment (capacity: gaseous hydrogen 200,000 kg/day) and 3% of the capital investment for maintenance [1].

Because of the high costs, alternatives such as high-pressure electrolizer output to provide high-pressure hydrogen directly to transmission pipelines are being explored [5]. Technologies may include proton exchange membrane (PEM), alkaline (KOH), high-temperature ceramic, or a combination thereof. Figure 10.23 shows the Norsk hydro electrolyzer, liquid KOH type (the gas cleanup equipment is not shown) with an input of 560 kW. For a megawatt (MW)-scale and 100 bars output, further R&D effort will be required.

10.2.6 Future Outlook

Economical analysis has shown that a pipeline network infrastructure is the most cost-effective and energy-efficient method to transport hydrogen in large quantities from the point of production to the point of use over a long distance. A complete hydrogen pipeline infrastructure will include both transmission and distribution to minimize the overall hydrogen transport cost.

Transporting gaseous hydrogen via a new pipeline requires high initial capital investment cost; however, only a small fraction of the total cost of the pipeline infrastructure. Comprehensive R&D efforts are needed and are being conducted for cost reduction,

TABLE 10.3

Compressor Technology Candidates for Giga Watt (GW)-Scale Hydrogen Pipeline Transmission

Compressor Type	Description	Benefits/Problems
Reciprocating	A well-proven technology for many different types of gases. Hydrogen, with its low molecular weight, becomes very difficult to be contained in compression cylinders and therefore requires very specialized designs.	Positive displacement, capable of high-pressure output. Oiled and oil-less lubrication types available. Oil lubrication may be necessary to contain GH2, but will contaminate the GH2. Usually electric motor driven. Well proven in 1000 kW size range for NG service. Few other advantages for GW-scale GH2 transmission pipeline service, where both the physical size and number of machines required make this type of compression a very remote possibility.
Centrifugal	A well-proven technology for many different types of gases, in sizes up to over 30,000 kW. Hydrogen, with its low molecular weight, is very difficult to compress. Approximately 60 compression stages would be required for boost from 30 to 100 bars, which would be a very costly machine.	The mechanical tolerances between a large number of stages become very difficult to maintain, which compromises interstage sealing, with very detrimental effects on compression efficiency and equipment capital cost.
Diaphragm	Compatible with all types of gases. Triple diaphragm construction isolates GH2 from hydraulic oil, preventing any contamination of the process gas or leakage into the atmosphere. High output pressure (>700 bars) available.	High compression ratio per stage, compared to reciprocating compressors, may reduce capital costs. Not available in large sizes for GH2 pipeline transmission; may be difficult and costly to scale up. Largest, currently available—2000 Nm^3/h.
Electrochemical	Multistate electrochemical compression uses a series of membrane electrode assemblies (MEAs), similar to those used in PEM fuel cells. An electrical potential across the MEA causes the transfer of hydrogen with increasing output pressure from state to stage.	This nonmechanical compressor addresses a number of shortfalls inherent in traditional compressors. It has no moving parts, so reduces the wear and tear, noise, and energy intensity problems of mechanical compressors. It is also significantly more compact, thus adaptable to a range of applications. The challenges of building units large enough for GH2 transmission pipeline service and preventing GH2 contamination have yet to be addressed.
Hydride	Compressors using reversible metal hydride alloys offer an economical alternative to traditional mechanical compressors for GH2. The simplicity and passive operation of the hydride compression process offers many advantages over mechanical compressors.	Hydride compressors are compact, silent, do not have dynamic seals, require very little maintenance, and can operate unattended for long periods. However, they are a very new and may be difficult to be built at the scale required for GH2 transmission pipeline service.

technical justification, and ensuring longevity of such a pipeline infrastructure. Transporting gaseous hydrogen via the existing natural gas pipelines is the low initial cost option for delivering large volume of hydrogen. However, the cost saving is marginal because, as mentioned earlier, the saving from utilizing the existing transmission pipeline is only a small portion of the total cost of the pipeline infrastructure. A strategy that new natural gas pipeline systems to be built with line pipe capable of transmitting gaseous hydrogen for

FIGURE 10.23
A photo of the Norsk hydro electrolyzer system that is capable of generating 4.2–4.3 kWh/Nm^3 output at 30 bars.

future conversion to renewable hydrogen services as and after the natural gas is depleted may be justified for future capital investment saving.

Lack of knowledge of HE of pipeline steels in high-pressure hydrogen environment is one of the major technical barriers for using the present natural gas pipeline steel grades. The damage will be caused due to high internal pressure and pressure fluctuations in the presence of hydrogen. Quantification of the effect of high-pressure hydrogen on line pipe steels and seam welds degradation and development of methods for preventing and controlling hydrogen degradation are strategically important. In addition, a material database, mainly, fracture toughness K_{IH} and threshold stress intensity factors K_{th} and ΔK_{th} for subcritical crack growth, and growth rate da/dt and da/dN for damage tolerance assessment and life prediction need to be established. Steel grades (tubing) for sour service may be well suited for hydrogen delivery and may be a temporary solution for meeting the today's need. Alternative materials, for example, CRLP™ or FRP are good candidates to replace steel line pipe, but require full characterization before being commercialized.

Sensor technologies for leak detection and integrity monitoring are important. Major sensor technologies have been identified and are being developed. The new pipeline systems should be equipped with leak and integrity monitoring sensors to ensure safe operation and meet the code requirements.

Hydrogen compression is another key technology that needs to be resolved. All the candidate compression technologies identified are costly in capital, energy, and O&M. Advances have centered on the optimization of subsystems for cost reduction and improvement of the energy efficiency of gaseous hydrogen compression. Centrifugal compression and electrochemical hydrogen compression technologies are the major focus for the development.

Because of the consequence of a failure in a transmission pipeline system, a pilot-scale R&D and demonstration pipeline system may be needed before designing commercial large-scale hydrogen pipeline systems. An International Hydrogen Transmission Test Facility (IHTTF) or International Hydrogen Renewable Transmission Demonstration Facility (IHRTDF) has been proposed for this purpose [4,5].

10.3 Bulk Storage

Following the transmission of hydrogen gas through pipelines, there will be requirements for bulk storage at various local distribution centers. The quantity of hydrogen will be dictated by the application.

Currently, all such large quantities of hydrogen are stored as liquid hydrogen (LH2) in insulated vessels, and such storage is expensive and more importantly product loss due to evaporation makes this process inefficient. The application of such LH2 is confined to high-purity applications in the chemical industry and aerospace applications. The external tank on the space shuttle has an aluminum LH2 tank with an operating pressure of 35 psia and a volume capacity in excess of 400,000 gal. These tanks have foam insulators, and also have thermal isolators to preclude any heat flow. Hydrogen refueling stations, which are primarily LH2 with boil-off protection, exist at the Munich Airport and Berlin among other places. Combining liquid storage with boil-off protection will be critical, and is feasible with compressed hydrogen (CH2) storage.

Innovative bulk storage options include mined salt caverns, depleted natural gas reservoirs, and liquid reservoirs [41]. Such storage methodologies will have to be developed to manage a hydrogen infrastructure, and may present some intermediate storage options. Issues such as efficiency cost and accessibility will determine applicability of any storage system. It is essential to recognize that any older infrastructure that is redirected toward hydrogen storage has to recognize the carbon steel interaction with moist hydrogen, dry hydrogen, or otherwise contaminated hydrogen.

Wind towers have been proposed as an alternate option for the storage of hydrogen. Electrolyzers would be the source of hydrogen in such towers, and can produce hydrogen up to 15 atm [42]. There are some practical design issues around fatigue, burst, corrosion, and HIC. Kottenstette and Cotrell [42] have reviewed all the design issues in detail and have identified pressure and some fatigue limitations versus cost on such storage options. The issue of moist hydrogen enhancing fatigue cracking has to be reviewed in further detail to establish some design options, which could include internal coatings and other loading considerations. However, the 30% reduction in potential cost as compared to pressure vessels has to drive further development work for wind turbine storage.

10.4 Vehicular Hydrogen Storage

Vehicular hydrogen storage, quantities ranging from 3 to 13 kg, represents the biggest R&D challenge. The options for bulk storage of hydrogen exist, and the present few economic and engineering challenges, however, will still face cost and efficiency constraints. A hydrogen infrastructure will require a minimum of two types of storage, which will include a local distribution (bulk) storage and vehicular storage. There are multiple hydrogen pathway options, and one possible pathway is a central hydrogen production plant that transports to local refueling locations. Another option is decentralized hydrogen production to replenish local distribution centers. Either way there will be requirements for local bulk storage, where the volumetric and gravimetric requirements will be determined by the transportation and distribution models that are finally adopted.

On a weight basis, hydrogen has three times the energy content when compared to gasoline (120 MJ/kg for hydrogen versus 44 MJ/kg for gasoline). However, hydrogen has four

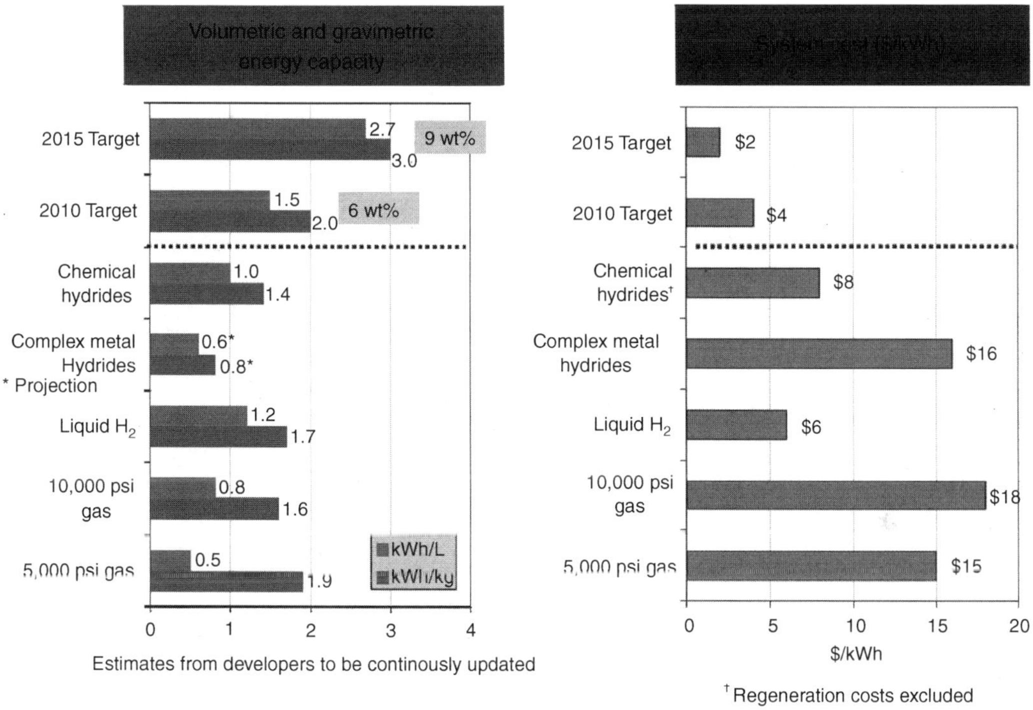

FIGURE 10.24
Comparison of volumetric, gravimetric energy capacity, and costs of various storage options.

times lower energy content when compared on a volume basis (8 MJ/L for hydrogen versus 32 MJ/L for gasoline). The storage options adopted for the application will have to reflect these weight and volume constraints for hydrogen. To compete with vehicles in use currently, the DOE has set a target driving range of >300 mi. for hydrogen-powered vehicles [43]. Further constraints and performance requirements were set on hydrogen storage systems to be competitive with current gasoline systems. The storage system targets set by the DOE are shown in Table 9.2. These are system-driven targets for the hydrogen storage options to be competitive with the current gasoline options. As shown in Figure 10.24 [44], the well-developed technologies of high-pressure tanks and LH2 meet some of the criteria. The focus of the DOE-sponsored research is on the materials-based storage systems with the objective to meet the goals in 2010 and 2015.

10.4.1 Compressed Hydrogen Storage

One kilogram of hydrogen occupies 11 m^3 of volume at ambient pressure and temperature and consequently storage requires enormous compression. High-strength steel or other metals are an option from a strength perspective; however, diffusivity of hydrogen through the steel and weight of the steel are major issues for vehicular storage. Other metallic options such as austenitic steels are an option; however, weight considerations would eliminate them as an option. For compressed natural gas storage applications, type-3 and -4 tanks are appropriate when weight considerations are critical. Type-3 tanks are metal lined with a load-bearing composite outer wrap. Type 4 is a lightweight tank with an internal liner to prevent gas diffusion and load-bearing composite over wrap. The external wrap on both type-3 and -4 tanks are a carbon fiber, a glass fiber, or hybridization of the two with epoxy

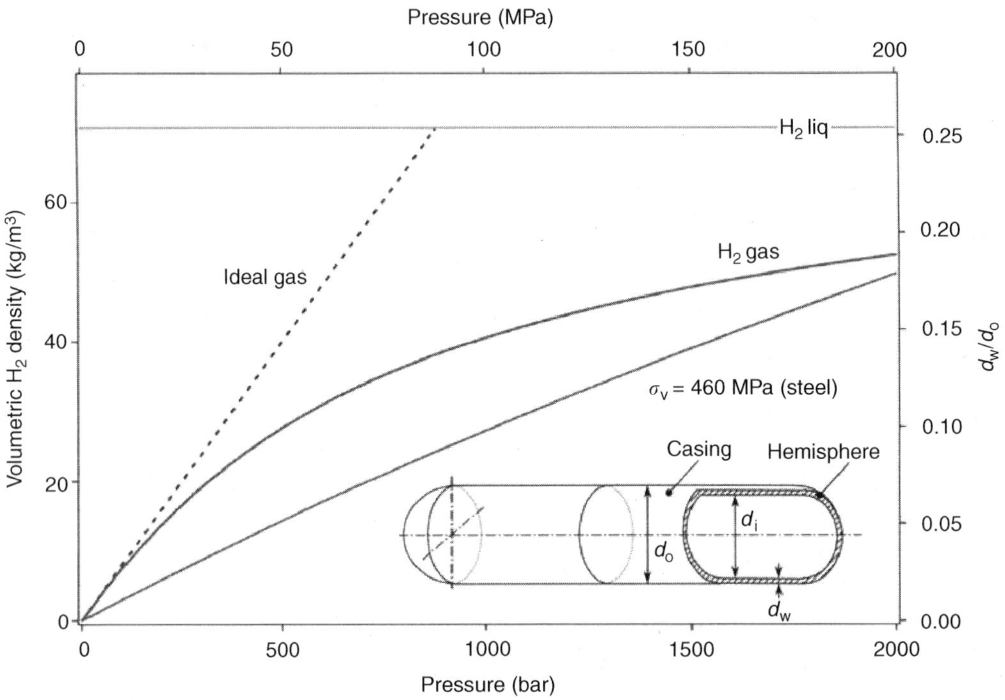

FIGURE 10.25
Nonideal volumetric density variation versus pressure.

resin as a binding material. There is one manufacturer of the 5,000 and 10,000 psi CH2 vessels. These have an internal liner that is a high-density polymer to prevent hydrogen diffusion. The next layer is a carbon fiber–epoxy that is load bearing, and finally an outer shell for impact and abrasion resistance. Several standards, groups, and institutes have certified and approved up to 10,000 psi storage tanks for fuel cell vehicles (FCVs).

A storage of 5 kg of CH2 at 5000 psi would require a volume of 212 L (or 56 gal.), which is substantially higher than conventional vehicles. And only 8.5% of the energy content of the hydrogen is used in compression [45]. The fill times are not an issue; however, the rate should be monitored for temperature increases. As shown in Figure 10.25 [46,47], since hydrogen is a nonideal gas, the volume density increases nonlinearly with increase in pressure, and beyond 700–800 bar the slope is smaller and further increases in compression will not yield substantially higher gravimetric density of hydrogen. The current challenge in CH2 is to develop conformable pressure vessels that can optimize space in FCVs. In addition, continuing reduction in cost and lower weight will be the focus in future developments. There was a theoretical study of the application of FCVs in urban India that demonstrated on an overall basis the net energy utilized for CH2 storage is the lowest as compared to LH2, metal hydride, and all other storage options [48]. This is a mature option for vehicular applications.

10.4.2 Liquid Hydrogen Storage

The energy density of hydrogen can be doubled as compared to the 10,000 psi CH2 to 70 g/L by liquefaction to 20 K [49]. Five kilogram of hydrogen only requires 71 L of volume, and this is equivalent on a volumetric basis to current vehicles. The phase diagram in Figure 10.26

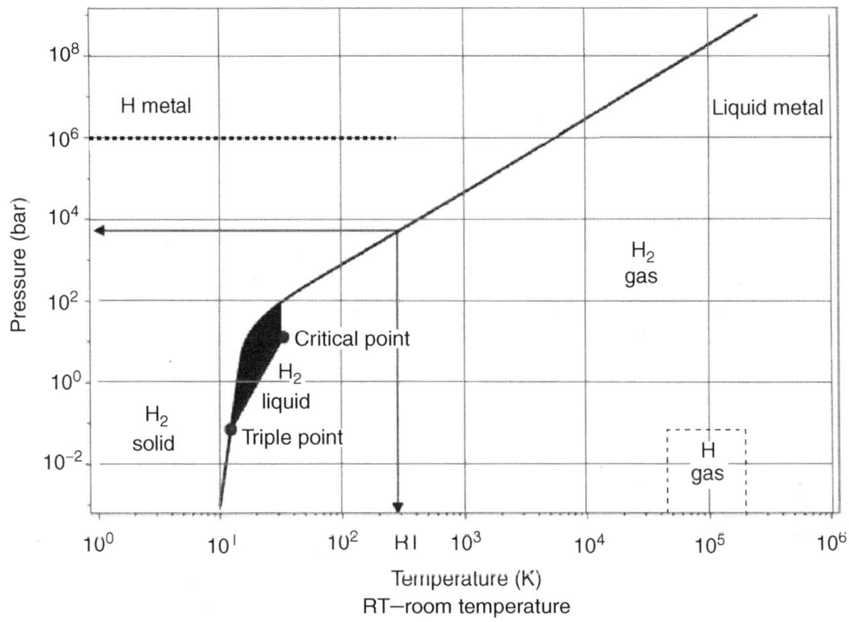

FIGURE 10.26
Hydrogen phase diagram (pressure versus temperature).

for hydrogen molecule shows the various forms of hydrogen as a function of pressure and temperature [46,47]. There is a small zone starting at the triple point, and ending at the critical point where hydrogen exists as a liquid with a density of 70.8 kg/m^3 at −253°C.

The energy utilized for cooling to LH2 requires an estimated 30% of the lower heating value (LHV) of hydrogen, which is substantially higher as compared to CH2. There is a period over which hydrogen will warm up in a storage vessel and convert to gas and will require venting from the LH2 tank. This period is the dormancy period, and there is one estimate of 4% per day for a 4.6 kg tank, and here the hydrogen will last for about 25 days [45,50]. The higher capacity of LH2 tanks will provide longer travel distances without refueling; however, the dormancy period and energy requirements for liquefaction are major issues.

The cryo-CH2 storage was a response to these issues, and insulated pressure vessels have been developed to address the dormancy issues [51,52]. On a weight basis, these will be heavier than the CH2 storage; however, lighter than the hydride storage options. One option that has been proposed and tested is an insulated pressure vessel that has the capability to operate at 20 K and also at a pressure of 240 atm, with higher pressures being an option. Aceves et al. [51,52] have designed and prototype tested such vessels. A type-3 vessel with an aluminum liner and composite outer wrap was utilized with a multilayer vacuum insulation and also an outer vacuum vessel. These have been extensively tested and also used in a demonstration vehicle. These vessels can be filled with LH2 or CH2 depending on the range of driving required. There is a three-fold increase in the dormancy period as the service pressure increases from 6 to 200 atm; consequently, such vessels can tolerate larger periods of inactivity with LH2. Additionally, evaporative losses are substantially decreased with increasing service pressures. A substantial practical advantage with cryocompressed storage is flexible fueling using CH2 and LH2. This hybrid solution will require further refinement of the vessel design although still addressing the weight and also ensuring conformable design.

Some preliminary results have been reported in the literature where a mixture of hydrogen with liquid nitrogen demonstrated substantially high densities at higher temperatures [53]. For example, 70 mol% H_2 mixtures had a density of 45 kg H_2/m^3 at 77 K at a pressure of 6000 psi. The mixtures had a higher compressibility compared to pure hydrogen. These are very encouraging results because they demonstrate some simpler solutions to vehicular hydrogen storage. Such mixtures combined with a cryocompressed vessel design may provide an optimal solution.

10.4.3 Hydride Storage

Hydrides absorb hydrogen and release heat during this reaction, and then require heat to release hydrogen when required as a fuel. Currently, there are commercially available hydride storage units. As a system, they have a substantial advantage on a volumetric basis (0.13–0.15 kg H_2/L) as compared to CH2 and LH2; however, have a very high system weight percent. The metal and intermetallic hydrides such as LaNi and FeTi—the simplest hydrides—operate at lower temperatures of 100°C and lower pressures. The weight percent of such hydrides range from 1.5% to 3%, and the volumetric density range from 0.05 to 0.12 kg H_2/L [45]. Higher gravimetric densities can be achieved by the higher-temperature hydrides such as MgH_2 and $MgNiH_2$ that operate at 300°C and at pressures ranging from 5 to 10 atm. The sodium alanate doped with Ti has a hydrogen weight percent twice as that of the traditional low-temperature hydrides. However, the slow reaction kinetics of the hydrides continues to be a major issue for vehicular applications. The alanates have approached a gravimetric density of 5%, which is still lower than the DOE system target of 6%. There is substantial R&D that is targeting some complex hydrides that exceed the DOE target; however, much development work is necessary. There are also chemical hydrides that release hydrogen by a chemical reaction. However, such reactions are not reversible. Storage in metal hydrides, carbons, and as chemical hydrides is discussed in Chapters 11 through 13.

There was a pilot test conducted on a city transit vehicle in the city of Augusta, Georgia that demonstrated the potential applicability of metal hydride storage [54]. Using a lanthanum-nickel-aluminum-type low-temperature hydride material, this supplied 6 kg/h of hydrogen to power an internal combustion (IC) engine to provide a driving range of 300 km [54]. This project conducted in 1998 is a demonstration of the possibilities with this technology.

10.4.4 Future Outlook

The CH2 with 25 g/L has the lowest density; however, the metal hydride option has the highest density with 125 g/L. These are material densities; however, in actual applications, the tank, piping, valves, and thermal management systems will substantially reduce the density and storage efficiency range from about 16% to 40% [55]. The CH2 because of its simplicity has a storage efficiency of around 40%, and the metal hydrides because of the heat exchanges and other associated systems have the lowest efficiency of 16% [55]. The efficiency for LH2 is 37%, which is comparable to the CH2 option. CH2 storage is considered appropriate for vehicles with an engine power $<$200 hp, and intermittent consumer usage. However, for vehicles with power $>$200 hp including taxis, buses, and trucks, the LH2 or cryocompressed storage option is necessary.

Other options such as carbon nanotubes and conformable materials are being researched for hydrogen storage [43,44]. These are funded through the DOE hydrogen storage program; however, the commercial applicability of these options are at least 10 years away.

10.5 Conclusion

Pipelines represent the primary option for the most efficient transportation mode in a hydrogen energy environment. Pipeline is the most economical option for transportation; however, the embrittlement of steels and diffusion losses represents a technical challenge. Sensor technologies for continued integrity management of such a pipeline network will be critical in the development of the hydrogen economy.

Vehicular storage of hydrogen has to become more efficient on a volumetric and gravimetric basis to enable the true application of hydrogen, and deliver the environmental advantages compared to fossil fuels. The DOE is furiously sponsoring research on materials development for efficient hydrogen storage with very aggressive quantitative goals by 2015.

References

1. U.S. Department of Energy: Hydrogen, fuel cells and infrastructure technologies program multi-year research, development and demonstration plan, Section 3.2, Hydrogen Delivery, January 21, 2005.
2. HFCIT. *Hydrogen Delivery: Current Technology*, 2007.
3. Keith, G. and Leighty, W. Transmitting 4,000 MW of New Windpower From North Dakota to Chicago: New HVDC Electric Lines or Hydrogen Pipeline, Draft Report, September 28, 2002.
4. Leighty, W., Hirata, M., O'Hashi, K., Asahi, H., Benoit, J., and Keith, G. Large renewables—hydrogen energy systems: Gathering and transmission pipelines for windpower and other diffuse, dispersed sources, *World Gas Conference*, Tokyo, Japan, June 1–5, 2003.
5. Leighty, W., Holloay, J., Merer, R., Somerday, B., Marchi, C.S., Keith, G., and White, D. Compressorless hydrogen transmission pipelines deliver large-scale stranded renewable energy at competitive cost, *The 23rd World Gas Conference*, Amsterdam, 2006.
6. FreedomCar FuelPartnership. Hydrogen Delivery Technology Roadmap, Final Report, 2007.
7. Ringer, M., Mintz, M., Gillette, J., Elgowairy, A., Brown, D., and Paster, M., Cost installing and operating hydrogen pipelines, *Second Panel Forum—Hydrogen Pipeline Transmission*, IPC2006, Calgary, September 2006.
8. DOE. Hydrogen Safety Fact Sheet, DOE Hydrogen Program, November 2006.
9. Feng, Z. Hydrogen Permeability and Integrity of Hydrogen Transfer Pipelines, FY 2006 Annual Progress Report, III.A.1, DOE Hydrogen Program, 2006.
10. Ringer, M. H2A Delivery Components Model Version 1.1: User Guide, April 2006.
11. Mintz, M., Gillette, J., and Elgowainy, A. Hydrogen distribution options analysis using a generalized model of hydrogen delivery, *Transportation Technologies and Fuels Forum*, Toronto, June 6, 2006.
12. Chen, T.P. Hydrogen Delivery Infrastructure Option Analysis, FY 2006 Annual Progress Report, DOE Hydrogen Program, 2006.
13. Leighty, W., Holloay, J., Merer, R., Somerday, B., Marchi, C.S., Keith, G., and White, D. Compressorless hydrogen transmission pipelines deliver large-scale stranded renewable energy at competitive cost, *Power-Gen Renewable Energy and Fuels*, Las Vegas, NV, April 10–12, 2006.
14. Pangborn, J., Scott, M., and Sharer, J. Technical prospects for commercial and residential distribution and utilization of hydrogen, *International Journal of Hydrogen Energy*, 2, 431–445, 1977.
15. Jasionowski, W., Johnson, D.G., and Pangborn, J.B. Suitability of gas distribution equipment in hydrogen service, *Proceedings of the 14th Intersociety Energy Conversion Engineering Conference*, T.N. Veriroglu and J.B. Taylor (Eds), vol. 1, Boston, MA, August 5–10, 1979, 759–762.

16. Jasionowski, W. and Huang, H. Gas distribution equipment in hydrogen service, IGT Report, Paper presented at the *15th Intersociety Energy Conversion Engineering Conference*, Seattle, WA, August 18–22, 1980.
17. Blazek, C.F., Biederman, R.T., Foh, S.E., and Jasionowski, W. Underground storage and transmission of hydrogen, *Proceedings of the 3rd Annual US Hydrogen Meeting*, Washington, DC, pp. 4-203–4-221, March 18–20, 1992.
18. Das, S. Material solutions for hydrogen delivery in pipelines, FY 2006 Annual Report, III.A.3, DOE Hydrogen Program, 2006.
19. Sofronis, P., Robertson, I.M., and Johnson, D.D. Hydrogen embrittlement of pipeline steels: Cause and remediation, 2006 Annual Progress Report, III.A.5, DOE Hydrogen Program, 2006.
20. Smith, B., Eberle, C., Frame, B., Blencoe, J., Anovitz, L., and Mays, J. FRP Hydrogen Pipelines, 2006
21. Frank, J. *The Mathematics of Diffusion*, Second Edition, Oxford University Press, Oxford, 1975, p. 52.
22. Alefeld, G., and Volkl, J. *Hydrogen in Metals I—Basic Properties*, Springer, New York, 1978, p. 328.
23. Gangloff, R. Hydrogen assisted cracking of high strength alloys, in *Comprehensive Structural Integrity*, I. Milne, R.O. Ritchie, and B. Karihaloo (Eds), vol. 6, Elsevier, New York, 2003.
24. Birnbaum, H.K., Roberson, I.M., Sofronis, P., and Teter, D. Mechanisms of Hydrogen Related Fracture—A Review, in *Corrosion-Deformation Interactions, CDI '96*, T. Magnin (Ed), The Institute of Mining, Materials, London, 1997, pp. 172–195.
25. Wei, R.P. and Gao, M. Chemistry, microstructure and crack growth response, in *Hydrogen Degradation of Ferrous Alloys*, R.A. Oriani, J.P. Hirth, and S. Smialowska (Eds), Noyes Publications, Park Ridge, NJ, 1985, pp. 579–607.
26. Loginow, A.W. and Phelps, E.H. *Steels for Seamless Hydrogen Pressure Vessels, Corrosion*, vol. 31, 1975, pp. 404–412
27. Walter, R.J. and Chandler, W.T. Cyclic-load crack growth in ASME SA-105 Grade II steel in high-pressure hydrogen at ambient temperature, in *Effect of Hydrogen on Behavior of Materials*, A.W. Thompson and I.M. Berstein (Eds), American Institute of Mining, Metallurgical, and Petroleum Engineers, New York, 1976, pp. 273–286.
28. Clarke, G.A. and Landes, J.D. *Journal of Testing and Evaluation*, 7(5), 251, 1979.
29. Anon. *Guide on Methods for Assessing the Acceptability of Flaws in Metallic Structures*, BS7910: 1999, British Standard, 1999.
30. Anon. *Finess-For-Service, API Recommended Practice 579*, First Edition, API Publishing Services, Washington, DC, 2000.
31. Adams, T. Evaluation of natural gas pipeline materials and infrastructure for hydrogen/hythane service, FY 2006 Annual Progress Report, III.A.4, 2006.
32. Mohitpour, M., Golshan, H., and Murray, A. *Pipeline Design and Construction: A Practical Approach*, ASME Press, New York, 2000.
33. Mohitpour, M., Pierce, C., and Hooper, R. The design and engineering of cross-country hydrogen pipelins. *ASME Journal of Energy Resources Technology*, 110, 203–207, 1988.
34. Pottier, J. Hydrogen transmission for future energy system, *Hydrogen Energy System*, Kluwer Academic Publishers, Netherlands, 1995, pp. 181–193.
35. Smith, B., Eberle, C., Frame, B., Blencoe, J., Anovtz, L., and Mays, J. FPR Hydrogen Pipelines, FY 2006 Annual Progress Report, III.A.2, 2006.
36. Chisholm, B.J., Moore, R.B., Barber, G., Khouri, F., Hempstead, A., Larsen, M., Olson, E., Kelley, J., Balch, G., and Caraher, J. Nanocomposites derived from sulfonated poly(butylene terephthalate), *Marcromolecules*, 35, 5508, 2002.
37. Zimmerman, T., Stephen, G., and Glover, A. Composite reinforced line pipe (CRLP^{TM}) for on shore gas pipelines, IPC2002, Paper No. 27125, *Proceedings of IPC02*, Calgary, October 2002.
38. Pitts, R.R., Smith, D., Lee, S., and Tracy, E. Interfacial Stability of Thin Film Hydrogen Sensors, FY 2004 Annual Progress Report, VI.3, DOE Hydrogen Program, 2004.
39. Tennyson, R.C., Morison, D., Colpitts, B., and Brownn, A. Application of Brillouin fiber optic sensors to monitor pipeline integrity IPC2004, Paper# 0711, Calgary, October 2004.

40. Zou, L., Ferrier, G.A., Afshar, S., Yu, Q., Chen, L., and Bao, X. Distributed Brillouin scattering sensor for discrimination of wall-thinning defects in steel pipe under internal pressure, *Applied Optics*, 43(7), 1583–1588, 2004.
41. Venter, R.D. and Pucher, G. Modelling of stationary bulk hydrogen storage systems. *International Journal of Hydrogen Energy*, 22(8), 791–798, 1997.
42. Kottenstette, R. and Cotrell, J. *Hydrogen Storage in Wind Turbine Towers*, National Renewable Energy Laboratory, NREL/TP-500-34656, 2003.
43. Read, C., Petrovic, J., Ordaz, G., and Satryapal, S. The DOE national hydrogen storage project: Recent progress in on-board vehicular hydrogen storage, *Material Research Society Symposium Proceedings*, vol. 885E, 2006.
44. Hydrogen, Fuel Cells and Infrastructure Technologies Program Multi-Year Research, Development and Demonstration Plan, www.eere.energy.gov/hydrogenandfuelcells/mypp, and the DOE Hydrogen Program Plan, www.hydrogen.energy.gov/pdfs/hydrogen_posture_plan.pdf.
45. Burke, A.F. and Gardiner, M. Hydrogen Storage Options: Technologies and Comparisons for Light-duty Vehicle Applications, Research Report UCD-ITS-RR-05-01, Institute of Transportation Studies, University of California, Davis, 2005.
46. Zuttel, A. Materials for hydrogen storage, *Materials Today*, September 2003.
47. Zuttel, A. Hydrogen storage methods, *Naturwissenschaften*, 91, 157–172.
48. Sarkar, A. and Banerjee, R. Net energy analysis of hydrogen storage options, *International Journal of Hydrogen Energy*, 30, 867–877, 2004.
49. Arnold, G. and Wolf, J. Liquid hydrogen for automotive application next generation fuel for FC and ICE vehicles. *Journal of Cryogenic Society (TEION KOGAKU) Japan*, 40(6), 221–230, 2005.
50. Arnold, G. *Advanced Hydrogen Storage Technologies, General Motors Hydrogen Storage and Refueling*, Global Alternative Propulsion Center, Mainz-Kastel, Germany.
51. Aceves, S.M., Berry, G.D., Martinez-Frias, J., and Espinosa-Loza, F. Vehicular storage of hydrogen in insulated pressure vessels. *International Journal of Hydrogen Energy*, 31, 2274–2283, 2006.
52. Aceves, S.M., Martinez-Frias, J., and Garcia-Villazana, O. Analytical and experimental evaluation of insulated pressure vessels for cryogenic hydrogen storage. *International Journal of Hydrogen Energy*, 25, 1075–1085, 2000.
53. Berry, G.D. *Hydrogen Absorption in Fluids: An Unexpected Solution for Onboard Hydrogen Storage*, UCRL-TR-209650.
54. Heung, L.K. *On-Board Hydrogen Storage for a City Transit Bus*, WSRC-MS-98-00107, 1998.
55. Miller, A., Hess, K., and Barnes, D. (2003). *Comparison of Practical Hydrogen Storage Volumetric Densities*, DE-FC36-99GO10458.
56. Mintz, M. and Ringer, M. H2A Delivery Analysis, FY 2006 Annual Progress Report, DOE Hydrogen Program, 2006.
57. Kesten, M. and Windgassen K.F. Hydrogen-assisted fatigue of periodically pressurized steel cylinders, in *Hydrogen Effects in Metals*, I.M. Bernstein and T.W. Thompson (Eds), American Institute of Mining, Metallurgical, and Petroleum Engineers, New York, 1981, pp. 1017–1025.
58. Hoover, W.R., Robinson, S.L., Stoltz, R.E., and Spingarm, J.R. *Hydrogen Compatibility of Structural Materials for Energy Storage and Transmission Final Report*, SAND81-8006, Livemore, CA, 1981.
59. Asahi, H. and Ishitsuka, T. *Applicability of Line Pipe to Transportation of Hydrogen-Containing Gas*, Steel Research Laboratories, Nippon Steel Corporation, Japan.
60. Asahi, H., Hirata, M., and O'Hashi, K. *Hydrogen Related Damage in Line Pipe*, Nippon Steel Corporation, Japan.
61. Leighty, W. Personal communication with H. Asahi, Nippon Steel, Jan 15, 2003.
62. Vosikovsky, O., Macecek, M., and Ross, D.J. Allowable defect sizes in a sour crude oil pipeline for corrosion fatigue conditions, *International Journal of Pressure Vessels and Piping*, 13, 197–226, 1983.

11

Hydrogen Storage in Metal Hydrides

K. K. Pant and Ram B. Gupta

CONTENTS

11.1 Introduction

The development of a hydrogen-based economy is contingent on the development of safe and cost-effective hydrogen storage systems. Several studies have been published on the storage of hydrogen by different techniques [1–27]. Hydrogen storage is a key element in the utilization of hydrogen fuel (Figure 11.1) [18], and the energy consumed in each step of the process—production, multiple storages, and transport should be minimized for a high energy efficiency. Today, the annual consumption of hydrogen in the United States alone is more than 90 billion m^3 [28,29], mostly in refineries and chemical plants where it is produced. For such an application, storage has not been critical. However, for fuel usage in vehicles, storage has been the main technical challenge. Present storage techniques for hydrogen include compressed gas, cryogenic liquid, and absorption on solids. This chapter deals with the hydrogen storage using solid metal hydrides. For optimum hydrogen storage, the metal should have following desirable properties:

- High hydrogen capacity per unit mass and unit volume
- Low dissociation temperature

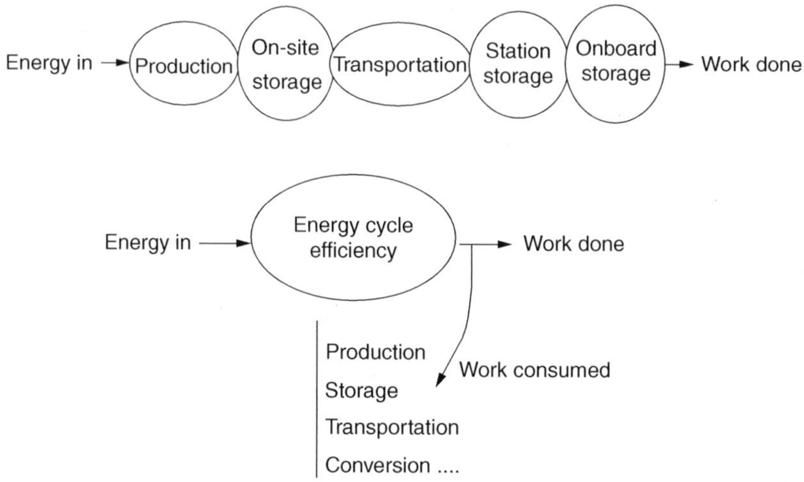

FIGURE 11.1
Importance of hydrogen storage. (Adapted from Heung, L.K., Using Metal Hydride to Store Hydrogen, U.S. Department of Commerce, National Technical Information Service Springfield, VA, 2003, available at http://sti.srs.gov/fulltext/ms2003172/ms2003172.pdf, May 2007.)

- Moderate dissociation pressure
- Low heat of formation to minimize the energy necessary for hydrogen release and low heat dissipation during the exothermic hydride formation
- Reversibility for limited energy loss during charge and discharge of hydrogen
- Fast kinetics
- High stability against O_2 and moisture for a long cycle life
- Cycleability
- Low cost of recycling and charging infrastructures
- High safety

Current metal hydrides, suitable for automotive applications, typically store 0.5–2 wt% hydrogen. Hydrides release heat when charged with pressurized hydrogen, and absorb heat to release hydrogen. More than 50 metallic elements of the periodic table can absorb hydrogen in good quantities; hence, the possible choices of hydride materials are enormous. But, only some of them are suitable for hydrogen storage at moderate temperatures and pressures. Practically, the temperature should not go above 100°C and below 10°C during charging and discharging, respectively. The choice is further narrowed by the need for a moderate charging pressure and a suitable discharging pressure. For example, a hydrogen supply pressure of 2.7 MPa is used, and the discharge pressure >0.2 MPa is required. To tailor metal properties to meet the requirements, alloys of various compositions are being utilized. These alloys are classified as AB_5 (e.g., $LaNi_5$), AB (e.g., FeTi), A_2B (e.g., Mg_2Ni), and AB_2 (e.g., ZrV_2).

Other key aspects in selecting a hydride include hydrogen capacity, energy density, and enthalpy of discharging (Table 11.1). The heat of hydrogenation (charging) and dehydrogenation (discharging) vary between 30 and 70 kJ/mol H_2. The typical intermetallic alloys shown in Table 11.1 are capable of storing hydrogen at pressures from <0.1 to 1.5 MPa. The special group includes materials from the AB_5, AB_2, and AB types, for example, $LaNi_{5-x}Al_x$, $TiV_{2-x}Mn_x$, and $FeTi_{1-x}Mn_x$, where x is a variable for adjusting the equilibrium pressure

TABLE 11.1

Energy Content of Various Types of Metal Hydrides

Media Type	Material	Hydride Form	H_2 Capacity (wt%)	Energy Density (kJ/kg Hydride)	Enthalpy of Hydrogenation (kJ/mol H_2)
AB_5	$LaNi_5$	$LaNi_5H_6$	1.3	1,850	30.1
AB	TiFe	$TiFeH_{1.95}$	1.7	2,560	28.1
AB_2	$ZrMn_2$	$ZrMn_2H_4$	1.7	2,419	53.2
A_2B	Mg_2Ni	Mg_2NiH_4	7.0	10,000	64.5
Metal	Mg	MgH_2	7.7	11,000	74.2

Source: Adapted from Heung, L.K., Using Metal Hydride to Store Hydrogen, U.S. Department of Commerce, National Techinical Information Service, Springfield, VA, 2003, available at http://sti.srs.gov/fulltext/ms2003172/ms2003172.pdf, May 2007.

TABLE 11.2

Metal Hydride Requirements for Hydrogen Storage for Mobile Applications

Property	Target
Gravimetric storage density	>6.5%
Volumetric storage density	>6.5%
De/rehydrogenation kinetics	Dehydrogenation <3 h Rehydrogenation <5 min
Rehydrogenation pressure	<5 MPa
Equilibrium pressure	About 0.1 MPa near room temperature
Specific energy (kWh/kg)	>2.0
Refueling rate (Kg H_2/min)	>1.5
Safety	As high as possible, that is, no ignition on exposure to air or water
Cycle life (cycles)	>500
Loss of usable hydrogen (g/h/Kg hydrogen stored)	<0.1

Source: Adapted from Schuth, F., Bogdanović, B., and Felderhoff, M., *Chem. Commun.*, 20, 2249–2258, 2004.

and stability of the material. Low levels of other elements may also be added to improve the overall performance.

Hydrides that need high temperatures and pressures to operate are undesirable due to the increased equipment and energy requirements. Schuth et al. [30] have summarized specific requirements of metal hydrides for mobile applications (Table 11.2), which exclude many of the materials. A compromise may be possible, depending on the values of various parameters. For example, if the storage capacity is sufficiently high, then slower hydrogenation kinetics may be accepted, since in that case material loading to the full capacity may not be required. The most stringent requirements with respect to gravimetric and volumetric storage capacities are posed by automobile applications, where the acceptable weight and volume of a storage tank has to be about 100 kg and 100 L, respectively. Rehydrogenation is more time crucial than dehydrogenation, as the rehydrogenation time is the time spent at the filling station and should be <5 min.

A comparison between different metal hydrides used for hydrogen storage is given in Figure 11.2 [31].

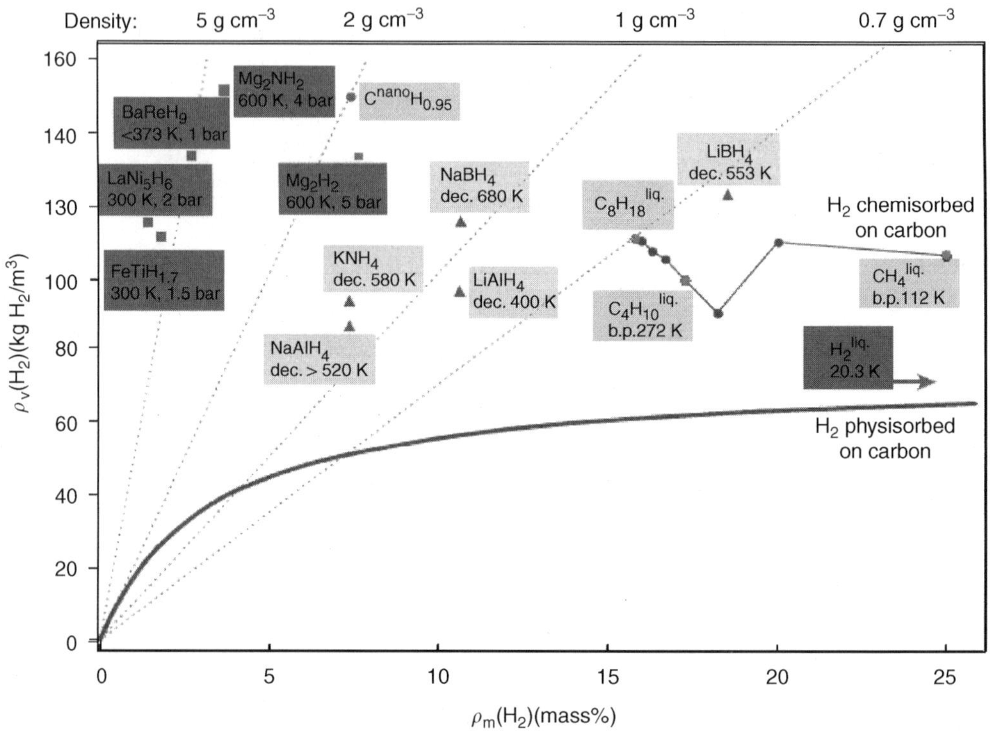

FIGURE 11.2
Stored hydrogen per mass and per volume. Comparison of metal hydrides, carbon nanotubes, petrol, and other hydrocarbons. (Reproduced with permission from Schlapbach, L., and Zuttel, A., *Nature*, 414, 353–358, 2001.)

11.2 Types of Metal Hydrides

Metals can combine with hydrogen to form metal hydrides by three different bond types: (1) ionic, (2) covalent, and (3) metallic. Based on the strength of the bonds, metal hydrides are broadly classified into two categories: (1) reversible and (2) irreversible. Examples of reversible metal hydrides are as follows:

- AB_5 is most common (e.g., nickel–metal hydride batteries with 1–1.25 reversible wt% H_2 capacity).
- AB_2 very common (1.3 reversible wt% H_2 capacity).
- AB (e.g., TiFe with 1.5 reversible wt% H_2 capacity).
- A_2B (e.g., Mg_2NiH_4 with 3.3–7.0 reversible wt% H_2 capacity).
- AB_3, A_2B_7.

The reversible metal hydrides can undergo a reversible equilibrium reaction as follows:

$$MH_{n,\text{solid}} + \text{Heat} \leftrightarrow M_{\text{solid}} + (n/2)H_2$$

where the forward reaction is the release of hydrogen or desorption and the backward reaction is the adsorption of hydrogen or charging. Reversible hydrides have been used for a long time as hydride-based batteries. These hydrides, such as $LaNi_5H_6$ or FeTiH, are typical low-temperature hydrides, that is, they have a dissociation pressure higher than 0.1 MPa at room temperature. In contrast, most of the light metal binary hydrides work at high temperatures. Although MgH_2 is an excellent material for reversible heat storage and can be used in solar thermal systems, it is the least stable one, with an equilibrium pressure of 0.1 MPa slightly below 300°C [32,33]. The light alkali metal hydrides such as LiH and NaH are more stable, with decomposition temperatures exceeding 500°C.

In an irreversible metal hydride, hydrogen is generated through a chemical reaction, which is not easily reversible onboard a vehicle. Common reactant is water or an alcohol. For example, sodium borohydride can produce hydrogen as

$$NaBH_4 + 2H_2O \rightarrow NaBO_2 + 4H_2$$

This reaction is not easily reversible onboard a vehicle. Hence, the by-products must be removed from the vehicle and regenerated offboard. The irreversible metal hydrides can include a transition metal, for example, Mg_2FeH_6 (5.5 wt% hydrogen capacity), a nontransition metal such as $Be(BH_4)_2$ (20.8 wt% hydrogen capacity), or $NaAlH_4$. Various promising metal hydrides are discussed in the following sections.

11.3 Sodium Alanates

Alanates are complex metal hydrides, which have a potential for a high hydrogen capacity, out of which sodium alanate ($NaAlH_4$) is a promising material for hydrogen storage with a theoretical reversible storage capacity of about 5.5 wt%. It is one of the known complex hydride with favorable thermodynamics and an acceptable gravimetric storage capacity for use in conjunction with a polymer electrolyte membrane (PEM) fuel cell system, which operates at about 90°C. Reversibility kinetics of $NaAlH_4$ is improved upon addition of titanium and or zirconium dopants, which causes a dramatic decrease in the primary crystal size as shown in Figure 11.3a and 11.3b. The elemental segregation of Na and Al found in particles after multiple absorption/desorption cycles may be a factor in the relatively low absorption capacity (Figure 11.3c) [34]. Sc-doped $NaAlH_4$ also exhibits

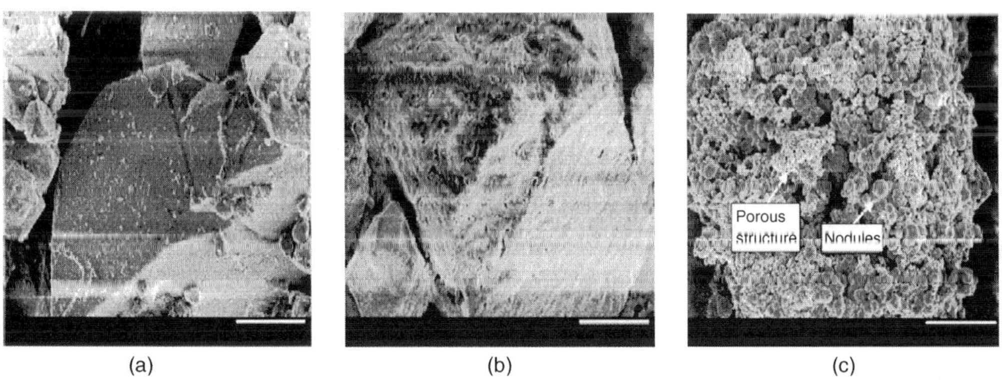

FIGURE 11.3
Scanning electron micrographs of (a) $NaAlH_4$ before the catalyst doping, (b) $NaAlH_4$ after Ti/Zr catalyst doping using a wet alkoxide precursor and mechanical milling, and (c) fully desorbed material after five absorption/desorption cycles. (Reproduced with permission from Thomas, G.J., Gross, K.J., Yang, N.Y.C., and Jensex, C. *J. Alloys Compd.*, 330–332, 702–707, 2002.)

comparable kinetics and a higher hydrogen storage capacity compared to Ti-doped $NaAlH_4$, depending on the temperature, metal dopant level, and cycling.

Hydrogen is released from $NaAlH_4$ in two steps:

$$3NaAlH_4 \leftrightarrow Na_3AlH_6 + 2Al + 3H_2 \quad \text{(3.7 wt\% hydrogen released)}$$

$$Na_3AlH_6 \leftrightarrow 3NaH + Al + 1.5H_2 \quad \text{(1.8 wt\% hydrogen released)}$$

Thus, $NaAlH_4$ first decomposes to evolve molecular hydrogen and forms an intermediate compound Na_3AlH_6, which further decomposes to NaH with the formation of additional metallic aluminum and hydrogen. The first reaction is thermodynamically favorable at ambient pressure and above 33°C and can release about 3.7 wt% hydrogen, whereas the second reaction takes place above 110°C and can release an additional 1.8 wt% hydrogen. Rehydrogenation is preferably done at 10 MPa and slightly above 100°C. The kinetics is faster at higher temperatures, but is offset by an increase in the hydrogen equilibrium pressure with increasing temperature. Therefore, a compromise between fast rate and proximity to equilibrium need to be determined. Recently, Wang et al. [35] have reported a synergistic effect of graphite cocatalyst with Ti on both the dehydrogenation and the hydrogenation kinetics of $NaAlH_4$ over a wide range of temperature (90–250°C) [35]. For example, codoping of 2 mol% Ti-doped $NaAlH_4$ with 10 wt% graphite significantly improved the kinetics of the first and the second hydrogen evolution reactions by lowering the dehydrogenation temperature over the crucial range 90–150°C by as much as 15°C compared to that without graphite. Similarly, the addition of graphite was shown to improve the dehydrogenation kinetics at 90°C and 110°C, respectively, by factors of 6.5 and 3.0 compared to a sample doped only with Ti. The typical hydrogenation/dehydrogenation behavior of various sodium alanates is given in Table 11.3.

TABLE 11.3

Hydrogen Absorption and Desorption Properties of Sodium Alanates

Material	Temperature (°C)	Pressure (bar)	Kinetics (min)	H_2 Storage (wt%)
Na_3AlH_6	T_{des}: 200	P_{des}: 1	t_{des}: 150	2.5
Na_2LiAlH_6	T_{abs}: 211	P_{abs}: 45	t_{abs}: 100	2.5
$NaAlH_4$ + 2 mol% $Ti(OBu^n)_4$ −2 mol% $Zr(OPr^i)_4$	T_{abs} and T_{des}: 125–165	P_{abs} and P_{des}: 101–202	t_{abs}: 60	3.0
$NaAlH_4$ + 4 mol% Ti	T_{abs}: 120	P_{abs}: 120	t_{abs}: 60	3.3
	T_{des}: 150	P_{des}: 1	t_{des}: 600	
$NaAlH_4$ + 2 mol% Ti	T_{abs}: 25–193	P_{abs}: 20–120	t_{abs}: 300–720	3.8
	T_{des}: 25–160	P_{des}: 1	t_{des}: 40	
$NaAlH_4$ + 2 mol% $Ti(OBu)_4$		P_{abs} and P_{des}:	t_{abs}: 1020	4.0
	T_{abs}: 120	60–150		
	T_{des}: 180–260		t_{des}: 120–300	
$NaAlH_4$ + 2 mol% $Ti(OBu)_4$		P_{abs} and P_{des}:		4.0
	T_{abs}: 135–120	150–130	t_{abs}: 330	
	T_{des}: 180–160		t_{des}: 90	
$NaAlH_4$ + 2 mol% $TiCl_3$	T_{des}: 125–100	P_{des}: 83–91	t_{des}: 20	4.0
$NaAlH_4$ + 2 mol% $Ti(OBu)_4$	T_{abs}: 104	P_{abs}: 88	t_{abs}: 1020	4.0
$NaAlH_4$	T_{abs} and T_{des}: 80–180	P_{abs} and P_{des}: 76–91	t_{abs}: 120–300	5.0

Note: Bu is C_4H_9, and Pr is C_3H_7.

Source: Adapted from Sakintuna, B., Lamari-Darkrim, F., Hirscher, M., and Dogan, B., *Int. J. Hydrogen Energ.*, 32(9), 1121–1140, 2007.

To explain the preceding results, it is assumed that $TiCl_3$ reduces to highly reactive Ti(0), which reacts with the hydrogen to form titanium hydride. Alternatively, there might be a Ti-catalyzed hydrogenation of the present NaH/Al mixture. Ball milling (b.m.) of $NaAlH_4$ with doping agents $TiCl_2$ or $TiCl_3{\cdot}1/3\ AlCl_3$ significantly enhanced the evolution of hydrogen from 1.5 to 4 and 8 mol H_2/mole dopant, respectively, according to the following reactions [33,34]:

$$x\text{NaAlH}_4 + \text{TiCl}_2^{(\text{b.m})} \rightarrow (x-2)\text{NaAlH}_4 + \text{Ti} + 2\text{Al} + 2\text{NaCl} + 4\text{H}_2 \uparrow$$

$$x\text{NaAlH}_4 + (\text{TiCl}_3)\,\tfrac{1}{3}\cdot \text{AlCl}_3^{(\text{b.m})} \rightarrow (x-4)\text{NaAlH}_4 + \text{Ti} + \tfrac{13}{3}\text{Al} + 4\text{NaCl} + 8\text{H}_2 \uparrow$$

The effects of liquid alkoxides, $Ti(OBu_n)_4 + Zr(OPr_i)_4$ (where, Bu: C_4H_9, Pr: C_3H_7), on the hydriding properties of $NaAlH_4$ were investigated by Sandrock et al. [19–21]. The liquid alkoxides contribute to hydrocarbon contamination of the released hydrogen. Moreover, the surface of $NaAlH_4$ is damaged by the catalyst, which decreases the cyclic capacity. The dry doping with $TiCl_3$ catalyst eliminates the high catalyst weight, low capacity, and contamination problems for the alkoxide catalysts. The hydriding and dehydriding rates increase by doping [19–21,33]. Under the same rehydrogenation conditions, the amount of evolved hydrogen from the Ti-doped $NaAlH_4$ is about 50 times as much as that of the undoped $NaAlH_4$. This indicates that the addition of titanium species enhances kinetically not only the dehydrogenation, but also the rehydrogenation reaction of $NaAlH_4$ [33–35].

11.3.1 Other Alanates

Among the well-studied alanates, only $NaAlH_4$ has favorable thermodynamic properties for hydrogen storage. For example, kinetically stabilized $LiAlH_4$ decomposes immediately below room temperature due to thermodynamic instability. Theoretically, lithium alanates also seem to be attractive for hydrogen storage because of the high hydrogen content. The total hydrogen content is 10.5 and 11.2 wt% for $LiAlH_4$ and Li_3AlH_6, respectively [36–39]. Unfortunately, $LiAlH_4$ has an extremely high equilibrium pressure of hydrogen, even at room temperature. $LiAlH_4$ is an unstable hydride, which decomposes easily, but cannot be rehydrogenated [19–21]. The desorption of $LiAlH_4$ occurs in two steps:

$$3\text{LiAlH}_4 \rightarrow \text{Li}_3\text{AlH}_6 + 2\text{Al} + 3\text{H}_2$$

$$\text{Li}_3\text{AlH}_6 \rightarrow 3\text{LiH} + \text{Al} + \tfrac{3}{2}\text{H}_2$$

The total hydrogen released in the two reactions is 5.3 and 2.65 wt% from the decomposition of $LiAlH_4$ and Li_3AlH_6, respectively, at temperatures between 160°C and 200°C. About 2.65 wt% of the total hydrogen content in $LiAlH_4$ remains unreleased in the form of LiH after the completion of the reaction, which can be desorbed only at very high temperatures (680°C). Therefore, the commercialization of lithium-based compounds is hindered by their slow kinetics and high-temperature requirement [33]. Ball milling of $LiAlH_4$ in the presence of titanium, zirconium, or scandium compounds releases part of the hydrogen [35–37], which suggests that these act primarily on the $[AlH_4]_2$ unit that is present in most alanates. Reversible storage up to 1.8 wt% has been reported for titanium catalyzed $LiAlH_4$, although no full dehydrogenation/rehydrogenation cycles have been carried out. Of the alkaline earth metal alanates, only $Mg(AlH_4)_2$ has an attractive hydrogen storage capacity (about 7 wt%) [36], however, the thermodynamic properties of this material are not favorable for reversible hydrogen storage under practical conditions as the hydrogenation enthalpy is near zero [21,38]. In addition to $Mg(AlH_4)_2$, $Ca(AlH_4)_2$ was recently obtained by ball milling of calcium hydride with $AlCl_3$ [35–40].

The alanates of the other heavier alkaline earth metals also have a low gravimetric storage density for practical applications. Instead of the hexahydrides M_3AlH_6 (where M is Li, Na, K, etc.), the pentahydrides $MAlH_5$ are normally formed as intermediate phases having interesting structures as seen in the case of $BaAlH_5$ [40]. The pentahydrides can be considered as H-bridged hexahydrides, and the decomposition pathways may be related. However, in the ball-milling process, material properties, such as chemical homogeneity, grain sizes and surface areas, and impurity levels, are difficult to control. Furthermore, nanostructured metal hydrides with high surface areas are desirable to improve the thermodynamics and kinetic properties of the hydrogen storage materials. Such nanoscale materials are difficult to obtain through ball milling. Thus, chemical routes for the bulk production of a variety of nanostructured, nanoporous metal hydrides, metal alanates ($MAlH_x$), and amides (MNH_2) for hydrogen storage applications need to be explored [41,42].

11.4 Boron Hydrides

All the elements of the boron group form polymeric hydrides $(MH_3)_x$ because the MH_3 monomers are strong Lewis acids and are unstable. Borane (BH_3) achieves electronic saturation by dimerization to form diborane (B_2H_6). Lithium tetrahydroborate, $LiBH_4$, has a hydrogen storage capacity of 18.4 wt%, which is the highest gravimetric storage capacity among the complex hydrides [31,43]. Its decomposition reaction to produce hydrogen is as follows:

$$Li(BH_4) \rightarrow LiBH_2 + H_2 \rightarrow LiH + B + \tfrac{3}{2}H_2$$

or

$$Li(BH_4) \rightarrow LiB + 2H_2$$

However, the thermodynamic properties of complex boron hydrides are unfavorable for use in mobile hydrogen storage applications except in hydrolytic storage, where hydrides are much more stable than the corresponding alanates. Another problem is that small amounts of volatile boranes can form that are toxic and can damage subsequent aggregates in a fuel cell system, which over time would lead to a loss of storage capacity. $LiBH_4$ starts to decompose thermally above 320°C, with the major decomposition occurring above 400°C. The total amount released is typically only about 50% of the theoretical capacity. Zuttel et al. [31,43,44] have reported that the addition of SiO_2 powder can reduce the onset temperature of the decomposition and that 9 wt% of hydrogen can be released below 400°C. In principle, the decomposition of the stable hydrides could be made easier if compounds that form either stable complex with the boron generated from the boron hydride decomposition or with the generated LiH would be added, thus providing an additional thermodynamic driving force.

11.5 Magnesium Hydride

In addition to the weight of the storage system, cost needs to be kept low. Magnesium hydride (MgH_2) is inexpensive and has a potential storage capacity of 7.6 wt% H_2. However, utilization of MgH_2 has some drawbacks [31,44,45]. For example, for hydrogenation, Mg needs to be exposed to a high temperature and pressure of hydrogen. Dehydrogenation, at ambient temperature gives a low pressure of about 0.1 Pa, mainly due to the very high dissociation enthalpy of 74.5 kJ/mol MgH_2. Thus, dehydrogenation needs to be carried out at 300°C to achieve a hydrogen pressure of more than 0.1 MPa [46]. Another difficulty is that MgH_2, similar to other light metal hydrides, has slow hydrogen absorption/desorption kinetics. A fast absorption/desorption kinetics is important especially for vehicular application to have a short refueling time and a sufficient rate of hydrogen release to fulfill the power requirement while driving [44–46].

The desired temperature range for hydrogenation and dehydrogenation steps is between 50°C and 100°C. The temperatures for MgH_2 can be reduced by forming a composite of small Mg crystals loosely agglomerated together as opposed to having a large chunk of Mg crystals. Channels with a lower binding energy and smaller migration barriers for the hydrogen are apparently formed and maintained throughout the hydride, thus enabling fast loading and unloading. The kinetics can be enhanced by adding a small amount of a catalytically active metal on the surface of the Mg clusters, or by forming alloys of Mg with small amounts (few percentage) of catalytically active components such as Ni, Fe, V, or Pd. The kinetics has been significantly accelerated by mechanically alloying nanoscaled Mg-based materials with small quantities of transition metals such as V, Ti, Fe, Ni, and Mn or their oxides [45–51]. However, the heavy transition metal atoms are only needed on the surface of the clusters, where the hydrogen gas dissociates and it is preferable not to have them in the interior of the grains. A catalytic transition metal that naturally migrates to the surface of the clusters and coats the surface can mitigate the problem. The high enthalpy of formation of the hydride is also a serious problem, which can be improved by adding Ni or Cu to Mg [45]. For example, the desorption temperature of the Mg_2Ni alloy is about 50°C lower than that for pure Mg, but this is still not low enough. Unfortunately, the addition of the heavy transition metals greatly reduces the storage capacity from 7.6 to 3.6 wt% [28].

11.6 Miscellaneous Hydrides

Several other complex hydrides also have useful properties for hydrogen storage. With respect to storage capacity, lithium beryllium hydrides are excellent. Ball milling of LiH with Be and subsequent hydrogenation at temperatures above 250°C leads to different lithium beryllium hydrides of the general formula—$Li_nBe_mH_{n12m}$ [52]. The material with n:m ratio near 3:2, has an equilibrium pressure of 0.1 MPa at about 250°C and can reversibly store about 8 wt% of hydrogen. As interesting as these properties may be, the toxicity of Be will prevent large-scale utilization.

Another interesting lithium-based system is Li_3N/Li_2NH [53]. Lithium nitride can be hydrogenated to lithium imide and lithium hydride (5.4 wt% H_2). The latter reaction can be used for reversible storage at 250°C. The formation of ammonia can be completely avoided by the addition of 1% $TiCl_3$ to the system, which has the positive additional effect to improve the kinetics [54]. Very fast kinetics has been reported for a partially oxidized lithium nitride [55].

With respect to gravimetric storage capacity, the materials formed from the light elements are most attractive. For example, boron nitrite (BN) nanotubes has an uptake of 4.2 wt% at 10 MPa hydrogen pressure [56]. Majority of the hydrogen is released at high

TABLE 11.4

Storage Capacity of Light Metal Hydrides

Hydride	Reaction	H_2 Storage Capacity (wt%)
LiH	$LiH + H_2O \rightarrow LiOH + H_2$	7.7
$LiBH_4$	$LiBH_4 + 4H_2O \rightarrow LiOH + H_3BO_3 + 4H_2$	8.6
$LiAlH_4$	$LiAlH_4 + 4H_2O \rightarrow LiOH + Al(OH)_3 + 4H_2$	7.3
NaH	$NaH + H_2O \rightarrow NaOH + H_2$	4.9
$NaBH_4$	$NaBH_4 + 4H_2O \rightarrow NaOH + H_3BO_3 + 4H_2$	7.3
$NaAlH_4$	$NaAlH_4 + 4H_2O \rightarrow NaOH + Al(OH)_3 + 4H_2$	6.4
MgH_2	$MgH_2 + 2H_2O \rightarrow Mg(OH)_2 + 2H_2$	6.3

temperatures between 300°C and 450°C, indicating that it is bound to dangling bond sites, which were formed during the collapse of the nanotubes.

A series of transition metal complex hydrides have been synthesized and studied structurally by a group of researchers [38,55–60]. However, for practical large volume applications these systems are either too expensive or have an insufficient gravimetric storage density. For example, Mg_2FeH_6 has shown a high hydrogen storage capacity of about 5.5 wt% with an exceptionally high volumetric hydrogen density (150 kg/m^3) and excellent stability in cycle tests. It can be a promising candidate for high-temperature (400–500°C) thermal energy storage; however, the thermodynamic properties do not allow the use of Mg_2FeH_6 for hydrogen storage in mobile applications [45,61].

The storage capacity of some lithium hydrides are compared with other promising hydrides in Table 11.4. While considering a recyclable process, one of the important issues is the ability to regenerate the chemical hydride. Lithium hydroxide forms a monohydrate, which is easily decomposed when heated to about 100°C. In contrast, many of the bimetal hydrides form multihydrates on reacting with water, and require much higher temperatures to decompose [59].

11.7 Nanoscale and Nanocomposite Materials

By reducing the particle size, the thermodynamic properties of material may be changed to a point where surface free energy can govern the reaction. For example, the improvement in the sorption properties of metal hydrides can be achieved by ball milling. Nanostructured metal hydrides address one of the key problems of hydrogen sorption kinetics: the diffusion of hydrogen within the metal leading to the formation of the hydride. Although the detailed mechanism of hydrogen diffusion in nanostructured metals needs to be determined for each specific metal–hydrogen system, numerous experiments indicate that in practically all metal hydrides, the kinetics of both absorption and desorption can be improved by an order of magnitude simply by reducing the grain size of the metal [62]. Nanocomposites, where both nanostructure and nanocatalysis can be incorporated into a more complex system, are the next step in designing high-performance metal hydrides. The key feature of such nanocomposites is that the metal hydrides are intermixed on the nanoscale in such a way that they preserve their basic thermodynamic characteristics, but at the same time interact. Figure 11.4 illustrates the nanostructure of Mg-11.3 atomic% Ni alloy that has been studied for hydrogen storage [63]. At this composition, the alloy is

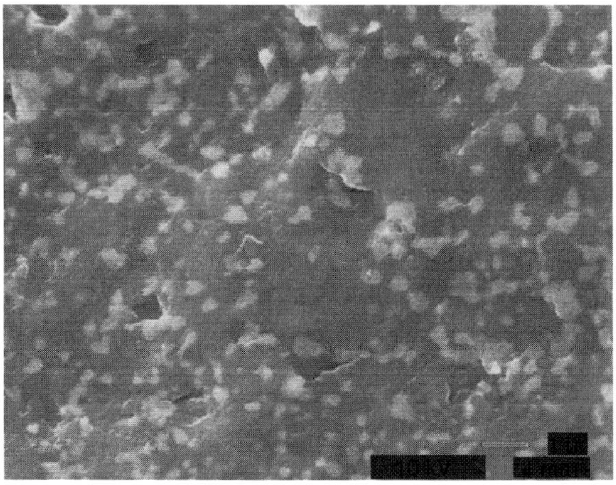

FIGURE 11.4
Scanning electron micrograph of magnesium–nickel alloy, containing 11.3 atomic% Ni, used for hydrogen storage. (Reproduced with permission from Schwarz, R.B., Storage of hydrogen in powders with nanosized crystalline domains, *Mater. Res. Bull.*, 24, 40, 1999, available at http://www.wtec.org/loyola/nano/US.Review/04_06.htm, May 2007.)

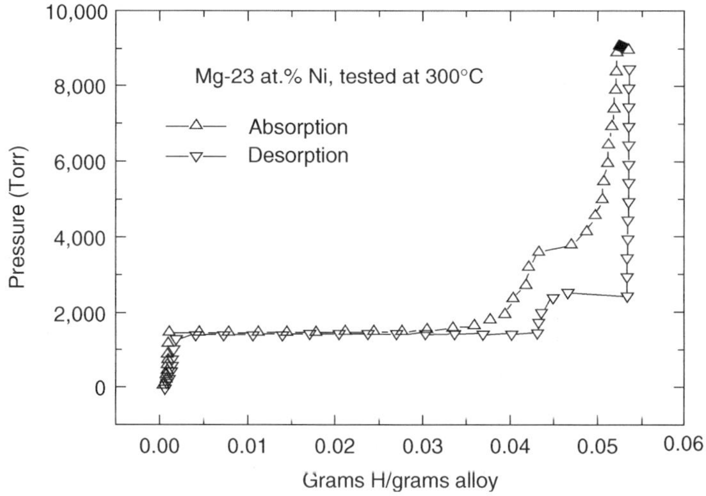

FIGURE 11.5
Hydrogen absorption/desorption characteristics of magnesium–nickel alloy containing 23 atomic% Ni. (Reproduced with permission from Schwarz, R.B., Storage of hydrogen in powders with nanosized crystalline domains, *Mater. Res. Bull.*, 24, 40, 1999, available at http://www.wtec.org/loyola/nano/US.Review/04_06.htm, May 2007.)

a two-phase mixture of pure magnesium (majority dark phase) and Mg_2Ni (small light inclusions). Most of the hydrogen is stored in the low-density magnesium phase. The small Mg_2Ni inclusions catalyze the decomposition of the molecular hydrogen, increasing the hydrogen absorption/desorption kinetics significantly. The catalyst is quite effective since it is finely divided and is in intimate contact with the pure magnesium phase [63]. The hydrogen absorption/desorption characteristics of this alloy is given in Figure 11.5, in which the first plateau corresponds to the absorption of hydrogen into pure magnesium, and the second plateau denotes the absorption of hydrogen into the Mg_2Ni phase. Owing to the presence of the nanostructure, the kinetics of hydrogen absorption/desorption into this alloy is relatively fast.

Another example of a nanocomposite is a nanomixture of a high-temperature Mg-based hydride with low-temperature FeTi-based hydrides, which can be used for cold starting. Beginning from the fully hydrogenated state, desorption may be initiated from the FeTi hydride at room temperature, when the consumption of hydrogen reduces the hydrogen pressure to below the equilibrium pressure of FeTi. The desorbed hydrogen can then be used, in part, for heating the system up to the temperature at which desorption from the high-temperature metal hydride can proceed. During the system cooling down, the temperature/pressure conditions become favorable for the low-temperature metal hydride (FeTi) to reabsorb the hydrogen that is being released by magnesium [62].

11.8 Thermodynamics

An important feature in hydrogen storage by metal hydrides is the "pressure plateau," which is a pressure at which the material reversibly absorbs/desorbs large quantities of hydrogen. For many applications, the plateau pressure should be close to ambient pressure, which allows the use of lightweight storage containers. Such pressure–concentration isotherms (PCI) are generally utilized for assessing the ability of metal hydrides to reversibly store hydrogen [64]. PCI diagrams indicate the maximum attainable storage capacity and thermodynamic conditions for the reversible reaction. PCI behaviors are shown in Figures 11.6 and 11.7 for Ti-doped $NaAlH_4$ [65] and unmilled and ball-milled MgH_2,

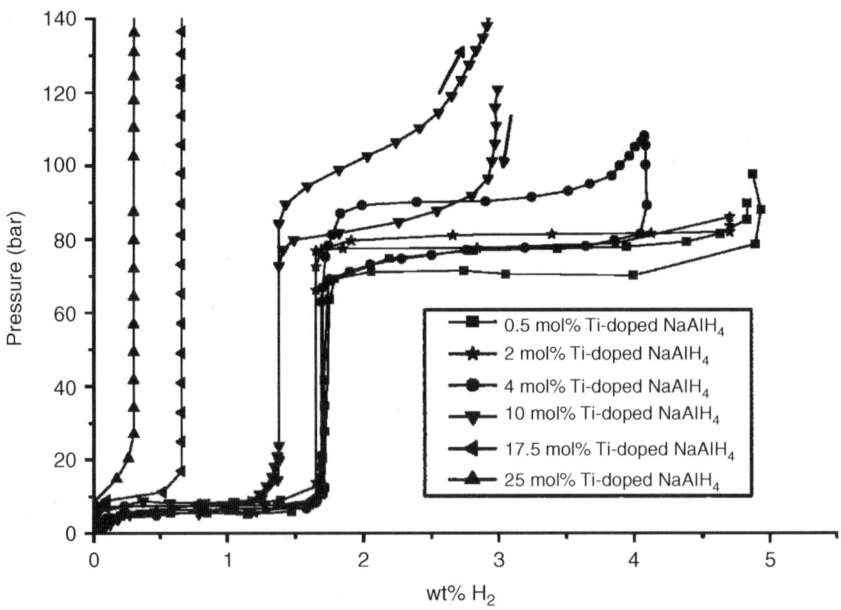

FIGURE 11.6
Effect of Ti doping on the PCIs for $NaAlH_4$ at 160°C. (Reproduced with permission from Streukens, G., Bogdanović, B., Felderhoff, M., and Schuth, F., *Phys. Chem. Chem. Phys.*, 8(24), 2889–2892, 2006.)

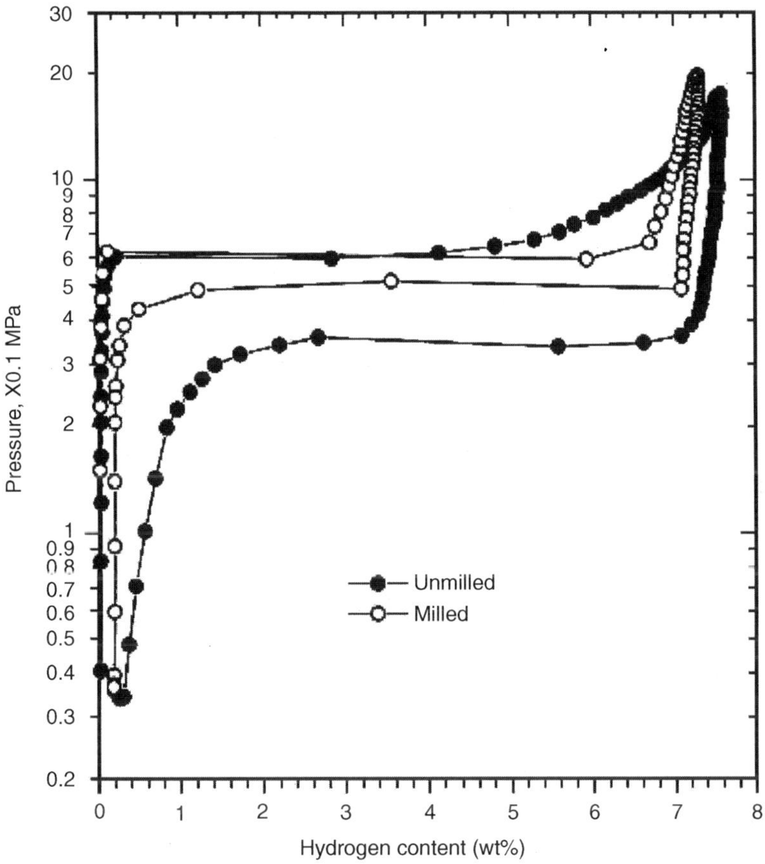

FIGURE 11.7
Pressure–concentration isotherms for unmilled and ball-milled MgH_2 at 350°C. (Reproduced with permission from Huot, J., Liang, G., Boily, S., Neste, A.V., and Schulz, R., *J. Alloys Compd.*, 293–295, 495–500, 1999.)

respectively [66]. The plateau pressure can be tailored to a specific application through alloying. As the hydrogen pressure together with the concentration of hydrogen in the metal is increased, interactions among the hydrogen atoms become locally important, and nucleation and growth of the hydride phase starts. The isotherms show a flat plateau and its length determines the amount of hydrogen that can be reversibly stored with small pressure variations. The equilibrium pressure depends strongly on temperature and is related to the changes in enthalpy and entropy. The entropy change corresponds mostly to the change from molecular hydrogen gas to dissolved hydrogen, which is roughly 130 J/K/mol for most metal–hydrogen systems under consideration. Lower values of entropy change (ΔS) can occur for complex metal hydrides, for example, consider the following reaction [57]:

$$LiBH_4 \rightarrow LiH + B + \tfrac{3}{2}H_2 \qquad \Delta S = 97\ \text{J/K/mol}$$

The presence of two pressure plateaus in Figure 11.6 corresponds to the two-step decomposition of the hydride complex given by the following equation:

$$6NaAlH_4 \leftrightarrow 2Na_3AlH_6 + 4Al + 6H_2 \leftrightarrow 6NaH + 6Al + 9H_2$$

Hydrogenation is sensitive to the surface modifications. For example, the ball-milled powders require less activation compared to the conventional powders. For nanocrystalline hydrides, the grain boundary does not dramatically affect the PCI, which describes the thermodynamic aspects of hydride formation. However, the pressure for hydrogen desorption of the unmilled MgH_2 is lower than that of the milled one as seen in Figure 11.7 [66].

The operating temperature of a metal hydride system is fixed by the plateau pressure in thermodynamic equilibrium and the overall reaction kinetics. Hydrogen is located in the form of atoms in the interstitial sites of the host metal lattice. The lattice expands during hydrogen sorption, often losing some of its high symmetry. As a consequence of the coexistence of the nonexpanded α phase and the anisotropically expanded β phase, lattice defects and internal strain fields are formed, which end in a decrepitation of the brittle host metals. The absorption or desorption of hydrogen by a metal hydride is associated with substantial heat (~30 W/g mol of hydrogen), which must be removed or supplied, respectively. As hydrogen initially diffuses into the pure metal interstices, the equilibrium pressure rises sharply, analogous to physical adsorption. When the physical sites are saturated, a second phase of the metal hydride (if applicable, as in Figure 11.6), begins to form. A plateau is observed on the pressure versus hydrogen/metal ratio for fixed temperature. When the second reaction is complete, another sharp pressure rise is seen if more hydrogen is added. During dehydrogenation, a lower pressure versus concentration profile is observed, representing a hysteresis effect. At a given temperature, the plateau pressure represents one point on a pressure–temperature (van't Hoff) plot (Figure 11.8), where the log of pressure versus reciprocal temperature is linear with a negative slope [30]. The slope is related to the heat of reaction, that is, the heat of hydrogenation. From Figure 11.8, it is clear that there are only a few metal hydrides that show about 0.1 MPa equilibrium pressures at slightly above room temperature—a highly desirable aspect for storage in mobile applications [30].

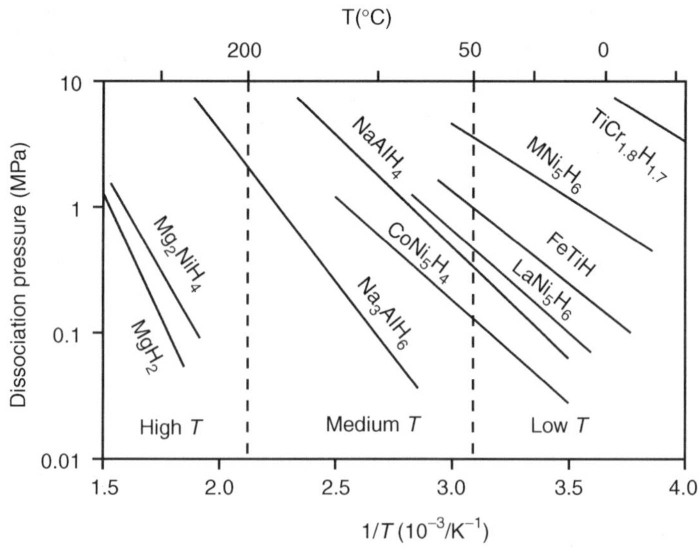

FIGURE 11.8
Van't Hoff plots for different metal hydrides. (Reproduced with permission from Schuth, F., Bogdanović, B., and Felderhoff, M., *Chem. Commun.*, 20, 2249–2258, 2004.)

Many elemental metals such as $PdH_{0.6}$, rare earth REH_2, REH_3, or MgH_2 form hydrides outside the pressure and temperature ranges attractive for mobile storage. In contrast, alloys derived from $LaNi_5$ show some promising properties including fast and reversible sorption with small hysteresis, plateau pressure of a few bars at room temperature, and good cycling life. The volumetric hydrogen density (measured crystallographically) of $LaNi_5H_{6.5}$ at 2 bar equals that of gaseous molecular hydrogen at 1800 bar; but the main advantage is that all the hydrogen desorbs at a pressure of 2 bar. The density for practical purposes can be reduced by the packing fraction of $LaNi_5$ powder, but is still higher than that of the liquid hydrogen. Storage in these intermetallics allows very safe hydrogen handling. But due to their high weight, the hydrogen storage remains below 2 wt%. However, it is attractive for electrochemical hydrogen storage in rechargeable metal hydride batteries, reaching a capacity of 330 mA h/g.

We have plenty of experience with manufacturing of AB_5-type metal hydrides, as more than one billion AB_5-type metal hydride batteries are produced every year.

11.9 Kinetics and Heat Transfer

Kinetics refers to the rates of hydrogen absorption and desorption; a faster kinetics at near-ambient temperatures and pressures corresponds to less time at the fueling station while operating under safe conditions. Reaction kinetics of light metal hydrides depends on the surface area of the material exposed to hydrogen, the dimensions and homogeneity of the microstructure, and the types of additives and catalysts. For example, fine microstructures with a high density of interfaces and phase boundaries provide short diffusion distances, fast diffusion pathways, and a high number of nucleation sites for the phase transformation in the solid-state reaction [45].

In addition to maintaining low weight and volume of hydride system, it is critical to maintain short cycle times for loading and unloading hydrogen. Usually, the rate-limiting step is the heat transfer (removal or supply of heat of hydrogenation) due to the low conductivity of the fine metal particles and a poor conduction between the particles and the container wall. Hence, it is necessary to improve the heat transfer and the hydrogen flow rate to achieve high hydrogenation/dehydrogenation rates [57]. Some additives in small quantities can greatly affect the transport. For example, addition of an organotransition metal catalyst to magnesium hydride provides excellent kinetics, high surface area, and high storage capacity (7 wt%), but it suffers from being pyrophoric. To improve the diffusivity of hydrogen in the metal hydride, various techniques such as ball milling can be applied, which in turn reduces the grain size, increases the defect concentration, and shortens the diffusion path [45].

11.10 Applications

11.10.1 Stationary Applications

Metal hydrides are excellent in storing hydrogen when space is limited. When compared to the conventional high-pressure cylinders, hydrogen stored in a metal hydride occupies only one-third of the volume. Also, the typical pressure of absorption/desorption is only 1–25 bar, as opposed to 200 bar typically in cylinder storage. Metal hydride fuel candles are being used in handheld burners. Using couples of metal hydride alloys with varying plateau pressure, it is now possible to realize thermal engineering applications such as refrigeration and air conditioning, heat pump, heat transformer, and solar heat harvest system. The hydrogen evolved from metal hydrides is of high purity (typically >99.9%), which can be used in a variety of applications. In large-scale stationary applications, metal hydrides can be used to load-level electric demand and supply; at low demand, charging of hydride is done, and at high demand, extra electricity is generated by utilizing the stored hydrogen. Availability of a wide range of alloys with varying plateau pressures and proper exploitation of metal hydride properties by advanced engineering can promote several engineering applications for metal hydrides.

The most common application of metal hydrides for hydrogen storage is the batteries. These hydride-forming alloys are of the nickel–metal hydride (Ni-MH) type, the negative electrode in the battery cell, which is activated, or charged, with the hydrogen. During the battery's lifetime, it proceeds through many hydriding/dehydriding cycles. The electrochemistry of the negative electrode can be represented as follows [67]:

$$\text{Alloy} + n\text{H}_2\text{O} + n\text{e}^- \leftrightarrow \text{Alloy(H}_n) + n\text{OH}^-$$

(Charging) (Discharging)

where n is the number of water molecules taking part in the reaction. Ni-MH batteries are typically used in portable computers and electronics—cell phones and power tools. They are also being used in the newly emerging hybrid vehicles. These batteries supply energy to and draw energy from the gas-powered motor, eliminating the need for a separate recharging.

11.10.2 Automotive Applications

Although metal hydrides are more suitable for stationary hydrogen storage, these have been successfully tested for automotive applications as well. For automotive applications, hydrogen loading needs to be done at a near-ambient temperature in a short duration. The storage can be designed for hydrogen use in an internal combustion (IC) engine and a fuel cell. A fuel cell–powered vehicle is about 2.5 times more energy efficient than the IC engine and, generally, requires a solid-state hydride with a gravimetric storage capacity of at least 9.1 wt% for a driving range 500 km [32,50,52]. This indicates that only hydrides with the highest possible gravimetric hydrogen storage capacities are viable for vehicular applications. The energy efficiency of an IC engine is about 25% when compared to a fuel cell with a direct current (DC) motor, which is up to 80%. The fuel tank with a capacity of 5 kg H_2 has to be refilled within 5 min using a hydrogenation rate of 17 g/s and be capable of supplying about 2 g/s H_2 to the propulsion system under full throttle conditions [50]. A typical stainless-steel cylinder filled with solid metal hydride powder is shown in Figure 11.9a [68]. In another one, Lasher [69] designed a carbon–fiber (CF)-wound composite tank (Figure 11.9b) with a metal liner and glass–fiber (GF) winding for penetration protection. The 2 mm thick stainless-steel (304 SS) metal liner is considered impermeable to hydrogen. The thickness of the CF winding was calculated assuming that it alone carries the hoop stresses at a safety factor of 2.25 at 10 Mpa maximum storage pressure. The metal hydride is in the form of a powder packed inside. The tube-sheet plenums (304 SS, 0.9 mm sheet thickness) for distributing and collecting the heat transfer fluid (HTF) are

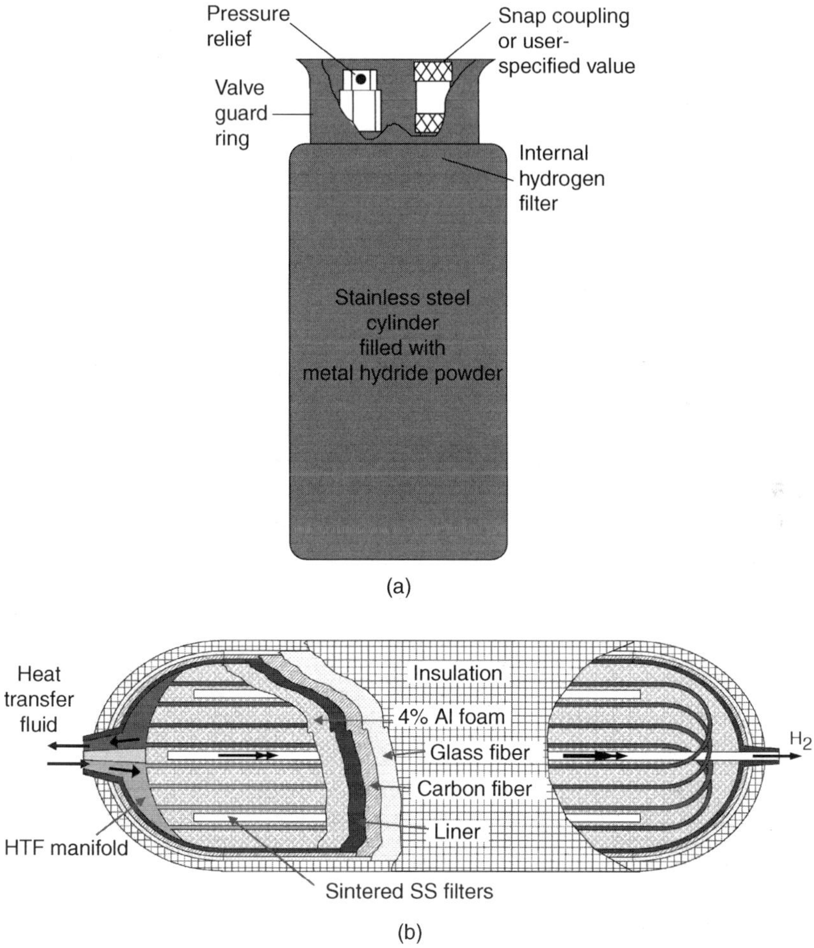

FIGURE 11.9
(a) Typical solid metal hydride hydrogen storage container. (Reproduced from SOLID-H™, Metal Hydride Hydrogen Storage Containers, available at http://www.hydrogencomponents.com/stor.html, January 2007. With permission.) (b) CF wound composite tank. (Reproduced with permission from Lasher, S., Comparison of hydrogen storage options, *NHA Annual Hydrogen Conference 2005*, Washington, March 29–April 1, 2005.)

inside the headspace of the tank, joined by U-tubes embedded in the foam. The U-tubes, also made from 304 SS, are 9.5 mm in outer diameter and 1.2 mm thick. Hydrogen enters and leaves the alanate tank from the end opposite to where the HTF communicates with the system. The central hydrogen tube is packed with a sintered metal filter medium to capture fine powders that may be emitted from the metal hydride. Hydrogen is collected or distributed through the four tubes inside the tank. These tubes also incorporate sintered metal filter media [69,70].

At present, practical metal hydrides can store a maximum of 3.5 wt% H_2, which is grossly inadequate for automotive application due to the burden of carrying the heavyweight metal during transit. Unless a hydrogen storage level of 6 wt% H_2 is achieved, it is difficult to realize commercial uses of metal hydrides for automotive applications. Therefore, complex materials such as alanates of magnesium or sodium are being tested. However, the Mg-based alloys require a high temperature (~300°C) to release hydrogen [30,38,46], which in turn, requires a heating device and an associated energy expenditure. An energy analysis for the release of hydrogen from metal hydrides was done by Maddalena et al. [46]. Depending on the combustion efficiency, a heat of 120–142 kJ/g can be obtained from hydrogen. For a release of 6 g of hydrogen from 100 g of magnesium hydride, 39.6 kJ is needed for heating from 25°C to 300°C (using heat capacity, 1.44 J/K g metal hydride), and 228 kJ is needed for endothermic heat of decomposition (using 38 kJ/g hydrogen). Hence, a total of 267.6 kJ is needed, which can be obtained from burning 2.2 g of hydrogen (based on 122 kJ/g H_2 combustion energy). The remaining 3.8 g of hydrogen can be effectively used in the fuel cell for driving, which will yield 452 kJ or 126 Wh energy (based on 120 kJ/g energy conversion). Therefore, an electric engine of 1 kW, at 50% efficiency, requires 1.6 kg of metal hydride powder to run for 1 h. The 1.6 kg of magnesium hydride may be distributed into 10 vessels of 160 g each, which will be heated at 300°C one by one every 360 s.

Metal hydride vessels for hydrogen storage are more like heat exchangers. They must have thermal jackets or conduits for heat transfer medium to cool and heat the material. During discharge, waste heat from the fuel cell coolant or the engine is used to supply the heat. The heat desorbs the hydrogen and generates a pressure to feed the fuel cell. During refueling, the bed is connected to a supply of hydrogen gas and is cooled by the internal or an external coolant. High pressure and quick cooling are needed for fast absorption and quick refueling. A typical schematic design of a prototype of a metal hydride reservoir is shown in Figure 11.10 [46]. The main stage is composed of a number of vessels filled with doped MgH_2, whereas the primer stage is filled by $LaNi_5$. The flux of hydrogen from the main-stage reservoir to feed the fuel cell and continuously refill the primer stage is electronically controlled. The hydrogen burner stage, located into the main-stage part, has an inlet for hydrogen and another for air supply. To initiate the hydride decomposition from the first container, a primer stage of the reservoir with about 0.5 kg of $LaNi_5$ (a hydride

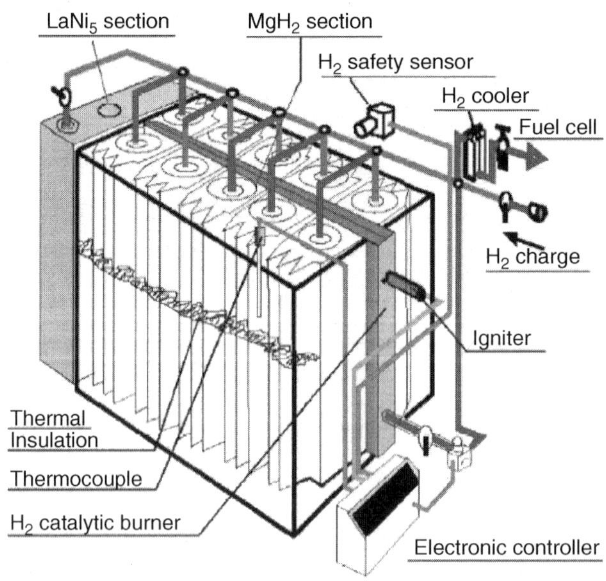

FIGURE 11.10
Schematic of a two-stage metal hydride reservoir. (Reproduced with permission from Maddalena, A., Petrisa, M., Paladea, P., Sartoria, S., Principia, G., Settimob, E., Molinasc, B., and Lo Russo, S., *Int. J. Hydrogen Energ.*, 31, 2097–2103, 2006.)

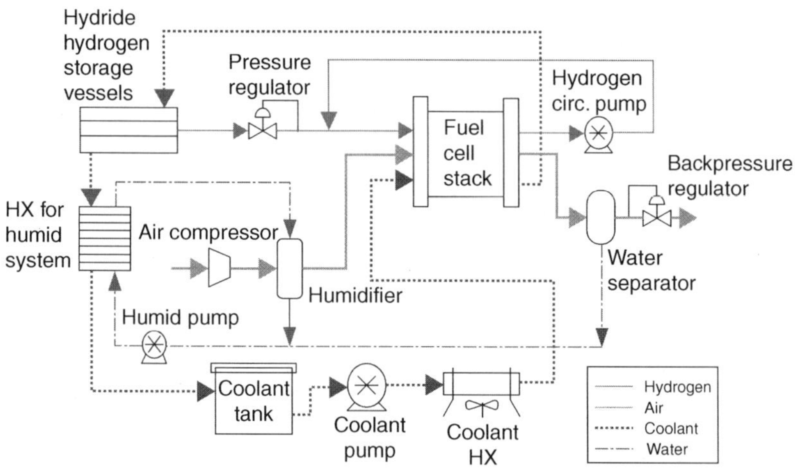

FIGURE 11.11
Metal hydride hydrogen-fuel-cell-powered system. (Reproduced with permission from Heung, L.K., Using Metal Hydride to Store Hydrogen, U.S. Department of Commerce, National Technical Information Service Springfield, VA, 2003, available at http://sti.srs.gov/fulltext/ms2003172/ms2003172.pdf, May 2007.)

which contains only about 1.3 wt% hydrogen, but desorbs it at room temperature with a plateau pressure of 0.18 MPa) may be used.

A metal hydride–based hydrogen fuel cell–powered system is shown in Figure 11.11 [18,70], where the original gasoline engine power train of the vehicle has been replaced by a hydrogen fuel cell and an electric motor. The system consisted of the fuel cell stack, the hydrogen storage system, the coolant loop, the water loop, and the air loop. During operation, the coolant of the fuel cell stack heats the metal hydride system to generate a discharge pressure between 0.2 and 2 MPa. Ahluwalia [70] studied the suitability of Ti-catalyzed sodium alanate as a reversible hydrogen storage medium for automotive polymer–electrolyte fuel cells under the constraint that it must meet certain performance criteria with respect to the amount of recoverable H_2 needed for vehicle driving range, minimum full flow of H_2 needed to satisfy vehicular power demands, maximum refueling time, and minimum H_2-delivery pressure. The results revealed that a 10-fold enhancement in published desorption kinetics is needed to attain 90% reversible H_2 storage capacity while satisfying the minimum full-flow requirement of 0.02 g H_2/kW. The faster the refueling rate, the higher is the peak heat transfer rate and the lower is the amount of recoverable H_2. For a system with 5.6 kg recoverable H_2, the peak heat transfer rate during refueling can exceed 1 MW.

11.11 Challenges

The main challenge in developing hydrogen as an automotive fuel is to develop and demonstrate viable hydrogen storage technologies. Low cost, offboard hydrogen storage systems, as required for hydrogen infrastructure needs to support transportation, stationary, and portable power markets by 2015. Unfortunately, none of the current hydrogen storage system

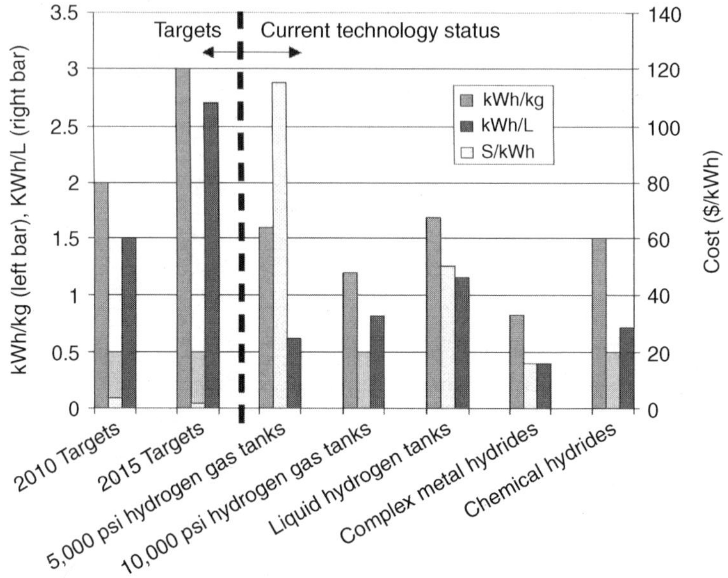

FIGURE 11.12
U.S. DOE targets specified for onboard hydrogen storage systems. (Adapted from Milliken, J., Cataquiz, L., Petrovick, J., Read, C., and Satyapal, S., Grand Challenge for Basic and Applied Research in Hydrogen Storage U.S. DOE Report, 2003.)

meets the Department of Energy (DOE) targets (Figure 11.12; for more details see Chapter 9). The major technical barriers in the use of hydrogen storage in metal hydrides are as follows:

1. *Cost.* It is desired to design low-cost materials and components for hydrogen storage systems as well as low-cost, high-volume manufacturing methods. The targeted storage system cost specified by DOE is \$4/kWh (~\$133/kg H_2) for the year 2010, whereas it should be further reduced to \$2/kWh (~\$67/kg H_2) by the year 2015.
2. *Weight and volume.* Materials and components that allow compact, lightweight, hydrogen storage while enabling 500 km driving range for light-duty vehicles are needed.
3. *Efficiency.* The energy required to get hydrogen in and out of the metal hydride material is an issue. Thermal management for charging and releasing hydrogen from the storage system needs to be optimized to increase the overall efficiency.
4. *Durability.* Materials and components that allow hydrogen storage systems with a lifetime of 1500 cycles and tolerance to fuel contaminants are needed.
5. *Refueling time.* There is a need to develop hydrogen storage systems with refueling times of <3 min for 5 kg of hydrogen.

Metal hydrides can potentially hold more hydrogen than any other solid storage system up to 12 wt%. But these metal hydrides tend to be heavy, and require high energy and high temperatures to break the ionic bonds to release the hydrogen. Current usable metal hydrides can hold no more than 1.8 wt% of hydrogen. Typical hydrogen content that can be achieved by various methods is shown in Figure 11.13 [18]. The present metal hydrides contain about 60 kg/m^3 of hydrogen compared with 70 kg/m^3 for liquid hydrogen and about 100 kg/m^3 for the liquid hydrocarbon fuels. Also, the light solids such as magnesium hydride and the

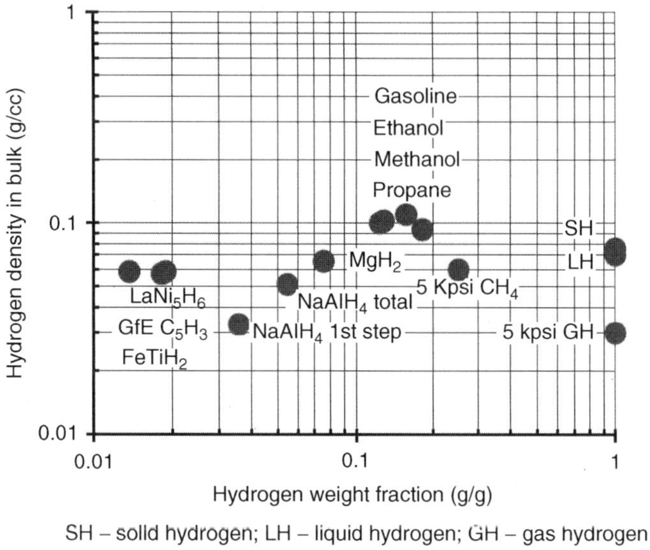

FIGURE 11.13
Hydrogen content in various storage options. (Adapted from Heung, L.K., Using Metal Hydride to Store Hydrogen, U.S. Department of Commerce, National Technical Information Service Springfield, VA, 2003, available at http://sti.srs.gov/fulltext/ms2003172/ms2003172.pdf, May 2007.)

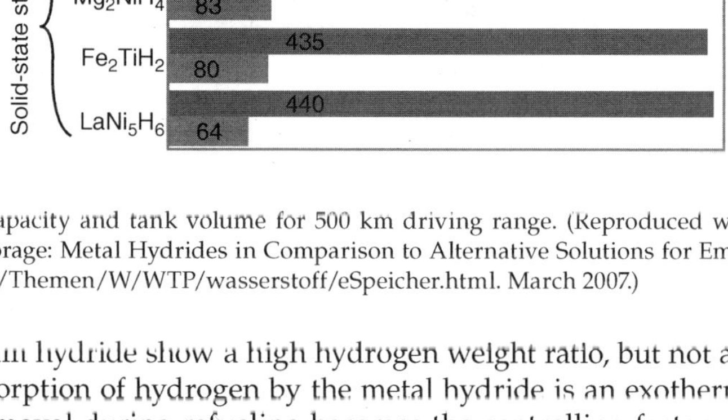

FIGURE 11.14
Hydrogen storage capacity and tank volume for 500 km driving range. (Reproduced with permission from GKSS, Hydrogen Storage: Metal Hydrides in Comparison to Alternative Solutions for Emission Free Vehicles. http://www.gkss.de/Themen/W/WTP/wasserstoff/eSpeicher.html. March 2007.)

sodium aluminum hydride show a high hydrogen weight ratio, but not a high volumetric density. The absorption of hydrogen by the metal hydride is an exothermic reaction, and the slow heat removal during refueling becomes the controlling factor. Another problem with the use of metal hydride is the size of the container. For every 100 km, approximately 1.2 kg of hydrogen is consumed in a hydrogen-based vehicle. Figure 11.14 shows the tank weight and volume required for various options for a 500 km driving distance [71].

Metal hydrides expand during the hydriding cycle and generate a large stress on the alloy and the container. The alloys tend to fragment into smaller particles as a result of this stress causing material movement and segregation, which create a large internal stress/strain. These strains can lead to container failure if left uncontrolled. Also, as the particles get smaller, there is an increased tendency for entrainment of the hydride fines in the gaseous hydrogen streams. Therefore, the container system must provide for these small particles and the resulting expansion. A container, in which the alloy is placed between the radially aligned lamellae of aluminum was tested [72] and found to provide a good performance. It has spring-loaded chambers that can accommodate charging rates of 0.6 m^3/h of hydrogen. For an 80 kg hydride system, it is possible to charge and discharge about 20,000 m^3 hydrogen in 1 h.

The main drawback of hydrogen stored in metals is that the gravimetric storage density is very low. The best possible storage can be achieved by forming light and reactive metal hydrides such as NaH, LiH, or MgH_2. But these pure metal hydrides are very stable and require higher energy input and higher temperature to release hydrogen. Magnesium is an excellent hydrogen storage medium since it can store about 7.6 wt% hydrogen. However, the absorption/desorption of hydrogen in magnesium is very slow. The materials are being modified by creating amorphous structures by rapid solidification, mechanical alloying, and catalytic additions to overcome the limitations.

Issues with complex metal hydrides include low hydrogen capacity, slow uptake and release kinetics, and cost. Even a promising candidate, sodium alanate (storage capacity of 5.5 wt%), is even below the year 2010 DOE target. In fact, practically, only 4 wt% reversible hydrogen storage has been demonstrated. Furthermore, the packing density of sodium alanate powders is low (e.g., about 50%); hence, a high system volume is needed. Although the currently available metal alanates are behind the target, it is envisioned that their continued study will lead to a fundamental understanding, which can be applied to the design and develop much improved complex metal hydrides. Another critical issue with complex metal hydride materials, due to their high hydrogenation enthalpies, is thermal management during refueling. Reversibility of these and new materials needs to be demonstrated for over a thousand cycles. Safety of a hydride-based storage system is an issue with respect to practical implementation as well. $NaAlH_4$, especially when loaded with titanium, is a very reactive solid, which can ignite with air or moisture. Strategies that reduce this problem have to be developed. Recently, numerous attempts have been made to use nanosized metal hydrides for improving storage properties, which may bring manifold impacts including improved cyclibility, better stability, ease of fabrication and handling, higher electronic conductivity, and higher power density.

A breakthrough in reaction kinetics of light metal hydrides has been achieved by preparing nanocrystalline materials using high energy milling (Figure 11.15) [71]. The absorption of hydrogen in conventional coarse material (size >1 μm) was slow, and refueling required hours. On the contrary, the nanocrystalline (size ~10 nm) material prepared by milling absorbed hydrogen much faster than the conventional particles up to full capacity. By addition of catalysts, absorption is enabled at room temperature and desorption at 200°C, although at lower pressures.

Favorable thermodynamics, kinetics, and storage capacity alone are not sufficient for practical applicability of metal hydrides as additional engineering problems need to be solved. Of primary importance is the pressure plateau at which the material reversibly absorbs/desorbs large quantities of hydrogen. For ease of applications, the plateau pressure should be close to ambient pressure because this allows the use of lightweight storage containers. The optimum operating pressure and temperature windows for PEMFC applications are 0.1–1 MPa and 25–120°C, respectively. The plateau pressure can be tailored to a specific application through alloying. Additionally, the hysteresis has to be minimized to increase the efficiency during the cyclic operation.

In view of the large-scale implementation of metal hydrides in automobiles, economic large-scale production of the materials would be necessary. Assuming a storage capacity of 10 wt% on a materials level, the storage of 5 kg of hydrogen for a medium-sized car would require 50 kg of storage material per car. This translates to 25 million t of storage

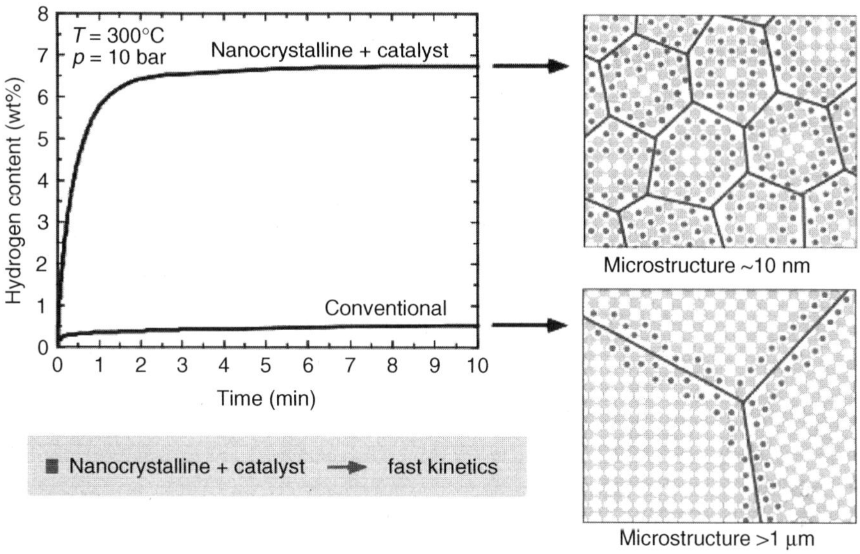

FIGURE 11.15
Hydrogen storage in reversible light metal hydrides. (Reproduced with permission from GKSS, Hydrogen Storage: Metal Hydrides in Comparison to Alternative Solutions for Emission Free Vehicles. http://www.gkss.de/Themen/W/WTP/wasserstoff/eSpeicher.html. March 2007.)

material for the roughly 500 million cars on the road today. Such a scenario necessitates the selection of suitable cost-effective large-scale processing routes for the production of metal hydride materials [73].

11.12 Conclusions

Hydrogen storage for vehicular applications is a major concern for the implementation of a hydrogen-based economy. The most important properties needed for a metal hydride–based hydrogen storage material are high hydrogen capacity, low hydrogenation/dehydrogenation enthalpies, reversibility, and low desorption temperature, which are very difficult to satisfy simultaneously because of thermodynamic and kinetic constraints. Out of a number of metal hydrides available for hydrogen storage, magnesium hydride with a 7.6 wt% storage capacity and sodium alanate with a 5.5 wt% storage capacity are emerging as materials of interest. In addition, four different classes of alloys are of interest: AB_5 (e.g., $LaNi_5$), AB (e.g., FeTi), A_2B (e.g., Mg_2Ni), and AB_2 (e.g., ZrV_2). Metal hydrides have a higher hydrogen storage density (6.5 H_{atoms}/cm^3 for MgH_2) than hydrogen gas (0.99 H_{atoms}/cm^3) or liquid hydrogen (4.2 H_{atoms}/cm^3). When compared to storage in the pressurized tank, metal hydrides occupy much less volume. However, the weight of metal hydrides systems are still high and further improvement in the material will be needed. Once the weight is reduced by the discovery of new materials, their use in mobile applications will be practical. Owing to a high heat of hydrogen absorption/desorption, metal hydrides need an effective heat management system, which adds to the weight. The storage properties such as pressure, temperature, rate of absorption/desorption are being enhanced by adding catalyst and creating nanostructures.

References

1. James, B.D., G.N. Baum, F.D. Lomax, C.E. Thomas, and I.F. Kuhn, Comparison Onboard Hydrogen Storage for Fuel Cell Vehicles, Directed Technologies, Inc., prepared for Ford Motor Company, under prime contract DE-AC02-94CE50389 to the United States Department of Energy, May 1996.
2. Hord, J., Is hydrogen a safe fuel? *Int. J. Hydrogen Energ.*, 3, 157–176, 1976.
3. Kelley, J.H., and R. Hagler, Storage, transmission and distribution of hydrogen, *Int. J. Hydrogen Energ.*, 5, 35–54, 1980.
4. Ogden, J.M., E. Dennis, M. Steinbugler, and J. Strohbehn, Hydrogen Energy Systems Studies, Final Report to U.S. DOE for Contract No. XR-11265-2, January 18, 1995.
5. Ogden, J.M., Prospectus for building a hydrogen energy infrastructure, *Annu. Rev. Energy Environ.*, 24, 227–279, 1999.
6. Winter, C.J., and J. Nitsch, *Hydrogen as an Energy Carrier*, Springer, New York, 1988.
7. Cadwallader, L.C., and J.S. Herring, Safety Issues with Hydrogen as a Vehicle fuel, September 1999. http://www.inl.gov/hydrogenfuels/projects/docs/h2safetyreport.pdf, April 2007.
8. Mitlitsky, F., A.H. Weisberg, and B. Myers, Vehicular hydrogen storage using lightweight tanks, *US DOE Hydrogen Program Annual Review Meeting*, Lakewood, CO, May 4–6, 1999.
9. Little, A.D., Multi-fuel Reformers for Fuel Cells Used in Transportation. Assessment of Hydrogen Storage Technologies. Phase I. Final Report, United States Department of Energy, Office of Transportation Technologies, March 1994.
10. Audus, H., O. Karstaad, and M. Koval, Decarbonization of fossil fuels: Hydrogen as an energy carrier, Hydrogen Energy Progress XI, *Proceedings of the 11th World Hydrogen Energy Conference*, Stuttgart, pp. 525–534, June 23–28, 1996.
11. Barnes, R.G., Ed., Materials Science Forum, *Hydrogen Storage Materials*, vol. 31, Trans-Tech Publications, Brookfield, VT, 31, 1988.
12. Benard, P., and R. Chahine, Modeling of hydrogen adsorption on activated carbon over the 77 K to 273 K Range, Hydrogen Energy Progress XII, *Proceedings of the 12th World Hydrogen Energy Conference*, Buenos Aires, Argentina, pp. 1121–1130, June 21–25, 1998.
13. Carpetis, C., Estimation of storage costs for large hydrogen storage facilities, *Int. J. Hydrogen Energ.*, 7(2), 191–203, 1981.
14. DeLuchi, M., Hydrogen vehicles: An evaluation of fuel storage, performance, safety, environmental impacts and cost, *Int. J. Hydrogen Energ.*, 14(2), 88–130, 1989.
15. Verbetsky, V.N., S.P. Malyshenko, S. Mitrokhin, V.W. Solovei, and Y. Shmalko, Metal hydrides: Properties and practical applications: Review of works in the CIS Countries, *Int. J. Hydrogen Energ.*, 23(12), 1165–1178, 1998.
16. Heung, L.K., On-board hydrogen storage system using metal hydride, in T.O. Saetre, Ed., *Hydrogen Power: Theoretical and Engineering Solutions*, U.S. Department of Commerce, National Technical Information Service, Springfield, VA, pp. 251–256, 1998.
17. Heung, L.K., T. Motyka, and W.A. Summers, Hydrogen storage development for utility vehicles, *Proceedings of HYPOTHESIS IV*, Stralsund, pp. 336–340, September 9–14, 2001.
18. Heung, L.K., Using Metal Hydride to Store Hydrogen, U.S. Department of Commerce, National Technical Information Service, Springfield, VA, 2003, available at http://sti.srs.gov/fulltext/ms2003172/ms2003172.pdf, May 2007.
19. Sandrock, G., and E.L. Huston, Rare earths-resources, science, technology and applications, in R.G. Bautista, N. Jackson Eds., *The Minerals Metals and Materials Society*, Warrendale, PA, p. 423, 1991.
20. Sandrock, G., Hydrogen energy system: production and utilization of hydrogen and future aspects, in Y. Yurum, Ed., *Application of Hydrides*, Kluwer Academic Publishers, Dordrecht, p. 253, 1995.
21. Sandrock, G., S. Suda, and L. Schlapbach, *Hydrogen in Intermetallic Compounds II. Surface and Dynamic Properties, Applications*, in L. Schlapbach, Ed., Springer, Berlin, 1992, p. 179.

22. Shi, J., T. Makoto, T. Hiroyuki, N. Kuriyama, and T. Sakai, Influence of carbon on electrode properties of V-Ti-Ni type hydrogen storage alloy, *J. Alloys Compd.*, 293–295, 716–720, 1999.
23. Eigen, N., M. Kunowsky, T. Klassen, and R. Bormann, Synthesis of $NaAlH_4$-based hydrogen storage material using milling under low pressure hydrogen atmosphere. *J. Alloys Compd.*, 430(1–2), 350–355, 2007.
24. Seo, C.Y., S.-J. Choi, J. Choi, C.N. Park, P.S. Lee, and J.Y. Lee, The influence of titanium on the electrochemical properties of the AB_5-type metal hydrides, *J. Alloys Compd.*, 350, 324–331, 2003.
25. Braess, H.H., and W. Strobel, Hydrogen as a fuel for road transport of the future: Possibilities and prerequisites, *Proceedings of the 11th World Hydrogen Energy Conference, Hydrogen Energy Prog.*, Stuttgart, pp. 1378–404.
26. Hydrogen Storage Alloy and Hydrogen Storage Systems, http://www.jmcusa.com/mh1.html, Japan Metal and Chemical Company (USA), Inc, April 2007.
27. Wasz, M.L., R.B. Schwarz, S. Srinivasan, and M.P.S. Kumar, Sn-substituted $LaNi_5$ alloys for metal hydride electrodes, *Mater. Res. Soc. Symp. Proc.*, 393, 237–242, 1995.
28. Abdel-Aala, H.K., M. Sadika, M. Bassyounia, and M. Shalabib, A new approach to utilize hydrogen as a safe fuel, *Int. J. Hydrogen Energ.*, 30, 1511–1514, 2005.
29. Satyapal, S., J. Petrovic, C. Read, G. Thomas, and G. Ordaz, The U.S. Department of Energy's National Hydrogen Storage Project: Progress towards meeting hydrogen-powered vehicle requirements, *Catal. Today*, 120, 246–256, 2007.
30. Schuth, F., B. Bogdanović, and M. Felderhoff, Light metal hydrides and complex hydrides for hydrogen storage, *Chem. Commun.*, 20, 2249–2258, 2004.
31. Schlapbach, L., and A. Zuttel, Hydrogen-storage materials for mobile applications, *Nature*, 414, 353–358, 2001.
32. Schulz, R., J. Strom-Olsen, L. Zaluski, and A. Zaluska, Nanocrystalline Mg or Be-based materials and use thereof for the transportation and storage of hydrogen, United States Patent 5,964,965, October 12, 1999.
33. Sakintuna, B., F. Lamari-Darkrim, M. Hirscher, and B. Dogan, Metal hydride materials for solid hydrogen storage: A review, *Int. J. Hydrogen Energ.*, 32(9), 1121–1140, 2007.
34. Thomas, G.J., K.J. Gross, N.Y.C. Yang, and C. Jensen, Microstructural characterization of catalyzed $NaAlH_4$, *J. Alloys Compd.*, 330–332, 702–707, 2002.
35. Wang, J., A.D. Ebner, and J.A. Ritter, Kinetic behavior of Ti-doped $NaAlH_4$ when cocatalyzed with carbon nanostructures, *J. Phys. Chem. B*, 110, 17353–17358, 2006.
36. Bellosta von Colbe, J.M., B. Bogdanović, M. Felderhoff, A. Pommerin, and F. Schüth, Recording of hydrogen evolution—a way for controlling the doping process of sodium alanate by ball milling, *J. Alloys Compd.*, 370, 104–109, 2004.
37. Reiser, A., B. Bogdanović, and K. Schlichte, The application of Mg-based metal hydrides as heat energy storage systems, *Int. J. Hydrogen Energ.*, 25(5), 425–430, 2000.
38. Bogdanović, B., and B. Spliethoff, Active magnesium hydride-magnesium systems for hydrogen storage, *Int. J. Hydrogen Energ.*, 12(12), 863–873, 1987.
39. Balema, V.P., K.W. Dennis, and V.K. Pecharsky, Rapid solid-state transformation of tetrahedral $[AlH_4]^-$ into octahedral $[AlH_6]_3^-$ in lithium aluminohydride, *Chem. Commun.*, 17, 1665, 2000.
40. Balema, V.P., K.W. Dennis, V.K. Pecharsky, V.P. Balema, V.K. Pecharsky, and K.W. Dennis, Solid state phase transformations in $LiAlH_4$ during high-energy ball-milling, *J. Alloys Compd.*, 313, 69, 2000.
41. Wang, J., and E. Ronnebro, Hydride developments for hydrogen storage, *Proceedings of the 2005 Spring TMS Conference*, DOE 2005 Hydrogen Program Annual Review, Washington, DC, May 23, p. 19, 2005.
42. Balema, V.P., J.W. Wiench, K.W.M. Dennis, M. Pruski, and V.K. Pecharsky, Titanium catalyzed solid-state transformations in $LiAlH_4$ during high-energy ball-milling, *J. Alloys Compd.*, 329, 108, 2001.
43. Züttel, A., S. Rentsch, P. Fisher, P. Wenger, P. Sudan, Ph. Mauron, and Ch. Emmenegger, Hydrogen storage properties of $LiBH_4$, *J. Alloys Compd.*, 356, 515, 2003.

44. Zuttel, A., C. Nutzenadel, P. Sudan, P. Mauron, C. Emmenegger, S. Rentsch, L. Schlapbach, A. Weidenkaff, and T. Kiyobayashi, Hydrogen sorption by carbon nanotubes and other carbon nanostructures, *J. Alloys Compd.*, 2002.
45. Ritchson, E., Nanotechnology and hydrogen storage in complex metal hydrides, available at http://engineering.purdue.edu/ChE/webpublications/ericjri/physics and chemistry of the alanate, March 2007.
46. Maddalena, A., M. Petrisa, P. Paladea, S. Sartoria, G. Principia, E. Settimob, B. Molinasc, and S. Lo Russo, Study of Mg-based materials to be used in a functional solid state hydrogen reservoir for vehicular applications, *Int. J. Hydrogen Energ.*, 31, 2097–2103, 2006.
47. Züttel, A., A. Borgschulte, and S.-I. Orimo, Tetrahydroborates as new hydrogen storage materials, *Scri. Mater.*, 56, 823–828, 2007.
48. Andresen, A.F., and A.J. Maeland, Hydrides for energy storage, *Proceedings of an International Symposium*, Geilo, Pergamon Press, New York, August 14–19, 1977.
49. Hufton, J., S. Mayorga, T. Gaffney, S. Naturaj, M. Roa, and S. Sircar, Sorption enhanced reaction process for production of hydrogen, *Proceedings of 1998 USDOE Hydrogen Program Review Meeting*, Arlington, VA, NREL/CP-570-25315, pp. 693–706, April 28–30, 1998.
50. Bogdanović, B., U. Eberle, M. Felderhoff, and F. Schutha, Complex aluminum hydrides, *Scr. Mater.*, 56, 813–816, 2007.
51. Baum, L., M. Meyer, and L. Mendoza-Zélis, Hydrogen storage properties of the Mg/Fe system, *Physica B*, 389, 189–192, 2007.
52. Zaluska, A., L. Zaluski, and J.O. Ström-Olsen, Lithium-beryllium hydrides: The lightest reversible metal hydrides, *J. Alloys Compd.*, 307, 157, 2000.
53. Chen, P., Z. Xiong, J. Luo, J. Lin, and K.-L. Tan, Interaction of hydrogen with metal nitrides and imides. *Nature*, 420, 302, 2002.
54. Ichikawa, T., S. Isobe, N. Hanada, and H. Fujii, Lithium nitride for reversible hydrogen storage, *J. Alloys Compd.*, 365, 271, 2004.
55. Hu, Y.H. and E. Ruckenstein, Highly effective Li_2O/Li_3N with ultrafast kinetics for H_2 storage, *Ind. Eng. Chem. Res.*, 43, 2464, 2004.
56. Tang, C.C., X.X. Ding, Z. Gan, J. Gao, X. Huang, S. Qi, and S. Fan, Synthesis of carbon nanotubes using supported catalysts modified by lanthanum species, *Carbon*, 40(13), 2497–2502, 2002.
57. Sudhakar, A.V., J. Karl Johnson, and S. David, Sholl identification of destabilized metal hydrides for hydrogen storage using first principles calculations, *J. Phys. Chem. B*, 110, 8769–8776, 2006.
58. Bronger, W. and G. Auffermann, New ternary alkali-metal-transition-metal hydrides synthesized at high pressures: Characterization and properties, *Chem. Mater.*, 10, 2723, 1998.
59. Olofsson-Martensson, M., U. Häussermann, J. Tomkinson, and D. Noreus, Stabilization of electron-dense palladium-hydrido complexes in solid-state hydrides, *J. Am. Chem. Soc.*, 122, 6960, 2000.
60. Zölliker, P., K. Yvon, P. Fischer, and J. Schefer, Dimagnesium cobalt(I) pentahydride, Mg_2CoH_5, containing square-pyramidal pentahydrocobaltate(4−) ($CoH_5$4−) anions, *Inorg. Chem.*, 24, 4177, 1985.
61. Zaluska, A., L. Zaluski, and J.O. Strom-Olsen, Lithium-beryllium hydrides: The lightest reversible metal hydrides, *J. Alloys Compd.*, 307(1–2), 157–166, 2000.
62. Zaluska, A., and L. Zaluski, New catalytic complexes for metal hydride systems, *J. Alloys Compd.*, 404–406, 706–711, 2005.
63. Schwarz, R.B., Storage of hydrogen in powders with nanosized crystalline domains, *Mater. Res. Bull.*, 24, 40, 1999, available at http://www.wtec.org/loyola/nano/US.Review/04_06.htm, May 2007.
64. Bogdanović, B. and M. Schwickardi, Ti-doped alkali metal aluminum hydrides as potential novel reversible hydrogen storage materials, *J. Alloys Compd.*, 1–9, 253–254, 1997.
65. Streukens, G., B. Bogdanović, M. Felderhoff, F. Schuth, Dependence of dissociation pressure upon doping level of Ti-doped sodium alanate—a possibility for "thermodynamic tailoring" of the system, *Phys. Chem. Chem. Phys.*, 8(24), 2889–2892, 2006.
66. Huot, J., G. Liang, S. Boily, A.V. Neste, and R. Schulz, Structural study and hydrogen sorption kinetics of ball-milled magnesium hydride, *J. Alloys Compd.*, 293–295, 495–500, 1999.

67. Becker, L., *Hydrogen Storage*, available at http://www.csa.com/discoveryguides/hydrogen/overview.php, March 2007.
68. SOLID-H™, Metal Hydride Hydrogen Storage Containers, available at http://www.hydrogencomponents.com, January 2007.
69. Lasher, S., Comparison of hydrogen storage options, *NHA Annual Hydrogen Conference 2005*, Washington, March 29–April 1, 2005.
70. Ahluwalia, R.K., Sodium alanate hydrogen storage system for automotive fuel cells, *Int. J. Hydrogen Energ.*, 32(9), 1251–1261, 2007.
71. GKSS, Hydrogen Storage: Metal Hydrides in Comparison to Alternative Solutions for Emission Free Vehicles, March 2007.
72. Milliken, J., L. Cataquiz, J. Petrovick, C. Read, and S. Satyapal, Grand Challenge for Basic and Applied Research in Hydrogen Storage, U.S. DOE Report, 2003.
73. Eigen, N., C. Keller, M. Dornheim, T. Klassen, and R. Bormann, Industrial production of light metal hydrides for hydrogen storage, *Scr. Mater.*, 56, 847–851, 2007.

12

Hydrogen Storage in Carbon Materials

K. K. Pant and Ram B. Gupta

CONTENTS

12.1 Introduction

Hydrogen is considered a key fuel for the future, but the problems related to its storage is a major concern in further development. In fact, hydrogen storage is the key enabling technology that must be significantly advanced in terms of performance and cost-effectiveness, if hydrogen is to become an important part of the world's energy economy. Significant research and development activities are being carried out to increase the efficiency of hydrogen-storage systems to make them competitive with the current fossil fuels for transportation and stationary applications [1–31]. To develop and demonstrate viable hydrogen-storage technologies, a set of objectives have been proposed by the U.S. Department of Energy (DOE) based on a driving range of 500 km (see Chapter 9). These objectives fix a target of 2 kWh/kg (6 wt%) and \$4/kWh for 2010, and of 3 kWh/kg (9 wt%) and \$2/kWh for 2015 [7]. Possible current approaches to vehicular hydrogen storage include (a) physical storage via compression or liquefaction, (b) chemical storage in irreversible hydrogen carriers, (c) reversible metal and chemical hydrides, and (d) gas-on

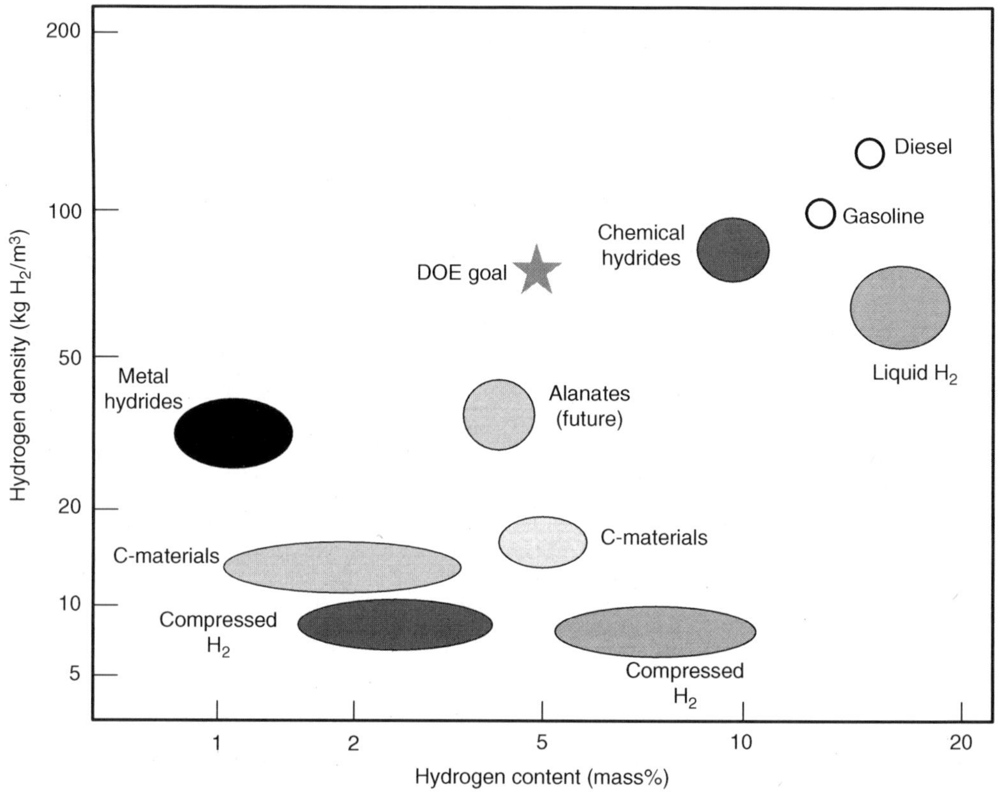

FIGURE 12.1
Volumetric versus mass densities achieved by various hydrogen-storage technologies. (Modified from Conte, M., Prosini, P., and Passerini, S., *Mater. Sci. Eng. B: Solid-State Mater. Adv. Technol.*, B108(1–2), 2–8, 2004.)

solid adsorption. Although each storage method possesses some desirable attributes, no approach satisfies all of the efficiency, size, weight, cost, and safety requirements for automobile applications. Volumetric and mass performances of various storage methods are shown in Figure 12.1 [31].

This chapter describes the storage of hydrogen in several forms of carbon, ranging from amorphous activated carbon (AC) to the ordered forms such as graphite and carbon nanotubes (CNTs). Carbon materials can be utilized for hydrogen storage in the following different ways:

i. Carbon forms covalent compounds with hydrogen that could be used to store and then reversibly recover hydrogen. Uses of reversible reaction pairs, such as decalin dehydrogenation/naphthalene hydrogenation, methylcyclohexane dehydrogenation/toluene hydrogenation, and tetralin dehydrogenation/naphthalene hydrogenation, have been proposed, which are discussed in detail in Chapter 13.
ii. Carbon can be made into fibers that are 10-fold stronger than steel. Carbon fiber reinforced tanks can store hydrogen up to 650 bars and are available in lightweight composite designs. These tanks can withstand explosive blast and a surface temperature of up to 700°C. Such tanks have been discussed in Chapter 10.

iii. Hydrogen can be adsorbed on a carbon surface. The maximum amount of hydrogen adsorption on any particular material can be estimated as the quantity of hydrogen that can be accommodated on a surface monolayer. With the assumption that the structure of the adsorbed hydrogen is closed-packed face centered, the minimum surface area required for the adsorption of 1 mol of hydrogen is 85.9 m^2. Based on the surface area of a single graphene sheet (1315 m^2/g), the storage capacity of hydrogen adsorbed on graphite is about 3 wt% that can be typically achieved at cryogenic temperatures. The interlayer distance between the graphene layers can be optimized by adding spacers to achieve a high binding energy to enhance physisorption. The additives not only act geometrically but also via a strong interaction with hydrogen, which can increase the adsorption energy leading to storage at a higher temperature. However, there is an increase in the material density. Binding can also be enhanced by making molecular hydrogen complexes of transition metals via Kubas binding [26,32]. In these organometallic complexes, the chemical bond in the hydrogen molecule interacts with the metal and leads to a relatively strong adsorption of molecular hydrogen. Use of additives can also increase the binding energy by hydrogen spillover. These additives act as a catalytic active center for the dissociation of molecular hydrogen to hydrogen atom that can easily spillover from the additive sites to the carbon network and finally become bound to carbon atoms [26].

iv. Carbon nanomaterials possess high surface area and porosity and low density. Generally, the storage capacity for different carbon materials depends linearly on their surface area and is independent of the individual nanostructure. Surface areas up to 3000 m^2/g can be achieved in AC, which provide a hydrogen-storage capacity of 2.4 times more than the same volume of compressed gas at 200 bars [33]. For example, at 200 bars, a specific adsorption capacity of approximately 8 g of hydrogen per 100 g of carbon can be obtained. Volumetric capacity of such AC pressurized storage is comparable to the density of cryogenic liquid hydrogen of about 0.7 g/cm^3. Besides the high polarizability and the high surface areas, adsorption can also be increased by curving the structure of the carbon surface. This curved structure leads to an overlap of the potential fields of carbon atoms, which increases the adsorption energy up to 30 kJ/mol [34].

v. The inherent lightness and a good hydrogen uptake by CNTs and graphite nanofibers (GNFs) are attractive as prospective hydrogen-storage materials. A comparison of the storage weight of hydrogen for different alternative hydrogen-storage concepts is presented in Table 12.1 and is compared with gasoline [1]. For all others except methane, a hydrogen-storage capacity of 5 kg is assumed that enough to run an advanced fuel-cell car for about 500 km. Methane (CH_4) as a compressed natural gas also represents a sort of chemical hydrogen storage representing a 25 wt% capacity. The data presented in Table 12.1 for methane are based on the storage weight and volume for different alternative hydrogen-storage concepts with 60% conversion efficiency [35]. Carbon nanotubes have the potential to compete with metal hydride, being comparable in size and weight. Hydrogen storage in inorganic nanotubes also has potential as the van der Waals gaps between the nanotube layers are potential candidates for hydrogen uptake. Furthermore, the inorganic nanotubes combine two elements, which is different from the pure CNTs [36].

TABLE 12.1

Storage Weight and Volume for Different Alternative Hydrogen-Storage Concepts (for Comparison, 5 kg Stored Hydrogen Are Assumed, Enough to Drive a Car for Approximately 500 km with 0.3 kWh/km Consumption)

System	Technical Parameters	Total Volume (L)	Weight (kg)	Weight Including Containment (kg)
Gasoline	Liquid	40	30	33
Compressed hydrogen	300 bars	310	5	38
	500 bars	220	5	47
	700 bars	180	5	55
Liquid hydrogen	25°C	150	5	36
CH_4	300 bars	82	20	46
Metal hydride	2% weight	128	258	284
	3% weight	85	172	189
Carbon materials	5% weight	100–300	105	115
	10% weight	50–150	55	60
	20% weight	25–75	30	33
	50% weight	10–30	15	17

Source: Adapted from Bünger, U. and Zittel, W. Hydrogen storage in carbon nanostructures—still a long road from science to commerce? *Appl. Phys. A-mater.* 72(2), 147–151, 2001.

12.2 Fundamentals of Adsorption

The use of hydrogen physisorption on porous materials is one of the main methods being considered for automobile applications. The objective is to store large amounts of hydrogen at a near-ambient temperature and safe pressures. In addition to the high sorption capacity, an ideal hydrogen adsorbent should have uniform and small nanopores in as high a density as possible, with minimal macroporosity and high thermal conductivity. While the first two characteristics together ensure that the internal volume of the adsorbent is not wasted, the third characteristic provides for managing the heat flux that is relatively large due to a high heat of adsorption.

The weak interaction of hydrogen physisorbed within the pores needs to be understood to optimize porous materials for hydrogen adsorption. The important measures are (a) amount adsorbed as a function of pressure, (b) temperature dependence of adsorption, (c) the enthalpies of adsorption, and (d) the characteristics of adsorption/desorption cycle. Finding the best adsorbents for hydrogen storage involves optimizing three parameters: (1) the characteristic binding energy of the hydrogen molecule with the material, (2) the available surface for adsorption processes, and (3) the bulk density of the adsorbate. The binding energy determines the operating temperature of a hydrogen storage system based on the solid [7]. The variation of attractive surface forces as a function of the distance from the surface decides whether van der Waals-type weak physisorption of molecular hydrogen occurs, or a dissociation and chemisorption of atomic hydrogen takes place. Owing to the attractive forces, the most stable position for an adsorbed molecule is with its center at about 1 molecular radius from the surface, and the attractive field rapidly diminishes at greater distances [26]. Once a monolayer of hydrogen adsorbate molecules or atoms has been formed, the gaseous species interacts with the liquid or solid adsorbate. The binding energy of the second layer of adsorbates is similar to the latent heat of sublimation or vaporization of the adsorbate. Consequently, adsorption at a temperature at or above the boiling point of the adsorbate at a given pressure leads to the adsorption of a single monolayer. The extent of adsorption, or the fraction surface coverage (θ), as a function of pressure (p) can be obtained from Langmuir isotherm as

$$\theta = \frac{Kp}{1 + Kp}$$

where K is the adsorption-equilibrium constant. As shown in Figure 12.2 for AC, hydrogen adsorbed decreases with increasing temperature and decreasing pressure [28]. With increasing pressure, hydrogen adsorption increases linearly up to a certain pressure, but at higher pressures the increase becomes gradual and eventually levels off at very high pressures [28,29].

The enthalpy of adsorption (ΔH_{ads}) can be obtained by

$$\Delta H_{ads} = -RT^2\left(\frac{d\ln p}{dT}\right)_\theta$$

where T is the temperature in Kelvin and R the gas constant (8.314 J/mol/K). Typical adsorption enthalpies range from 4 to 8 kJ/mol, which signifies a weak interaction between a poorly polarizable H_2 molecule and the pore surface. This is why, at ambient pressure, hydrogen does not significantly adsorb on many materials even with a high porosity. Interestingly, this interaction is enhanced in nano-sized pores resulting in high adsorption as shown in Figure 12.3 for various nanostructured carbons [37]. The quasiperiodic icosahedral nanoporous carbon material can reach 6 wt% adsorption at 3.8 MPa and 77 K. High gravimetric uptake of hydrogen at moderate pressures can be attributed to the significant physical adsorption of hydrogen on the external surface of this nanostructure.

In addition to the desirable thermodynamics (enthalpy of adsorption), a fast adsorption/desorption kinetics is required for a fast recharging and discharging of hydrogen. Hence, the activation energies of the adsorption and desorption steps play an important role

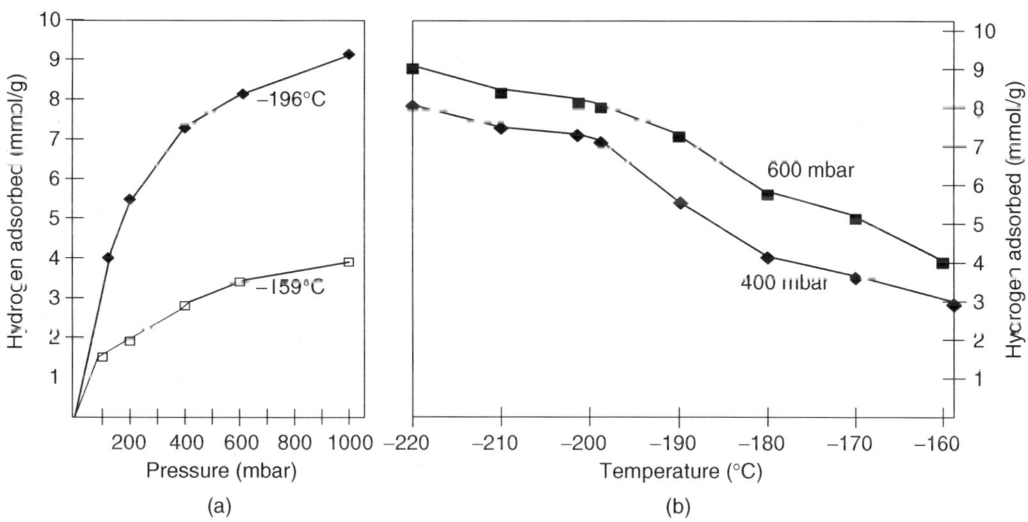

FIGURE 12.2
Isotherms (a) and isobars (b) for hydrogen adsorption on activated carbon. (Modified from Zhao, X., Xiao, Bo., Fletcher, A.J., Thomas, K.M., Bradshaw, D., and Rosseinsky, M.J., *Science*, 306, 1012, 2004.)

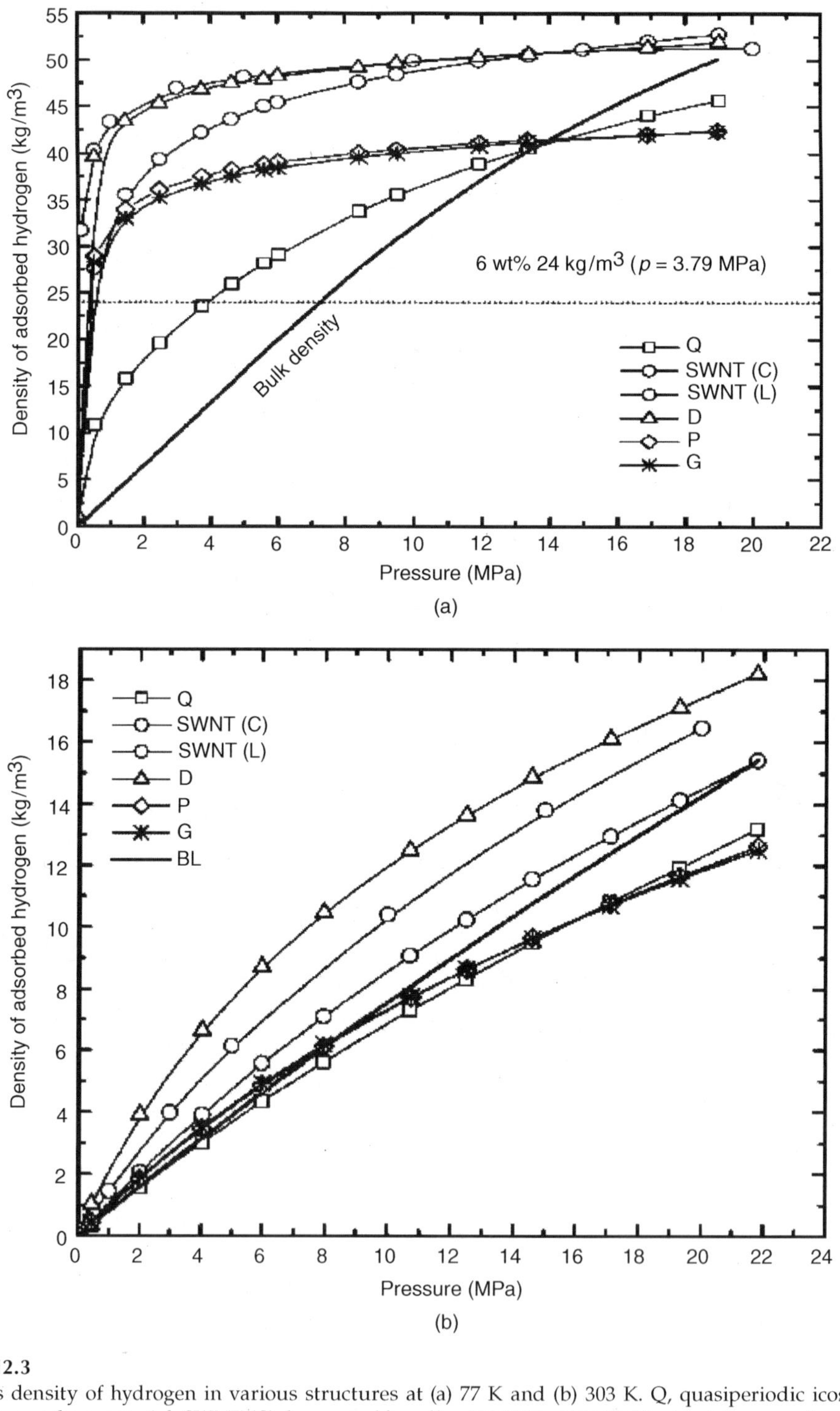

FIGURE 12.3
The mass density of hydrogen in various structures at (a) 77 K and (b) 303 K. Q, quasiperiodic icosahedral nanoporous carbon material; SWNT (C), hexagonal bundle of SWNTs; SWNT (L), simulation results of hydrogen storage in SWNTs; D, diamond nanoporous carbon material; P, primitive nanoporous carbon material; G, gyroid nanoporous carbon material; and BL, bulk density of hydrogen. (Reproduced from Kowalczyk, P., Hoyst, R., Terronesb, M., and Terrones, H., *Phys. Chem. Chem. Phys.*, 9, 1786–1792, 2007. With permission from the PCCP Owner Societies, http://dx.doi.org/10.1039/b110420a, the DOI suffix is 'b110420a'.)

in system kinetics. Usually the activation energy for adsorption is low, providing a fast adsorption. However, there is a significant barrier for the desorption process, making it slow.

To enhance the H_2 uptake due to physical adsorption, it is necessary to decrease the thermal motion of H_2 molecules on the solid surface, as low temperatures are ideal for physical adsorption. However, an alternative way to increase the amount of H_2 physisorbed is by optimizing the morphology of nanostructures of the adsorbant. In the search for a high storage capacity, a range of new materials have recently been proposed: superactivated carbons (ACs), activated carbon fibers (ACFs), single-walled carbon nanotubes (SWNTs), single-walled carbon nanohorns (SWNHs), multiwalled carbon nanotubes (MWNTs), GNFs, and organic–inorganic hybrid porous complexes. The unifying feature of these materials is that they all have pores of molecular dimensions. Such nanomaterials can produce high packing density of supercritical hydrogen, although not high enough to satisfy the DOE target. Various types of carbon materials that may be used for hydrogen storage are discussed in the following sections.

12.3 Activated Carbon

Activated carbon is a low-cost material and is available at the industrial level with a typical pore diameter of <1 nm and can have a specific surface as high as 3000 m^2/g. AC is made of amorphous carbon containing very small graphite crystallites with nanoporosity and some meso- and macro-porosity. AC is prepared from carbon-rich organic precursors by dry distillation to form carbonized organic precursors, which can be activated to increase the pore volume by thermal or chemical treatments. In the thermal activation process, carbon is heated at a temperature ranging between 700°C and 1000°C in the presence of oxidizing gases such as steam, CO_2, steam/CO_2 mixtures, or air. In the chemical activation process, carbon is heated between temperatures 500°C and 800°C in the presence of dehydrating substances such as $ZnCl_2$, H_3PO_4, or KOH, which are leached out afterward. A typical molecular structure of AC is shown in Figure 12.4 [38].

The surface area in ACs correlates well with the amounts of hydrogen adsorbed, which is low at room temperature and moderate pressures. For example, the hydrogen adsorption is <1 wt% at 100 bars and 298 K, despite a high (2800 m^2/g) surface area [38]. Although hydrogen is mainly adsorbed by physical interaction, surface groups and other reactive sites can also be incorporated into the AC to promote hydrogen sorption. However, such processes tend to operate mostly via chemisorption and are not a useful means of enhancing the reversible storage of hydrogen.

12.4 Graphite

Graphite occurs naturally, but can also be made by heating petroleum coke or coal pitch to 2200°C. Graphite is one of the four ordered carbon allotropes with sp^2 bonding, and the other carbon allotropes are diamond with sp^3 bonding, fullerene with perturbed sp^2 bonding, and carbyne with sp^1 bonding. As shown in Figure 12.5a, each carbon atom in graphite is covalently bonded to three other surrounding carbon atoms [3]. The flat sheets (layers)

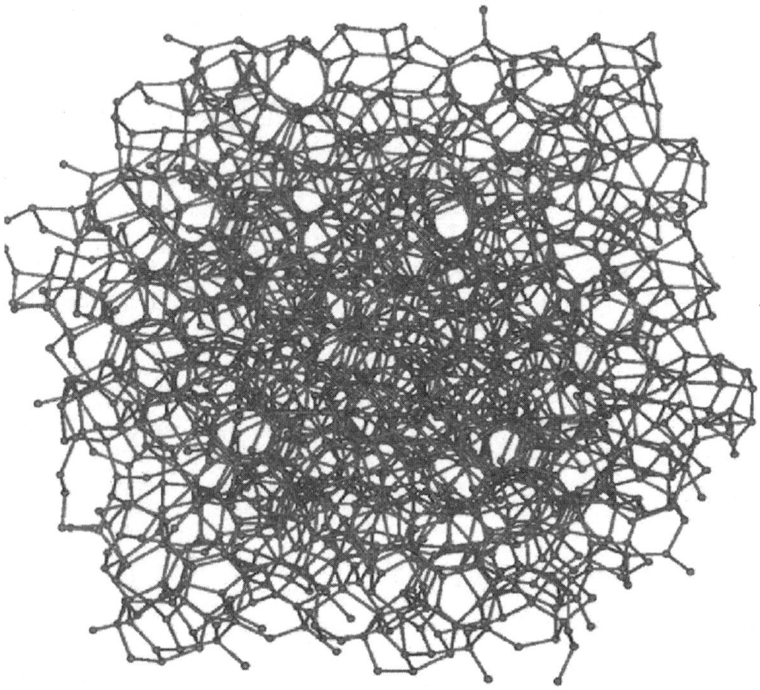

FIGURE 12.4
Molecular structure of activated carbon. (Modified from Jin, H., Lee, Y.S., and Hong, I., *Catal. Today*, 120, 399–406, 2007.)

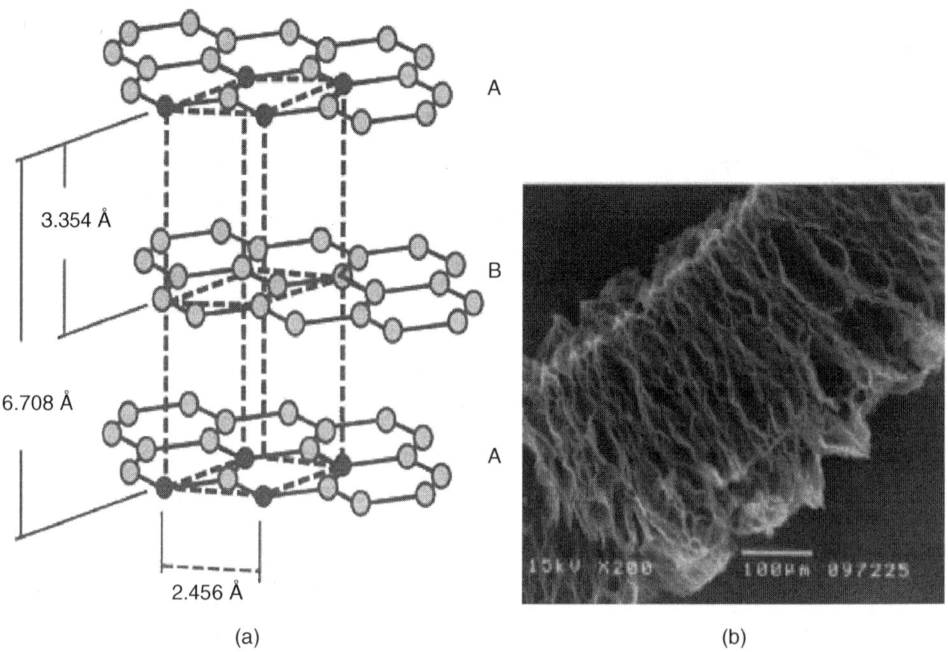

FIGURE 12.5
Graphite (a) molecular structure and (b) SEM micrographs after exfoliation. (Figure 12.5b reproduced from website www.electronics-cooling.com/assets/images/, February 2007; Inagaki, M., and Suwa, T., *Carbon*, 39, 915–920, 2001. With permission.)

of carbon atoms are bonded into hexagonal structures. The layers are not covalently connected to the surrounding layers; instead, these are connected together by weak van der Waals forces much like that of mica. The unit cell dimensions are $a = b = 2.456$ Å and $c = 6.694$ Å. The carbon–carbon bond length in the bulk form is 1.418 Å and the interlayer spacing is $c/2 = 3.347$ Å [3].

In its pure glassy (isotropic) synthetic forms, pyrolytic carbon, carbon fiber, and graphite are extremely strong and heat-resistant (up to 3000°C) materials, and are used in reentry shields for missile nose cones, solid rocket engines, pebble bed reactors, brake shoes, electric motor brushes, and as electrodes in electrical discharge machines. Expandable graphites are used in firestops, particularly plastic pipes and gaskets, fitted around the perimeter of a fire door. During a fire, the graphite expands and chars to resist fire penetration reducing the likelihood of the spread of fire and fumes. Exfoliated graphite (Figure 12.5b), which is an important raw material for the industrial production of flexible graphite sheets, is attractive because of its very high sorption capacity. Usually it is prepared by the rapid heating of the residue compound of natural graphite flakes with sulfuric acid at 1200°C [39,40].

Despite the relatively high polarizability of graphite, its hydrogen-storage capability is rather low due to its small interlayer distances and its low specific surface area (SSA). However, the surface area can be increased by ball milling, for example, it is reported that after 80 h of ball milling in a 10 bars hydrogen atmosphere, graphite nanostructures contain up to 0.95 hydrogen atoms per carbon atom, or 7.4 wt% from which 80% could be desorbed at temperatures above 300°C [41]. In addition, the intercalation of hydrogen into graphite at room temperature and moderate pressures can be done by slightly expanding the interlayer spacing.

12.5 Graphene

Graphene is a single planar sheet consisting of sp^2 bonded carbon atoms, which is identical to one of the layers of the graphite structure but isolated as shown in Figure 12.6 [26]. The sheet has a finite size and noncarbon elements can appear at the edge of the sheet. Graphenes are aromatic in nature and are generally formed only of hexagonal cells. A typical sheet has the chemical formula $C_{62}H_{20}$, where hydrogen is used to terminate the dangling bonds. If a pentagonal cell is present then the plane warps into a conical shape.

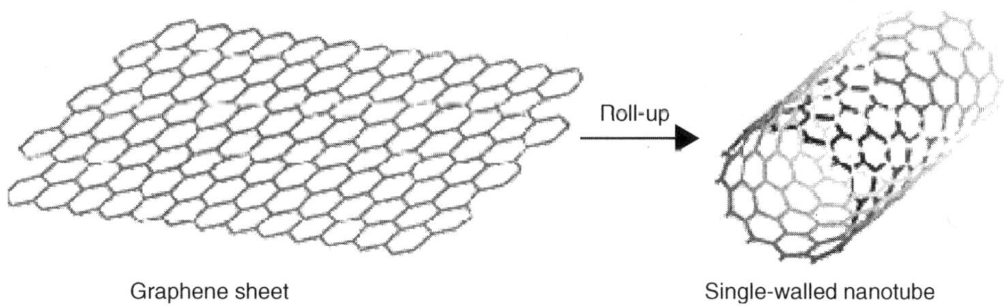

FIGURE 12.6
Molecular structure of graphene. (Reproduced from Strobel, R.J., Garche, J., Moseley, P.T., Jorissen, L., and Wolf, G., *J. Power Sources*, 159, 781–801, 2006. With permission.)

Graphene, a single atomic layer of graphite, is a flat sheet of benzene rings with zig-zag and arm-chair configurations. An interesting characteristic of these two structures is that the zig-zag structure has metallic transport properties. CNTs are empty cylinders, which can be viewed as rolled up sheets of a single or several concentric layers of graphene. Planar graphene does not exist in the free state, as it is unstable with respect to the formation of curved structures such as soot, fullerenes, and nanotubes. For a graphene sheet with an SSA of 1315 m^2/g as adsorbent, the maximum theoretical concentration is 0.4 hydrogen atoms per surface carbon atom or 3.3 wt% hydrogen on sheets with hydrogen atoms on one side. Attempts have been made to isolate a single graphene sheet by mechanical exfoliation of small fragments of highly oriented pyrolytic graphite to study the electrical properties of graphene [26].

12.6 Fullerene

Fullerenes are closed-cage carbon molecules C_{2n} ($n > 10$) composed of a combination of n carbon atoms of sp^2 (trigonal) hybridization arranged into 12 pentagons and $(n-20)/2$ hexagons. Fullerenes are soluble in certain solvents such as toluene and carbon disulfide. Spherical fullerenes are often referred to as "buckyballs," and cylindrical fullerenes as "buckytubes" or "nanotubes" that are formed in the presence of metal catalysts. The smallest possible complete fullerene molecule could be built up from 32 carbon atoms. The shape of spherical fullerene that is similar to a soccer ball is called a truncated icosahedron with 20 hexagons and 12 pentagons [42,43]. Both the spherical and the cylindrical fullerenes are hollow. The most common fullerene is C_{60} that consists of 60 carbon atoms, which are bonded in a nearly spherical configuration. Besides this many other forms of fullerenes are known, for example, C_{70}, C_{76}, and C_{84} or even C_{240} and C_{540}. The typical structures of two common fullerenes are shown in Figure 12.7. Fullerenes are generally formed by

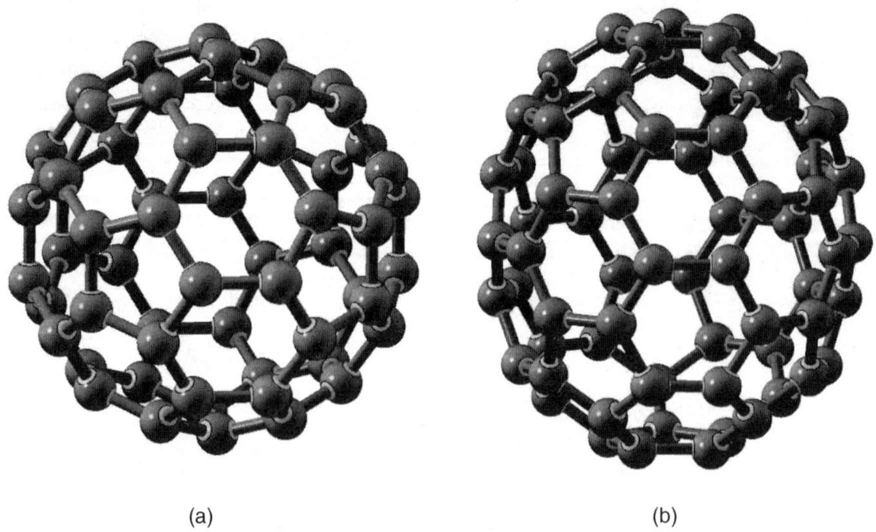

FIGURE 12.7
Molecular structures of fullerenes: (a) C_{60} and (b) C_{70}. (Modified from Peera, A.A., Alemany, L.B., and Billups, W.E., *Appl. Phys.* A, 78, 995–1000, 2004.)

vaporizing carbon by a pulsed laser, or an arc discharge, and mixed with an inert gas, followed by gradual condensation.

Owing to their unique molecular structures, fullerenes such as C_{60} are the only form of carbon that can be hydrogenated and dehydrogenated reversibly [15]. Reduced fullerenes can be prepared by dissolving-metal reductions or by transfer of hydrogenation using dihydroanthracene as the hydrogen donor that leads to $C_{60}H_{36}$ isomers containing approximately 5 wt% hydrogen. The hydrides from the transfer hydrogenation route are preferred to store hydrogen since the $C_{60}H_{36}$ isomers prepared by this route are more stable than those obtained from dissolving-metal reductions. Fullerenes are the simplest hydrofulleride in which the hydrogen atoms are bonded vicinally. Stable fullerenes cannot have adjacent pentagons because this introduces too much strain into the cages. It serves as a model for the thermally induced removal of hydrogen from a fullerene and are a class of carbon molecules in which the carbon atoms are arranged into 12 pentagonal faces and 2 or more hexagonal faces [26,39,42–46].

C_{60} and C_{70} fullerites without a catalyst interact with gaseous hydrogen (deuterium) at 373–773 K and pressures over 10 MPa. $C_{60}D_{8-36}$ deuterides have been produced at 373–773 K and 15 MPa. Hydrogenation of the mixture from fullerite (C_{60}, C_{70}) and metal phases (Pd, V, LaNi5; LaNi4:65Mn0:35; $CeCO_3$) at 1–5 MPa and 570–670 K result in the formation of the mixture from metalhydrides and hydrofullerenes. At 400–700 K, $C_{60}Pt$ and $C_{60}Pd_{4.9}$ metalfullerides interact with hydrogen to form hydrogen compounds: metalfullerides or hydrofullerenes. In both cases, hydrogen evolves during heating at 800 K [5].

Possibility of hydrogen localization within the fullerene frame as a method for hydrogen capacity enhancement is of great interest. At 77 K, about 3 mol H_2/mol C_{60} can be present in the interstitial sites of $C_{60}H_x$ as H_2 molecules and evolve from the sample during its heating up to 293 K [5]. According to theoretical calculations, a C–H bond is weaker than a C–C bond (68 versus 83 kcal/mol bond energy) [15]. Therefore, on heating, the C–H bond will break faster than the C–C bond, giving off hydrogen and preserving the fullerene structure [15], thereby making such a structure useful for hydrogen storage.

Mechanical and chemical treatments of C_{60} can result in a structurally strong material. However, the details of the structural changes are not yet fully established. It is known from studies of C_{60} polymers obtained by chemical methods that linking of C_{60} molecules with rigid covalent bonds results in an increase in density and decrease in cell volume compared to the pristine C_{60} [43]. However, in the case of milling-driven modifications, the reverse effect was observed. Hydrogen adsorption by C_{60} increases due to the milling with maximum at the milling time, which corresponds to complete amorphization of material.

12.7 Carbon and Graphite Nanofibers

Various carbon nanostructures have been investigated to obtain higher storage densities. Carbon nanofibers (CNFs) are layered graphitic nanostructures that initially generated significant interest and great controversy. The reported values of the hydrogen-storage capacity of such structures ranged from <1 wt% to several tens of weight percentages at moderate pressures and temperatures. CNFs are formed at high temperatures from a mixture of carbon-containing gases and hydrogen, or from hydrocarbons alone, on catalysts comprising mostly nickel- and iron-based alloys. Depending on the nature and shape of the catalyst, the ratio of the hydrocarbon/hydrogen reactant mixture, and the reaction conditions, different types of CNFs can be formed that differ in morphology, crystallinity, and shape. The fibers are usually 5–100 μm long having a diameter of 5–100 nm [47]. The SSAs (BET) of these ordered crystalline solids are normally between 100 and 300 m^2/g, but can sometimes reach as high as 700 m^2/g.

GNFs consisting of the stacks from graphene plates and cones, with a length of 10–100 μm, a section of 10–50 nm, and an interlayer space of 0.337 nm, have plenty of open edges that can favor hydrogen sorption. GNFs produced by pyrolysis of C_2H_2 on Fe:Ni:Cu = 85:10:5 sorb 6.5 wt% H_2. GNFs from C_6H_6 or CH_4 absorb 4–5.7 wt% H_2. GNFs having a parallel-bladed structure sorb up to 10–13 wt% H_2. In a recent study, Xu et al. [48] synthesized the GNF by decomposition of C_2H_4:H_2 = 1:4 gas flowing at 873 K using NiCu catalyst. These GNFs (Figure 12.8) showed a relatively high surface area and pore volume and were found to adsorb hydrogen comparable with the SWNHs. The structure of the platelets provides many edges on a nanofiber that represent a large surface, with many interaction sites for adsorbates [6], hence the interlayer distance can play an important role as shown in Table 12.2. The hydrogen-storage capacity of CNFs depend on their structure, structure defects, pretreatment, purification, geometry (surface area, tube diameter, and length), arrangement of tubes in bundles or "ropes," storage pressure, temperature, etc.

The proposed configurations for hydrogen adsorbed on graphite in different phases indicate a commensurate phase with a nearest-neighbor spacing of 4.26 Å for low coverage of hydrogen; a denser and an incommensurate close-packed structure with a nearest-neighbor distance of 3.51 Å at higher pressure, and the monolayer that would be required to achieve the densities if five layers were present between every two sheets of graphite [6]. Consequently, a theoretical storage density of 60 wt% H_2 would require a nearest-neighbor distance within one of these five monolayers of only 0.98 Å [2]. Hydrogen-adsorption studies have been conducted on similarly fabricated fibers at temperatures between 77 and 300 K. The absolute level of hydrogen desorption measured for the graphite nanofiber is typically <0.08 wt%.

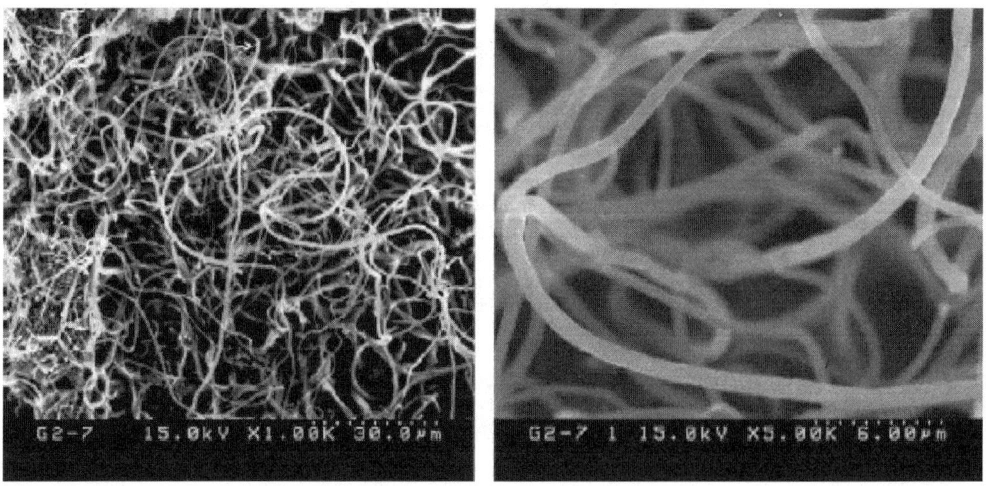

FIGURE 12.8
SEM of graphitic nanofiber prepared by decomposition of C_2H_4:H_2 = 1:4 gas flow over NiCu alloy powder. (Reproduced from Xu, W.-C., Takahashi, K., Matsuo, Y., Hattori, Y., Kumagai, M., Ishiyama, S., Kanekoc, K., and Iijima, S., *Int. J. Hydrogen Energ.* 32(12), 2504–2512, 2007. With permission.)

TABLE 12.2

Hydrogen Storage in Graphitic Nanofibers as a Function of Temperature and Interlayer Distance, at 100 bars

Temperature (K)	Interlayer Distance (Å)	Hydrogen Storage (wt%)
77	6	3
	9	7.5
	12	9
	20	12.5
298	6	0.8
	9	1.4
	12	1.6
	20	2.4
77	7	4
	10	7
200	7	2.4
	10	2.6
300	7	1.2
	10	1.2

Source: Adapted from Colomer, J.-F., Piedigrosso, P., Wilems, I., Journet, C., Bernier, P., Van Tendeloo, G., Fonseca, A., and Nagy, J.B., *J. Chem. Soc. Faraday Trans.*, 94, 3753, 1998.

12.8 Multiwalled Carbon Nanotubes

The CNTs are formed by rolled graphite sheets, with an inner diameter starting from 0.7 nm up to several nanometers and a length of 10–100 μm. The CNTs are described as usually closed on both sides by a hemisphere, that is, half of a fullerene. Typical transmission electron microscopy (TEM) and scanning electron microscopy (SEM) of CNTs are shown in Figure 12.9 [14]. Tubes formed by only one single layer are called SWNTs, whereas those consisting of multiple concentric layers are called MWNTs. The interlayer distance in MWNTs is close to the interlayer distance in graphite, which is equal to half of the unit cell parameter *c* (0.671 nm) [2]. The diameter of SWNTs varies from 0.7 to 3 nm, whereas MWNTs show typical diameters of 30–50 nm.

MWNTs consist of layers of nested concentric cylinders of graphite with a hollow center. The spacing between each cylinder is similar to the interplanar spacing in graphite, with the number of shells varying from 2 to 50. MWNTs have inner and outer diameters that are typically 2–10 and 15–30 nm, respectively, and are generally 5–100 microns in length. Large bundles of MWNTs with diameters up to 200 μm [2,49] and long individual tubes, over 2 mm, have been reported. MWNTs can be produced by the vaporization of carbon in an electric arc [50,51], via benzene pyrolysis [30], or the decomposition of acetylene [52,53] with metal catalysts. Most of the shells of MWNTs are inactive for hydrogen adsorption and only serve to increase the carbon mass and therefore decrease the hydrogen-storage capacity [27].

Chen and Wu [36] reported remarkable hydrogen-storage capacities for alkali-metal-doped MWNTs formed by the catalytic decomposition of CH_4. The walls of the nanotubes formed via this process are conical in shape and for this reason these nanotubes have been sometimes classified as graphitic nanofibers [45]. Lee et al. [54,55] addressed the problem of hydrogen storage in MWNTs and reported that hydrogen atom chemisorption is favored between the tube walls where hydrogen-storage capacity is independent of diameter, and they estimated it to be on the order of 7.7 wt%. The intercalation of

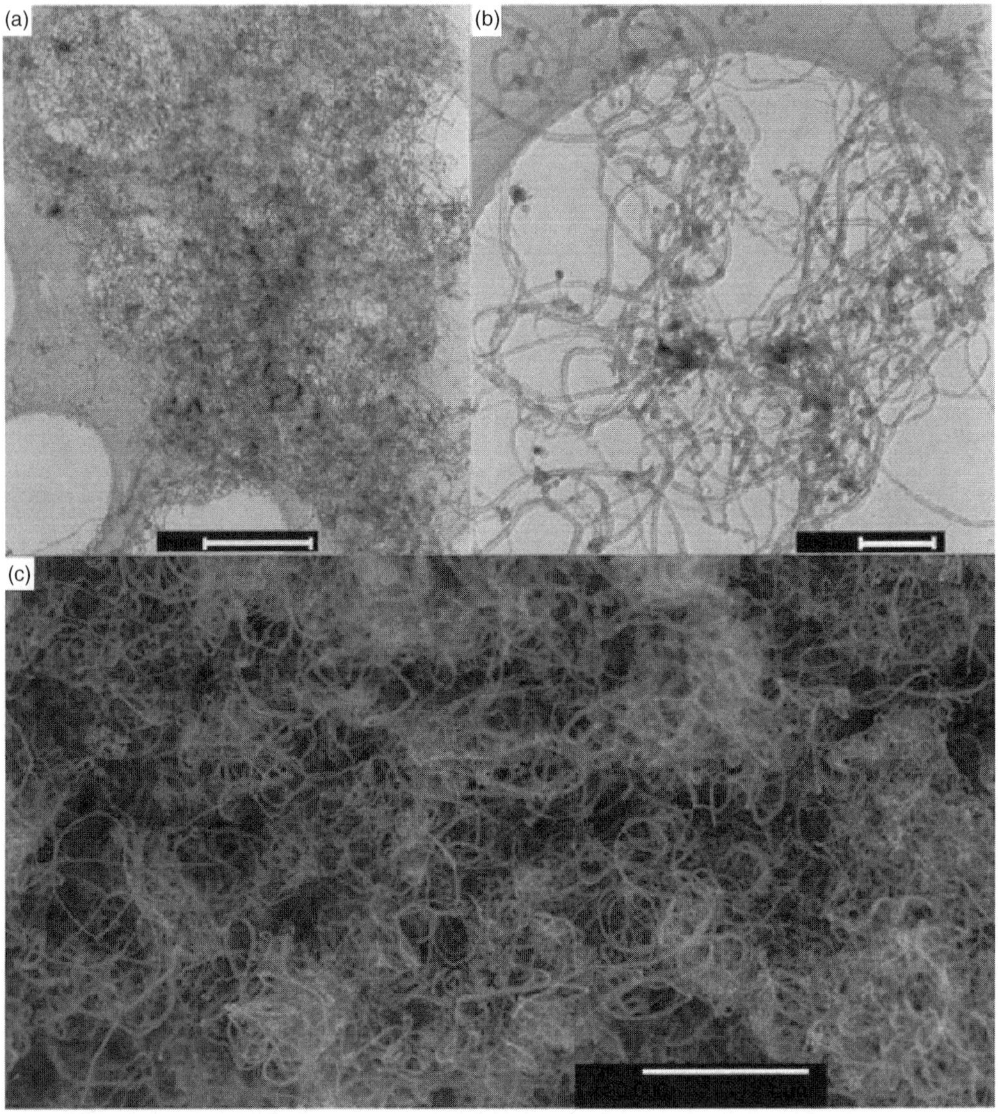

FIGURE 12.9
Typical TEM (a,b) and SEM (c) images showing the morphologies of the MWNTs (CNT). (Reproduced from Ning, G.Q., Wei, F., Luo, G.H., Wang, Q.X., Wu, Y.L., and Yu, H., *Appl. Phys. A*, 78, 955–959, 2004. With permission.)

alkali metal increases the spacing between the graphitic layers, which facilitates hydrogen diffusion into the material. For example, Li and K have been incorporated in the MWNTs through solid-state reactions with the metal carbonates or nitrates, with H_2 uptake of 20 wt% for Li-doping at 653 K and 14 wt% for K-doping at ambient temperature. Unfortunately, the K-doped nanotubes were reported to combust on exposure to air. The hydrogen adsorption was believed to proceed by a dissociative mechanism, and an infrared spectrum of the hydrogen-charged Li-doped material was interpreted to indicate the presence of both LiH and CH species [56]. Interestingly, Li-doped MWNTs formed under identical conditions were shown to exhibit a weight increase of 12 wt%

when exposed to "wet H_2" and only 2.5 wt% in the presence of "dry H_2" [57]. When K-doped nanotubes were exposed to dry H_2, a weight increase of only 1.8% was observed [52]. It has been reported [58] that a single titanium atom coated on a single-walled nanotube can bind up to four hydrogen molecules, which can have up to 8 wt% of hydrogen storage. However, all these results need to be validated before their practical application on a commercial scale.

12.9 Single-Walled Carbon Nanotubes

The SWNTs have a remarkable form of elemental carbon giving unique properties, for example, its diameter and chirality may impact the hydrogen adsorption [2,7,21,26]. The narrow pore-size distribution of SWNTs makes them attractive candidates as adsorbents for hydrogen, and their cylindrical geometry can lead to a deeper potential well. Calculations, based on the minimization of the energy of a hydrogen molecule located at the center of a 6.4 Å internal-diameter SWNT using a continuum cylindrical Lennard–Jones potential, show that the minimum sorption binding energy inside the SWNT would be 12.3 kJ/mol. This binding energy is two to three times larger than that for AC. Very small radii of the SWNT, however, can lead to repulsive interactions inside the nanotube. Early experiments on SWNTs reported adsorbed amounts between 0 and 10 wt%. The spread of values was mainly due to the experimental errors due to the small amounts of samples available for sorption measurements [7].

SWNTs can be synthesized by the coevaporation of a cobalt catalyst and graphite in an electric arc furnace. Fibers typically consisted of 7–14 bundled SWNTs, 10–15 Å in diameter. The hydrogen-adsorption properties of the early SWNT materials has been investigated by several researchers [2,12,13,18,48,59–61]. In addition, SWNTs have been reportedly produced at a high yield by the laser vaporization method [59]. The calculated amounts of hydrogen absorbed at the surface and in the cavity of the nanotubes are shown in Figure 12.10 [41]. The surface adsorption is highest for the CNT with the largest SSA, that is, the SWNT. The bulk absorption in the cavity is proportional to the diameter of the tubes and is also highest for SWNT. Furthermore, the weight percentage of adsorbed hydrogen in the cavity of the tube increases with the tube diameter considering that hydrogen would condense in the nanotube. Hydrogen uptake varies linearly with tube diameter although the tube volume increases as the square of the radius. This is because the uptake is proportional to the surface area (one layer adsorption), that is, the number of carbon atoms, which increases linearly with the increasing tube diameter.

Xu et al. [48] investigated the hydrogen-storage capacities of various carbon materials, including AC, SWNHs, SWNTs, and GNFs at 303 and 77 K. The hydrogen-storage capacity of carbon materials was <1 wt% at 303 K, and a ACs had the highest capacity (0.67 wt%). By lowering the adsorption temperature to 77 K, hydrogen-storage capacity of carbon materials increased significantly to 5.7 wt% at a relatively low pressure of 3 MPa. Hydrogen-storage capacity of carbon materials was proportional to their SSA and the volume of micropores, and the narrow micropores was preferred to adsorption of hydrogen, indicating that all carbon materials adsorbed hydrogen gas through physical adsorption on the surface.

Most measurements show that under ambient conditions, the amount of hydrogen adsorbed in SWNTs is small (<1 wt%), whereas under cryogenic conditions, hydrogen

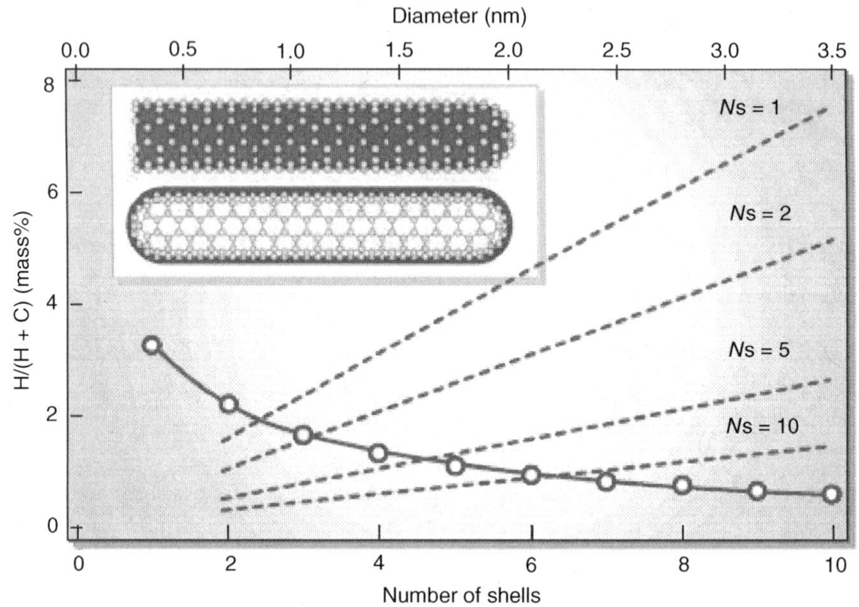

FIGURE 12.10
Calculated amount of adsorbed hydrogen in weight percent on CNTs assuming condensation of hydrogen: (a) Monolayer adsorbed at the surface of nanotubes as a function of the number of shells (NS) (markers and line, axis: left and bottom) and (b) hydrogen condensed in the cavity of nanotubes as a function of the diameter (dashed lines, axis: left and top) for various number of shells, NS = 1 (SWNT), NS = 2, NS = 5, NS = 10. (Reproduced from Schlapbach, L. (ed.), Hydrogen in intermetallic compounds II. Surface and dynamic properties, applications, *Topics Appl. Phys.* 67, Springer, Berlin, 67, 15–95, 1992. With permission.)

uptake ranges from 1 to 2.4 wt% [62]. An adsorption enthalpy of 4.3–4.5 kJ/mol was reported, which is typical of carbon adsorbents. Thermal treatments under high vacuum and high temperature can enhance the adsorption properties of SWNTs by removing guest species to open the access to adsorption sites and increasing the specific surface available to the hydrogen molecules.

Kowalczyk et al. [37] used grand canonical Monte Carlo simulation to model the hydrogen storage in the primitive, gyroid, diamond, and quasiperiodic icosahedral nanoporous carbon and CNTs. At 3.8 MPa and 77 K, the idealized model of a hexagonal bundle of SWNTs adsorbed 4.6 wt%, as shown in Figure 12.11, whereas the primitive, diamond, and gyroid nanoporous carbon materials adsorbed around 3 wt%. At this condition, the volumetric density of hydrogen for diamond nanoporous carbon material was 47 H_2 kg/m, whereas for the primitive and gyroid ones it was below 37 kg/m. Unfortunately, the hydrogen uptake drops by threefold when temperature is raised to 303 K. All investigated nanoporous carbon materials were not effective against compression above 20 MPa at 77 K since adsorbed density approached the density of the bulk fluid. At 3.8 MPa pressure, they did not observe complete saturation; however, the increment of gravimetric uptake and volumetric density of hydrogen was very slow. Further increase in the bulk pressure of hydrogen slightly enhanced the volumetric density and gravimetric uptake of hydrogen due to compression of adsorbed hydrogen. Adsorbed hydrogen molecules fit the internal and external channels of the diamond structure reasonably well, as shown in Figure 12.12a through 12.12d. For all other nanoporous carbon materials studied, the gravimetric uptake of hydrogen at 303 K and pressure up to 22 MPa was <2 wt%. The efficiency of hydrogen storage by

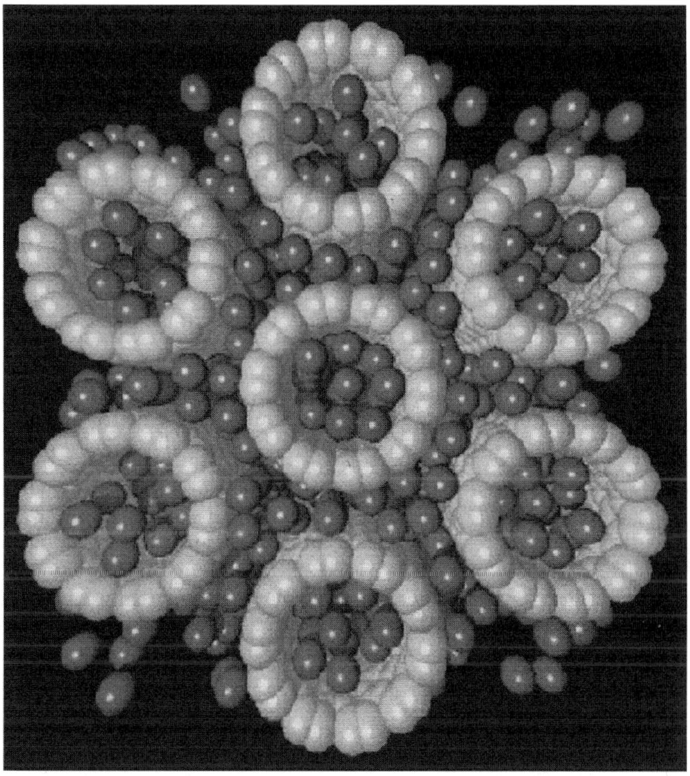

FIGURE 12.11
A snapshot of hydrogen adsorbed in the idealized hexagonal bundle of the (10,10) CNTs at 3.8 MPa and 77 K. (Reproduced from Kowalczyk, P., Hoyst, R., Terronesb, M., and Terrones, H., *Phys. Chem. Chem. Phys.*, 9, 1786–1792, 2007. With permission.)

physisorption in diamond nanoporous carbon materials and bundle of SWNTs was better than that based on pure compression in a tank. Unfortunately, the volumetric density of hydrogen in all investigated nanoporous carbon materials around room temperature was below 18 kg/m. The lower volumetric density of hydrogen in primitive carbon nanoporous material resulted from the complexity of the pore shapes of the mentioned nanoporous carbon materials (Figure 12.12c) [37,63].

Iijima et al. [64] developed a new type of carbon particle produced by CO_2 laser ablation of graphite at room temperature without any metal. In this material, an individual particle was composed of an aggregate of many horn-shaped sheaths of single-walled graphene sheets, which was named SWNH. SWNH is a new material that is similar to SWNT with high purity (>95%) without any metal catalyst. SWNH has been thought to be an ideal candidate for hydrogen-storage study, without any possible effect of metallic particles on hydrogen-storage capacity [48] because two strong adsorption sites of the internal and external nanopores of the assembly are present in SWNH. Because SWNH as prepared has no open intrananohorn space, hydrogen gas could be adsorbed only in the external nanopores of the assembly, leading to the low amount of hydrogen adsorption. However, the intraparticle pores become open by oxidation pretreatment, and hydrogen could be adsorbed in these open internal nanopores. Oxidation of SWNH at 623 K doubled the SSA, which resulted in the increase in hydrogen adsorption [48].

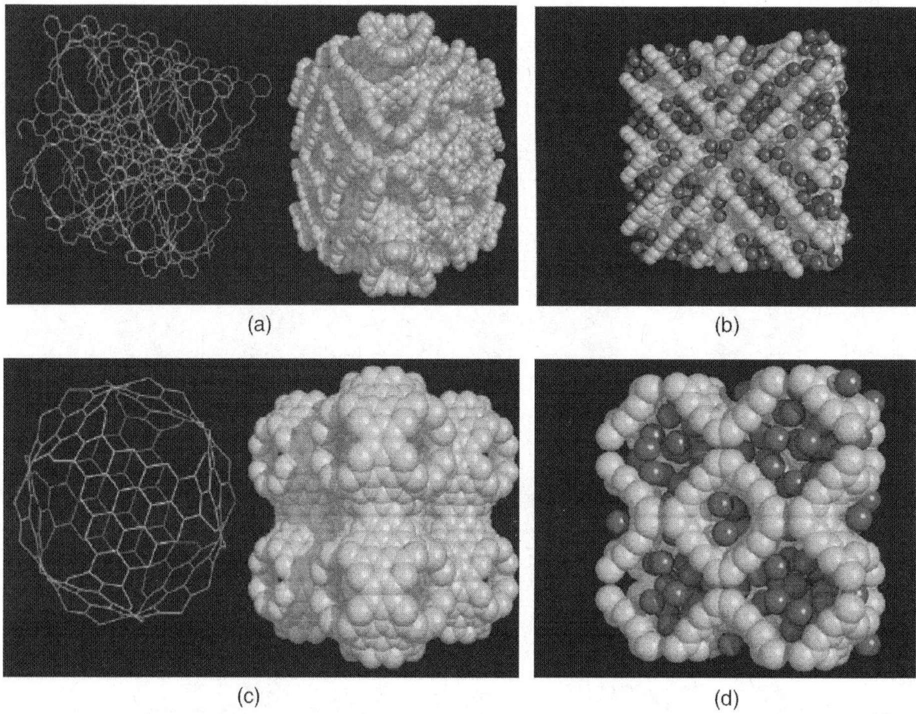

FIGURE 12.12
(a) Diamond carbon nanoporous material. The left image shows carbon rings on the surface and the right image shows eight unit cells of the structure decorated by carbon atoms. (b) Hydrogen adsorbed in the diamond nanoporous carbon material at 3.8 MPa and 77 K collected from the simulation (cross section is shown). (c) Primitive carbon nanoporous material. The left image shows carbon rings on the surface and the right image shows eight unit cells of the structure decorated by carbon atoms. (d) A snapshot of hydrogen adsorbed in the primitive nanoporous carbon material at 3.8 MPa and 77 K collected from the simulation (cross section is shown). (Reproduced from Kowalczyk, P., Hoyst, R., Terronesb, M., and Terrones, H., *Phys. Chem. Chem. Phys.*, 9, 1786–1792, 2007. With permission.)

12.10 Hybrid Materials

The CNTs can be used in combination with other hydrogen-storage materials to utilize the synergetic effects based on changes in crystal and electronic structures. For example, metal organic frameworks (MOFs) are nanostructures of transition metals bridged by carbon ligands that have recently been proposed as sorbent for hydrogen storage. The main interest in these materials lie in their very large SSAs (1000–6000 m^2/g) and the fact that they may be customized to specific storage applications by varying ligands, transition metals, and doping them with metals. It has been reported that titanium alloys, in combination with SWNT, have a hydrogen uptake of 2–8 wt% [30,58,64–67], which is much better than 2.5 wt% in the alloy alone. In these hybrid materials, it is not clear whether hydrogen is stored as molecules or as atoms. Since the storage capacity is significantly increased in the presence of metals, it is assumed that the metal also acts as a catalyst, which promotes the formation of hydride (H_{ad}) via a spillover (from the metal to the carbon surface) as follows:

$$H_2O + e^- + M \rightarrow M{\cdot}H_{ad} + OH^-$$

$$M{\cdot}H_{ad} + C \rightarrow C{\cdot}H_{ad} + M$$

Most of the hydrogen uptake in nanoporous materials such as carbon nanostructures and MOFs is mostly due to physisorption. Studies of micropore filling of ACs show that the density of hydrogen in the pores approaches that of liquid hydrogen. Any further improvement of the storage density would entail storing hydrogen in the porous structure at densities beyond liquid hydrogen. Increasing the operating temperature of a physisorption storage system will require going beyond current carbon and MOF nanostructures. Doping of carbon nanostructures with certain metals could lead to nondissociative binding of hydrogen molecules and enable them to be used to reversibly store hydrogen with enhanced sorption capacity. The maximum gravimetric excess adsorption density correlates linearly with the specific surface and the pore volume of the adsorbent. This suggests that the best adsorbents for practical storage applications will generally be the one with the largest specific surface per unit volume. The maximum gravimetric excess adsorption density correlates linearly with the specific surface and the pore volume of the adsorbent. The simulation studies carried out by Bénard et al. [68] showed at low pressure, the gravimetric adsorbed density of hydrogen increased with the density of the carbon nanostructure. This is because the greater nanostructure density reflects narrower pores and increased surface-hydrogen interactions due to surface potential overlaps. The reverse trend was observed at high pressures, where the specific surface available for adsorption plays a more important role. Results also conclude that the high-pressure trend is reversed when considering the volumetric adsorbed density: per unit volume, the high-density materials always have the largest adsorbed density. The linear correlation between adsorbed density and specific surface was also observed for these structures at 35 bars and 77 K.

For physisorbed hydrogen, these results confirmed that maximizing the specific surface of the adsorbent seems to remain the optimal strategy for the design of materials for hydrogen storage if volumetric considerations are not critical. For MOF adsorbents and in hydrogen sorption simulations on carbon nanostructures, it was observed that the low-pressure region of the adsorbent isotherms depends on the adsorbate–adsorbent interaction, leading to a reverse trend with respect to the higher-pressure region. In addition, the excess hydrogen density adsorbed on single-wall nanotubes is strongly dependent on sample preparation. Experiments show that at room temperature, reversible hydrogen uptakes of <1 wt% have been obtained. Whereas, at 77 K, the excess adsorption of 2–7.3 wt% at 50–80 bars have been reported [62,68].

Typical frameworks consist of rigid organic spacer molecules that are bonded into a three-dimensional structure by metal complexes that fix the direction of the spacers. The minimized unit cell for MOF5, one of the first MOF materials that showed promising hydrogen adsorption, is shown in Figure 12.13 [69]. The spacer molecules contain aromatic rings that make them rigid building units. The MOFs have very large internal surfaces; therefore, if by chemical modification the adsorption interaction strength could be increased, this would provide a pathway toward successful hydrogen storage [69].

So far, the excess density of hydrogen stored on carbon and MOF nanostructures has been limited to about 6–7.5 wt% at 77 K and 30–40 bars. A study of micropore filling of ACs shows that, at these temperatures, the density of hydrogen in the pores approaches that of liquid hydrogen [69]. Any further improvement of the storage density would entail storing hydrogen in the porous structure at densities beyond liquid hydrogen.

Increasing the operating temperature of a physisorption storage system will require going beyond current carbon and MOF nanostructures. A possible strategy is to enhance the sorption capacity of carbon nanostructures by doping them with certain metals, which

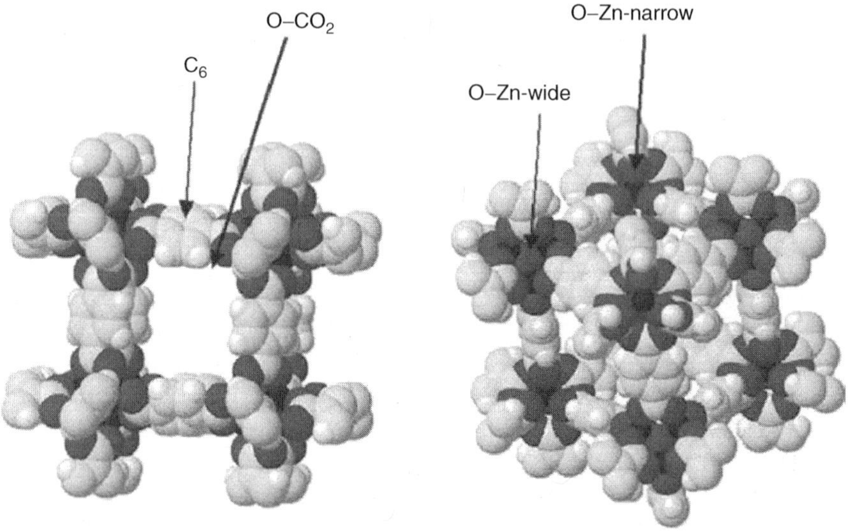

FIGURE 12.13
The minimized unit cell of MOF5 viewed from two different directions. In the left image the open pore structure is visible. White spheres, hydrogen; light gray, carbon; dark gray, oxygen; black, zinc. The hydrogen-adsorption sites are indicated by arrows. (Reproduced from Mulder, F.M., Dingemans, T.J., Wagemaker, M., and Kearley, G.J., *Chem. Phys.*, 317, 113–118, 2005. With permission.)

could lead to nondissociative binding of hydrogen molecules and enable them to be used to reversibly store hydrogen. The enthalpy of adsorption is largest for SWNTs and smallest for MOFs, suggesting stronger adsorbent–adsorbate interactions for SWNTs. At high coverage, both values converge to below 5 kJ/mol that is consistent with physisorption.

Hydrogen storage by spillover has been proposed as a mechanism to enhance the storage density of carbon-based and MOF nanostructures. For example, the doping of carbon nanomaterials significantly increased the hydrogen uptake compared to the commercially available materials. This approach relies on the use of a supported metallic catalyst to dissociate molecular hydrogen, relying on surface diffusion through a bridge to store atomic hydrogen in a receptor. An increase by a factor of 2.9 for the AC and by 1.6 for SWNTs at 298 K and 1 bar was obtained using a Pd/carbon catalyst [7]. Palladium can capture hydrogen into its bulk structure or chemically react to form hydrides of low stability that can enhance the intrinsic sorption capacity of the carbon support by spillover. Since the hydrogen sorption process in Pd nanoparticles has been described as reversible, it is expected that Pd-loaded carbon materials give mainly reversible H_2 sorption isotherms. For raw carbon, the maximum hydrogen sorption at room temperature in the Pd-loaded samples has been found to be of 0.5 wt% at atmospheric pressure, which is 10-fold more than without palladium loading [30].

12.11 Electrochemical Hydrogen Storage

Analogous to the formation of metal hydrides, a technique has been proposed in which carbons are used to sorb hydrogen produced *in situ* by electrolysis of water. Here, the carbon is used as a negatively polarized electrode. The process itself is attractive considering energy efficiency since it associates, in a unique step, the hydrogen production from water electrolysis and its simultaneous storage in the carbon substrate, with the help of the driving force provided by the negative polarization. The charge/discharge phenomenon can be represented as follows:

$$<C> + xH_2O + xe^- \rightarrow <CH_x> + xOH^-$$

where $<CH_x>$ stands for the hydrogen inserted into the carbon during charging and oxidized during discharging. Considering the high hysteresis between the sorption and the desorption (reduction and oxidation) processes, it was assumed that hydrogen, after adsorption, diffuses in the bulk of the carbon material occupying sites of high energy. The hydrogen-storage capacity is influenced by the electrolyte, the carbon material, and the charge and discharge conditions. Because the charge reaction proceeds normally via an adsorbed atom, hydrogen could also be built into the carbon materials, atomically without the need to overcome the dissociation barrier. Results indicate that electrochemical hydrogen storage on highly pure SWNTs may be attractive. Incorporation of conductive materials such as gold, nickel, copper, or palladium in the form of micrometer-sized powders to the nanocarbon electrode improves the electronic properties of CNTs. The electrochemical storage of hydrogen generally gives 0.3–3 wt% capacity; however, capacities up to 7.7 wt% has also been reported [26,70,71], which needs to be further verified.

12.12 Challenges

12.12.1 Technical

The hydrogen-storage capacities of the carbon materials are still low. Out of various carbons, it seems that CNFs and nanotubes should be further explored. An efficient development of carbon materials for hydrogen storage requires (a) an understanding of the hydrogen-adsorption mechanism, so that the adsorption energy can be increased to 20–40 kJ/mol, (b) an appropriate method to fabricate materials having a high gravimetric density of sites displaying such interactions, and (c) approaches to arrange these sites in space to enable high-volumetric storage densities.

Pure and hybrid SWNT materials remain prime candidates for development because these materials can be formed into well-packed nanoporous solids and have properties that may be controlled through nanotube geometry including the introduction of defects, attachment of electronic species, elemental substitution on the nanotube lattice, and the introduction of adventitious dopants or catalytic species. Other nanostructured carbon materials such as nanohorns, MWNTs, and metal oxide frameworks are also of interest because they provide additional systems in which adsorption interactions with hydrogen may be investigated and controlled. The hydrogen-storage device presumably would contain the nanotubes in a loose mixture. Since these are tough structures, no special care for their stability is required. However, the outer containment and especially the interfaces to the user and the filling system must prevent losses of the material. In addition, some space factor must be taken into account to allow a certain "breathing" of the tube mixture during loading and unloading. Thus, rigid carbon structures may be advantageous from this point of view.

The mechanism of hydrogen storage in carbon materials is not clear yet. A physical adsorption of hydrogen in carbon materials including SWNT, MWNT, AC, and amorphous carbon has been suggested because the hydrogen-storage capacity of carbon materials is proportional to their SSA [29,48,68] and the volume of micropores. The diameter of the micropores also affects the hydrogen-storage capacity with the optimum pore diameter for adsorbing hydrogen being 0.7 nm, which corresponds to adsorbing two layers of hydrogen. However, the unusual high hydrogen-storage capacity observed experimentally could not be interpreted only in terms of physisorption. Panella et al. [72] reported a linear relation between hydrogen uptake and SSA for all samples including ACs and CNTs with a slope of 1.91×10^{-3} wt%/m^{2}g at 77 K, smaller than the theoretical value of 2.28×10^{-3} wt%/m^{2}g, which is an upper limit of hydrogen adsorbed in a monolayer by assuming the density of the liquid adsorbate. However, the micropores seem to play an important role for storing hydrogen. The hydrogen-storage capacity of carbon materials is proportional to their micropore volume at both temperatures of 303 and 77 K. Results suggest that a carbon material with higher volume of micropores as well as a larger SSA is suitable as hydrogen-storage materials.

Hou et al. [73] considered small "carbon islands" as the main hydrogen-adsorption sites in an MWNT. The hydrogen-storage capacity of a CNT varies widely and the reason for such a variation is not clear, possibly caused by the impurity such as metal catalysts or amorphous carbon. It is not clear yet how the metallic catalyst particles, which are used during the preparation of nanotube samples, affect the hydrogen-storage capacity of nanotubes.

Li et al. [74] proposed that hydrogen is stored at two sites in carbon materials. First, hydrogen is adsorbed at the surface of the material and second, the hydrogen molecules permeate into the space between carbon layers. Therefore, the sum of the two types is the total adsorption. The first type of storage was calculated based on the surface area and the second type based on the storage in graphited samples. The SSA plays the most important role, which forms the main part of the storage. The first type of adsorption is described accurately by the same parameters despite their different morphologies, which demonstrates that the interaction between hydrogen and carbon surfaces is similar. The second type of storage is not affected much by morphology. Therefore, the hydrogen storage is determined by the surface state of the materials.

Computational work [7,9,68,70] on the adsorption of atomic hydrogen on model graphite sheets reveal that considerable substrate surface reconstruction is needed for stable chemisorption of hydrogen on a basal plane, with the closest carbon puckering out of the previously flat graphene sheet. Georgakis et al. [75] theoretically investigated hydrogen adsorption in carbonaceous microporous materials by exploiting molecular dynamics simulations combined with geometry optimization calculations of the solid structures. Insertion of oxygen functional groups causes a decrease in hydrogen adsorption and this effect is more evident in narrow pores. The amount of hydrogen adsorbed reached 4.4 wt% for the purely carbonaceous materials, whereas for the oxygenated materials adsorption was limited to 3.3 wt%. Investigation revealed that on entering the pores, hydrogen molecules are organized in layers parallel to the solid planes, following the planes' distortion. In small 5.5 Å pores, H_2 molecules orient themselves parallel to the slit walls, whereas in larger pores they align in groups perpendicular to the wall surface. And then the adsorption progresses with the formation of multiple layers of adsorbed molecules at exact pores sizes. In the layer formation process, the hydrogen–hydrogen distance gradually increases with the pore size, mainly due to the growth of the available space inside the pores. Adsorbed hydrogen molecules are in a compressed state inside the solid pores, and the hydrogen density is higher than the liquid hydrogen density of 0.071 g/cm^3. A better understanding of the unique interactions of hydrogen with carbon-adsorbent surfaces is required so that the optimal storage materials may be identified. An increase in the SSA and an expansion of the interlayer of graphite are possible via the following arrangement as suggested in the literature [76–80]:

i. Enlargement of the distance between the graphene sheets by intercalation of, for example, alkali metals and metal halides.
ii. Producing GNFs, which is an arrangement of graphene platelets in a parallel, perpendicular, or angled orientation with respect to the fiber axis. The many edges of the platelets provide a large surface with many interaction sites for adsorbates.
iii. Producing graphite nanotubes, which can be pictured as rolled-up graphene sheets, with enhanced adsorption sites between pseudographitic layers, and on interstitial sites, both inside and outside the tube.
iv. Ball milling graphite leads to an enormous increase in SSA by reducing the size of the fundamental particles.
v. Producing carbon foams with an unusual structure based on rigidly interconnected segments of graphite. Furthermore, metal nanoparticles can be relatively easily incorporated to catalyze the formation of graphitic structures.

Fast adsorption/desorption kinetics and relatively small (<10 kJ/mol) adsorption enthalpies are observed for hydrogen adsorption on many porous materials, which indicates that physisorption on porous materials is suitable for fast recharging with hydrogen [81,82]. The narrowest pores make the biggest contribution to hydrogen-adsorption capacity, whereas mesopores contribute to total pore volume, but little to hydrogen capacity, and are detrimental for the overall volumetric capacity. Hence, porous materials with very narrow pores or pore-size distributions are required for enhanced hydrogen capacity at low pressures.

The temperature dependence of hydrogen adsorption is also a major consideration for storage applications. From the results available to date, it is difficult to predict whether increasing pressure will be sufficient to increase the adsorption capacity at ambient temperatures to a similar level to that achieved at 77 K. Kinetic trapping of hydrogen occurs in some MOF materials and this is a method where there are possibilities for improving the temperature dependence of hydrogen adsorption. Design strategies are necessary for the synthesis of MOF materials with improved adsorption capacity, enhanced H_2 surface site interactions with higher enthalpies of adsorption, and improved kinetic trapping characteristics. Improvements in hydrogen-adsorption capacity of MOFs require an improved understanding of optimization of adsorption sites, whereas secondary interactions, which allow framework flexibility and window opening in MOFs, are important in kinetic trapping characteristics [82].

Adsorption can also be increased by curving the carbon surface. In microporous solids with capillaries that have a width not exceeding a few molecular diameters, the potential fields from opposite walls overlap so that the attractive force acting on hydrogen molecules is greater than that on an open flat surface. Hence, at low temperatures, far more hydrogen can be adsorbed in the tube than on a flat surface, but the ratio decreases strongly with increasing temperature, from 55 at 50 K to 11 at 77 K [2,26,83]. It has been speculated [80,81] that for CNTs, besides the overlapping of the potential fields, there might also be an increase in the hydrogen-storage capacity due to capillary storage, where several layers of hydrogen could condense inside the tube and even between carbon crystal platelets. Under the influence of capillary forces, hydrogen will shrink from an effective gas diameter (square root of the product of gas collision and kinetic diameter) of 2.6 Å to about 1.28 Å. Therefore, the idea is to draw on the capillary forces at ambient temperatures to achieve a hydrogen-storage capacity equivalent to filling the CNTs with liquid hydrogen. Based on this liquidlike state of hydrogen at room temperature no refrigeration would be necessary.

12.12.2 Economic and Health

Besides hydrogen storage, nanostructured carbons are potentially useful in a wide range of applications including microelectronics, micromechanics, and batteries. Therefore, a high demand for these materials is expected, which will necessitate breakthroughs for the economic production in large quantities. The production of CNTs (e.g., with 10 wt% H_2 capacity) for one storage tank of 5 kg hydrogen would require an electricity consumption of about 1.2 GW. This means that the production of the storage system would consume as much energy as the total vehicle would consume to run for about 3.6 million km. This is 20 times its life span and at least a factor of 400 above an acceptable level [2,26,83]. The drawbacks of the current production technology are highlighted by the fact that it would take about 27 years to produce the carbon material for a single storage tank, if produced with one production line. Currently, SWNTs are commercially available at a cost of about \$60/g, which would end up with a tank cost of about \$3 million per vehicle [8]. Neglecting other process steps and investment costs of the laboratories, the electricity consumption for a laser-grown storage tank adds up to a cost of about \$60,000 per storage tank assuming an approximate electricity costs at \$0.05/kWh.

Generally, carbon is a harmless nontoxic substance; however, inhaling nanotubes is likely to be harmful. The potential hazards of inhaled CNTs have not been sufficiently evaluated, although the results of several *in vitro* and non-aerosol-associated *in vivo* studies suggest that exposure to these materials could represent a significant hazard. The hazard studies in rats indicate that inhaled, nano-sized (i.e., <100 nm) carbon black particles may produce significant lung toxicity. The smaller nano-sized carbon black particles produce even greater toxicity [84–88]. Exposures to a high dose (5 mg/kg) of SWNT produced mortality in 15% of the instilled rats within 24 h, resulting from the mechanical blockage of the large airways by the instilled aggregate and not toxicity *per se*. Instillation of MWNT in the lungs of rats produced pulmonary lesions characterized by the formation of collagen-rich granulomas as well as inflammatory and fibrotic effects [84–88]. Furthermore, the production processes still possess health risks resulting from the toxicity of the materials used for the preparation. However, many of the toxicity hazards can be reduced if nanostructured carbons are produced as monolithic bulk materials.

12.13 Conclusion

The major bottleneck in the use of hydrogen-fueled vehicles is the onboard hydrogen storage and the carbon materials that are potential candidates for storing hydrogen. Despite high surface area, especially that of nanostructured carbons, the storage capacity is <3 wt%. Many conventional carbon materials are inexpensive and lightweight, but the nanotubes and related materials are prohibitively expensive for building storage devices on a mass scale. Carbon materials do possess a fast adsorption/desorption kinetics that is needed in practical applications. Carbon/metal hybrid materials are emerging with promising performance; however, further research is needed to reach acceptable storage capacity.

References

1. Bünger, U., W. Zittel, Hydrogen storage in carbon nanostructures—still a long road from science to commerce, 72(2), 147–151, 2001.
2. Dillon, A.C., M.J. Heben, Hydrogen storage using carbon adsorbents: Past, present and future. *Appl. Phys. A* 72, 133–142, 2001.
3. Heben, M., Minutes of Meeting of the Hydrogen Technical Advisory Panel, February 28–29, 2000, http://www.eren.doe.gov/hydrogen/spring00.html.
4. Zuttel, A., P. Sudan, P. Mauron, P. Wenger, Model for the hydrogen adsorption on carbon nanostructures. *Mater. Sci. Process. Appl. Phys. A* 78, 941–946, 2004.
5. Schur, D.V., B.P. Tarasov, S.Yu. Zaginaichenko, V.K. Pishuka, T.N. Veziroglu, Yu.M. Shul'g, A.G. Dubovoi, N.S. Anikina, A.P. Pomytkin, A.D. Zolotarenko, The prospects for using of carbon nanomaterials as hydrogen storage systems. *Int. J. Hydrogen Energ.* 27, 1063–1069, 2002.
6. Chambers, A., C. Park, R.T.K. Baker, N.M. Rodriguez, Hydrogen storage in graphite nanofibers. *Phys. Chem. B* 102, 4253–4256, 1998.
7. Bénard, P., R. Chahine, Storage of hydrogen by physisorption on carbon and nanostructured materials. *Scr. Mater.* 56, 803–808, 2007.
8. Hynek, S., W. Fuller, J. Bentley, Hydrogen storage by carbon sorption. *Int. J. Hydrogen Energ.* 22(6), 601, 1997.
9. Chahine, R., T.K. Bose, Hydrogen adsorption and cohesive energy of single-walled carbon nanotubes. *Int. J. Hydrogen Energ.* 19, 161, 1994.
10. Orimo, S., G. Majer, T. Fukunaga, A. Zuttel, L. Schlapbach, H. Fujii, Hydrogen in the mechanically prepared nanostructured graphite. *Appl. Phys. Lett.* 75(20), 3093, 1999.
11. Heben, M.J., I.M. Robertson, R. Stumpf, T. Vogt (eds), Materials for hydrogen storage—2004. Proceedings of the Symposium held on 1–2 December 2004 in Boston, MA, *Mater. Res. Soc. Symp. Proc.* 837, 153pp., 2005.
12. Wu, X.B., P. Chen, J. Lin, K.L. Tan, Hydrogen uptake by carbon nanotubes. *Int. J. Hydrogen Energ.* 25(3), 261–265, 2000.
13. Dillon, A.C., B.P. Nelson, Y. Zhao, Y.-H. Kim, C.E. Tracy, S.B. Zhang, Importance of turning to renewable energy resources with hydrogen as a promising candidate and on-board storage a critical barrier. *Mater. Res. Soc. Symp. Proc.* 895, 115–122, 2006.
14. Ning, G.Q., F. Wei, G.H. Luo, Q.X. Wang, Y.L. Wu, H. Yu, Hydrogen storage in multi-wall carbon nanotubes using samples up to 85 g, *Appl. Phys. A*, 78, 955–959, 2004.
15. Peera, A.A., L.B. Alemany, W.E. Billups, Hydrogen storage in hydrofullerides. *Appl. Phys. A* 78, 995–1000, 2004.
16. Rodriguez, N.M., A. Chambers, R.T.K. Baker, Catalytic engineering of carbon nanostructures. *Langmuir* 11, 3862, 1995.
17. Heben, M.J., VI.C.1 DOE Carbon-Based Hydrogen Storage Center of Excellence DOE Hydrogen Program FY 2005 Progress Report National Renewable Energy Laboratory.
18. Dillon, A.C., J.L. Blackburn, P.A. Parilla, Y. Zhao, Y.-H. Kim, S.B. Zhang, A.H. Mahan, J.L. Alleman, K.M. Jones, K.E.H. Gilbert, M.J. Heben, Discovering the mechanism of H_2 adsorption on aromatic carbon nanostructures to develop adsorbents for vehicular applications. *Mat. Res. Soc. Proc.* 837, 117–123, 2005.
19. Zhao, Y., Y.-H. Kim, A.C. Dillon, M.J. Heben, S.B. Zhang, Interaction of hydrogen with carbon based molecules through transition metal atoms. *Mat. Res. Soc. Proc.* 19(4), 2004.
20. Lee, Y.-W., R. Deshpande, A.C. Dillon, M.J. Heben, H. Dai, B.M. Clemens, The role of metal catalyst in near ambient hydrogen adsorption on multi-walled nanotubes. *Mat. Res. Soc. Proc.* 2004.

21. Züttel, A., P. Wenger, P. Sudan, P. Maurona, S.-i. Orimob, Hydrogen density in nanostructured carbon, metals and complex materials. *Mater. Sci. Eng.* B108, 9–18, 2004.
22. Schulz, R., J. Huot, G. Liang, S. Boily, G. Lalande, M.C. Denis, J.P. Dodelet, Recent developments in the applications of nanocrystalline materials to hydrogen technologies. *Mater. Sci. Eng. A* 267, 240–245, 1999.
24. Dillon, A.C., P.A. Parilla, T. Gennett, K.E.H. Gilbert, J.L. Blackburn, Y.-H. Ki Zhao, S.B. Zhang, J.L. Alleman, K.M. Jones, T. McDonald, M. Heben, Hydrogen Storage in Carbon-Based Materials DOE FY, 2004.
25. Meregalli, V., M. Parrinello, Review of theoretical calculations of hydrogen storage in carbon-based materials. *Appl. Phys. A* 72, 143–146, 2001.
26. Strobel, R.J., J. Garche, P.T. Moseley, L. Jorissen, G. Wolf, Hydrogen storage by carbon materials. *J. Power Sources* 159, 781–801, 2006.
27. Satyapal, S., J. Petrovic, C. Read, G. Thomas, G. Ordaz, The U.S. Department of Energy's National Hydrogen Storage Project: Progress towards meeting hydrogen-powered vehicle requirements. *Catal. Today* 120(3–4), 246–256, 2007.
28. Zhao, X., Bo. Xiao, A.J. Fletcher, K.M. Thomas, D. Bradshaw, M.J. Rosseinsky, Hysteretic adsorption and desorption of hydrogen by nanoporous metal-organic frameworks. *Science* 306, 1012, 2004.
29. Texier-Mandoki, N., J. Dentzer, T. Piquero, S. Saadallah, P. David, C. Vix-Guterl, Hydrogen storage in activated carbon materials: Role of the nanoporous texture. *Carbon* 42(12–13), 2744–2747, 2004.
30. Lueking, A.D., R.T. Yang, N.M. Rodriguez, R. Baker, K. Terry, Hydrogen storage in graphite nanofibers: Effect of synthesis catalyst and pretreatment conditions. *Langmuir* 20(3), 714–721, 2004.
31. Conte, M., P. Prosini, S. Passerini, Overview of energy/hydrogen storage: State-of-the-art of the technologies and prospects for nanomaterials. *Mater. Sci. Eng. B: Solid-State Mater. Adv. Technol.* B108(1–2), 2–8, 2004.
32. Kubas, G.J., R.R. Ryan, B.N. Swanson, P.J. Vergatini, J. Wasswerman, Molecular hydrogen complexes: Coordination of a sigma bond to transition metals. *J. Am. Chem. Soc.* 120, 1988.
33. Orimo, S., A. Zuttel, L. Schlapbach, G. Majer, T. Fukunaga, H. Fujii, Hydrogen interaction with carbon nanostructures: Current situation and future prospects. *J. Alloys Compd.* 356–357, 716–719, 2003.
34. Okamoto, Y., Y. Miyamoto, *Ab initio* investigation of physisorption of molecular hydrogen on planar and curved graphene, *J. Phys. Chem. B* 105, 3470, 2001.
35. Züttel, A., C. Nützenadel, P. Sudan, P. Mauron, C. Emmenegger, S. Rentsch, L. Schlapbach, A. Weidenkaff, T. Kiyobayashi, Hydrogen sorption by carbon nanotubes and other carbon nanostructures. *J. Alloys Compd.* 1, 676, 2002.
36. Chen, J., F. Wu, Review of hydrogen storage in inorganic fullerene-like nanotubes. *Appl. Phys. A* 78, 989–994, 2004.
37. Kowalczyk, P., R. Hoyst, M. Terronesb, H. Terrones, Hydrogen storage in nanoporous carbon materials: Myth and facts. *Phys. Chem. Chem. Phys.* 9, 1786–1792, 2007.
38. Jin, H., Y.S. Lee, I. Hong, Hydrogen adsorption characteristics of activated carbon. *Catal. Today* 120, 399–406, 2007.
39. Available on website, www.electronics-cooling.com/assets/images/, February 2007.
40. Inagaki, M., Suwa, T., Pore structure analysis of exfoliated graphite using image processing of scanning electron micrographs, *Carbon* 39, 915–920, 2001.
41. Schlapbach, L. (ed.), Hydrogen in intermetallic compounds II. Surface and dynamic properties, applications, *Topics Appl. Phys.* 67, Springer, Berlin, 67, 15–95, 1992.
42. Dillon, A.C., P.A. Parilla, Y. Zhao, T. Genett, K.E.H. Gilbert, J.L. Blackburn, Y.-H. Kim, Y. Zhao, S.B. Zhang, J.L. Alleman, M.J. Blackburn, M.J. Heben, *Proceedings of US DOE Hydrogen Program*, Review 2005.
43. Lavin, J.G., S. Subramoney, R.S. Ruoff, S. Berber, D. Tománek, Scrolls and nested tubes in multiwall carbon nanotubes. *Carbon* 40, 1123–1130, 2002.
44. Kroto, H.W., J.R. Heath, S.C. O'Brien, R.F. Curl, R.E. Smalley, C_{60}: Buckminsterfullerene. *Nature* 318, 162, 1985.

45. Sandrock, G., G. Thomas, The IEA/DOC/SNL on-line hydride databases. *Appl. Phys. A* 72, 153–155, 2001.
46. Fantini, C., A. Jorio, M. Souza, L.O. Ladeira, M.A. Pimenta, A.G. Souza Filho, R. Saito, Ge.G. Samsonidze, G. Dresselhaus, M.S. Dresselhaus, Step-like dispersive Raman modes in carbon nanotubes. *Phys. Rev. Lett.* 93, 2004, http://link.aps.org/abstract/PRL/V93/e147406.
47. Quinn, D.F., J.A. MacDonald, Natural gas storage. *Carbon* 30(7), 1097–1103, 1992.
48. Xu, W.-C., K. Takahashi, Y. Matsuo, Y. Hattori, M. Kumagai, S. Ishiyama, K. Kanekoc, S. Iijima, Investigation of hydrogen storage capacity of various carbon materials. *Int. J. Hydrogen Energ.* 32(13), 2504–2512, 2007 (available online).
49. Wang, X.K., X.W. Lin, V.P. Dravid, J.B. Ketterson, R.P.H. Chang, Carbon nanotubes synthesized in a hydrogen arc discharge. *Appl. Phys. Lett.* 66, 427, 1995.
50. Iijima, S., Helical microtubules of graphitic carbon. *Nature* 354, 56, 1991.
51. Iijima, S., T. Ichihashi, Single-shell carbon nanotubes of 1-nm diameter. *Nature* 363(6430), 603–605, 1993.
52. Ivanov, V., J.B. Nagy, P. Lambin, A. Lucas, X.B. Zhang, The study of carbon nanotubules produced by catalytic method. *Chem. Phys. Lett.* 223, 329, 1994.
53. Colomer, J.-F., P. Piedigrosso, I. Wilems, C. Journet, P. Bernier, G. Van Tendeloo, A. Fonseca, J.B. Nagy, Purification of catalytically produced multi-wall nanotubes. *J. Chem. Soc. Faraday Trans.* 94, 3753, 1998.
54. Lee, S.M., S. Mi, K.H. An, Y.H. Lee, G. Seifert, T. Frauenheim, A hydrogen storage mechanism in single-walled carbon nanotubes. *J. Am. Chem. Soc.* 123(21), 5059–5063, 2001.
55. Lee, S.M., Y.H. Lee, S.M. Lee, Y.H. Lee, Hydrogen storage in single-walled carbon nanotubes. *Appl. Phys. Lett.* 76(20), 2877–2879, 2000.
56. Chen, P., X. Wu, J. Lin, K.L. Tan, High H_2 uptake by alkali-doped carbon nanotubes under ambient pressure and moderate temperatures. *Science* (Washington, DC) 285(5424), 91–93, 1999.
57. Yang, R.T., Hydrogen storage by alkali-doped carbon nanotubes. *Carbon* 38(4), 623–626, 2000.
58. Yildirim, T., J. Iniguez, S. Ciraci, Molecular and dissociative adsorption of multiple hydrogen molecules on transition metal decorated C60. *Phys. Rev. B: Condens. Matter Mater. Phys.* 72(15), 153403 (4 pages), 2005.
59. Guo, T., P. Nikolaev, A. Thess, D.T. Colbert, R.E. Smalley; Catalytic growth of single-walled nanotubes by laser vaporization. *Chem. Phys. Lett.* 243(1,2), 49–54, 1995.
60. Miller, M.A., M.A. Gorte, Carbon Carbon Nanotube Sorption Science—External Carbon Nanotube Sorption Science—External Peer Review of NREL Activities. U.S. Department of Energy Office of Hydrogen, Fuel Cells and Infrastructure Technologies, 2004.
61. Bethune, D.S., C.-H. Kiang, M.S. de Vries, G. Gorman, R. Savoy, J. Vasquez, R. Beyers, Cobalt-catalyzed growth of carbon nanotubes with single-atomic-layer walls. *Nature* (London), 363(6430), 605–607, 1993.
62. Poirier, E., R. Chahine, P. Bénard, G. Dorval Douville, L. Lafi, P.A. Chandonia, Hydrogen adsorption measurements and modeling on metal-organic frameworks and single-walled carbon nanotubes. *Langmuir* 22(21), 8784–8789, 2006.
63. Jordá-Beneyto, M., F. Suárez-Garcí, D. Lozano-Castelló, D. Cazorla-Amoró, A. Linares-Solano, Hydrogen storage on chemically activated carbons and carbon nanomaterials at high pressures. *Carbon* 45, 293–303, 2007.
64. Iijima, S., M. Yudasaka, R. Yamada, S. Bandow, K. Suenaga, F. Kokai, Nano-aggregates of single-walled graphitic carbon nano-horns. *Chem. Phys. Lett.* 309, 165–170, 1999.
65. Guay, P., B.L. Stansfield, A. Rochefort, On the control of carbon nanostructures for hydrogen storage applications. *Carbon* 42, 2187, 2004.
66. Ye, Y., C.C. Ahn, C. Witham, B. Fultz, J. Liu, A.G. Rinzler, D. Colbert, K.A. Smith, R.E. Smalley, Hydrogen adsorption and cohesive energy of single-walled carbon nanotubes. *Appl. Phys. Lett.* 74(16), 2307–2309, 1999.
67. Hirscher, M., M. Becher, M. Haluska, U. Detlaff-Weglikowska, A. Quintel, G.S. Duesberg, Y.M. Coi, P. Downes, M. Hulman, S. Roth, I. Stepanek, P. Bernier, Hydrogen storage in sonicated carbon materials. *Appl. Phys. A: Mater. Sci. Process.* 72(2), 129–132, 2001.

68. Bénard, P., R. Chahine, P.A. Chandonia, D. Cossement, G. Dorval-Douville, L. Lafia, P. Lachance, R. Paggiaro, E. Poirier, Comparison of hydrogen adsorption on nanoporous materials. *J. Alloys Compd.* 446–447, 380–384, 2007.
69. Mulder, F.M., T.J. Dingemans, M. Wagemaker, G.J. Kearley, Modelling of hydrogen adsorption in the metal organic framework MOF5. *Chem. Phys.* 317, 113–118, 2005.
70. Hynek, S., W. Fuller, J. Bentley, J. McCullough, Hydrogen storage using carbon adsorbents: past, present and future, 32(12), 1999–2004, In *Hydrogen Energy Progress X*, ed. D.L. Block, T.N. Veziroglu, International Association for Hydrogen Energy, Coral Gables, FL, 1994, pp. 985–1000.
71. Cheng, H.M., F. Li, G. Su, H.Y. Pan, L.L. He, X. Sun, M.S. Dresselhaus, Large-scale and low-cost synthesis of single-walled carbon nanotubes by the catalytic pyrolysis of hydrocarbons. *Appl. Phys. Lett.* 72(25), 3282–3284, 1998.
72. Panella, B., M. Hirscher, S. Roth, Hydrogen adsorption in different carbon nanostructures. *Carbon* 43, 2209–2214, 2005.
73. Hou, P.X., S.T. Xu, Z. Ying, Q.H. Yang, C. Liu, H.M. Cheng, Hydrogen adsorption/desorption behavior of multi-walled carbon nanotubes with different diameters. *Carbon* 41, 2471–2476, 2003.
74. Li, Y., D. Zhao, Y. Wanga, R. Xue, Z. Shen, X. Li, The mechanism of hydrogen storage in carbon materials. *Int. J. Hydrogen Energ.* 32(13) 2513–2517, 2007.
75. Georgakis, M., G. Stavropoulos, G.P. Sakellaropoulos, Molecular dynamics study of hydrogen adsorption in carbonaceous microporous materials and the effect of oxygen functional groups. *Int. J. Hydrogen Energ.* 32(12), 1999–2004, 2007.
76. Pinkerton, F.E., B.G. Wicke, C.H. Olk, G.G. Tibbetts, G.P. Meisner, M.S. Meyer, J.F. Herbst, Thermo gravimetric measurement of hydrogen absorption in alkali-modified carbon materials. *J. Phys. Chem.* 104(40), 9460–9467, 2000.
77. Wu, X.B., P. Chen, J. Lin, K.L., Hydrogen uptake by carbon nanotubes. *Int. J. Hydrogen Energ.* 25(3), 261–265, 1999.
78. Pederson, M.R., J.Q. Broughton, Nanocapillarity in fullerene tubules. *Phys. Rev. Lett.* 69(18), 2689–2692, 1992.
79. Patchkovskii, S.J.S. Tse, S.N. Yurchenko, L. Zhechkov, Th. Heine, G. Seifert, Graphene nanostructures as tunable storage media for molecular hydrogen. *Proc. Natl. Acad. Sci. USA* 102(30), 10439–10444, 2005.
80. Dillon, A.C., J.L. Blackburn, P.A. Parilla, Y. Zhao, Y.-H. Kim, S.B. Zhang, A.H. Mahan, J.L. Alleman, K.M. Jones, K.E.H. Gilbert, M.J. Heben. Discovering the mechanism of H2 adsorption on aromatic carbon nanostructures to develop adsorbents for vehicular applications. *Materials Research Society Symposium Proceedings*, 837 (Materials for Hydrogen Storage, 2004), 2005, pp. 117–123.
81. Schlapbach, L., A. Züttel, Hydrogen-storage materials for mobile applications. *Nature* 414, 353–358, 2001.
82. Thomas, K.M., Hydrogen adsorption and storage on porous materials. *Catal. Today* 120, 389–398, 2007.
83. Warheit, D.B., What is currently known about the health risks related to carbon nanotube exposures? *Carbon* 44, 1064–1069, 2006.
84. Warheit, D.B., B.R. Laurence, K.L. Reed, D.H. Roach, G.A. Reynolds, T.R. Webb, Comparative pulmonary toxicity assessment of single-wall carbon nanotubes in rats. *Toxicol. Sci.* 77(1), 117–125, 2004.
85. Jia, G., H. Wang, L. Yan, X. Wang, R. Pei, T. Yan, Y. Zhao, X. Guo, Cytotoxicity of carbon nanomaterials: Single-wall nanotube, multi-wall nanotube, and fullerene. *Environ. Sci. Technol.* 39(5), 1378–1383, 2005.
86. Warheit, D.B., B.R. Laurence, K.L. Reed, D.H. Roach, G.A. Reynolds, T.R. Webb, Lung toxicity bioassay study in rats with single-wall carbon nanotubes. *Proceedings of the ACS Symposium Series*, 890 (Nanotechnology and the Environment), 2005, pp. 67–90.
87. Muller, J., F. Huaux, N. Moreau, P. Misson, J.F. Heilier, M. Delos, Respiratory toxicity of multi-wall carbon nanotubes. *Toxicol. Appl. Pharmacol.* 207, 221–231, 2005.
88. Maynard, A.D., P.A. Baron, M. Foley, A.A. Shvedova, E.R. Kisin, V. Castranova, Exposure to carbon nanotube material: Aerosol release during the handling of unrefined single-walled carbon nanotube material. *J. Toxicol. Environ. Health A*, 2004.

13

Hydrogen Storage in Organic Chemical Hydrides on the Basis of Superheated Liquid-Film Concept

Shinya Hodoshima and Yasukazu Saito

CONTENTS

13.1 Introduction

13.1.1 Organic Chemical Hydrides for Hydrogen Storage

Hydrogen is regarded as an ideal energy carrier because of its unique properties such as high efficiency and pollution free nature. However, to popularize hydrogen energy systems, for example, stationary fuel cells, hydrogen combustion turbines and hydrogen-fueled vehicles powered by onboard fuel cells or internal combustion engines (ICEs), various technological issues regarding production, transportation, storage, supply, and safety must be solved. In particular, development of the most suitable material for hydrogen storage and the most efficient method for absorbing and desorbing hydrogen through the storage material under mild conditions are urgent tasks.

With respect to the hydrogen storage, storage materials require the following characteristics: (1) high storage capacities in weight and volume, (2) facile reversibility, and (3) safe and cost competitive. On the basis of these viewpoints, organic chemical hydrides [1–15], consisting of reversible catalysis pairs such as decalin dehydrogenation/naphthalene hydrogenation (Equation 13.1), methylcyclohexane dehydrogenation/toluene hydrogenation (Equation 13.2), and tetralin dehydrogenation/naphthalene hydrogenation (Equation 13.3) have been proposed as the storage material of hydrogen by our research group.

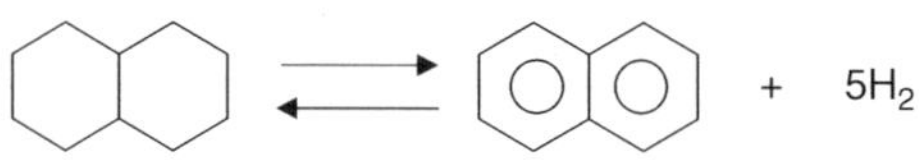

(13.1)

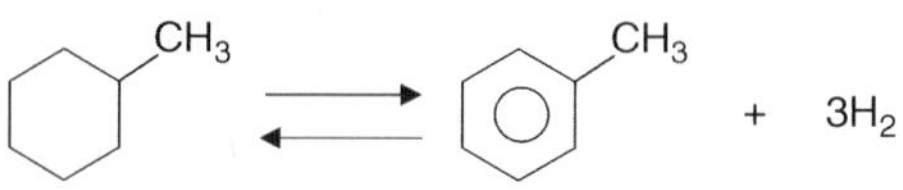

(13.2)

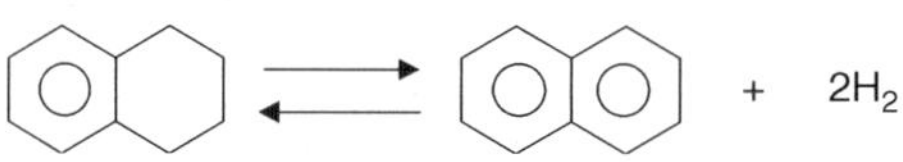

(13.3)

Figure 13.1 compares hydrogen storage in weight and volume capacities. Decalin can store hydrogen with high capacities (7.3 wt%, 64.8 kg-H_2/m^3), which satisfy the target values (6.5 wt%, 62.0 kg-H_2/m^3) presented by the U.S. Department of Energy (DOE) [16] and those regulated by the U.S. Council for Automotive Research (USCAR). Decalin and naphthalene are accepted socially as safe commodity chemicals since they have been utilized as a solvent and an insect killer at home, respectively, for long periods. A catalytic process of naphthalene hydrogenation to produce decalin has been commercialized since the 1940s [17]; its reaction mechanism is investigated in detail later [17–19]. The tetralin-intermediate path for catalytic hydrogenation of naphthalene to decalin was revealed [17], in which the step from naphthalene to tetralin proceeds much faster than that from tetralin to decalin. Furthermore, existing infrastructures (e.g., gas station and tank lorry) are available for storage, transportation, and supply of hydrogen because properties of these organic chemical hydrides are similar to kerosene. A pair of methylcyclohexane dehydrogenation/toluene hydrogenation with comparatively high storage capacities (6.2 wt%, 46.5 kg-H_2/m^3) is also usable as a hydrogen carrier for the same reasons as the decalin/naphthalene pair. These advantages suggest that both decalin and methylcyclohexane are suitable as an onboard storage material for operating hydrogen vehicles.

In contrast, a catalytic reaction pair of tetralin dehydrogenation/naphthalene hydrogenation (Equation 13.3) is another choice for stationary applications. Although the storage densities of tetralin are relatively low (3.0 wt%, 28.2 kg-H_2/m^3), rates of absorbing and desorbing hydrogen

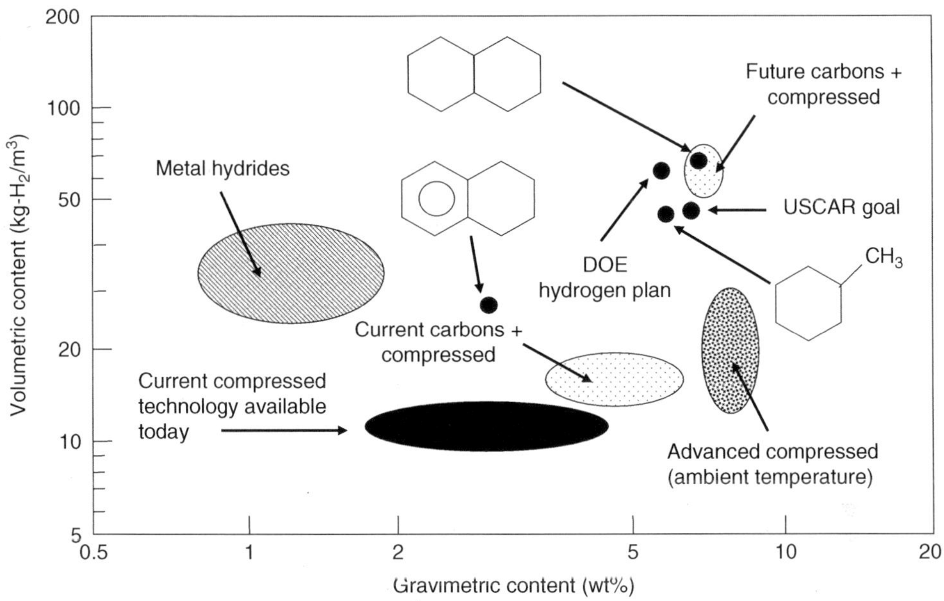

FIGURE 13.1
Comparison of capacities of hydrogen storage materials.

are much faster than the others. With respect to stationary fuel cells with large space for hydrogen storage, tetralin is, therefore, more suitable than decalin or methylcyclohexane in terms of rapid hydrogen supply.

Figure 13.2 illustrates a new system for storage and transportation of hydrogen by utilizing organic chemical hydrides and existing infrastructures. Repeated catalysis between dehydrogenation of naphthenic hydrocarbons and hydrogenation of aromatic hydrocarbons would make it possible to combine electrolysis hydrogen produced from renewable energy (e.g., wind power) and by-product hydrogen recovered from various industrial processes (e.g., coal-oven gas) with the hydrogen demand practically for stationary fuel cells and hydrogen vehicles. Unlike the steam reforming of methane, moreover, it is to be noted that organic chemical hydrides provide pure hydrogen without emission of carbon dioxide (CO_2) and carbon monoxide (CO) during the catalytic dehydrogenation.

Naphthalene itself is solid at ambient temperatures (m.p.: 80.5°C) but is dissolved easily in aromatic compounds such as toluene (refer Table 13.1) [10,12], so that the oily mixture can be handled as a "naphthalene oil." The naphthalene oil is catalytically hydrogenated to decalin and methylcyclohexane simultaneously. Decalin and methylcyclohexane are converted into hydrogen and naphthalene oil again by dehydrogenation catalysis. From the handling viewpoint, the naphthalene oil may be deemed as a preferential and practical material for hydrogen storage and transportation.

To establish the hydrogen storage and transportation system as shown in Figure 13.2, it is necessary to realize efficient hydrogen generation from organic chemical hydrides under mild reaction conditions. However, dehydrogenation of naphthenic hydrocarbons is highly endothermic (e.g., the enthalpy change of decalin dehydrogenation: +297.3 kJ/mol at 25°C) and restricted thermodynamically at low temperatures as indicated from the relationship between the Gibbs energy change and the temperature (Figure 13.3) [20]. As well

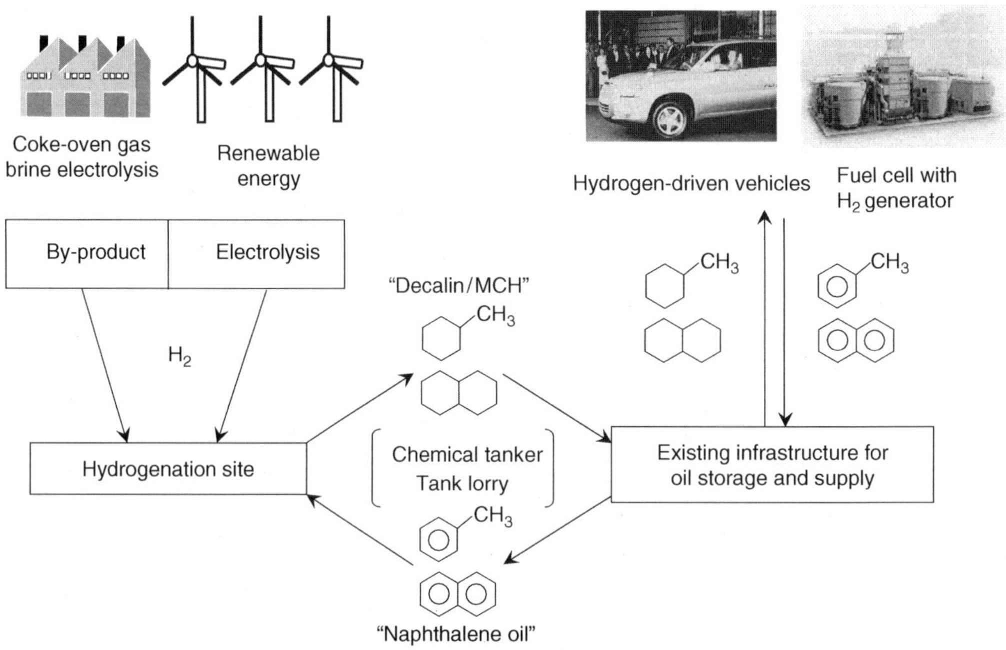

FIGURE 13.2
Hydrogen energy systems by use of organic chemical hydrides.

TABLE 13.1
Candidate Component of Naphthalene Oil as Hydrogen Storage Material

			Storage Capacities of Hydrogen	
Candidate Compound	**Melting Point (°C)**	**Boiling Point (°C)**	**(wt%)**	**(L/5 kg-H_2)**
Toluene	−94.5	110.6	6.2	104.1
o-Xylene	−25.2	144.4	5.4	116.3
m-Xylene	−47.4	139.3	5.4	121.4
Naphthalene	80.2	218.0	7.3	76.5–78.9
1-Methylnaphthalene	−30.6	244.8	6.6	84.9–87.2

Source: Reproduced from Hodoshima, S., Takaiwa, S., Shono, A., Satoh, K., and Saito, Y., *Appl. Catal. A*, 283(1–2), 235–242, 2005. With permission from Elsevier.

known in the process of petroleum reforming [21], conventional heterogeneous catalysis in the solid–gas phase for dehydrogenation of cycloalkanes requires higher heating temperatures than 450°C because of their thermodynamic and kinetic restrictions, resulting in large consumption of exergy and serious coke formation over the catalyst surface. In the Euro-Quebec Hydrogen Project in the 1980s [22,23], the catalytic reaction pair of methylcyclohexane dehydrogenation/toluene hydrogenation (Equation 13.2) was once proposed but a serious disadvantage was pointed out because the European demand side had to afford the thermal energy for reaction at high temperatures. Furthermore, there was no dehydrogenation catalyst with high catalytic performance. Newson et al. [24,25] carried out methylcyclohexane or decalin dehydrogenation by heterogeneous catalysis at a heating temperature as high as 440°C to attain sufficient conversions (90–98%). The existing

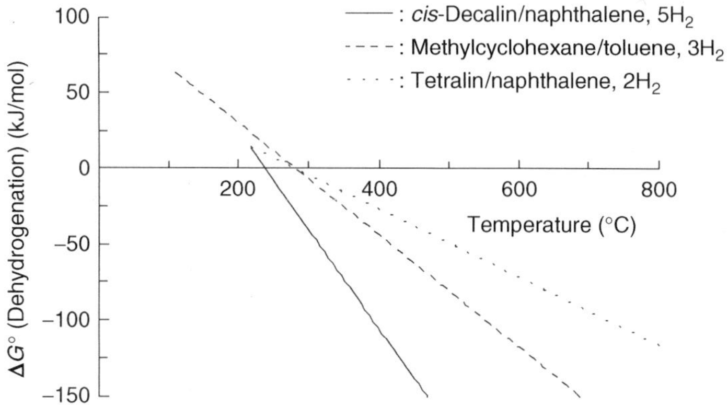

FIGURE 13.3
Gibbs energy change of reversible reaction pairs of dehydrogenation/hydrogenation as a function of temperature.

heterogeneous catalysis for dehydrogenation is, therefore, unfavorable as a method for supplying hydrogen to fuel cells or hydrogen vehicles.

13.1.2 Superheated Liquid-Film-Type Catalysis for Hydrogen Supply

Catalytic dehydrogenation of cycloalkanes has been well developed in petroleum industries as a part of the reforming process [26], where high temperatures were imposed to solid catalysts on account of thermodynamic and kinetic requirements. It was elucidated, however, that hydrogen was yielded even at 151°C catalytically from cyclooctane with the aid of Wilkinson complexes under boiling and refluxing conditions [27]. A catalytic cycle was constituted, with turnover numbers exceeding unity, since reactive distillation was operated for separating product hydrogen from condensable components of the other product cyclooctene and residual substrate cyclooctane, which could remove the equilibrium restriction. Moreover, it was conceivable for organometallic catalysis to assume that carbon–hydrogen bond activation would proceed easily at either the α position of coordinated alkane or the β position of ligand alkyl as suggested from quantum chemical calculations on catalytic alkene hydrogenation with Wilkinson complexes [28]. Indeed, thermal alkane dehydrogenation at a sufficiently low temperature was successfully attained for the first time in this attempt of hydrogen generation from the cyclooctane solution of Wilkinson complexes under boiling and refluxing conditions.

As its natural consequence, suspended metallic catalysts were applied to alkyl-substituted cyclohexanes since dehydrogeno-aromatization reactions were well known to be difficult with homogeneous catalysis. Under boiling and refluxing conditions, carbon- or alumina-supported platinum, palladium, rhodium, and ruthenium catalysts were found to be active for hydrogen generation from their suspensions of ethylcyclohexane, 1,2-dimethylcyclohexane, 1,3-dimethylcyclohexane, 1,4-dimethylcyclohexane, and methylcyclohexane, yielding ethylbenzene, *o*-xylene, *m*-xylene, *p*-xylene, and toluene, respectively, besides hydrogen [29].

Although the suspended catalysts under boiling and refluxing conditions were active for dehydrogeno-aromatization of alkyl-substituted cyclohexanes, the levels of reaction rates were unsatisfactorily low. A breakthrough toward low catalytic activities was accomplished by our introduction of "superheated liquid-film concept" (*vide infra*) [30].

Removal of thermodynamic restriction through reactive distillation and enhancement of hydrogen generation reactivity due to this concept made it possible to utilize organic chemical hydrides in the field of hydrogen storage from a novel standpoint.

The superheated liquid-film-type catalysis [1–15,31], a novel concept developed by our group, made it possible for the first time to generate hydrogen quite efficiently from organic chemical hydrides at moderate heating temperatures of 210–280°C without coke formation. By charging an adequate amount of reactant solution (e.g., decalin, methylcyclohexane, and tetralin) to a constant amount of catalyst consisting of metallic nanoparticles (e.g., platinum, platinum–tungsten, platinum–rhenium, and nickel–ruthenium) supported on activated carbon, the liquid-film state is formed (Figure 13.4). Because its surface of the catalyst layer is covered with a thin liquid-film of solution but not suspended, this catalyst state has been designated as a "liquid-film state." This liquid-film state was confirmed to exhibit much higher catalytic performance than that in the ordinary heterogeneous catalysis in the solid–gas phase (sand-bath state) or the solid–liquid phase (suspension state). Furthermore, the superheated liquid-film-type catalysis enables us to utilize waste heat of below 300°C for endothermic dehydrogenation. Waste heat recovered from solid oxide fuel cells, various industrial processes, hydrogen-fueled combustion turbines, and ICEs would be available as heat sources. As a consequence, the total efficiency of hydrogen energy systems by use of these organic chemical hydrides would be improved.

In the present chapter, the fundamental investigations for realizing hydrogen supply from organic chemical hydrides under superheated liquid-film conditions and the thermodynamic analysis on the superheated liquid-film-type catalysis are described on the experimental basis in batchwise and continuous operations. In addition, a new proposal for automotive application by use of organic chemical hydrides and its energetic analysis of the hydrogen storage and transportation system are introduced.

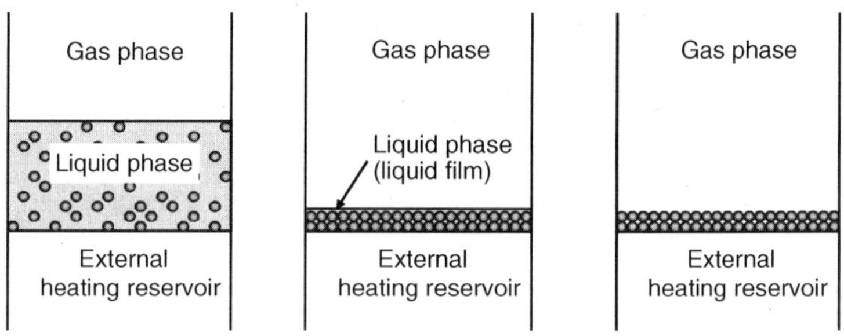

	Suspended state	Superheated liquid-film state	Solid–gas contact state
Amount ratio of catalyst to solution	Small	Medium	Large
Temperature gradient	$T_{ext} > T_{cat} = T_{liq}$	$T_{ext} > T_{cat} > T_{liq}$	$T_{ext} > T_{cat} = T_{gas}$
Reaction rate	Small	Large	Small
Conversion	Small	Large	Small

T_{ext}: external heating temperature, T_{cat}: catalyst-layer temperature, T_{liq}: boiling point, T_{gas}: gas-phase temperature

FIGURE 13.4
Characteristics of dehydrogenation catalyst immersed with liquid substrate under boiling conditions.

13.2 Efficient Hydrogen Generation from Organic Chemical Hydrides with Superheated Liquid-Film-Type Catalysis in Batchwise Operation

13.2.1 Catalytic Dehydrogenation of Decalin and Methylcyclohexane over Carbon-Supported Platinum-Based Nanoparticles under Superheated Liquid-Film Conditions

Platinum-based nanoparticles (e.g., Pt [1–15], Pt–Re [10,15], and Pt–W [5,6,15]) supported on granular activated carbon (KOH-activation, BET specific surface area: 3100 m^2/g, pore volume: 1.78 cm^3/g, average particle size: 13 μm, average pore size: 2.0 nm, Kansai Netsukagaku Co. Ltd. [32]) were mainly used as the dehydrogenation catalysts in the present study.

Before preparing these carbon-supported Pt-based catalysts, a support pretreatment toward granular activated carbon with an aqueous solution of NaOH (pH 14) was carried out by immersing for 24 h to promote the anion exchange between the ligand chloride of impregnated metal precursers (K_2PtCl_4) and the aqueous hydroxide ion (OH^-) inside the micropores of the activated carbon [33].

A platinum catalyst supported on the granular activated carbon (Pt/C, 5 wt-metal%) was prepared by a conventional impregnation method in the following manner [1–15]. First, granular powders of the base-pretreated activated carbon were stirred with a K_2PtCl_4 aqueous solution at room temperature for 48 h, which was followed by dropwise addition of a $NaBH_4$ aqueous solution at 90°C for 30 min. After standing for another 30 min, the catalysts were filtered and then washed for removing residual ions with a large amount of water. Finally, the Pt/C catalyst was evacuated at 70°C for 10 h.

A platinum–rhenium composite catalyst supported on the granular activated carbon (Pt–Re/C, 5 wt-Pt%, mixed molar ratio of Pt/Re = 2) [10] was prepared by a "dry-migration method" [33,34] as follows: (1) The Pt/C catalyst prepared earlier (5 wt-metal%) was evacuated at 180°C for 1 h; (2) The mixture (molar ratio of Pt/Re = 2) of the Pt/C catalyst and a cyclopentadienylrhenium tricarbonyl complex ($Re(Cp)(CO)_3$) were stirred under nitrogen atmosphere at room temperature for 1 h and then heated at 100° for 1 h, with the temperature kept at a constant; (3) This mixture was further stirred under hydrogen atmosphere at 240°C for 3 h; and finally (4) the Pt–Re/C composite catalyst was evacuated at 180°C for 1 h. A platinum–tungsten composite catalyst supported on the granular activated carbon (Pt–W/C, 5 wt-Pt%, mixed molar ratio of Pt/W = 1) [5,6] was also prepared similarly by the dry migration method. All the catalysts were evacuated inside the reactor at 150°C for 1 h before use.

Figure 13.5a shows an experimental apparatus for catalytic dehydrogenation under atmospheric pressure in a batchwise mode [1–15]. Catalytic dehydrogenation of decalin to naphthalene using a batchwise reactor (eggplant-type-flask, volume: 50 mL) was performed under boiling and refluxing conditions by heating at 210°C and cooling at 5°C. As illustrated in Figure 13.5b, only hydrogen produced from decalin is separated from the boiling liquid phase by repeating evaporation and condensation. Continuous removal of the generated hydrogen can shift the chemical equilibrium to the dehydrogenation side as the consequence of boiling and refluxing conditions, that is, reactive distillation conditions [35], so that dehydrogenation proceeds stationarily in spite of thermodynamically unfavorable temperatures.

Commercial decalin (mixture of *cis*- and *trans*-decalin, volume ratio of cis-form/trans-form: 1/1) was used for the catalytic reaction test in the present study. Reaction conditions were optimized by varying charged amounts of decalin (0.5, 0.75, 1.0, 1.25, 1.5, and 3.0 mL)

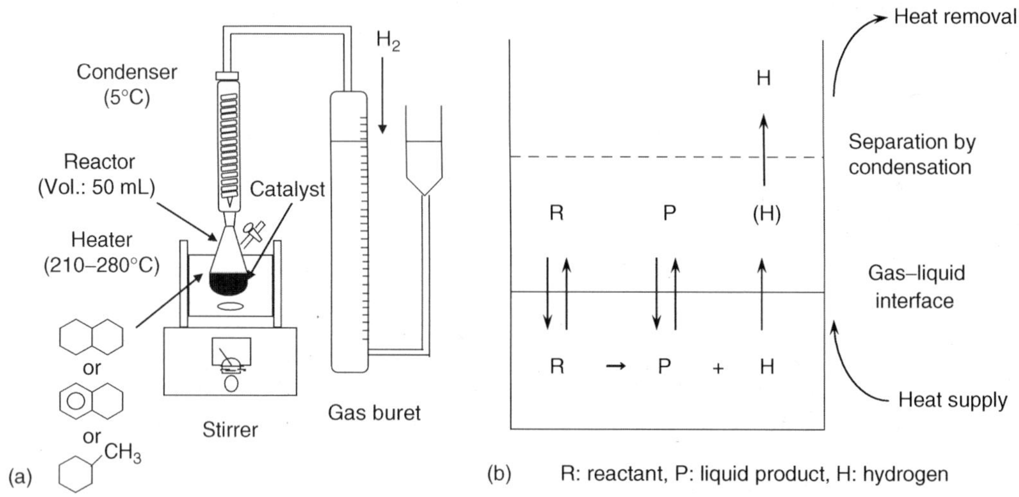

FIGURE 13.5
Experimental apparatus for catalytic dehydrogenation under reactive distillation conditions in batchwise operation. (a) Batchwise apparatus and (b) liquid-phase dehydrogenation under reactive distillation conditions.

relative to a constant amount of the platinum catalyst supported on a granular activated carbon (Pt/C, 0.30 g). The hydrogen generated from decalin was collected in a gas buret and pursued volumetrically for 2.5 h. Reaction rates were obtained from the gaseous amount due to hydrogen continuously, while the final extent of conversion was checked with gas chromatographic analysis on the accumulated condensates at the final stage of reaction. Keeping the best amount ratio of catalyst to-decalin optimized at the heating temperature of 210°C, catalytic decalin dehydrogenation at 240°C and 280°C was carried out over both the Pt/C and the Pt–Re/C catalysts [12].

The time courses of hydrogen generated for 2.5 h from various charged amounts of decalin (0.5, 0.75, 1.0, 1.25, 1.5, and 3.0 mL) with a constant amount of the carbon-supported platinum catalyst (Pt/C, 0.30 g) were compared with each other under common boiling, externally heated at 210°C, and refluxing conditions (Figure 13.6). In addition, a characteristic correlation of the charged amounts of decalin with both catalytic conversions after 2.5 h and initial rates of hydrogen generation was given on the basis of these time courses (Figure 13.7). It is apparent from Figures 13.6 and 13.7 that the amount ratios of catalyst/decalin are quite sensitive to the catalytic performance for dehydrogenation.

When the catalyst was suspended uniformly in 3.0 mL decalin, not only low initial rates of hydrogen generation but also poor 2.5 h conversions were obtained in spite of the excess amounts of substrate in the reaction medium.

In the case of 1.0 mL decalin, the catalyst could not be suspended, but its surface layer just became wet with the substrate solution. The present state of the catalyst was designated as a "superheated liquid-film state" [1–15,31] since the carbon-supported catalyst was covered with a thin film of liquid substrate. Here, the directly measured temperature was higher than the boiling point as the consequence of "superheating" [36] at the interface of thermal conduction from the external heat reservoir to the boiling liquid bulk inside the reactor. The highest initial rate of hydrogen generation and the highest decalin conversion (2.5 h) were accomplished simultaneously only by adopting this superheated liquid-film-type catalysis (Figure 13.7). Indeed, the chemical equilibrium can be shifted toward the dehydrogenation side even at 210°C within the unfavorable temperature range for decalin under the conditions of reactive

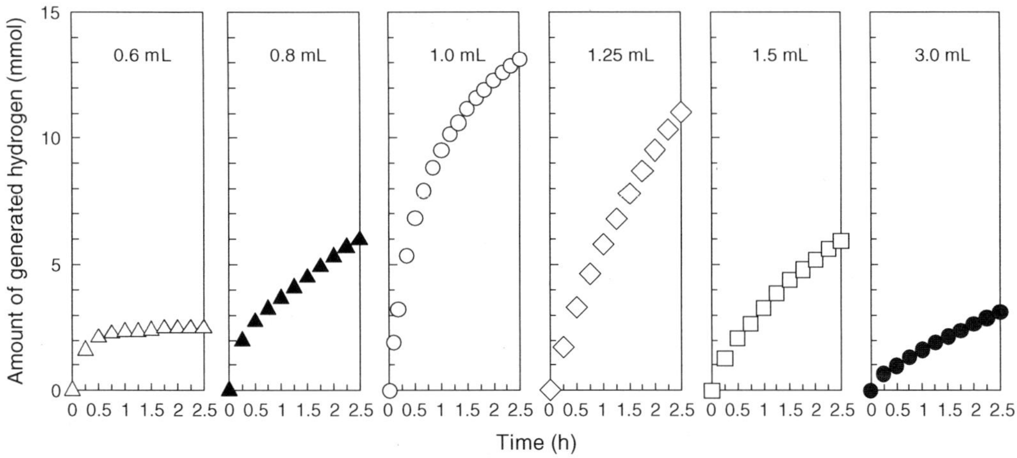

FIGURE 13.6
Time courses of hydrogen generated from decalin with carbon-supported platinum catalyst at various charged amounts of decalin in a batchwise operation. Catalyst: platinum nanoparticles supported on the granular activated carbon (Pt/C, 5 wt-metal%). Charged amount of decalin: 0.6 (△), 0.8 (▲), 1.0 (○), 1.25 (◇), 1.5 (□), and 3.0 mL (●), 0.30 g. Reaction conditions: boiling and refluxing by heating at 210°C and cooling at 5°C.

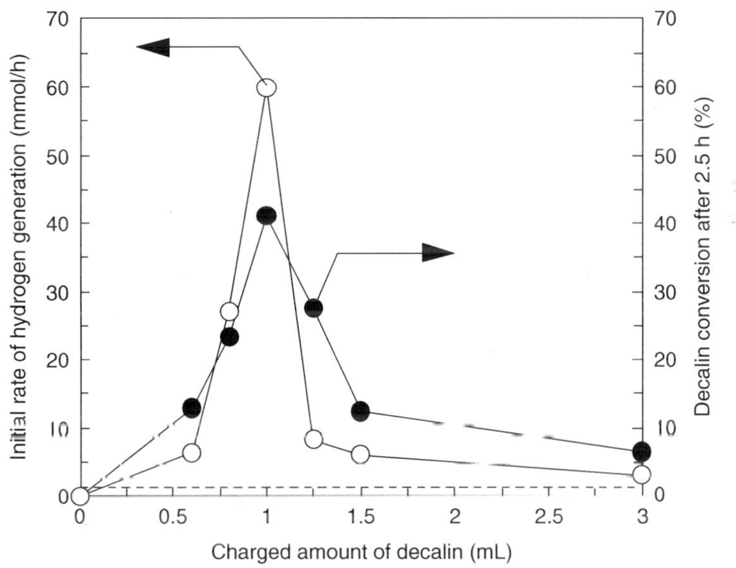

FIGURE 13.7
Relationship between catalytic dehydrogenation activities and charged amount of decalin. ○: Initial rate of hydrogen generation; ●: decalin conversion after 2.5 h; equilibrium conversion for decalin dehydrogenation. 1.2% at 210°C and 1 atm (a broken line). Catalyst: platinum nanoparticles supported on granular activated carbon (Pt/C, 5 wt-metal%), 0.30 g. Reaction conditions: boiling and refluxing by heating at 210°C and cooling at 5°C.

distillation. But kinetic control is operating here to the superheated liquid-film states since a temperature gradient at the superheated active site accelerates product desorption from the catalyst surface.

In the case when the initially charged amount of decalin was as small as 0.5 mL, the catalyst surface became dry within an hour as if the catalyst behaved as a sand-bath.

Because the gaseous part of the reactor space was large enough for such a tiny amount of liquid decalin, the substrate supply to the catalyst surface was apt to become deficient especially after the reaction time elapsed. As a consequence, the reaction rate decreased rather rapidly in this sand-bath state.

Dependence of the hydrogen generation rate from decalin on the naphthalene concentration using the catalysts in the superheated liquid film (0.30 g/1.0 mL) and the suspended (0.30 g/3.0 mL) states is shown in Figure 13.8a, and 13.8b [6,10]. In a typical batch-type operation under boiling and refluxing conditions, the naphthalene concentration in the decalin solution increased gradually as the result of its condensation into the liquid droplets formed in the refluxing condenser. Gradual decrease in the reaction rate was consequently ascribed to the adsorption of naphthalene on the catalyst surface. However, the kinetic behaviors of these two states were in sharp contrast with each other. As apparent from Figure 13.8b, the suspended state gave a small initial- and rapid rate decrease even with a small increment of naphthalene concentration, whereas the superheated liquid-film state resulted in high generation rates not only at the initial stage but also at a wide range of naphthalene concentrations. These kinetic behaviors in dehydrogenation catalysis had been observed for decalin at the corresponding catalyst states.

Kinetic analysis with a Langmuir-type rate equation (Equation 13.4) [37] gave us the magnitudes of reaction rate constant (k) and retardation constant due to product naphthalene (K) for the superheated liquid film (0.30 g/1.0 mL) and the suspended states (0.30 g/3.0 mL) with the same Pt/C catalyst as summarized in Table 13.2. It is apparent that excellent performance with carbon-supported platinum nanoparticles in the superheated liquid-film state is realized in dehydrogenation catalysis on the basis of reaction rate and retardation constants.

$$v_{\mathrm{h}} = \frac{k}{(1 + K\,[\mathrm{naphthalene}])} \qquad (13.4)$$

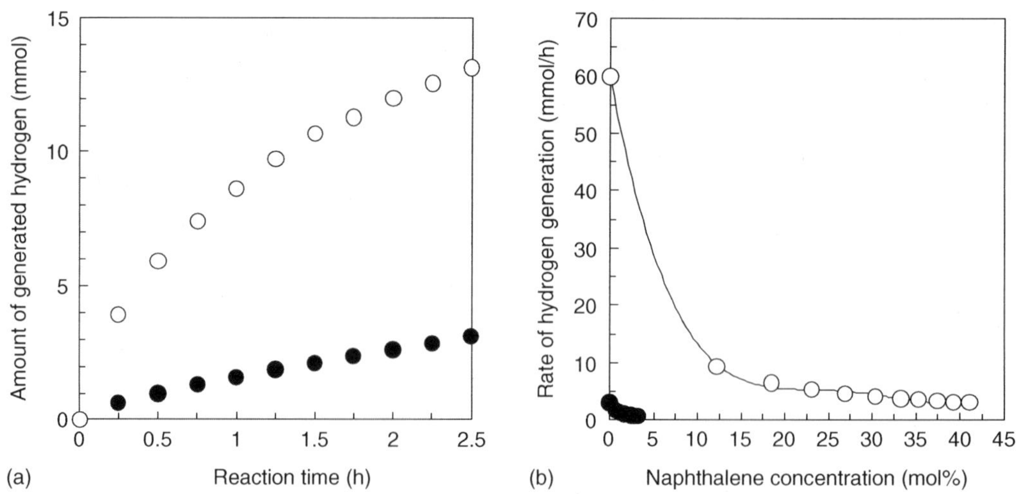

FIGURE 13.8
Time courses of hydrogen generated from decalin and rate of hydrogen generation as a function of naphthalene concentration in superheated liquid-film and suspended states. (a) Time course of hydrogen generation for 2.5 h and (b) relationship of hydrogen generation rate with naphthalene concentration. Catalyst: platinum nanoparticles supported on granular activated carbon (Pt/C, 5 wt-metal%), 0.30 g. Charged amount of decalin: 1.0 mL (○, superheated liquid-film state), 3.0 mL (●, suspended state). Reaction conditions: boiling and refluxing by heating at 210°C and cooling at 5°C.

TABLE 13.2

Comparison of Kinetic Parameters for Decalin Dehydrogenation in Superheated Liquid-Film and Suspended States

	Superheated Liquid-Film State	Suspended State
Catalyst/substrate ratio (g/mL)	0.30/1.0	0.30/3.0
Rate constant: *k* (mol/h/Pt-g)	3.3	0.21
Retardation constant: *K* (mL/mmol)	5.2	16.4

Note: Catalyst = platinum nanoparticles supported on granular activated carbon (Pt/C, 5 wt-metal%). Reaction conditions = boiling and refluxing by heating at 210°C and cooling at 5°C; *k* and *K* calculated from the equation: $v = k/(1 + K\,[\text{naphthalene}])$.

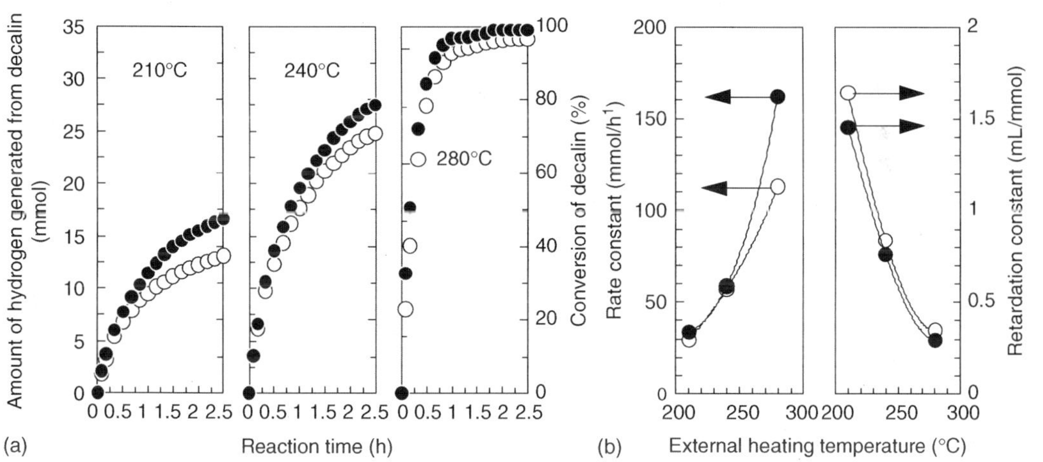

FIGURE 13.9
Hydrogen generation from decalin with carbon-supported platinum-based catalysts at various heating temperatures under superheated liquid-film conditions. (a) Time courses of dehydrogenation activities with Pt-based catalysts and (b) kinetic parameters of Pt-based catalysts as a function of heating temperature. Catalyst: platinum nanoparticles supported on granular activated carbon (Pt/C, 5 wt-metal%, ○) and platinum–rhenium composite nanoparticles supported on granular activated carbon (Pt–Re/C, 5 wt-Pt%, Pt/Re = 2, ●), 0.30 g. Catalyst/solution ratio: 0.30 g/1.0 mL (superheated liquid-film state). Reaction conditions: boiling and refluxing by heating at 210°C, 240°C, and 280°C and cooling at 5°C. (Reproduced from Hodoshima, S., Takaiwa, S., Shono, A., Satoh, K., and Saito, Y., *Appl. Catal. A*, 283(1–2), 235–242, 2005. With permission from Elsevier.)

where v_h is the generation rate of hydrogen (mmol/h), k the reaction rate constant (mmol/h), K the retardation constant due to naphthalene (mL/mmol), and [naphthalene] the molar concentration (mmol/mL).

At the amount ratio of the Pt/C catalyst to decalin solution of 0.30 g/1.0 mL under boiling conditions, the catalyst surface became just wet, resulting in the superheated liquid-film state and the highest catalytic activities. Furthermore, the fact that the external heating temperature rose from 210°C to 280°C could bring about enormously enhanced dehydrogenation activities with the superheated Pt-based catalysts (Figure 13.9a) [10]. In addition, synergetic effects were evident for platinum–rhenium composite catalysts (Pt–Re/C), with which the decalin conversion of almost 95% was attained within 1 h by heating at 280°C (Figure 13.9a) [10]. The higher the heating temperature, the larger the rate constant and the lower the product retardation, as evident with the Pt/C and the Pt–Re/C catalysts from Figure 13.9b.

Under common boiling and refluxing conditions by heating externally at 210°C and refluxing at 5°C, the temperatures of the catalyst layers in various charged amounts of decalin (0, 1.0, and 3.0 mL) with a constant amount of the granular activated carbon (0.285 g) were measured by inserting a thermocouple directly into the center part of each carbon support. The catalyst-layer temperature in each immersed state varied with the charged amount of decalin (Figure 13.10). In the suspension state (3.0 mL), the catalyst-layer temperature was almost equal to the boiling point of decalin (192°C), whereas its temperature in the liquid-film state (1.0 mL) was kept at the level just between the external heating temperature (210°C) and the boiling point (192°C). In case of charging no reactant to the catalyst (0 mL), its temperature became increasingly near to the heating temperature of 210°C. In addition, by raising the external heating temperature from 210°C to 280°C, the temperatures of catalyst layers in the liquid-film states also increased with the heating temperature (Figure 13.11).

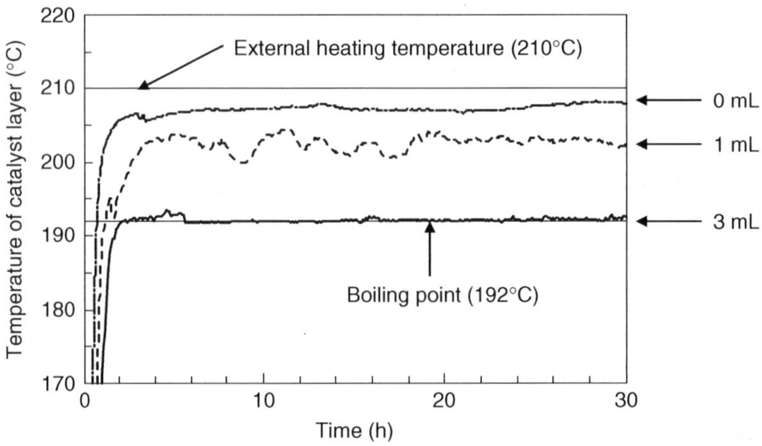

FIGURE 13.10
Relationship of charged amount of decalin with catalyst-layer temperature. Catalyst support: granular activated carbon, 0.285 g. Charged amount of decalin: 0, 1.0, and 3.0 mL. Reaction conditions: boiling and refluxing by heating at 210°C and cooling at 5°C.

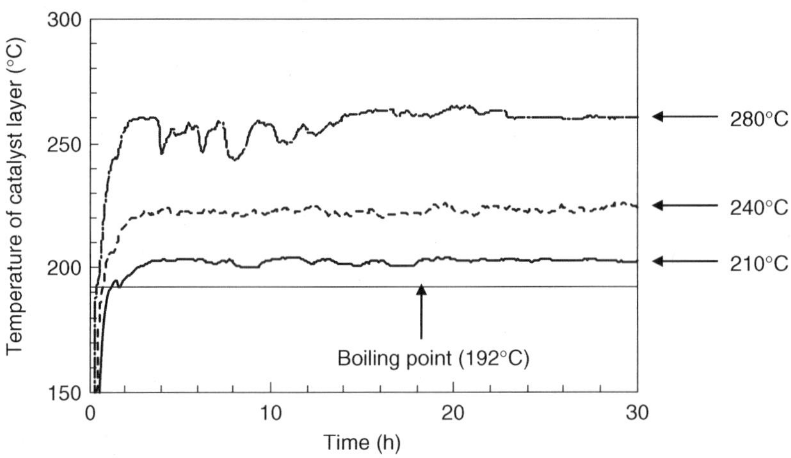

FIGURE 13.11
Relationship of external heating temperature with catalyst-layer temperature. Catalyst support: granular activated carbon, 0.285 g. Charged amount of decalin: 1.0 mL. Reaction conditions: boiling and refluxing by heating at 210°C, 240°C, and 280°C and cooling at 5°C.

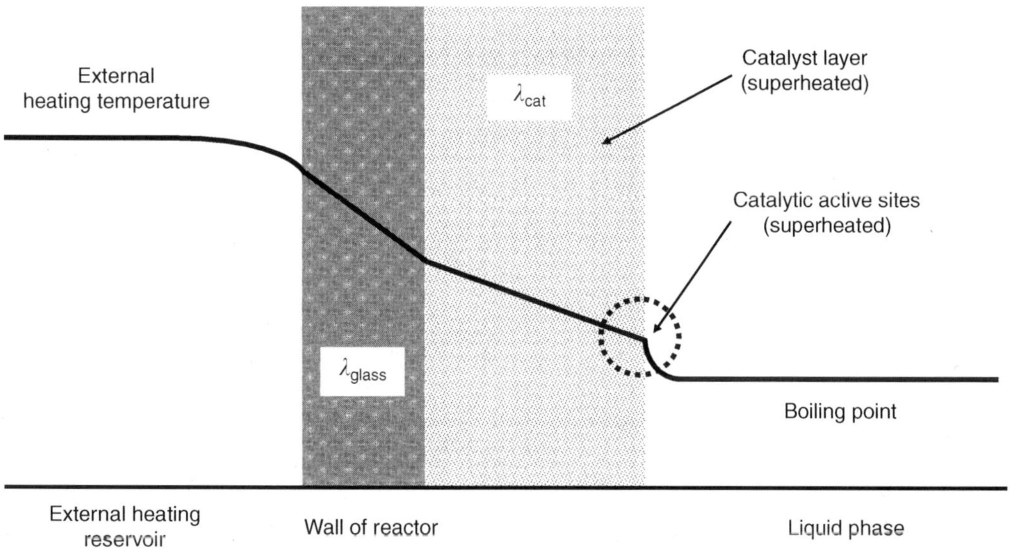

FIGURE 13.12
Temperature gradient among reactor wall, catalyst-layer, and reactant solution under superheated liquid-film conditions.

As shown in Figures 13.10 and 13.11, it was experimentally confirmed that the temperatures of the catalyst layers in the liquid-film states (1.0 mL) were higher than the boiling point of the reactant solution. This is a consequence of "superheating" [36] because the catalyst layer in the liquid-film state, being directly on the reactor bottom, is just wetted with a thin liquid film of the solution under boiling and refluxing conditions. Stationary thermal flow from the external heating reservoir to the catalyst layer results in its temperature remaining higher than the boiling point. Therefore, as illustrated in Figure 13.12, a temperature gradient exists between the catalyst layer and the liquid phase stationarily, which facilitates product desorption (e.g., hydrogen and naphthalene) from the surface, and generates a lot of vacant active sites. Consequently, the reaction rates were increased and the adsorption equilibrium of naphthalene to the catalyst surface was shifted to the side of desorption, resulting in high rate constant (k) and low retardation constant (K), simultaneously. In addition to the superheated liquid-film conditions, continuous hydrogen removal from the reaction medium makes it possible to shift the chemical equilibrium toward the dehydrogenation side and gives high catalytic conversions.

In the suspended state, however, no temperature gradient arises at the catalyst surface since the active site temperature is the same as the boiling point of the solution. Only small magnitudes of reaction rates and conversions were obtained in the suspended state because of diminished rate constant k and enlarged retardation constant K (Table 13.2).

As shown in Table 13.1, toluene is a candidate compound to form the naphthalene oil. To utilize the reaction pair of methylcyclohexane dehydrogenation/toluene hydrogenation as an additive component, it is, thus, necessary to generate hydrogen efficiently from methylcyclohexane under mild reaction conditions.

In the same manner as the decalin dehydrogenation (see Figure 13.5), catalytic dehydrogenation of methylcyclohexane to toluene over the carbon-supported platinum nanoparticles (Pt/C, 5 wt-metal%) was investigated by comparing dehydrogenation catalysis for decalin under boiling and refluxing conditions by heating at 180–240°C and cooling at 5°C. Figure 13.13 shows the time courses of catalytic conversions in the dehydrogenation

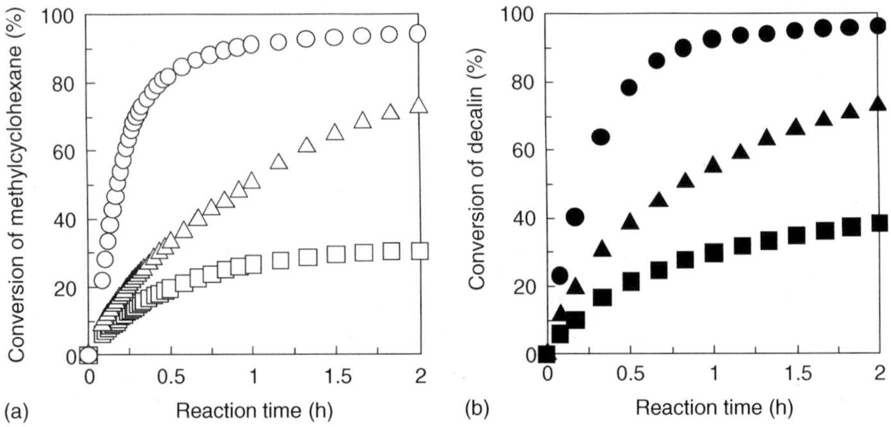

FIGURE 13.13
Time courses of catalytic conversion for dehydrogenation of naphthenic hydrocarbon under superheated liquid-film conditions. (a) Methylcyclohexane dehydrogenation and (b) decalin dehydrogenation. Catalyst: platinum nanoparticles supported on granular activated carbon (Pt/C, 5 wt-metal%), 0.30 g. Catalyst/solution ratio: 0.30 g/1.0 mL (superheated liquid-film state). Reaction conditions: boiling and refluxing by heating at 180–280°C and cooling at 5°C (□: 180°C; △: 210°C, ○: 240°C, ■: 210°C, ▲: 240°C, ●: 280°C).

of methylcyclohexane and decalin under the same superheated liquid-film conditions (amount ratio of catalyst to reactant: 0.30 g/1.0 mL). Under the same superheated liquid-film conditions by heating at 210°C, methylcyclohexane and decalin gave catalytic conversions of 50.8% and 29.7% for 1 h, respectively. At 240°C, catalytic methylcyclohexane conversion above 90% was attained within 1 h, whereas decalin conversion resulted in 55.1% for 1 h. The pair of methylcyclohexane/toluene was found to be suitable as an additive compound from the experimental results that catalytic dehydrogenation of methylcyclohexane is facile compared to decalin dehydrogenation under the common reaction conditions. Moreover, this result suggests that methylcyclohexane itself could be a practical medium for hydrogen storage due to high energy density and fast hydrogen supply (high power density).

13.2.2 Catalytic Dehydrogenation of Tetralin over Carbon-Supported Platinum Nanoparticles under Superheated Liquid-Film Conditions

Under boiling and refluxing conditions by heating at 210°C–240°C and cooling at 5°C, catalytic dehydrogenation of tetralin to naphthalene over the carbon-supported platinum catalyst was performed in the same batchwise reactor as the decalin dehydrogenation (see Figure 13.5) [12]. Figure 13.14 shows the relationship of the dehydrogenation activities of the Pt/C catalysts (0.75 or 0.30 g) with the charged amount of tetralin. It was revealed that the catalyst-to-tetralin amount ratios were also quite sensitive to initial rates of hydrogen generation (Figure 13.14a) and tetralin conversions (Figure 13.14b) at the initial stage (0.5 h). An adequate amount of tetralin was definitely determined for a certain catalyst amount (0.75 or 0.30 g), at which the catalyst displayed the highest ability.

In the adequate case (1.0 or 2.0 mL tetralin), the catalyst appeared to be wet differently from dry sand-bath or suspension states. As in the case of decalin dehydrogenation under the superheated liquid-film conditions, the catalyst temperature is higher than the boiling point, exhibiting a temperature gradient, and the substrate liquid is limited in amount to

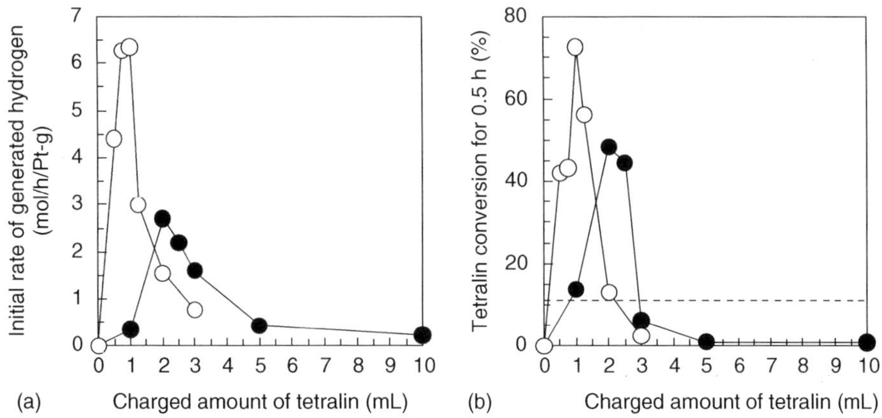

FIGURE 13.14
Relationship of catalytic dehydrogenation activities with charged amount of tetralin. (a) Initial rate of hydrogen generation and (b) tetralin conversion for 0.5 h (equilibrium conversion for teralin dehydrogenation: 11.2% at 210°C and 1 atm (a broken line)). Catalyst: platinum nanoparticles supported on granular activated carbon (Pt/C, 5 wt-metal%), 0.30 (○) and 0.75 g (●). Reaction conditions: boiling and refluxing by heating at 210°C and cooling at 5°C. (Reproduced from Hodoshima, S., Nagata, H., and Saito, Y., *Appl. Catal. A*, 292, 90–96, 2005, With permission from Elsevier.)

a thin film. The Pt/C catalyst in the superheated liquid-film state gave the highest initial rates of hydrogen generation and tetralin conversions, irrespective of the catalyst amount (Figure 13.14a and 13.14b), when the catalyst surface was covered with a thin liquid film of the tetralin solution under boiling and refluxing conditions. The highest tetralin conversions were 48.3% at the amount ratio of 0.75 g/2.0 mL and 72.5% at 0.30 g/1.0 mL, respectively (Figure 13.14b). It is noteworthy that the unfavorable limitation of chemical equilibrium (the dotted line in Figure 13.14b) was surmounted with endothermic dehydrogenation catalysis in the superheated liquid-film states quite efficiently at the temperature range as low as the boiling point of tetralin (207°C). In addition, the initial rate per 1 g of platinum and the tetralin conversion were much higher at the amount ratio of 0.30 g/1.0 mL than those at the ratio of 0.75 g/2.0 mL (Figure 13.14a and 13.14b). These experimental results attract our interests because they suggest to us that the superheated liquid-film state is controllable by the extent of superheating. To attain high reaction rates and conversions, sharp temperature gradients at the catalyst–solution interface are important; in such cases, efficient desorption of reaction products from the active sites and enhanced rates of the dehydrogenation processes are guaranteed. A compact reactor is, therefore, possible with improved catalytic performance for dehydrogenation [12].

Table 13.3 shows dehydrogenation activities for tetralin with the Pt/C catalyst in the superheated liquid-film states at the external heating temperature of 240°C as well as 210°C. Here the optimum amount ratio, giving the highest dehydrogenation activities, was found to be 0.30 g/1.0 mL irrespective of the heating temperature. The external heating temperature, raised from 210°C to 240°C, enormously enhanced reaction rates and conversions by use of the same Pt/C catalyst in the superheated liquid-film states. At 240°C, tetralin conversion higher than 95% was attained within 25 min, whereas the equilibrium conversion (22.5% at 240°C and atmospheric pressure) exceeded within only 2 min [12]. It is noteworthy that even the thermodynamically unfavorable endothermic dehydrogenation proceeds at this heating temperature.

TABLE 13.3

Dehydrogenation Activities for Decalin and Tetralin under Superheated Liquid-Film Conditions

	Initial Rate of Reaction (mol/h/Pt-g)		Conversion for 30 min (%)	
	210°C	240°C	210°C	240°C
Decalin ($C_{10}H_{18}$)	0.80	1.2	18.5	33.7
Tetralin ($C_{10}H_{12}$)	3.1	7.5	72.5	96.0

Note: Catalyst = platinum nanoparticles supported on granular activated carbon (Pt/C, 5 wt-metal%), 0.30 g; charged amount of decalin and tetralin = 1.0 mL (superheated liquid-film state). Reaction conditions = boiling and refluxing by heating at 210°C and 240°C and cooling at 5°C.

Source: Reproduced from Hodoshima, S., Nagata, H., and Saito, Y., *Appl. Catal. A*, 292, 90–96, 2005, With permission from Elsevier.

Dehydrogenation activities, compared for tetralin and decalin [5,12] under the same superheated liquid-film conditions over the same Pt/C catalyst, exhibited around 3.9–6.3 times preference of tetralin (Table 13.3), which can certainly be ascribed to advantageous adsorption due to the π-bonding capability of its aromatic part [17–19]. It was, thus, confirmed experimentally that tetralin is superior to decalin as the organic hydrogen carrier for stationary applications in terms of rapid hydrogen supply or power density, provided that the density of fuel storage is unimportant.

13.2.3 Catalytic Dehydrogenation of Organic Chemical Hydrides over Carbon-Supported Nickel-Based Nanoparticles under Superheated Liquid-Film Conditions

The feasibility of carbon-supported nickel-based catalysts as the alternative to the platinum catalyst is studied in this chapter. Carbon-supported nickel (Ni/C, 10 wt-metal% [12]), ruthenium (Ru/C, 10 wt-metal% [12]), and nickel–ruthenium composite (Ni–Ru/C, 10 wt-metal%, mixed molar ratio of Ni/Ru: 0.25, 1, 4, 8, and 16 [12]) catalysts were prepared similarly by the impregnation method. Granular powders of the activated carbon without the base pretreatment were stirred with the $NiCl_2$, $RuCl_3$, and $NiCl_2$–$RuCl_3$ aqueous solutions at room temperature for 24 h, respectively. Reduction and washing were carried out in the same way as done for the Pt/C catalyst. Finally, these nickel-based catalysts were evacuated at 70°C for 10 h.

Dehydrogenation activities of the carbon-supported nickel (Ni/C, 10 wt-metal%), ruthenium (Ru/C, 10 wt-metal%), and nickel–ruthenium composite (Ni–Ru/C, 10 wt-metal%, mixed molar ratio of Ni/Ru: 0.25, 1, 4, and 8) catalysts in the superheated liquid-film states (amount ratio of catalyst to reactant: 0.317 g/0.75 mL) were examined for decalin under boiling and refluxing conditions by heating at 280°C and cooling at 5°C. As shown in Figure 13.15, dehydrogenation activities of the single-component ruthenium or nickel metal catalysts supported on activated carbon were poor even in the superheated liquid-film state. However, the Ni–Ru composite catalyst was proved to be excellent especially at the mixed molar ratio of Ni/Ru: 1 (total amount of hydrogen generated for 2.5 h: 12.2 mmol; catalytic conversion of decalin for 2.5 h: 51.0%) by heating at 280°C. As for the single-component metal catalyst, the dehydrogenation activity of ruthenium was superior to nickel. However, Ni–Ru composite catalysts with adequate mixed ratio were ascertained to include more Ru metal rather than Ni metal. The Ni–Ru/C catalyst (10 wt-metal%, Ni/Ru = 1) with the best performance, prepared by the conventional coimpregnation method, could generate hydrogen from decalin at 280°C with a reaction rate comparable to the Pt/C catalyst (5 wt-metal%) at 210°C.

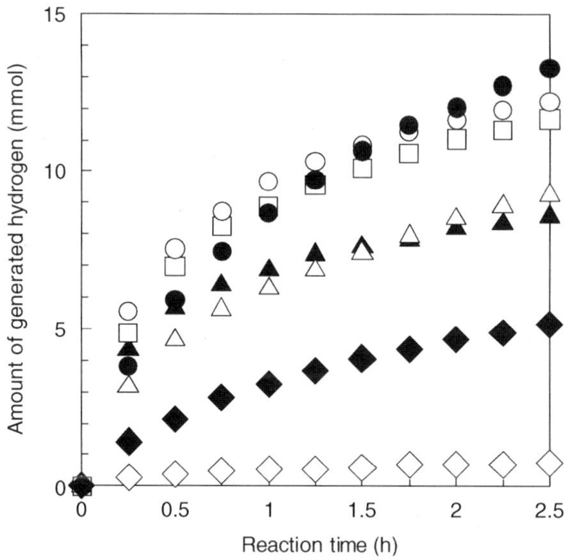

FIGURE 13.15
Dehydrogenation activities for decalin dehydrogenation over carbon-supported nickel-based and platinum catalysts under superheated liquid-film conditions. Catalyst: nickel nanoparticles supported on granular activated carbon (Ni/C (◇), 10 wt-metal%), 0.317 g; ruthenium nanoparticles supported on granular activated carbon (Ru/C (◆), 10 wt-metal%), 0.317 g; nickel–ruthenium composite nanoparticles supported on granular activated carbon (Ni–Ru/C, 10 wt-metal%, mixed molar ratio of Ni/Ru: 0.25 (△), 1 (○), 4 (□), and 8 (▲)), 0.317 g; and platinum nanoparticles supported on granular activated carbon (Pt/C (●), 5 wt-metal%), 0.30 g. Charged amount of decalin: 0.75 mL (Ni/C, Ru/C, and Ni–Ru/C catalysts) and 1.0 mL (Pt/C catalyst). Reaction conditions: boiling and refluxing by heating at 210°C (Pt/C catalyst) and 280°C (Ni/C, Ru/C, and Ni–Ru/C catalysts) and cooling at 5°C.

On the experimental basis of the reaction tests (Figure 13.15) using commercial decalin (volume ratio of cis-/trans-form: 1), dehydrogenation activities for geometrical isomers of decalin (*cis-* and *trans*-decalin) were investigated by using the Ni–Ru/C catalyst (mixed molar ratio of Ni/Ru: 1) and the Pt/C catalyst. It was found that decalin dehydrogenation proceeded more rapidly by use of cis- rather than trans-isomer or commercial decalin (Figure 13.16). Catalytic dehydrogenation activities of the Ni–Ru/C catalyst and the Pt/C catalyst were generally in the order of *cis*-decalin > the mixture > *trans*-decalin.

Because the *cis*-decalin molecule extends its two methine carbon–hydrogen bonds on the same side in contrast to *trans*-decalin, the carbon–hydrogen bond dissociation of adsorbed decalin would be advantageous to the cis-isomer on the catalyst surface (Figure 13.17). A possible reaction path by octalin to naphthalene in dehydrogeno-aromatization of decalin will be favored to the cis-isomer, since its alkyl intermediate provides the second hydrogen atom from the methine group to the surface active site easily.

Actually, *cis*-decalin dehydrogenation proceeded facilely as the amount of generated hydrogen of 15.2 mmol and the conversion of 62.4% were obtained for 2.5 h (Figure 13.16a). Because this activity surpassed the level of the Pt/C catalyst, *cis*-decalin seems to be more suitable than *trans*-decalin as the storage medium of hydrogen from the viewpoints of both catalytic activity and molecular compactness. Aiming at a promising method for supplying hydrogen for onboard applications, it is, therefore, very important for catalytic naphthalene hydrogenation to attain any high selectivity to *cis*-decalin.

Tetralin dehydrogenation over the Ni/C-, the Ru/C-, and the Ni–Ru/C catalysts (mixed molar ratio of Ni/Ru: 0.25, 1, 4, 8, and 16) in the superheated liquid-film states (amount

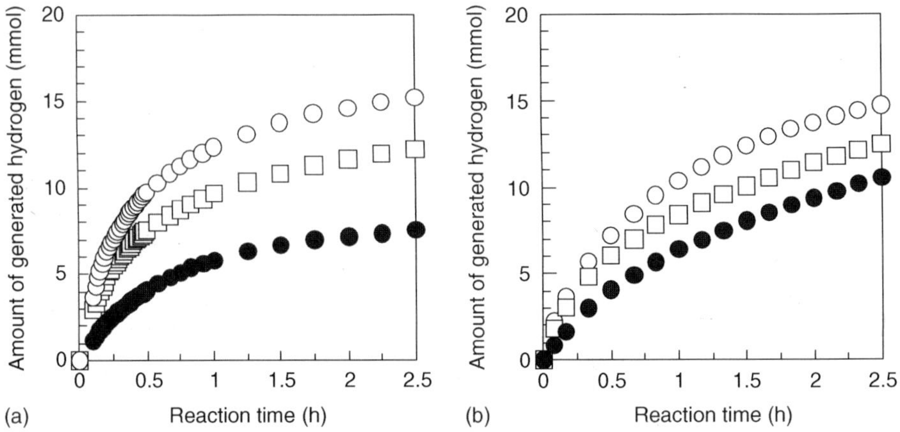

FIGURE 13.16
Decalin isomer effect of dehydrogenation activity by use of carbon-supported platinum catalyst or nickel–ruthenium composite catalyst. (a) Carbon-supported nickel–ruthenium composite catalyst and (b) Carbon-supported platinum catalyst. ●: *trans*-decalin; □: decalin (*cis* : *trans* = 1:1); ○: *cis*-decalin. Catalyst: nickel–ruthenium composite nanoparticles supported on granular activated carbon (Ni–Ru/C, 10 wt-metal%, mixed molar ratio of Ni/Ru: 1), 0.317 g, and platinum nanoparticles supported on granular activated carbon (Pt/C, 5 wt-metal%), 0.30 g. Charged amount of decalin: 0.75 mL (Ni–Ru/C catalyst) and 1.0 mL (Pt/C catalyst). Reaction conditions: boiling and refluxing by heating at 210°C (Pt/C catalyst) and 280°C (Ni–Ru/C catalyst) and cooling at 5°C.

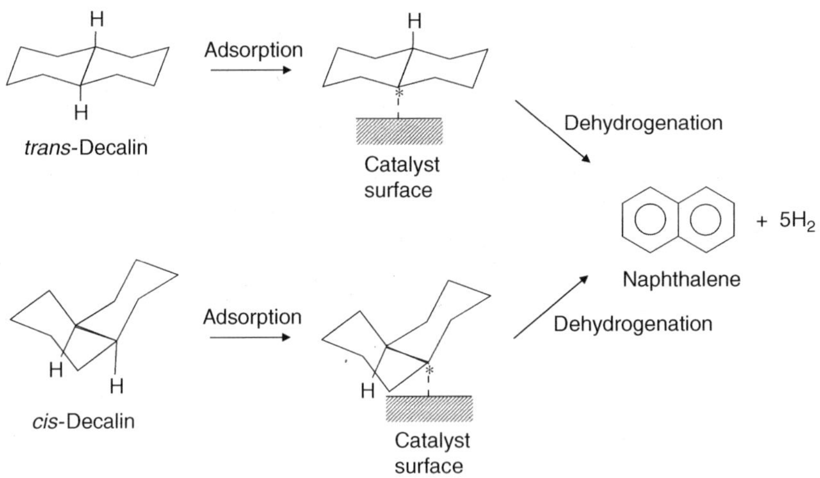

FIGURE 13.17
Contrast between decalin isomers on the molecular structures of adsorption resulting in different dehydrogenation activities.

ratio: 0.317 g/1.0 mL) was also examined by heating at 240°C [12]. Catalytic performance of the Ni–Ru/C composite catalysts for tetralin dehydrogenation exhibited a significant synergetic effect, where the mixed ratios of Ni/Ru within the range 4–16 gave high initial reaction rates and conversions [12]. These nickel-based composite catalysts generated hydrogen from tetralin by heating at 240°C in catalytic activities, comparable with those of the Pt/C catalyst heated at 210°C [12].

13.3 Efficient Hydrogen Generation from Organic Chemical Hydrides with Superheated Liquid-Film-Type Catalysis in Continuous Operation

13.3.1 Catalytic Hydrogen Generation from Organic Chemical Hydrides under Superheated Liquid-Film Conditions by Use of Laboratory-Scale Continuous Reactor

To realize efficient hydrogen generation from organic chemical hydrides with the superheated liquid-film-type catalysis in a continuous operation, catalytic dehydrogenation by use of a continuous reactor was investigated on a laboratory scale [13,14].

Figure 13.18 shows experimental apparatus for catalytic dehydrogenation in a continuous mode [13,14]. In this continuous operation on a laboratory scale (reactor: round-bottom flask, volume: 100 mL), catalytic dehydrogenation of decalin to naphthalene was performed with a constant amount of the Pt/C catalyst (1.1 g) by heating at 230–270°C and cooling at 25°C under atmospheric pressure. Here liquid-phase reactant was supplied continuously to the catalyst at various feed rates (0.28, 0.5, 1.1, 1.7, and 2.3 mL/min at 230°C; 0.26, 0.6, 1.1, 2.3, and 3.4 mL/min at 270°C). All of the vapors were removed from the reactor through a thermally protected part to a condenser cooled at 25°C (see Figure 13.18). After confirming the stationary reaction rates by a flow meter, the rates of hydrogen generated from decalin were recorded every 5 min for 30 min by the flow meter, and the liquid composition, consisting of naphthalene and unconverted decalin, was analyzed with gas chromatographic analysis. The temperature of each catalyst layer during reaction was measured by directly inserting a thermocouple into the center part of the catalyst layer. In this way, a relationship of catalytic activities with the feed rate of decalin and the temperature of the catalyst layer was investigated.

Figure 13.19a shows a relationship of the catalyst-layer temperature during the reaction with the feed rate of decalin in the continuous operation. The smaller the feed rate of decalin, the higher the catalyst-layer temperature. It was also revealed that dehydrogenation activities for decalin were dramatically changed in accordance with the feed rate of decalin (Figure 13.19b and 13.19c).

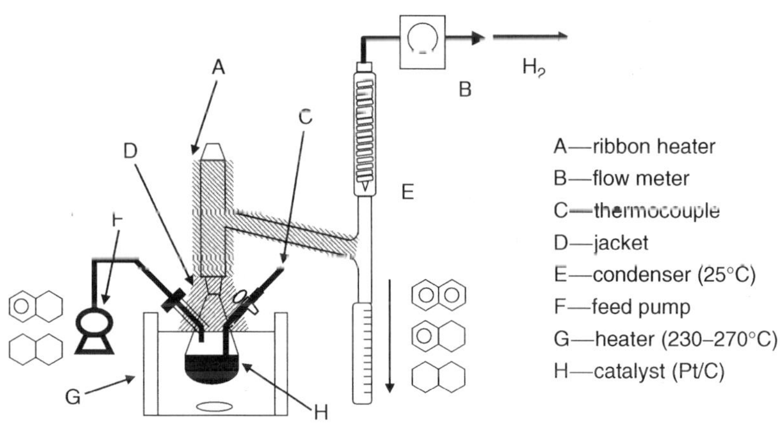

FIGURE 13.18
Experimental apparatus for catalytic dehydrogenation under reactive distillation conditions in laboratory-scale continuous operation. (Reproduced from Hodoshima, S., Shono, A., Satoh, K., and Saito, Y., *Chem. Eng. Trans.*, 8, 183–188, 2005. With permission.)

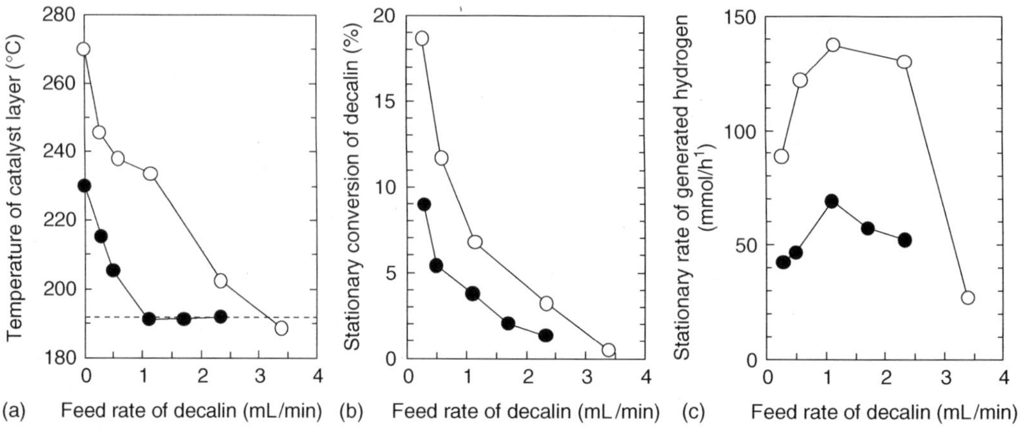

FIGURE 13.19
Relationship of catalyst-layer temperature and dehydrogenation activities with feed rate of decalin in laboratory-scale continuous operation. (a) Catalyst-layer temperature versus feed rate of decalin (boiling point of decalin: 192°C (a broken line)), (b) stationary conversion versus feed rate of decalin, and (c) stationary hydrogen-generation rate versus feed rate of decalin. Catalyst: platinum nanoparticles supported on granular activated carbon (Pt/C, 5 wt-metal%), 1.1 g. Reaction conditions: boiling and refluxing by heating at 230°C (●) and 270°C (○) and cooling at 25°C.

At large feed rates (1.1–2.3 mL/min at 230°C, 3.4 mL/min at 270°C), the catalyst temperature in each immersed state was almost equal to the boiling point of decalin (192°C). These catalysts were suspended uniformly in sufficient amounts of decalin solution, resulting in low catalytic conversions comparatively (Figure 13.19b).

Below a certain feed rate (1.1 mL/min at 230°C, 3.4 mL/min at 270°C), the temperatures were higher than the boiling point (Figure 13.19a) and catalytic conversions of decalin were remarkably enhanced [14]. In this range of feed rate, the extent of superheating increased with the decrease of the feed rate, so that the temperature gradient at the catalyst–solution interface became sharp in accordance with the superheating extent and catalytic conversions were improved as shown in Figure 13.19a and 13.19b. From these experimental results, it was confirmed that the superheated liquid-film states, giving excellent catalytic performance, were realized in the continuous operation [13,14]. In contrast, the highest reaction rate was attained in a certain range of feed rate as shown in Figure 13.19c. In this continuous operation, a stationary reaction rate is determined from the feed rate and the stationary conversion. The highest magnitude was apparently obtained at about 1.1 mL/min consequently. However, it was revealed that high reaction rates were given at a rather wide range of the feed rate unlike the batchwise operation. As perceived in Figures 13.7 and 13.14, the optimum range for realizing the superheated liquid-film state in the batchwise operation was sharp and sensitive with respect to the amount ratio of catalyst to reactant solution, whereas the superheated active sites that give high catalytic performance were allowed at a rather wide range in the continuous operation. Reactor design for keeping stationary high reaction rates, thus, will be rather easy due to the wide thickness of the superheated liquid-film layer.

By using the same experimental apparatus as in the decalin dehydrogenation (Figure 13.18), catalytic dehydrogenation of tetralin to naphthalene was performed with continuous supply of tetralin solution at various feed rates (0.26, 0.5, 1.2, 1.7, and 2.1 mL/min) into the reactor, containing a constant amount of the Pt/C catalyst (1.1 g), under boiling and refluxing conditions by heating at 230–240°C and cooling at 25°C [13]. The same tendency

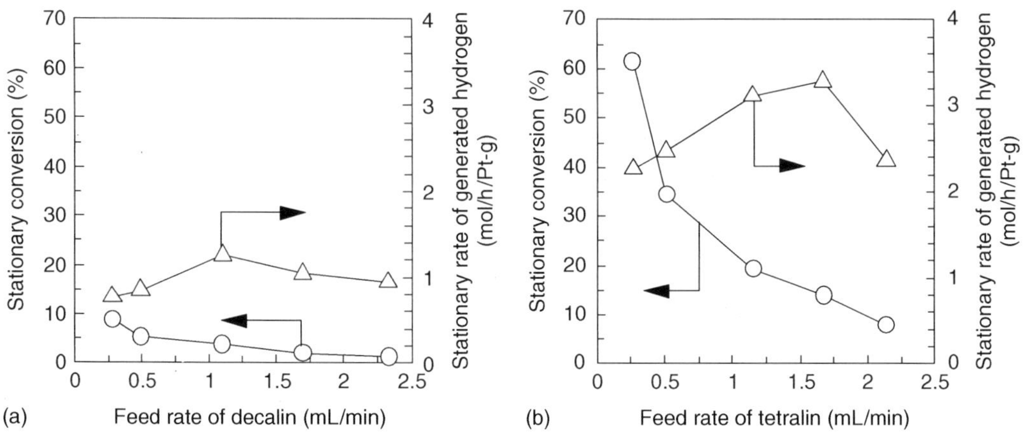

FIGURE 13.20
Comparison of dehydrogenation activities for decalin and tetralin. (a) Decalin dehydrogenation and (b) tetralin dehydrogenation. Catalyst: platinum nanoparticles supported on granular activated carbon (Pt/C, 5 wt-metal%), 1.1 g. Reaction conditions: boiling and refluxing by heating at 230°C and cooling at 25°C. (Reproduced from Hodoshima, S., Shono, A., Satoh, K., and Saito, Y., *Chem. Eng. Trans.*, 8, 183–188, 2005. With permission.)

as the decalin dehydrogenation was observed with respect to the relationship of catalytic activities with the feed rate of tetralin and the temperature of the catalyst layer. It was confirmed that tetralin dehydrogenation in the continuous mode also proceeded quite efficiently under superheated liquid-film conditions (e.g., 61.5% at 0.26 mL/min) [13].

Figure 13.20a and 13.20b shows a comparison of dehydrogenation activities for decalin and tetralin under the common reaction conditions by heating at 230°C with the same continuous reactor [13]. The rates of generated hydrogen and tetralin conversions were much higher than those for decalin at all the feed levels. As for the stationary one-pass conversion, tetralin exhibited around 6–7 times larger conversions than the decalin under the same superheated conditions (0.26–0.5 mL/min). It is noted that catalytic conversion higher than 60% was attained at 0.26 mL/min, although the heating temperature (230°C) is restricted thermodynamically for endothermic tetralin dehydrogenation. Facile dehydrogenation of tetralin would certainly be ascribed to advantageous adsorption due to the π-bonding capability of its aromatic part. It was, therefore, confirmed that tetralin as a storage medium for stationary use is superior to decalin in terms of rapid hydrogen supply.

To examine the catalyst deterioration under the present superheated liquid-film conditions, a long-term tetralin dehydrogenation over the carbon-supported platinum catalyst (Pt/C) was carried out under the superheated liquid-film conditions (heating temperature: 240°C) at the amount ratio of 1.1 g/0.5 mL/min [13]. Figure 13.21 shows the time courses of reaction rate and conversion in the tetralin dehydrogenation. High conversion (around 50%) was maintained for longer than 5 h.

The catalyst deactivation is generally ascribed to coke formation over the catalyst surface in addition to sintering or aggregation of metal particles. Exothermic catalysis often suffers from catalyst sintering caused by the rise in temperature during the reaction. In contrast, the present endothermic dehydrogenation was performed at a mild temperature of 240°C, which is low enough to avoid the coke formation. Moreover, the products (e.g., naphthalene and hydrogen) or any poisoning impurities adsorbed on the catalyst surface were easily removable by vigorous bubbling due to the temperature gradient in the superheated liquid-film state under boiling conditions, resulting in preventing catalysts from deterioration.

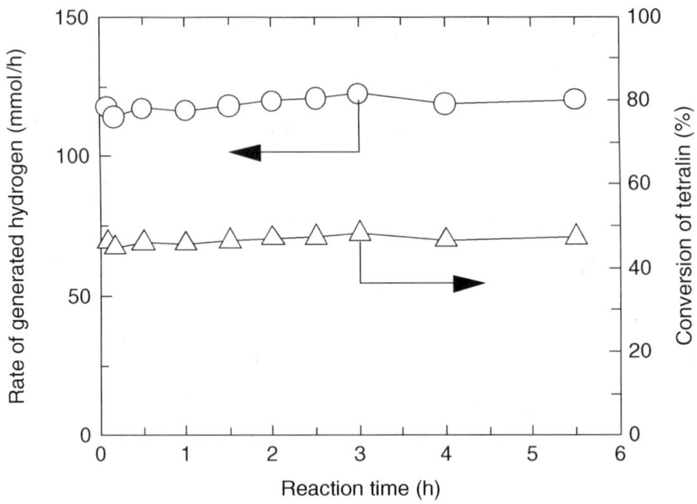

FIGURE 13.21
Time courses of dehydrogenation activities with carbon-supported platinum catalyst under superheated liquid-film conditions in laboratory-scale continuous operation. Catalyst: platinum nanoparticles supported on granular activated carbon (Pt/C, 5 wt-metal%), 1.1 g. Feed rate of tetralin: 0.5 mL/min (superheated liquid-film conditions). Reaction conditions: boiling and refluxing by heating at 240°C and cooling at 25°C. (Reproduced from Hodoshima, S., Shono, A., Satoh, K., and Saito, Y., *Chem. Eng. Trans.*, 8, 183–188, 2005. With permission.)

13.3.2 Demonstration Test for Efficient Hydrogen Supply from Decalin with Superheated Liquid-Film-Type Catalysis by Use of Bench-Scale Continuous Reactor

For the purpose of demonstrating the superheated liquid-film-type catalysis in a continuous operation on a bench scale, catalytic dehydrogenation of decalin using a continuous reactor together with a new supporting material was investigated [10,11].

Instead of the granular activated carbon, the activated carbon cloth (ACC) (BET specific surface area: 1834 m^2/g, average pore size: 1.5 nm, Kuraray Chemical Co.) was adopted as the catalyst support in this continuous reaction because of its practical advantages. A platinum catalyst supported on this ACC was prepared by an impregnation method [10]. A piece of the ACC was cut into a disk with a diameter of 5 cm and a thickness of around 0.8 mm. The disk-shaped ACC was stirred with a K_2PtCl_4 aqueous solution at room temperature for 24 h. This step was followed by dropwise addition of the $NaBH_4$ aqueous solution at room temperature for 10 min. After the mixture was kept standing for another 30 min, the catalysts were filtered and then washed with a large amount of water to remove the residual ions. Finally, the Pt/ACC catalyst (5 wt-metal%) was evacuated at 200°C for 5 h.

In a continuous reactor on a bench scale (stainless reactor, volume: 1.2 L) (Figure 13.22), catalytic dehydrogenation of decalin was performed with constant amounts of platinum catalyst supported on the ACC (disk-shaped Pt/ACC catalyst (5 wt-metal%) with a diameter of 5 cm, 0.29 g (one layer), $0.29 \times 2 = 0.58$ g (two layers), and $0.29 \times 3 = 0.87$ g (three layers)) under boiling and refluxing conditions by heating at 280°C and cooling at 25°C. Decalin solution was supplied continuously to the catalyst bed at the various feed rates (1.5, 2.0, 2.5, 3.0, and 5.0 mL/min). The stationary generation rates of hydrogen were recorded every 5 min for 30 min by the flow meter and the liquid composition was analyzed gas chromatographically. The relationship of catalytic activities for dehydrogenation with the feed rate of decalin was examined similarly.

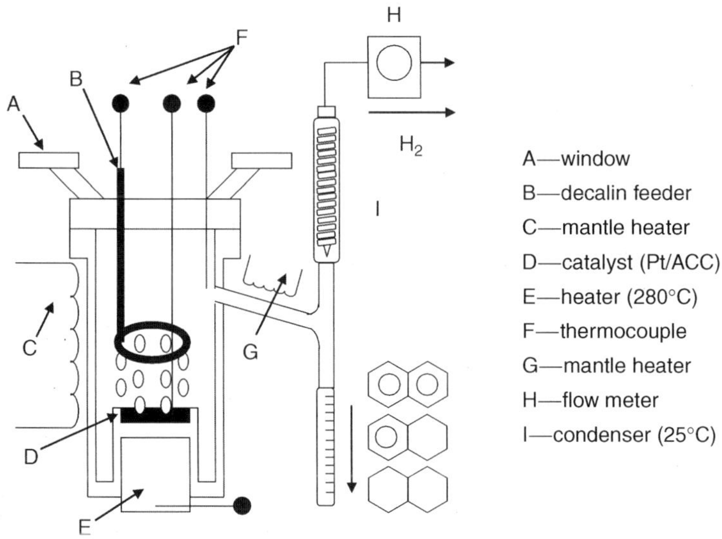

FIGURE 13.22
Experimental apparatus for catalytic dehydrogenation under reactive distillation conditions in bench-scale continuous operation.

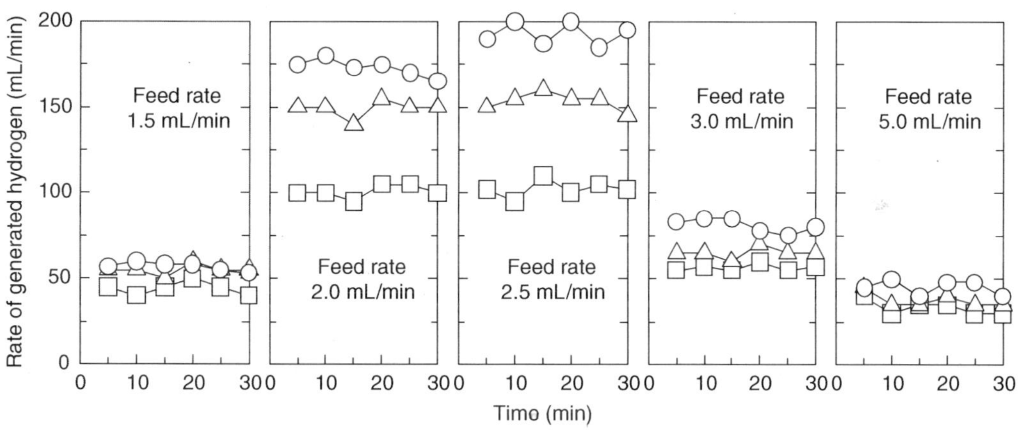

FIGURE 13.23
Time courses of rate of hydrogen generated from decalin with carbon-supported platinum catalyst at various feed rates in bench-scale continuous operation. Catalyst: platinum nanoparticles supported on ACC (5 wt-metal%), 0.29 g (one layer, □), 0.58 g (two layers, △), and 0.87 g (three layers, ○). Feed rate of decalin: 1.5, 2.0, 2.5, 3.0, and 5.0 mL/min. Reaction conditions: boiling and refluxing by heating at 280°C and cooling at 25°C.

The time courses of hydrogen generation were pursued in the continuous operation at various feed rates (Figure 13.23). In the batchwise operation, a gradual decrease of reaction rates was observed. This was due to the product accumulation over the active sites, as shown in Figure 13.8b. In contrast, the rates of hydrogen generated from decalin were almost constant in this continuous operation at all the feed levels (Figure 13.23). Direct visual observation of immersed catalysts could discriminate the states among suspension (5.0 mL/min), liquid-film (2.0–2.5 mL/min), and sand-bath (solid–gas phase catalysis; 1.5 ml/min) states in accordance with the feed rates of decalin. In the liquid-film state, the highest magnitudes of stationary hydrogen generation rate and catalytic conversion were accomplished simultaneously (Figure 13.24).

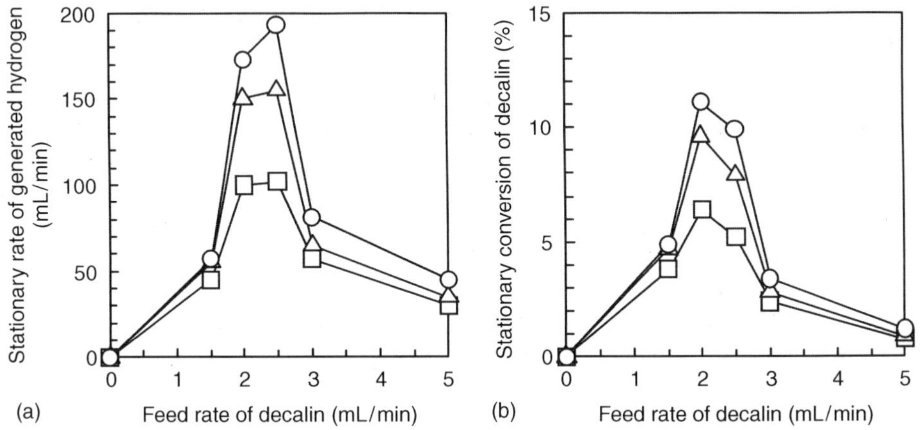

FIGURE 13.24
Relationship of dehydrogenation activities with feed rate of decalin in bench-scale continuous operation. (a) Stationary rate of generated hydrogen and (b) stationary conversion. Catalyst: platinum nanoparticles supported on ACC (5 wt-metal%), 0.29 g (one layer, □), 0.58 g (two layers, △), and 0.87 g (three layers, ○). Reaction conditions: boiling and refluxing by heating at 280°C and cooling at 25°C.

In addition, the more the number of piled catalyst sheet, the larger the dehydrogenation activities in the liquid-film state especially, as evident in Figures 13.23 and 13.24. It is easy for the liquid reactant to penetrate into the catalyst layer consisting of the ACC with lots of air space. By piling the ACC, therefore, decalin would penetrate into catalyst layer further and be kept in it for a long time, resulting in an enhanced catalytic performance.

The catalysis pair of decalin dehydrogenation/naphthalene hydrogenation is usable as a hydrogen storage medium loaded in hydrogen vehicles. Here, the power density obtainable from decalin is the most significant factor for the onboard application.

Sufficient power of 50–100 kW for fuel-cell-vehicle operation requires high rates of hydrogen generation from the onboard dehydrogenation reactor [38]. Because the heating areas in this endothermic reactor are the most important for compact design, the following estimation of the catalyst-layer area needed for the targeted power (50 kW) is attempted on the basis of experimental results obtained in the bench-scale operation [10].

In the present estimation, a continuous dehydrogenation reactor in which decalin is supplied to the catalyst at various feed rates without internal refluxing is assumed. Here all the condensable products and unconverted decalin were removed from the reactor to the condensation part (see Figure 13.22). Now, the stationary rates of hydrogen generation (V_H), naphthalene formation (V_N), and evaporation of unconverted decalin (V_D) are defined as the magnitudes per area of the catalyst layer (mol/m^{2}h). All these rates are expressed from mass balance as follows:.

$$V_H = 5 \times \left(\frac{A}{100}\right) \times D \tag{13.5}$$

$$V_N = 5 \times \left(\frac{A}{100}\right) \times D \tag{13.6}$$

$$V_D = \left[1 - \left(\frac{A}{100}\right)\right] \times D \tag{13.7}$$

where A (%) is the stationary one-pass conversion of decalin and D (mol/m^{2}h) the feed rate of decalin per area of the catalyst layer made of a disk-shaped ACC for loading of decalin, with a diameter of 5 cm in this study. Therefore, the catalyst-layer area needed (S (m^2)) is correlated with the desired stationary rate of hydrogen generation (L_H (mol/h)) given as follows:

$$S = \frac{L_H}{V_H} = \frac{L_H}{[5 \times (A/100) \times D]} \tag{13.8}$$

In the case of 50 kW power, the rate of hydrogen supply needed (L_H) is around 1.69×10^3 (mol/h) at the energy-conversion-efficiency level of 45% for the proton exchange membrane fuel cell (PEM-FC) [38].

As is evident from Figure 13.25a, the area loading of decalin is closely related with the stationary conversion. Because the highest conversion was attained at around the area loading of 70 (L/m^{2}h) (= 4.48×10^2 (mol/m^{2}h)) with the catalyst in the superheated liquid-film state, Equation 13.8 is transformed in to Equation 13.9, giving a reciprocal correlation curve between the one-pass conversion (A) and the reaction area needed to a 50 kW power (S) (Figure 13.25b).

$$S = \frac{L_H}{V_H} - \frac{1.69 \times 10^3}{[5 \times (A/100) \times 4.48 \times 10^2]} = \frac{75.3}{A} \tag{13.9}$$

Although the criteria of a one-pass decalin conversion above 70% and the reaction area of around 1.0 m^2 are fulfilled for fuel cell vehicles, increment of contact time between decalin and catalyst in the continuous reactor are possible by adding a function of internal refluxing. A piston-flow-type reactor with the function of internal refluxing will make it possible to lengthen the contact time and improve the dehydrogenation activities per unit heating area. Figure 13.26 shows a concept for the basic design of a compact reactor in a continuous mode to realize a high one-pass conversion under superheated liquid-film conditions [10,15]. Liquid decalin is continuously supplied from the inlet of this reactor to the superheated catalyst layer, followed by repeating evaporation and

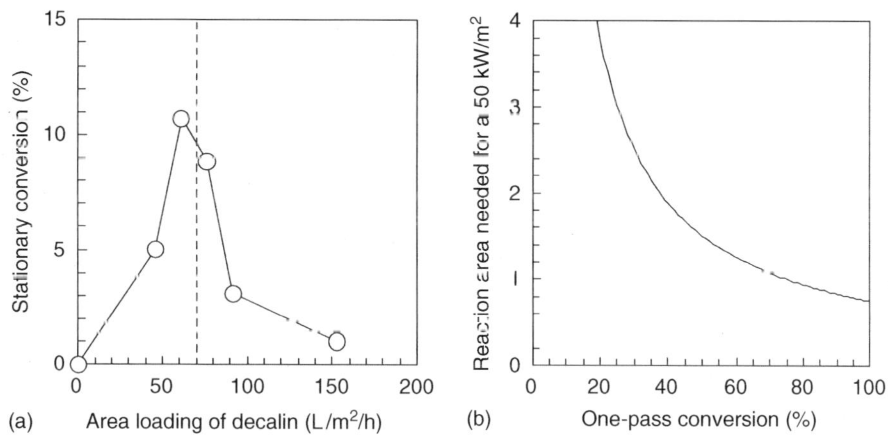

FIGURE 13.25
Relationship of stationary conversion with area loading of decalin and a correlation curve between reaction area needed for 50 kW power and stationary conversion. (a) Stationary conversion versus area loading of decalin and (b) reaction area needed for 50 kW versus one-pass conversion. Catalyst: platinum nanoparticles supported on ACC (5 wt-metal%), 0.87 g (three layers). Reaction conditions: boiling and refluxing by heating at 280°C and cooling at 25°C.

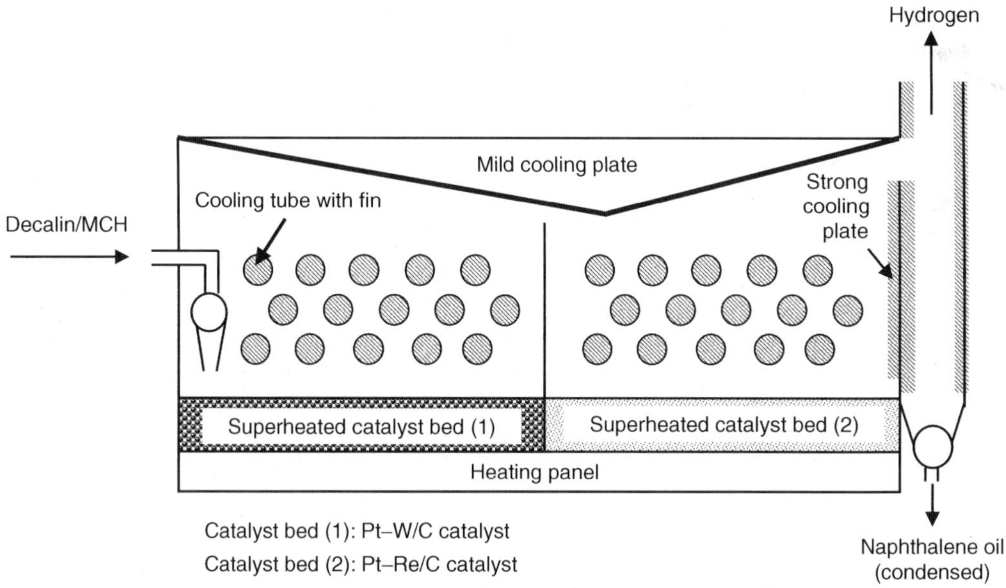

FIGURE 13.26
A piston-flow-type reactor for dehydrogenation in continuous mode under superheated liquid-film conditions.

condensation through internal refluxing. Before getting rid of the reactor, unconverted decalin and intermediate tetralin will be converted completely into naphthalene. Hydrogen and condensates are separated finally by the procedure of total condensation. Moreover, adequate choice of composite catalysts can modulate the activities in accordance with the extent of conversions or naphthalene concentrations. In this piston-flow-type reactor, naphthalene should become rich gradually in proportion to the distance from the reactor inlet. Adoption of different composite catalysts is suggested here, since the catalytic activity on carbon–hydrogen bond dissociation for alkane and the sensitivity toward retardation due to naphthalene adsorption can be selected on the experimental basis [15]. As reported previously [15], enhancement of dehydrogenation activities for decalin was realized by using carbon-supported platinum-based composite catalysts (e.g., Pt–W/C and Pt–Re/C catalysts). The Pt–W/C composite catalyst exhibited the highest reaction rate at the initial stage, whereas the Pt–Re/C composite catalyst showed the second highest reaction rate in addition to a low sensitivity to retardation due to naphthalene adsorbed on catalytic active sites [15]. Consequently, the Pt–W/C and the Pt–Re/C catalysts should be, thus, set close to and distant from the inlet (catalyst beds (1) and (2) in Figure 13.26), respectively [10,15].

13.4 Organic Chemical Hydrides for Storage, Transportation, and Utilization of Hydrogen

13.4.1 Onboard Hydrogen Supply from Organic Chemical Hydrides

Onboard hydrogen required for 500 km cruise is estimated to be 5 kg, amounting to 77.7 L of commercial decalin (cis/trans = 1) and 104.1 L of methylcyclohexane in volume, which corresponds to 1.09 and 1.52 times of the required amount of liquefied hydrogen (70.6 L), respectively.

As for onboard supply of hydrogen from its liquefied or compressed state, no external input of energy is needed additively. It is to be noted that endothermic dehydrogenation reactions are able to proceed with heat supply at proper temperatures for organic chemical hydrides. Moreover, evaporation heat is required for the carbon-supported catalysts in the superheated liquid-film states. Onboard hydrogen supply from organic chemical hydrides might not be so practical for PEM-FC-driven vehicles. With respect to hydrogen vehicles driven by ICEs, however, waste heat lower than 300° exhausted from the engine is now useful for providing heat of both reaction and evaporation. In other words, "chemical recuperation of low-quality waste heat" [39] is possible by utilizing endothermic dehydrogenation reactions of organic chemical hydrides, which will improve the total efficiency in the vehicular operation.

The onboard hydrogen supplied from organic chemical hydrides in the ICE vehicles should be used multipurposely: (a) not only as the single-component fuel but also as the mixed-phase addend to gasoline or light oil fuel to enlarge the limit of lean burn efficiently, keeping the heat contents of carbonaceous fuels high enough and (b) as the reducing reagent toward nitrogen oxides (NOx) to assist the de-NOx catalyst without adding urea aqueous solutions or sacrificing a part of the loaded fuel in intermittent small pulses for NOx reduction with steam-reformed hydrogen

13.4.2 Chemical Recuperation of Waste Heat by Utilizing Organic Chemical Hydrides

At present, waste heat exhausted from the ICE is removed with any efficient radiator system through direct apparent heat exchanging. On the contrary, organic chemical hydrides can recuperate the chemical energy of endothermic reaction heat during exhausted heat removal. Heat transfers accompanying the phase change of evaporation and condensation of aromatic products and unconverted reactants will certainly facilitate the removal of heat from the ICE parts, with adoption of any new radiator system compelled.

As an onboard reactor equipped with a function of heat exchange, a piston-flow-type reactor for catalytic dehydrogenation under superheated liquid-film conditions (Figure 13.26) will replace conventional radiator systems. For the purpose of recuperating exhausted waste heat as chemical energy of hydrogen, this radiator system works not only as a catalytic hydrogen generator but also as a heat exchanger recovering the waste heat from the engine for endothermic dehydrogenation. At the same time, the aromatic products separated from hydrogen through condensation are stored by refueling the evacuated space of the same cassette-type container with a mobile partition, which will be replaced later at the gas station with the one full of organic chemical hydrides. The overall storage system including the dehydrogenation unit, thus, will be compact in weight and volume. In addition, the combination of catalysts highly active for dehydrogenation such as the Pt–W/C and of catalysts highly resistant for aromatic adsorbates such as the Pt–Re/C will be an adequate choice for enhancing catalytic performance per area of catalyst bed in the reactor.

A new evaluation standard for the dehydrogenation catalysts in the superheated liquid-film states is introduced here. This standard is called as the "ratio of heat recuperation" [39], being defined as the ratio of endothermic reaction heat to the denominator of heat supplied from the external thermo-reservoir to the catalyst layer shown as follows (Equations 13.10 and 13.11):

$$[\text{Heat supplied to the catalyst layer}] = [\text{Reaction heat}] + [\text{Evaporation heat}] \quad (13.10)$$

$$[\text{Ratio of heat recuperation}] = \frac{[\text{Reaction heat}]}{[\text{Supplied heat}]} = \frac{[\text{Reaction heat}]}{[\text{Reaction heat}] + [\text{Evaporation heat}]} \quad (13.11)$$

TABLE 13.4

Chemical Recuperation of Waste Heat by Catalytic Dehydrogenation of Methylcyclohexane under Superheated Liquid-Film Conditions

Temperature of Catalyst Layer (°C)	One-Pass Conversion (%)	Ratio of Heat Recuperation[a] (%)
210	6.6	15.6
245	37.0	47.5
285	82.6	63.3

Note: Catalyst = platinum nanoparticles supported on ACC (Pt/ACC, 5 wt-metal%) 3.0 g. Feed rate of methylcyclohexane = 3.5 mmol/min (superheated liquid-film state).

[a] [Ratio of heat recuperation] = [reaction heat]/([reaction heat] + [evaporation heat]).

Source: Based on Saito, Y., Hodoshima, S., Shono, A., and Otake, K., Hydrogen production from cyclohexane and hydrogen production. *Proceedings of 9th Asian Hydrogen Energy Conference*, Tokyo, 2007.

One-pass conversion obtained in the continuous dehydrogenation reactor equipped with an external condenser (Figures 13.18 and 13.22) gives the extent of energy recuperation from the ICE waste heat through hydrogen generation, because induction of heat from the external thermo-reservoir to the catalyst layer must be consumed stationarily to allot the supplied heat to the endothermic reaction heat as well as the evaporation heat.

A series of experimental results on the ratio of heat recuperation is shown in Table 13.4. Here temperatures of the catalyst layer, composed of a sheet of ACC dispersed with platinum nanoparticles (5 wt-metal%, 3.0 g), were varied at the constant feed ratio (3.5 mmol/min) of liquid methylcyclohexane (MCH) [39]. At the catalyst-layer temperature of 285°C, the heat recuperation ratio of 63.3% was attained, being much larger than the magnitudes of 15.6 and 47.5% for the catalyst-layer temperatures of 210°C and 245°C, respectively.

The theoretical limit of heat recuperation ratio for methylcyclohexane at 285°C is given as 67.2% on the basis of the reaction heat 202.2 kJ/mol-MCH for immediate 100% conversion of methylcyclohexane to toluene and hydrogen together with 98.6 kJ/mol-MCH of the evaporation heat (=apparent heat + latent heat) for toluene, hydrogen, and methylcyclohexane. From the dependence of the one-pass conversion for methylcyclohexane dehydrogenation over the carbon-supported platinum catalyst in the superheated liquid-film state on the catalyst-layer temperature, it was revealed that the heat consumption for reaction became more predominant than for evaporation, resulting in higher magnitudes of thermo-recuperation ratio at higher catalyst-layer temperatures. In the flow-type catalytic reactor, the thermo-recuperation ratio ought to be enhanced by immersing the carbon-supported platinum nanoparticles with the liquid film of methylcyclohexane. Reactants can stay as long as possible inside the catalyst layer under the superheated conditions with a steep temperature gradient. This is the reason why activated carbon with a high surface area was chosen as the catalyst support. It is well anticipated for this support material to retain the organic chemical hydrides within the catalyst layer and perform the reaction further due to large amounts of micropores, offering high affinity with hydrocarbon reactants.

13.4.3 Exergetic Analysis of Organic Chemical Hydrides as the Media for Hydrogen Storage and Transportation

Hydrogen-driven vehicles are nowadays managed with either compressed hydrogen or liquefied hydrogen supplied from the hydrogen station facilities. From the viewpoint of exergy consumption, however, a large amount of mechanical energy is consumed in vain without any chance of recovering low-temperature or high-pressure exergy. In contrast, organic chemical hydrides are also potential storage media for operating the ICE vehicles as described earlier. To compare each storage method for automotive applications, exergetic analysis of hydrogen storage and transportation systems was attempted with respect to organic chemical hydrides and compressed and liquefied hydrogen.

The electrolysis hydrogen from renewable energy (e.g., wind- and solar power) and the by-product hydrogen from industrial processes were targeted in the present estimation, irrespective of the distance between the hydrogen supply site and the demand site or the scale of both sites. As the common basis for comparison, the following process was assumed: 1 kg hydrogen in the standard state (298.15 K and 0.1 MPa) obtained at the supply site is stored inside vehicles by each method (organic chemical hydrides and compressed and liquefied hydrogen) and becomes the standard state again when the hydrogen is supplied to ICE vehicles or onboard fuel cells at the demand site. The exergy consumed during hydrogen storage and transportation in this process was calculated with regard to each method.

Exergetic analysis is based on the first and second law of thermodynamics. The exergy content of a certain amount of energy equals the total amount of work obtainable from that energy. Exergy is, therefore, a measure for the quality of energy. It is possible to perform work with energy when this energy is not in equilibrium with the natural environment. This lack of equilibrium can be the result of not only a temperature or pressure difference with the natural environment (physical exergy), but also the result of a difference in the chemical potential (chemical exergy) relative to the natural environment. By defining a reference temperature, pressure, and chemical composition of the natural environment, it is possible to define the universal exergy content for every substance and subsequently for all process streams. According to this definition, the standard chemical exergy of hydrogen is estimated to be 238.38 kJ/mol on the basis that the difference of the standard chemical exergy between the product and the reactant in Equation 13.12 is equal to the standard Gibbs energy change in this reaction [40].

$$H_2(g) + \tfrac{1}{2}O_2(g) \rightarrow H_2O(g) \tag{13.12}$$

Provided that hydrogen in the standard state is produced at the supply site and recovered at the demand site, the chemical exergy of hydrogen can be conserved. However, exergy loss originated from irreversible processes is inevitable for hydrogen storage and transportation systems. Because hydrogen used for operating hydrogen vehicles or fuel cells is considered to be in the standard state, a large amount of exergy provided for compression or liquefaction of hydrogen is released to the natural environment and consumed without its recovery. The exergetic analysis in this chapter is, therefore, to assess the exergy loss in each irreversible process with regard to organic chemical hydrides and compressed and liquefied hydrogen.

Exergy (E) is defined by the following equation, where the enthalpy (H) and the entropy (S) at the temperature (T) are given with subscript zero referring to the standard state (298.15 K and 0.1 MPa). Mechanical energy required for compression or liquefaction of hydrogen is equal to the magnitude corresponding to each state exergy.

$$E = (H - H_0) - T_0(S - S_0) \tag{13.13}$$

Meanwhile, thermal exergy (E_T) as the maximum work obtainable from the thermal energy (calorific value: Q) at the temperature (T) is expressed using the Carnot coefficient as follows:

$$E_T = (1 - T_0/T)Q \tag{13.14}$$

Under constant pressure, exergy change (ΔE) in case of raising temperatures from T_1 to T_2 is given as follows:

$$\Delta E = \int_{T_1}^{T_2} (1 - T_0/T)C_P \, dT$$
$$= C_P[(T_2 - T_1) - T_0 \ln(T_2/T_1)] \tag{13.15}$$

where C_P is the specific heat at constant pressure. T_1 represents the temperature of the ambient for liquid-phase reactant or of the boiling point for the product, whereas T_2 represents the temperature of the boiling point for the reactant or of the external heating temperature for the product, with various values of C_P used as the function of substances, temperatures, and the state of its liquid or gaseous phase.

Exergy change (ΔE) in phase change from state 1 to state 2 at the phase equilibrium temperature (T_S) is given as follows:

$$\Delta E = \Delta h_{12}(1 - T_0/T_S) \tag{13.16}$$

where Δh_{12} is the enthalpy change as the latent heat in this phase change.

Catalytic hydrogenation of aromatic compounds proceeds exothermically under atmospheric pressure, without addendum required for keeping the reaction conditions. In contrast to the other media such as compressed hydrogen and liquefied hydrogen, external heat should be provided toward organic chemical hydrides at the site of hydrogen utilization.

Heat flow from any external thermo-source into the dehydrogenation reactor should take the role of affording the endothermic reaction heat and the evaporation heat of both reactant and product in addition to the apparent heat for raising their temperatures from the ambient up to the external heating one. Under assumptions of the sufficient amounts of active catalyst and the adequate feed rates of organic chemical hydride, the minimum required heat is obtained as shown in the example of methylcyclohexane at 285°C on the basis of 100% conversion of methylcyclohexane to toluene and hydrogen (Table 13.5).

Organic chemical hydrides demand thermal energy at the stage of hydrogen generation in the cyclic reversible system. Notably, the temperature range required for this purpose are so moderate that new roles would be allotted to low-quality thermo-sources such as ICE vehicles. Consumed heat must be larger than an amount of 202.2 kJ/mol-MCH calculated

TABLE 13.5

External Supplied Heat and its Exergy Consumption in Methylcyclohexane Dehydrogenation under Superheated Liquid-Film Conditions

	Supplied Heat (kJ/mol-MCH)	Exergy Consumption (kJ/mol-MCH)
Apparent heat in liquid phase	16.82	2.01
Reaction heat[a]	202.20	41.12
Latent heat of evaporation	38.00	8.48
Apparent heat in gas phase	43.76	15.48
Total	300.78	67.09[b]

[a] C_7H_{14} (*l*) → C_7H_8 (*l*) + 3 H_2 (g) (external heating temperature: 285°C, conversion: 100%).
[b] Total amount of dissipated exergy consumption: 67.09 − 41.12 = 25.97 (kJ/mol-MCH) = 4.33 (MJ/kg-H_2).

for the methylcyclohexane case (Table 13.5), however, since irreversible processes of heat dissipation (e.g., bubble formation in the liquid film or heat leakage from the reactor) are inevitable. Here, it is to be noted that the endothermic dehydrogenation heat is recuperated as the chemical energy, being different in nature from the thermal energy used for performing this reaction, as explained in the previous section with reference to the concept of "thermo-recuperation" [39]. The thermal exergy recuperated as the chemical energy at the boiling point of methylcyclohexane (101.1°C) was calculated as 41.12 kJ/mol-MCH in terms of exergy; $\Delta E_R = \Delta E_R (1 - T_0/T_R)$, where ΔE_R is the enthalpy change as the reaction heat of dehydrogenation and T_R the boiling point (Table 13.5).

From the viewpoint of exergy loss, organic chemical hydrides are now compared to the other media for hydrogen storage and transportation as shown in Table 13.6. Hydrogen compression up to 35 and 70 MPa requires mechanical energy, amounting to 17.5 and 22.1 MJ/kg-H_2, respectively [41]. Hydrogen liquefaction demands 42.0 MJ/kg-H_2 at least [42], where the largest scale of production is adopted. On the contrary, hydrogen storage and regeneration by organic chemical hydrides require much less extent of exergy than the others as far as the gaseous hydrogen in the standard state at both sites of its supply and demand is adopted as the common basis for comparison. Exothermic catalysis affords heat necessary to keep the reaction temperatures for hydrogenation of aromatic compounds, whereas endothermic catalysis for the dehydrogenation of organic chemical hydrides proceeds stationarily at such temperatures as afforded with low-quality waste heat. Although exothermic reaction heat is dissipated at the hydrogen supply site, the energy material, that is, hydrogen, is regenerated as the result of chemical recuperation at the hydrogen demand site. The heat for recovering hydrogen from organic chemical hydrides is recuperated but a certain extent of thermal energy is dissipated here for heating up to the reaction temperature and evaporation under superheated liquid-film conditions for liquid-phase dehydrogenation. In the case of methylcyclohexane dehydrogenation, thermal processes of temperature raise and evaporation consume thermal exergy, consisting of the apparent heat in the liquid (2.01 kJ/mol-MCH) and gaseous (15.49 kJ/mol-MCH) phases and the latent heat of evaporation (8.48 kJ/mol-MCH). Therefore, the overall exergy loss in the dehydrogenation process amounts to 25.97 kJ/mol-MCH (=4.33 MJ/kg-H_2) in total (Tables 13.5 and 13.6).

It is to be noted that the difference in dissipation of energy between the thermal energy demanded by organic chemical hydrides and the mechanical energy by compression or liquefaction of hydrogen is quite significant because the former is supplied as waste heat, whereas the latter is lost at the site of hydrogen utilization for fuel cells or ICEs.

The onboard hydrogen supplied from organic chemical hydrides will be utilized well in the ICE vehicles. Even for stationary use of hydrogen, distribution of organic chemical hydrides will play an important role at stations or sites, where waste heat at modest temperatures is dissipated in vain without chemical recuperation.

TABLE 13.6

Comparison of Exergy Loss during Storage and Transportation of Hydrogen

	Pressure (MPa)	Temperature (K)	Exergy Loss (MJ/kg-H_2)
Compressed hydrogen	70	298.15	22.1 [41]
Liquefied hydrogen	0.1	20.4	42.0 [42]
Organic chemical hydride[a]	0.1	298.15	4.33

[a] C_7H_{14} (l) $\rightarrow$ C_7H_8 (l) + $3H_2$ (g) (external heating temperature: 285°C, conversion: 100%).

Required conditions for organic chemical hydrides to supply hydrogen will be summarized as follows:

1. Liquid reactants should be fed into the dehydrogenation reactor so as to keep the catalysts immersing slightly and not dipping deeply in the superheated liquid-film states. Otherwise, temperature gradients throughout the catalyst layer are not kept under boiling conditions. Moreover, direct gas–solid contact must be avoided even intermittently because adsorbates on the dry catalyst have no force for desorption other than equilibrium driving due to circumstance temperature and partial pressure. It is to be noted that aromatic compounds on the catalyst surface are apt to change into polymerized molecules, resulting in catalyst deactivation. Enforced desorption of the aromatic product should be performed as soon as possible, for which temperature gradients at the active sites play an important role to elongate the life of catalysts.
2. Functional catalysts should be prepared and loaded skillfully so as to dissociate the C–H bonds of cyclic alkanes efficiently for attaining extensive generation rates of hydrogen or vehicle power and to convert substrates into products at high conversions for ensuring long distances to vehicle cruise. Properly prepared catalysts for highly active or product-tolerant usage are recommended to place in the dehydrogenation reactor at the initial and rear positions, respectively. To remove unwillingly deposited carbon from the active sites, special care must be taken during the procedure of catalyst preparation since a certain metallic species is well oriented to the catalysis function of hydrogenolysis.
3. Low-quality waste heat can be recovered from various sources. One of the answers to such a question as how to deal with exhaust heat at temperatures lower than 300°C is certainly catalytic recuperation through hydrogen regeneration from organic chemical hydrides. Different reactivity among them will enlarge the choice in accordance with heating temperatures.

Some comments should be added here, however, on waste heat at higher temperatures than 300°C and on their unsteady modes with respect to quantity and quality. Pressurized hydrogen is taken out as the gaseous product from the dehydrogenation reactor under boiling and refluxing conditions even at high external heating temperatures since the boiling point of the substrate solution is controllable through the regulated outlet valve. Unsteady periods and amounts of waste heat will find their good targets in energy conversion from thermal to chemical. Recuperated hydrogen will accommodate to versatile usage, taking the proper role as secondary energy.

13.5 Thermodynamic Backgrounds of the Superheated Liquid-Film Concept

Figure 13.27 illustrates a schematic view of a carbon-supported metallic catalyst in the superheated liquid-film state.

The dehydrogenation catalyst is covered with a thin liquid film of the substrate solution so as to prevent the catalyst inside the liquid phase and the substrate vapor in the gas phase from direct gas–solid contact. Under boiling and refluxing conditions, vigorously formed

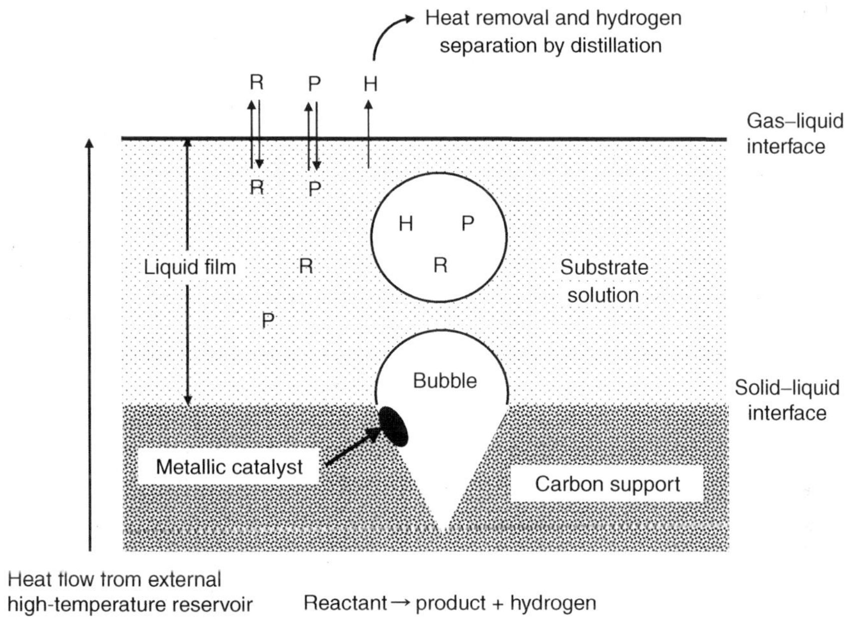

FIGURE 13.27
Schematic view of the carbon-supported dehydrogenation catalyst under superheated liquid-film conditions.

bubbles are able to convey the gaseous mixture of hydrogen and condensable components of reactant and product efficiently from the catalyst surface to the gas–liquid interface, with hydrogen separated from the mixture by fractional distillation.

This kind of mass flow is accompanied with heat flow driven from the external thermoreservoir at a higher temperature than the boiling point of the substrate solution, resulting in a temperature gradient throughout the catalyst layer, to the vapor phase under boiling and refluxing conditions.

Our "superheated liquid-film concept" stands on the thermodynamic basis of (1) equilibrium shifts due to reactive separation under boiling and refluxing conditions and (2) irreversible processes of heat flows through the catalyst layer as well as bubble formation from the catalyst surface.

Chemical equilibrium in the formation reaction of hydrogen from cycloalkane is shifted toward the dehydrogenation side under boiling and refluxing conditions because hydrogen as one of the reaction products is separated through the refluxing condenser out from the catalysis system, which consists of the catalyst, the substrate solution, and the condensates, that is, the condensable aromatic compound and the residual cycloalkane. Moreover, adsorption equilibrium of the aromatic product between the catalyst surface and the substrate solution is shifted toward the desorption side under boiling conditions as far as bubbles are formed vigorously.

As stated earlier, the retardation term owing to the product adsorption or the constant K in the Langmuir-type equation, $v = k/(1 + K[P])$ (Equation 13.4), where P is the condensable product), is related to its adsorption equilibrium driven by the concentration term in solutions. Under boiling conditions, the reaction products are desorbed not only to the solution bulk but also into the bubble formed at the catalyst surface. Because the molecular translational entropy is far larger in the gas phase than in the liquid phase, the product molecules are more stable in the bubble than in the solution. Different extents of bubble formation are, therefore, sensitive to the retardation term, contributing to the predominance of the superheated liquid-film state to the suspended state in the liquid–solid heterogeneous catalysis for cycloalkane dehydrogenation.

Another theoretical basis of the superheated liquid-film concept lies on the irreversible thermodynamics developed by Prigogine [43]. According to this theory, irreversible chemical processes would be described (Equation 13.17) by extending the equation of De Donder, provided that simultaneous reactions were coupled in a certain thermodynamic model, as follows:

$$A_1V_1 + A_2V_2 > 0 \tag{13.17}$$

where A is the chemical affinity and V the reaction rate. In a certain coupling mode between the chemical process 1 and the other process 2, one of the terms is allowed to become negative as far as the magnitude of the other is extensively negative or the equation $A_1V_1 \gg 0$ allows a new relationship between A and V of $A_2V_2 < 0$.

Even if one of the processes is not chemical but is categorized as a phase change, for example, evaporation, the extended De Donder's equation (Equation 13.17), is known to be valid. Any large magnitude of entropy production rate (*diS/dt*) due to evaporation might give a correlation such as

$$A_1V_1 = T\left(\frac{diS}{dt}\right) \gg 0 \tag{13.18}$$

resulting in the negative product for the chemical process 2. Notwithstanding the chemical affinity of a reaction to be negative or thermodynamically unfavorable ($A_2 < 0$), it is, therefore, possible for this reaction to proceed ($V_2 > 0$) in a certain coupling mode between these irreversible processes.

A famous example of the same category in irreversible coupling phenomena is "active transportation" [44], in which K^+ ions are transported through a membrane from a diluted side to the other concentrated side against entropy increase, with the expense of another entropy increase induced by H^+ transfer through the same membrane in a countercurrent.

Under boiling and refluxing conditions, heat flows from the high-temperature external heating reservoir through the catalyst layer to the boiling solution. The catalyst in the superheated liquid-film state is set inside the temperature gradient zone. The temperature gradient was induced from the difference of temperatures between the heat reservoir and the solution at the boiling point. In case of suspension, the catalyst temperature is the same as the boiling point. In the superheated liquid-film state, however, the catalyst layer is set inside the temperature gradient zone, where the layer temperature is in-between with respect to the heat source and the boiling solution.

Under thermally nonequilibrium conditions, where heat flows stationarily through the catalyst in the temperature gradient zone, the Gibbs energy change (ΔG^*) is given as

$$\Delta G^* = \Delta H - \Delta(TS) \tag{13.19}$$

in contrast to the familiar equation in thermal equilibrium

$$\Delta G = \Delta H - T\Delta S \tag{13.20}$$

By taking the difference of these equations,

$$\Delta G - \Delta G^{*} = S\Delta T \tag{13.21}$$

where ΔT is smaller than the temperature difference between T_{ext} and T_{boil}, the following inequality equation is derived:

$$T_{ext} - T_{boil} > \Delta T > 0 \tag{13.22}$$

Therefore,

$$\Delta G - \Delta G^{*} > 0 \text{ or } \Delta G^{*} < \Delta G \tag{13.23}$$

As far as the Prigogine-type coupling between heat- and mass flow is operative from the catalytic active sites to the vigorously generating bubbles, the adsorbed products will be readily taken out and a large amount of vacant sites will be generated stationarily. Therefore, the restriction of equilibrium conversion ($\Delta G < 0$) even becomes removable as a consequence of $\Delta G^{*} < \Delta G$ by adopting the superheated liquid-film catalysis.

The characteristics of equilibrium shifts and the irreversible nature in the superheated liquid-film concept are summarized in Figure 13.28.

At the beginning stage of dehydrogenation, the substrate organic hydride is adsorbed onto the catalyst surface from the liquid phase directly and easily. Catalytic reaction processes will succeed it, until the surface sites are filled with the adsorbed reactant and products. Once product desorption starts to form and grow a bubble, product readsorption becomes unfavorable due to the increment of translational entropy of the product molecule in the bubble, if compared with that in the solution, shifting the adsorption equilibrium for the product and suppressing its effect of rate retardation.

Steep temperature gradients inside the catalyst layer will enhance the bubble formation and bring about efficient product desorption and effective regeneration of vacant active sites consequently. There irreversible processes are followed by another irreversible act of bubble detachment from the surface.

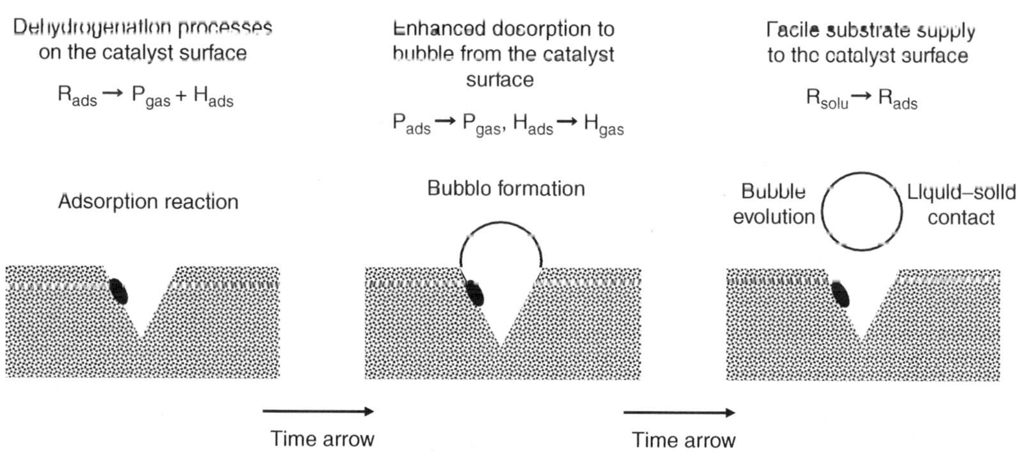

FIGURE 13.28
Sequential irreversible processes on dehydrogenation catalyst under superheated liquid-film conditions.

The time arrow is irreversible. The dehydrogenation reaction under superheated liquid-film conditions is initiated again with the facile adsorption of substrate from the solution to the catalyst surface.

13.6 Conclusions

Any technological breakthrough on storage and supply of hydrogen is urgently required to popularize hydrogen energy systems. In the present study, organic chemical hydrides with appropriate characteristics have been proposed as the storage material of hydrogen. With respect to the hydrogen supply, efficient hydrogen generation from organic chemical hydrides under moderate heating conditions of 210°C–285°C was accomplished only by adopting the superheated liquid-film-type catalysis developed by our research group, so that excessive exergy loss and coke formation over the catalyst surface could be prevented simultaneously.

The use of organic chemical hydrides on the basis of superheated liquid-film concept would, thus, make it possible to combine electrolysis hydrogen produced from renewable energy and by-product hydrogen recovered from various industrial processes, with the hydrogen demand practically for stationary fuel cells and hydrogen vehicles.

References

1. Liu, C., M. Sakaguchi, and Y. Saito, Long-term storage and long-distance transportation of hydrogen by use of catalyst-assisted decalin dehydrogenation/naphthalene hydrogenation pair. *J. Hydrogen Energy Syst. Soc. Jpn.*, 22(1), 27–32 (1997).
2. Hodoshima, S., and Y. Saito, Characteristics of decalin dehydrogeno-aromatization catalyst driven by waste heats for long-term storage and long-distance transportation of hydrogen. *J. Hydrogen Energy Syst. Soc. Jpn.*, 24(1), 13–18 (1999).
3. Hodoshima, S., and Y. Saito, Long-term storage and long-distance transportation of hydrogen by use of liquid organic compounds. *J. Hydrogen Energy Syst. Soc. Jpn.*, 25(2), 36–43 (2000).
4. Takaiwa, S., S. Hodoshima, H. Arai, and Y. Saito, Liquid-phase dehydrogeno-aromatization catalysis of decalin for pure hydrogen/air fuel cell. *J. Hydrogen Energy Syst. Soc. Jpn.*, 26(2), 44–49 (2001).
5. Hodoshima, S., H. Arai, and Y. Saito, Liquid-film-type catalytic decalin dehydrogeno-aromatization for long-term storage and long-distance transportation of hydrogen. *Int. J. Hydrogen Energ.*, 28(2), 197–204 (2003).
6. Hodoshima, S., H. Arai, S. Takaiwa, and Y. Saito, Catalytic decalin dehydrogenation/naphthalene hydrogenation pair as a hydrogen source for fuel-cell vehicle. *Int. J. Hydrogen Energy.*, 28(11), 1255–1262 (2003).
7. Takeuchi, H., S. Takaiwa, S. Hodoshima, and Y. Saito, Hydrogen formation with use of a flow-type pulse reactor for catalytic decalin dehydrogenation. *J. Hydrogen Energy Syst. Soc. Jpn.*, 28(1), 61–66 (2003).
8. Hodoshima, S., and Y. Saito, Characteristics of decalin dehydrogenation catalysis in the superheated liquid-film state for mobile storage of hydrogen. *J. Chem. Eng. Jpn.*, 37(3), 391–398 (2004).
9. Shinohara, C., S. Kawakami, T. Moriga, H. Hayashi, S. Hodoshima, Y. Saito, and S. Sugiyama, Local structure around platinum in Pt/C catalysts employed for liquid-phase dehydrogenation of decalin in the liquid-film state under reactive distillation conditions. *Appl. Catal. A*, 266(2), 251–255 (2004).

10. Hodoshima, S., S. Takaiwa, A. Shono, K. Satoh, and Y. Saito, Hydrogen storage by decalin/naphthalene pair and hydrogen supply to fuel cells by use of superheated liquid-film-type catalysis. *Appl. Catal. A*, 283(1–2), 235–242 (2005).
11. Hodoshima, S., H. Nakamura, A. Shono, K. Satoh, and Y. Saito, Hydrogen supply from organic chemical hydride with carbon-supported metallic catalyst under superheated liquid-film conditions. *J. Hydrogen Energy Syst. Soc. Jpn.*, 30(1), 36–41 (2005).
12. Hodoshima, S., H. Nagata, and Y. Saito, Efficient hydrogen supply from tetralin with superheated liquid-film-type catalysis for operating fuel cells. *Appl. Catal. A*, 292, 90–96 (2005).
13. Hodoshima, S., A. Shono, K. Satoh, and Y. Saito, Hydrogen storage by organic chemical hydrides and hydrogen supply to fuel cells with superheated liquid-film-type catalysis. *Chem. Eng. Trans.*, 8, 183–188 (2005).
14. Shono, A., T. Hashimoto, S. Hodoshima, K. Satoh, and Y. Saito, Continuous catalytic dehydrogenation of decalin under mild conditions. *J. Chem. Eng. Jpn.*, 39(2), 211–215 (2006).
15. Hodoshima, S., A. Shono, K. Satoh, and Y. Saito, Catalytic hydrogen supply from a decalin-based chemical hydride under superheated liquid-film conditions. *Stud. Surf. Sci. Catal.*, 159, 177–180 (2006).
16. http://www.energy.gov/
17. Weitkamp, A.W., Stereochemistry and mechanism of hydrogenation of naphthalene on transition metal catalysts and conformational analysis of the products. *Adv. Catal.*, 18, 1–110 (1968).
18. Schmitz, A.D., G. Bowers, and C. Song, Shape-selective hydrogenation of naphthalene over zeolite-supported Pt and Pd catalysts. *Catal. Today*, 31, 45–56 (1996).
19. Lin, S.D., and C. Song, Noble metal catalysts for low-temperature naphthalene hydrogenation in the presence of benzothiophene. *Catal. Today*, 31, 93–104 (1996).
20. Stull, O.R., E.F. Westrum Jr., and G.C. Sinke, *The Chemical Thermodynamics of Organic Compounds*, Wiley, New York, 1969, pp. 15–20.
21. Wiseman, P., *Petrochemicals*, Chichester, England, 1986, pp. 90–91.
22. Gretz, J., J.P. Baselt, O. Ullmann, and H. Wendt, The 100 MW Euro-Quebec hydro-hydrogen pilot project. *Int. J. Hydrogen Energ.*, 15, 419–424 (1990).
23. Gretz, J., B. Drolet, D. Kluyskens, F. Sandmann, and O. Ullmann, Status of the hydro-hydrogen pilot project (EQHHPP). *Int. J. Hydrogen Energ.*, 19, 169–174 (1994).
24. Newson, E., T.B. Truong, P. Hottinger, F.V. Roth, and T.H.H. Schucan, Optimization of seasonal energy storage in stationary systems with liquid hydrogen carriers, decalin and methylcyclohexane. *Proceedings of 12th World Hydrogen Energy Conference, Buenos Aires*, Argentina, 1998, pp. 935–942.
25. Newson, E., T.H. Haueter, P. Hottinger, F. Vonrof, G.W.H. Scherer, and T.H.H. Schucan, Seasonal storage of hydrogen in stationary systems with liquid organic hydrides. *Int. J. Hydrogen Energ.*, 23, 905–909 (1998).
26. Gates, B.C., J.R. Katzes, and G.C.A. Schmit, *Chemistry of Catalytic Processes*, McGraw-Hill, 1979, pp. 184–324.
27. Fujii, T., and Y. Saito, Thermocatalytic formation of molecular hydrogen and cyclo-octene from cyclo-octane by rhodium complexes. *J. Chem. Soc., Chem. Commun.*, 757–758 (1990).
28. Koga, N., and K. Morokuma, Ab initio molecular orbital studies of catalytic elementary reactions and catalytic cycles of transition-metal complexes. *Chem. Rev.*, 91, 823–842 (1991).
29. Yukawa, K., X. Li, T. Fujii, T. Kawamura, and Y. Saito, Dehydrogeno-aromatization of cyclohexanes with suspended noble-metal catalysts. *J. Chem. Soc. Faraday Trans.*, 89, 3641–3644 (1993).
30. Saito, Y., and N. Yamamoto, Hydrogen production from cyclohexane and hydrogen production. *J. Hydrogen Energy Syst. Soc. Jpn.*, 19(2), 17–24 (1994).
31. Meng, N., S. Shinoda, and Y. Saito, Improvements on thermal efficiency of chemical heat pump involving the reaction couple of 2-propanol dehydrogenation and acetone hydrogenation. *Int. J. Hydrogen Energ.*, 22, 361–367 (1997).
32. Otowa, T., Development and application of high surface area activated carbon. *Hyoumen (Surface)*, 34(2), 62–65 (1996).
33. Hodoshima, S., T. Kubono, S. Asano, H. Arai, and Y. Saito, Preparation of nanosize bimetallic particles on activated carbon. *Stud. Surf. Sci. Catal.*, 132, 323–326 (2001).

34. Nakabayashi, M., M. Yamashita, and Y. Saito, Preparation of size-controlled ruthenium metal particles on carbon from hydro-carbonyl cluster complex. *Chem. Lett.*, 1275–1278 (1994).
35. Agreda, V.H., L.R. Partin, W.H. Heise, High-purity methyl acetate via reactive distillation. *Chem. Eng. Progr.*, 86, 40–46 (1990).
36. Heidrich, A., and M. Jakob, *Heat Transfer*, Wiley, New York, 1957, Vol. 1, p. 618.
37. Mears, D.E., M. Boudart, The dehydrogenation of isopropanol on catalysts prepared by sodium borohydride reduction. *A.I.Ch.E. Journal*, 12, 313–321 (1966).
38. Appleby, A.J., The electrochemical engine for vehicles. *Sci. Am.*, 281(1), 74–79 (1999).
39. Saito, Y., S. Hodoshima, A. Shono, and K. Otake, Chemical recuperation of low-quality heats with use of dehydrogenation catalysts for organic hydrides. *Proceedings of 9th Asian Hydrogen Energy Conference*, Tokyo, 2007.
40. Kojima, K., *Principle of Efficient Energy Utilization*, Chapter 8, Baifukan, Tokyo, 2004.
41. Bossel, U., The physics of the hydrogen economy. *European Fuel Cell News*, 10(2), 1–16 (2003), available at http://www.efcf.com/reports/
42. Saito, T., E. Tobihara, and A. Hanzu, *Energy Conversion*, University of Tokyo Press, Tokyo, 1995, p. 176.
43. Glansdorff, G., and I. Prigogine, *Thermodynamic Theory of Structure, Stability and Fluctuations*, Chapter 3, Wiley, London, 1971.
44. Abrahams, J.R., A.-G.W. Leslie, R. Lutter, and J.-E. Walker, Structure at 2.8 Å resolution of F1-ATPase from bovine heart mitochondria. *Nature*, 370, 621 (1994).

Section III

Safety and Environmental Aspects of Hydrogen

14

Hydrogen Codes and Standards

James M. Ohi

CONTENTS

14.1 Introduction

Large quantities of hydrogen have been used safely as a chemical feedstock and industrial gas for many years, and standards, codes, and regulations governing its storage, distribution, and use at industrial sites are well established. Hydrogen is used as a feedstock in the petroleum refining, chemical, petrochemical, and synthetic fuels industries. Examples include making ammonia for fertilizer and removing sulfur in petroleum refining for products such as "reformulated" gasoline. Hydrogen is also used in the food processing, semiconductor, glass, and steel industries as well as by electricity utilities to cool the rotor and stator coils in large turbine generators. The use of hydrogen as an energy carrier for consumer markets is expected to grow over the next decade, and the development and promulgation of codes and standards for this use are essential to establish a market-receptive environment for commercial hydrogen products and systems.

Standards provide a base for a uniform system of accepted performance such as those found in engineering practice standards, material standards, and test standards. Hydrogen standards are typically written under a consensus process by technical committees (TC) representing a cross section of interested parties and issued in the United States, for example, by organizations such as the American Society of Mechanical Engineers (ASME) for pressure vessels, pipelines, and piping; the Compressed Gas Association (CGA) for pressure vessel operation and maintenance; and the Underwriters Laboratories (UL) for product certification.

In the United States, the American National Standards Institute (ANSI) facilitates the development of national standards by accrediting the procedures of standards developing organizations (SDOs) such as those mentioned earlier [1]. ANSI administers and coordinates the voluntary standardization system of the private sector in the United States. ANSI does not itself develop standards; rather, it facilitates the development of standards by establishing consensus among qualified groups. These groups work cooperatively to develop voluntary national consensus standards. Accreditation by ANSI signifies that the procedures used by the SDOs meet ANSI's essential requirements for openness, balance, consensus, and due process. Standards as consensus documents can be issued jointly with ANSI as long as the ANSI procedures are followed.

Standards address issues such as equipment design, plant layout, and operating procedures. In themselves, standards are not mandatory. However, standards are widely included in regulations and in such cases are mandatory. Model codes typically address the safe construction, operation, and maintenance of buildings and facilities. Model codes often incorporate standards by reference and are, in turn, often adopted as part of regulations. For example, a state or local jurisdiction can adopt and enforce a model building code, with or without modifications, to regulate building construction and occupancy within its jurisdiction.

For hydrogen energy use in the United States, the International Code Council (ICC) and the National Fire Protection Association (NFPA) are the two principal model code development organizations. The ICC was formed in 1994 through the merger of three regional building code organizations: the Building Officials and Code Administrators International (BOCA), International Conference of Building Officials (ICBO), and Southern Building Code Congress International (SBCCI) [2]. The ICC develops and publishes a family of model codes; the most relevant for hydrogen energy are the International Fire Code (IFC), International Fuel Gas Code, International Building Code (IBC), and International Mechanical Code (IMC). Many of these codes have been adopted by state and local jurisdictions. The NFPA develops and publishes both standards and codes [3]. For hydrogen energy, the most widely used of these are NFPA 55 (Standard for the Storage, Use, and Handling of Compressed Gases and Cryogenic Fluids in Portable and Stationary Containers, Cylinders, and Tanks), NFPA 52 (Vehicular Fuel Systems Code), and NFPA 30A (Code for Motor Fuel Dispensing Facilities and Repair Garages). The NFPA is in the process of incorporating all of its provisions for hydrogen into a single document to be called NFPA 2, Hydrogen Technologies Code. NFPA 2, scheduled for publication in 2010, has the potential to become a national hydrogen code analogous to the National Electric Code (NFPA 70).

The adoption and enforcement of codes and standards in the United States take place under the jurisdiction of some 44,000 entities that include city, county, and state governments as well as special districts such as port and tunnel authorities. Regulations make use of the existing standards, either by incorporating appropriate sections of the standards (incorporation by transcription) or by referring to those sections (incorporation by reference). Many of these entities adopt model codes by reference, but states and municipalities can impose more stringent or additional requirements. In either case, the incorporated content of the standard is enforceable by law. As with almost all projects, including those that use hydrogen, it is necessary to obtain the approval of an authority having jurisdiction (AHJ) at multiple levels of government. The extremely decentralized enforcement of codes and standards means that the permitting process for hydrogen fueling stations (HFS) can be very time-consuming with different jurisdictions having different requirements or with developers having to undergo a lengthy approval process each time a facility is proposed.

The federal government plays a limited role in the development, adoption, and enforcement of codes and standards, but federal safety regulations are incorporated in the Code of Federal Regulations (CFR). Those that apply to hydrogen are embodied primarily in 49 CFR (1995) and 29 CFR (1996), under the jurisdictions of the Department of Transportation (DOT) and Occupational Safety and Health Administration (OSHA), respectively. The DOT regulates the transportation of hydrogen and classifies and treats gaseous and liquefied hydrogen as a compressed, flammable gas. The OSHA regulates the safe handling of hydrogen in the workplace. The DOT chairs an interagency committee to coordinate the regulatory environment for hydrogen applications and maintains a Web site on the current U.S. statutes and regulations that may be applicable to hydrogen [4]. An agency within the DOT, the National Highway Transportation Safety Administration (NHTSA) regulates the safety of automobiles, including hydrogen vehicles, under the Federal Motor Vehicle Safety Standards (FMVSS), whereas the Office of Pipeline Safety (OPS) within the DOT's Research and Innovative Technology Administration (RITA) regulates pipelines, including those that transport hydrogen. The OSHA regulations are intended to protect worker safety that for hydrogen includes separation distances, or "setbacks," between hydrogen components and building features, such as air intakes [5].

Although most industrialized countries have adopted regulations, codes, and standards (RCS) that govern the use of hydrogen, many of these countries also support the development of international standards to facilitate international trade and commerce. For hydrogen energy, the key international SDOs are the International Organization for Standardization (ISO) and the International Electrotechnical Commission (IEC). Each of these organizations has specific TC and subcommittees (SC) devoted to developing consensus international standards for hydrogen applications. For ISO, the key TC are TC197 (Hydrogen Technologies), TC22/SC21 (Electric Vehicles/Hydrogen Vehicles), TC58 (Gas Cylinders), and TC220 (Cryogenic Vessels). The key activities under ISO TC197 include standards and technical specifications for hydrogen fuel quality, gaseous hydrogen fueling stations, hydrogen generators, and hydrogen detectors. In participating countries, TC197 activities are coordinated by a Technical Advisory Group (TAG) of subject matter experts. In the United States, the TC197 TAG is managed by the CGA, which maintains a Web site to facilitate coordination of U.S. experts and stakeholders [6]. The key TC for IEC is TC105, Stationary Fuel Cells, and the U.S. TAG for TC105 is managed by CSA America (CSA) [7]. Information about domestic and international hydrogen RCS, current activities of relevant ISO and IEC TC, including draft standards under preparation or review, can be found at www.fuelcellstandards.com, a Web site supported by the U.S. Department of Energy (DOE). Another useful source of information on hydrogen safety, codes, and standards is the Hydrogen Safety Report, a monthly newsletter published by the National Hydrogen Association (NHA) at www.hydrogensafety.info also supported by the DOE.

This chapter provides an overview of hydrogen code and standards with an emphasis on the national effort supported and managed by the DOE.

14.2 The DOE Program for Hydrogen Codes and Standards

For the past decade, the Office of Hydrogen, Fuel Cells, and Infrastructure Technologies in the DOE has sponsored a collaborative national effort by government and industry to prepare, review, and promulgate codes and standards needed to expedite hydrogen infrastructure development and to help enable the emergence of hydrogen as a significant energy carrier. In addition, the DOE has worked to harmonize national and international standards, codes, and regulations that are essential for the safe use of hydrogen by consumers

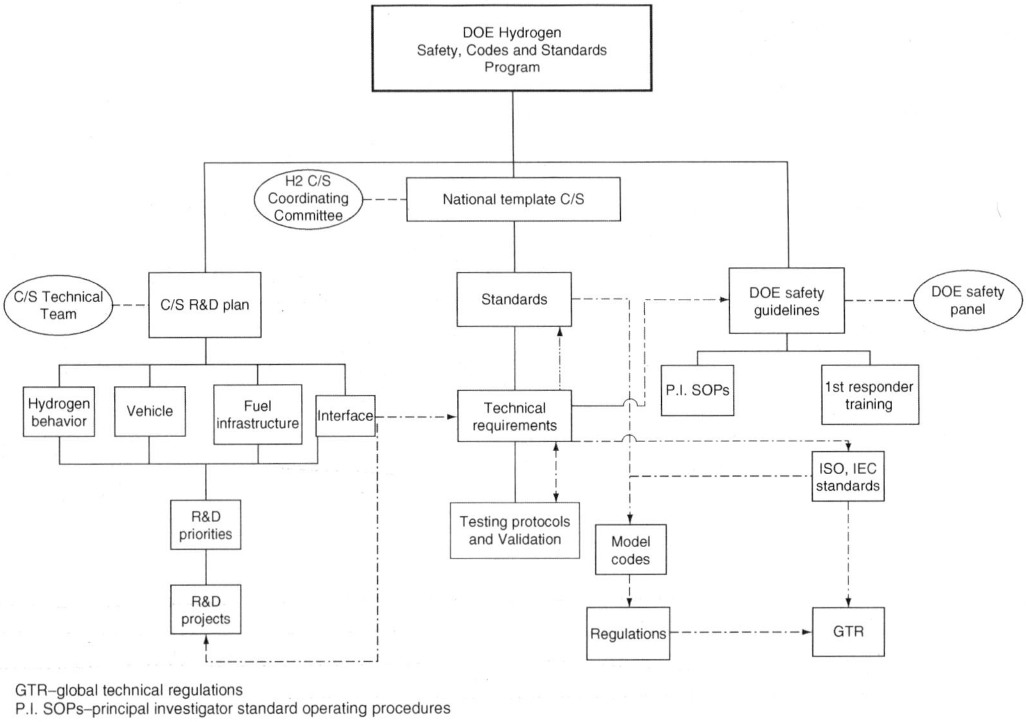

FIGURE 14.1
DOE Hydrogen Safety, Codes and Standards Program.

in the United States and throughout the world. The National Renewable Energy Laboratory (NREL) provides technical and programmatic support to the DOE for this effort.

The DOE has also launched a comprehensive research, development, and demonstration (RD&D) effort to obtain the data needed to establish a scientific basis for requirements incorporated in hydrogen codes and standards. This RD&D is planned, conducted, and evaluated in collaboration with an industry through the U.S. FreedomCAR and Fuel Partnership formed to examine and advance precompetitive research and development (R&D) of technologies to enable high-volume production of affordable hydrogen fuel cell vehicles and the national hydrogen infrastructure necessary to support them. The codes and standards activities of the Partnership are conducted through the Codes and Standards Technical Team (CSTT) that adopted a roadmap to guide the RD&D.

The structure of this effort is shown in Figure 14.1. This chapter focuses on domestic codes and standards shown in the center of Figure 14.1 and on the RD&D effort shown on the left-hand side of Figure 14.1.

14.2.1 National Templates

Over the past several years, a coordinated national agenda for hydrogen and fuel cell codes and standards has emerged through the DOE leadership and the support and collaboration of industry and key standards and model code development organizations (SDOs and CDOs). For example, hydrogen is recognized as a fuel gas, and hydrogen applications are incorporated in the 2003 and 2006 editions of the ICC model codes. Provisions for the safe use of hydrogen are included in ICC's International Building, Residential, Fire, Mechanical,

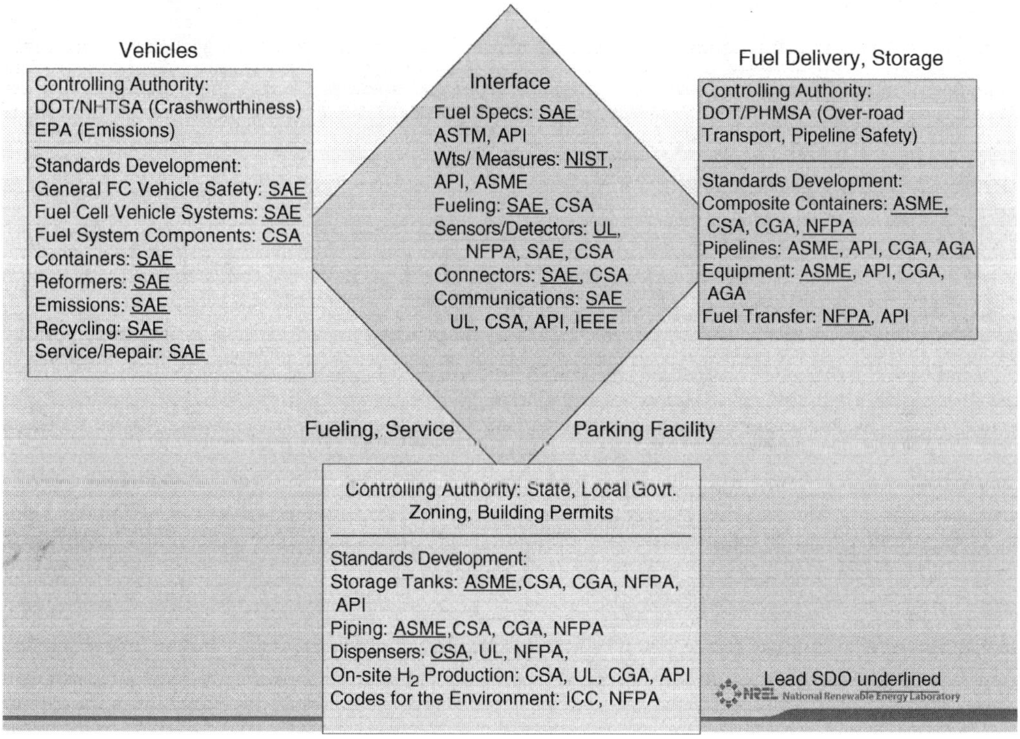

FIGURE 14.2
National template for hydrogen vehicle systems and refueling facilities.

and Fuel Gas Codes. Also, NFPA has incorporated hydrogen safety requirements into its family of codes and standards, as noted earlier. The consolidation of all hydrogen safety requirements into a single document (NFPA 2) will be a major step toward the development of a national hydrogen code.

A key to the success of the national hydrogen and fuel cell codes and standards development efforts to date has been the creation and implementation of national templates through which the DOE, NREL, and major SDOs and CDOs coordinate the preparation of critical standards and codes for hydrogen and fuel cell technologies and applications. The national templates for vehicle systems and fueling and refueling facilities and for stationary and portable systems are shown in Figures 14.2 and 14.3, respectively. The national templates have helped the DOE to create and maintain a coordinated national agenda for hydrogen and fuel cell codes and standards. The DOE leadership has coincided with the emergence of heightened national and international interest in hydrogen energy in general and in codes and standards in particular.

The national templates have been accepted by the major SDOs and CDOs in the United States, the FreedomCAR and Fuel Partnership, key industry associations, and many state and local governments as the guideposts for the coordinated development of standards and model codes. All of the relevant major SDOs and CDOs in the United States are part of this national effort: the ANSI, ASME, American Society of Testing and Materials (ASTM),

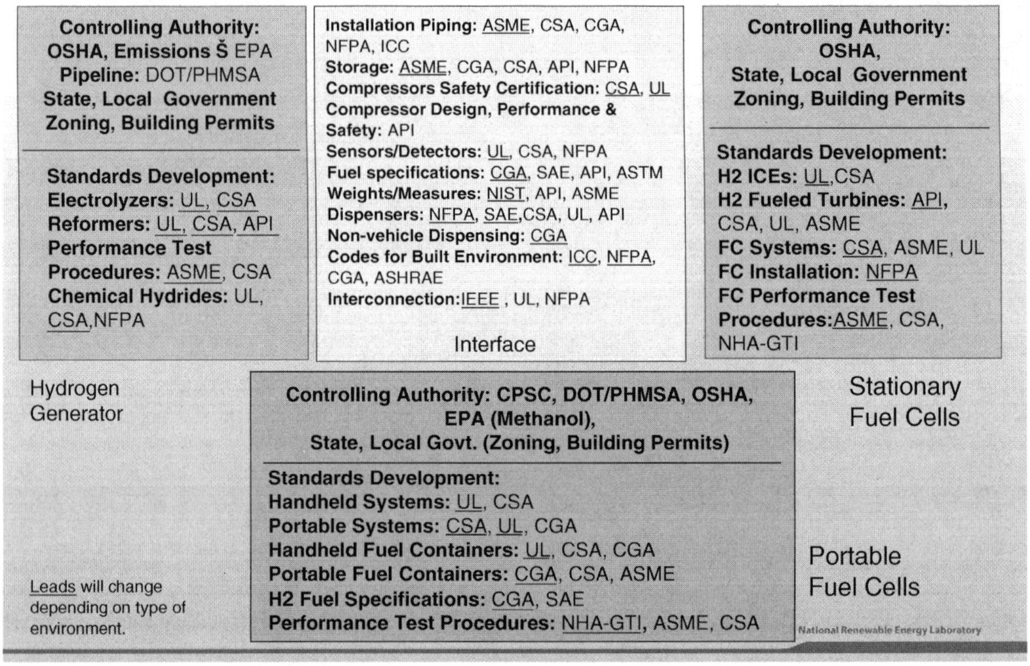

FIGURE 14.3
National template for stationary and portable hydrogen energy systems.

CGA, CSA, ICC, NFPA, Society of Automotive Engineers (SAE), and UL. Industry participants include the FreedomCAR Fuel Partnership (Chrysler, Ford Motor Company, General Motors, BP, Chevron, ConocoPhillips, ExxonMobil, and Shell Hydrogen); other industry members such as Ballard Power Systems, General Electric, Plug Power, Hydrogenics, UTC Power; and industry associations such as the American Petroleum Institute (API), NHA, and the U.S. Fuel Cell Council (USFCC). Other federal agencies involved include the DOT and the National Institute of Standards and Technology (NIST). Other organizations participate on an as-need basis.

The objectives of the national templates are to

- Establish by a consensus of the national codes and standards development organizations such as the CDO or SDO that will have the lead in the development of codes and standards for establishing safety requirements for specific components, subsystems, and systems (as shown in the templates) and the organizations that will work collaboratively with (or in support of) the lead organization.
- Minimize duplication of efforts in the codes and standards development.
- Establish "boundaries" and interfaces among standards for components, subsystems, and systems and identify harmonization requirements *across* such standards.
- Identify codes and standards development needs and gaps and identify the organizations that should have responsibility for addressing the gaps.

Implementation of the national templates is coordinated through the National Hydrogen and Fuel Cells Codes and Standards Coordinating Committee, created by the DOE,

NREL, NHA, and USFCC. The committee conducts monthly conference calls to update participants on the current activities and to discuss key issues. In addition, the committee meets quarterly to coordinate codes and standards development and prevent duplication of effort, identify critical deficiencies and gaps in hydrogen codes and standards development that could have an adverse impact on market acceptance, determine a collaborative strategy and action plan to address critical gaps and deficiencies, and identify specific opportunities for organizations to work together in developing codes and standards. The minutes of conference calls and proceedings of meetings are posted at www.hydrogenandfuelcellsafety.info.

The DOE supports the implementation of templates through subcontracts with a number of SDOs and CDOs designated for lead roles on the templates. It should also be noted that significant work to implement the templates is being carried out by organizations not funded by the DOE. While the templates were not intended to specify which organizations should receive the DOE funding, they have helped to solidify the roles of the organizations identified as having a lead role in developing a particular standard.

To summarize, the templates continue to function as the seminal documents that help to create a more unified national approach to the development of hydrogen and fuel cell codes and standards. The templates and the National Hydrogen and Fuel Cells Codes and Standards Coordinating Committee that was formed to manage the templates have created a "virtual national forum" for SDOs, CDOs, industry, government, and other interested parties to address codes and standards issues, both immediate and long-term.

14.2.2 Selected Highlights of National Templates

The ICC conducted its initial Code Development Hearings (CDH) for the 2006/2007 code cycle in September 2006, and its Final Action Hearings were held on May 22–25, 2007, in Rochester, New York. The Final Action Hearings will result in the 2008 editions of the ICC family of model codes. Before 2006, most of the hydrogen safety requirements incorporated into the ICC family of codes were developed by the *Ad Hoc* Committee for Hydrogen Gas (AHC), which was established by the ICC and supported by the DOE. During 2004 and 2005, the AHC's efforts resulted in the development of new hydrogen safety requirements (and modifications to the existing requirements) that were incorporated into the 2006 editions of the IBC, IFC, and International Fuel Gas Code.

After the completion of its mission with the ICC 2005/2006 code development cycle, the ICC disbanded the AHC as planned. During 2006, the DOE facilitated the creation of the Hydrogen Industry Panel on Codes (HIPOC) by the ICC, NFPA, and NHA to continue the development of hydrogen and fuel cell–related safety requirements for incorporation not only into ICC codes but also into NFPA codes and standards. The objectives of HIPOC are to work collaboratively to coordinate proposed changes and additions to the ICC and NFPA documents and to harmonize technical requirements in these. The activities of HIPOC are not intended to substitute for the codes and standards development procedures and processes for either of the organizations. Rather, HIPOC is a group of industry experts with experience in getting consensus code modifications into documents of both organizations.

In late 2006, the NFPA initiated the development of NFPA 2 with the establishment and staffing of the NFPA Hydrogen Technology Technical Committee. The Committee held its first meeting in November 2006 during which it developed the overall structure of NFPA 2 and established task groups to draft major sections. The scheduled completion date for NFPA 2 is 2010. During 2006, NFPA also worked on revision cycles for NFPA 52, NFPA 30A, and NFPA 55.

The NFPA Research Foundation, in a collaborative project with the DOE and the telecommunications industry, completed a draft report on diffusion of hydrogen leaks from cabinet-enclosed hydrogen storage tanks. The purpose of this research is to establish a better scientific foundation for setback requirements for hydrogen fuel cell systems used in telecommunication applications.

14.2.3 RD&D for Codes and Standards

The RD&D Roadmap helps to guide the DOE activities that will provide data required for SDOs to develop performance-based codes and standards for a commercial hydrogen-fueled transportation sector in the United States. The Roadmap reflects the experience and priorities of the members of the FreedomCAR and Fuel Partnership, which include the DOE, energy companies (BP, Chevron, ConocoPhillips, ExxonMobil, and Shell Hydrogen), and the automotive companies (Chrysler, Ford, General Motors) belonging to the U.S. Consortium for Automotive Research (USCAR). The contents of the Roadmap are reviewed and revised by the Partnership as needed to reflect the changing needs and opportunities.

By evaluating specific needs for RD&D, assessing the status of ongoing RD&D, and revising the Roadmap as needed, the Partnership will ensure that new U.S. projects are efficiently leveraged and coordinated with those undertaken internationally. Through the International Partnership for the Hydrogen Economy (IPHE), the DOE works with individual countries as well as contributes to global RD&D efforts. Information requirements of international SDOs are considered to help align RD&D projects with a need for code and standard development.

The Roadmap includes an assessment of the existing hydrogen and fuel cell codes and standards and those that are in the process of being established domestically and internationally and identifies information needs and gaps related to those codes and standards for a hydrogen-based transportation system. The CSTT of the Partnership reviews RD&D projects to address gaps and to provide documented research to the SDOs on a continuing basis.

The Roadmap is organized into four focus areas:

1. Hydrogen behavior
2. Hydrogen-fueled vehicles
3. Hydrogen fuel infrastructure
4. Fuel–vehicle interface

The technical goal for each of these focus areas is to gather sufficient information and validating experience on technology applications so that the responsible SDO or CDO can proceed with better data on which to base requirements incorporated in its codes and standards. Each focus area is subdivided into key target areas, which identify important information needs for which information is required by the SDOs and CDOs to fully develop codes and standards (Figure 14.4). Completion of RD&D for the individual technical target areas, in conjunction with information distribution, is expected to result in the subsequent development of safe, performance-based codes and standards. The Roadmap will be implemented over the next 5 years assuming required budget and resources are available.

14.2.4 Selected Highlights of RD&D Roadmap

Sandia National Laboratory (SNL) completed a literature survey on flammability limits for mixtures of hydrogen and air. Nearly 80 investigations of hydrogen flammability

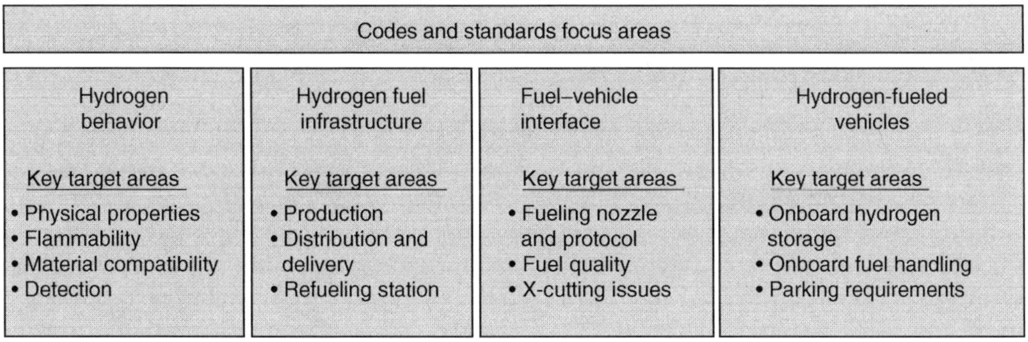

FIGURE 14.4
RD&D Roadmap focus areas.

limits between 1920 and 1960 were identified. The flammability limits measured in these studies were found to be very consistent once differences in test apparatus were accounted for. Based on this review, SNL concluded that the flammability limits of hydrogen are well established and do not require further research. A unique aspect of hydrogen is that the lean flammability limit is significantly different for up-, down-, and sideward propagating flames. Although the generally accepted value for the upward-propagating lower flammability limit of hydrogen in air is 4% mole fraction, experimental data in the literature indicate that the limit may be as high as 7.2% for horizontal propagating flames and 8.8% for downward propagating flames. It is noteworthy that the value of about 8% agrees well with the 8% value for hydrogen in air observed in experiments by the University of Miami as a requirement to achieve ignition of turbulent hydrogen jet flows.

A technical reference Web site on materials compatibility for hydrogen applications was opened to public access by SNL in January 2005 (http://www.ca.sandia.gov/matlsTechRef). Technical reference chapters can be downloaded for the following materials: high-strength steel 9Ni-4Co, type 304 and type 316 stainless steels, nitrogen-strengthened stainless steels 21-6-9 and 22-13-5, A-286 precipitation-strengthened stainless-steel low-alloy ferritic steels Fe–Cr–Mo and Fe–Cr–Mo–Ni as well as austenitic steel Fe–Ni–Co, copper, and duplex stainless steels. The chapters provide a comprehensive summary of the materials compatibility of each specific alloy in high-pressure hydrogen gas (or with internal hydrogen precharged from high-pressure hydrogen gas), including hydrogen transport, mechanical properties with emphasis on fracture properties, and the effects of materials processing on hydrogen embrittlement, such as how forging and welding affect the microstructure and materials behavior in high-pressure hydrogen gas. Data for the low-alloy ferritic steels are relevant for seamless pressure vessels used in hydrogen gas storage. Copper is important for seals and ancillary devices in engineering systems. Pure coppers are embrittled by hydrogen if oxygen is present in the alloy. Duplex stainless steel is an important high-strength corrosion-resistant alloy used in chemical processing and is of stated interest for piping and valving components for hydrogen service.

Defining hydrogen fuel quality specifications is a high priority for the Roadmap. Such specifications must be quantified at the vehicle–station interface and must consider how the presence of small amounts of contaminants affects the performance and durability of fuel cell and balance of plant; material compatibility of onboard and stationary hydrogen storage systems; and the operation and maintenance of hydrogen production, purification, and delivery systems. Most importantly, fuel quality specification must consider trade-offs between the cost of hydrogen fuel and the performance and complexity of fuel cell and hydrogen infrastructure technologies. Preliminary fuel quality guidelines based on the available data and information have been prepared by the ISO Technical Committee 197 and by the SAE [8]. The ISO guidelines apply only to proton exchange membrane (PEM) fuel cells for road vehicles. The guidelines are conservative and are intended for the precommercial demonstration phase of fuel cell vehicle and hydrogen fuel infrastructure development. Before these guidelines can become international and national standards, respectively, a comprehensive, structured R&D and testing effort is needed to determine the effects, especially degradation mechanisms, of various contaminants on fuel cell electrodes and membranes. Hydrogen production, purification, and delivery technologies must also be evaluated to determine the technical and economic drivers that affect fuel quality and cost. Standard analytical procedures, including sampling and instrumentation to measure nonhydrogen fuel constituents to the levels required (sub-ppm for selected constituents), are also critical needs. Finally, implications of hydrogen fuel quality for the complexity, performance, and durability of fuel cell systems and upstream hydrogen infrastructure and the cost of hydrogen fuel must be understood so that critical trade-offs can be assessed.

Facilitating the permitting of HFS is a high priority for both the Partnership and the Secretary of Energy's Hydrogen Technical Advisory Panel. The DOE and NREL launched an initiative to address this need by organizing and conducting a kick-off workshop in February 2007 that brought together HFS developers, state and local code officials, and other interested parties. The key output from the workshop was a set of findings and recommendations to the DOE on how it could help to facilitate the permitting of HFS:

- Develop an information repository for HFS with validated, "third-party" data and information
  - Identify applicable codes and standards (specific safety requirements) and make them more accessible to permitting officials
  - Develop a detailed process flowchart for permitting HFS
- Develop a template for code officials to navigate the permitting process
- Note best practices for application of codes and standards for HFS
  - Develop fact sheets on hydrogen technologies and HFS equipment for permitting officials
  - Develop a permitting pathway from "behind-the-fence" stations to retail stations

The NREL is developing a web-based information repository as recommended. A concept for an HFS information repository is shown in Figure 14.5, with potential outputs noted in brackets. Items included under "permitting process" and "retail hydrogen station" will be linked to specific sections of applicable codes and standards in the database shown at the bottom of Figure 14.5.

The DOE and NREL conducted a second workshop with the National Association of State Fire Marshals (NASFM) in July 2007 with key fire and building code officials to review and apply available codes and standards to five "case studies" of HFS site plans prepared by two leading energy companies. These case studies covered the likely near-term configurations of HFS. The key outcomes of this workshop were the identification of the required modifications in the primary codes and standards used for permitting HFS, the delineation of the level of detail of critical information required in the permitting process, and the linkage of the permitting process to land-use planning and zoning

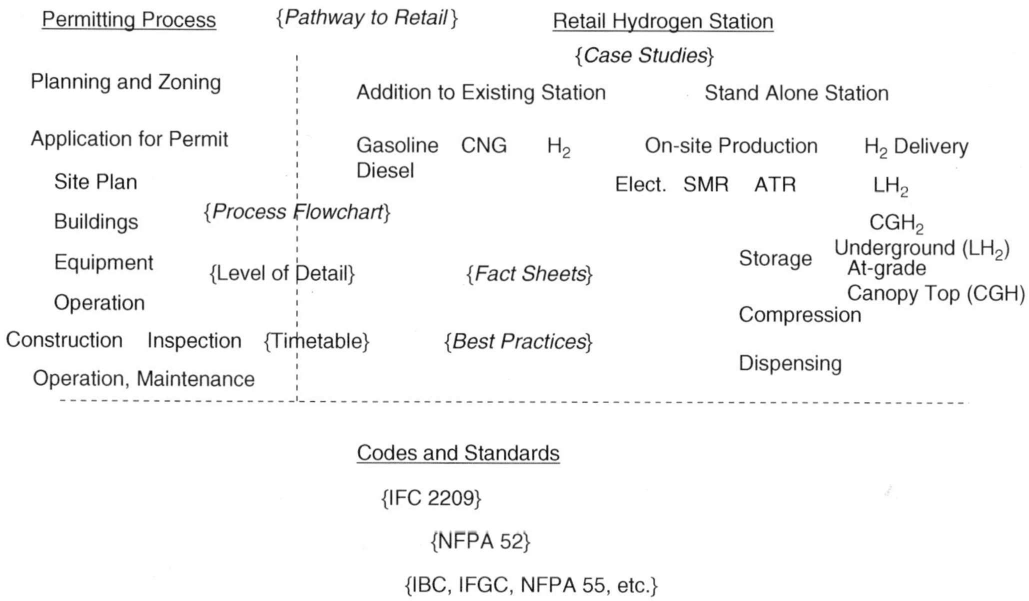

FIGURE 14.5
Concept for an information repository for permitting hydrogen fueling stations.

issues. The modifications needed in codes and standards were referred to HIPOC, which will address these needs, in turn, prepare and propose changes to the ICC and NFPA as appropriate. The DOE and NREL will continue this initiative by conducting additional national and regional workshops and by launching the information repository by the end of the calendar year.

14.3 Key Issues

Three key issues for codes and standards include better synchronization of codes and standards development with RD&D, performance- versus design-based (or prescriptive) standards, and coordinating national codes and standards development with international standards development. For all the three issues, the critical need is to establish a scientific foundation for requirements incorporated in codes and standards. If the science underlying these requirements can be established, harmonization of requirements and consensus among international experts can be more readily achieved, and the DOE can focus on the critical RD&D needs as identified in the Roadmap to facilitate harmonization and consensus.

14.3.1 Synchronizing RD&D and Development of Codes and Standards

The first issue concerns better synchronization of codes and standards development by the SDOs and model code organizations with the RD&D needed to establish a sound technical and scientific basis for requirements incorporated in codes and standards.

A major barrier to incorporating recent scientific data into the determination, as an example, of separation distances for hydrogen storage is the different and, to a large extent, incompatible time requirements of SDOs and researchers. For example, under the ICC's established procedures, new editions to its family of model codes are published every 3 years, and ICC committees must necessarily adhere to strict deadlines governing submittal of code change proposals so that adequate public review can be allowed and public hearings can be arranged and scheduled. The revision cycle takes place over an 18-month period that encompasses the deadline for the receipt of proposed code proposals and the publication of amendments to an existing edition or a new edition of a model code. In contrast, R&D, while not open-ended, cannot be scheduled to provide definitive results by a certain deadline. For most R&D projects, time and schedule are highly dependent on the nature of the problem being assessed and the resources available.

A current example of an effort to incorporate both hydrogen behavior R&D and quantitative risk assessment (QRA) in the codes and standards development process is the work underway by the NFPA to compile all of its hydrogen provisions in a single document, NFPA 2, to be published in 2010. The TC responsible for drafting NFPA 2 has created task groups to develop specific sections or to address key issues that confront such a new comprehensive code for hydrogen as an energy carrier. Given the importance of defining adequate but safe separation distances and the relatively long period of time available, Task Group 6 is addressing separation distances that will be incorporated in NFPA 2 from the ground up, that is, beginning with available data, models, and QRA tools. The task group has defined an approach that will allow it to identify, assess, and select a set of measures and criteria that will enable it to specify separation distances for HRS that are based on a foundation of scientific and technical data, analysis, modeling, and QRA.

Although the TC for NFPA 2 held its first meeting in November 2006 and its work is in a very preliminary stage, the approach taken by Task Group 6 may provide a template for how R&D can be better integrated into the codes and standards development process. The objective of Task Group 6 is to validate or revise the existing prescriptive hydrogen separation distances in NFPA 55 using new research and investigating the use of risk-informed processes. One of the first steps that the task group took was to create a joint effort with the TC of NFPA 55, thus integrating technical experts on the task group with other experts responsible for establishing the existing separation distances for hydrogen storage. The task group has delineated a step-by-step process to address separation distances and has started with outdoor, aboveground gaseous storage systems. For such systems, the task group is considering separate tables for different pressure regimes, for example, 15–250 psi, >250–3,000 psi, >3,000–7,500 psi, and >7,500–15,000 psi, as storage pressure is a key factor for hydrogen behavior during unintended releases.

In addition to storage pressure, another key factor that affects separation distance is the size of the leak. Leak size, expressed as a diameter for the simplest case, depends on the hazard scenario, such as a fueling hose breaking or being disconnected from the dispenser. The task group is examining data that relate leak size to frequency of leaks to determine what leak sizes it should use in constructing a table of separation distances. Such data, unfortunately, are not readily available for hydrogen systems and surrogates for such data must be used. For example, the size of piping typically decreases with the pressure of the system in which it is used, and leak size can be defined in terms of the flow area as a percentage of the pipe diameter. In the worst case, a pipe is severed, and the flow area will be equal to the pipe diameter, whereas for a "small leak," the flow area may be 10% of the pipe diameter.

For a given pressure and leak size, separation distance tables are being constructed around exposure categories. These categories include components or elements of a facility, for example, air intakes for HVAC systems, lot lines, openings in buildings, structures and buildings, and equipment, that require protection from potential hazard scenarios due to unintended hydrogen releases and other events that could create hazards at the facility. These hazard scenarios include hydrogen gas release and subsequent entrainment or accumulation, fire spreading to or from adjacent equipment or structures, and ignition of a release or venting of hydrogen. Exposures are then matched with hazard scenarios against which the exposure must be protected.

For each exposure category and its accompanying hazard scenario or scenarios, the task group has identified consequence parameters that provide measurable criteria on which separation distances can be based. These parameters are the decay distance of an unignited plume of hydrogen to 4% volume fraction in air, the radiative heat flux level of an ignited hydrogen jet, and the flame length of an ignited jet. These parameters will vary with pressure and leak size as well as the physical and environmental conditions at the site when the event occurs. The radiative heat flux parameter provides a way to measure the severity of a hazard against which a minimum separation distance is intended to protect people exposed to this hazard. A radiative flux of 1.6 kW/m^2 can be considered as a "no-harm" level as it is essentially equivalent to exposure to the sun on a clear day, whereas a flux 4–5 kW/m^2 can cause second-degree burns within 20 s of exposure, and 25 kW/m^2 can cause structural damage, significant injury within 10 s, and death within 60 s of exposure [9]. If a given level of radiative heat flux is the most appropriate consequence parameter to measure the degree of protection that a separation distance should provide for a given exposure category and hazard scenario, the distance at which the heat flux subsides to that level will be equivalent to the separation distance. For example, at lot lines the heat flux should not cause harm to people and should be set at no more than 1.6 kW/m^2, whereas for buildings and structures made of noncombustible materials, a heat flux as high as 25 kW/m^2 could be allowed. However, for the exposure category of air intakes, the decay distance of an unignited hydrogen plume to 4% volume fraction is the most appropriate consequence parameter. The separation distance from a potential leak source to an air intake can be determined by assessing the flammable vapor cloud formation characteristics of the potential leak through modeling and experiments based on the volume and pressure of the leak source, the leak size and location, and the physical and environmental conditions at the site.

Separation distances for aboveground gaseous hydrogen systems based solely on consequence parameters can be reduced through mitigation measures such as barrier walls and noncombustible enclosures. The task group will evaluate the effects of mitigation measures through experiments, modeling, and analysis. The task group will also apply QRA techniques to meet its objective. These techniques and their application to separation distances for HFS are described in detail elsewhere [10].

The approach described in the preceding paragraphs may be more successful in incorporating recent R&D results and state-of-the-art modeling and analysis in the code development process because of several factors. First, the timeline to develop and publish NFPA 2 was set from the beginning to allow sufficient time for supporting R&D, modeling, and analysis to be performed. Thus, R&D and code development are better synchronized than would be the case when R&D has to fit into a code revision cycle already underway. Second, the creation of task groups composed of TC members and outside experts to address key issues allows for better interaction between researchers and code development experts, particularly in the case of Task Group 6, which has formed a joint effort with

TC members of NFPA 55. Third, the DOE has supported both critical R&D for safety, codes, and standards and the work of key SDOs such as the NFPA for a number of years and this support has fostered a good working relationship at both institutional and personal levels among researchers and codes and standards experts.

14.3.2 Performance-Based versus Prescriptive Standards

The key difference between performance-based and prescriptive, or design-based, standards is that the former specifies tests that a component, subsystem, or system must pass without regard to the materials they are made of or how they are designed. An example of such a standard is crash-testing requirements in the FMVSS. These standards establish the parameters of crash testing that vehicles must undergo and the criteria they must meet to pass the crash tests. These criteria specify how the vehicle must perform and not how they should be designed or built to meet the criteria. For example, Section 301 of the FMVSS (49 CFR 571.301) requires the protection of occupants in vehicles from exposure to fire that result from fuel spillage during and after crashes and specifies performance requirements for the fuel systems of vehicles. The standard limits the amount of fuel spillage from vehicles during and after frontal, rear, and side impact tests. The standard, however, does not prescribe how the requirements are to be met, and manufacturers must certify that their vehicles meet these requirements. The standard is enforced by the DOT's NHTSA through crash testing of vehicles purchased randomly from dealer showrooms.

Design-based, or prescriptive, standards, however, specify the materials and design for components, subsystems, and systems that are covered by the standard. An example of such standards are those found in the ASME's Boiler and Pressure Vessel Code, which specify the materials, design, fabrication, installation, testing and inspection, and operation and maintenance that boilers and pressure vessels must meet. Such prescriptive standards are common for stationary components and systems that must be approved by state and local code officials. Most of these requirements are based on historical practice and safety experience and have served the affected industries well.

In most circumstances, performance-based standards are preferred by industry, as such standards tend not to limit innovative solutions or inhibit technological progress. The trend among SDOs and CDOs seems to favor performance-based requirements when feasible and when there is sufficient industry support for them during the development of a particular standard. Development of such requirements, in turn, often requires additional RD&D to obtain data and conduct modeling and analysis. The DOE supports RD&D needed to determine requirements that are to the extent possible performance-based.

14.3.3 Coordination of International and Domestic Codes and Standards

The DOE through the CSTT continued to fund hydrogen behavior RD&D to provide better scientific data to underpin requirements in codes and standards. The DOE also funded work on risk assessment to support better risk-informed requirements. There is also a significant collaboration between the DOE and organizations in Europe and Japan to share information on hydrogen behavior research and siting requirements for HFS. For example, the DOE is participating in HyPER (Installation Permitting Guidance for Hydrogen and Fuel Cell Stationary Applications), which is an European Commission (EC) effort to develop fast-track approval for small stationary hydrogen and fuel cell systems and safe procedures to enable a comprehensive installation process for developers, design engineers, manufacturers, and installers across the European Union. The DOE is also actively participating in two other EC efforts, HyApproval that focuses on harmonizing permitting requirements for HFS and HySafe to coordinate hydrogen safety RD&D and to sponsor and organize the biennial International Conference on Hydrogen Safety (ICHS). In addition, the DOE participates in and provides support to the convener of the International Energy Agency's Annex 19 work on hydrogen safety, which is focused on applying risk assessment techniques to help define hydrogen safety requirements. By supporting both domestic RD&D and international collaboration, the DOE is helping to improve synchronization of codes and standards development with RD&D.

Providing consistent and sustained support of international standards development remains a key issue. Most of the meetings of TC responsible for drafting international standards are held outside the United States and require significant investment of time and money on the part of U.S. stakeholders. The DOE supports this activity to facilitate the development of consensus international standards that are essential for a market receptiveness of hydrogen and fuel cell technologies. This consensus can be facilitated through continued efforts of international experts to develop performance-based standards based on collaborative RD&D among key agencies and organizations in the United States, EC, and Asia.

14.4 Conclusion

Two key needs for hydrogen safety in consumer applications are the incorporation of data and analysis from RD&D into the codes and standards development process and the adoption and enforcement of these codes and standards by state and local code officials. The DOE supports a comprehensive program to address both these aspects of hydrogen safety. For the first, the DOE is working with the automobile and energy industries to identify and address high-priority RD&D to establish a sound scientific basis for requirements that are incorporated in hydrogen codes and standards. The high-priority RD&D needs are incorporated and tracked in a Roadmap adopted by the CSTT of the FreedomCAR and Fuel Partnership. The DOE and its national laboratories conduct critical RD&D and work with key standards and model code development organizations to help incorporate RD&D results into the codes and standards process. To address the second aspect, the DOE has, for example, launched an initiative to facilitate the permitting process for HFS. A key element of this initiative will be a web-based information repository, a toolkit that includes information fact sheets, networking charts to encourage information exchange among code officials who have permitted or are in the process of permitting HFS, case studies of likely near-term HFS configurations, and a database of requirements incorporated in key codes and standards. The information repository will be augmented by workshops for code officials and station developers in jurisdictions that are likely to have HFS in the near future.

The national templates have guided the DOE's effort to accelerate the development of key standards and model codes for hydrogen and fuel cell applications. With the help and cooperation of SDOs, CDOs, industry, and other interested parties, the DOE has established a coordinated national agenda for hydrogen and fuel cell codes and standards. With the adoption of an RD&D Roadmap by the Partnership and through its implementation by the CSTT, the DOE will help strengthen the scientific basis for requirements incorporated in these codes and standards that, in turn, will facilitate international market receptivity for hydrogen and fuel cell technologies.

List of Acronyms

AHC	*Ad Hoc* Committee for Hydrogen Gas
AHJ	authority having jurisdiction
ANSI	American National Standards Institute
API	American Petroleum Institute
ASME	American Society of Mechanical Engineers
ASTM	American Society of Testing and Materials
BOCA	Building Officials and Code Administrators International
CDH	Code Development Hearings
CDO	Code Developing Organization
CFR	Code of Federal Regulations
CGA	Compressed Gas Association
CSTT	Codes and Standards Technical Team
DOE	Department of Energy
DOT	Department of Transportation
FMVSS	Federal Motor Vehicle Safety Standards
HFS	hydrogen fueling stations
HIPOC	Hydrogen Industry Panel on Codes
ICBO	International Conference of Building Officials
ICC	International Code Council
ICHS	International Conference on Hydrogen Safety
IEC	International Electrotechnical Commission
IMC	International Mechanical Code
IPHE	International Partnership for the Hydrogen Economy
ISO	International Organization for Standardization
NASFM	National Association of State Fire Marshals
NFPA	National Fire Protection Association
NHA	National Hydrogen Association
NHTSA	National Highway Traffic Safety Administration
NIST	National Institute of Standards and Technology
NREL	National Renewable Energy Laboratory
OPS	Office of Pipeline Safety
OSHA	Occupational Safety and Health Administration
PEM	proton exchange membrane
QRA	quantitative risk assessment
RCS	regulations, codes, and standards
RD&D	research, development, and demonstration
RITA	Research and Innovative Technology Administration
SAE	Society of Automotive Engineers
SBCCI	Southern Building Code Congress International
SC	subcommittees
SDOs	standards developing organizations
SNL	Sandia National Laboratory
TAG	Technical Advisory Group
TC	technical committees
UL	Underwriters Laboratories
USCAR	U.S. Consortium for Automotive Research
USFCC	U.S. Fuel Cell Council

References

1. ANSI, *Overview of the U.S. Standardization System*, 2nd edition, ANSI, NY, 2007.
2. http://www.iccsafe.org (accessed August 8, 2007).
3. http://www.nfpa.org (accessed August 8, 2007).
4. http://www.hydrogen.gov/regulations (accessed August 8, 2007).
5. 29CFR Part 1910.103 and http://www.osha.gov (accessed August 8, 2007).
6. http://www.cganet.com/isotc197 (accessed August 8, 2007).
7. http://www.csa-america.org (accessed August 8, 2007).
8. ISO TS14687-2, Hydrogen Fuel Product Specification-Part 2; SAE J2719, Information Report on the Development of a Hydrogen Quality Guideline for Fuel Cell Vehicles.
9. LaChance, J., Risk-Informed Separation Distances for an Example Hydrogen Fueling Station, *Presentation at IEA Task 19 Meeting*, Tsukuba, Japan, January 31–February 2, 2007.
10. Ohi, J.M., Moen, C., Keller, J., Cox, R., Risk Assessment for Hydrogen Codes and Standards, *Proceedings of the HySafe International Conference on Hydrogen Safety*, Pisa, Italy, September 8–10, 2005.

15

Hydrogen Sensing and Detection

Prabhu Soundarrajan and Frank Schweighardt

CONTENTS

15.1 Introduction

Hydrogen is a highly flammable gas and will burn at concentrations as low as 4% in air. The lower explosive limit (LEL) and upper explosive limit (UEL) are the two most common terminologies used to indicate the flammable levels for many fuels including hydrogen. As indicated in Table 15.1, hydrogen is one of the least flammable materials at 4% but has a larger window (4–75% v/v H_2) of flammability in comparison to natural gas, gasoline, propane, ethane, methane, propylene, etc. The flammability limit of hydrogen is seven times wider than methane (Figure 15.1 [1]). It is, therefore, critical for a hydrogen sensor to have a wider measurement range (1–99% v/v H_2) for safety applications than most common fuels. Hydrogen is the lightest of elements and the smallest molecule; it, therefore, has the greatest tendency to leak. Thus, for a process safety application, a hydrogen leak can be more dangerous and its detection becomes more challenging than other gases. This chapter provides a general understanding of different hydrogen sensing technologies, their importance, applications, and limitations. Section 15.2 provides an overview of the hydrogen measuring principles including traditional industrial methods and discusses their limitations. Section 15.3 focuses on the operation mechanisms and recent advancements of low-cost solid-state sensors for hydrogen detection. Section 15.4 outlines the applications of hydrogen sensors in industrial processes. Section 15.5 provides an overview of sensors in hydrogen fuel applications. Section 15.6 provides a brief market overview with a list of manufacturers of hydrogen sensors. The purpose of this chapter is to provide a reference guide on hydrogen sensors for the readers and expert hydrogen community.

TABLE 15.1

Explosive Limits for Common Constituents in Process Industries

Fuel	LEL (%)	UEL (%)
Gasoline	1.4	7.6
Propane	2.1	10.1
Ethane	3	12.4
Hydrogen	4	75
Methane	5	15
Propylene	2.0	11.1

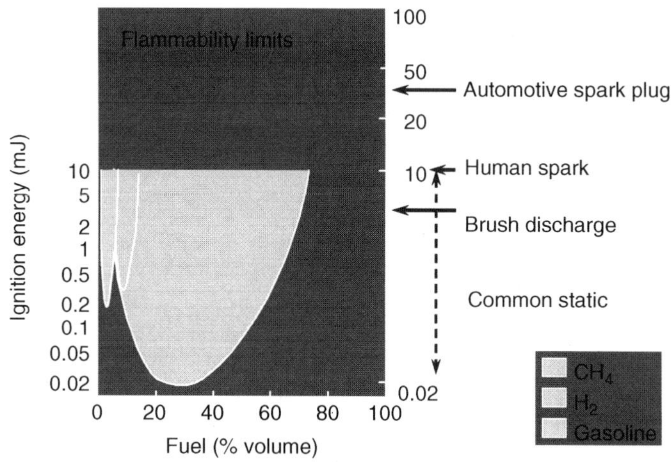

FIGURE 15.1
Flammability limits for methane, hydrogen, and gasoline. (Courtesy of Air Products and Chemicals, Inc., Allentown, PA.)

15.2 Hydrogen Measuring Principles: An Overview

The four key functional parameters to be satisfied by any hydrogen sensor to gain wide acceptance for use within the hydrogen infrastructure [2] of production, storage, transportation, and utilization are

- *Performance*. Sensors must have a broad operating range in air, nitrogen, and inert backgrounds (2–99% v/v) and good sensitivity well before the explosive limit (4% H_2) in air backgrounds to meet National Fire Protection Association (NFPA) safety guidelines. Continuous operation in gas streams containing hydrocarbons, sulfur compounds, and carbon monoxide is required for petrochemical applications. Operation in wet chlorine backgrounds is necessary in chloralkali process plants. Fast response time to line out (<60 s) in high humidity is desired for diagnostic study and optimal operation of hydrogen transportation systems (vehicles, electrolyzers, battery rooms, fuel cells, storage containers, etc.). Response time is measured as the time taken for 90% response (t90) to a particular hydrogen concentration. In a process safety system, the gas delivery flow rate and total volume of the lines to the sensor must be considered. Transport for on-line or at-line process analysers such as gas chromatographs and mass spectrometers involve line purging. It takes 15 volumes to exchange a gas from 100% to <1 ppm with a new gas in a continuous flow stream. In a process plant, this can take several minutes using even the newest sampling systems to the point of use due to the reduction in mole fraction of a gas with subsequent flush volumes (Figure 15.2). The best configuration to record accurate response times in process streams is to use a recommended depth to insert the sensor directly in the sample stream. A recommended installation point for the sensor is 35–40% into the tube diameter and not at a point near the wall. In some cases, the sensor may be pointed away from the moving gas stream to avoid particles and reduce condensation.
- *Lifetime*. Sensors must have a usable lifetime consistent with the application for which it is intended. Sensors should also have a well-defined replacement,

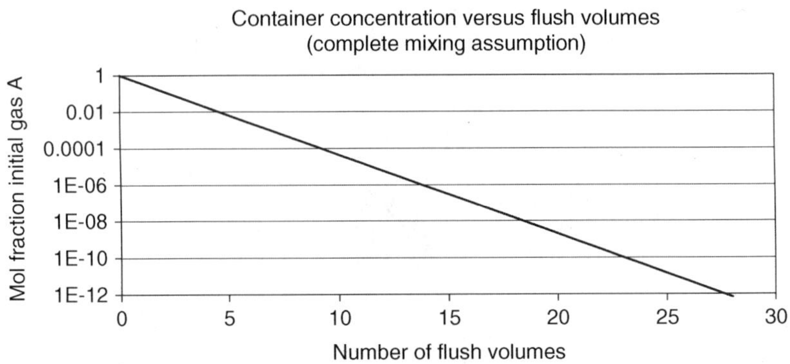

FIGURE 15.2
Chart showing the reduction in mole fraction of a gas with flush volumes.

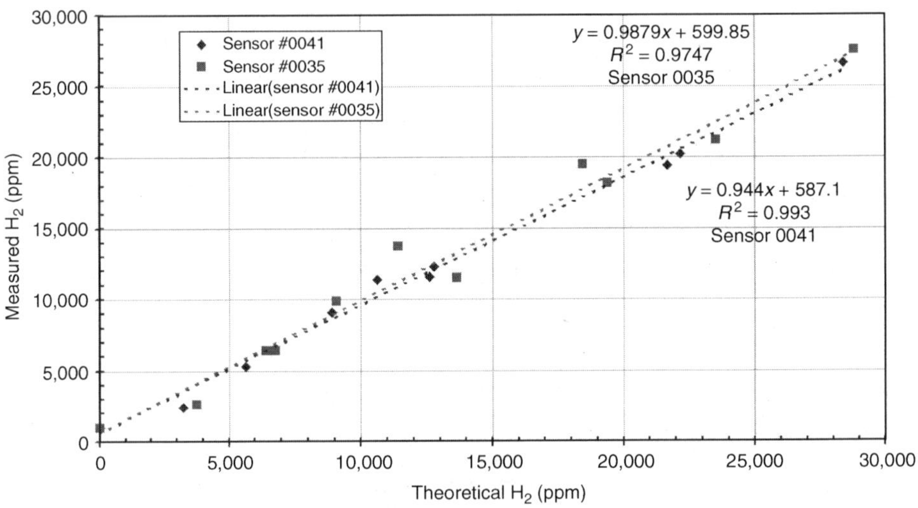

FIGURE 15.3
Chart showing an example comparison of the theoretical versus measured hydrogen reading from a sensor. (Courtesy of Air Products and Chemicals, Inc., Allentown, PA.)

maintenance, and operational costs associated with them. Hydrogen analytical systems (gas chromatograph or mass spectrometer) for process monitoring and refinery applications are typically used for <10 years and require daily to weekly calibration. A low-cost in-line solid-state sensor with minimal annual calibration is more cost effective in the lifetime of a process. For transportation and automotive applications, the desired lifetime is greater than 10 years with minimal (annual) calibration. During this period, the hydrogen sensor must be operational without cleaning, frequent calibration, or replacements while exposed to harsh environmental conditions.

- *Reliability*. Sensors must indicate the presence of hydrogen specifically, and not provide false alarms. The sensors should have consistent reproducibility (e.g., ± 5% over 1 year at 2% v/v H_2 in air), long-term stability and minimal drift rates. Response must not drift outside acceptable limits over that lifetime without providing an alarm. Figure 15.3 [3] shows an example comparison of the theoretical versus measured

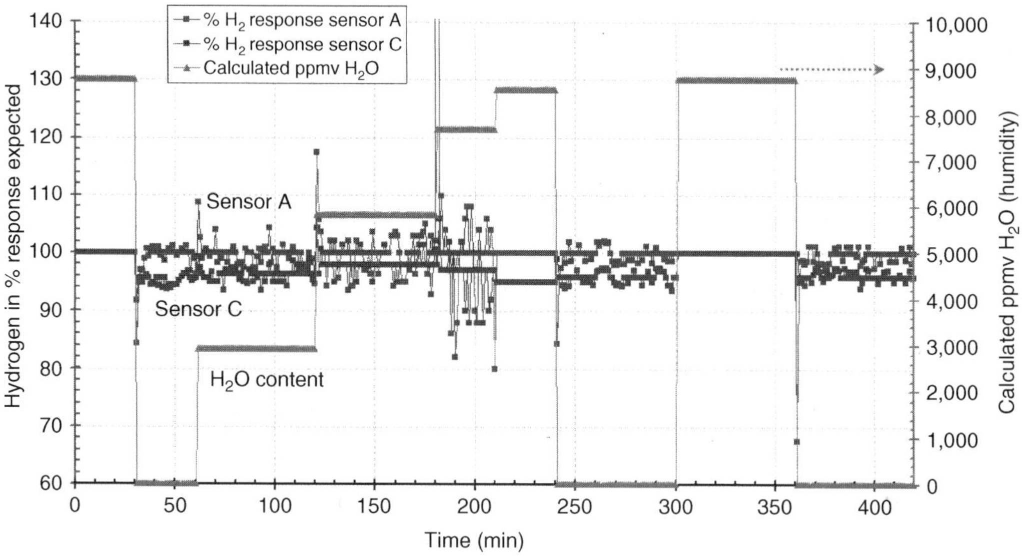

FIGURE 15.4
Chart showing the comparison of two sensors detecting hydrogen for 400 min in the presence of humidity within acceptable limits. Actual acceptable limits may differ based on applications. (Courtesy of Air Products and Chemicals, Inc., Allentown, PA.)

hydrogen reading from a sensor. Such a comparison can be used by users to define acceptable drift rates. The functionality of the sensors should be easily verifiable and there should be a low tolerance for false alarms. Material selectivity to hydrogen is necessary to the design of a hydrogen sensor to prevent false alarms. Sensors should be able to survive multiple excursions to hydrogen concentrations without damage. Hydrogen specificity in the sensor material is highly desired for sensor operation in contaminant backgrounds. As an example, Figure 15.4 shows the comparison of two sensors detecting hydrogen for 400 min in the presence of humidity within acceptable limits. Such performance is a must for a hydrogen sensor in automotive applications.

- *Cost.* Sensors and their controllers must justify their cost (purchase, installation, and maintenance) versus current technology by a factor of four (4) or more to warrant purchase. In industrial process applications today, a unit cost of <\$10,000 is highly desired for a sensor with minimal maintenance and installation costs for multiple-point installations. The process industry is currently using Class I/Div II Gp B gas chromatographs that cost \$100,000 to buy/install, and require additional skilled labor to maintain. The typical gas chromatography (GC) process analyser has a limited number of sample points (e.g., <6) in process facilities due to gas transfer times. In some cases, the distance from the analyser to the sample point is limited to <500 ft. (tubing runs). In all cases, a process analyser must then switch sample stream (purge lines) to initiate the next sample and bring the data to the distributed control system (DCS). Overall, this can take 3–8 min per sample (sample/purge/test/report). If six sample points were interfaced to one process analyser a full sample cycle can take ~45 min, even with continuous bypass purge. Therefore, six independent hydrogen sensors, at the same cost-to-buy, can operate simultaneously gathering data every minute. For automotive, fuel cell applications, a low-cost sensor package is needed with minimal maintenance costs, such as exchanging the sensor element (<\$50). The cost alternative is better for most applications as long as performance, lifetime, and reliability are not compromised.

15.2.1 Mechanisms for Hydrogen Sensing

Several mechanisms for hydrogen sensing and detection have been studied since the early 1900s. Several traditional sensing mechanisms that are still widely used in the industry include GC, mass spectrometry (MS), catalytic bead (CB), and thermal conductivity. Semiconducting metal oxide and CB sensors are popular solid-state technologies, which employ heated catalysts to sense hydrogen. These sensors require heating to about 300°C (a potential source of ignition!) to enable surface reactions that promote hydrogen sensing. Electrochemical sensors are based on known electrolytic reactions of hydrogen. Sensors based on catalytic combustion are generally nonspecific, electrochemical hydrogen sensors with liquid- or solid-type electrolytes having leakage issues. The hydrogen sensors based on thermal conductivity, CB, metal oxide, and electrochemical technologies require the presence of oxygen for sensor operation. Oxygen plays a crucial role in promoting the grain boundary formation in metal oxide sensors and electron transfer reactions in electrochemical sensors. The most promising solid-state technology is based on a hydrogen-specific material, palladium, which does not require oxygen for operation. Palladium-based sensors are gaining wide popularity in the industry due to their reliability and high specificity to hydrogen. Any hydrogen sensor technology needs to satisfy the three basic requirements—sensitivity, selectivity, and specificity (i.e., the sensor should be functional over the range of interest and be free of cross-interference from other gases present in the environment). This chapter will introduce the recent developments in solid-state hydrogen sensors and outline their industrial applications.

15.2.2 Traditional Approaches to Hydrogen Sensing

15.2.2.1 Thermal Conductivity

Thermal conductivity is the most widely applied measuring principle for the determination of hydrogen. The measuring principle is based on the differences in thermal conductivity of the gases to be measured. A thermal conductivity detector (TCD) measures the concentration of a gas in a binary gas mixture by measuring the thermal conductivity of the sample gas and comparing it to the thermal conductivity of a selected reference gas (Figure 15.5). Two or four ultrastable, precision glass-coated thermistors are used, one (or two) in contact with the sample gas and the other (or two) in contact with the reference gas (such as chlorine in a sealed chamber). The thermistors are mounted so that they are in close proximity to the walls of the gas chambers. The entire cell is temperature-controlled, and the thermistors are heated to an elevated temperature in a constant current Wheatstone bridge. The thermistors lose heat to the walls of the gas chambers at a rate that is proportional to the thermal conductivity of the gas surrounding them. Thus, each thermistor will reach a different equilibrium temperature. The temperature difference between the two (or four) thermistors is detected in the Wheatstone bridge, and the resulting bridge voltage is amplified and converted to a linear 4–20 mA output proportional to the concentration of hydrogen in the binary gas mixture. The TCD will detect any gas or vapor that has a thermal conductivity that differs significantly from the high thermal conductivity of helium. The general accepted lower detection limit for the TCD is 50 ppm v/v.

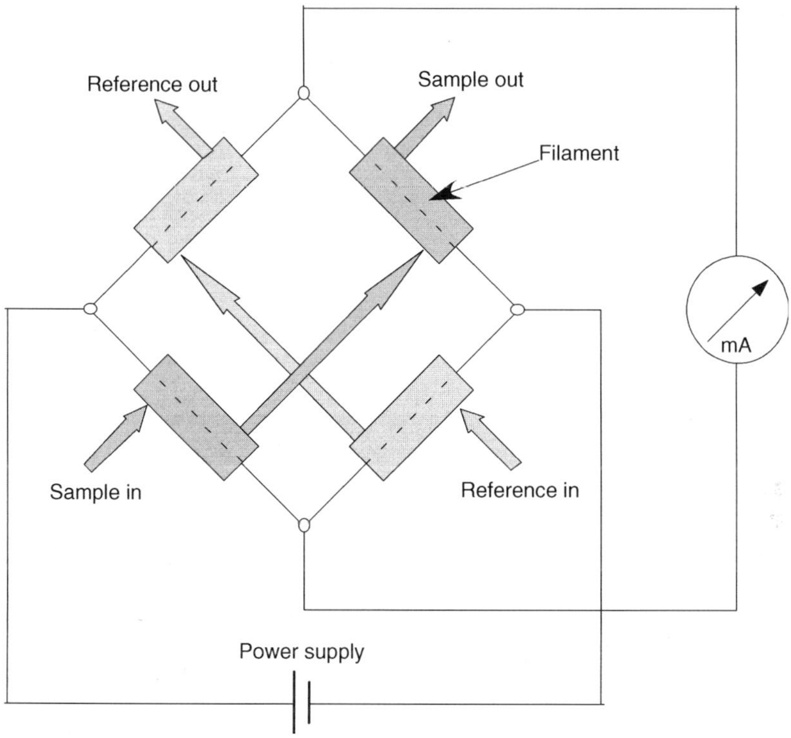

FIGURE 15.5
Schematic drawing of a TCD.

15.2.2.2 Gas Chromatography

Gas chromatography is the second most applied measuring principle for hydrogen detection. The disadvantages of GC are long response times (minutes) due to the chromatography, time-intensive sample preparation, consumable (carrier and calibration gases), and labor-intensive handling procedures. An advantage, however, is the ability to measure other gases such as nitrogen, oxygen, and carbon dioxide in the presence of hydrogen. But, this adds time to the total analysis.

Process gas chromatographs must be configured as Class I/Div II Gp B, with "Z" or "X" purge; or be intrinsically safe when situated within a plant analyser shed (not in a lab environment) to meet NFPA and must incorporate safety reviews. Such systems are not manned and provide 4–20 mA signals of the species in question to the DCS in the control room. That is, they are set to respond to the concentration range expected by precalibration using a series of costly gas reference standards. The use of a dedicated hydrogen sensor avoids these issues and costs. It is true that the GC and mass spectrometer can identify and quantify other species in the sample stream, but if the need is to control the process hydrogen content a sensor system can do the task at the lowest cost, in the shortest time frame, and at the highest precision.

15.2.2.3 Mass Spectrometry

Mass spectrometry is another widely used technology for hydrogen sensing in the industry. Process mass spectrometers such as gas chromatographs must be safety certified to operate in the plant environment. Mass spectrometers require special air-conditioned shelters, safety monitors, many calibration gases (e.g., fragmentation gases), and routine (weekly) maintenance by a skilled operator. The mass spectrometer can be configured to identify a host of species in a sample stream as long as there is no overlap in signal. Similar to GC, the mass spectrometer requires extensive sample systems and continuous bypass flow of sample gases that add to the complexity and cost to perform.

15.2.2.4 Laser Gas Analysis

A laser gas analyser exploits the phenomenon that hydrogen gas molecule struck by laser light which absorbs it and reemits light at a different frequency. The molecular bonds of hydrogen molecule uniquely determine the resulting "Raman shift" frequency of the reemitted light. The shifts are discrete and precise such that the intensity of light observed at various shifted frequencies is directly proportional to the concentration of hydrogen in the measured gas mixture. A standard setup for laser Raman spectroscopy consists of a very stable laser and precise optics such as fibers, lenses, holographic gratings, notch filters, a measuring cell, a very stable diode array detector, and a computer with dedicated software. These instruments measure the full Raman spectrum and are very expensive (~$100,000 and upward).

15.2.3 Solid-State Approaches to Hydrogen Sensing

A wide variety of solid-state sensors based on hydrogen-specific palladium, metal oxide semiconductor (MOS), CB, electrochemical, and surface acoustic wave (SAW) technology are used in the industry for several years. Microelectromechanical systems (MEMS), and nanotechnology-based devices for the measurement of hydrogen are the recent developments. These developments are mainly driven by the demands of the fuel cell industry. Solid-state approaches are gaining rapid popularity within the industry due to their low cost, low maintenance, replacements, and flexibility of multiple installations with minimal labor.

15.2.3.1 Hydrogen-Specific Palladium-Based Sensors

There are three major classes of palladium-based hydrogen sensors [4]. The most popular class of palladium-based sensors is based on palladium resistors. A thin film of palladium deposited between two metal contacts shows a change in conductivity on exposure to hydrogen due to the phase transition in palladium. The palladium field-effect transistors (FETs) or capacitors constitute the second class, wherein the sensor architecture is in a transistor mode or capacitor configuration. The third class of palladium sensors includes optical sensors consisting of a layer of palladium coated on an optically active material that transforms the hydrogen concentration to an optical signal.

15.2.3.2 Metal Oxide and Catalytic Bead Sensors

Metal oxide sensors [5] are also known as semiconductor-based sensors since they use a semiconducting film as the sensing element. The MOS have to be heated up to high temperatures and require the presence of oxygen. At high temperatures (300–500°C), there is a grain boundary formation in the MOS that enable detection of a number of gases. The MOS consume high power and have high false alarm issues. CB sensors are a variation of this type and have an active sensing element with a coated catalyst and a passivated reference element for ambient temperature and pressure compensation.

15.2.3.3 Electrochemical-Based Sensors

Electrochemical sensors [6] represent the most commercially successful sensor type due to the simplicity of their operation, high sensitivity, and proven sensor mechanisms. The sensor architecture mimics a battery containing an anode, cathode, and a supporting electrolyte. These sensors can operate well at room temperature in the presence of oxygen. Hydrogen is detected using a specific electrochemical reaction between the anode and the cathode. The resulting signal is either a current or a voltage change in the external electrical circuit.

15.2.3.4 Surface Acoustic Wave and Microresonance-Based Sensors

A SAW is an acoustic wave traveling along the surface of a material having some elasticity, with amplitude that typically decays exponentially with the depth of the substrate. The SAW is commonly used in piezoelectric devices in electronic circuits. SAW devices measure the change in frequency and propagation of the SAW waves across a known medium as a function of hydrogen concentration. SAW-based hydrogen sensors may include a hydrogen-sensitive material [7,8].

15.2.3.5 Nanotechnology Approaches for Hydrogen Sensing

Nanotechnology involves the manipulation of materials in the molecular level to provide significant material property enhancements. Carbon nanotubes, nanoparticles, nanowires, nanowhiskers, metallic nanotubes, metal oxide nanostructures, and nanoclusters are some of the nanomaterials investigated for hydrogen sensing. Most of these materials are nanoscale components of bulk materials, which have been investigated for hydrogen sensing for several years. Nanomaterials have high surface to bulk ratio resulting in higher sensitivity, faster response time, and high selectivity in comparison with their bulk counterparts.

15.3 Operation Mechanisms of Solid-State Sensors

15.3.1 Theory of Palladium-Based Hydrogen Sensors

Thin-film palladium has been the leading material investigated for sensing hydrogen due to its unique material response to hydrogen. The sensing of hydrogen using palladium is a two-step process (Figure 15.6). In the first step, molecular hydrogen dissociates on the surface of palladium and is converted into elemental hydrogen, which is the rate-limiting step for the detection. The elemental hydrogen then diffuses into the palladium lattice causing an expansion in the lattice and a phase transition. The main principle of hydrogen sensing is the conversion of the α phase (conductive metallic phase) of palladium, to the β phase (less conductive phase). A film of palladium assembled between two conductive electrodes detects hydrogen by a change in electrical conductivity. Upon exposure to hydrogen, the conductivity of the thin-film palladium decreases due to the change from the predominant α phase to the less conductive β phase.

It is known that the transition from the α state to the β state is accompanied by a lattice increase [9] of about 5%. This transition occurs at about 18 mm Hg (or about 2.37% H_2 concentration). It takes a higher increasing level of hydrogen to transition from α phase to β phase

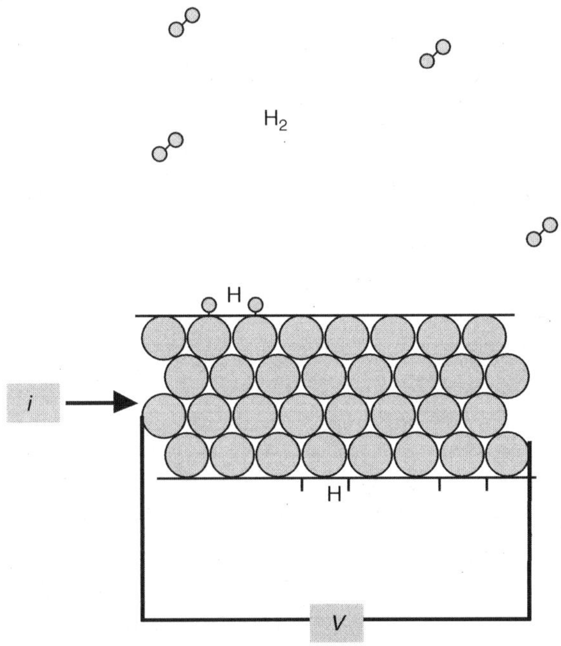

FIGURE 15.6
Mechanism of hydrogen sensing with palladium thin films.

as the temperature increases. Above the critical temperature in the Pd/H system, two solid hydride phases of differing composition cannot exist together in the solid under steady state at specified hydrogen content. It is also clear that the α- to β-composition varies continually with temperature and that, generally, there is no simple ratio of hydrogen atoms to palladium during this process. Because there is a large crystal lattice change due to the phase transition in palladium (the ratio of H to Pd atoms can jump from 0.01 to 0.6 in the β phase), polycrystalline palladium films crack and the electrical resistance of the film develops hysteresis over time. Therefore, the material property of palladium should be improved to fabricate a sensor with robust performance over a wide temperature range of operation.

An addition of a second metal (alloy) to palladium was shown to extend the temperature range of operation of palladium-based sensors. For example, the addition of nickel (Ni) has shown a significant change in the resistance and sensing behavior as hydrogen sensors due to the reversal in the phase transition in palladium films. The α- to β-phase transition was found in the 5.3% Ni alloy at about 250 Torr and alloys of Ni $>$ 8% showed no phase change behavior up to 630 Torr in pure H_2 at ambient temperatures [10]. The temperature range of operation, speed of response, and sensitivity has been established as a function of nickel percentages in the sensor. The addition of higher percentages of nickel to the palladium films increases the stability of the film at high concentrations of hydrogen and prevents sensor hysteresis.

15.3.2 Types of Palladium Hydrogen Sensors

Several types of palladium-based hydrogen sensors have been reported in the literature. The most notable ones are based on Pd thin-film resistors, FETs, Pd nanowires, Pd nanoparticle networks, Pd nanoclusters, and Pd nanotubes as shown in Table 15.2.

TABLE 15.2

Technology Approaches to Palladium-Based Hydrogen Detection and Sensing

Technology	Pros	Cons
Pd field effect devices [9]	Good sensitivity Fast response time	Sensor hysteresis and fouling Not selective to hydrogen
Pd-based resistors [8]	High range of detection (0.5% H_2 concentrations and above) Cost is relatively low	Limited operation range Heater required High power
Pd-coated fiber optic detectors	Remote detection Good sensitivity Can network to multiple sites	High cost Temperature dependence
Pd mesowire detector	Very low power (below milliWatt) on/off detector	Development required Fabrication process not repeatable
Pd nanoparticle network sensor	Broad dynamic range	Possible interference from CO and humidity (characteristics of all Pd-based hydrogen sensors)
Pd nanoclusters and nanotubes	High sensitivity Very low Fast response time	Research under development

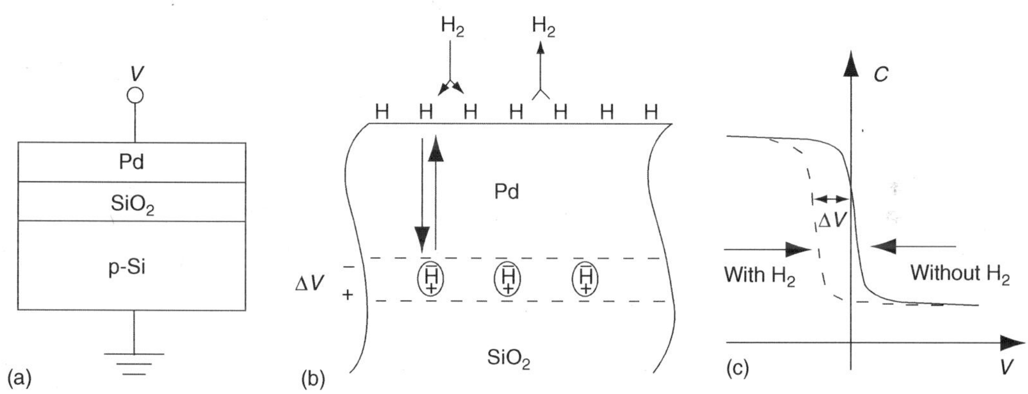

FIGURE 15.7
(a) Device architecture, (b) mechanism of operation, and (c) electrical response of Pd-MIS hydrogen sensor.

15.3.2.1 Palladium Field-Effect Sensors

Hydrogen sensors based on the "field effect" of palladium have been investigated extensively in the literature [11]. The field effect results due to the rapid dissolution of hydrogen in the palladium surface arranged in a Pd–SiO_2–Si configuration (Figure 15.7). The sensor relies on an electric field resulting from the charge transfer between palladium and hydrogen on its surface. The FETs [12] and metal–insulator–semiconductor (MIS) [13] are the two major types of device structures that have been studied for palladium-based hydrogen sensing. Palladium is catalytically active, permeable to hydrogen, and can be readily used in FET and MIS devices. A MIS hydrogen sensor (essentially a capacitor) was fabricated based on the palladium thin film as the sensing layer. The device architecture is shown schematically in Figure 15.7a and consists of three elements: (1) a layer of catalytically active metal, palladium, whose outer surface is exposed to hydrogen; (2) a layer of dielectric material, typically, silicon dioxide; and (3) a doped silicon substrate. The configuration of these layers and their electronic assembly can be altered to fabricate hydrogen sensors that function as resistors, capacitors, or transistors.

Figure 15.7b illustrates the operating principle of the capacitor-type hydrogen sensor. A hydrogen molecule is adsorbed (i.e., chemically bound) to the metal surface, dissociated into two hydrogen atoms, which then further diffuse rapidly through the palladium and bind at interface sites. The binding process creates a change in the electronic properties of the device (a change in the work function), which causes a voltage shift in the capacitance versus voltage curve (Figure 15.7c). The change in the voltage shift was measured and quantified to the concentration of hydrogen. Capacitance-based hydrogen sensors such as the MIS devices, however, are affected by cross-interference from other gases. The sensor is cross-sensitive to oxygen, which prevents hydrogen from adsorbing to the sensor surface since it binds more tightly to metals than hydrogen. The sensor is also cross-sensitive to gases, such as acetylene that can dissociatively adsorb and form atomic hydrogen at the surface, which then travels to the interface and registers as a signal.

Palladium has been used as the sensitive material to fabricate FET devices for hydrogen sensing. A FET was fabricated with a palladium thin film and a sacrificial oxide layer [14] (Figure 15.8). The presence of the interfacial oxide layer allows larger turn-on voltage when compared with conventional metal–semiconductor FETs. The device architecture as shown in Figure 15.8a has a thin-film palladium as the gate layer on a gallium arsenide semiconductor. The hydrogen atoms diffuse and pass through the Pd metal and form a dipole layer at the metal and semiconductor interface. This dipole layer generated by hydrogen atoms at the Pd/oxide interface decreases the depletion width and lowers the metal–semiconductor Schottky barrier height. The dipole layer between the metal and semiconductor corresponds to a voltage drop that in turn corresponds to the concentration of hydrogen. The sensor is expensive to manufacture and consumes high power because of the heating requirement. Thin-film palladium also suffers from hysteresis and delamination at high temperatures. Sandia National Laboratory developed a hydrogen sensor by integrating palladium–nickel (Pd–Ni) alloy films into complementary metal oxide semiconductor (CMOS) microelectronic technology [15].

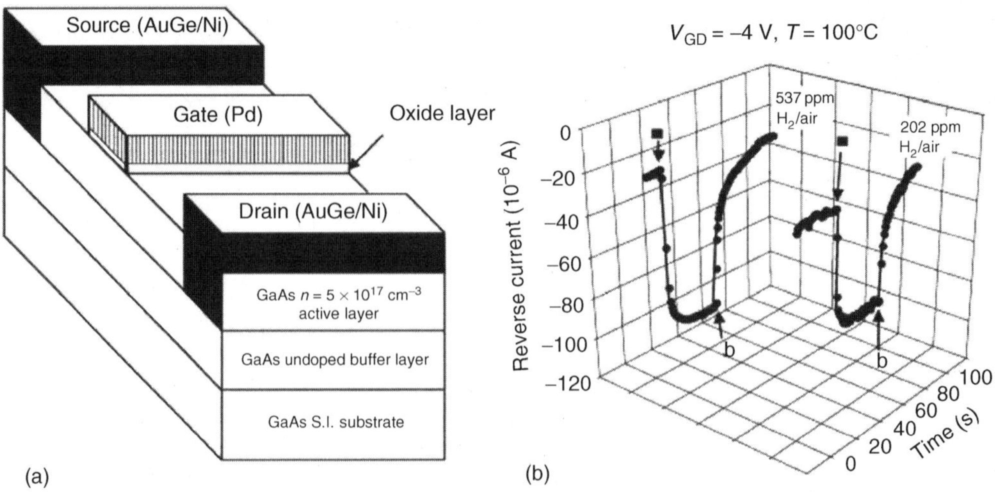

FIGURE 15.8
(a) Schematic representation of the palladium-based FET, and (b) response of palladium-based FET sensor to hydrogen at 100°C.

15.3.2.2 *Palladium-Based Resistors*

Thick- and thin-film palladium-based resistors have been reported for hydrogen sensing [16]. The thick-film device uses printed palladium paste on a ceramic substrate in a four-resistor network (Wheatstone bridge). Two opposed resistors are covered to isolate them from the ambient atmosphere. The exposure of the uncovered resistors to hydrogen results in a change in resistivity of the thick-film material and a shift in the balance point of the bridge, which can be scaled to the hydrogen concentration. The thin-film device is equivalent in design to the thick film; here, much thinner films (typically vacuum deposited) are used as the resistors. Thin-film palladium detectors have been prepared by depositing palladium through electron beam evaporation [17], RF magnetron sputtering [18], micro contact printing [19], and wet electrochemistry. Most palladium resistors have fouling issues on the palladium surface due to impurities and pollutants in, or reaction with the air. The fouling on the palladium surface can be reduced by the addition of a second metal (alloy) to palladium. For example, palladium–nickel (PdNi) resistors (Figure 15.9b) fabricated for detecting higher concentrations of hydrogen (up to 100%) have showed limited fouling on the sensor surface. The PdNi resistor sensors have on-chip microthermometers and microheaters for maintaining constant chip temperature. The addition of nickel increases the surface stability and durability of the sensor at high hydrogen concentrations. PdNi resistor technology is most suited in hydrogen-specific (0–100% v/v) measurement for industrial process applications.

15.3.2.3 *Palladium-Coated Fiber Optic Sensors*

A fiber optic hydrogen sensor consists of a palladium coating at the end of an optical fiber that senses the presence of hydrogen in air. When the coating reacts with the hydrogen, its optical properties are changed. Light from a central electro-optic control unit is projected down the optical fiber where it is either reflected from the sensor coating back to central

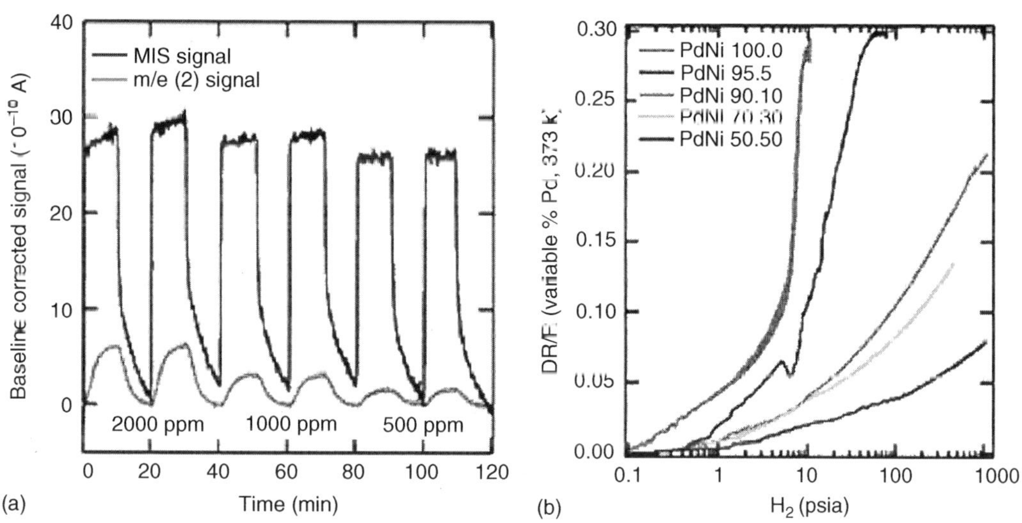

FIGURE 15.9
(a) Response of a thin-film palladium-based MIS microsensor (ppm range) and (b) response of palladium-based chemical resistor to hydrogen pressure illustrating the effects of Ni composition in the film.

optical detector or is transmitted to another fiber leading to the central optical detector. A change in the reflected or transmitted intensity indicates the presence of hydrogen. The fiber optic detector offers inherent safety by removing all electrical power from the test sites and reduces signal-processing problems by minimizing electromagnetic interference. The fiber optic hydrogen sensors can be fabricated using a palladium-coated single-mode tapered optical fiber [20]. The attenuation change of the fiber-mode when the device was exposed to hydrogen is used to detect and measure hydrogen concentration in gaseous atmospheres. A wavelength multiplexed hydrogen sensor that employs a Pd-coated tapered fiber combined with a fiber Bragg grating [21] has been demonstrated. The fiber optic hydrogen sensor required heating the palladium layer optically with a diode laser at 980 nm to obtain fast response times. The heating process consumes high power ~60 mW to provide the maximum value of the initial response velocity for a sensor exposed to 4% hydrogen concentration at −30°C. Butler [22,23] suggested the use of a palladium thin film as a mirror at the end of a fiber optic cable. A hydrogen sensor based on a diode laser and beam splitters was reported [24]. Ito [25] proposed a palladium-catalyzed reaction of amorphous tungsten oxide with hydrogen in a fiber optic hydrogen detector. The reaction causes partial reduction of the tungsten oxide and introduces a strong optical absorption band near 800 nm. The increase in absorption reduces the intensity of the light beam reflected by the coated optical fiber.

A fiber optic hydrogen sensor that is commercially available is the micromirror sensor as shown in Figure 15.10a. The sensor consists of a thin film of palladium evaporated at the cleaved end. In an alternative approach, a core-exposed fiber optic device [26] consisting of a palladium thin film around the fiber core (Figure 15.10b) excites the light traveling in the fiber. In the presence of the Pd coating, the evanescent fields are altered (Figure 15.10c). As the Pd film absorbs the hydrogen, it expands; because of the formation of the palladium-hydride, the real and imaginary part of its refractive index change. The expansion of the Pd thin film induces a strain in the fiber and through the elasto-optic effects, it influences the phase of the guided light. The change in the imaginary part of the refractive index results in a change in the absorption of the guided light and, hence, it can be detected by monitoring the light intensity. The change in the real part of the palladium's refractive index results in effective phase changes in the guided light that can be detected using

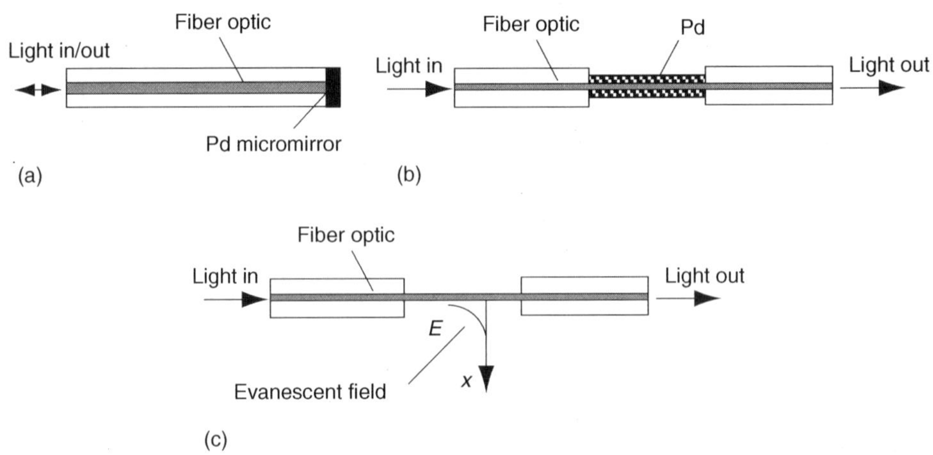

FIGURE 15.10
(a) Fiber optic hydrogen sensor with Pd micromirror, (b) evanescent field fiber optic hydrogen sensor, and (c) evanescent field profile near a core-exposed fiber optic.

interferometric techniques. The fiber optic sensor's sensitivity should be independently optimized at different temperatures since the fiber optic material is temperature sensitive. The sensor had a lower-level sensitivity of 2000 ppm with a response time of 20–30 s.

15.3.2.4 Palladium Mesowire and Nanoparticle Detectors

The sensitivity of hydrogen sensors can be improved by increasing the surface atoms of palladium available for reaction with hydrogen. The nanoscale structures of palladium provide a high specific surface area for the catalytic conversion of hydrogen. A laboratory-scale palladium mesowire hydrogen detector was demonstrated on a flexible substrate using a transfer process [27]. The mesowire sensor is based on resistivity changes caused not by surface adsorption, but on changes in the structure of the mesowires. Each sensor comprises up to 100 mesoscopic palladium wires arranged in a parallel configuration. In contrast to palladium-based resistive hydrogen sensors, the resistance of mesowire sensor decreases instead of increasing in the presence of hydrogen. The palladium mesowires were fabricated on a step-edge graphite substrate by a two-step electrochemical process. The palladium mesowires as fabricated had nanoscopic gaps and high resistance. On exposure to hydrogen, the nanowires expanded in volume creating electrical shorts and effectively forming a highly conductive circuit. The palladium nanowire had open break junctions (Figure 15.11a) before exposure to hydrogen. The break junctions close due to the expansion of the nanowires resulting from the phase transition in palladium and the corresponding lattice expansion. The mesowire sensor had a lower detection limit of 5000 ppm and a response time of milliseconds. The sensor fabrication process involves the transfer of palladium mesowires to an insulating substrate. The mesowire technology holds promise, but the fabrication process has to be improved for manufacturing repeatable hydrogen sensors. Recently, a palladium nanoparticle network–based sensor has been developed by precisely controlling the size and the distance between the nanoparticles [28]. The sensor contains a variety of nanoparticle networks that provide a broad dynamic sensitivity range for hydrogen (10–40,000 ppm) [29]. As shown in Figure 15.12, the nanoparticles are apart from one another in the network in air (open circuit), in the absence of hydrogen. On exposure to hydrogen, the palladium nanoparticles expand in size and volume due to the phase transition and lattice expansion, and physically contact each other. The contact between adjacent nanoparticles results in the effective decrease of the network resistance (close circuit). Unlike the mesowire detector, the nanoparticles are fabricated in a conductive circuit

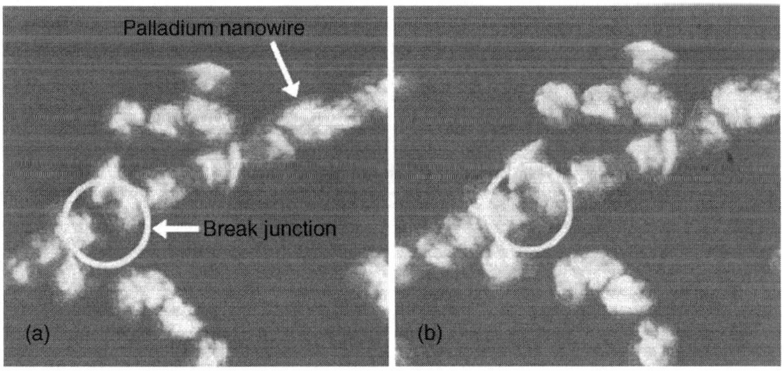

FIGURE 15.11
Atomic force microscopy of palladium nanowires showing the (a) open break junction before exposure to hydrogen and (b) closed break junction after exposure to hydrogen.

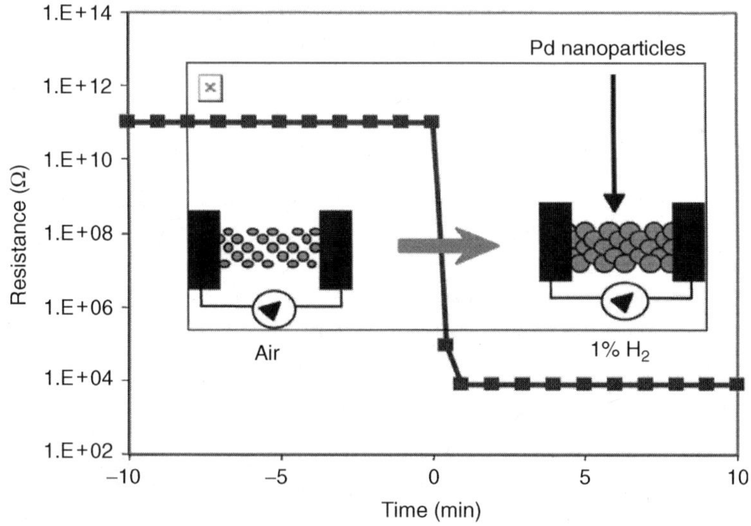

FIGURE 15.12
Schematic representation of the operation of the nanoparticle network hydrogen sensor.

to measure small increments in hydrogen concentrations. The sensor can detect as low as 1 ppm hydrogen. The palladium mesowire sensors have to be developed into a process-hardened package to be suitable for process applications.

15.3.2.5 Palladium Nanoclusters and Nanotube Detectors

Most of the research in nanostructures of palladium used for hydrogen sensors has been in the laboratory scale. Some of the promising research is mentioned here to outline recent advances in palladium nanostructures for hydrogen sensing. Hydrogen sensors have been fabricated using palladium nanotubes, nanoclusters, nanoparticles, and nanowires. Single-walled carbon nanotubes (SWNTs) decorated with Pd nanoparticles were used as molecular hydrogen sensors [30]. The Pd-modified SWNT (Pd-SWNT) showed significant electrical conductance modulation on exposure to small concentrations of H_2 in air, and the response reversible on removal of hydrogen. Researchers in Penn State University reported a self-cleaning hydrogen nanosensor based on an array of TiO_2-nanotubes [31] that clean the sensor surface from contaminants through a photocatalytic oxidation process.

A novel coating based on a self-assembled monolayer of siloxane was used to develop a hydrogen nanosensor by researchers in Argonne National Laboratory [32]. A monolayer of siloxane was precoated on a glass substrate followed by thermal evaporation of a thin layer of palladium. The siloxane coating on glass altered the size, nanometric gaps, and spatial distribution of the evaporated palladium nanoclusters. The gaps between neighboring palladium nanoclusters on the siloxane-coated glass were 10 times smaller on average than the gaps between the much larger, spread out clusters on the bare glass. The sensing mechanism of palladium nanoclusters is based on the hydrogen-induced lattice expansion (HILE) in Pd, which causes the swelling of the Pd nanobeads, thus narrowing/closing the gaps between the beads in the Pd chains. Unlike bulk PdH_x, the lattice constant pressure isotherms of Pd nanoparticles and small nanoclusters exhibit an even increase in the lattice constant as a function of hydrogen gas pressure. A slight narrowing of the intercluster gaps can still produce an appreciable increase in film conductance since the electron tunneling current between the intercluster gaps depends exponentially on the gap spacing.

The sensors showed a response time of few milliseconds for 2% H_2 and a lower detection limit of 25 ppm.

Palladium nanotubes have been fabricated using electroless deposition within the pores of track-etched polycarbonate membranes [33]. For direct palladium electroless deposition, the activated membrane was transferred to a fresh solution containing 1 mM $PdCl_2$, 13 mM disodium EDTA, and 18 mM hydrazine (N_2H_4). In this solution, $PdCl_2$ is the palladium precursor with hydrazine as the reducing agent. The complexing agent (disodium EDTA) chelated with the palladium ions forming $[Pd(EDTA)]^{2-}$, which allows the following reaction:

$$2[Pd\ (EDTA)]^{2-}_{(aq)} + N_2H_{4(aq)} + 4OH^-_{(aq)} \rightarrow 2Pd_{(s)} + 2EDTA^{4-}_{(aq)} + N_{2(g)} + 4H_2O_{(l)} \qquad (15.1)$$

The reaction can be carried out at room temperature on the pore walls and the surface of the membrane. The palladium nanotubes span through the entire thickness of the membranes. The hydrogen sensor prepared by this process showed a lower detection limit of 500 ppm with good resolution. The sensitivity can be increased by optimizing the pore diameter and density. This technique can be extended to other gases such as ammonia, oxygen, and methane. Palladium nanostructures such as nanotubes and nanoclusters have potential applications in hydrogen sensing, but much development is needed to commercialize these materials for the process monitoring industry.

15.3.3 Metal Oxide and Catalytic Bead Hydrogen Sensors

Metal oxides are semiconductor-type materials that constitute a major category of gas sensors. The most widely researched metal oxide used for sensing applications is tin oxide (SnO_2). Typical processes for manufacturing SnO_2-based sensors involve a thick semiconducting film made of micron size particles. Electron transport between particles is limited by the huge energy barriers at their grain boundaries (Figure 15.13a). The most common

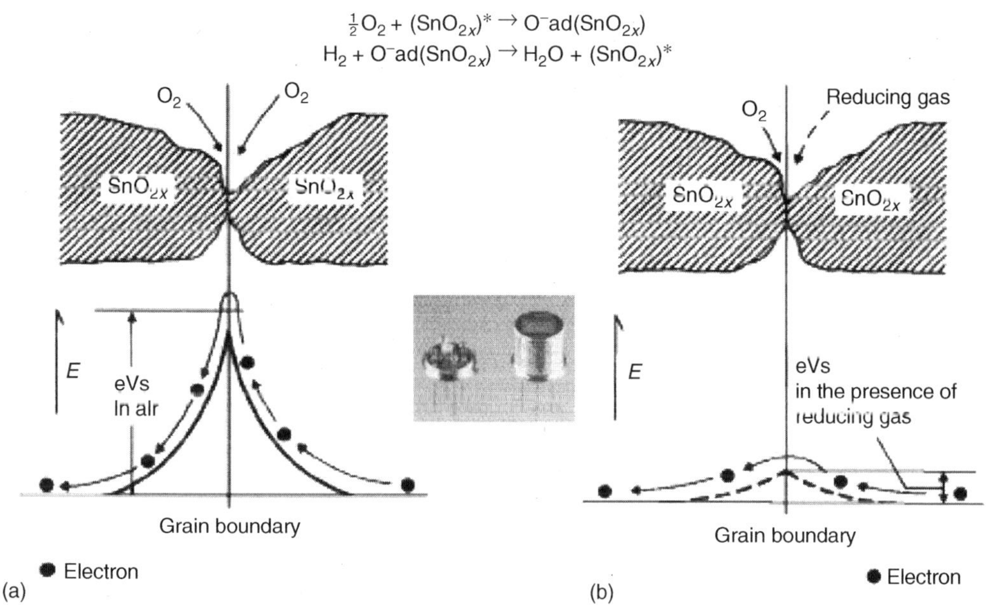

FIGURE 15.13
(a) Model of intergrain potential barrier (in the absence of gases) and (b) model of intergrain potential barrier in the presence of reducing gas (hydrogen).

method of overcoming the energy barrier is by thermal activation (i.e., heating the metal oxide to high temperatures in air).

Hydrogen sensing using metal oxide sensors involves a two-step process [34]. In the first step, ambient oxygen is adsorbed on the defect sites of the metal oxide surface (SnO_2). Hydrogen gas diffuses to the metal oxide surface in the second step to react with the adsorbed oxygen. This oxidation reaction transfers an electron into the metal oxide, which can be detected as a change in conductivity. In addition, this change in conductivity is due to the decrease in energy barrier of the metal oxide (Figure 15.13b). MOS have very fast response times and a broad gas detection capability, but they have high power consumption and are not selective to hydrogen. The surrounding gas environment will have an effect on the concentration of hydrogen recorded by the detector. The heated metal oxide sensors are attractive for low-cost hydrogen alarms for leak detection applications. The sensors have a lower detection limit of 30 ppm but are cross-sensitive to methane and other hydrocarbons. Recently, Citizen Watch Company Ltd., Japan, introduced a hydrogen sensor [35] based on a coil (double helix) structure. Palladium oxide has also been investigated for hydrogen sensing. A hydrogen-sensitive MOS detector based on a palladium oxide device was reported [36]. CB technology, commonly used for combustible hydrocarbon gases is also used for hydrogen sensing. A coil of wire is coated with glass or ceramic material, which is, in turn, coated with a catalyst such as platinum (Figure 15.14). The coil is electrically heated to a temperature that will allow it to burn (catalyze) hydrogen. The liberated heat is proportional to the concentration of the hydrogen present. This heat increases the temperature of the wire coil, increasing its resistance. The increase in resistance is measured electrically and is the source of the signal. CB sensors are not hydrogen

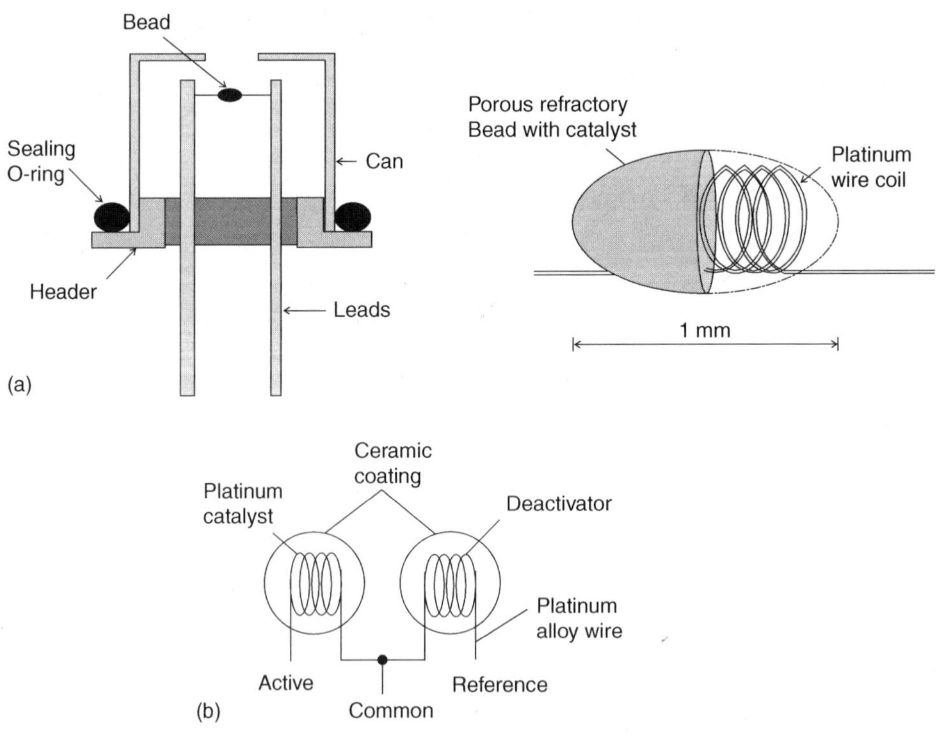

FIGURE 15.14
(a) Schematic illustration of CB sensor technology and (b) sensor configuration showing coils coated with catalyst.

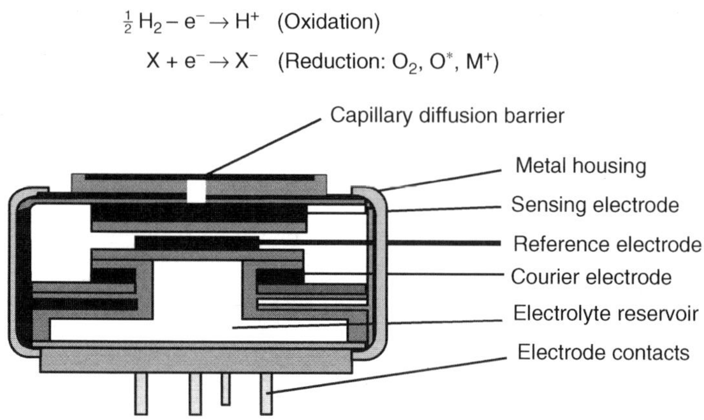

FIGURE 15.15
Electrochemical reaction scheme in a hydrogen sensor with a membrane for gas transport and a solid housing to prevent electrolyte leakage.

specific and can cause false alarms from common hydrocarbon and organic compounds. CB sensors are inexpensive and are available in different packages.

15.3.4 Electrochemical Hydrogen Sensors

Electrochemical sensors have been used for several decades for monitoring several toxic and volatile organic compounds. There has been little research in the electrochemical hydrogen sensors in comparison to the palladium-based sensors. Electrochemical sensors operate similar to a battery with a liquid or solid electrolyte. The electrochemical sensors generally have a two or three electrode configuration with a membrane for gas transport and a solid housing to prevent electrolyte leakage (Figure 15.15). The advantages of electrochemical sensors are low power, high sensitivity, and good selectivity. Most electrochemical hydrogen sensors are amperometric, that is, the electrochemical reaction of hydrogen causes a change in the circuit current. Polymer electrolyte membranes (PEM) have been used as hydrogen sensors at room temperatures [37]. Amperometric solid-state hydrogen sensors [38] have been fabricated with yttria-stabilized zirconia (YSZ) disk sandwiched between two platinum films. The electrochemical sensor technology is well established for the hydrogen sensor but is not practical for process monitoring applications due to their limited temperature range of operation and easy contamination by process gases.

15.3.5 Piezoelectric-Based Hydrogen Sensors

The principle of piezoelectric or quartz crystal sorption hydrogen detectors is based on a change in frequency of the crystal due to the absorption of H_2 molecules on a thin layer of hydrogen-sensitive coating. The decrease in the resonance frequency is in proportion to the amount of dissolved H_2 gas molecules and, hence, the gas concentration. The application of an electric field to the quartz crystal gives rise to a resonant frequency. The frequency changes in the quartz crystal due to the absorption of hydrogen in the coating. This technique has limited success due to the dependency of the sensor on the thickness of the coating and the interference from temperature and pressure.

A piezoelectric-sound-resonance cavity (PSRC) has been used as a sensing element for hydrogen [39]. The sensing mechanism was based on the acoustic property differences between H_2 and air. The concentration of hydrogen is indicated by the shift of the sound-resonance state of the PSRC. A lower detection limit of 8 ppm with a response time of 1.5 s and a broad detection range was demonstrated. Another type of piezoelectric hydrogen sensor is based on the SAW technology [40]. These sensors are fabricated on substrates such as quartz, lithium niobate, or zinc oxide using processes similar to those used in the production of large-scale integrated electronic circuits. The principle of sensing is based on the absorption of gas molecules on a functional hydrogen-sensitive layer. The gas absorption perturbs the properties of the SAWs that travel on a piezoelectric substrate. The energy of the SAW is localized within one or two acoustic wavelengths of the surface and interacts strongly with the medium adjacent to the surface. The SAW sensors are generally characterized by their high sensitivity, good linearity, and wide versatility.

15.4 Hydrogen Sensors for Industrial Process Applications

The need to quantify the absolute concentration (partial pressure) of hydrogen in process operations exists in a diversity of industrial applications [41] ranging from hydrogenating cooking oil (edible oils) to hydrotreating petroleum crude into heating oil, gasoline, diesel, and jet fuels, specifically to know that the process is safe and operating at peak efficiency is critical.

15.4.1 New Catalytic Technology in Foods

New catalytic hydrogenation technologies enable the production of partially hydrogenated edible oils with the desired low amount of TFAs (<10 wt%) without the subsequent elevation of solid fat levels in the oil mixture. Efforts to produce an affordable catalytic technology with high activity and trans-selectivity that could be used as a drop-in substitute in existing hydrogenation reactors are in development. The use of hydrogen sensors during these upgrades is under consideration as a cost-effective means to quantify and control the process. In the past, the hydrogen partial pressure was kept "on the high side" to ensure complete reaction and then excess gases were vented/flared.

When hydrogen is produced on-site, purified, and then passed into the process, hydrogen sensors can be placed at the feed line, in the hydrogenation zone, and at the outlet of the off-gas to provide more precise control of the quality of the product.

15.4.2 Metals Processing

15.4.2.1 Hydrogen in Aluminum Production

During molten aluminum processing, the most important dissolved gas is hydrogen, which is produced when aluminum reacts with moisture to form aluminum oxide and hydrogen. The solubility of hydrogen in the liquid aluminum is much higher than that in solid aluminum, thus dissolved hydrogen can lead to porosity during solidification. To produce sound castings, molten aluminum must often be degassed to remove hydrogen. The efficiency of these degassing processes can be improved using real-time measurement of the hydrogen content during the process.

Currently, systems for measuring the hydrogen content in molten aluminum are commercially available. In the most common systems, an inert gas (usually nitrogen) flows through a probe and over the molten metal, such that chemical equilibrium between the hydrogen partial pressure in the gas and the concentration of hydrogen in the molten aluminum can be established (i.e., according to Sievert's law). The sample gas is then transported to an off-line analyser (GC or MS), where its hydrogen content is measured and then related to the amount of hydrogen dissolved in the alloy. The gas chromatograph (Figure 15.16 [42]) process takes 6–10 min to perform, whereas the mass spectrometer (Figure 15.17 [43]) can respond in <2 min, from the time of sample. Although these nitrogen carrier gas systems are used in industry, the cost to buy/install/maintain of the system (GC or MS) is far too high for some applications. In response to the need for lower-cost hydrogen sensors, research has been performed on developing low-cost solid-state sensors. Solid-state in-line hydrogen sensors (Figure 15.18) are compact and can be located online and can line out in seconds to provide real-time process control at the lowest-cost per data point.

The newest process hydrogen sensors can quantify the hydrogen content directly from the nitrogen gas, and may be able to respond to the hot off-gas closer to the process to improve the spend of response for closed-loop control.

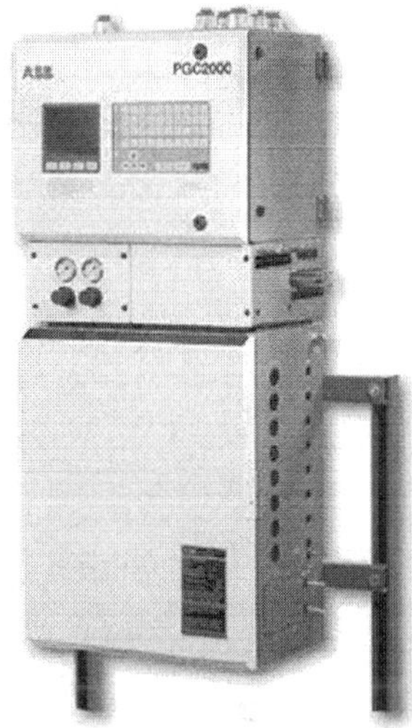

FIGURE 15.16
Picture of process gas chromatograph. (From ABB Instruments, ABB Inc., Lewisburg, WV. With permission.)

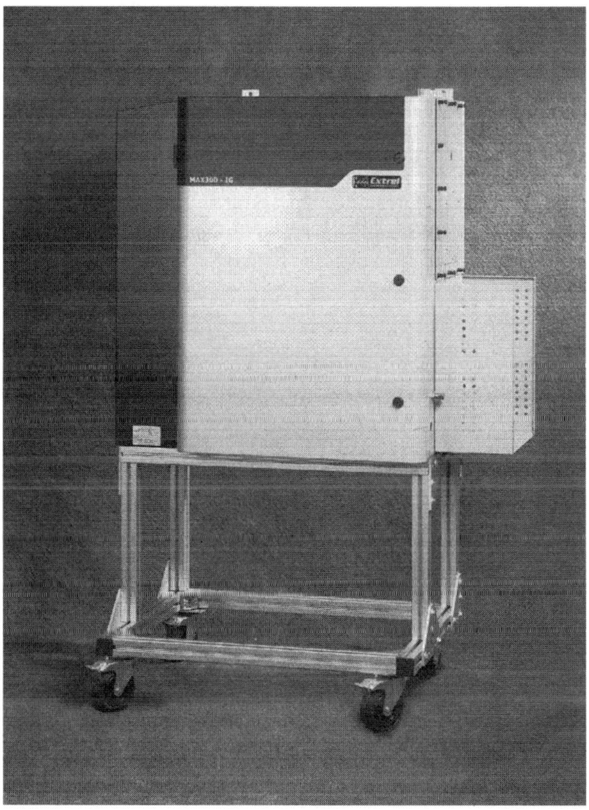

FIGURE 15.17
Picture of process mass spectrometer. (From EXTEL, Extrel CMS, Pittsburgh, PA. With permission.)

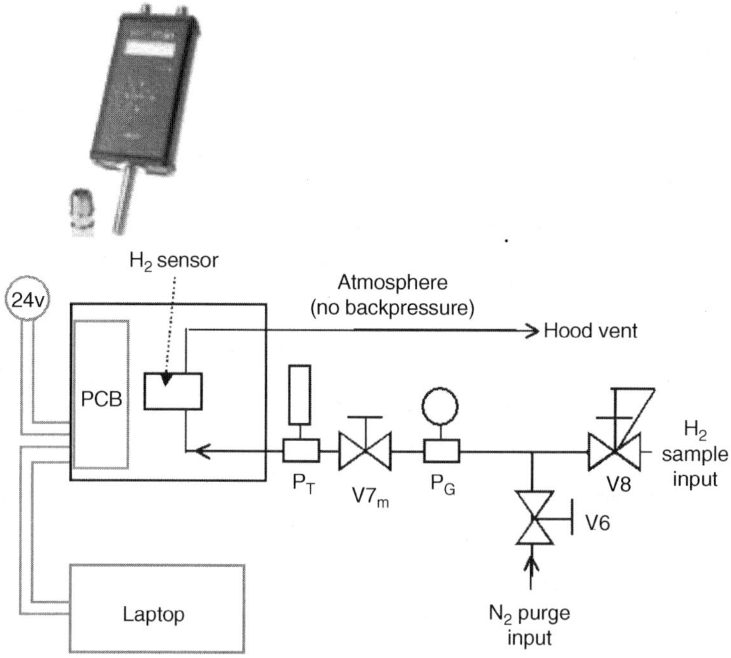

FIGURE 15.18
Picture of solid-state process hydrogen sensor implemented in a process stream. (PCB—printed circuit board.)

15.4.2.2 Hydrogen in Atmosphere Control

A hydrogen–nitrogen protective atmosphere in heat treating operations provides a safer, nontoxic alternative to dissociated ammonia in metal and material processing applications. In addition to more consistent atmospheres and improved reliability, hydrogen offers better regulatory compliance, improved safety, lower maintenance requirements, and lower capital costs. Flexible atmosphere systems from hydrogen concentrations of 1–100% can be used for these applications with no heating elements or catalysts. Bulk gas suppliers can use these atmosphere systems more reliably than dissociated ammonia generators. The system can further be conveniently shut down for planned maintenance and can be started up without losing valuable production time. An on-line hydrogen sensor can be applied to meter the hydrogen gas composition to within ±0.2%, to increase the efficiency of hydrogen use.

Hydrogen is used for annealing atmospheres. For carbon steel annealing, atmosphere compositions of 5–10% hydrogen with the balance nitrogen are commonly used in both continuous and batch furnace types. For stainless and other high alloy products, the hydrogen content is typically in the range 75–100%. Typical hydrogen purity is 99.995%. These atmospheres serve to transfer heat, remove residual rolling oils, prevent surface oxidation, and reduce any surface oxides that may be present, resulting in a bright annealed surface free from oxides, deposits, and discoloration. The ability to control the hydrogen content for consistent product quality is obvious. Process hydrogen sensors can be directly applied to this industry to monitor and provide feedback control for the hydrogen required. The newest hydrogen sensors can provide direct and real-time concentrations of hydrogen feed to the process and vent gas composition for recovery.

15.4.3 Pharmaceutical Applications

Multiphase hydrogenation reactions involving solid catalysts play a critical role in the pharmaceutical industry. Nearly 20% of all reaction steps in a typical fine chemical synthesis are catalytic hydrogenation. Mass and heat transport resistances are obstacles in such reaction systems. The microscale geometry of the microreactor technology offers gains against these hurdles; hence, their use would greatly benefit chemical processing in the pharmaceutical and other industries. This is indeed possible with a combination of reducible functions. Hydrogen sensor can assist the specialty chemical and pharmaceutical chemist (research and development [R&D] and pilot plant) to develop the most effective means to produce the final medical product. Nearly every major pharmaceutical firm has such reactors. The concentration of hydrogen in the feed and vent streams are painstakingly determined by GC or MS. An in-line process hydrogen sensor will reduce cost, save time, and increase the total number of data points.

15.4.4 Petrochemical Refinery Applications

15.4.4.1 Hydrogenation Applications

Hydrogenation is the chemical reaction of hydrogen with an unsaturated organic compound under the influence of temperature, pressure, and catalysts. There are several types of hydrogenation reactions (see Table 15.3). Hydrogenation is synonymous with reduction in which oxygen or some other elements (most commonly nitrogen, sulfur, carbon, or halogen) is withdrawn from, or hydrogen is added to, a molecule. When hydrogenation is capable of producing the desired reduction product, it is generally the simplest and most efficient procedure. The typical process is usually overpressurized in the reactor with hydrogen because an accurate partial pressure cannot be measured in real time. New hydrogen sensors can now work under high pressure (1–100 atm) and temperature (50–300°F) conditions to meet this challenge. Hydrogenation is used extensively in industrial processes. Important examples are the synthesis of methanol, liquid fuels, hydrogenated vegetable oils, fatty alcohols from the corresponding carboxylic acids, alcohols from aldehydes prepared by the aldol reaction, cyclohexanol and cyclohexane from phenol and benzene, respectively, and hexamethylenediamine for the synthesis of nylon from adiponitrile.

TABLE 15.3
Types of Hydrogenation Reactions

Substrate	Reaction	Metal	Solvent
Azides	$RN_3 \rightarrow RNH_2$	Pd, Pt, Ni	Polar
Aromatic nitro groups	$ArNO_2 \rightarrow ArNH_2$	Ni, Pd, Pt	Various
Benzyl derivatives (debenzylation)	$ArCH_2X \rightarrow ArCH_3 + HX X =$ OR, NR_2	Pd	Protic, acidic, or basic
Alkenes	$R_2C{=}CR_2 \rightarrow R_2HC{-}CHR_2$	Pd, Pt, Rh,	Various
Alkynes	$RC{=}CR \rightarrow RHC{=}CHR$	Pd/Pb	Low polarity
Aliphatic C=O groups	$R_2CO \rightarrow R_2CHOH$	Ni, Ru,	Polar
Aromatic C=O groups	$ArCOR \rightarrow ArCH(OH)R$	Pd, Pt, Cu	Polar
Aryl halides	$ArX \rightarrow ArH$ X = Cl, Br, I	Pd	Basic
Nitriles	$RCN \rightarrow RCH_2NH_2$	Ni, Ph/Pd, Pt	Basic/acidic
Imines	$R_2C{=}NR \rightarrow R_2CHNHR$	Pd, Pt	Various
Oximes	$R_2C{=}NOR$ R_2CHNH_2	Ni/Pt, Pd	Basic/acidic

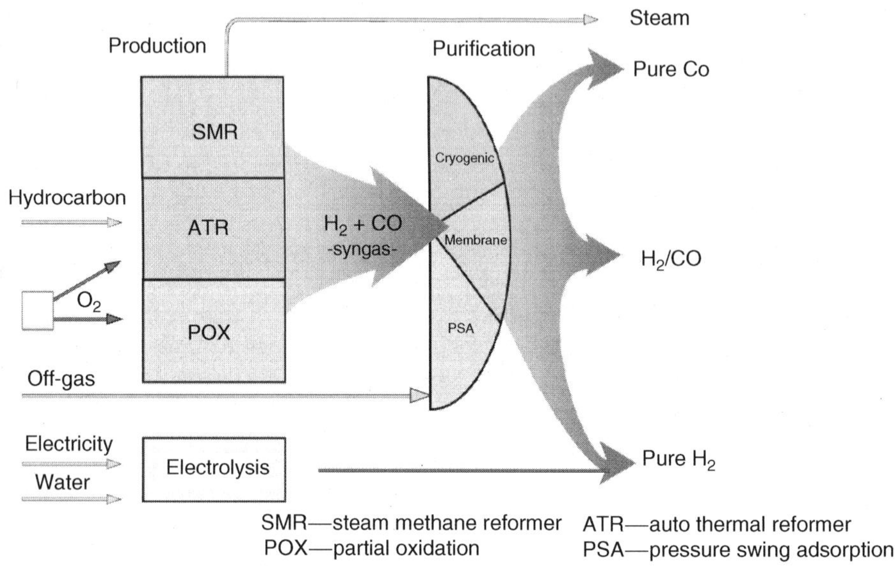

FIGURE 15.19
Types of hydrogen production techniques.

15.4.4.2 Hydrogen Production

There are several types of hydrogen production techniques. Most of the techniques were described in the previous chapters. In all these processes (Figure 15.19), continuous hydrogen monitoring is highly desired.

15.4.4.3 Hydrodesulfurization

Hydrodesulfurization (HDS) is a catalytic chemical process widely used to remove sulfur (S) from natural gas and refined petroleum products such as gasoline or petrol, jet fuel, kerosene, diesel fuel, and fuel oils (Figure 15.20 [44]). The purpose of removing the sulfur is to reduce the sulfur dioxide (SO_2) emissions that result from using fuels in automotive vehicles, aircraft, railroad locomotives, ships, gas or oil burning power plants, residential and industrial furnaces, and other forms of fuel combustion. Another reason for removing sulfur from the naphtha streams within a petroleum refinery is that sulfur, even in extremely low concentrations, poisons the noble metal catalysts (platinum and rhodium) in the catalytic reforming units that are subsequently used to upgrade the octane rating of the naphtha streams. In an industrial HDS unit, such as in a refinery (Figure 15.20), the HDS reaction takes place in a fixed-bed reactor at elevated temperatures ranging from 300°C to 400°C and elevated pressures ranging from 30 to 130 atm, typically in the presence of a catalyst consisting of an alumina base impregnated with cobalt and molybdenum. The application of hydrogen sensors throughout the plant can better define process control loops and improve the overall process efficiency.

A typical refinery application of the hydrogen sensor is to measure the total hydrogen content in a mixed gas matrix as a function of time. The development of the process hydrogen sensor now allows direct point-of-use analysis of hydrogen every 12 s to within ±0.2% in a range 50–100% v/v. Figure 15.21 shows a typical output of a process hydrogen sensor. Placement of four to six process hydrogen sensors in a refinery operation can improve gas management and process efficiency.

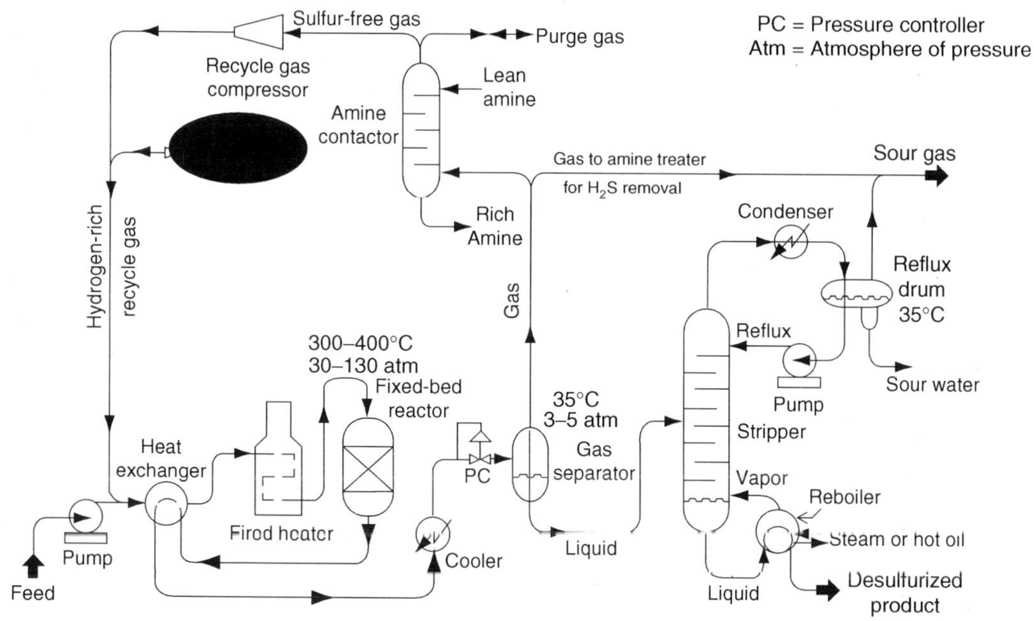

FIGURE 15.20
Schematic showing the hydrodesulfurization process in refineries. (From Yamaguchi, N., Trans-Energy Research Associates, Seattle, WA. With permission.)

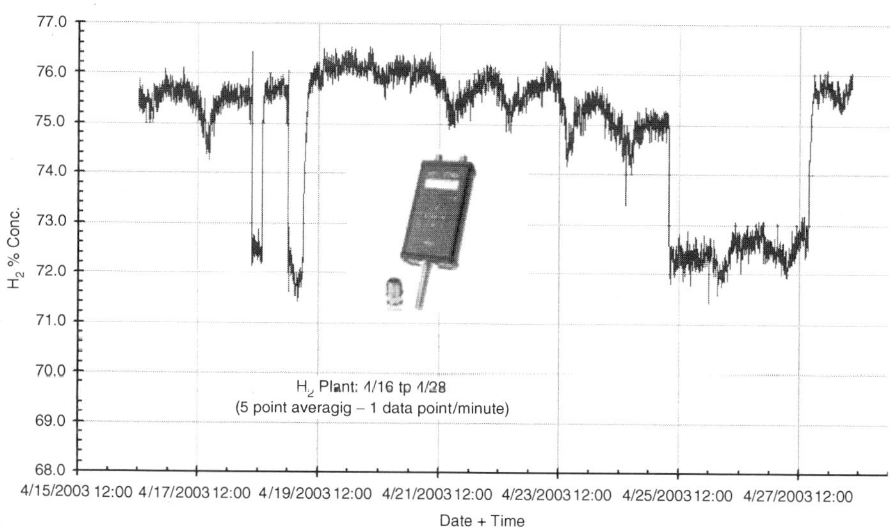

FIGURE 15.21
Real-time output of a solid-state hydrogen sensor that can be used in process refineries to improve efficiency. (Courtesy of Air Products and Chemicals, Inc., Allentown, PA.)

15.4.5 Chlorine Manufacturing

To ensure a safe, uninterrupted, and smooth operation of chlorine production plants, it is vital to prevent the formation of explosive gas mixtures. Continuous monitoring of the hydrogen concentration in chlorine gas in several stages of the chlorine production process can help in identifying the formation of such explosive mixtures in an early stage.

A mixture of hydrogen and chlorine gas, eventually in combination with air, can be very explosive if one of the components exceeds certain limits. In chlorine production plants, based on the electrolysis of sodium chloride solutions, there is always a production of hydrogen. It is, therefore, essential to be aware of the actual hydrogen content of chlorine gas process streams at any time. There are several places in the chlorine production process where the hydrogen content in the chlorine gas can accumulate above the explosion limits. Within the chloralkali industry, mainly two types of processes are used for the production of chlorine—the mercury- and the membrane-based electrolysis of sodium chloride solutions (brine).

One of the main differences between these two process types is the place where hydrogen is produced. In the mercury-based process, the hydrogen is produced in a separate reactor where the sodium amalgam is decomposed. Therefore, in this type of process no direct contact between the concentrated chlorine and hydrogen process streams is possible. Whereas in the membrane-type process, production of chlorine and hydrogen takes place in the same cell, separated only by a thin sheet of ion exchange membrane. In mercury-based electrolysis, the main source of hydrogen is unwanted side reactions at the electrodes. Normally, the produced hydrogen from this source is continuous but at a low level. A sudden rise of hydrogen from this source is not to be expected because hydrogen is produced in a different cell. However, a slow rise is possible. Therefore, the response time for hydrogen measurements in the outlet of mercury-based electrolyzers can be in the order of minutes.

In the membrane-based electrolysis, a continuous low-level "leakage" of hydrogen through small pores in the ion exchange membrane occurs. But when the membrane gets damaged (cracked), a sudden rise in hydrogen concentration above the explosion limits can occur within seconds. Therefore, the response time for hydrogen measurements in the outlet of membrane-type electrolyzers is set to <4 s.

After leaving the electrolyzers, the chlorine gas is dried and compressed (liquefied), during which the hydrogen content in the chlorine gas will rise; no sudden rises are to be expected in this stage of the process, hence, hydrogen measurements can have a response time of minutes here.

For applications where a fast response (<10 s) is needed, thermal conductivity and infrared (IR) spectrometry as measuring principles are the only choices for the measurement of hydrogen in chlorine gas. GC is a good alternative only if other gases also have to be measured and a slower response (>6 min) time is acceptable. When concentration variations of other components in the gas mixture are expected, differential thermal conductivity or IR spectrometry (conversion of hydrogen to hydrogen chloride) has to be applied. From the other discussed techniques, Raman gas laser, the H2scan solid-state sensor, and a semiconductor or electrochemical device behind a chlorine-resistant membrane appear to be the most promising for the chlorine manufacturing industry.

15.4.6 Power Generation and Nuclear Waste Monitoring

Hydrogen monitoring in power transformers and nuclear power plants are critical to prevent failures and catastrophic explosions. Hydrogen is the incipient fault marker in oil-cooled power transformers and is generated due to the degradation of oil–cellulose insulation systems. The life of the electrical transformer depends on the life of the internal insulation system, which can be shortened by a number of events: exposure to extreme conditions, aging, and wear and tear. The oil–cellulose insulation system in a transformer cannot be replaced and due to the omnipresence of oxygen and water, insulating oil deterioration is normal. A sudden increase in the levels in electrical insulating oils can indicate an imminent explosion; closely monitoring hydrogen is an effective tool in predicting and preventing transformer failure. Currently, analytical techniques such as gas chromatographs are used with cumbersome sampling procedures to measure hydrogen levels in insulating oils of power transformers.

Accurate hydrogen monitoring in nuclear power reactors is also critical for reactor efficiency and safety applications. Existing nuclear power plants are applying for and receiving plant life extension (20 years). Baseline demand for power has caught up with capacity. The nuclear industry has concluded that the only viable option to meet the increasing base load demand is to begin the construction of new generation nuclear power plants. Demand for hydrogen monitoring will continue for existing plants and increase in support of the new plants. Hydrogen specificity, long-term stability, ease of operation, and installation will be key considerations for hydrogen sensors in nuclear power plants.

Real-time measurement of hydrogen inside nuclear waste barrels is critical for safety in the nuclear industry. It is estimated that there are over 500,000 low-level waste containers in the United States. The presence of hydrogen in the container is an early indicator of the problem. There is a great need for a sensor that would continually monitor the condition inside the waste barrel and feed real-time data to an operator. The nuclear waste monitoring system would measure five critical components within the storage container that could result in a container leaking radiation, or in an extreme case cause a container to explode with the potential for catastrophic damage. The five sensors used in the device are temperature, moisture, pressure, radiation, and hydrogen. Today, there is no real economic way to continuously monitor the condition inside a nuclear waste barrel except for taking a sample of the air inside the container and then having an expensive diagnostic test performed. Currently, tests are performed to determine if hydrogen levels exceed the levels of safety and are early indictors of a future problem (i.e., an explosion with devastating consequences). The Department of Energy (DOE) takes statistical samples of only 10% of the containers in the United States every year because of the high cost ($1000.00 per sample per barrel) and the time it takes to extract the sample and send it off to a laboratory for diagnosis. The big issue in the United States and worldwide is that there could be containers that are in a near-explosive state because only a small number of containers are tested; thus, the risk of an explosion is increased. It is a priority of the energy department to clean up the nuclear waste barrels and monitor hydrogen accurately in storage containers for safe operation. The DOE owns and operates a major nuclear facility near Idaho Falls where hydrogen specific monitoring is highly desired to protect the surrounding area from hydrogen leaks. The facility currently uses a solid-state hydrogen-specific sensor supplied by H2scan Corporation to measure and remotely indicate the level of hydrogen.

15.4.7 New In-Line Sensor for Industrial Process Applications

The use of *in situ* hydrogen sensors to replace the need to safely sample and conduct off-line laboratory analysis or to use costly on-line GC or MS is most attractive for the process industry. Recently, a new solid-state hydrogen sensor has been developed and tested that can be installed at multiple points in a process plant and directly linked to the DCS or a program logic control (PLC). The technology is unique in the marketplace and continuously monitors hydrogen levels from 15 ppm to 100% by volume in the presence or absence of oxygen. H2scan Corporation offers a wide range sensor (WRS) based on a novel "chip on a flex" technology, wherein a hydrogen-specific chip is mounted on a flexible substrate.

The sensor element constitutes a palladium–nickel alloy resistor with a temperature sensor and a proprietary coating. The sensor has a broad operating temperature range and a sophisticated temperature control loop that includes a heater and a temperature sensor, which controls the die temperature within ±0.1°C.

A molecular-level coating enables the sensor to operate continuously in most process streams. The Model 700 HY-OPTIMA in-line process monitor (Figure 15.21) offered by H2scan has been tested in gas streams of 1000 ppm H_2S, 100 ppm CO, include CO_2, CH_4, alkanes, alkenes, alkynes etc. for refinery applications, in 35% Cl_2, and 95% relative humidity for corrosive applications. The ruggedness of the coating extends the sensor's operational life in process streams, including complex hydrocarbons, to over 3 years, with low sensor replacement costs. The sensor has a robust mechanical design with a 3 in. × ½ in. corrosion-resistant sensor probe for installation into a gas stream operating in a wide temperature (−20°C to 100°C) and pressure (up to 100 psig) range using VCR or compression fittings that make it transferable to the New Sample System Initiative (NeSSI) platform led by Honeywell, and integrated into the consortium for process analytics. The sensor is ATEX certified for operation in Class I/Div II Gp B hazardous locations. The output (4–20 mA) from the sensor is readily integrated into existing DCS. The product has a life expectancy of 10 years and the recommended calibration frequency is 3 months. When in-process sensors are combined with hydrogen sensors for safety monitoring (leak detection), a plant using hydrogenation technology could apply 20–30 such sensors if the network connection can be made common to reduce cost to install. The broad range of hydrogen sensor technology and the leak detection capability of the H2scan technology lends itself to hydrogen transport in existing natural gas pipelines. The current natural gas pipeline infrastructure will need hydrogen-specific sensors at critical joints, valves, and around transfer stations to alert for hydrogen leaks. In a similar manner, the ability of the H2scan technology to sense high concentrations (5–99%) of hydrogen in the presence of other hydrocarbons provides a special market opportunity and offers an alternative to cumbersome analytical techniques for the process industry (Figure 15.22).

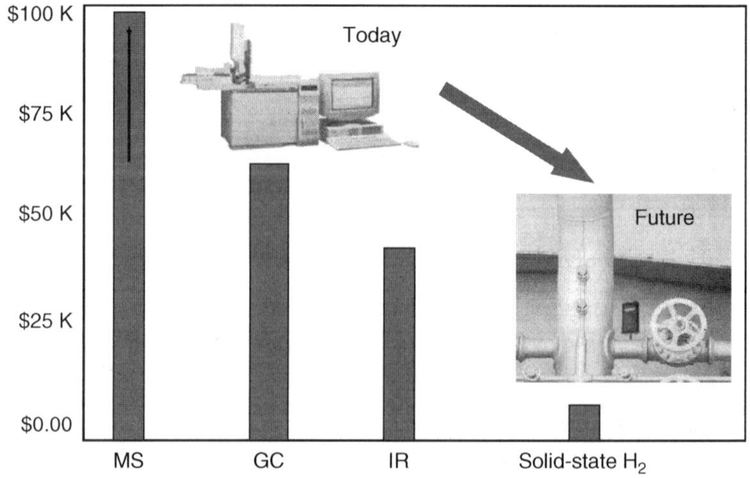

FIGURE 15.22
Graph indicating the cost per process analyser used in the industry today and in the future.

15.5 Sensors in Hydrogen Fuel Applications

15.5.1 Hydrogen Refueling Stations

Using a flammable material as a fuel is not new. Gasoline, as reviewed earlier has a lower flammability point than hydrogen, and if spilled is a source of fuel for a fire. The one point that makes hydrogen different is that one can smell the fume of gasoline, whereas hydrogen is odorless. Adding an odorant to hydrogen or causing an odor to occur has been patented [45]. Such additives can be deleterious to fuel cells, and in some cases would add to sulfur off-gases. This area of research is important since it helps the user to smell hydrogen before the explosive limits. The automotive industry would prefer to have an intrinsically safe fueling system so that customers are not afraid to ride in the vehicle.

As the current concepts to fuel hydrogen-powered vehicles range from high pressure (650–1000 atm) storage tanks of gaseous hydrogen to chemically bound hydrogen that can be released on demand. The amount of hydrogen needed per full charge to archive the 250–300 mi. of run time is the goal. The need for safety sensors has been reviewed by the automotive industry and the U.S. DOE. The number of sensors suggested per vehicle range from two to six. These sensors have different working environments, but if one type of sensor can meet all environments the cost per unit drops, and the cost to maintain the system can be better controlled.

In the high-pressure hydrogen vessel cars and the chemically bound hydrogen service, hydrogen sensors are needed at key connection points that are not welded: at the storage vessel valve/seal/interface to the delivery tubing to the carburetor, above the storage vessel, in the truck compartment where gases can be trapped, within the engine compartment at points of seals/fittings, and in the passenger compartment. All such sensors will have to be interfaced to the vehicles' computer system to initiate a fail-safe shutdown or close out the ignition system from starting the car if a leak is present above a threshold.

A more critical function is the fueling of the hydrogen car by the general public. The use of high-pressure fueling nozzles and equipment to load up to 10,000 psig (666 atm) of hydrogen is under development. External leak detection devices are generally used by that alert personnel once hydrogen has escaped to the open environment from a closed system. Recent U.S. patents [46,47] describe the use of solid-state hydrogen sensors that can be embedded in the hydrogen-filling system nozzle for refueling applications (Figure 15.23). As shown in Figure 15.23 (*top*), hydrogen is transferred from a source container to a destination container through a transfer line and the fueling nozzle that forms a part of a monitoring and control system comprising a controller and an alarm. The hydrogen sensors positioned inside the nozzle housing to allow for the direct immediate knowledge of presence of hydrogen in transit and the controller can activate an alarm and stop the flow of hydrogen if necessary. A conceptual hydrogen fueling station is shown in Figure 15.23 (*bottom*), using multiple fueling nozzles, each of which is coupled to a common hydrogen source by control valves. Each hydrogen sensor is independently coupled to the controller and can be used to shut down a particular hydrogen transfer to a fuel port of the vehicle's hydrogen tank.

The sensors are used for providing an alert while hydrogen gas is transferred through a refueling nozzle from a source to a destination container, detect possible leaks at the point of greatest concern, and provide a means to shut down the filling process. Similar hydrogen sensors can be located outside the service pump area and at the compressor to establish an alert, should hydrogen be leaking above a threshold value.

Such hydrogen sensors must be rugged, hydrogen specific, must not saturate at hydrogen safety levels, and must have a long and stable lifetime (>12 months); they must not

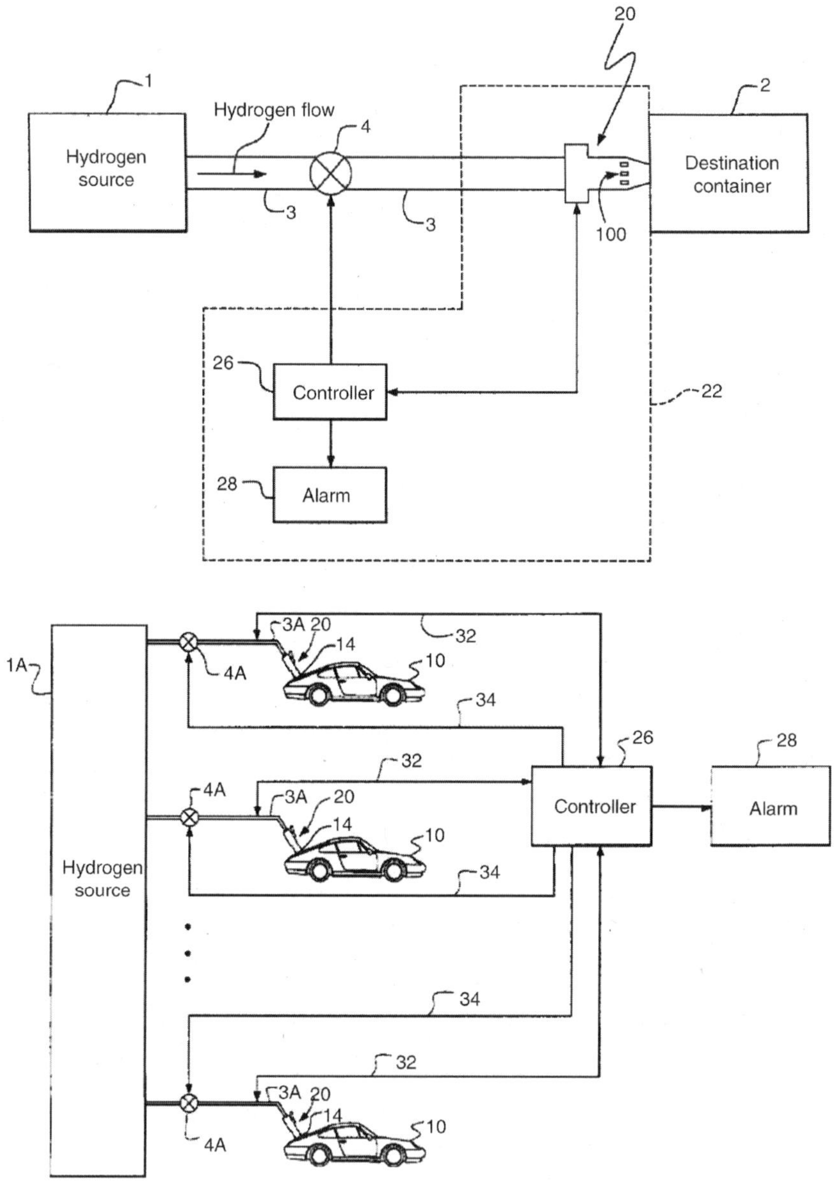

FIGURE 15.23
A hydrogen fueling nozzle with a monitoring and control system for transferring hydrogen between containers (*top*). Schematic illustrating a hydrogen fueling station (*bottom*).

require oxygen to function; they should be able to respond to >0.2% hydrogen within seconds; and they must not be affected by the environment (temperatures from −10°C to +50°C and under humid (>99% relative humidity) conditions).

15.5.2 Hydrogen Transport Pipelines

The infrastructure for gaseous hydrogen delivery by pipelines must include transmission and distribution. In conventional terminology, *transmission* lines generally use relatively large-diameter, high-pressure (35–100 bar or 500–1500 psi) pipelines for moving large

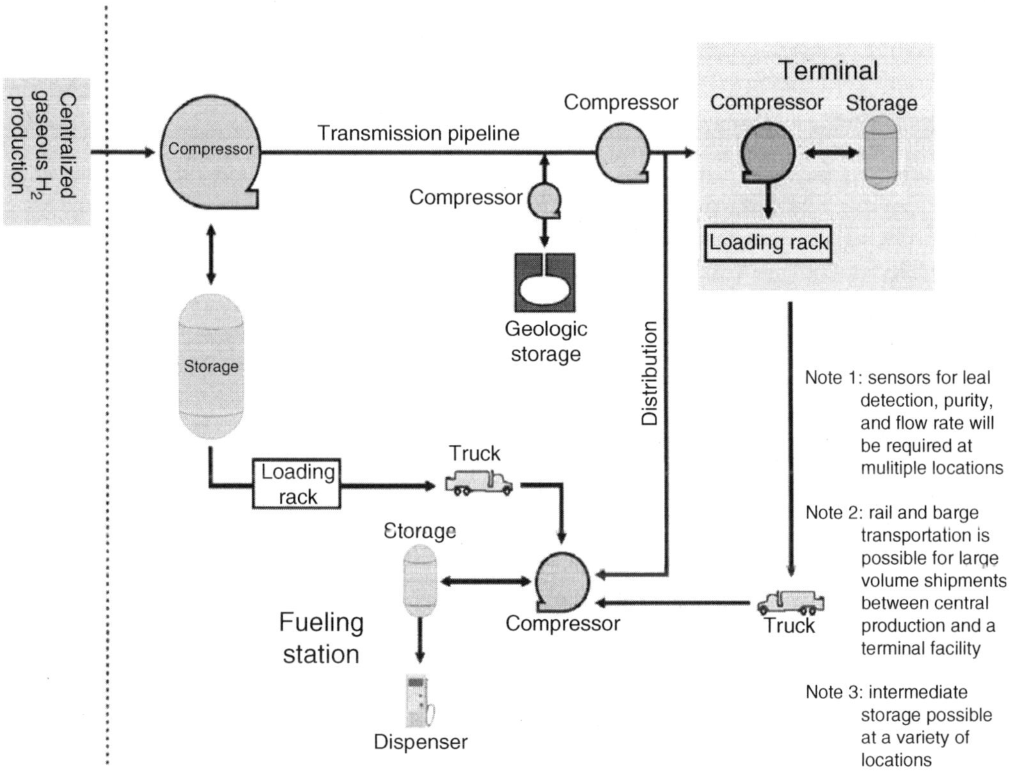

FIGURE 15.24
Gaseous hydrogen delivery system.

volumes of gas over long distances (Figure 15.24). In contrast, *distribution* lines typically provide more localized delivery of smaller volumes of gas through smaller-diameter, lower-pressure (0.3–14 bar or 5–200 psi) pipelines. For hydrogen delivery, pressures in distribution lines are likely to be higher (14–28 bar or 200–400 psi) than in natural gas distribution lines due to the need for high pressures at refueling stations and power sites. Furthermore, with appropriate separations technology included, the issues discussed in the following text apply to pure hydrogen gas as well as to mixtures of natural gas containing a substantial fraction (10–30%) of hydrogen gas.

The United States currently has >1000 km (630 mi.) of dedicated steel hydrogen transmission pipelines, mostly in Texas, Louisiana, and California at refinery operations. There are significant technical questions that must be addressed before establishing a major hydrogen pipeline infrastructure. The chief technical concern is hydrogen embrittlement of metallic pipelines and welds. In the simplest sense, hydrogen embrittlement describes the decrease in ductility or toughness of materials as a result of interaction with atomic hydrogen. Because welds are particularly susceptible to embrittlement, pipeline materials that eliminate the need for welding together pipeline sections (e.g., "spoolable" pipeline materials) may also help solve the embrittlement problem.

Currently, no commercial pipelines for liquid hydrogen exist. Without breakthrough technologies, liquid hydrogen delivery in pipelines is considered impractical and cost prohibitive. In addition to the high cost and energy inefficiency of current liquefaction technologies, the engineering requirements for constructing a pipeline with appropriate materials and codes are problematic [48].

The application of hydrogen sensors to current and future gas-filled (or liquid filled) pipelines is a major area requiring development. Multiple microsensors that can be "wrapped" around pipeline joints and valves to detect the first level of leakage would significantly aid in improving the safety. These sensors would then have to be monitored through radio frequency from low-flying aircraft that can scan the many miles of pipeline at frequent intervals, similar to natural gas pipelines monitored today using ground sensing analytical systems (hydrocarbon sensors/gas chromatographs—TCD) from vehicles.

The hydrogen sensors required for such operations must be stable over a large range of environmental conditions, respond to 0–0.5–30% hydrogen in humid air, be powered by solar cells or long (>2 years) life batteries, and can be easily replaced as a "belt" and slipped around or pulled through a channel. Such pipelines would be underground (1–3 m), even under water, allowing for pockets of hydrogen or mixed gases to concentrate, sometimes above the 4% flammability level.

Pipeline safety management is one aspect; the other is to quantify the hydrogen content within the pipeline. In some cases under consideration, hydrogen will be mixed with natural gas for transport, and separated by membrane technology at the point of use. This would take advantage of the current natural gas pipeline infrastructure and reduce time of costs. In this case, hydrogen sensors would be placed inside the pipeline to monitor the "absolute" concentrate (partial pressure). These sensors would be connected through wire or radio frequency to control stations to confirm mixing is complete or that the concentration of hydrogen being delivered is correct. In these cases, the hydrogen sensors must be a full-range device (0.1–99.5%) and operate at changing pressures (1–30 atm) and temperatures from −20°C to +50°C.

15.5.3 Hydrogen Storage Tanks: Cryogenic/Noncryogenic

Today, in the United States there are only 10 liquid hydrogen facilities making 5,400 to 32,000 kg/day. All of these are managed by the major gas producers, Air Products, Air Liquide, Praxair, and BOC-Linde. The transportation of liquid hydrogen is limited by distance (boil-off) and roadway access as defined by the U.S. Department of Transportation. Liquid hydrogen is of the highest purity being cryodistilled at −253°C, and has a limited market (NASA, electronics, specialty stainless steel and fuel cell R&D).

Storage of liquid hydrogen is achieved in large well-insulated tanks from which it is dispensed to liquid tankers for transport over the road. The use of hydrogen sensors (mostly flammable sensors) in and around such facilities is a common safety practice today. As hydrogen-specific sensors and solid-state sensors become more reliable, and if their cost is reduced they will replace the more common flammable gas sensors.

Such storage tanks are constantly "boiling off" hydrogen, as are liquid hydrogen tankers, otherwise the pressure within the tank or tanker will rise well above the nominal 10–15 psig above the liquid hydrogen. Therefore, the use of hydrogen sensors is to monitor the consistent build-up of hydrogen in specific areas. Here, the precision of the sensor and its accuracy to report the changing partial pressure are the most important characteristics.

15.5.4 Hydrogen Fuel Cells and Automobiles

Hydrogen is used as a fuel in proton exchange membrane fuel cells. Hydrogen concentrations of 60–100% are used as feed gas to fuel cells. The hydrogen is converted to water and electrons through a proton exchange mechanism in the polymer membrane. Hydrogen-powered fuel cells are used to drive automotives and can be used as backup for generators and small power plants. Hydrogen is a cleaner fuel for fuel cells compared to natural gas since it has lower emissions. Hydrogen, when used as a fuel source in an automobile or in a generator room has to be accurately monitored to prevent the risks of explosion.

Accurate monitoring of hydrogen in fuel cell stacks are required for proton exchange membrane (PEM) fuel cells. Hydrogen is used as a feed in a fuel cell (Figure 15.25) [49] in the anode side, and air is fed into the cathode side. A catalyst causes the hydrogen to split into protons and electrons. The PEM membrane allows the protons to pass through the cathode. The electrons travel along the circuit to the cathode to cause an electric current. Hydrogen detection is required in the fuel cell anode loop (~60–100% H_2 in nitrogen background). There is a high concentration of water vapor at the cathode side as the electrons and protons combine with oxygen to form water. The hydrogen sensor installed in the anode loop of the fuel cell should not have a response time delay or lower sensitivity in the presence of water vapor (H_2O). The emissions from fuel oxidant outlet of a fuel cell has to be monitored for hydrogen concentrations to prevent possible explosions since the concentration of hydrogen is in the range 0–4% H_2 in a oxygen-reduced background. Hydrogen

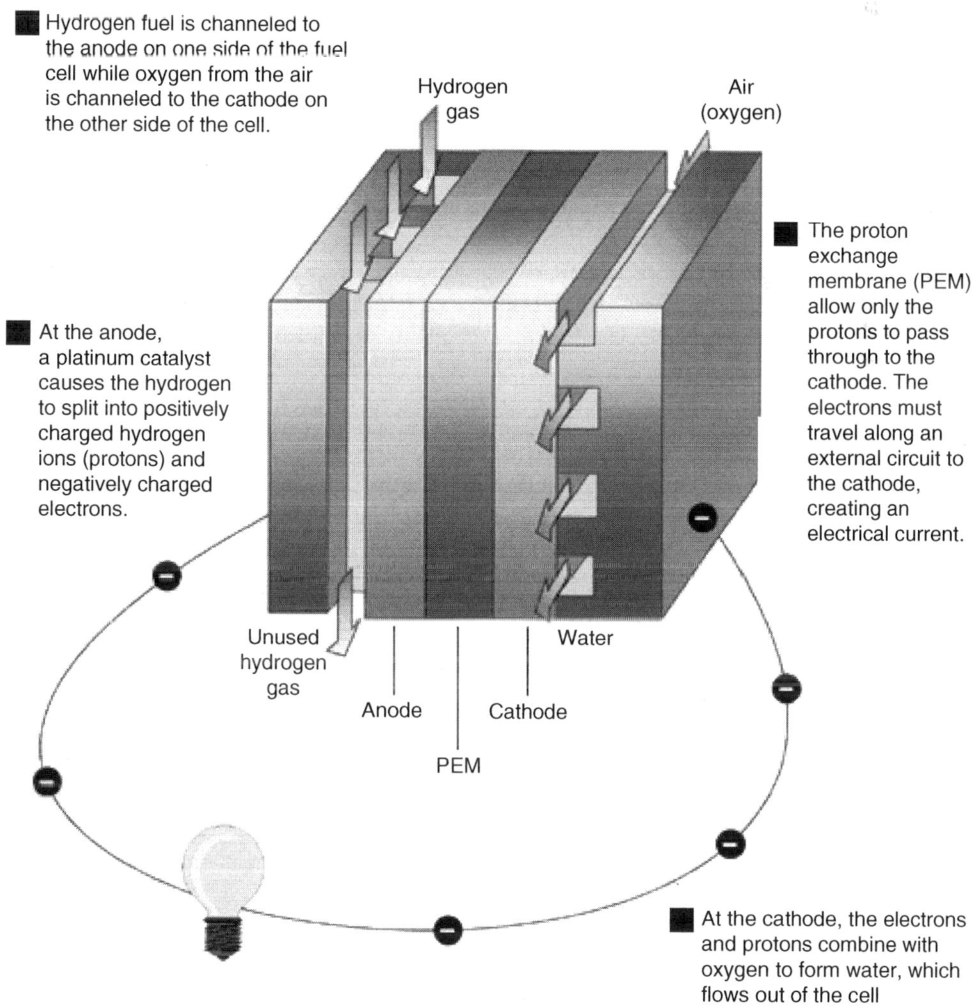

FIGURE 15.25
A typical PEM hydrogen fuel cell.

leak detection is also required in the fuel cell ventilation areas. An accurate leak detector should be able to detect 0–4% H_2 in air backgrounds without any cross-sensitivity to common gases in exhausts such as methane (CH_4), carbon dioxide (CO_2), carbon monoxide (CO), and nitrous oxides (NO, NO_2).

The major requirement for a reliable hydrogen sensor operation in the fuel cell environment is in 100% condensing humidity. Most of the fuel cells have abundant humidity and the sensor needs to operate continuously in humid environments. In some cases, the hydrogen sensor can also be operated at very low temperatures (as low as −40°C). The fuel cells regularly have a cold start, when operated from a very low ambient temperature; the sensor needs to attain ambient temperature quickly (<30 s) and continue operation well below ambient temperature before the fuel cell itself reaches the ambient temperature.

Figure 15.26 shows the hydrogen sensor locations in a fuel cell vehicle. In a hydrogen-powered vehicle, gas mileage is important, therefore, the "unused hydrogen" is recirculated. Hydrogen area monitors are needed in garages to ensure that there is no leak from a hydrogen-powered vehicle. The passenger compartment should be monitored for hydrogen leaks to ensure the safety of passengers. The hydrogen is generally stored in the compartment and the ambient hydrogen has to be monitored in the fuel storage area and high pressure leaks. Accurate hydrogen monitoring is also required in fuel cell ventilation systems, oxidant outlets, and fuel cell anode loop as described in Figure 15.26. In a fuel cell loop, a fast responsive sensor is necessary to monitor the recirculated hydrogen during start up and shut down and to control electrical load appropriately to avoid catalyst degradation.

Most solid-state sensors are heated to well above 100°C and can operate in the "cold start" condition in a fuel cell. Another important performance parameter for a hydrogen sensor in a fuel cell is its resistance to water entry. Most fuel cells have excess of liquids including water during operation. It is highly possible that water will splash or penetrate into the hydrogen sensor mounted in the ventilation or outlet of a fuel cell. Hydrophobic

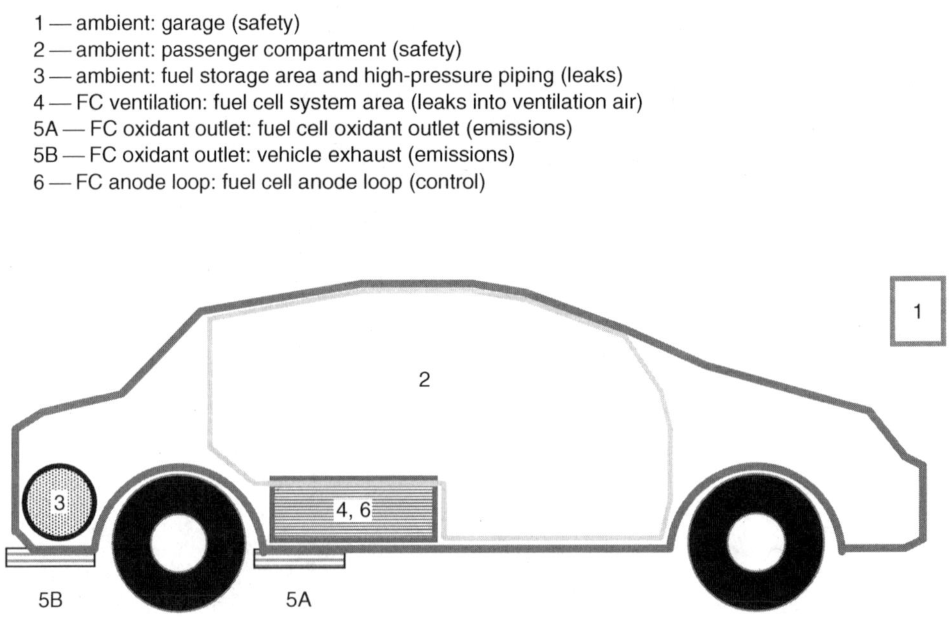

FIGURE 15.26
Hydrogen sensor locations in fuel cell vehicle.

membranes are generally used in hydrogen sensors to prevent water entry into the sensors. Recent developments in solid-state hydrogen sensors and molecular-level inorganic coating technologies enable sensor operations without interference from condensing water and high relative humidity. Hydrogen sensors for fuel cell applications also need to operate in a wide operating temperature range (−40 to +100°C) and should have fast response times.

15.6 Hydrogen Sensors: A Market Overview

Several hydrogen sensing technologies and detection schemes were discussed in this chapter. The success of each technology is determined by its performance in the target application domain.

- The application of hydrogen sensors to the broad production processes of petrochemical hydrogenation is a multitrillion dollar business throughout the world. The main goal of the process engineer in each of the hydrogen production facility is to increase production and maintain consistent product at the lowest possible cost. An in-line process monitor (cost <$10,000) is highly desired as a replacement to traditional gas sampling techniques such as GC.
- There is a need for accurate in-line hydrogen monitor in individual electrolyzers in chloralkali industry that will provide huge cost savings in chlorine manufacturing processes. Again, an in-line process monitor that can be installed in individual electrolyzers is highly desired.
- Hydrogen specificity is highly desired in hydrogen refueling stations, lead acid battery rooms, and nuclear power plants in radioactive waste recycling processes. There is a need for hydrogen-specific sensors that will replace CB technologies, eliminate false alarms, and prevent plant shut downs.
- The fuel cell and automotive markets seem to be a large market for low-cost (<$100/sensor) hydrogen sensors. Low power sensors are in high demand for multiple-point operations in fuel cells and internal combustion engines.
- Hydrogen monitoring in power transformers is highly desired. Accurate hydrogen sensors for transformer headspace and low cost hydrogen sensors immersed in oil have considerable market potential.
- MEMS-based hydrogen sensors have been used in NASA shuttle missions STS–95 and STS–96. Hydrogen sensors were also a part of an "Integrated Vehicle Health Monitoring HEDS Technology Demonstration" series conducted at the NASA Kennedy Space Center.
- Hydrogen detection (100 ± 1 ppm H_2 v/v) has been used in medical diagnostics to detect the gastrointestinal disorders. Hydrogen is one of the gases produced in the intestinal lumen by bacteriological breakdown of carbohydrates. Accurate hydrogen detection is useful to determine the carbohydrate breakdown deficiencies, excessive bacterial growth in the gut, lactose intolerance screening, and time of passage through the small intestine. At present, there is no sensor in the marketplace that can satisfy this requirement.

Table 15.4 summarizes the commercially available hydrogen sensor products for leak detection and process monitoring. Hydrogen sensors are available with different

TABLE 15.4

Commercial Hydrogen Sensor Technologies for Leak Detection and Process Monitoring

Type	Property Measured	Vendor	Range	Needs O_2?	Selectivity (Environmental Effects)	Approximate Cost
			Leak Detection			
Palladium based					Hydrogen specific, response may be slowed by CO, H_2S, and humidity	\$500–\$3,000
Pd–Ni alloy	Resistance	H2 scan Corporation	0.5–100%	No	No interference from CO (30%), H_2S (1000 ppm), CH_4, humidity (95% RH), and chlorine (35%)	
	Capacitance		15 ppm to 1%			
Pd–Ag alloy	Resistance	Makel Engineering	1,000 ppm to 100%	Yes	Humidity: 0–95% RH at 50°C	
Pd nanoparticles	Resistance	Applied Nanotech	100 ppm to 5%	Yes	Can be affected by pressure	
Catalytic bead					False alarms from hydrocarbons and volatile organic compounds	\$500–\$4,000
Catalytic bead	Resistance	RKI Instruments	0–5%	Yes	Uses special molecular sieve filter to select H_2 + exclude VOCs, CO	
Catalytic bead	Resistance	Fuel Cell Safety Inc.	1–5%	Yes	Can respond to reducibles (VOC, CO)	
Catalytic bead	Resistance	Analytical Technology Inc.	0–100% LEL	Yes		
Semiconductor metal oxides					Response influenced by humidity, pressure, oxygen, and flow rate	\$100–\$500
Semiconducting oxides	Resistance	Dupont	0–1,000 ppm		Can have environmental, O_2 effects	
Semiconducting oxides	Resistance	RKI Instruments	0–2,000 ppm	Yes	Uses special molecular sieve filter to select H_2 + exclude VOCs, CO	
			0–10,000 ppm			
Tin oxide (SnO_2)	Resistance	Figaro Engineering, Inc.	50–1,000 ppm	Yes	Model TGS862 is "more" selective to H_2 than VOCs	
					Model TGS2620 is general purpose combustible gas analyser	
Tin oxide	Resistance	Neodym Systems	0–20,000 ppm	Yes	Uses Figaro sensors, but selectively	
Mixed metal oxide	Resistance	City Technologies	0–1,000 ppm		Uses a carbon filter to remove VOC interferences	
Electrochemical					Needs hydration	\$300–\$1,200
					Sensitivity to NO, H_2S, and CO	

Electrochemical	Current	Control Instruments	100–1,000 ppm	No	Needs hydration Sensitivity to NO, H_2S and is 10X less insensitive to CO and SO_2	
Electrochemical	Current	Draeger Safety Inc.	100–1,000 ppm	No	Interference from CO	
Electrochemical	Current	Toxalert, Inc.	0–1,000 ppm	No		
Nanotube technologies					Under development	\$100–\$500
Nanotube	Resistance	Nanomix		No		
Nanotube	Resistance	NanoMaterials Research	10–1,000 ppm		Responds well to ethanol (better than H_2)	
Process Monitoring						
Solid state					Low cost/sensor point. Attractive alternate to analytical techniques	<\$8,000
Pd–Ni alloy	Resistance	H2 scan Corporation	15 ppm to 100% H_2 v/v	No	No interference from CO (30%), H_2S (1,000 ppm), CH_4, humidity (95% RH), and chlorine (35%)	
Electrochemical	Capacitance Voltage	Unisense	0.02–25% H_2	No	Operates by diffusion of hydrogen through a silicone membrane to a hydrogen oxidizing platinum anode	
Thermal conductivity					Uses the thermal conductivity (TC) of H_2 versus much higher TC of HC, N_2, O_2, and CC to make linear curve for H_2	<\$25,000
Binary gas analyser		Applied Analytics, Gow-mac	0–100%	No	±0.5% s response time	
Gas chromatograph					Cumbersome sampling, calibration, and maintenance	\$50,000–\$80,000
Gas chromatograph		ABB, Siemens, Thermo-Fisher		No		Up to \$40,000 in additonal parts
Mass spectrometer					Cumbersome sampling, calibration, and maintenance	\$50,000–\$110,000
Process mass spectrometer		Extrel, Thermo-Fisher, Amtek		No		Up to \$80,000 in additonal sampling costs

Note: RH—relative humidity; VOCs—volatile organic compounds packaging configurations depending on the application requirement. Palladium-based resistors are the most hydrogen specific and should be used where a false alarm cannot be tolerated during hydrogen detection. The most commercially successful low-cost technology is CB, semiconductor metal oxide, and electrochemical sensors. Nanotube-based technologies are still under development by research laboratories for leak detection.

Process hydrogen sensing technologies include GC, MS, thermal conductivity, and emerging solid-state technologies. Process gas chromatographs and mass spectrometers have high unit costs, but are still most widely used in the industry. These instruments have to be enclosed in a closed room and have to be monitored continuously. Process-hardened in-line solid-state sensors will eventually replace costly single-point analysers (Figure 15.22). Networked point-of-use sensors for remote monitoring of hydrogen in process streams offer several advantages including rapid deployment—low cost to install; in-line operation—low cost of ownership; and operate with limited sample conditioning—no lab space/analyst required.

15.7 Conclusion

As the demand for hydrogen sensors grows, the accurate hydrogen detection is critical in several industries for everyday safety and process improvement. The user may consider the information presented in this chapter as a reference guide for hydrogen sensing technologies. However, the user should carefully consider the application and environmental parameters before selecting an optimum hydrogen sensor. Real-time, in-line hydrogen sensors are in immediate need for the process industry as replacements to analytical techniques such as GC. Hydrogen specificity is a must for almost any applications to prevent false alarms.

Hydrogen sensors will be ubiquitous in the next 5–10 years due to the advent of hydrogen economy and related infrastructure. This chapter presented the need for accurate hydrogen sensing and outlined the major sensor technologies, their merits and demerits for the targeted application. The palladium-based hydrogen-specific sensors are the most extensively researched and hold the most promise for the industry. Hydrogen sensors will find applications in petrochemical plants, petroleum refineries, hydrogen production facilities, chlorine plants, nuclear power plants, fuel cells, IC engines, power transformer monitoring, medical diagnostics, leak detection, and other applications.

References

1. Hansel, J., Hydrogen Safety Presentation, Air Products.
2. *Proceedings of the 2002 U.S. DOE Hydrogen Program Review*, NREL/CP-610-32405.
3. Schweighardt, F., Zatko, D., Air Products.
4. Lewis, F.A., *The Palladium Hydrogen System*, Academic Press, London, 1967.
5. Eranna, G., E., Joshi, Runthala and Gupta, Oxide materials for development for integrated gas sensors—a comprehensive review, *Critical Reviews in Solid State and Material Sciences*, 29, 111, 2004.
6. Bard, A.J., Stratmann, M. and Unwin, P.R., Eds., Instrumentation and electroanalytical chemistry, *Encyclopedia of Electrochemistry*, Vol. 3, 2003.
7. Pohl, A., A review of wireless SAW sensors, *IEEE Transactions on Ultrasonics, Ferroelectrics, and Frequency Control*, 47, 317, 2000.

8. Samuel, J.I., Sasikaran, K., Kourosh, K.-Z., Adrian, T. and Wojtek, W., A layered surface acoustic Wave $ZnO/LiTaO_3$ structure with a WO_3 selective layer for hydrogen sensing, *Sensor Letters*, 1, 33, 2003.
9. Lewis, F.A., *The Palladium Hydrogen System*, Academic Press, New York, 1967.
10. Huges, R.C., Schubert, W.K. and Buss, R.J., Solid-state hydrogen sensors using palladium-nickel alloys: Effect of alloy composition on sensor response, *Journal of Electrochemical Society*, 142, 249, 1995.
11. Robins, I., Ross, J.F. and Shaw, J.E.A., The logarithmic response of palladium-gate metal-insulator-silicon field-effect transistors to hydrogen, *Journal of Applied Physics*, 60(2), 843, 1986.
12. Cassidy, J., Pons, S. and Janata, J., Hydrogen response of palladium coated suspended gate field effect transistor, *Analytical Chemistry*, 58(8), 1757, 1986.
13. Salomonsson, A., Eriksson, M. and Dannetun, H., Hydrogen interaction with platinum and palladium metal–insulator–semiconductor devices, *Journal of Applied Physics*, 98, 014505, 2005.
14. Lin, K.-W., Cheng, C.-C., Cheng, S.-Y., Yu, K.-H., Wang, C.-K., Chuang, H.-M., Chen, J.-Y., Wu, C.-Z. and Liu, W.-C., A novel Pd/oxide/GaAs metal-insulator-semiconductor field-effect transistor (MISFET) hydrogen sensor, *Semiconductor Science and Technology*, 16, 997, 2001.
15. Hughes, R.C. and Schubert, W.T., Solid-state hydrogen sensors using palladium nickel alloys—effect of alloy composition on sensor response, *Journal of Electrochemical Society*, 142(1), 249, 1995.
16. Pitts, J.R., Liu, P., Lee, S.-H. and Tracy, C.E., Interfacial instability in hydrogen sensors, *Proceedings of the 2000 DOE Hydrogen Program Review*, Available at http://www1.eere.energy.gov/hydrogenandfuelcells/pdfs/28890g.pdf.
17. Thomas, R.C. and Hughes, R.C., Sensors for Detecting Molecular Hydrogen Based on Pd Metal Alloys, *Journal of Electrochemical Society*, 144(9), 3245, 1997.
18. Oh, Y.-s., Hamagami, J.-i., Watanabe, Y., Takata, M. and Yanagida, H., Palladium thin film hydrogen detector, *Journal of the Ceramic Society of Japan*, 101(1174), 618–620, 1993.
19. Wolfe, D.B., Love, J.C., Paul, K.E., Chabinyc, M.L. and Whitesides, G.M., Fabrication of palladium-based microelectronic devices by microcontact printing, *Applied Physics Letters*, 80, 2222, 2002.
20. Villatoro, J., Díez, A., Cruz, J.L. and Andrés, M.V., Highly sensitive optical hydrogen sensor using circular Pd-coated single mode tapered fibre, *Electron*, 37, 1011, 2001.
21. Zalvidea, D., Díez, A., Cruz, J.L. and Andrés, M.V., A wavelength multiplexed hydrogen sensor based on a palladium-coated fibre taper and a Bragg grating, *Electron*, 40, 301, 2004.
22. Butler, M.A., Fiber optic sensor for hydrogen concentrations near the explosive limit, *Journal of Electrochemical Society*, 46(138), L46, 1991.
23. Butler, M.A., Micromirror optical-fiber hydrogen sensor, *Sensors and Actuators, B*, 22, 155, 1994.
24. Garcia, J.A. and Mandelis, A., Study of the thin-film palladium/hydrogen system by an optical transmittance method, *Review Scientific Instruments*, 67(11), 3981, 1996.
25. Ito, K. and Kubo, T., Gas detection by hydrochromism, *Proceedings of the 4th Sensor Symposium*, Tsukuba, pp. 153–156, December 1984.
26. Tabib-Azar, M., Sutapun, B. and Petrick, P., *Proceedings of SPIE—Volume 3513 Microelectronic Structures and MEMS for Optical Processing IV*, 80, September 1998. Available at http://spiedl.aip.org/getabs/servlet/GetabsServlet?prog=normal&id=PSISDG00351300000100008000000l&idtype=cvips&gifs=yes
27. Favier, F., Walter, E.C., Zach, M.P., Benter, T. and Penner, R.M., Hydrogen sensors and switches from electrodeposited palladium mesowire arrays, *Science*, 293, 2227, 2001.
28. Pavlovsky, I., Soundarrajan, P. and Yaniv, Z., Palladium nanoparticle hydrogen sensor, *Gases and Technology*, 18(July/August), 18, 2006.
29. Yaniv, Z. and Soundarrajan, P., Achieving the potential of nano gas sensors, *Sensors*, September 2006.
30. Kong, J., Chapline, M.G. and Dai, H., Functionalized carbon nanotubes for molecular hydrogen sensors, *Advanced Materials*, 13(18), 1384, 2001.
31. Mor, G.K., Varghese, O.K., Pishko, M.V. and Grimes, C.A., A room-temperature TiO_2-nanotube hydrogen sensor able to self-clean photoactively from environmental contamination, *Journal of Materials Research*, 19(2), 628, 2004.

32. Xu, T., Zach, M.P., Xiao, Z.L., Rosenmann, D., Welp, U., Kwok, W.K. and Crabtree, G.W., Self-assembled monolayer-enhanced hydrogen sensing with ultra thin palladium films, *Applied Physics Letters*, 86, 203104, 2005.
33. Yu, S., Welp, U., Hua, L.Z., Rydh, A.R., Kwok, W.K. and Wang, H.H., Fabrication of palladium nanotubes and their application in hydrogen sensing. *Chemistry of Materials*, 17, 3445, 2005.
34. Barsan, N. and Weimar, U., Understanding the fundamental principles of metal oxide based gas sensors; the example of CO sensors with SnO_2 sensors in the presence of humidity, *Journal of Physics: Condensed Matter*, 15, R813–R839, 2003.
35. http://www.citizen.co.jp/english/release/06/060202gs.html (accessed July 15, 2006).
36. Fogelberg, J. and Petersson, L.-G., Kinetic modeling of the H sub 2-O sub 2 reaction on palladium and of its influence on the hydrogen response of a hydrogen sensitive palladium metal-oxide-semiconductor device, *Surface Science*, 350(1–3), 91, 1996.
37. Ramesh, C., Velayutham, G., Murugesan, N., Ganesan, V., Dhathathreyan, K.S. and Periaswami, G., An improved polymer electrolyte-based amperometric hydrogen sensor, *Journal of Solid State Electrochemistry*, 7(8), 511, 2003.
38. Tan, Y. and Tan, T.C., Sensing behavior of an amperometric hydrogen sensor theoretical modeling and experimental verification, *Journal of Electrochemical Society*, 142(6), 4001, 1995.
39. Dong, S., Bai, F., Li, J. and Viehland, D. Sound-resonance hydrogen sensor, *Applied Physics Letters*, 82(25), 4590, 2003.
40. http://www.sandia.gov/mstc/technologies/microsensors/hightempacoustic.html (accessed January 15, 2007).
41. http://www.praxair.com/praxair.nsf/AllContent/7E8A75E1CC87FA998525655E000B40DA?OpenDocument&URLMenuBranch=88FFF0AAE2BDB4948525706F00587825 (accessed August 17, 2007).
42. ABB Instruments, ABB, Inc., Lewisburg, WV.
43. EXTEL, Extrel CMS, Pittsburgh, PA.
44. Yamaguchi, N., Trans-Energy Research Associates, Seattle, WA.
45. Puri, P.S. and Pe, G.P., Addition of odorants to hydrogen by incorporating odorants with hydrogen storage materials, U.S. Patent Application, April 2006.
46. Cohen, J.P. and Schweighardt, F.K., Fueling nozzle with integral molecular leak sensor, U.S. patent Application, April 2006.
47. Cohen, J.P. and Schweighardt, F.K., Fueling nozzle with integral molecular leak sensor, U.S. patent Application, March 2007.
48. U.S. Department of Energy, The Freedom Car-Fuel Partnership, U.S. Department of Energy, 2005.
49. http://www.pbs.org/wgbh/nova/sciencenow/3210/01-fcw.html (accessed August 20, 2007).

16

Hydrogen Safety

Fotis Rigas and Spyros Sklavounos

CONTENTS

16.1 Introduction

Edwin Drake drilled the first oil well in Pennsylvania in 1859 to use extracted oil as a substitute for whale oil—the main lighting source and feedstock for consumer and chemical products then. This resource was hazardous to get and dwindling due to heavy exploitation, as it happens currently with oil. Petroleum presented many advantages over whale oil and solved a great deal of ecological and resource security problems associated with the old resource. Nevertheless, after a century and a half of use, petroleum has created new problems related to environmental pollution and energy security [1].

To address these problems and to face the oil wells dry up and the simultaneous rapid increase of energy demands all over the world, many efforts have been dedicated to the gradual substitution of fossil fuels by new energy vectors. These efforts have been dedicated to the gradual substitution of all ordinary fossil fuels (oil, coal, and natural gas) by new energy vectors, primarily because of the natural resources exhaustion and the simultaneous rapid increase of energy demands worldwide.

Hydrogen is considered as one of the most promising fuels for generalized use in the future, mainly because it is an energy-efficient, low-polluting, and renewable fuel. Hydrogen is versatile and clean, and aiming at environmental benefits, the production of hydrogen from renewable energy sources and nuclear is under consideration [2,3]. Yet, discussion of potential problems that widespread use of hydrogen might bring, such as whale oil and petroleum before it, is missing. The identification and control of potential adverse consequences is an ethical requirement whenever any new technology is introduced, and certainly in the hydrogen case with such world-influencing potential [1].

Hydrogen has been, for more than a century, produced and used with a high safety record for commercial and industrial purposes such as refinery and chemical processes and rocket propulsion. Industry has significant experience in safe handling of hazardous materials in chemical plants, where only well-trained personnel come in contact with hydrogen. Nevertheless, the wide use of hydrogen as an energy carrier will result in its use by laypersons necessitating different safety regulations and technologies that are now under development [4,5].

One of the major issues affecting the acceptance of hydrogen for public use is the safety of hydrogen installations (production and storage units) as well as its applications (i.e., as vehicle fuel or home use). The hazards associated with the use of hydrogen can be characterized as physiological (frostbite and asphyxiation), physical (embrittlement and component failures), and chemical (burning or explosion), the primary hazard being inadvertently producing a flammable or explosive mixture with air [6,7]. As far as European countries are concerned, hazardous chemicals installations come under the so-called SEVESO II Directive (96/82/EC) for the control of major-accident hazards involving dangerous substances. Hydrogen is included in this directive, indeed with stricter minimal quantity for the Directive's implementation than any other ordinary fuel [8,9].

In recent theoretical [10] and computational [11] works, safety comparisons between hydrogen and other fuels do not allow a clear viewpoint for the safest one. Indeed, in the past, certain hydrogen applications gave rise to severe accidents with significant economic and societal cost, affirming the need of augmented safety measures wherever hydrogen is handled [12]. Undoubtedly, the need for safety measures should be pointed out when loss prevention and public safety are concerned. This demands the knowledge of potential hazards of storing, transporting, and using hydrogen.

16.2 Historical Survey of Hydrogen Accidents

In the previous decades, severe accidents have happened involving hydrogen utilized in industrial and other applications. The determinant causes for a catastrophic event may be classified into the following [13]:

- Mechanical or material failure
- Corrosion attack
- Overpressurization
- Enhanced embrittlement of storage tanks at low temperatures
- Boiling liquid expanding vapor explosion (BLEVE)
- Rupture due to impact by shock waves and missiles from adjacent explosions
- Human error

A list of accidents related to hydrogen collected from some accident databases, such as UNEP, OECD, MHDAS, and BARPI, is shown in Table 16.1 [3,14]. This is not a complete list of all accidents involving hydrogen, and the databases should be extensively updated. Yet, these characteristic examples prove that, in practice, hydrogen is liable to major chemical accidents posing a considerable risk not only for onsite, but also for offsite damage as well.

Typical examples of such accidents with consequences on people and property are quoted in Sections 16.2.1 through 16.2.3 and may be found in detail in relevant literature and databases [12,15,16].

16.2.1 The Hindenburg Disaster

The disaster occurred at Lakehurst, New Jersey, on May 6, 1937 and was for many years under investigation to identify the reasons that caused the ignition of the hydrogen gas used for buoyancy of the giant airship "Hindenburg". In that accident, the ignition of hydrogen proceeded rapidly to fire toward the tail section of the craft. The fire was almost simultaneously succeeded by an explosion that engulfed the 240 t craft causing it to crash onto the ground killing 36 people. The overall results indicated that the outer shell and the paint of the airship were flammable and could be ignited from electrical sparks. Indeed, prevailing atmospheric conditions at the time the accident occurred could generate considerable electrostatic discharge activity on the airship.

16.2.2 Hydrogen Release during Maintenance Works

On October 23, 1989, at Pasadena, Houston, a massive and destructive hydrogen vapor cloud explosion occurred in a polyethylene plant causing loss of life to 22 people, injuries to 100 people, and extensive damage to the whole plant. The explosion strength was estimated to be equivalent to 2.4 t of the explosive trinitrotoluene (TNT). The accident resulted from a release of process gas during maintenance on a reaction loop line. The Occupational Safety and Health Administration (OSHA) detailed numerous defects in the management of the installation, among which were lack of a hazard assessment study, inadequate safety distances between the control room and the reactor, lack of an effective permit system for maintenance activities, and erroneous design of building ventilation intakes.

TABLE 16.1

Summary of Accidents Related to Hydrogen

Year	Date	Location	Origin of Accident	Death	Injury	Evacuated
2001	05.01	Oklahoma (trailer), OK, USA	Fire	1	1	15
2001	04.18	Labadie, MO, USA	Fire	NA	NA	NA
2000	09.14	Pardies	Fire and explosion	NA	NA	NA
2000	09.03	Gonfreville-Lorcher	Explosion	—	12	—
1999	05.07	Panipat, India	Fire	5	—	—
1999	04.08	Hillsborough, USA	Fire and explosion	3	50	38
1998	09.15	Torch, Canada	Fire	NA	NA	NA
1998	06.08	Auzouer, Touraine	Explosion and fire	—	1	200
1998	04.25	France	Fire	NA	NA	NA
1996	NA	United States	Explosion and fire of a transmission pipe	—	—	—
1994	NA	Japan	Fire in a desulfurization unit	—	2	—
1993	NA	Russia	Cloud explosion	—	4	—
1992	04.22	Jarrie	Fire	1	2	—
1992	01.18	Pennsylvania, USA	Fire	1	3	—
1992	NA	Japan	Explosion and fire in a refinery hydrogenation unit	9	8	—
1992	01.16	Sodegaura, Japan	Explosion	10	7	—
1992	01.08	Wilmington, USA	Leakage	—	16	—
1992	NA	Hong Kong	Hydrogen explosion in a power plant	2	19	—
1992	NA	U.S. nuclear power plant	Fire and explosion	NA	NA	NA
1991	06.09	Pardies	Fire	NA	NA	NA
1991	02.14	Daesan, Korea	Explosion	—	2	—
1991	10.–	Hamau-Frankfurt, Germany	Explosion	NA	NA	NA
1990	07.25	Birmingham, UK	Fire and gas cloud	—	>60	70,050
1990	04.29	Ottmarsheim	Fire	NA	NA	NA
1989	NA	United States	Pipeline failure and jet fire	7	8	—
1989	10.23	Pasadena, Houston, TX, USA	Explosion	22	100	NA
1988	06.15	Genoa, Italy	Explosion	3	2	15,000
1986	01.28	Challenger, USA	Explosion	7	—	—
1984	05.25	Tempelhof-feld, Germany	Explosion	NA	NA	NA
1984	10.13	Waziers, France	Fire	NA	NA	NA
1980	NA	EU chemical industries	Fire and explosion	NA	NA	NA
1978	06.12	United States	Fire	—	9	—
1975	NA	Ilford, Esses	Explosion	1	—	—
1937	05.06	Hindenburg, Lakehurst, NJ, USA	Fire	36	NA	NA

Source: After Rigas, F. and Sklavounos, S., *Int. J. Hydrogen Energ.*, 30, 1501, 2005; Rosyid, O.A., System-Analytic Safety Evaluation of the Hydrogen Cycle for Energetic Utilization, Dissertation, Otto-von-Guericke University, Magdeburg, Germany, 2006.

16.2.3 Pressurized Hydrogen Tank Rupture

In 1991 at Hanau, Frankfurt (Germany), a tank of 100 m^3 hydrogen pressurized at 45 bars burst when stored outdoors in an industrial plant. The shock wave and the missiles of the tank shell caused heavy damage on the plant units. Investigations showed that welding in the metallic shell suffered from extensive cracks from the inner side to the outside. Likely, the corners along the welding caused concentration of stresses so that the first cracks appeared in those parts. Under the influence of hydrogen, cracks grew much faster than normally leading to material failure and subsequent mechanical explosion. As a result, all similar tanks in the country were checked, manufacturing rules revised providing upper limits for corners, and new test methods for detecting cracks in early stages during operation were established. Undoubtedly, this accident contributed to improvements in hydrogen tanks safety.

16.3 Hydrogen Properties Associated with Hazards

16.3.1 General Consideration

Table 16.2 gives selected thermophysical, chemical, and combustion properties of hydrogen [17–22]. Some of the key overall properties of hydrogen that are relevant to its employment as an engine fuel are also listed in Table 16.2 together with the corresponding values of methane—the other promising gaseous fuel for engine applications—and those of gasoline [18]. In cases that property values could not be found, these are represented by isooctane vapor [19] or the arithmetic average of normal heptane and octane [20].

16.3.2 Hydrogen Gas Properties Related to Hazards

Hydrogen gas is colorless, odorless, and not detectable in any concentration by human senses in atmospheric conditions. Therefore, inability of *detection* renders hydrogen a potential asphyxiant and a latent fuel, ready to ignite. With regard to its *buoyancy*, hydrogen gas is about 14 times lighter than air in normal conditions and this is why any leak moves fast upward, thus reducing ignition hazards. Nevertheless, the cold dense fuel vapors produced by liquid hydrogen (LH_2) spills will initially remain close to the ground and then rise more slowly than standard temperature and pressure fuel gases [3].

The *limiting oxygen index* is the minimum concentration of oxygen that will support flame propagation in a mixture of fuel vapors and air. For hydrogen, no flame propagation is observed at normal temperature and pressure (NTP) conditions if the mixture contains $<5\%$ by volume oxygen. *Burning velocity* (subsonic) in NTP air for hydrogen depends on pressure, temperature, and mixture composition. This high burning velocity of hydrogen indicates the difficulty of confining or arresting hydrogen flames. Explosion confinement is much more difficult to achieve in detonations where velocities are three orders of magnitude faster.

16.3.3 Liquefied Hydrogen Properties Related to Hazards

Hazards in addition to those taken for hydrogen gas should be taken when handling or storing LH_2. Therefore, any LH_2 splashed onto the skin or into the eyes can cause frostbite burns or hypothermia. Vents and valves in storage vessels and dewars may be blocked by *accumulation of ice* formed from moisture in the air. Excessive pressure may then result in mechanical failure with jet release of hydrogen and potentially in BLEVE.

Moreover, storage vessels should be kept under positive pressure to prevent air from entering and producing flammable mixtures. Liquefied hydrogen may be contaminated with air condensed and solidified from the atmosphere. The quantity of *solidified air* can increase during repeated refilling or pressurization of storage vessels producing explosive mixtures with hydrogen.

TABLE 16.2

Selected Comparative Properties of Hydrogen, Methane, and Gasoline

	Value			
Property	**Hydrogen**	**Methane**	**Gasoline**	**References**
Heat of combustion (low) (kJ/g)	119.93	50.02	44.50	17,20
	119.70	46.72	44.79	10,19
Heat of combustion (high) (kJ/g)	141.86	55.53	48.00	17,20
	141.70	52.68	48.29	19
Flammability limits in NTP air (vol%)	4.0–75.0	5.3–15.0	1.0–7.6	17–20
			1.2–6.0	
Flammability limits in NTP oxygen (vol%)	4.1–94.0	—	—	17
Detonability limits in NTP air (vol%)	18.3–59.0	6.3–13.5	1.1–3.3	17,18,20,21
	13.5–70.0			
Detonability limits in NTP oxygen (vol%)	15–90	—	—	17
Stoichiometric composition in air (vol%)	29.53	9.48	1.76	17,20
Minimum ignition energy in air (mJ)	0.017	0.29	0.24	17,20
	0.02	0.28	0.25	8,19,20
	0.14		0.024	
Autoignition temperature (K)	858	813	501–744	17,19,20
			500–750	
Adiabatic flame temperature in air (K)	2318	2148	~2470	17–20
		2190		
Thermal energy radiated from flame to surroundings (%)	17–25	23–33	30–42	17,18,20
Burning velocity in NTP air (cm/s)	265–325	37–45	37–43	17,20
Detonation velocity in NTP air (km/s)	1.48–2.15	1.39–1.64	1.40–1.70	17,20
Energy of stoichiometric mixture in NTP air (MJ/m^3)	3.58	3.58	3.91	22
Diffusion coefficient in NTP air (cm^2/s)	0.61	0.16	0.05	17,18,20
Buoyant velocity in NTP air (m/s)	1.2–9.0	0.8–6.0	Nonbuoyant	17,20
Limiting oxygen index (vol%)	5.0	12.1	11.6	20
Maximum experimental safe gap in NTP air (cm)	0.008	0.12	0.07	17,20
Quenching gap in NTP air (cm)	0.064	0.203	0.200	17,19,20
Detonation induction distance in NTP air	*L/D* ~100	—	—	20

Note: STP (standard temperature and pressure): 273.15 K (0°C), 101.3 kPa (1 atm); NTP: 293.15 K (20°C), 101.3 kPa; NBP (normal boiling point): boiling point at 101.3 kPa.

Source: After NASA, *Safety Standard for Hydrogen and Hydrogen Systems*, Report NSS 1740.16, 1997, chap. 2; Adamson, K.A. and Pearson, P., *J. Power Sourc.* 86, 548, 2000; Karim, A., *Int. J. Hydrogen Energ.*, 28, 569, 2003; Hord, J., *Int. J. Hydrogen Energ.*, 3, 157, 1978; Baker, W.E. and Tang, M.J., *Gas, Dust and Hybrid Explosions*, Elsevier, Amsterdam, 1991; Rosyid, A. and Hauptmanns, U., System Analysis Safety Assessment of Hydrogen Cycle for Energetic Utilization, *Proc. Int. Cong. Hydrogen Energy and Exhibition*, Istanbul, 2005.

Liquefied hydrogen confined, for instance, in a pipe between two valves will eventually warm to ambient temperature resulting in a significant pressure rise. Standard storage system designs usually assume a heat leak equivalent to 0.5% per day of the liquid contents.

The high density of the saturated vapor resulting immediately after release from a liquefied hydrogen storage leaking vessel causes the hydrogen cloud to move horizontally or downward for some time as it has been shown experimentally by National Aeronautics and Space Administration (NASA) in the Langley Research Center at the White Sands Test Facility in 1980 and simulated effectively later using a computational fluid dynamics (CFD) approach [3].

Since electrical conductivity of liquefied hydrogen is about 10^{19} W cm at 25 V, the electric current–carrying capacity is small and more or less independent of the imposed voltage. Investigation has shown that *electric charge buildup* in flowing liquefied hydrogen is not a great concern [17].

16.4 Hydrogen Hazards

16.4.1 Physiological Hazards

Personnel present during leaks, fires, or explosions of hydrogen mixtures with air can incur several types of injury. *Asphyxiation* may occur when entering a region where hydrogen or another nontoxic gas has displaced the air, lowering the oxygen concentration below 19.5% by volume.

Blast waves from explosions may cause *overpressure injury* as a result of a combination of overpressure and duration at a given distance. For instance, the threshold of lung rupture may be 70 kPa (0.7 atm) overpressure for 50 ms or 140–210 kPa (1.4–2.0 atm) overpressure for 3 ms.

Thermal burns result from the radiant heat emitted by a hydrogen fire and absorbed by a person, which is directly proportional to many factors including exposure time, burning rate, heat of combustion, size of the burning surface, and atmospheric conditions (mainly wind and humidity). For instance, thermal radiation flux exposure level of 0.95 W/cm^2 may cause skin burns in 30 s. *Cryogenic burns* may result from contact with cold fluids or cold vessel surfaces. Exposure to large liquefied hydrogen spills could result in *hypothermia*, if proper precautions are not taken [17].

16.4.2 Physical Hazards

16.4.2.1 Hydrogen Embrittlement

Owing to hydrogen embrittlement, the mechanical properties of metallic and nonmetallic materials of containment systems may degrade and fail resulting in leaks. Hydrogen embrittlement depends on many factors such as environmental temperature and pressure, purity of metal, concentration and exposure time to hydrogen, stress state, physical and mechanical properties, microstructure, surface conditions, and the nature of the crack front of material [23].

The types of hydrogen embrittlement are *environmental hydrogen embrittlement* of metals and alloys resulting in increased surface cracks, *internal hydrogen embrittlement* caused by absorbed hydrogen resulting in premature failures with internally starting cracks, and *hydrogen reaction embrittlement* caused by the chemical reaction of the absorbed hydrogen with one or more of the constituents of the metal to form a brittle metal hydride or methane with the carbon in steels [24].

Studies of the *deterioration of mechanical properties* observed in hydrogen cracking [25–27] have shown that the susceptibility of a metal (or alloy) due to hydrogen increases with the strength of the metal. The internal and the environmental hydrogen embrittlement maximizes in the range 200–300 K (−73°C to 27°C), whereas hydrogen reaction embrittlement occurs at temperatures above room temperature. In addition, the susceptibility to embrittlement generally increases with the tensile stress.

The *main factors* governing hydrogen embrittlement are the metal surface films that can restrict hydrogen absorption, the effect of electrical discharge machining that may insert hydrogen (produced by the decomposition of a dielectric fluid) into a machined component, and the effect of trapping sites within the structure of the metal.

Control of hydrogen embrittlement can be obtained by oxide coatings, elimination of stress concentrations, additives to hydrogen, proper grain size, and careful alloy selection. Additional measures may be the use of aluminum as construction material or medium strength steel (for hydrogen gas) and stainless steel (for LH_2), the design of metal components assuming an up to fivefold safety factor, and the consideration of the fact that environmental and internal hydrogen embrittlement increases in the temperature range 200–300 K (−73°C to 27°C) [23].

16.4.2.2 Low-Temperature Mechanical Properties of Structural Materials

The main considerations of mechanical properties of metals and alloys at low temperatures taken into account for safety reasons are the transition from ductile-to-brittle behavior, certain unconventional modes of plastic deformation, and mechanical and elastic properties changes due to phase transformations in the crystalline structure.

With regard to *low temperature embrittlement*, this change may lead to an accidental failure of a hydrogen storage vessel or pipe and cause an accident. Such an accident occurred in Cleveland in 1944 due to low-temperature embrittlement of a liquefied natural gas (LNG) storage vessel made of 3.5% nickel steel with a capacity of 4248 m^3. The vessel ruptured and released 4163 m^3 of LNG, which ignited after having spread into nearby storm sewers. As a domino effect, a nearby storage vessel collapsed from the fire and spilled its contents, which also burned with flames up to about 850 m height. The accident resulted in 128 deaths and 200–400 injuries and an estimated U.S.$6,800,000 (1944 dollars) in property damage [23].

Different structural materials have different *thermal contraction coefficients*, meaning that accommodations should be made for their different dimensions at cryogenic temperatures. If not, problems associated with safety (e.g., leaks) may arise. Generally, the contraction of most metals from room temperature (300 K) to a temperature close to the liquefaction temperature of hydrogen (20 K) is <1%, whereas the contraction for most common structural plastics is from 1% to 2.5% [23].

16.4.3 Chemical Hazards

16.4.3.1 General Consideration and Accident Statistics

Safety measures in installations handling hydrogen include the elimination of all likely sources of ignition. An investigation of industrial accidents has shown that 53% of them occurred because of leaks, off-gassing, and equipment ruptures as listed in Table 16.3 [28]. Of the 145 incidents listed in Table 16.3, 84 (58%) were explosions, 53 (37%) were fires, and the remaining 8 (5%) were either combined fire explosions or unignited releases due to vessel rupture. Purging or vent-exhaust incidents account for 15% and the remaining 32% were other types of incidents. In ammonia plants, most of the accidents are due to leaking of gaskets and valve packing as shown in Table 16.4 [29].

Since LH_2 has the greatest energy per weight of fuel, NASA has used LH_2 to fuel the second and third stages of the Apollo moon rockets as well as the current fleet of space shuttles. Nevertheless, NASA has paid a great toll with 107 hydrogen incidents only in 1974 of which 87 involved the release of gaseous hydrogen (GH_2) or LH_2 as shown in Table 16.5. When incidents were not caused by equipment failure, they primarily occurred when procedures were not prescribed or when the prescribed procedures were not followed [30].

TABLE 16.3

Industrial Hydrogen Accidents in the United States for the Period 1968–1977

Category	Number of Incidents	Percentage of Total Accidents
Undetected leaks	32	22
Hydrogen–oxygen off-gassing explosions	25	17
Piping and pressure vessel ruptures	21	14
Inadequate inert gas purging	12	8
Vent and exhaust system incidents	10	7
Hydrogen–chlorine incidents	10	7
Others	35	25
Total	145	100

Source: After Zalosh, R.G. and Short T.P., *Compilation and Analysis of Hydrogen Accident Reports*, C00-4442-4, Factory Mutual Research Corp., Norwood, OH, 1978.

TABLE 16.4

Hydrogen Accidents in Ammonia Plants in North America for the Period 1969–1974

Classification	Number	Percentage of Total Accidents
Gaskets	46	27
Equipment flanges	23	14
Piping flanges	16	9
Valve flanges	7	4
Valve packing	10	6
Oil leaks	24	14
Transfer header	9	5
Auxiliary boiler	8	5
Primary reformer	7	4
Cooling tower	3	2
Electrical	2	1
Miscellaneous	16	9

Source: After Williams, G.P., *Chem. Eng. Prog*., 74, 9, 1978.

TABLE 16.5

Hydrogen Incidents in NASA for the Year 1974

Description	Accidents Involving the Release of Hydrogen	Percentage of Total Accidents
Accidents involving the release of LH_2 or GH_2	87	81
Location of hydrogen release		
To atmosphere	71[a]	66
To enclosures (piping, containers, etc.)	26[a]	24
Ignition of hydrogen releases		
To atmosphere	44	41
To enclosures	24	22

[a] Hydrogen was released to both locations in 10 accidents.

Source: After Ordin, P.M., A review of hydrogen accidents and incidents in NASA operations, *9th Intersociety Energy Conversion Engineering Conference, ASME*, New York, 1974.

16.4.3.2 Flammability of Hydrogen

The flammability limits of mixtures of hydrogen with air, oxygen, or other oxidizers depend on the ignition energy; temperature; pressure; presence of diluents; and size and configuration of the equipment, facility, or apparatus. Such a mixture may be diluted with either of its constituents until its concentration shifts below the lower flammability limit (LFL) or above the upper flammability limit (UFL). The flammability limits of hydrogen mixtures with either air or oxygen widen for upward flame propagation and narrow for downward flame propagation. Mixtures of LH_2 and liquid oxygen (LOX) or solid oxygen as oxidizer are not hypergolic. In accidental fires of these mixtures during the mixing process, the system caught fire because the required ignition energy is very small [17]. However, LH_2 and liquid or solid oxygen can detonate when initiated by a shock wave. Flammability limits move closer to one another when pressure reduces below 101.3 kPa (1 atm).

Hydrogen–air mixture. The lowest pressure for which a low-energy ignition source ignites a hydrogen–air mixture is approximately 6.9 kPa (0.07 atm) at a hydrogen concentration between 20% and 30% by volume [17]. Hydrogen–air mixtures ignited by a 45 mJ spark ignition source at 311 K (38°C) have an LFL equal to 4.5% by volume over the pressure range 34.5–101.3 kPa (0.34–1 atm). An increasingly higher LFL was required to obtain combustion below 34.5 kPa (0.34 atm). The lowest pressure for which a low-energy ignition source could inflame the mixture was 6.2 kPa (0.06 atm) at a hydrogen–air mixture of between 20% and 30% by volume hydrogen [31]. Yet, using a strong ignition source, the lowest pressure for which ignition can occur is 0.117 kPa (0.0012 atm). For downward propagation, the LFL of hydrogen–air decreases from 9.0% to 6.3% by volume hydrogen and the UFL increases from 75% to 81.5% by volume hydrogen when the temperature is increased from 290 to 673 K (17–400°C) at a pressure of 101.3 kPa (1 atm).

Hydrogen–oxygen mixtures. The flammability limits for hydrogen–oxygen mixtures at 101.3 kPa (1 atm) range from 4% to 94% by volume hydrogen for upward propagation in tubes. Reduced pressures increase the LFL [17]. The lowest pressure observed for ignition is 57 Pa (0.56 atm) at a hydrogen concentration of 50% by volume when a high-energy ignition source was used. At elevated pressures, the LFL does not change with pressure up to 12.4 MPa (122 atm), whereas at 1.52 MPa (15 atm) UFL is 95.7% by volume hydrogen [23]. The LFL decreases from 9.6% to 9.1% by volume hydrogen and the UFL increases from 90% to 94% by volume hydrogen, when the temperature rises from 288 to 573 K (15–300°C).

Effects of diluents and halocarbon inhibitors. The effect of equal concentrations of added diluents in wide tubes on flammability limits for hydrogen in air increases in the order: $CO_2 < N_2 < He < Ar$. Thus, argon is the least effective in reducing the flammable range of hydrogen in air. Measurements performed for the water vapor at 422 K (149°C) have shown that water is the most effective of all these diluents in reducing the flammability range of hydrogen in air [32,33]. The effect of the flame inhibitors N_2, CH_3Br, and $CBrF_3$ required to extinguish hydrogen diffusion flames in air has been compared [34]. The chlorinated inhibitors were more effective when added to the air stream, whereas nitrogen was more effective when added to the fuel stream.

Autoignition temperature. The autoignition temperature of hydrogen in air differs only slightly than that in oxygen. It is interesting to note that autoignition temperatures in both cases depend not only on GH_2 concentration and pressure, but also on the surface treatment of containers as well. Therefore, the autoignition temperatures found in the literature are very dependent on the system and values should be applied only to similar systems. At 101.3 kPa (1 atm), the range of reported autoignition temperatures for stoichiometric hydrogen in air is from 773 to 850 K (500–577°C), whereas for stoichiometric oxygen it is from 773 to 833 K (500–560°C). At pressures from 20 to 50 kPa (0.20–0.49 atm), GH_2–air ignitions have occurred at 620 K (347°C) [17].

Quenching gap in air. Quenching distance is the passage gap dimension required to prevent the propagation of an open flame through a flammable fuel–air mixture that fills the passage. In parallel-plate configuration, the quenching gap is defined as the spark gap between two flat electrodes at which ignition of combustible fuel–air mixtures is suppressed. The quenching gap for hydrogen is 0.6 mm at NTP air. Nevertheless, this value depends on the temperature, pressure, and composition of the combustible gas mixture and the electrode configuration. Generally, faster burning gases have smaller quenching gaps. Consequently, flame arresters for faster burning gases require smaller apertures. The lowest quenching distance found in the literature for hydrogen is 0.076 mm [35]. There are three major factors that determine the quenching distance for a gaseous fuel such as hydrogen: ignition energy, mixture composition, and pressure. Quenching gap depends on the *ignition energy*. For instance, low ignition energy of the order 0.001 mJ corresponds to a small gap of the order 0.01 cm. In contrast, high ignition energy of the order 10 mJ corresponds to a larger gap of the order 1 cm. *Pressure* and *composition* also affect the quenching distance, the quenching distance increasing rapidly at very low pressures. Since specific values for hydrogen–air mixtures are not available, the effect of pressure as a function of tube diameter for deflagration and detonation of acetylene–air mixtures can be used as a guideline [17].

16.4.3.3 Ignition Sources

All ignition sources, such as open flames, electrical equipment, or heating equipment, should be eliminated or safely isolated in buildings or special rooms containing hydrogen systems, and operations should be conducted as if unforeseen ignition sources could occur. The ignition of GH_2–air mixtures usually results in deflagration, whose explosive consequences are significantly less severe than if detonation results. Potential ignition sources ranked according to frequency of occurrence is as follows [36]: open flames > hot works > electrical (short circuits, sparks, and arcs) > hot surfaces > personnel smoking > friction > spontaneous ignition > autoignition > hot particles > static discharge > others (e.g., lightning).

Electrical sparks can be the result of electrical discharges between objects having different electrical potentials, such as breaking electrical circuits or discharges of static electricity. *Static electricity sparks* can ignite hydrogen–air or hydrogen–oxygen mixtures. Static electricity is caused by many common articles such as hair or fur when combed or stroked or an operating conveyor belt. People can generate high-voltage charges of static electricity on themselves when walking on a synthetic carpet or dry ground, moving while wearing synthetic clothing, sliding on automobile seats, or combing their hair. The flow of GH_2 or LH_2 in ducts or turbulence in containers can generate charges of static electricity, as any other nonconductive liquid or gas. In addition, static charges may be induced during electrical storms [37–39].

Hard objects coming into shearing contact with one another can cause *friction sparks,* as in the cases of metal striking metal, metal striking stone, or stone striking stone. Friction sparks are particles of burning material, initially heated by the mechanical energy of friction and impact that have been sheared off as a result of contact. Sparks from hand tools have normally low energy, whereas mechanical tools, such as drills and pneumatic chisels, can generate high-energy sparks.

Hard objects striking one another can also produce *impact sparks*. Impact sparks are usually produced by impact on a quartzitic rock, such as the sand in concrete, and then small particles of the impacted material are thrown off.

Minimum spark energy for ignition is defined as the minimum spark energy required to ignite the most easily ignitable concentration of fuel in air and oxygen. The minimum spark energies of hydrogen in air are 0.017 mJ at 101.3 kPa (1 atm), 0.09 mJ at 5.1 kPa (0.05 atm), and 0.56 mJ at 2.03 kPa (0.02 atm). The minimum spark energy required for ignition of hydrogen in air is considerably less than that for methane (0.29 mJ) or gasoline (0.24 mJ). Nevertheless, the ignition energy for all the three fuels is sufficiently low that ignition in air is relatively certain in the presence of any weak ignition source such as, sparks, matches, hot surfaces, open flames, or even a weak spark caused by the discharge of static electricity from a human body.

Hot objects and flames. Objects at temperatures from 773 to 854 K (500–581°C) can ignite hydrogen–air or hydrogen–oxygen mixtures at atmospheric pressure. Substantially cooler objects, about 590 K (317°C), can also ignite these mixtures after prolonged contact at less than atmospheric pressure. In cases that ignition sources are a required part of hydrogen use, provisions should be made to acceptably contain any resulting deflagration or detonation. For instance, a combustor or engine should not be operated in atmospheres containing hydrogen without well-dispersed water sprays in its exhaust. Experience has shown that multiple bank sprays will partially suppress the detonation pressures and reduce the number and temperature of ignition sources in an exhaust system. Water sprays should not be relied on as a means of avoiding detonations. Carbon dioxide may be used with the water spray to further reduce hazards.

16.4.3.4 Explosion Phenomena

The general term *explosion* corresponds to a rapid energy release and pressure rise and can occur in both reactive and nonreactive systems. Pressurized gas vessel failures are typical examples of nonreactive explosions, whereas premixed gases of fuel and oxidizer permit a rapid energy release by chemical reactions in reactive systems. Gas explosions do not necessarily require the transmission of a wave through the explosion source medium, although explosions in most cases involve some kind of wave such as a deflagration or detonation wave. Deflagration is the phenomenon in which the flame front moves through a flammable mixture in the form of a subsonic wave with respect to the unburned mixture. The terms *deflagration, combustion, flame,* and *burn* are found in the literature to be used interchangeably as synonymous. *Detonation* is the phenomenon in which the flame front coupled to a shock wave propagates through a detonable mixture in the form of a supersonic wave with respect to the unburned mixture. Generally, detonation propagates a thousand times faster than deflagration resulting in pressures at the detonation front 15–20 times higher than the initial ones. This explains why detonation has much greater potential for causing personnel injury or equipment damage than deflagration.

Especially, for gaseous mixtures exploding in the open the term *unconfined vapor cloud explosion (UVCE)* is used, whereas when it explodes in confined spaces the term *confined vapor cloud explosion (CVCE)* is used. In a very poor or very rich fuel mixture, but still within flammability limits, the flame front travels in the cloud at low velocity and insignificant pressure increase—a phenomenon known as *flash fire.*

Another explosion phenomenon encountered in the storage of liquefied hydrogen under pressure is the BLEVE, which occurs from the sudden release of a large mass of pressurized liquid into the atmosphere. The content of the vessel released into the atmosphere, if flammable, ignites forming a nearly spherical burning cloud, the so-called *fireball*. For more details on explosion phenomena see Ref. 21.

16.4.3.4.1 *Detonation Limits*

The maximum and minimum concentrations of a gas, vapor, mist, spray, or dust in the air or other gaseous oxidant for a stable detonation to occur are the so-called upper and lower *detonation limits*. These limits depend on the size and geometry of the surroundings as well as other factors. Therefore, detonation limits found in the literature should be used with caution. Detonation limits are sometimes confused with deflagration limits and the term *explosive limits* is then used inconsiderately [40].

As shown in Table 16.2, detonable concentrations are narrower than flammable concentrations for hydrogen as well as for all other gases. The worst case during an accidental release is reaching around stoichiometric concentration of a flammable gas for a detonation to occur provided an ignition source is present. These detonable concentrations are quite infrequent to occur in an enclosed area and rather unlikely in the open air for most gases possessing lower and higher detonation limits very close to one another. Unfortunately, in some gases, among which is hydrogen, this concentration span is quite broad and the detonability range is easily attained leading to much more serious accidents than those of the narrow spanned gases [41].

The higher diffusion coefficient of GH_2 compared to other gases (shown in Table 16.2) means that any release will quickly mix with the surrounding air, reach the dangerous range, and then come down to the safe region below the LFL. In an LH_2 leak, fast vaporization occurs first and molecular diffusion in the air follows. However, the positive buoyancy of GH_2 can move a release to higher floors in a building endangering these areas or to safe heights in the open air.

Detonation limits for any fuel–oxidizer mixture depend on the nature and degree of confinement among other factors [42].

16.4.3.4.2 *Ignition Energy*

Minimum ignition energy depends on hydrogen concentration in the detonable mixture. This is about 1 g of tetryl for hydrogen–air mixtures at the stoichiometric composition (possessing the highest sensitivity) increasing to several tenths of grams for very rich or very poor mixtures (near the detonation limits) [17]. Stable detonation parameters such as detonation temperature and pressure can be calculated by the Chapman–Jouguet approach using the Gordon–McBride computer code [43]. The dependence of detonation temperature and pressure on hydrogen concentration for hydrogen–air mixtures and hydrogen–oxygen mixtures is shown in Figures 16.1 and 16.2, respectively. Maximum values of approximately 3000 K and 1600 kPa for hydrogen–air mixtures near stoichiometric composition are observed, whereas these maxima rise to about 3800 K and 2000 kPa for stoichiometric hydrogen–oxygen mixtures.

16.4.3.4.3 *Catastrophic Effects*

With regard to the *catastrophic effects* of an accidental release and subsequent explosion, the energy of explosion is often taken into consideration. Calculations such as 1 g stoichiometric hydrogen in NTP air is equivalent to 24 g of TNT are usually found in the literature, but they are misleading since they do not take into account the weight of the oxidant (air in this case). By including air in the calculations, it becomes clear that the stoichiometric hydrogen–air mixture produces only two-thirds of the explosion energy of TNT. This difference can be easily explained by the dilution effect of nitrogen in the air. In addition, it should be noted

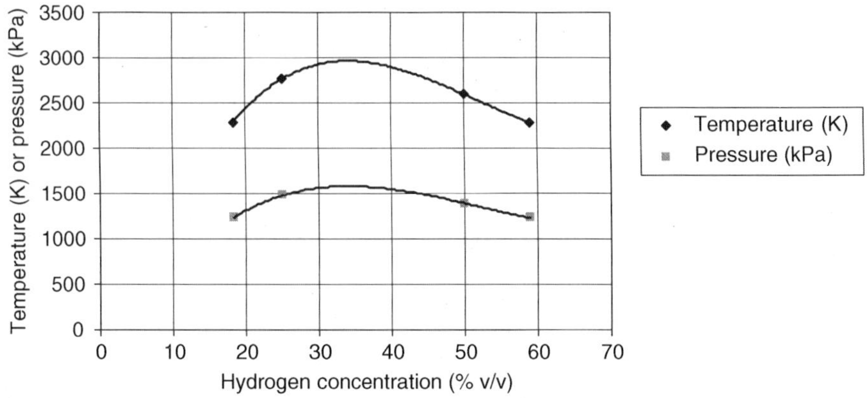

FIGURE 16.1
Detonation pressure and temperature of hydrogen–air mixtures starting from 101.3 kPa (1 atm) and 298 K (25°C). Chapman–Jouguet calculations using the Gordon–McBride code. (After Gordon, S. and McBride, B.J., *Computer Program for Calculation of Complex Chemical Equilibrium Compositions and Applications*, NASA Reference Publication, Cleveland, Ohio, 1994.)

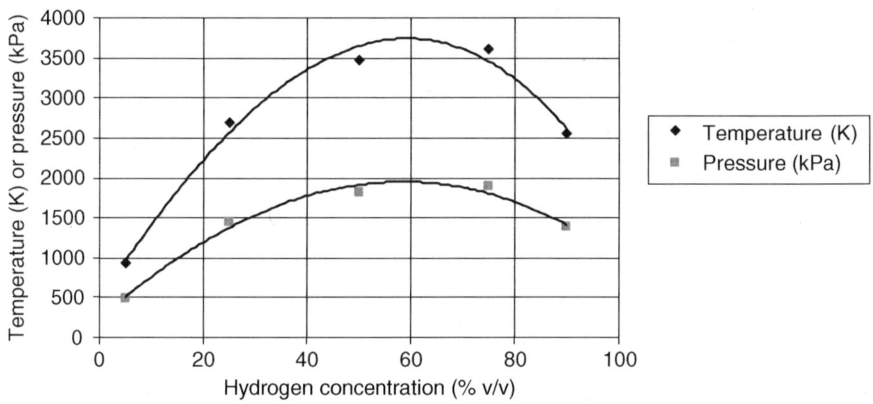

FIGURE 16.2
Detonation pressure and temperature of hydrogen–oxygen mixtures starting from 101.3 kPa (1 atm) and 298 K (25°C). Chapman–Jouguet calculations using the Gordon–McBride code. (After Gordon, S. and McBride, B.J., *Computer Program for Calculation of Complex Chemical Equilibrium Compositions and Applications*, NASA Reference Publication, Cleveland, Ohio, 1994.)

that in an accidental release mixing of hydrogen with air is not perfect; therefore, only a small fraction of the theoretical energy will be produced.

Other factors that determine the catastrophic effects of an explosion are the initial density of the explosive (which is more than three orders of magnitude higher for TNT than for hydrogen–air mixture) and detonation velocity (which is three to five times higher in TNT). Therefore, the resulting pressure wave from a hydrogen explosion is considerably flatter (longer duration and lower maximum overpressure) than TNT, and destruction effects are mainly caused by impulse rather than overpressure.

16.4.3.4.4 Deflagration-to-Detonation Transition and Overdriven Detonation

The oxidation (combustion) of hydrogen in the air or other oxidant (e.g., pure oxygen) can be subsonic (deflagration) or supersonic (detonation) depending on the strength of the ignition source. A high energetic source such as a shock wave can initiate a detonation, whereas a low-strength energy source such as a flame or a spark can initiate a deflagration. However, there is a possibility that a deflagration transits to a detonation after the flame has traveled for some time and distance when factors such as sufficient degree of confinement and the presence of obstacles inducing turbulence in the flame front, limit energy losses, and suitably mix the ingredients, thus triggering a detonation. This phenomenon, which is not a continuous process but rather a stepwise change, is known as *deflagration-to-detonation transition* (*DDT*) [44]. The theoretical approach is based on the work of Eichert [45], who modeled the transition phase using an acceleration parameter only.

Generally, the composition range in which a detonation can take place is narrower than that of deflagration. The range for detonation limits found in Table 16.2 for a hydrogen–air mixture is from 18.3% to 59% hydrogen. However, with sufficiently strong ignition sources, these limits can be extended and, in addition, an *overdriven detonation* can be obtained possessing a higher detonation velocity that gradually declines to a stable detonation at a thermodynamically determined lower velocity.

16.5 Hazards in Hydrogen Storage Facilities

16.5.1 Hazard Spotting

In atmospheric conditions, hydrogen is a colorless, odorless gas much lighter than air in normal conditions. The low density in conjunction with the small particle size allows the penetration of hydrogen molecules in some metals and alloys such as cast iron and high-carbon steel [46]. The penetration may end in small hydrogen leaks, or, in the presence of cracks within the wall, contribute to cracks spread, material strength decrease, and subsequent fracture.

Hydrogen reacts violently with oxidizers such as nitrous oxide, halogens (especially with fluorine and chlorine), and unsaturated hydrocarbons (i.e., acetylene) with intensely exothermic reactions. Hydrogen gas forms combustible or explosive mixtures with the atmospheric oxygen over a wide range of concentrations as mentioned earlier. Hydrogen is nontoxic; however, besides the hazards of flash fire, jet fire, and gas cloud explosion presented in the case of an accidental release, hydrogen may act as an asphyxiant at sufficiently high concentrations by depleting the oxygen available in atmospheric air.

16.5.1.1 Thermal Hazards

Hydrogen is stored either in the gaseous state under high pressure and normal temperature, or in the liquid state under low temperature and moderate pressure. In any case there is a significant hazard potential for mechanical explosion of hydrogen vessels (tanks and cylinders) when exposed to high temperature or thermal radiation. Usually, the reason for overheating is a neighboring fire (primary event) that causes temperature increase in the vessel shell and the content (Figure 16.3). The vessel finally bursts (secondary event) spilling its content, which (if flammable) is usually immediately ignited and burns in the form of a jet fire or a fireball. Typical ignition sources for the spilling fuel may be electrostatic sparks developed during content discharging or the flames of the neighboring fire. The situation is known as domino effect since an initial incident gives rise to another generating a chain of incidents with escalating impact on the surroundings [47,48].

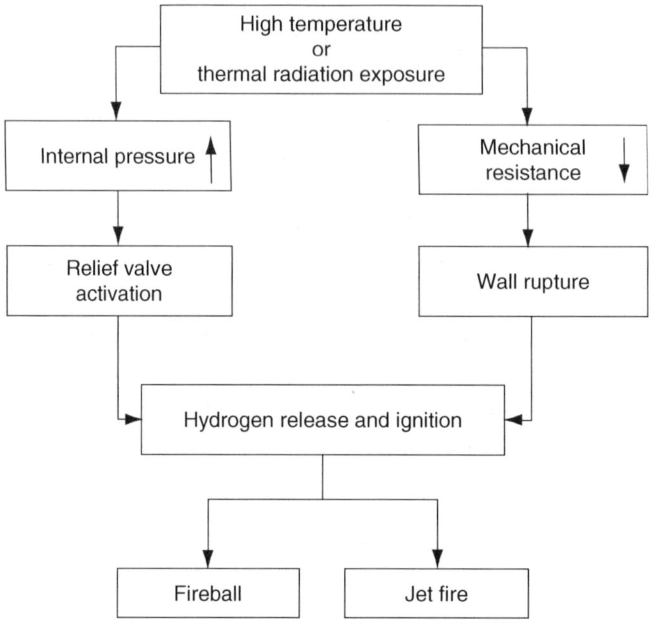

FIGURE 16.3
Thermal radiation effect on pressurized hydrogen vessel. (From Rigas, F. and Sklavounos, S., *Int. J. Hydrogen Energ.*, 30, 1501, 2005. With permission from International Association of Hydrogen Energy.)

In contrast to high-pressure GH_2 storage vessels, liquefied hydrogen vessels operate in moderate to low pressures that normally do not exceed 20 bars. As a result, it is reasonable for the walls to be designed with lower pressure resistance than that of hydrogen gas (whose operation pressure may be over than 400 bars) being liable to increased risk of failure. In case that the container is engulfed in a fire, its metal is heated and loses mechanical strength. While the liquid phase absorbs significant amounts of heat, vapors possess far lower specific heat capacity. Therefore, the heat supplied to that part of the container where the vapor phase exists will raise much more the local wall temperature thus weakening its strength [47].

Concerning liquefied gas storage, vessel overheating may result in internal temperatures higher than the boiling point of the content, indeed without vaporization initiation of the liquid phase, and then the liquid is superheated. The phenomenon is observed when there is a shortage of nucleation sites (i.e., impurities, crystals, or ions) in the bulk of the liquid. However, there is a temperature limit above which the fluid cannot remain in the liquid state anymore (homogeneous nucleation limit or superheat limit temperature). In this limit, random molecular density fluctuations within the bulk of a liquid produce holelike regions of such molecular dimensions that may act as bubbles [48]. The end result is the explosive flash of the liquid accompanied by a strong shock wave that propagates through the fluid and ruptures the container spilling the content into the atmosphere (BLEVE) [49]. Missiles of the ruptured walls may travel hundred of meters, whereas the flammable content ignites forming a sphere that burns from the outer to inner layers (fireball). The overall process is shown in Figure 16.4.

16.5.1.2 Cryogenic Storage Conditions

At hydrogen refilling stations and vehicles, hydrogen has to be compressed at high pressures (typical value: 400 bars) due to the low energy content per unit volume.

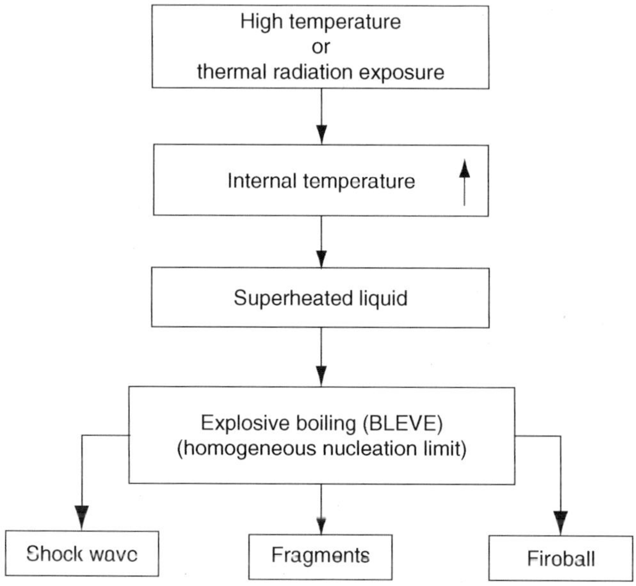

FIGURE 16.4
Thermal radiation effect on liquefied hydrogen vessel. (From Rigas, F. and Sklavounos, S., *Int. J. Hydrogen Energ.*, 30, 1501, 2005. With permission from International Association of Hydrogen Energy.)

Nevertheless, in some applications, hydrogen needs to be stored in the liquid state at very low temperatures for volume restriction. These temperatures are often lower than −73°C so that storage conditions are characterized as cryogenic distinguished from the refrigerated ones. These conditions are preferred, for example, in uses such as rocket propulsion as well as in warehousing for convenience and economy. A crucial factor for the liquefaction implementation is the critical temperature above which the gas cannot be liquefied by pressure application only. In practice, hydrogen is kept liquefied at extremely low temperatures below −240.2°C and moderate pressures (20–30 bars). Major hazards related to cryogenic storage stem from

- *Embrittlement of service materials.* The low temperatures inside storage tanks and transmission pipelines may cause significant susceptibleness of the structural material in vibrations and shocks. Mild steel and most iron alloys, at LH_2 temperatures, lose their ductility being liable to increased risk of mechanical failure [50].
- *LH_2 spills.* Liquefied hydrogen releases yield dramatically larger volumes of combustible clouds (1 L of liquid produces 851 L of gas on evaporation). Therefore, the consequences of a fire or explosion are more extensive than that of a pressurized hydrogen release.
- *Extremely low temperature.* It can cause severe tissue frostbite if the material comes into contact with human body. Flesh may stick fast to cold insufficiently insulated pipes or vessels and tear on attempting to withdraw it [51].
- *Hydrogen cloud dispersion.* As discussed hereinafter, hydrogen spill originated from cryogenic storage resulting in cloud formation that disperses with a way similar to a heavier-than-air gas increasing the risk of accidental fires and explosions.

16.5.2 Hazard Evaluation

16.5.2.1 Methodology

Hazard evaluation in hydrogen storage facilities deals with the determination of all credible accidental scenarios. To do this, a variety of methods (event tree, failure mode and effect analysis, what-if, fault tree) can be found in the relevant literature [11,12,52]. Of them, event tree analysis (ETA) is a formal technique and one of the standard approaches used for industrial incidents investigations. ETA is a logic model that graphically portrays the combination of events and circumstances in an accident sequence. It is an inductive method, which begins with an initiating undesirable event and works toward a final result (outcome). The general procedure for ETA involves the following steps:

1. Determination of the initiating events that can result in certain types of accident
2. Identification of the critical factors that may affect the initiating event evolution
3. Construction of the event tree taking into account the interaction between the critical factors and the initiating event
4. Designation and evaluation of resulting accidental events

Applying ETA in fuel gas releases, the critical factors that may affect substantially the final outcome are the time of ignition of the resulting cloud and the confinement provided by the surroundings. The former is related to the mixing of escaping fuel gas with air. When immediate ignition occurs, gas cloud mixing with atmospheric oxygen is still limited; thus, the ignition takes place on the outer layer that is between the flammable limits, whereas the inner core of the cloud is too rich in fuel to ignite. As buoyancy forces of the hot gases begin to dominate, the burning cloud rises and becomes more spherical in shape forming a ball into flames. This elevation gradually causes further mixing of the gas with oxygen, which brings new volumes of gas into flammable limits sustaining the fire. In contrast, when delayed ignition occurs, the fuel cloud may have been adequately mixed with air, so that after ignition it flashes back. It differs from fireball since it proceeds faster and can burn from inner to outer flammable layers provided a proper ignition source is found there.

16.5.2.2 ETA Method Application

The application of ETA regarding accidental hydrogen release is displayed in Figure 16.5. It is obvious that, unless an immediate ignition takes place, there is some time of dispersion that intervenes between release and ignition. Generally, if hydrogen flammability zones were known, it would be possible to take preventive measures and to prepare emergency response planning against fires and explosions. Consequently, a major issue arises regarding the computation of the dispersion succeeding an accidental hydrogen release.

Furthermore, even if no ignition takes place, escaping hydrogen may accumulate into closed spaces adjacent to the source posing asphyxiation hazard for the people being there. Hydrogen dispersion may be considered safe only when no ignition occurs and no space confinement exists.

16.5.3 Qualitative Prediction of Cloud Travel

Generally, the behavior of a gas (or vapor) cloud during dispersion can be either buoyant or gravity driven. The former is regarded with heavier-than-air gases and the later with lighter-than-air gases. Currently, dispersion estimations can be performed via the use of semiempirical one-dimensional models (the so-called box models) and the CFD codes.

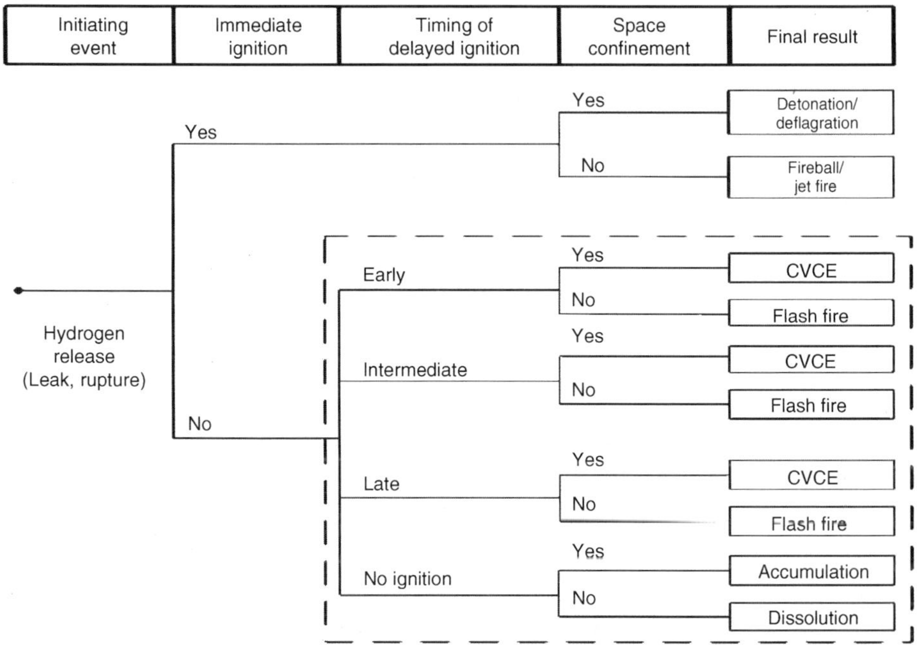

FIGURE 16.5
Event tree analysis adapted to accidental hydrogen releases. (From Rigas, F. and Sklavounos, S., *Int. J. Hydrogen Energ.*, 30, 1501, 2005. With permission from International Association of Hydrogen Energy.)

The selection of the appropriate dispersion model in an accidental release scenario requires the behavior of dispersing gas to be known since each model is specialized on one kind of release (buoyancy or gravity driven).

Hydrogen preserves the property of light gas behavior even when releases from high-pressure storage systems occur. But, once hydrogen escapes while stored in the liquid state, its dispersion behavior turns to that of a heavy rather than a light gas. The different way in which a gas disperses according to the storage conditions has been observed in field-scale trials with another light gas (natural gas) that was left to escape and disperse while stored in the liquid state [53]. In this case, instead of elevation the vapor cloud moved along the ground being under the LFL limit for several tenths of meters.

16.5.4 Gas Dispersion Simulation

16.5.4.1 CFD Codes

CFD codes have been successfully used for providing hydrogen gas dispersion estimations [54,56]. The CFD code CFX was first validated against the experimental data of large scale gas dispersion trials for isothermal heavy gas [55] as well as for nonisothermal dense gas releases [56]. The first step in a CFD simulation attempt includes the definition of the three-dimensional geometry (computational domain), in addition to the subdivision of the entire domain into a number of smaller control volumes (cells) that form a mesh. The geometry comprises a unified set of parametric surfaces built in an appropriate CAD interface. In CFX (a CFD code), the solution process initiates with discretizing the spatial domain into finite control volumes using the mesh constructed in the preprocessor stage. The subsequent balance equations integration over each control volume is based on the finite volume method (FVM) [57].

16.5.4.2 CFD Application in Gaseous and Liquefied Hydrogen Releases

The reliability of CFD simulation considering cryogenic gas release has been examined against field-scale experimental data including hydrogen and natural gas releases. A characteristic example of natural gas experiments is the Burro series trials of LNG spills. The project was sponsored by the U.S. Department of Energy (U.S. DOE) and was conducted by Lawrence Livermore National Laboratory (LLNL) and the Naval Weapons Center (NWC) at the China Lake site in California during the summer of 1980 to determine the transport and dispersion of vapor from spills of LNG on water. A detailed description and a result analysis of the experiments are given in the relevant report in Ref. 58. Hydrogen dispersion experiments with data available to the public [59] are the LH_2 spills conducted by NASA Langley Research Center at the White Sands Test Facility in 1980. The objective of the program was to study the phenomena associated with the rapid release of large amounts of LH_2, with emphasis on the generation and dispersion of the flammable hydrogen cloud.

Figures 16.6 and 16.7 show the experimental gas concentration histories versus the computational ones for the LNG Burro 8 trial and the LH_2 trial 6. The experimental observations of the concentration change are found statistically in good agreement with the computationally predicted histories showing that the CFD predictions are reasonable.

The behavior of hydrogen during its atmospheric dispersion has been examined considering two different cases: isothermal release under pressurization conditions and cryogenic (nonisothermal) release under liquefaction conditions. Three-dimensional simulations run for a typical accidental release scenario, which included hydrogen discharge under pressurization and liquefaction conditions in an imaginary computational domain with dimensions 160 m $\times$ 40 m $\times$ 30 m. Hydrogen release was modeled as a ground-level area source with inflow rate equal to 2 kg/s for 5 s, whereas typical atmospheric conditions (1 atm, 20°C), ground temperature equal to 15°C, and wind speed equal to 3 m/s were considered.

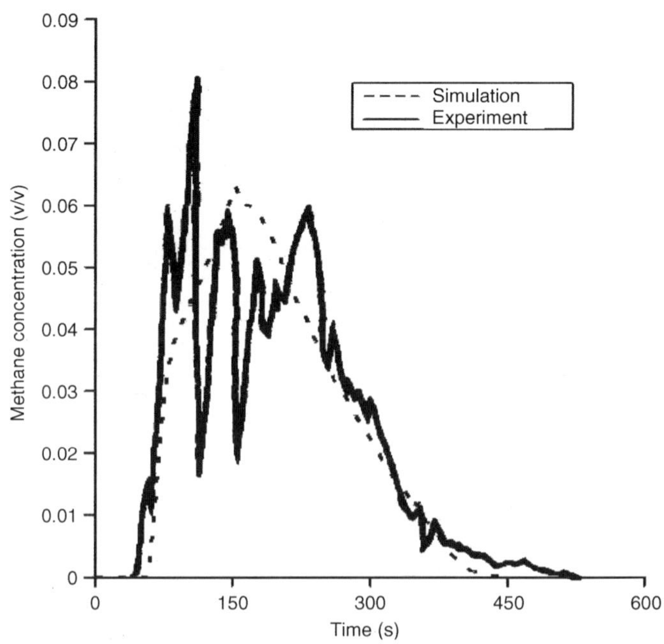

FIGURE 16.6
Computational and experimental methane concentration histories at point (140, 0, 1) for the Burro 8 trial. (From Sklavounos, S. and Rigas, F., *Energ. Fuel.*, 19, 2535, 2005. With permission.)

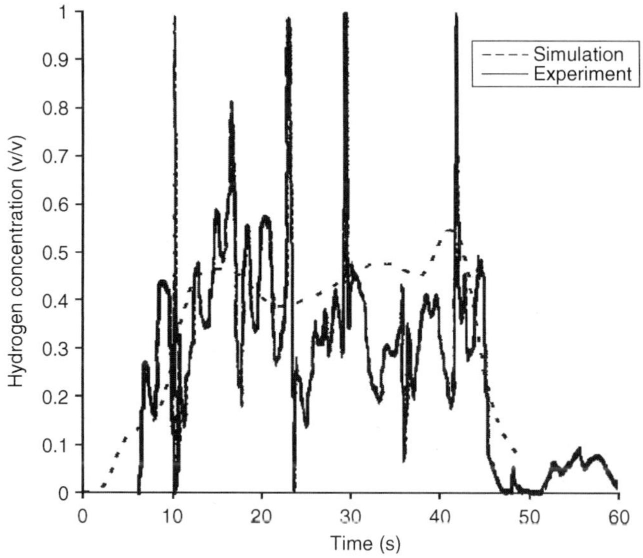

FIGURE 16.7
Computational and experimental hydrogen concentration histories at point (9.1, 0, 1) for the LH_2 trial 6. (From Sklavounos, S. and Rigas, F., *Energ. Fuel.*, 19, 2535, 2005. With permission.)

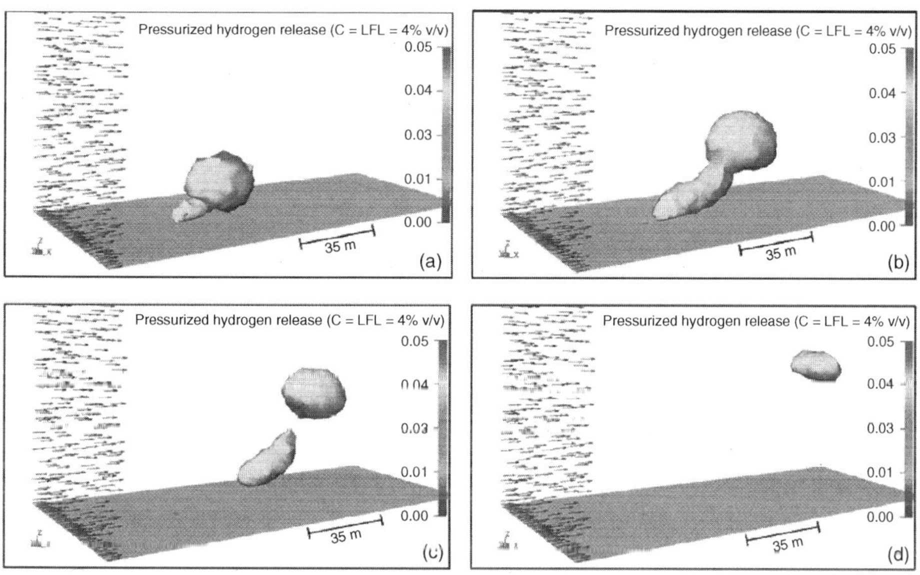

FIGURE 16.8
Simulated dispersion snapshots following pressurized hydrogen release at times (a) 2.4 s, (b) 6.2 s, (c) 9.6 s, and (d) 12.4 s after start of release. (From Rigas, F. and Sklavounos, S., *Int. J. Hydrogen Energ.*, 30, 1501, 2005. With permission from International Association of Hydrogen Energy.)

The results demonstrated substantial differentiation between pressurized and cryogenic hydrogen release as far as cloud dispersion is regarded. In the former case, the cloud lifts immediately moving almost vertically (Figure 16.8a through 16.8d), whereas in the latter it travels downwind, almost horizontally remaining close to the ground (Figure 16.9a through 16.9d).

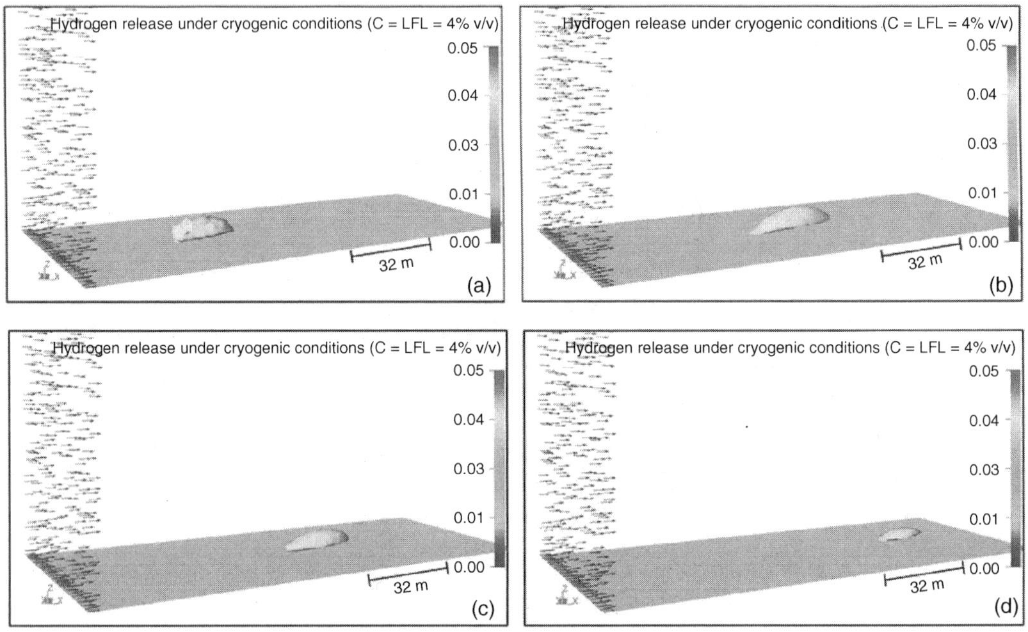

FIGURE 16.9
Simulated dispersion snapshots following cryogenic hydrogen release at times (a) 3.5 s, (b) 10 s, (c) 13 s, and (d) 18.5 s after start of release. (From Rigas, F. and Sklavounos, S., *Int. J. Hydrogen Energ.*, 30, 1501, 2005. With permission from International Association of Hydrogen Energy.)

It can be seen from Figure 16.10a and 16.10b that hydrogen cloud spreading is similar to that of LNG that has been experimentally observed in field-scale trials [53], namely horizontal cloud shift at low height [60]. In both cases, the greater inertia of the heavy cryogenic gas reduces the rate of turbulent mixing thus leading to delayed dissolution times, whereas the greater density of the cloud produces a gravity-driven flow that tends to reduce the cloud height and increase its width. In contrast to lighter-than-air gas, the stay of heavy cloud at low height significantly increases the risk of ignition.

Conclusively, the simulation of atmospheric dispersion of gaseous pressurized and liquefied *cryogenic hydrogen releases* showed that hydrogen disperses as a heavier-than-air gas when discharged in cryogenic conditions, in contrast with high-pressure gas release that produce lighter-than-air cloud. In the former case, hydrogen remains within the LFL range close to the ground for some time, increasing the risk for various types of fires and explosions. Consequently, accidental liquefied hydrogen release scenarios should be addressed as heavier-than-air releases using the appropriate dispersion models for calculating heavy and not light gas dispersion. Moreover, this change of the behavior of a light gas on the dispersion during cryogenic release should be kept in mind in hazard evaluation procedures and safety measures establishment.

In practical applications, dispersion calculations need to be performed in *topographically complex environment*. Thus, solid obstacles intervening in the area dispersion should be counting in the computations. In previous validation works, it has been proved that CFD codes constitute powerful tools for complex terrain dispersion simulation providing high accuracy results with excellent visualization capabilities, which can be helpful in quantitative risk analysis applications [55]. The dominating mixing mechanism between

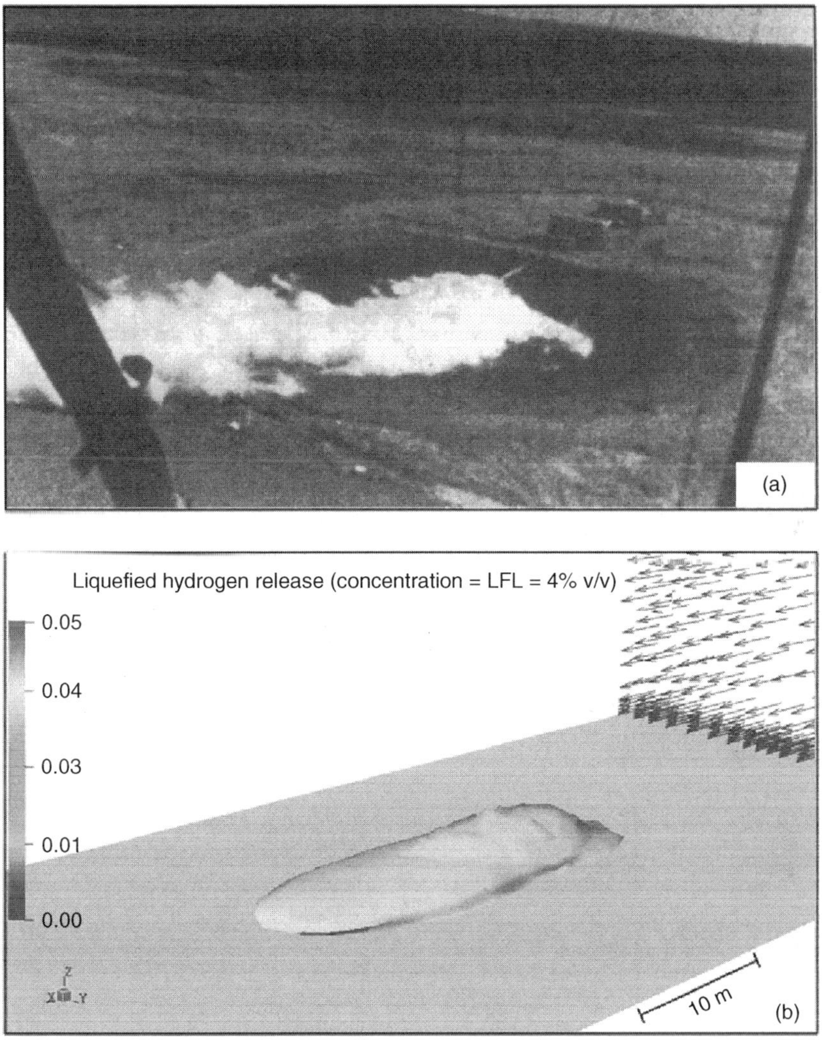

FIGURE 16.10
Qualitative comparison of dispersing cloud following (a) liquefied natural gas release; experimental record from Burro series trials (From Lees, F.P., *Loss Prevention in the Process Industries*, Butterworth Heinemann, Oxford, 1996.) and (b) liquefied hydrogen release; computational estimation 7 s after release initiation. (From Rigas, F. and Sklavounos, S., *Int. J. Hydrogen Energ.*, 30, 1501, 2005. With permission from International Association of Hydrogen Energy.)

gas and air, which is the turbulence flow developing in the vicinity of the solid obstacle, was then computationally approached satisfactorily by several turbulence models: the $k-\varepsilon$ and shear stress transport (SST) models showed improved robustness during solver processing. The SSG model entailed increased the central processing unit (CPU) time without significant enhancement of accuracy of results. The SSG, $k-\varepsilon$, and SST models appeared to overestimate maximal concentrations recorded in the trials, whereas the $k-\omega$ model to underestimate them.

Generally, the results obtained through the numerical simulation showed good agreement with the experimental data leading to the conclusion that CFD techniques can be effectively used in consequence assessment procedures concerning toxic/flammable dispersion scenarios in real terrains, where box models have limited capabilities.

Atmospheric releases of flammable gases such as hydrogen may lead to major fires with extensive effects on the surroundings. In activities where hazards are associated with cloud fires, there is the need of societal risk assessment that involves the estimation of hazardous zones due to the resulting thermal radiation. However, till now only limited work has been done on modeling the effects of flash fires, in a way that available techniques may be judged insufficient [47].

A methodology for modeling *flash fire radiation* is proposed by Raj and Emmons [47], who modeled a flash fire as a two-dimensional, turbulent flame propagating at constant speed. Yet, the model is liable to simplifying assumptions (i.e., stationary location of burning cloud) and presents certain deficiencies (i.e., ignorance of ignition site location). Moreover, a multipoint grid-based approach for radiation impact analysis was adopted by Pula et al. [61] providing estimations of radiation effects at different locations in the area of interest. A three-dimensional computational approach based on fluid dynamics techniques was recently attempted aiming at the estimation of resulting thermal radiation emissions and overpressure in large-scale cloud fires [62,63].

The experimentation in the field of gas cloud fires appears to be limited. The unique set of large-scale experiments that involve the release, dispersion, ignition, and combustion of flammable natural gas clouds in the open air is that with the code name Coyote. Coyote series trials conducted by LLNL in 1983 at California's Nevada Test Site, Nevada provided an integrated dataset for use in validation studies [64,65]. The objective of the experiments was to determine the transport and dispersion of vapors from LNG spills, and in addition to investigate the damage potential of vapor cloud fires. Transient simulations

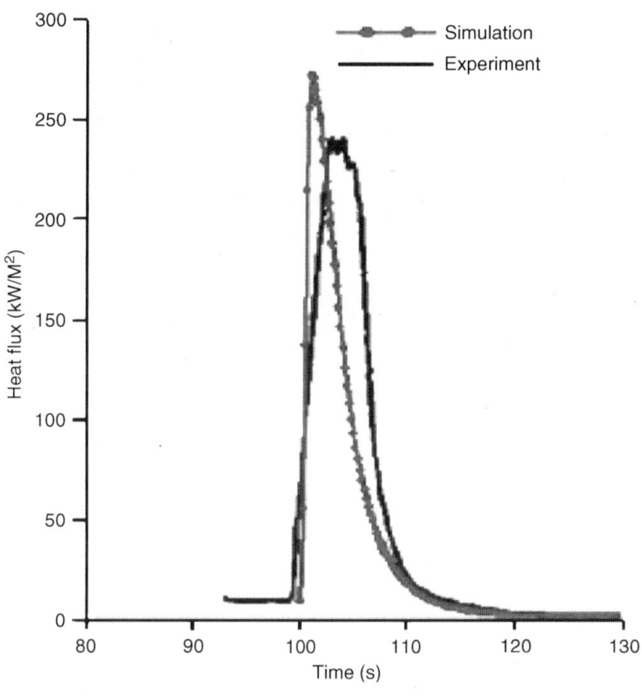

FIGURE 16.11
Experimental versus computational heat flux histories for Coyote trial 3 at position (65, 57, 1). (From Rigas, F. and Sklavounos, S., *Chem. Eng. Sci.*, 61, 1444, 2006. With permission from Elsevier.)

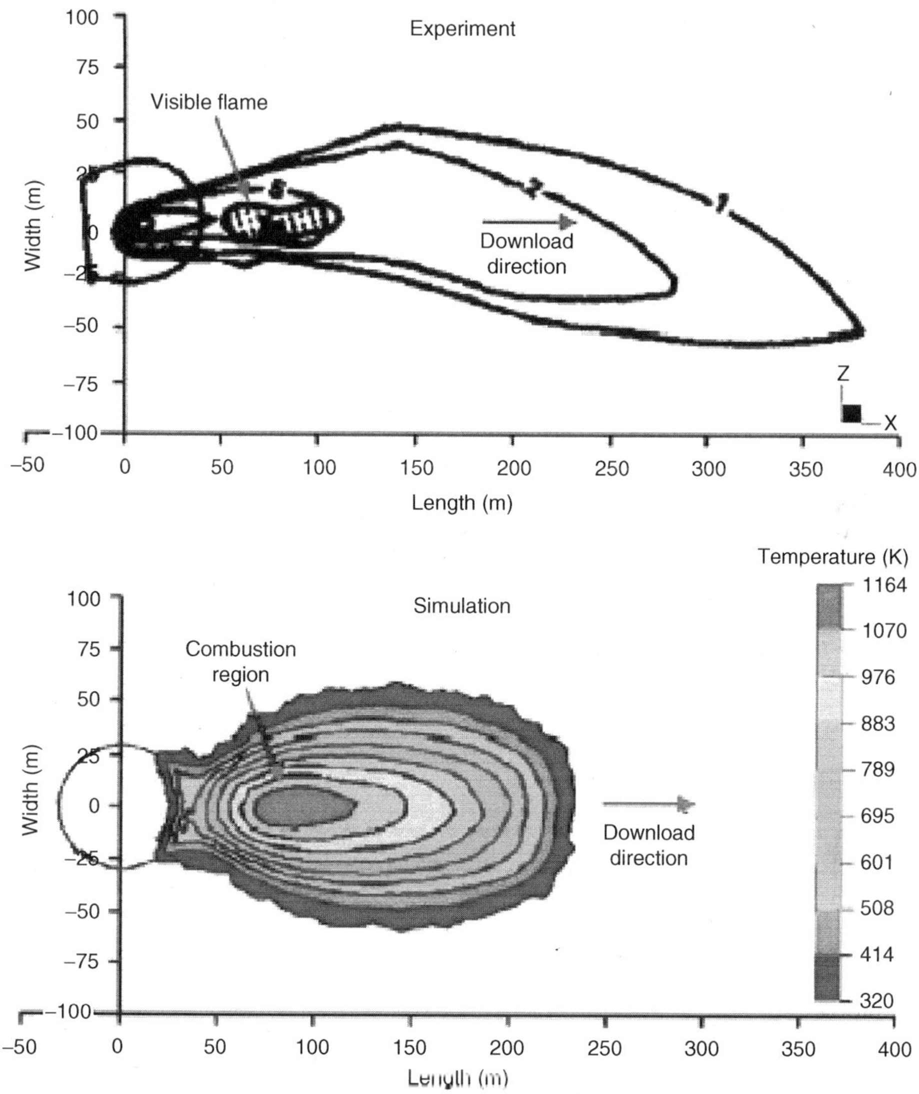

FIGURE 16.12
Experimental and simulated burning cloud snapshot for Coyote trial 3 (1 m height, 103 s). (From Rigas, F. and Sklavounos, S., *Chem. Eng. Sci.*, 61, 1444, 2006. With permission from Elsevier.)

of the flammable cloud combustion performed with CFX (a CFD code) provided reasonable estimations of the thermal flux histories (Figure 16.11), whereas the predictions of the thermal impulses were proved to be statistically valid based on the calculation of certain statistical measures [62].

In addition, postprocessing of the computational results allowed visualizations of the combustion progress within the cloud. Thus, comparisons between the computational and the experimentally observed combustion regions were obtained revealing good agreement between the corresponding snapshots (Figure 16.12). Generally, it was concluded that the evolution of a gas cloud fire and the oncoming consequences may be efficiently simulated utilizing CFD techniques.

16.6 Hazards in Hydrogen Use as a Fuel for Transport

16.6.1 Safety Comparisons of Hydrogen, Methane, and Gasoline

Thermophysical, chemical, and combustion properties of hydrogen, methane, and gasoline are compared in Table 16.2. Of them, gasoline is certainly the easiest and perhaps the safest fuel to store because of its higher boiling point, lower volatility, and narrower flammability and detonability limits. This generalized assumption is based on previous discussion concerning fire and explosion hazards. Nevertheless, hydrogen and methane (the principal ingredient of natural gas) can also be safely stored using current technology.

Despite its volumetric energy density, hydrogen has the highest energy-to-weight ratio of any fuel. Unfortunately, this weight advantage is usually overshadowed by the high weight of the hydrogen storage tanks and associated equipment. Thus, most hydrogen storage systems made for transport applications are considerably bulkier or heavier than those used for liquid fuels such as gasoline or diesel fuels. Attempting a comparison with regard to the hazards posed by hydrogen, methane, and gasoline, the following remarks can be made [3,20]:

- *Size of molecules.* Since the hydrogen molecule is smallest of all it will leak through permeable materials where methane and gasoline will not.
- *Fuel spills.* In the event of a fuel spill, it is expected that a fire hazard develops most rapidly in the descending order: hydrogen, methane, and gasoline. With regard to fire duration, gasoline fires last the longest and hydrogen fires are the shortest ones, while all the three fuels burn at nearly the same flame temperature. In fact, for spillage of identical liquid fuel volumes, hydrocarbon fires will endure 5–10 times longer than hydrogen fires.
- *Odorization.* Natural gas is odorized so that leaks can be detected, whereas gasoline is normally smelly. Odorization of natural gas is a prudent although not entirely effective safety measure. Hydrogen as an industrial gas or fuel cell vehicle fuel is not odorized because sulfur-containing substances (mercaptans) contaminate the catalysts of a fuel cell.
- *Buoyancy.* Hydrogen is 14.5 times lighter than air at NTP, whereas methane is 1.8 times lighter and gasoline vapor is heavier than air. Thus, hydrogen will rise much more quickly causing greater turbulent diffusion, which reduces its concentration below the LFL more rapidly. Moreover, hydrogen diffuses into air approximately 4 times faster than methane and 12 times faster than gasoline, thus causing rapid concentration decrease in safe levels.
- *Energy of explosion.* The energy of explosion values given in Table 16.2 should be considered as the theoretical maxima, and yield factors of 10% are considered reasonable for fuel–air explosions. For equivalent volume storage, hydrogen has the least theoretical explosive potential of the three fuels considered, albeit it has the highest heat of combustion and explosive potential on a mass basis.
- *Flammability and detonability limits.* The broader flammability and detonability limits of hydrogen coupled with its rapid burning velocity renders hydrogen a greater explosive threat than methane or gasoline.
- *Ignition energy.* The ignition energy of hydrogen is about 1/14 and 1/12 of that for methane and gasoline, respectively. However, the energy levels required for

the ignition of these fuels are so low that common ignition sources such as static electricity discharge from a human body will ignite any of these fuels in air.

- *Autoignition temperature.* Hydrogen and methane possess unusually high autoignition temperatures (585°C and 540°C, respectively), whereas gasoline with autoignition temperature ranging from 227°C to 477°C appears more hazardous.
- *Deflagrations.* A *confined deflagration* of hydrogen–air or methane–air will produce a static pressure rise ratio of less than 8:1. Explosion pressures for confined deflagrations of gasoline–air are about 70–80% of those for hydrogen–air. *Unconfined deflagration* overpressures are usually <7 kPa. However, a pressure of 3–4 kPa is sufficient to cause structural damage to buildings; therefore, unconfined large-volume gas-phase explosions can be destructive. Thus, it is apparent that confined deflagrations with up to 8 atm (811 kPa) of explosion pressure can be devastating and even unconfined deflagrations can cause slight-to-moderate structural damage, and injure people via fire and window-glass shrapnel.
- *Detonations.* Pressure rise ratios of ~15:1 for hydrogen–air or methane–air *detonations* and a ratio of ~12:1 for a gasoline–air detonation are normally expected. The impulse created by explosion pressure profile should be taken into account in evaluating explosion damage and in the design of barricades or structures to mitigate explosion consequences.
- *Shrapnel hazards* depend on explosion overpressures, and with ordinary enclosures ($L/D < 30$) are about the same for hydrogen–air and methane–air and somewhat less severe for gasoline–air mixtures. Nevertheless, in long structures, such as tunnels or pipes, hydrogen poses a greater explosion risk than the other two fuels due to its greater tendency for a DDT. Thus, hydrogen presents the greatest hazard to shrapnel damage.
- *Radiant heat.* Owing to the heat-absorbing water vapor created during hydrogen combustion and the absence of a carbon combustion reaction, the radiant heat from a hydrogen fire is significantly less than a hydrocarbon fire, and this reduces the risk of secondary fires. The reduced radiant heat entails reduced heating of adjacent equipment in case of a major fire and hence reduced probability for a domino effect that leads to escalating damage and losses.
- *Hazardous smoke.* The potential for smoke inhalation damage is judged to be most severe in the descending order: gasoline, methane, and hydrogen fires.
- *Flame visibility.* Unlike visible methane and gasoline flames, hydrogen burns with a near-invisible flame in daylight, but contaminants in air generally add some visibility. Hydrogen flames are visible at night and modern detection equipment can detect them even in daylight.
- *Fire fighting.* Normally, hydrogen and methane fires should be allowed to burn until gas flow is stopped or until liquid spills are consumed because of the potential explosive hazard resulting from extinguishing these fires. However, the fire should be controlled by cooling the storage tanks with water in all situations. Dry chemicals and high-expansion foams can be used to extinguish methane and gasoline fires.

Conclusively, hydrogen has been used and stored safely in the industry for quite a long time as compressed gas or liquefied hydrogen, and it seems that metal hydride storage will be equally safe or even safer. Consideration of future hydrogen applications reveals no safety problems in the industrial and commercial markets. Although hydrogen safety problems have been efficiently controlled in the industry, additional safety analyses are needed in the transportation and residential fuel markets.

16.6.2 Hydrogen Vehicle Hazards

Hydrogen is a promising fuel gas for transport; therefore, hazards associated with this application should be thoroughly investigated. These hazards should be considered in situations when a vehicle is inoperable, in normal operation, and in collisions.

Potential hazards are commonly related with fire, explosion, or toxicity as in any other case. *Toxicity* hazards can be ignored because neither hydrogen nor its combustion products are toxic, as mentioned earlier. *Fire and explosion* of hydrogen used for transport may come from the fuel storage on vehicle, the fuel supply lines, or from the fuel cell if such a system is used. Among them, fuel cell is the least hazardous, although hydrogen and oxygen in the existing technology are separated by a thin polymer membrane (20–30 μm). The initiating event in this case would be the membrane rupture resulting in the combination of hydrogen and oxygen, yet in such an event the fuel cell would lose its potential that can be easily detected by a control system. Then, the supply lines could be immediately disconnected. The operating temperature of a fuel cell (60–90°C) is too low to act as a thermal ignition source, but the two reactants may combine on the catalyst surface to create ignition conditions. Nevertheless, the damage potential is limited since the amount of hydrogen in the fuel cell and the fuel supply lines is small.

The greatest damage potential is located in the fuel gas tank due to the largest amount of hydrogen found there. The failure modes that should be considered in the case of normal operation, and in collision as well, are the following:

- *Catastrophic rupture of a tank.* It may be the result of a manufacturing defect, abusive handling or stress fracture, puncture by a sharp object, or external fire combined with failure of pressure relief device to open.
- *Massive leak.* It may be due to faulty operation of pressure relief valve, chemically induced failure of tank wall, puncture by a sharp object, or normal operation of pressure relief valve in case of fire.
- *Slow leak.* Probable reasons may be stress cracking of tank liner, faulty operation of pressure relief valve, faulty coupling from tank to the feed line, or failure of the fuel line connection.

Since a catastrophic rupture is rather unlikely, failure modes resulting in large or slow *hydrogen release* have been identified experimentally. Countermeasures to avoid the preceding failure modes could be for instance

- *Leak prevention* by a suitable safety design, allowing for shock and vibration tolerance of high-pressure lines.
- *Leak detection* by a proper detector or by adding an odorant in the fuel, which according to the existing technology is a problem for fuel cells.
- *Ignition prevention* by an automatic disconnection of the battery in case of an accident, separation of fuel supply lines and electrical systems, and design of proper systems for both active and passive ventilation, such as openings to allow hydrogen to escape upward.

Based on these failure modes, a detailed risk assessment of most probable or most severe hydrogen accident scenarios has been conducted in some studies [66] including fuel tank fire or explosion in unconfined spaces and in tunnels, fuel line leaks in unconfined spaces and in a garage, and refueling station accidents.

These studies have determined that

- A well-designed hydrogen fuel cell car should be safer than a natural gas or a gasoline car in collisions in open spaces.
- A hydrogen fuel cell car should be nearly as safe as a natural gas car, and both should be safer than a gasoline or propane car in a tunnel collision.
- The greatest risk appears in the scenario of hydrogen leak in a garage, which in the absence of passive or active ventilation may result in fire or explosion.

Conclusively, hydrogen fuel presents similarities and differences from other transportation fuels, with some parameters tending to make accidents more severe and others making accidents less severe. Thus, it is not clear whether hydrogen would be more or less risky in transportation. The good safety records of trucks carrying compressed and liquefied hydrogen over the road in the United States add confidence that there are not large, unknown risks associated with hydrogen fuel use and despite public perception, the use of hydrogen as a fuel for transport may be safer than that of gasoline or natural gas in some aspects [67].

16.7 Hydrogen Safety Guidelines

The following are general guidelines to ensure safety in the hydrogen storage and transfer areas. Good illumination, lightning protection, alarm systems, and gas detection and sampling systems should be provided in such a facility. More detailed safety guidelines concerning safety policy; safety in construction, operation, maintenance, and final disposition of a hydrogen facility; as well as safety measures in buildings and test chambers used in hydrogen service and emergency procedures can be found in *Safety Standard for Hydrogen and Hydrogen Systems* [68] from which some synoptic information is given in this section enriched with other references.

- *Electrical considerations.* Areas where flammable hydrogen mixtures are expected to occur or areas where hydrogen is stored, transferred, or used and where the hydrogen normally is contained are classified as highly hazardous zones according to internationally accepted regulations. All electrical sources of ignition should be prohibited in these areas, using approved explosion-proof equipment or selecting nonarcing approved equipment. *Explosion-proof equipment* has an enclosure strong enough to contain the pressure produced by igniting a flammable mixture inside the enclosure. Since it is not gas tight, the joints and threads must be tight enough and long enough to prevent issuance of flames or gases that would be hot enough to ignite a surrounding flammable mixture. Another method to prevent a gas explosion is to locate the equipment in an enclosure purged and maintained above ambient pressure with an inert gas. Intrinsically safe installations used in the facility should be approved for hydrogen service.
- *Hydrogen transmission lines* carrying hydrogen from trailers and storage vessels must be aboveground installations. Hydrogen lines crossing roadways should

be installed in concrete channels covered with an open grating and should not be located beneath electric power lines. Area surfaces located below liquefied hydrogen lines from which condensed liquid air may drop must be constructed of noncombustible materials such as concrete. A hydrogen leak, due to its very low density, can easily find the way to upper floors in a building traveling through connecting ducts and other openings and cause secondary explosions in other rooms.

- *Elimination of ignition sources.* Installations using hydrogen should be protected from *lightning* by lightning rods, aerial cable, and ground rods suitably connected. Lightning strikes may cause inducing sparks; therefore, all equipment in a building should be bonded and grounded to prevent sparks. *Static electricity* may be generated in moving machinery belts or in flowing fluids containing solid or liquid particles. The measures taken to limit electrostatic charge generation and accumulation include bonding and grounding of all metal parts within a system, use of conductive machinery belts, personnel clothes made of antistatic fibers, and conductive and nonsparking floors. Sparks may also be generated by other mechanisms such as *friction and impact.* Even spark-proof tools can cause ignitions because the energy required for ignition of flammable hydrogen–air mixtures is extremely small.
- *Hot objects, flames, and flame arrestors.* The measures taken to eliminate ignition by flames and hot objects and prevent proliferation of a fire to other areas (domino effect) include prohibition to use a flame, welding or cutting within the exclusion area around a hydrogen facility, equipment of internal combustion systems with exhaust system spark arrestors, and carburetor flame arrestors. Only *flame arrestors* specifically designed for hydrogen applications must be used taking into account the oxidant present. Flame arrestors can quench a flame on the basis of sufficient heat removal from the gas mixture. It is quite difficult to develop flame arrestors and explosion-proof equipment for hydrogen due to its small quenching distance (0.6 mm). Sintered-bronze flame arrestors may be effective in stopping hydrogen flames, whereas sintered stainless steel is not so effective.
- *Design and construction of buildings.* Buildings in which hydrogen is used must be constructed of light and noncombustible materials on a sufficiently strong frame. Windowpanes must be made of shatterproof glass or plastic. Floors, walls, and ceilings should be designed and installed to limit the generation and accumulation of static electricity and shall have a fire resistance rating of at least 2 h. *Explosion venting* must be provided only in the exterior walls or the roof. The venting area should be not $<0.11\ m^2/m^3$ of room volume. Vents, designed to relieve at a maximum internal pressure of 1.2 kPa, may consist of one or a combination of walls of light material, lightly fastened hatch covers or outward-opening swinging doors in exterior walls or roof. *Doors* should be hinged to swing outward in an explosion and must be readily accessible to personnel. *Walls* or partitions must be continuous from floor to ceiling and securely anchored. At least one wall must be an exterior wall, and the room must not be open to other parts of the building. Only indirect means, such as steam and hot water, should be used for *heating* in rooms containing hydrogen.
- *Barricades.* These are constructed to protect uncontrolled areas from the effects of a hydrogen system failure and to protect a hydrogen system from the hazards of adjacent or nearby operations. Barricades have been shown to be most effective against

fragments and only marginally effective in reducing overpressures. The housing of equipment provides partial protection in many cases. Shrapnel protection can be achieved by blast curtains or blast mats placed adjacent to the equipment to be protected. Barricades are commonly constructed as *mounds* and *single-revetted barricades.* A mound (earthworks) is an elevation of sloped dirt with a crest at least 0.91 m wide, whereas single-revetted barricades are mounds supported on a retaining wall on the side facing the hazard source. Simulations [69] and experimental works [70] have shown that

- Barricades should be designed to block the line of sight between equipment from which fragments can originate and protect items.
- Barricades should be placed adjacent to the fragment source for maximum protection.
- The efficiency of barricades depends on the height aboveground and barricade location, dimensions, and configuration.
- Barricades can reduce peak overpressures and impulses behind the barricades, but reflections on obstacles may amplify blast waves behind barricades.
- Single-revetted barricades are more efficient than mounds.

- *Liquid spills and vapor cloud dispersion.* An LH_2 spill from a storage vessel will result in a brief period of ground-level flammable cloud travel. The quick evaporation of the liquid causes the hydrogen vapors to mix quickly with air, dilute to nonflammable concentrations, warm up, and become positively buoyant [3]. If barricades are chosen, these should not confine excessively the vapor cloud formed because it may lead to detonation rather than simple burning of escaped hydrogen. LH_2 spill tests in an open-ended (U-shaped) bunker may produce detonation of the hydrogen–air mixture even without a roof. The use of dikes and barricades around storage tanks is not recommended for LH_2, although this is required for LNG because it may prolong evaporation and ground-level travel of the flammable cloud. Hydrogen detectors should be positioned to indicate the possible ground-level travel of flammable mixtures. No sewer drains must be located in an area in which a liquefied hydrogen spill could occur.
- *Shields and impoundment areas.* Impoundment areas and shields to control the extent of liquid and vapor travel caused by spills should be included in the design of a facility using hydrogen. The terrain below the transfer piping should be directed to a sump or impoundment area. Crushed stone should be used in the impoundment area to provide added surface area for liquefied hydrogen dissipation. Careful design of installations should eliminate possible confinement developed by the equipment or buildings because initial combustion or deflagration of hydrogen–air mixtures may evolve to DDT.

16.8 Conclusion

This chapter aims at highlighting physiological, physical, and chemical hazards associated with hydrogen storage and distribution systems as well as to reveal potential accidental scenarios that hydrogen may yield under certain conditions. Special attention is paid to hydrogen storage facilities because of the greater hazards encountered in them due to the increased quantities stored and handled, and on hazards resulting from the use of hydrogen as a fuel for transport. Hazard evaluation in these systems based on ETA shows that hydrogen may lead to a series of accident scenarios that can pose severe threats to properties and people. Computer simulations with the aid of CFD of both gaseous and liquefied hydrogen are verified by experiments. Especially, dispersion resulting from liquefied hydrogen spills agrees with experimental results showing that the resulting cloud behaves as a heavy rather than a light gas, remaining at flammable concentrations at low heights, thus significantly increasing the risk for accidental fires and explosions. A numerical approach to flash fires created by flammable gas clouds leads to effective simulation of combustion process and reasonable estimations of thermal fluxes. The main benefits obtained using CFD modeling is the control of cloud ignition timing and the precise definition of flammable limits and hence of combustible zones within the cloud volume. Safety guidelines are finally given for the prevention of accidents in hydrogen-handling installations and the protection of installations as well as the surroundings.

References

1. Cherry, R.S., A hydrogen utopia? *Int. J. Hydrogen Energ.*, 29, 125, 2004.
2. Akansu, S.O., Dulger, Z., Kahraman, N., and Veziroglu, T.N., Internal combustion engines fueled by natural gas–hydrogen mixtures, *Int. J. Hydrogen Energ.*, 29, 1527, 2004.
3. Rigas, F. and Sklavounos, S., Evaluation of hazards associated with hydrogen storage facilities, *Int. J. Hydrogen Energ.*, 30, 1501, 2005.
4. Momirlan, M. and Veziroglu, T.N., Current status of hydrogen energy, *Renew. Sust. Energ. Rev.*, 6, 141, 2002.
5. EUR 22002, *Introducing Hydrogen as an Energy Carrier*, EUROPEAN COMMISSION, Directorate-General for Research Sustainable Energy Systems, Brussels, 2006.
6. Schulte, I., Hart, D., and van der Vorst, R., Issues affecting the acceptance of hydrogen fuel, *Int. J. Hydrogen Energ.*, 29, 677, 2004.
7. Dincer, I., Technical, environmental and exergetic aspects of hydrogen energy systems, *Int. J. Hydrogen Energ.*, 27, 265, 2002.
8. EEC, On the Control of Major-Accident Hazards Involving Dangerous Substances, Directive 96/82/EC, Brussels, 1996.
9. Kirchsteiger, C., Availability of community level information on industrial risks in the EU. *Trans. IChemE Proc. Safety Environ.*, 78, 81, 2000.
10. IChemE, *Accident Database* (CD form), Loughborough, 1997.
11. Taylor, J.R., *Risk Analysis for Process Plant, Pipelines and Transport*, Chapman & Hall, London, 1994, p. 102.
12. CCPS, *Guidelines for Hazard Evaluation Procedures*, AIChE, New York, 1992, p. 69.
13. Federal Institute for Materials Research and Testing, *Hydrogen Safety*, German Hydrogen Association, Brussels, 2002, p. 18.
14. Rosyid, O.A., System-Analytic Safety Evaluation of the Hydrogen Cycle for Energetic Utilization, Dissertation, Otto-von-Guericke University, Magdeburg, Germany, 2006.
15. Warren, P., *Hazardous Gases and Fumes*, Butterworth-Heinemann, Oxford, 1997, p. 96.
16. Khan, F.I. and Abbasi, S.A., Major accidents in process industries and an analysis of causes and consequences, *J. Loss Prevent. Process Ind.*, 12, 361, 1999.
17. NASA, *Safety Standard for Hydrogen and Hydrogen Systems*, Report NSS 1740.16, Office of Safety and Mission Assurance, Washington, 1997, Chap. 2.
18. Adamson, K.A. and Pearson, P., Hydrogen and methanol: A comparison of safety, economics, efficiencies and emissions, *J. Power Sourc.* 86, 548, 2000.
19. Karim, A., Hydrogen as a spark ignition engine fuel, *Int. J. Hydrogen Energ.*, 28, 569, 2003.

20. Hord, J., Is hydrogen a safe fuel? *Int. J. Hydrogen Energ.*, 3, 157, 1978.
21. Baker, W.E. and Tang, M.J., *Gas, Dust and Hybrid Explosions*, Elsevier, Amsterdam, 1991, p. 42.
22. Rosyid, A. and Hauptmanns, U., System Analysis Safety Assessment of Hydrogen Cycle for Energetic Utilization, *Proc. Int. Cong. Hydrogen Energy and Exhibition*, Istanbul, 2005.
23. NASA, *Safety Standard for Hydrogen and Hydrogen Systems*, Report NSS 1740.16, Office of Safety and Mission Assurance, Washington, 1997, Chap. 3.
24. Gray, H.R., Testing for hydrogen environment embrittlement: experimental variables, in *Hydrogen Embrittlement Testing*, American Society for Testing and Materials, Special Technical Publication ASTM STP-543, Philadelphia, PA, 1974, p. 133.
25. Chandler, W.T. and Walter, R.J., Testing to determine the effect of high pressure hydrogen environments on the mechanical properties of metals, in *Hydrogen Embrittlement Testing ASTM 543*, American Society for Testing and Materials, Philadelphia, PA, 1974, p. 170.
26. Groenvald, T.D. and Elcea, A.D., Hydrogen stress cracking in natural gas transmission pipelines, in *Hydrogen in Metals*, Bernstein, I.M. and Thompson, A.W., Eds., ASM, 1974.
27. Rowe, M.D., Nelson, T.W., and Lippold, J.C., Hydrogen-induced cracking along the fusion boundary of dissimilar metal welds, *Welding Res.*, February, 1999, p. 31.
28. Zalosh, R.G. and Short, T.P., *Compilation and Analysis of Hydrogen Accident Reports*, C00-4442-4, Factory Mutual Research Corp., Norwood, OH, 1978.
29. Williams, G.P., Causes of ammonia plant shutdowns, *Chem. Eng. Prog.*, 74, 9, 1978.
30. Ordin, P.M., A review of hydrogen accidents and incidents in NASA operations, *9th Intersociety Energy Conversion Engineering Conference*, ASME, New York, 1974.
31. Thompson, J.D. and Enloe, J.D., Flammability limits of hydrogen–oxygen–nitrogen mixtures, *Combust. Flame*, 10, 393, 1966.
32. McHale, E.T., Geary, R.W., von Elbe, G., and Huggett, C., Flammability limits of H_2–O_2-fluorocarbon mixtures, *Combust. Flame*, 16, 167, 1971.
33. Coley, G.D. and Field, J.E., Role of cavities in the initiation and growth of explosions in liquids, *Proc. Roy. Soc. Lond. Ser. A*, 335, 67, 1973.
34. Creitz, E.C., Inhibition of diffusion flames by methyl bromide and trifluoromethyl-bromide applied to the fuel and oxygen sides of the reaction zone, *J. Res. App. Phys. Chem.*, 65, 389, 1961.
35. Wionsky, S.G., Predicting flammable material classifications, *Chem. Eng.*, 79, 81, 1972.
36. IchemE—Institution of Chemical Engineers, *Hazard Spotting User's Guide*, Multimedia Package, Rugby, Warwickshire, 2003.
37. Beach, R., Preventing static electricity fires, *Chem. Eng.*, 71, 73, 1964.
38. Beach, R., Preventing static electricity fires, *Chem. Eng.*, 72, 63, 1965.
39. Beach, R., Preventing static electricity fires, *Chem. Eng.*, 72, 85, 1965.
40. Bjerketvedt, D., Bakke, J.R., and Wingerden, K V, Gas explosion handbook, *J. Hazard. Mater.*, 52, 1, 1997.
41. Philips, H., *Explosions in the Process Industries*, IChemE, Warwickshire, 1994, p. 5.
42. Lee, J.H. et al., Hydrogen–air detonations, in *Proc. 2nd International Workshop on the Impact of Hydrogen on Water Reactor Safety*, M. Berman, Ed., SAND82-2456, Sandia National Laboratories, Albuquerque, 1982.
43. Gordon, S. and McBride, B.J., *Computer Program for Calculation of Complex Chemical Equilibrium Compositions and Applications*, NASA Reference Publication, Lewis Research Center, Cleveland, Ohio, 1994, p. 1311.
44. Lee, J., Initiation of gaseous detonation, *Annu. Rev. Phys. Chem.*, 28, 75, 1977.
45. Eichert, H., Hydrogen–air deflagrations and detonations: numerical calculation of 1-d tube combustion processes, *Int. J. Hydrogen Energ.*, 12, 171, 1987.
46. Warren, P., *Hazardous Gases and Fumes*, Butterworth-Heinemann, Oxford, 1997, p. 96.
47. CCPS, *Guidelines for Evaluating the Characteristics of Vapor Cloud Explosions, Flash Fires and BLEVEs*, AIChE, New York, 1994, p. 158.
48. NIST, *CRC Handbook of Chemistry and Physics*, CRC Press, Boca Raton, FL, 1994, p. 6.259.
49. Shebeko, Yu.N., Shevchuck, A.P., and Smolin, I.M., BLEVE prevention using vent devices, *J. Hazard. Mater.*, 50, 227, 1996.

50. Pohanish, R.P. and Green, S.A., *Rapid Guide to Chemical Incompatibilities*, Van Nostrand Reinhold, New York, 1997, p. 434.
51. Carson, P. and Mumford, C., *Hazardous Chemicals Handbook*, Butterworth-Heinemann, Oxford, 1994, pp. 181, 202.
52. HSE, Application of QRA in Operational Safety Studies, Health and Safety Executive report 025, 2002, p. 19.
53. LLNL, *Burro Series Data Report*, UCID: 19075 Vol. 1, LLNL/NWC, Livermore, 1982.
54. Swain, M.R., Filoso, P., Grilliot, E.S., and Swain, M.N., Hydrogen leakage into simple geometric enclosures, *Int. J. Hydrogen Energ.*, 28, 229, 2003.
55. Sklavounos, S. and Rigas, F., Validation of turbulence models in heavy gas dispersion over obstacles, *J. Hazard. Mater.*, 108, 9, 2004.
56. Sklavounos, S. and Rigas, F., Fuel gas dispersion under cryogenic release conditions, *Energ. Fuel.*, 19, 2535, 2005.
57. Versteeg, H.K. and Malalasekera, W., *An Introduction to Computational Fluid Dynamics—The Finite Volume Method*, Longman, New York, 1995, p. 85.
58. LLNL, *Description and Analysis of Burro Series 40-m3 LNG Spill Experiments*, LLNL/NWC report No. UCRL-53186, Lawrence Livermore National Laboratory, Berkeley, CA, 1981.
59. Witcofski, R.D. and Chirivella, J.E., Experimental and analytical analyses of the mechanisms governing the dispersion of flammable clouds formed by liquid hydrogen spills, *Int. J. Hydrogen Energ.*, 9, 425, 1984.
60. Lees, F.P., *Loss Prevention in the Process Industries*, Butterworth Heinemann, Oxford, 1996.
61. Pula, R., Khan, F., Veitch, B., and Amyotte, P., 2005. Revised fire consequence models for offshore quantitative risk assessment. Journal of Loss Prevention in the Process Industries 18, 443–454.
62. Rigas, F. and Sklavounos, S., Simulation of Coyote series trials—Part II: a computational approach to ignition and combustion of flammable vapor clouds, *Chem. Eng. Sci.*, 61, 1444, 2006.
63. Sklavounos, S. and Rigas, F., Simulation of Coyote series trials—Part I: CFD estimation of non-isothermal LNG releases and comparison with box-model predictions, *Chem. Eng. Sci.*, 61, 1434, 2006.
64. LLNL, *Coyote Series Data Report*, UCID: 19953 Vol. 1, 2. LLNL/NWC, Livermore, 1983.
65. LLNL, *Vapor Burn Analysis for the COYOTE Series LNG Spill Experiments*, UCID: 53530, LLNL/NWC, Livermore, 1984.
66. Swain, M.R., Shriber, J., and Swain, M.N., Comparison of hydrogen, natural gas, liquefied petroleum gas, and gasoline leakage in a residential garage, *Energ. Fuels*, 12, 83, 1998.
67. Farrell, A.E., Keith, D.W., and Corbett, J.J., A strategy for introducing hydrogen into transportation, *Energ. Policy*, 31, 1357, 2003.
68. NASA, *Safety Standard for Hydrogen and Hydrogen Systems*, Report NSS 1740.16, Office of Safety and Mission Assurance, Washington, 1997, Chap. 4.
69. Sklavounos, S. and Rigas, F., Computer simulation of shock waves transmission in obstructed terrains, *J. Loss Prevent. Process Ind.*, 17, 407, 2004.
70. HSE, *Explosion Hazard Assessment: A Study of the Feasibility and Benefits of Extending Current HSE Methodology to Take Account of Blast Sheltering*, Health and Safety Executive report HSL/2001/04, Sheffield, 2001.

17

Carbon Sequestration

Ah-Hyung Alissa Park, Klaus S. Lackner, and Liang-Shih Fan

CONTENTS

17.1 Introduction and Background

Fossil fuels, which include oil, natural gas, and coal, have largely been used as affordable sources of energy for transportation, electricity, and residential (heating and cooling) services. According to recent figures published by the U.S. Department of Energy (DOE) and Energy Information Administration (EIA), the uses of the fossil fuels are projected to increase from 420 quadrillion BTU in 2003 to as high as 720 quadrillion BTU by the year 2030 (Energy Information Administration, 2006). The continued reliance on fossil fuels is primarily due to their relatively low cost compared with renewable sources such as biomass, wind, geothermal, and solar power. Dependence on fossil fuels will continue even if the energy sector adapts to hydrogen, which in the foreseeable future will be mainly produced from carbonaceous materials. From the point of view of managing carbon in the environment, hydrogen is particularly an attractive fuel because it is environmental friendly and clean and can be consumed without concern over greenhouse gas (GHG) emissions at the point of use. As discussed in earlier chapters, there are various methods of producing hydrogen using different energy sources including coal and biomass. Hydrogen is likely to be produced from fossil energy resources as long as natural gas and coal provide hydrogen at a fraction of the cost of the electricity that is required by electrolysis. However, the use of these energy sources is limited by their environmental impacts.

One of the major concerns in using fossil fuels for hydrogen production is the emission of GHGs such as carbon dioxide (CO_2). As shown in Figure 17.1, the atmospheric concentration of CO_2 historically fluctuated naturally on the timescales of ice ages. Concerns, however, stem from the recent dramatic increase in the CO_2 concentration, which coincides with global industrial development. The atmospheric concentration of CO_2 rose from 280 ppm in 1800 to 370 ppm in 2000. This rise is mainly due to the high use of fossil fuels (Fankhauser et al., 2001; IPCC, 2005; Caldeira, 2006). The increase in the atmospheric CO_2 concentration has been most severe over the last few decades, and, unless actions are taken, the atmospheric CO_2 concentration could reach 580 ppm, a threshold value thought to trigger severe climate change, within a mere 50 years (Stix, 2006; IPCC, 2007).

Owing to their molecular structures, CO_2 and other GHGs including methane (CH_4), nitrous oxide (N_2O), sulfur hexafluoride (SF_6), hydrofluorocarbons (HFCs), and perfluorocarbons (PFCs) absorb a significant fraction of the thermal radiation reradiating the energy delivered by the sun as visible light to the surface of the earth at longer wavelength and, thus, prevents heat from escaping the earth's atmosphere. Trapping this heat in the atmosphere results in an increase of the earth's surface temperature. This phenomenon is known as the greenhouse effect. CO_2 is by far the most significant contributor to global

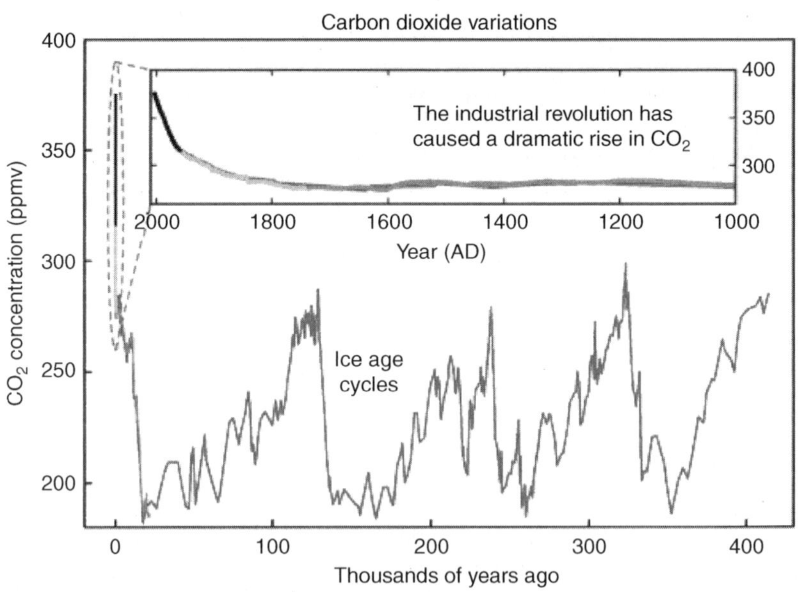

FIGURE 17.1
Carbon dioxide variations. (Prepared by Robert A. Rohde from publicly available data and is incorporated into the Global Warming Art project, 2005.)

warming and accounts for 64% of the increased greenhouse effect (Bryant, 1997). There is now sufficient scientific evidence to support the claim that human activities cause global warming (National Academy of Sciences Committee, 2001; Oreskes, 2004; IPCC, 2005; IPCC, 2007). International science and engineering communities are now actively investigating the various impacts of global warming on humans and environments. For example, researchers have found that increased warming of the earth's surface has lead to the melting of the polar ice caps. Some argue that this will result in a significant rise in the sea level and unpredictable climate changes.

In addition to climate change, the increased atmospheric concentration of CO_2 is likely to have wide-spread ecological effects in various environments, since CO_2 is a physiologically active gas, in plants as well as animals. The acidic nature of CO_2 will also lead to changes in the chemistry of the ocean's surface, which is in equilibrium with the atmosphere. Once the shift in the oceanic chemical balance becomes significant, it will affect ecosystems. It has been shown, for example, that doubling CO_2 concentration in the atmosphere will reduce the rate of calcium carbonate deposition in coral reefs by 30–40% (Langdon et al., 2000).

In the past, the ocean has been the largest sink for atmospheric CO_2, which can compensate for some emissions in the foreseeable future (Archer et al., 1997). However, as evidenced from rising CO_2 concentrations in the atmosphere, this natural oceanic uptake is not sufficient to compensate for the current and future CO_2 emission rates. Thus, the escalating concerns regarding global climate change led to the creation of the Kyoto Protocol. Initiated in 1997, the Kyoto Protocol is an international treaty established by the United Nation Framework Convention on Climate Change (UNFCCC) in an effort to globally reduce GHG emissions. This agreement sets specific and binding targets for reducing emissions of GHGs for each signatory nation. The Kyoto Protocol entered into force on February 16, 2005, and its overall goal is to reduce the collective average GHG emissions of the industrialized countries by 5% below the 1990 levels by 2008–2012 (United Nations, 1997). As of December 2006, 169 countries and other governmental entities have ratified the Kyoto Protocol, which represents a total of 61.6% of global GHG emissions. Although the United States has not ratified the Kyoto Protocol, there have been significant efforts on both the national- and local-level to reduce the emission of various GHGs (e.g., New York taxicabs to go hybrid, California emission restrictions, and increased governmental and industrial funding for carbon sequestration research and development).

To meet the ever-increasing global energy demands while stabilizing the CO_2 level in the atmosphere, current carbon emissions have to be reduced by at least a factor of three. In this chapter, the options for the capture of CO_2 and its subsequent disposal are discussed. Together, these technologies are known as "carbon sequestration." The term "sequestration" refers to "putting aside and locking up" carbon that would otherwise be released while processing fossil fuels. Among some researchers, the term sequestration has become synonymous with biological sequestration, that is, the growth of biomass to capture and store CO_2 (Harvey, 2004) or geological sequestration, that is, injection of CO_2 into various geological formations. In this chapter, the term is used more broadly: sequestration refers to any storage scheme that does not release CO_2 into the atmosphere. Carbon sequestration schemes can be classified by their capacity and permanence of storage, environmental consequences, and cost of implementation.

Any viable system for sequestering carbon must be (1) effective and cost competitive, (2) stable as long-term storage, and (3) environmentally benign (U.S. DOE, 2004). Because the application of CO_2 sequestration technologies or other carbon management strategies will raise the cost of energy, it is unlikely that they will be introduced without regulatory pressure. However, the trade-off between ignoring the risk of climate change and paying for a carbon neutral energy infrastructure is more easily managed if the cost of the carbon management can be held low. Sequestration as one of the options will only be considered if it is economically competitive with alternative energy sources with lower CO_2 emissions. Hence, alternative forms of energy including wind and nuclear energy limit the acceptable price of CO_2 sequestration.

Carbon sequestration generally involves three steps—capture, transportation, and disposal. The current estimated cost of capture and compression of CO_2 to pipeline pressure (~135 atm) is approximately \$30–50/ton CO_2. Transportation of CO_2 adds \$1–3/100 km (Wallace, 2000). Disposal costs vary significantly. In some cases they are quite low, for example, \$4/ton of CO_2 for underground injection into easily accessible geological formations. Based on the current technologies, the overall sequestration cost is estimated to be between \$20 and \$100/ton of CO_2 sequestered (U.S. DOE, 2004; Penner, et al., 2003; IPCC, 2005). In the following sections, the details of various carbon management technologies are discussed with respect to the hydrogen or hydrogen-rich fuel productions.

17.2 CO_2 Capture from Concentrated Sources

The CO_2 capture is generally the most expensive of the three steps in any sequestration scheme. CO_2 separation or capture is economically competitive at large, concentrated industrial sources such as power- or cement-plants. High concentration and large volumes of CO_2 allow for efficient separation of CO_2 from the gas mixture. Hydrogen production involving fossil fuels can also be performed in large centralized facilities. The centralization of hydrogen production would avoid the emission of CO_2 from distributed small points of emissions where hydrogen would displace carbon-based fuels. The carbon problem would be shifted upstream where carbon capture efficiency could be higher, whereas the consumption sectors would be decarbonized (Johansson et al., 1996; Ogden, 1999). For example, hydrogen-fueled automobiles do not emit CO_2 since carbon is removed from the fuel during the preparation and refining processes regardless of the initial forms of the energy source (e.g., natural gas, coal, and biomass). The use of hydrogen fuel in automobiles, therefore, decarbonizes the transportation sector.

Traditional research and development on the carbon capture technologies has concentrated on power plants, which is currently the largest source of CO_2 emission (30% globally). These technologies can be applied to other sources including cement plants (6%), steel mills (6%), and hydrogen production facilities (1%) (Lackner et al., 2005). With the increased interest in hydrogen, tailoring carbon capture strategies for hydrogen production facilities becomes important. Currently available CO_2 capture technologies include physical and chemical absorption and adsorption, sorbent injection, refrigeration, and membrane separation. In addition, the ongoing research and development on carbon capture strategies in terms of hydrogen production focuses on improving the energy-intensive low-temperature amine scrubbing process and employing reactive CO_2 separation using high-temperature sorbents such as limestone, potassium carbonates, lithium silicates, and sodium carbonates, which yield a sequestration-ready CO_2 gas stream (White et al., 2003).

17.2.1 Physical and Chemical Absorption and Adsorption

Typical CO_2 separation techniques that have been extensively used in chemical and petrochemical industries, particularly during the synthesis of ammonia and methanol, involve physical and chemical absorption of CO_2 at low-temperature conditions (Halman and Steinberg, 1999). The absorbed CO_2 is then released by applying heat or lowering the system pressure, which also regenerates the solvent. Amine solutions such as monoethanolamine (MEA) and methyldiethanolamine (MDEA) are examples of the commercial chemical absorption solvents. These amine scrubbing solvents exhibit good sorption capacity under low acid gas (e.g., CO_2) concentrations; however, when the acid gas concentration is high, they can react with the acid gas and form heat-stable salts that plug the system. Thus, MEA and MDEA are generally used for capturing CO_2 at low partial pressures.

Other organic solvent-based scrubbing technologies include the Rectisol process using refrigerated methanol and the Selexol process using a solvent made of dimethyl ether of polyethylene glycol. Different from amine-based scrubbing systems, which rely on chemical reactions with the acid gases, the Rectisol and Selexol processes are based on physical absorption and are suitable for systems with high CO_2 partial pressures. Therefore, these processes are more commonly used in the gasification and hydrogen production facilities where the CO_2 concentration is high. However, these processes are also energy intensive due to the need for solvent refrigeration (Breckenridge et al., 2000). The operating cost of the Rectisol process is higher than the Selexol process (Heaven et al., 2004) due to the lower solvent temperature in the absorption step of Rectisol (−60°C) when compared to Selexol (~5°C), where only mild refrigeration is necessary (Higman, 1995).

The absorption technique using hot potassium carbonate has also been developed to capture CO_2 (Probstein and Hicks, 1990). The chilled ammonia process is another solvent-based CO_2 capture technology where ammonia carbonate slurries are used to capture 90% of the CO_2 in the gas stream mixture gas forming ammonia bicarbonate in the process. A pilot-scale chilled ammonia unit for 5 MW equivalent flue gas capture is under construction by ALSTOM and EPRI. Although this process is developed for a combustion system, the results will provide valuable information for the future development of such a process for hydrogen production. According to ALSTOM, commercial products on chilled ammonia process will be available by 2010 (Alstom, 2007).

Generally, the main impediment with these absorption/stripping technologies is the low operating temperature and the high operating pressure. Because the gas mixture has to be cooled before the separation, a part of the energy penalty involves cooling of the plant exhaust and it becomes significant if the CO_2 concentration in the exhaust is low. The parasitic energy consumption of the conventional amine scrubbing process is as high as 30% of the plant's electric power output, reducing the total plant efficiency or capacity (Châtel-Pélage et al., 2005). Recently, some processes including the MEA process and the chilled ammonia process claim that they have solved these issues and their processes are efficient and economically favorable. The MEA process is currently considered to be the most attractive choice for separating CO_2 from the gas mixture (e.g., flue gas) since this system can absorb CO_2 at low partial pressures. For the application of the MEA technology, the energy penalty is significantly reduced and it is as low as 15% of the power plant's electric power output (Mimura et al., 1997)

Once a high concentration hydrogen stream has been produced, adsorption on solids is another well-known method of selectively removing gases from a gas stream. A fixed bed of high surface area materials such as molecular sieves, activated carbon, silica gel, and zeolite can adsorb CO_2 and other gases from the stream (Kikkinides et al., 1993). These adsorbents are selective toward gases with high polarity and low volatility. Owing to its high volatility and low polarity, hydrogen does not adsorb on the adsorbent bed, and thus, a high-purity hydrogen gas stream can be obtained by capturing the other gas components including CO_2 (Stocker et al., 1998). Various techniques are employed for CO_2 separation including pressure swing adsorption (PSA), thermal (or temperature) swing adsorption (TSA), and vacuum pressure swing adsorption (VPSA). In PSA and TSA, similar to absorption techniques, adsorbed CO_2 is released by either lowering the bed pressure or heating the saturated bed, respectively. However, this method is usually low in capacity and often not very selective in the adsorption process. In VPSA, pressure difference is also used to carry out the CO_2 desorption, but a vacuum is applied to further pull additional CO_2 out of the bed. Generally, the regeneration cycles of the VPSA process require only a few seconds.

17.2.2 Cryogenic Separation

Cryogenic separation techniques require an operating temperature even lower than that of absorption or adsorption processes. In this method, CO_2 is separated from other gases by condensation at cryogenic temperatures (Halman and Steinberg, 1999). To minimize the energy requirement for refrigeration, the gases may be pressurized to increase the partial pressure of CO_2 allowing its condensation at higher temperatures. Recent development of the cryogenic separation technologies aims mainly at integrated gasification combined cycle (IGCC) configurations or oxyfuel combustion plants. Cryogenic methods, while possible, are usually too expensive and are often limited to purifying CO_2 for specialty applications.

17.2.3 Membrane Separation

To overcome the energy penalty of the aforementioned separation methods operating at low-temperature conditions, there have been efforts to develop CO_2 capture techniques that operate under higher-temperature conditions (e.g., gasification temperature). A particular advantage of these methods is that the hot gas can be recycled without undergoing large temperature swings. Membrane separation processes are one particular implementation of this concept. As a mixture of gases diffuse through a membrane system, only the desired gas constituent, in this case CO_2 or hydrogen, would either pass or be blocked by the pores of the membrane. Because membrane technology separates CO_2 based on its molecular size or chemical characteristics, there is no need for phase changes or refrigeration of the gas mixture. Furthermore, depending on the materials used in the membrane fabrication, the membrane surface and each component in the gas mixture interact differently and this results in the selective affinity for one gas. Hence, the selected gas permeates faster resulting in a partial separation of the different gas species (Wallace, 2000). Membranes made out of various types of permeable materials such as polymers, metals, and rubber composites have been investigated (McFarlan et al., 2004; Bugante et al., 1989). Palladium and molecular sieves are also currently under study (McFarlan et al., 2004). In some cases, the membrane technology is incorporated with the absorption method to provide a larger surface area for contact while preventing mixing between the gas and the liquid phases. Energy penalties in membrane systems are driven by the selectivity of the membrane and the pressure drop required to pass the gas through the separation membrane.

17.2.4 Reaction-Based Sorbent Injection

High-temperature reactants that can absorb CO_2 at elevated temperatures in a chemical reaction offer another way to avoid the energy penalties that are associated with creating low operating temperatures or high pressures in the absorption step. In such a system, the heat of absorption is delivered at a temperature that makes it economically attractive. A successful implementation of the injection of reaction-based sorbents will, therefore, require the incorporation of effective process integration schemes to minimize the parasitic energy consumption for CO_2 capture. The reaction-based processes using solid sorbents can separate CO_2 while operating at the elevated temperatures suitable for the gasifier application. Moreover, the sorbent injection process can be retrofitted to an existing gasifier without any significant modifications to the plant, and hence, reduces the potential capital investment. Such methods utilize the carbonation/calcination reaction (CCR) process of a solid metal oxide (MO) to capture CO_2 by the following chemical reaction (Iyer et al., 2004; Balasubramanian et al., 1999; Fink et al., 1974):

$$MO\ (s) + CO_2\ (g) \leftrightarrow MCO_3\ (s) + \text{heat} \tag{17.1}$$

At low temperatures or high partial pressure of CO_2, the reaction proceeds to the right. At high temperatures or low partial pressure of CO_2, the reaction proceeds to the left. Gaseous CO_2 is bound into the solid matrix during the carbonation process yielding metal carbonate. The metal oxide is then regenerated by either temperature or pressure swings. At temperatures beyond the calcination temperature that is specific for the given partial pressure of CO_2 and the specific choice of the metal oxide, a pure CO_2 stream is produced as the metal oxide is regenerated. For the $CaO/CaCO_3$ system, the transition between carbonate and oxide at 1 atm of CO_2 occurs at 898°C (calcination temperature), and, as the partial pressure of CO_2 increases, the transition temperature of the system also rises.

The advantages of the reaction-based separation process are (a) CO_2 capture under wide range of mixture gas conditions (i.e., 150–700°C and <1 atm), (b) high equilibrium capacity, (c) low equilibrium CO_2 concentration, and (d) generation of a pure CO_2 stream. Often the initial metal oxide sorbents are prepared by calcining raw carbonate materials. $CaCO_3$, $MgCO_3$, $ZnCO_3$, $PbCO_3$, $CuCO_3$, and $MnCO_3$ are the potential candidates based on the range of carbonation and calcination temperatures and reactivity (Iyer et al., 2004).

Among the metal oxides, calcium-based sorbents have been extensively investigated. The CCR process, which uses the reengineered limestone sorbent has efficiently and economically shown to capture CO_2 at 600–700°C. Although limestone sorbent is inexpensive, its effective sorption capacity is generally low due to its small surface area and pore sizes (~30% conversion). Hence, the modification of limestone sorbent to form calcium-based sorbents possessing mesopores has been the main focus of the development of CCR process. The conversion of the engineered calcium-based sorbents in carbonation reaction is >90%.

Another important advantage of the CCR process is that this process can be designed to capture both CO_2 and H_2S simultaneously, rendering it a multipollutant control technology for hydrogen production in gasification systems. CO_2 and H_2S are first captured in the carbonator/sulfator system. The carbonated sorbent, $CaCO_3$, is then sent to the calciner to regenerate the calcium oxide sorbent for subsequent cycles while yielding a sequestration-ready, pure CO_2 stream. Sulfated sorbent, which is not regenerated, since the calcination temperature is significantly higher than for calcium carbonate, and fly ash are removed from the system by means of a purge stream. Carbonation is a valuable heat source, which can provide as much as one-third of the total thermal capacity of a plant. In the CCR process, steam can be generated from three high-quality heat sources: (1) the reaction heat of the carbonation or sulfation, (2) sensible heat left in the hot gas after supplying energy to the calciner (between 600°C and 850°C), and (3) sensible heat in the pure CO_2 stream from the calciner at 800–850°C. The steam can be used in a turbine system for additional electricity generation and in driving various feed water pumps in the plant. Hence, the total parasitic energy consumption of the plant is immensely reduced to 15% including CO_2 compression (10%), which is half of that of the conventional CO_2 separation technologies such as amine scrubbing process (Fan and Iyer, 2006). These ongoing efforts suggest a very promising future for the reaction-based sorbent injection technology.

17.3 Integrated Carbon Capture Technologies

A number of carbon capture technologies described earlier are now being directly integrated with hydrogen production processes to produce carbon-free alternative fuels while achieving process intensification. Hydrogen production facilities including IGCC plants can be designed to capture carbon dioxide and make it available for disposal. This is already done at gasification plants operating in refineries that remove CO_2 to obtain pure hydrogen. To obtain a pure hydrogen stream, the carbon monoxide (CO) contained in the syngas is catalytically reacted with steam (H_2O) to produce carbon dioxide (CO_2) and hydrogen (H_2).

$$CO + H_2O \leftrightarrow H_2 + CO_2 \tag{17.2}$$

This reaction is known as the water–gas shift (WGS). The resulting carbon dioxide is then separated from hydrogen by conventional cleaning operations such as chemical or physical sorbents or traditional membrane separation as described in the previous section. As discussed previously, these processes are often costly and highly energy intensive. The efficiency of the hydrogen production in the above gasification–WGS system with a conventional CO_2 separation unit is reported to be about 64% (Simbeck and Chang, 2002). If hydrogen is to be used as a pollution-reduced energy source in the future, better gasification and hydrogen purification processes that are more economically favorable and technically efficient need to be developed. Hence, there is a need to develop an integrated carbon capture capability.

Novel zero-emission plants that eliminate all emissions to the atmosphere are designed to improve their efficiency by capturing CO_2 at high temperatures. There are a number of new carbon capture concepts under development and promising results are being generated at various stages of demonstration. One of the methods of improving the yield of hydrogen from coal is to use limestone sorbents. Based on the CCR process described earlier, limestone sorbent is calcined to form calcium oxide, which is then used to react and continuously remove the CO_2 product formed during the WGS reaction. There have been several reports on sorption-enhanced hydrogen production by coupling steam methane reforming (SMR) and *in situ* CO_2 capture using a sorbent (Hufton et al., 1999; Balasubramanian et al., 1999; Lopez-Ortiz and Harrison, 2001). Calcium oxide–assisted SMR was attempted in earlier studies (Balasubramanian et al., 1999; Lopez-Ortiz and Harrison, 2001), which describe the performance of a single-step sorption-enhanced process using a Ni-based catalyst to produce hydrogen. When the Ni-based catalyst was mixed with dolomite–CaO powder, the H_2 concentration reached 97% while capturing CO_2. By applying this carbonation–calcination loop, a number of integrated carbon capture processes that enhance hydrogen production using calcium-based sorbents such as the Zero Emission Coal Allianc (ZECA)- (Lackner et al., 2004), HyPr-Ring-, ALSTOM-, GE- (Lin et al., 2002), and calcium looping process are being developed.

Another approach to produce carbon-free hydrogen employed in coal-direct chemical looping and syngas redox (SGR) processes use iron-based intermediate looping particles as oxygen carriers, which inherently produce a sequestration-ready CO_2 stream while producing hydrogen. Generally, chemical looping processes involve a sorbent, typically a metal or more probably a low oxidation state metal oxide that can be oxidized in air (Leithner, 2004). The oxide is reduced by fuel gases in a subsequent step. A variation of this approach oxidizes the metal not in air but in a chemical reaction with steam to produce a pure stream of hydrogen. Some advantages of this approach to hydrogen production are that it eliminates the separation of hydrogen from the fuel gas, the use of a separate WGS reactor, and the cleanup of fuel gas before hydrogen production (Mattisson et al., 2004).

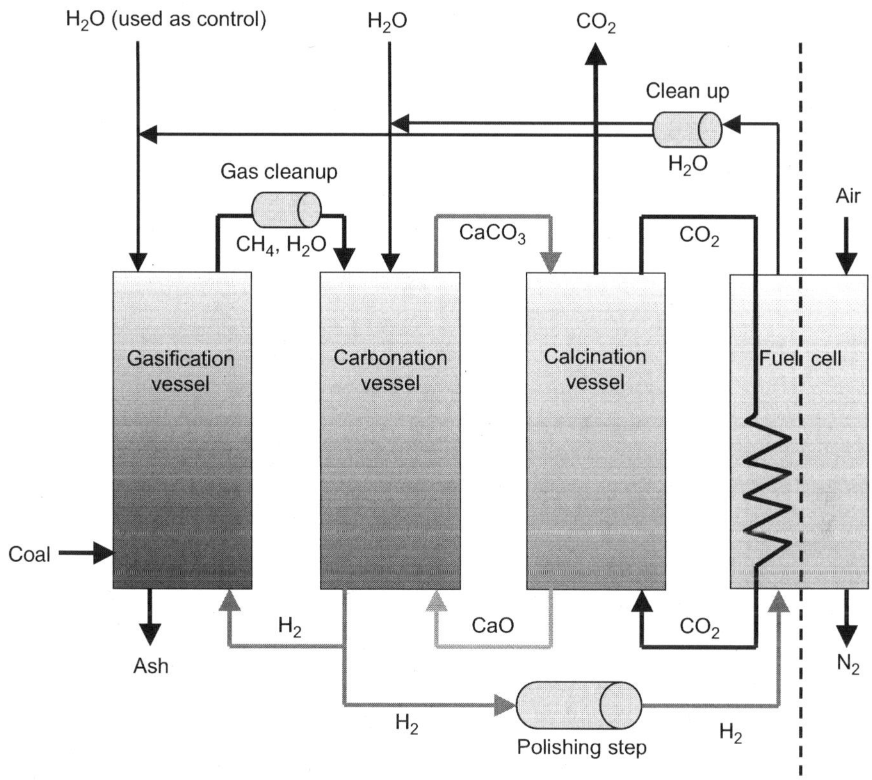

FIGURE 17.2
Schematic of ZECA process: the anaerobic hydrogen production and fuel cell system. Material flows are idealized to predominant components.

Moreover, from chemical looping processes, the sequestration-ready CO_2 stream can also be produced at higher pressures.

In the following sections, a number of processes developed by incorporating CO_2 capture with hydrogen production are discussed. These processes are good examples of process intensification with a goal to reduce the excess steam requirement while simultaneously removing CO_2 as well as sulfur, providing a sequestration-ready CO_2 stream while producing a high-purity H_2 stream at high temperatures and pressures.

17.3.1 ZECA Process

The ZECA process illustrated in Figure 17.2 is developed by Los Alamos National Laboratory in the United States. In this process, hydrogen is first used to gasify coal to produce a methane-rich stream (reaction 17.3), which is then reformed using steam in the presence of calcium-based sorbents to produce hydrogen (reaction 17.5). As shown in reaction 17.6, the sorbent removes the CO_2 produced during the reforming reaction, thereby shifting the equilibrium of reaction 17.5 in the forward direction (Ziock et al., 2001). The major reactions are as follows:

Hydrogasifier:

$$C + 2H_2 \rightarrow CH_4 \tag{17.3}$$

$$H_2O\text{ (liquid)} \rightarrow H_2O\text{ (gas)} \tag{17.4}$$

CaO reformer:

$$CH_4 + 2H_2O \rightarrow CO_2 + 4H_2 \quad (17.5)$$

$$CaO + CO_2 \rightarrow CaCO_3 \quad (17.6)$$

Because the carbonation reaction is exothermic, the carbon capture process also provides energy for the reforming reaction (Ziock et al., 2001). Hydrogen is then used in various chemical processes or converted to electricity using an energy conversion system such as a solid oxide fuel cell. Solid oxide fuel cells are attractive because they integrate electricity production with the separation of oxygen from air. The fuel for the solid oxide fuel cell need not be hydrogen, it could be any mixture of CO and H_2, or may even include methane that undergoes steam reforming within the fuel cell chamber (George, 2000). Similar to coal, other carbonaceous materials including biomass can be used to produce the fuel gas (H_2 and CO) in an upstream gasification step. The fuel cell generates heat while separating the oxygen from the air and generates electricity. By transferring the heat to the upstream gasification process, it is possible to achieve extremely high efficiency. The ZECA plant design aims at 70% conversion efficiency while delivering CO_2 in a concentrated stream (Lackner and Ziock, 2001).

17.3.2 HyPr-Ring Process

The HyPr-Ring process was developed in Japan and involves coal gasification using pure oxygen and high-pressure steam with simultaneous CO_2 capture. This process involves three principal reactions occurring in two reactors. As illustrated in Figure 17.3, the HyPr-Ring process consists of two fluidized-bed reactors: the gasifier and the regenerator.

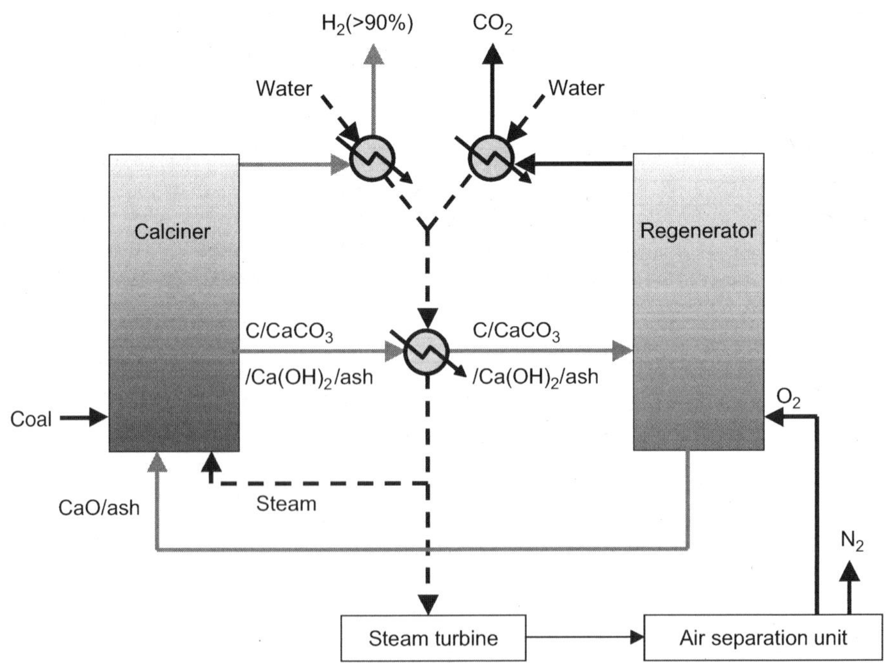

FIGURE 17.3
Schematic diagram of HyPr-Ring process.

In the gasifier, coal is fed along with calcium oxide, steam, and oxygen. The excess steam in the gasifier drives the following reaction toward the formation of H_2:

Gasifier:

$$C + 2H_2O + CaO \rightarrow CaCO_3 + 2H_2 \tag{17.7}$$

A high partial pressure of steam is used to hydrate the CaO so that the exothermic reactions of carbonation supply a sufficient amount of energy for the endothermic gasification reaction. The calcium-based sorbent used in the HyPr-Ring process is either CaO or $CaO/Ca(OH)_2$.

Calcium oxide also captures CO_2 generated in the WGS reaction, resulting in a product gas stream of 91% hydrogen mixed with methane. The solids from the gasifier consist of mostly reacted CaO sorbents ($CaCO_3$) and some unconverted carbon that is to be introduced to a regenerator along with oxygen. In the regenerator, unconverted carbon is combusted with O_2 to liberate heat. The heat is then used in the calcination of $CaCO_3$ to recover CaO sorbent. The reactions occurring in the regenerator can be seen as

Regenerator:

$$C + O_2 \rightarrow CO_2 + \text{heat} \tag{17.8}$$

$$\text{Heat} + CaCO_3 \rightarrow CO_2 + CaO \tag{17.9}$$

It is estimated that a H_2 production efficiency of 77% can be achieved using the HyPr-Ring process, while producing a sequestration-ready, pure CO_2 stream (Lin et al., 2005). The distribution of carbon or coal between the two reactors is determined by the amount of heat required in the second reactor for the calcination.

The main issues related to the further development of the HyPr-Ring process are the process high operating pressure (~70 bar), the use of excess steam, and the low purity of the produced hydrogen (Lin et al., 2005). Furthermore, at high temperatures and long holding times, there is considerable solid–solid interaction between the calcium-based sorbent particles and the coal-included minerals, which reduces the CO_2 sorption capacity of the sorbent (Lin et al., 2005). Steam also increases the interaction between the sorbent and the coal minerals at high temperatures and pressures (Lin et al., 2002).

17.3.3 ALSTOM Process

The ALSTOM's hybrid combustion gasification chemical looping coal power technology also uses solids as a heat transfer media. The ALSTOM process is designed to operate in three different configurations depending on the desired products: (1) indirect coal combustion for heat generation, (2) coal gasification for producing syngas, and (3) coal gasification for producing hydrogen (Andrus et al., 2006). For the first and second configurations, a single chemical loop of $CaS–CaSO_4$ is used, whereas for the third configuration two chemical loops of $CaS–CaSO_4$ and $CaO–CaCO_3$ are used.

In the first configuration, two reactors are used with calcium sulfate ($CaSO_4$) as a looping medium. Coal reduces the calcium sulfate to calcium sulfide in the first reactor while also forming a high-purity CO_2 stream. The calcium sulfide that is formed is then combusted in the second reactor with air. A portion of the heat generated from the combustor is used to compensate for the heat required for coal gasification in the first reactor while the rest is used to produce high-temperature, high-pressure steam for electricity generation.

Although similar to the first one, the second configuration uses a much higher coal to $CaSO_4$ ratio and a higher steam feed rate for the first reactor. Hence, the reduction of $CaSO_4$

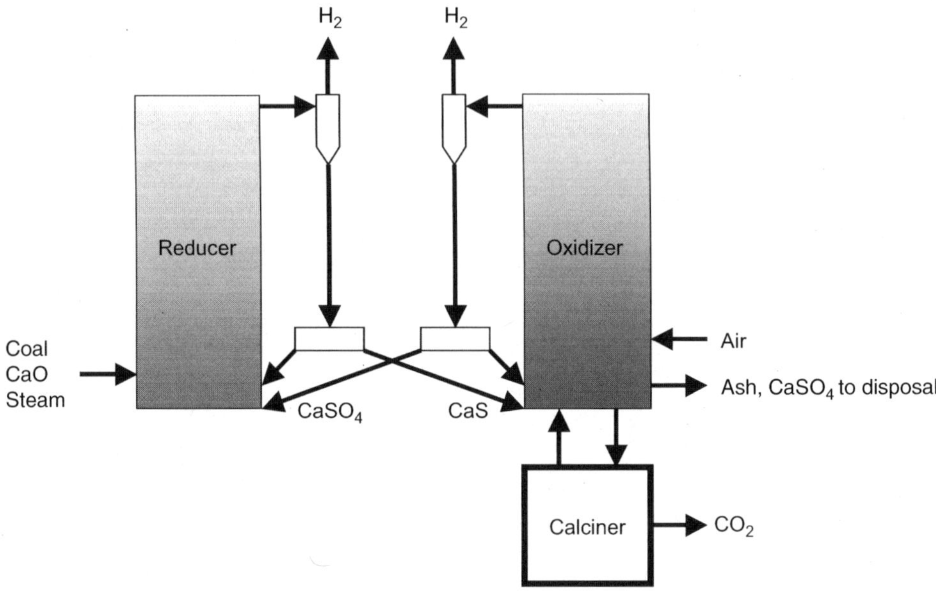

FIGURE 17.4
ALSTOM's hybrid combustion-gasification process for hydrogen production.

is accompanied by the formation of syngas (CO and H_2) due to the presence of an excessive amount of carbon and steam. In this configuration, most of the heat generated in the combustor is used to compensate for the heat required for coal gasification in the first reactor. The product for this configuration is syngas and most of the carbon in coal is converted to gaseous CO and, thus, there is no need for carbon capture as the carbon still remains in the fuel.

Only in the third configuration, shown in Figure 17.4, pure hydrogen is produced. This is achieved by introducing the third reactor (calciner) and an additional chemical loop (CaO–$CaCO_3$ loop). In this scheme, even more steam than the second configuration is fed into the system to further produce hydrogen by the WGS reaction in addition to the reduction reaction of $CaSO_4$. Calcium oxide is used in the first reactor of the second chemical loop to capture the CO_2 generated by the WGS reaction, which drives the reaction toward the formation of pure H_2 as the final product.

Reducer reactor:

$$4C + CaSO_4 \rightarrow CaS + 4CO \tag{17.10}$$

$$CO + H_2O \leftrightarrow CO_2 + H_2 \tag{17.11}$$

$$CaO + CO_2 \rightarrow CaCO_3 \tag{17.12}$$

Oxidizer reactor:

$$CaS + 2O_2 \rightarrow CaSO_4 \tag{17.13}$$

Calciner:

$$CaCO_3 \rightarrow CaO + CO_2 \tag{17.14}$$

The heat integration of this configuration includes the utilization of the heat generated from the combustion of calcium sulfide in the first chemical loop in the regeneration of the calcium-based sorbent in the second chemical loop (Andrus et al., 2005). The output gas

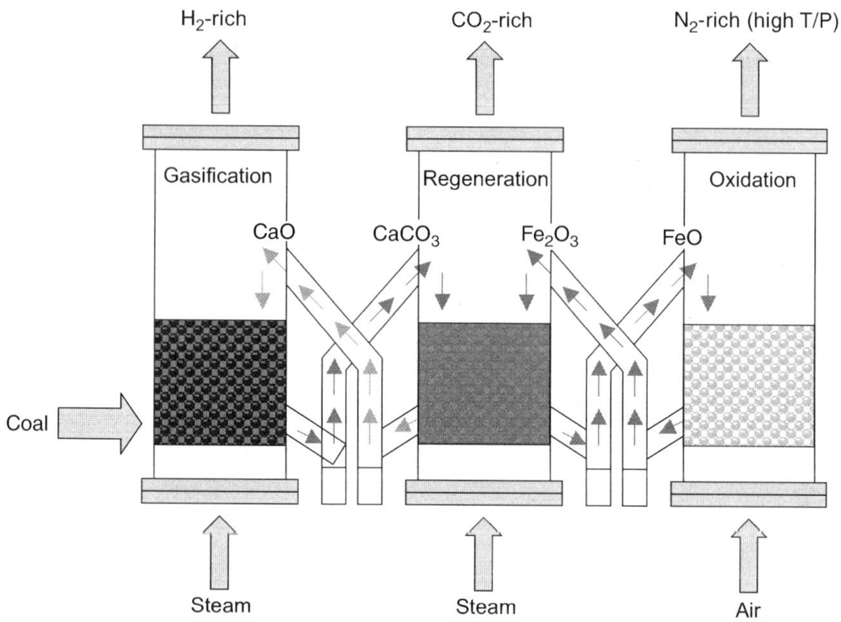

FIGURE 17.5
Schematic diagram of fuel-flexible gasification–combustion process.

stream from the calciner is a pure, concentrated CO_2 stream, which can later be processed and sequestered for the ultimate carbon management.

17.3.4 General Electric Fuel-Flexible Process

Another integrated carbon capture technology is called the fuel-flexible process developed by General Electric (GE). This process takes different feedstocks such as coal and biomass and produces hydrogen and electricity in adjustable ratios (Rizeq et al., 2002). The reaction scheme for this process involves two chemical loops operated using three fluidized-bed reactors as shown in Figure 17.5.

In the first reactor, coal is partially gasified with steam to form a mixture of H_2, CO, and CO_2. The CO_2 is then captured by the calcium-based sorbents by carbonation reaction. Similar to ALSTOM's process, the depletion of CO_2 results in the WGS reaction toward the formation of hydrogen. Moreover, sulfur in the coal is also captured by the same calcium-based sorbent forming $CaSO_3$. As a result, a high-purity hydrogen stream is obtained from the first reactor. The solids in the first reactor, which mainly consist of reacted sorbents (mostly $CaCO_3$ with a small quantity of $CaSO_3$) and unconverted carbon, are introduced to the second reactor where high-temperature steam is injected. In this reactor, the unconverted carbon reacts with an oxygen carrier metal, mainly Fe_2O_3. The heat of reaction and the high-temperature steam provide heat for the regeneration of the spent sorbents transferred from the first reactor. The gaseous product of the second reactor is a mixture of CO_2 and SO_2. Finally, the third reactor is used to regenerate the reduced oxygen carrier obtained from the second reactor by reacting it with air (Rizeq et al., 2002).

Gasifier:

$$C + 2H_2O \leftrightarrow CO_2 + 2H_2 \tag{17.15}$$

$$CaO + CO_2 \rightarrow CaCO_3 \tag{17.16}$$

Regenerator:

$$C + Fe_2O_3 \rightarrow CO_2 + FeO \quad (17.17)$$

Oxidation reactor:

$$4FeO + O_2 \rightarrow 2Fe_2O_3 \quad (17.18)$$

Heat from all the hot exhaust gas streams is used for steam generation to drive the steam turbine. Thus, the final products from the GE fuel-flexible process are pure hydrogen from the first reactor, CO_2 from the second reactor, and heat for electricity production from the third reactor. A portion of the solids in the chemical loops needs to be purged to avoid ash accumulation and maintain solid reactivity (Rizeq et al., 2002). The overall energy conversion efficiency for the GE fuel-flexible process is estimated to be 60% (Rizeq et al., 2003).

17.3.5 Calcium Looping Process

The calcium looping process developed by the Ohio State University (OSU) also uses high-temperature sorbent to enhance hydrogen production while capturing CO_2. Although there are a number of carbon capture technologies that utilize calcium-based sorbents, this process is particularly promising due to the specific design of the sorbent particle properties. High sorption capacity and long-term stability of the particles are crucial to the viability of the chemical looping process for hydrogen production. OSU's patented calcium sorbent is tailored using surface modifiers to provide mesoporous morphological properties that allow for high CO_2 capture as well as sulfur removal capacity at high temperatures while producing hydrogen with purity >95% without the WGS catalyst. The purity of the hydrogen stream increases beyond 99% if OSU's sorbent technology is combined with the WGS catalyst. In a lab scale, OSU's super sorbent has shown recyclability over 100 or more cycles while maintaining its reactivity.

The calcium looping technology provides a "one box" mode of operation for production of high-purity hydrogen with integrated CO_2, sulfur, and chloride capture, which integrates WGS reaction, hydrogen separation, CO_2 capture, and sulfur removal in one consolidated unit. Another advantage is that in addition to generating a pure hydrogen stream, it is also capable of adjusting the H_2 to CO ratio in the outlet to the required level while removing sulfur down to very low levels, thus making it an attractive scheme for various chemical processes (e.g., ethanol and Fischer–Tropsch syntheses). The process has the potential to achieve a higher system efficiency with a lower overall footprint by combining different process units in a single stage.

Figure 17.6 illustrates a gasification process integrated with the calcium looping process. Once the water gas mixture is formed at the exit of the gasifier, calcium oxide fines are injected into the fuel gas stream. As the fuel gas flows past the WGS catalyst, the WGS reaction takes place and forms additional CO_2. The injected CaO sorbent particles react with CO_2 and H_2S in the gas stream, thereby allowing further catalytic WGS reaction to occur. The reactions involved in the calcium looping scheme are

WGS reaction (Equation 17.11):

$$CO + H_2O \leftrightarrow CO_2 + H_2$$

CO_2 capture:

$$CaO + CO_2 \rightarrow CaCO_3 \quad (17.19)$$

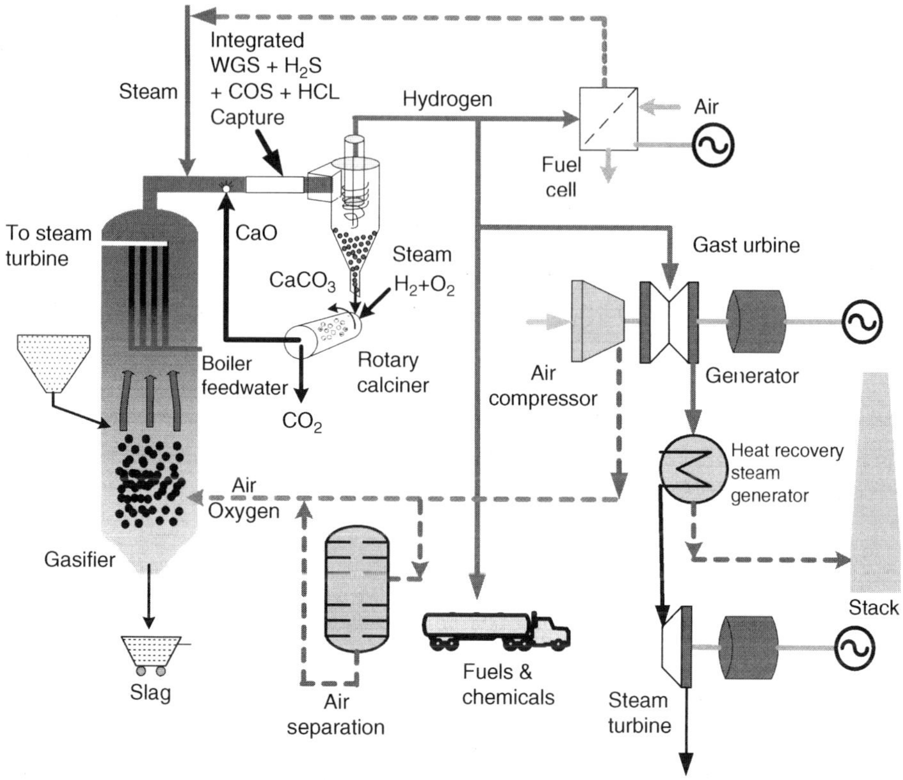

FIGURE 17.6
Conceptual flowsheet depicting integration of various units in the calcium looping processes for H_2 generation in typical coal-gasifier facility.

Sulfur capture:

$$CaO + H_2S \rightarrow CaS + H_2O \tag{17.20}$$

This process can be tailored to maximize the hydrogen production while capturing CO_2, or the ratio between CO and H_2 in the product stream can be adjusted for different applications. For example, the optimal ratio of H_2 to CO is ~2 for the Fischer–Tropsch synthesis.

At the exit of the WGS reactor, the reacted $CaCO_3$ particles are captured by a high temperature cyclone and the spent solids are then sent to a rotary calciner to thermally decompose the $CaCO_3$ back to CaO at temperatures between 660°C and 900°C with pure CO_2 formed in the outlet gas stream. The pure CO_2 gas can be economically compressed, cooled, liquefied, and transported for its safe sequestration. Hydrogen can be used for various energy conversion systems or hydrogen-enriched fuel production without any further low-temperature cleanup requirement.

17.3.6 Coal-Direct Chemical Looping Reforming Process and Syngas Redox Process

The coal-direct chemical looping process developed by the OSU is a chemical looping reforming (CLR) process that drastically simplifies the reaction scheme by utilizing only one chemical loop. The simplified flow diagram of the coal-direct CLR process is shown in Figure 17.7. Here, specially tailored highly reactive Fe_2O_3 particles that can undergo multiple (>100) reduction–oxidation cycles are used to convert coal into hydrogen. In this

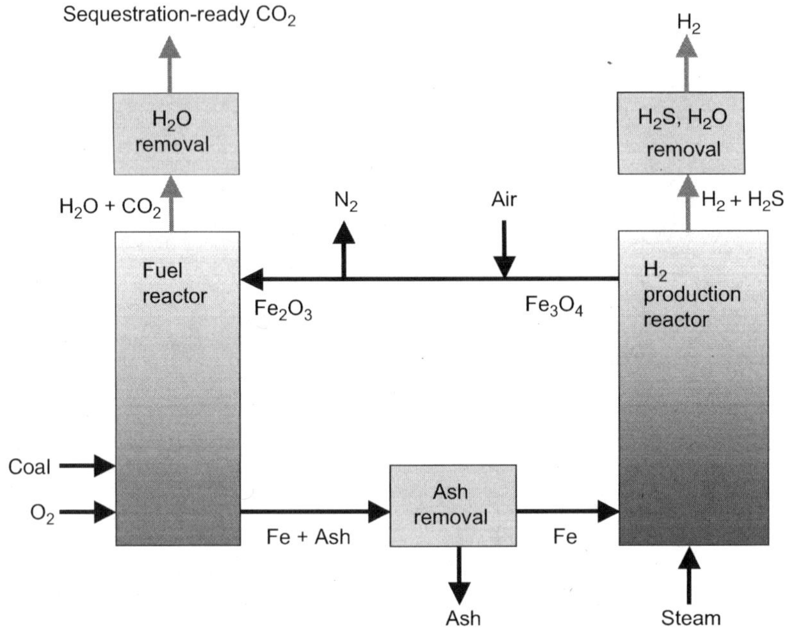

FIGURE 17.7
Simplified schematic of the CLR system.

process, Fe_2O_3 particles are introduced into the fuel reactor together with fine coal powder. By using a suitable gas–solid contacting pattern, coal will be gasified into CO and H_2. In the following reaction, the elemental composition of coal is given based on Pittsburgh #8 coal (Stultz and Kitto, 1992).

Fuel reactor:

$$C_{11}H_{10}O\text{ (coal)} + 8.67Fe_2O_3 \rightarrow 11CO_2 + 5H_2O + 17.33Fe \tag{17.21}$$

H_2 production reactor:

$$3Fe + 4H_2O \leftrightarrow Fe_3O_4 + 4H_2 \tag{17.22}$$

The reductive gas would react with Fe_2O_3 particles to form pure Fe while producing a high-concentration CO_2 and H_2O flue gas stream. H_2O in the flue gas can be readily condensed leaving a sequestration-ready CO_2 stream. The reduced Fe particle from the fuel reactor will enter the next reactor to react with steam to generate hydrogen while being oxidized to Fe_3O_4. The excess water is then condensed out to obtain a high-purity hydrogen steam. The resulting Fe_3O_4 exiting from the hydrogen production reactor will then be pneumatically conveyed back to the fuel reactor. During the particle conveying, the Fe_3O_4 particle will be oxidized with air to its original state, that is, Fe_2O_3, and recycled back to the fuel reactor. The regeneration of Fe back to Fe_2O_3 may be achieved in a number of ways. It is possible to react the Fe with air or oxygen while generating heat in the process. The oxidation of different metals, including iron, has been extensively investigated in previous studies (Lyngfelt et al., 2001; Jin and Ishida, 2004; Cho et al., 2004; Ishida et al., 2005; Corbella et al., 2006; Mattisson et al., 2006). Compared to other looping processes, coal-direct CLR process represents a much simpler process scheme with expected hydrogen production efficiency of more than 80% (Gupta et al., 2006).

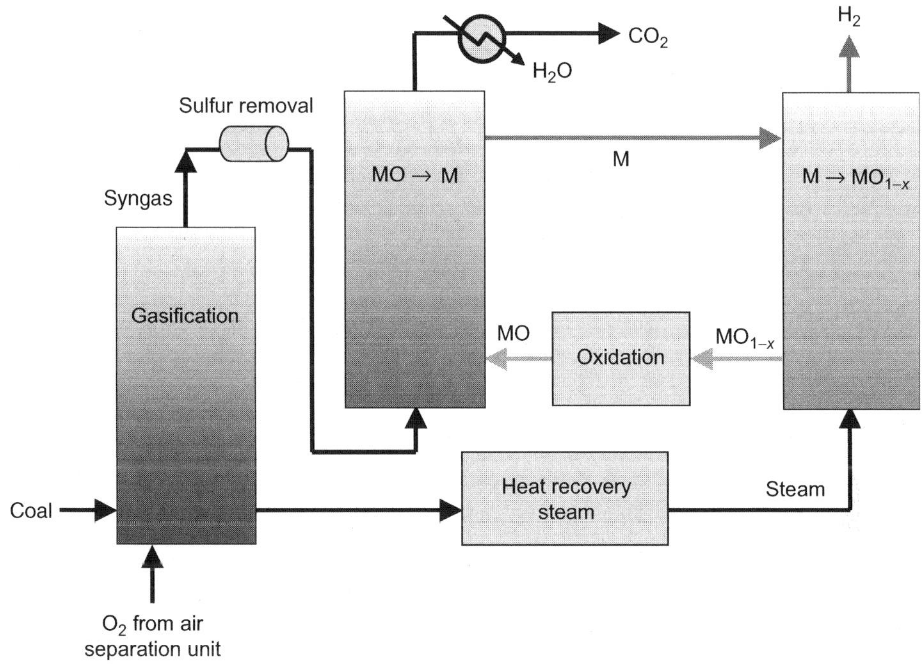

FIGURE 17.8
Simplified schematic of the syngas chemical looping process for hydrogen production from coal.

Based on a similar concept, SGR process shown in Figure 17.8 is developed where syngas, rather than coal, is fed into the fuel reactor to reduce the Fe_2O_3 particles and then produce H_2 in the hydrogen production reactor.

Fuel reactor:

$$3CO + Fe_2O_3 \rightarrow 3CO_2 + 2Fe \quad (17.23)$$

$$3H_2 + Fe_2O_3 \rightarrow 3H_2O + 2Fe \quad (17.24)$$

H_2 production reactor:

$$3Fe + 4H_2O \rightarrow Fe_3O_4 + 4H_2 \quad (17.25)$$

$$4Fe_3O_4 + O_2 \rightarrow 6Fe_2O_3 \quad (17.26)$$

As shown in the preceding reactions, the SGR process is based on similar reactions as the steam–iron process developed in the 1970s for production of hydrogen from producer gas (Benson, 1969; Charles, 1962; Gasior, 1961; Katell et al., 1962). However, the steam–iron process has a very low per pass conversion of producer gas and steam. Moreover, the iron oxide particles had poor recyclability and oxygen carrying capacity. As a result, the economic feasibility of the steam–iron process was found to be low. Hence, steam reforming of natural gas took over the market for hydrogen production.

With the unique reactor and smart particle designs, the oxygen carrying capacity and recyclability of the iron-based chemical looping particles have been radically improved, and OSU's SGR process has shown a great potential for commercialization. Its pilot-scale demonstration is currently under way. In addition, the sulfur present in coal is expected to react with Fe and form FeS. Hence, the SGR process has the capability of removing pollutants such as H_2S and mercury to low- to moderate-levels (<25 ppm) at high temperatures. The energy conversion efficiency for the SGR process is estimated to be 74% (Thomas et al., 2004; Velazquez-Vargas et al., 2006).

Besides using coal or syngas as feedstock to generate high-purity hydrogen, OSU's chemical looping processes can also process other carbonaceous fuels (e.g., coal and biomass), which makes the utilization of by-products from fossil fuel processing industry possible. Furthermore, the generation of hydrogen using steam makes the process compatible to any existing system that is a net producer of low-grade heat (such as low-pressure low-temperature steam) since water can be used to extract the heat of any exhaust stream with temperatures higher than 100°C to produce steam. The key to the high efficiency of this process stems from the staged reactions in the looping reactors and the design of highly reactive and highly recyclable particles.

17.3.7 Membrane Process

A number of techniques described earlier have achieved a high conversion efficiency for hydrogen by shifting the equilibrium of the WGS reaction. To shift the equilibrium toward hydrogen production, one of the reaction products, CO_2 is captured from the product stream through reaction-based technologies. This can also be achieved by employing membranes. Membrane technology, when applied to the WGS reaction with *in situ* removal of either CO_2 or H_2 from the reacting product gases, can significantly enhance the performance of the WGS reactor, and hence, the efficiency or economics of the process. However, currently existing membranes are not capable of completely separating CO_2 or hydrogen from the mixture, and a considerable pressure drop across the membrane leads to costly operation and poor process economics (Roark et al., 2002). In addition, any hydrogen contribution to the CO_2 stream would create substantial loss of valuable fuel.

Because of the current limitations, the development of advanced membranes has been one of the key areas of research. A high-temperature CO_2 membrane that operates in the same temperature range as that of the WGS reaction has been developed and tested (Chung et al., 2005), and hydrogen membranes are also being developed. These membrane technologies are now being implemented to the gasification systems at the laboratory or pilot stage. For example, in 2006, the National Energy Technology Laboratory in the United States showed that membrane technologies have the potential to decrease the cost of producing hydrogen through coal gasification by more than 25% (Stiegel and Ramezan, 2006). The Oak Ridge National Laboratory is developing a porous, ceramic membrane that simultaneously removes hydrogen and, thereby, enhances the conversion of the WGS reaction (Stiegel and Ramezan, 2006). In a prototype system being developed at the Oak Ridge National Laboratory, syngas produced in a conventional gasifier and cleanup system is sent to a membrane reactor that has two compartments: (1) a syngas reaction compartment, in which the CO in the syngas reacts with steam to form H_2 and CO_2 and (2) a hydrogen-collection compartment. The two are separated by a membrane that only allows for the passage of hydrogen, which generates a high-purity hydrogen stream without needing traditional separation. The product gas from the reaction compartment, which comprises mostly of CO_2 plus some CO and hydrogen residues, is combusted with oxygen in a combined cycle system to generate electricity for parasitic energy consuming processes.

Another example of the high-temperature membrane technologies is a dual-phase membrane developed at Columbia University, which consists of solid oxide and molten carbonate. This membrane technology takes advantage of oxide and carbonate ionic transport. At the face of the membrane exposed to the high concentration of CO_2, carbonation dioxide combines with an oxide anion to become carbonate ion. However, at the opposite face, a carbonate ion drops an oxide ion back into the solid oxide phase since the partial pressure of CO_2 is lower (Lackner et al., 2006). This membrane technology is capable of separating CO_2 at temperatures about 600–800°C.

Membrane technology can also be used to reduce the concentration of hydrogen sulfide to the parts per billion levels suitable for fuel cell applications (Stiegel and Ramezan, 2006). The challenges of using membranes, however, lie in the material (organic, inorganic, or hybrid) used in connection with its permeability and selectivity properties, particularly at high-temperature conditions. Concerns are cost and reliability in usage. Membranes can be effectively used in separation when coupled with reactor operation. For H_2 production, a process simulation indicates that utilizing hot gas cleanup units and advanced H_2 separation membranes with the ability to operate at 600°C can increase the coal gasification energy conversion efficiency for electricity and H_2 coproduction by ~11% when compared to SMR processes (Stiegel and Ramezan, 2006).

17.4 CO_2 Capture from Diffuse Sources

As mentioned earlier, separation of CO_2 at concentrated sources is easier than from the environment, and carbon capture at upstream decarbonizes many subsequent economic sectors. However, it does require significant changes in the existing infrastructure of power and chemical plants. Furthermore, approximately half of all emissions arise from small, distributed sources. Many of these emitters are vehicles for which onboard capture is not practical. Thus, unless all the existing automobiles are replaced by either hydrogen-powered fuel cell cars or electric cars, the capture of CO_2 from the air provides another alternative for small mobile emitters.

The atmosphere represents a large and well-mixed buffer of CO_2. Emissions over the course of a year are about 1% of the total CO_2 in the atmosphere. As a result, it is not necessary to capture a specific CO_2 emission; it suffices to capture an amount of CO_2 that cancels an emission that has happened or is about to happen. In general, CO_2 capture from air would occur at the disposal site, and thus, avoid the CO_2 transportation infrastructure that is necessary for capture at centralized sources.

Biomass sequestration, which has been embraced by coal companies as the least expensive way to start carbon management, relies on capture from air. Cofiring with biomass followed by sequestration also would lead to a net CO_2 reduction in the atmosphere (Keith, 2001). On a per ton basis, this option is quite cost effective, but the amount of land area that would be required is extremely large to be practical. Consequently, it is worthwhile to consider other options. Currently, studies that consider capture of CO_2 directly from the air by chemical means are under way. These processes will require good chemical sorbents that do not pose any environmental concerns in their own right and can be regenerated cost effectively.

The scale of an air capture system is surprisingly small. At a wind speed of 6 m/s, the U.S. per capita emissions of 22 t/year flow through an opening of 0.2 m^2—the size of a television screen (Lackner et al., 1999). Consequently, an air contacting apparatus, whether it just stands in the wind or utilizes wind, flows through towerlike structures that can be very compact. The major challenge being addressed by current research is the design of a sorbent recycle system. Strong alkali solutions including NaOH and $Ca(OH)_2$ readily pull CO_2 out of the air, but the recovery is energy intensive. $Ca(OH)_2$ in contact with air turns into $CaCO_3$, whereas NaOH turns into Na_2CO_3. The latter is converted back into sodium hydroxide in a process used in the pulp and paper industry known as the Kraft process (Bujanovic et al., 2003) as

$$Na_2CO_3 + Ca(OH)_2 \rightarrow 2NaOH + CaCO_3 \tag{17.27}$$

In the end, both approaches require the calcination of limestone to lime. The energy penalty in this process is about 4.5–5 GJ/ton of CO_2, which amounts to a 30–40% energy penalty of a transportation sector that uses air extraction for managing its own CO_2 emissions. This is comparable to the energy penalty incurred in the conversion of fossil fuels into hydrogen as a transportation fuel (Zeman and Lackner, 2004).

17.5 Transport of CO_2

Once the CO_2 is captured and compressed, it needs to be transported to the sequestration or utilization locations, unless the capture and sequestration processes are located at the same site. A CO_2 transportation infrastructure could be done with a rather conventional approach. On land, pipelines for long-distance CO_2 transport already exist. For example, a pipeline system more than 500 mi. long connects CO_2 fields in Southern Colorado to oil fields in West Texas. The CO_2 is purchased at about \$15/ton for tertiary oil recovery. The cost of CO_2 transportation is a function of distance, whereas the costs of pipeline construction vary significantly by region (Doctor et al., 1997). The construction and operation of pipelines for ocean would be quite different from land-based pipelines. Generally, CO_2 is transported at supercritical pressures (~2000 psi). If CO_2 is sequestered at geological formations, the transferred CO_2 may require additional compression at the injection site depending on the specifics of the reservoir (Doctor et al., 1997).

In addition to pipelines, high-pressure tankers can transport CO_2. Small CO_2 tankers already exist (Veder, 2004), but there is currently no use of larger vessels. Nevertheless the technology for large vessels seems less daunting than for liquefied natural gas tankers, which is a well-established technology. The use of large tankers is attractive if they are transported over water because the cost of CO_2 transportation would be rather insensitive to the distance (Möllersten et al., 2004). Furthermore, many large CO_2 point sources such as power plants and pulp mills are located near water. One study showed that assuming a CO_2 production rate of 20 kg/s, the combined cost of CO_2 transportation over the distance of 700 km and geological injection is approximately \$17/t CO_2 (Ekstrom et al., 1997).

17.6 CO_2 Storage

Disposal or long-term storage of CO_2 is the final step in the carbon management. Although the cost of compression and final storage of CO_2 account for a relatively small portion of the total carbon management cost (Mattisson et al., 2001), it is necessary to achieve permanent and efficient disposal of the CO_2. For carbon storage, ocean, geological, and mineral-based sequestration exist. With each capable of storing well over ~10^3 Gt, a significant reduction in the current anthropogenic carbon emission levels, which is estimated to be 6–GtC per year, could be sustained for several hundred years (Herzog et al., 1997). At current rates, the total CO_2 production for a century would be 650 GtC or 2500 Gt of CO_2. For comparison, Lake Michigan contains 5000 Gt of water. Regardless of the method of sequestration, the permanent disposal of carbon dioxide is difficult due to its large scale.

While the sequestration technology is still too new to settle on one specific approach, any storage option that would be useful beyond small niche markets must possess the following properties: have a large capacity, be safe, be environmentally benign, and guarantee for the long-term stability of storage. Storage lifetime constraints will become more stringent over time. The total leakage of all storage must be less than environmentally tolerable emissions. Leakage emissions $L(t)$ are a function of time t. They are related to total stored carbon $S(t)$ and the representative lifetime of all storage τ through

$$L(t) = \frac{S(t)}{\tau} \tag{17.28}$$

Leakage rates start small, because S initially is small. As S increases, lifetime constraints grow more stringent. Until there is a comprehensive understanding of the global carbon cycle, the maximum allowable leakage rate will remain uncertain. It is, however, unlikely to surpass 1 GtC/y. This suggests that τ will have to be in excess of 1000 years even before $S(t)$ reaches a 1000 GtC. Unless fossil fuels are phased out, such amounts will be stored long before the end of the twentieth century. This is a serious constraint: biomass turnover is measured in years to decades and ocean turnover times are measured in centuries. The various options of carbon sequestration technologies are discussed next.

17.6.1 Ocean Sequestration

The ocean is a well-known natural sink for carbon and is already absorbing a large fraction of the emitted CO_2 through natural processes. Therefore, nonaction equates to partial ocean sequestration. In equilibrium, the partitioning of CO_2 between ocean and atmosphere is roughly 4 to 1 (Butler, 1991). As the partial pressure of CO_2 in the air rises, the ratio gradually moves toward a higher fraction of CO_2 remaining in the atmosphere. The CO_2 carrying capacity of the ocean is determined by its carbonate alkalinity, which is the charge weighted sum of ion concentrations excluding protons, hydroxyl ions, and bicarbonate ions. Charge neutrality demands that the alkalinity, A, is equal to the sum of the ionic charge in these remaining ions,

$$A = [OH^-] + [HCO_3^-] + 2[CO_3^{2-}] - [H^+] \tag{17.29}$$

For ocean water with a pH near 8, one can neglect protons and hydroxyl ions, thus

$$A \approx [HCO_3^-] + 2[CO_3^{2-}] \tag{17.30}$$

At constant alkalinity, the ocean can absorb more CO_2 by replacing one $[CO_3^{2-}]$ with two $[HCO_3^-]$. Because 90% of all the inorganic CO_2 in the ocean is in the form of $[HCO_3^-]$, the uptake capacity of the ocean is far less than the 39,000 GtC present in the ocean. Eliminating all carbonate ions would reduce the pH to about 4 and store no more than 3500 GtC. In equilibrium with 550 ppm of CO_2 in the air, the pH of the ocean would drop by 0.3 and if stirred into the entire ocean, it could pick up about 1000–1500 GtC. Without active interference, the approach to this equilibrium would take millennia (Lackner, 2002).

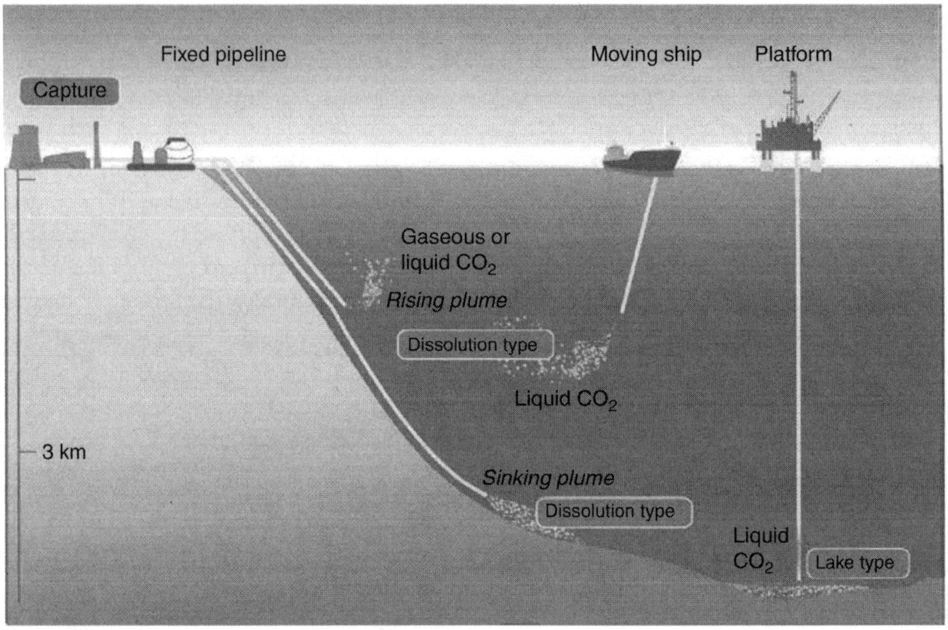

FIGURE 17.9
Ocean disposal of CO_2: five injection scenarios. (From Butler, J.N., *Carbon Dioxide Equilibria and Their Applications*, Lewis Publisher, Michigan, 1991. With permission.)

In spite of these limitations, estimates for ocean storage capacity in excess of 100,000 billion GtC have been cited (Herzog et al., 1997). Such numbers would only be feasible if alkalinity (i.e., NaOH) is added to the ocean to neutralize the corresponding amounts of carbonic acid formed. Over thousands of years, the dissolution of calcareous oozes at the bottom of the ocean could provide such alkalinity (Archer et al., 1997; Broecker and Takahashi, 1978).

Various CO_2 injection methods have been investigated to enhance the ocean's CO_2 uptake. Ocean disposal options fall into several categories as illustrated in Figure 17.9. One is to dilute the dissolved CO_2 at a depth below the mixed layer. CO_2 can be stored using this method for decades to centuries, although capacities are limited. Another approach is to form lakes of CO_2 at the bottom of the ocean. Below 2700 m, compressed CO_2 is denser than seawater and sinks to the bottom. In addition, CO_2 will react with seawater to form a clathrate, a cagelike structure with approximately six water molecules per CO_2 (Dendy and Sloan, 2003). Clathrates are solids that can form at temperatures slightly higher than the melting point of water and, thus, form spontaneously in the presence of liquid CO_2 near the bottom of the ocean. However, they are not stable and dissolve into ocean water once the CO_2 concentration in the water drops back to normal levels. Since 1977, when the first concept of direct ocean sequestration of CO_2 was introduced (Marchetti, 1977), significant research efforts have taken place. Some estimate the cost for the direct disposal of CO_2 in the oceans to be as low as \$1–6/ton of CO_2 (Freund and Ormerod, 1997). However, others argue that a more realistic estimate is \$5–15/ton of CO_2 (Herzog et al., 1995).

Although ocean sequestration is promising in terms of carbon sequestration capacity, the environmental impact may limit the acceptability of ocean storage since the strategy is predicated on the notion that the impacts on the ocean will be less than the avoided impact of these emissions to the atmosphere. Environmental concerns in ocean sequestration focus on long-term chronic issues of altering the ocean chemistry as well as on the local effects of low pH (as low as 4) and its effect on marine organisms. Although it is true that the oceans would also get acidified if the problem is ignored, the need for dealing with elevated levels of CO_2 in the atmosphere may not only be limited to climate change concerns but also to changes in ocean water chemistry. It does not help to replace one environmental problem with another.

17.6.2 Geological Sequestration

Underground storage in geological formations is a major option for disposing of CO_2. Geological sequestration is attractive due to existing technology and demonstrated low injection cost. Hence, geologic storage of CO_2 has rapidly advanced and is currently being investigated at various scales around the world. Ideal geological formations that can be used for geological CO_2 sequestration include depleted oil and gas reservoirs (200–500 GtC), deep unmineable coal seams (100–300 GtC), and deep saline aquifers (10^2–10^3 GtC). Mined salt domes and rock caverns have also been considered as viable options (Herzog et al., 1993, 1997). These estimated capacities at this point are very uncertain and may prove very large.

First tried in 1972 in Scurry County, Texas, supercritical CO_2 has been used as an effective fluid for enhanced oil and gas recovery (EOR) throughout the Permian Basin of West Texas and eastern New Mexico, and is now being pursued to a limited extent in Kansas, Mississippi, Wyoming, Oklahoma, Colorado, Utah, Montana, Alaska, and Pennsylvania (U.S. DOE, 2007). These injections do not qualify as sequestration as the CO_2 has been extracted from naturally existing underground wells about 500 mi. away (Ruether et al., 2002). However, since the injection of CO_2 into depleted oil and gas reservoirs for the purpose of carbon sequestration is similar to EOR, the existing injection technology of EOR is being applied to develop an efficient geological sequestration process. The current ongoing CO_2 sequestration–EOR projects include EnCana's CO_2–EOR process being carried out at the Weyburn Oil Field in southern Saskatchewan, Canada. It is expected that 18 million t of CO_2 will be injected while recovering additional 130 million barrels of oil from the field (U.S. DOE, 2007).

However, these projects and similar projects in Canada demonstrate that the disposal of CO_2 is practically feasible. Nevertheless, the capacity of tertiary oil and gas recovery is quite limited. Estimates range from 20 to 60 GtC (Holloway, 2001 and IPCC, 2005). The uncertainty of the practice for CO_2 storage, however, lies in the long-term stability of the reservoirs (Blunt et al., 1993).

Displacing the methane tied up in deep unmineable coal adds another small carbon sink to the portfolio of options and this process is called enhanced coal bed methane recovery. When injected into a coal bed, CO_2 can replace adsorbed methane. By doing so, coal beds can serve as a CO_2 reservoir and a source for methane production (Parson and Keith, 1998). This method is attractive in the sense that most of the injected CO_2 will be immobilized by either physical or chemical adsorption on the coal surface.

Beyond the use of CO_2 in the recovery of oil and gas, large underground reservoirs such as deep aquifers represent the best long-term underground geologic storage option. Such aquifers are generally saline and are separated from shallower aquifers and surface water supplies. The estimated storage capacities of these aquifers in the United States alone are 5–500 billion t of CO_2 (Bergman and Winter, 1995). This technology has been demonstrated by Statoil at the Sleipner West gas reservoir in the North Sea, where CO_2 is injected into an aquifer under the seafloor as illustrated in Figure 17.10 (American Geological Institute, 2003). This project shows that injection into deep aquifers is economically

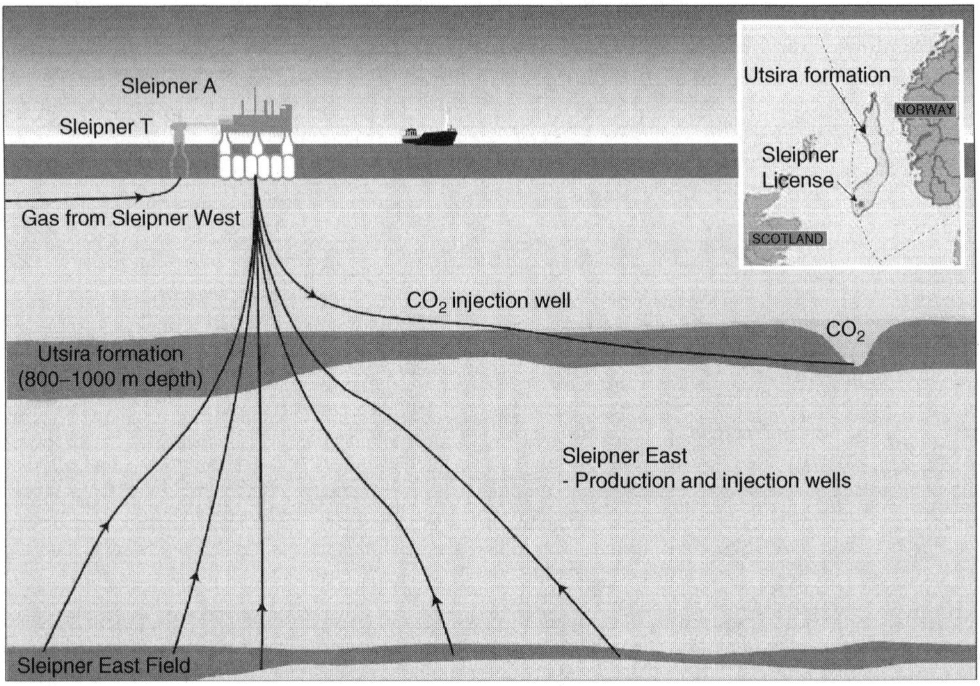

FIGURE 17.10
Saline aquifers CO_2 sequestration in Statoil's Sleipner West gas reservoir. (From American Geological Institute (AGI), Demonstrating Carbon Sequestration, *Geotimes,* http://www.agiweb.org/geotimes/mar03/feature_demonstrating.html#netl, 2003. With permission.)

feasible (Herzog, 2000). Figure 17.11 shows the location of deep saline aquifers in the United States. Unlike in the enhanced coal bed methane recovery where CO_2 is physically or chemically adsorbed, in aquifer sequestration, the injected CO_2 will probably exist as a less-stable phase (i.e., a supercritical CO_2) without being fully dissolved. Hence, there may be a greater potential for a higher leakage rate (Gale, 2004; IPCC, 2005).

There are a number of additional projects under way or on the drawing boards that will further advance knowledge in the critical areas of storage security and integrity, storage optimization, monitoring and verification, and risk assessment and mitigation. It is likely that there will be a large number of sites where CO_2 can be stored safely and for all practical purposes permanently, and the geological sequestration will make great contributions to solving global climate change issues (Apps, 2006). The central question is long-term safety, long-term leakage rate, and ultimately the capacity of these technologies. Further, the actual cost may be higher than the estimated value due to the cost associated with future monitoring of the injected CO_2. Thus, near-surface and in-depth monitoring technologies will play a key role in the success of the geological sequestration method.

17.6.3 Mineral Sequestration

As the amount of CO_2 stored increases, it becomes progressively more difficult to guarantee a physical barrier that prevents CO_2 from returning to the atmosphere. A chemical conversion to a thermodynamically lower state, thus, would be desirable and is indeed possible. CO_2 is the anhydrous form of carbonic acid and, therefore, can be used to displace

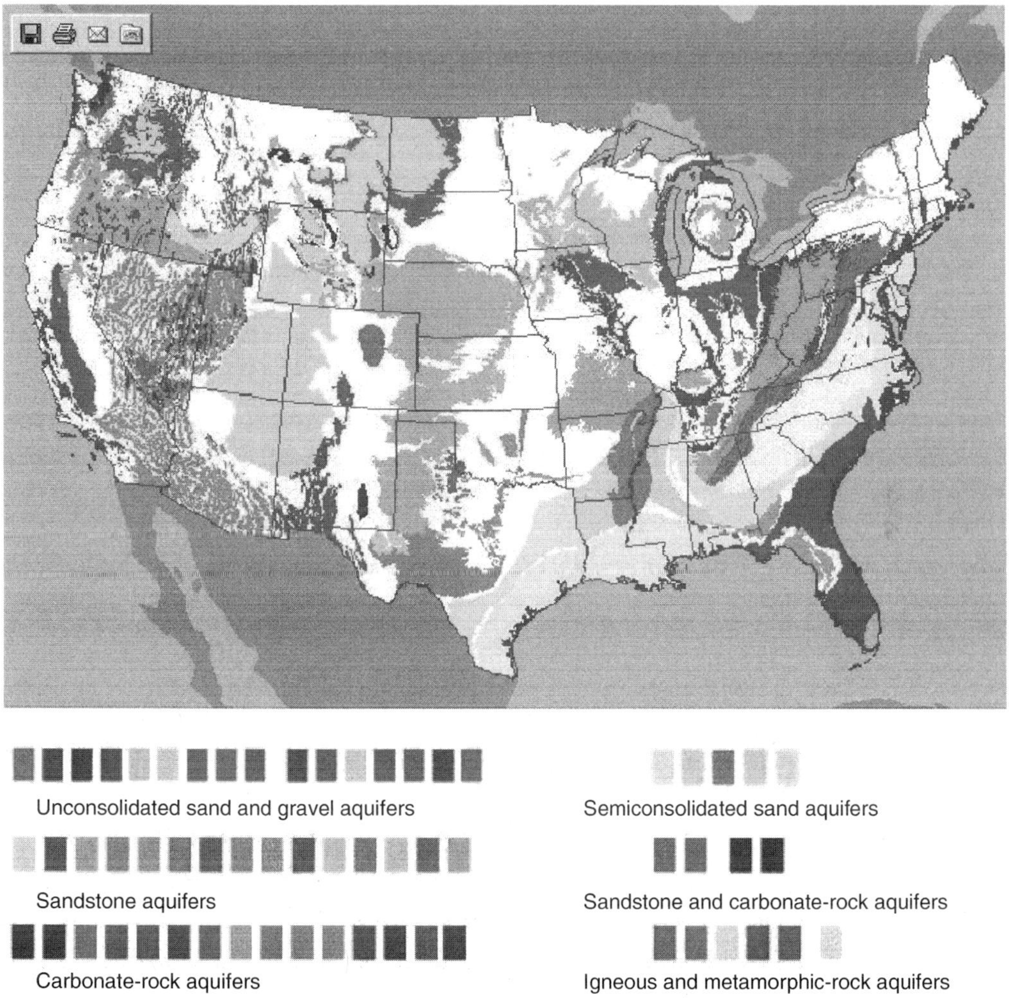

FIGURE 17.11
Saline aquifers in the United States based on the U.S. Geological Survey. (Adapted from the USGS map—*Principal Aquifers of the United States*. http://store.usgs.gov/scripts/wgate/ZWW20/!? language=en&~theme=GP&OSTORE=USGSGP&~OKCODE=START.)

weaker acids such as silicic acid. The formation of carbonates from silicates is well known as geological weathering. Thermodynamically, CO_2 can be bound as a carbonate. In many instances these carbonates dissolve in water, but some, such as magnesium or calcium carbonates, are remarkably stable as solids.

Even during geological sequestration described in the previous section, CO_2 can react with surrounding minerals and ions in ground water to form various carbonates including magnesium carbonate. Because magnesium carbonate is thermodynamically stable, immobilization of carbon in a mineral matrix truly provides a permanent disposal of CO_2 in addition to the amount stored as supercritical CO_2. Unfortunately, the reaction between CO_2 and mineral is rather slow in nature, and thus, the portion of carbon storage by mineralization is very limited in the geological sequestration. To achieve a high reaction rate it has been proposed to perform the carbonation reaction in a chemical reactor (Lackner et al., 1995).

Mineral carbonation is a new and, consequently, less studied method of sequestration. Mineral resources are plentiful for storing all the carbon that could ever be released in the consumption of fossil fuels. This sequestration process offers a safe and permanent method of CO_2 disposal, since there is almost no possibility of accidental release of CO_2 from the disposal site as CO_2 is chemically incorporated into the mineral and immobilized. Furthermore, the reactions that bind CO_2 to the mineral are exothermic in nature leading to the formation of thermodynamically stable carbonate forms (IPCC, 2005).

To sequester a significant quantity of CO_2, it is important to choose minerals that are abundantly available in the earth. Ultramafic igneous rocks (primarily peridotites and serpentinites) contain large amounts of MgO bound into a silicate structure. Dunite is a rock made up entirely of olivine (Mg_2SiO_4). More abundant ultramafic rock is serpentine ($Mg_3Si_2O_5(OH)_4$), which is hydrolyzed olivine. Peridotite and serpentinite rocks containing olivine and serpentine exist in amounts far exceeding fossil carbon resources (Goldberg, 2001). Figure 17.12 shows the map showing the worldwide locations of those deposits. There are large deposits in the United States and Puerto Rico (Goff and Lackner, 1998). There are also large deposits of these minerals in Canada and parts of Europe.

Mineral sequestration can proceed in industrial processes or underground. Industrial processes could be based on gas–solid reactions, aqueous processes, or combinations of the two. As shown in Figure 17.13, most of the current mineral sequestration processes use aqueous systems due to better reaction kinetics. In aqueous mineral sequestration, Mg in olivine or serpentine is dissolved in solution and reacts with dissolved CO_2 to form $MgCO_3$. Although the cost of mining, mineral preparation, and tailing disposal is well known and relatively inexpensive, the chemical reaction process is very expensive for a practical implementation (Lackner et al., 1997; IPCC, 2005). The bottleneck is the slow kinetics of the process and the high cost of accelerating the process by various means. Heat treatment of serpentine, external attrition grinding, and acid dissolution have all been shown to work, but the cost of the process will have to be reduced from current estimates of $69/ton of CO_2 (Penner et al., 2003) to more acceptable numbers such as $30/ton (Lackner et al., 1995). Because carbonic acid has proven to be too weak to dissolve serpentine or olivine with sufficient speed, some of the more promising approaches are based on the use of mixtures of weak acids that enhance the dissolution of serpentine and olivine while still allowing for their recovery at minimal energy cost (Park et al., 2003; Park and Fan, 2004). Particularly, a pH swing process that dissolves Mg at a low pH and precipitates $MgCO_3$ under a high pH offers a viable approach for mineral sequestration CO_2, while producing value-added solid products—silica, iron oxide, and magnesium carbonate (Park and Fan, 2004). The utilization of these solid products will lead to the cutback in the overall cost of the carbon mineral sequestration process and an additional reduction in the carbon emission related to the production of silica, iron oxide, and magnesium carbonate.

In an alternative scheme, CO_2 could be directly injected into selected underground mineral deposits for carbonation (*in situ* carbonation). This process envisions pumping CO_2 into an underground deposit of porous magnesium or calcium-bearing rock. In contrast to other underground storage reservoirs of CO_2 (as in aquifers and depleted oil/gas reservoirs), this process would result in chemically stable carbonates. Therefore, it poses a substantially lower long-term risk. The reaction would be aided by naturally high pressures (overburden) and could proceed more rapidly than mineral weathering on the earth's surface. Research efforts are being carried out to enhance the kinetics of the *in situ* chemical reaction between CO_2 and the mineral surface. The cost of the carbon mineral sequestration process is still higher than that of the geological sequestration process. Thus, further optimization of the carbon mineral sequestration method is needed to improve its economic feasibility.

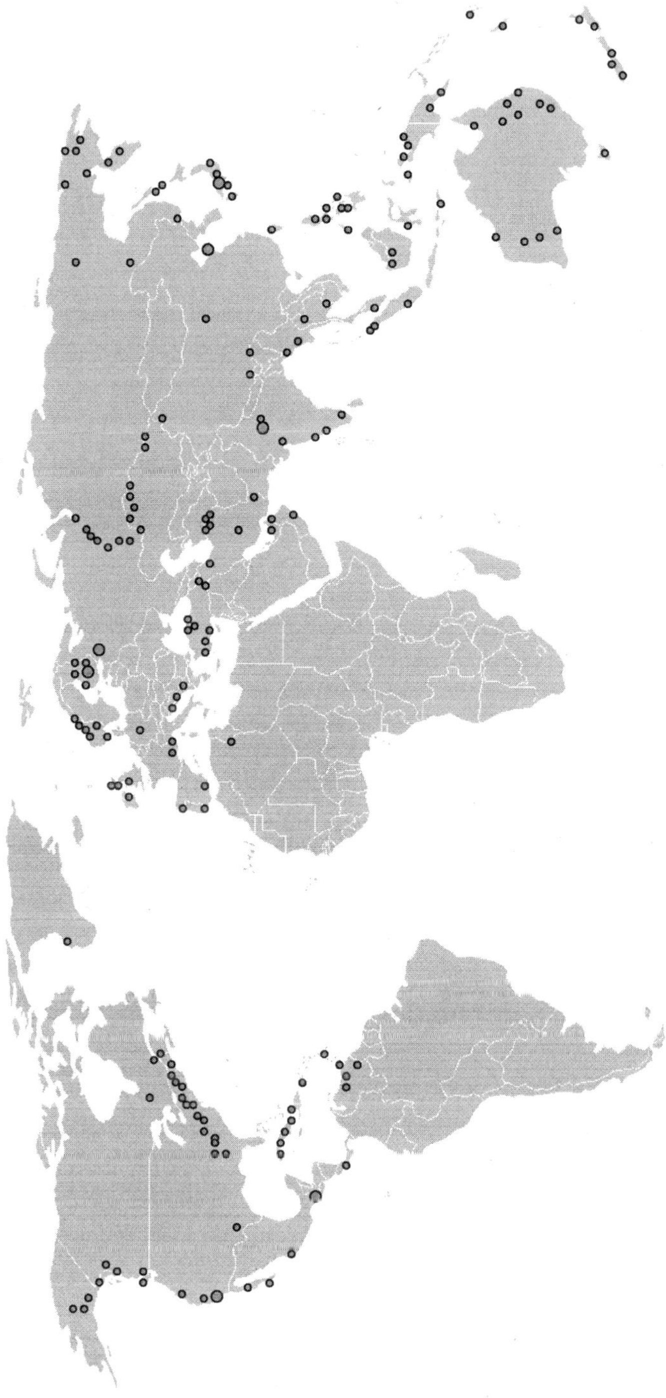

FIGURE 17.12
Worldwide distribution of magnesium-rich ultramafic rocks.

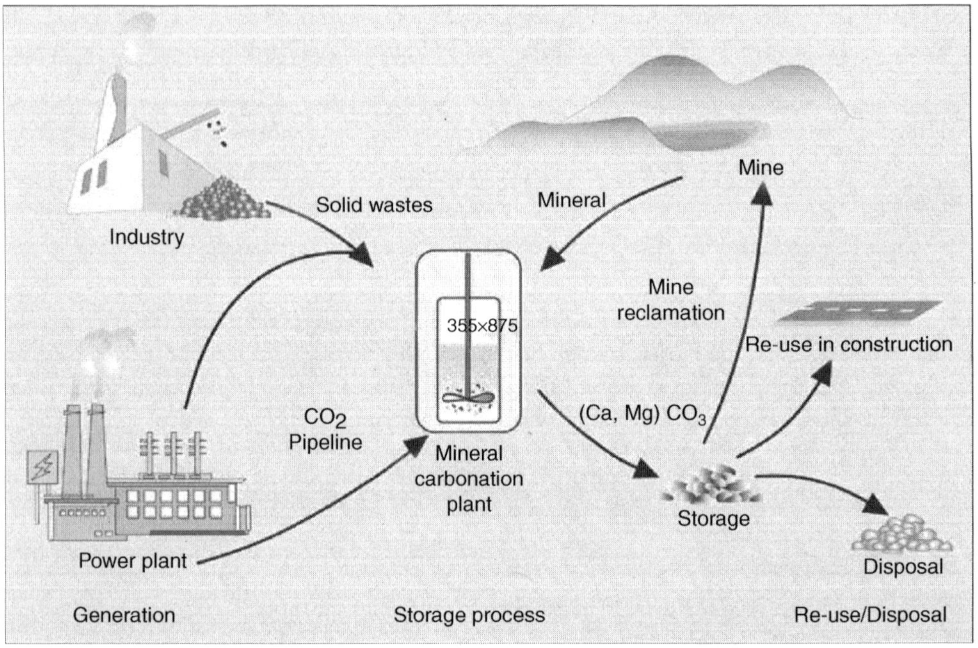

FIGURE 17.13
CO_2 sequestration by mineral carbonation. (From Goldberg, P., Mineral Sequestration Team Activities, Introduction, Issues & Plans, *Powerpoint Presentation for Mineral Carbonation Workshop August 8, 2001 sponsored by NETL*. http://www.netl.doe.gov/publications/proceedings/01/minecarb/goldberg.pdf, 2001. With permission.)

17.6.4 Biological Sequestration

Another important sequestration method is biobased. This method also mimics the natural capture of carbon in the forms of aboveground biomass (e.g., plants and trees), underground biomass (e.g., roots and microorganisms), and (organic and inorganic) carbon in solids. Worldwide biomass amounts to 610 GtC and soil carbon to 1580 GtC (Griffin and Seemann, 1996). One of the largest natural carbon cycles found in the environment is driven by photosynthesis. Trees and other plants sequester CO_2 during periods of growth. Plants take up CO_2 and water and convert it into reduced carbon compounds such as starch and cellulose. Photosynthesis fixes about 100 Gt of carbon per year. Most of this is returned through respiration and decomposition, but it appears that over the last decade, midlatitude forests have been net sinks for CO_2 (Houghton, 2003). The net uptake of CO_2 in a midlatitude forest is approximately 3.7 t of C per hectare per year (Wofsy et al., 1993). Thus, forestation and agricultural fixation of carbon, either in biomass or soil carbon, can play a role in carbon sequestration. Biomass or biological sequestration could make up for past deforestation, which is responsible for a small fraction of the increased atmospheric CO_2. Biomass sequestration is most useful when it is the by-product of other societal goals. Environmental concerns would arise where monocultures for purposes of biological fixation of carbon would replace less carbon efficient natural systems and change the natural food chain.

The cost of biological sequestration is often quite low. The estimated cost of planting forests in developing countries could be as low as \$3–\$10/ton of CO_2 captured (Paustian et al., 1998). In addition, the sequestration product, biomass, can be used as a carbon neutral energy source. However, the biological sequestration technologies are often limited by space requirements. Thus, researchers are currently investigating various biosystems including algae production to capture CO_2 at a rapid rate to minimize the space requirement.

17.7 Conclusions

As discussed in this chapter, there are a number of carbon capture and sequestration methods being investigated and developed at present. In fact, it is important to concurrently develop these sequestration methods because there will not be a single solution that can cure the carbon emission problem due to its massive emission rate. Rather, optimal solutions for the carbon sequestration will regionally be different. Before technologies can be implemented, it is important to gain public acceptance through public outreach and education. It is also important to establish the net reduction in CO_2 emissions that can be achieved by the various sequestration methods through life cycle analysis.

The costs of the carbon capture and sequestration processes described here are rather high for immediate applications. However, based on the current rate of the technology development and the emerging strong global commitment toward solving the climate change problem, the future for the carbon management is promising. Fossil fuels provide a large resource base for low-cost energy carriers in the form of hydrogen or electricity, and without carbon capture and storage technology, the vast majority of this resource cannot be utilized.

Although in the short term, the cost of CO_2 capture and transport are likely to dominate the cost of carbon sequestration, in the end, the cost of disposal will become more important. There are two major reasons. First, as technologies improve and hydrogen production facilities are designed around the concept of carbon capture, the incremental cost of capturing carbon dioxide will become smaller. Indeed in some of the zero-emission designs, the capture processes separate CO_2 while capturing other pollutants, and thus, the net costs of carbon capture may be close to zero. Pipeline costs will decrease as the pipeline network gets larger. At the same time, however, the demands on CO_2 storage will gradually increase. As the reservoirs get larger, concerns over environmental consequences will increase and the concerns over leakage will become larger. Meanwhile, the easily accessible sites will be used up and the cost of monitoring will increase.

CO_2 sequestration technology is a viable contender among a set of options to stabilize the atmospheric CO_2 level over the next few decades (Lackner, 2002). The technical ability to start such a process exists today, but the institutional structures required to reduce CO_2 emissions are still missing. For hydrogen production, carbon sequestration at the very least will buy time for alternatives to fossil fuels to become competitive. It is, however, equally possible that carbon sequestration removes the major environmental obstacles to the use of fossil fuels in which case it may prove competitive with alternatives for a long time to come.

References

Alstom, Press Releases, Alstom and Statoil to jointly develop project for chilled ammonia-based CO_2 capture for natural gas in Norway, http://www.alstom.com/pr_corp/2007/corp/43066.EN.php?languageId=EN&dir=/pr_corp/2007/corp/&idRubriqueCourante=15445, June 21, 2007.

American Geological Institute (AGI), Demonstrating Carbon Sequestration, *Geotimes*, http://www.agiweb.org/geotimes/mar03/feature_demonstrating.html#netl, 2003.

Andrus, H.A.E. Jr. et al., ALSTOM's Hybrid Combustion-Gasification Chemical Looping Technology Development, *Proceedings of the 22nd Annual International Pittsburgh Coal Conference*, Pittsburgh, PA, September 2005.

Andrus, H.E. et al., *Hybrid Combustion-Gasification Chemical Looping Coal Power Technology Development*, ALSTOM Technical Report for Project DE-FC26-03NT41866, 2006.

Apps, J.A., *A Review of Hazardous Chemical Species Associated with CO_2 Capture from Coal-Fired Power Plants and Their Potential Fate in CO_2 Geologic Storage*, LBNL, Berkeley, CA, 2006.

Archer, D., Kheshgi, H., and Maier-Reimer, E., Multiple timescales for neutralization of fossil fuel CO_2, *Geophys. Res. Lett.*, 24(4), 405, 1997.

Balasubramanian, B. et al., Hydrogen from methane in a single-step process, *Chem. Eng. Sci.*, 54, 3543, 1999.

Benson, E.H., Method for the Production of a Mixture of Hydrogen and Steam, U.S. Patent No. 3,421,869, 1969.

Bergman, P.D. and Winter, E.M., Disposal of CO_2 in aquifers in the U.S., *Energ. Convers. Manag.*, 36, 523, 1995.

Blunt, M., Fayers, F.J., and Orr, F.M., Carbon dioxide in enhanced oil recovery, *Energ. Convers. Manag.*, 34, 1197, 1993.

Breckenridge, W. et al., Use of SELEXOL Process in Coke Gasification to Ammonia Project, *Proceedings of Laurance Reid Gas Conditioning Conference*, http://www.uop.com/objects/92SelexCokeGasifAmm.pdf, March 2000.

Broecker, W.S. and Takahashi, T., Neutralization of fossil fuel CO_2 by marine calcium carbonate, in *The Fate of Fossil Fuel CO_2 in the Oceans*, Andersen, N.R., Malahoff, A. Eds., Plenum Publishing Corporation, New York, 1978.

Bryant, E., *Climate Process and Change*, Cambridge University Press, Cambridge, 1997.

Bugante, E.C. et al., Methane production from hydrogen and carbon dioxide and monoxide in a column bioreactor of thermophilic methanogens by gas recirculation, *J. Ferm. Bioeng.*, 67(6), 419, 1989.

Bujanovic, B., Cameron, J.H., and Yilgor, N., Comparative studies of kraft and kraft–borate pulping of black spruce, *J. Pulp Paper Sci.*, 29(6), 190, 2003.

Butler, J.N., *Carbon Dioxide Equilibria and Their Applications*, Lewis Publisher, Chelsea, MI, 1991.

Caldeira, K., Forests, climate, and silicate rock weathering, *J. Geochem. Exploration*, 88(1–3), 419 (Special Issue), 2006.

Charles, H.W., Hydrogen Manufacture, U.S. Patent No. 3,027,238, 1962.

Châtel-Pélage, F. et al., Applications of oxygen for NO_x control and CO_2 capture in coal-fired power plants, *Second International Conference on Clean Coal Technologies for our Future*, Sardinia, May 10–12, 2005.

Cho, P., Mattisson, T., and Lyngfelt, A., Comparison of iron-, nickel-, copper- and manganese-based oxygen carriers for chemical-looping combustion, *Fuel*, 83(9), 1215, 2004.

Chung, S.J. et al., Dual-phase inorganic metal-carbonate membrane for high temperature carbon dioxide separation, *Ind. Eng. Chem. Res.*, 44, 7999, 2005.

Corbella, B.M. et al., Performance in a fixed-bed reactor of titania-supported nickel oxide as oxygen carriers for the chemical-looping combustion of methane in multicycle tests, *I&EC Res.*, 45(1), 157, 2006.

Dendy, E. and Sloan, J., Fundamental principles and applications of natural gas hydrates. *Nature*, 426, 353, 2003.

Doctor, R.D., Molburg, J.C., and Thimmapuram, P.R., Oxygen-blown gasification combined cycle, carbon dioxide recovery, transport and disposal, *Energ. Convers. Manag.*, 38, (Suppl.), S575, 1997.

Ekstrom, C. et al., *Technologies and Costs in Sweden for Capture and Storage of CO_2 from Combustion of Fossil Fuels for Production of Power, Heat, and/or Transportation Fuels*, Vattenfall Utveckling AB, Stockholm, 1997.

Energy Information Administration (EIA), *International Energy Outlook-2006*, Office of Integrated Analysis and Forecasting, Report # DOE/EIA-0484, U.S. Department of Energy, Washington, June 2006.

Fan, L.-S. and Iyer, M., Coal cleans up its act, *TCE*, 36, October 2006.

Fankhauser, S. et al., Vulnerability to climate change and reasons for concern, A synthesis, in *Climate Change 2001, Impacts, Adaptation, and Vulnerability—A Report of Working Group II of the Intergovernmental Panel on Climate Change*, McCarthy, J.J., Canziani, O.F., Leary, N.A., Dokken, D.J, and White, K.S., Eds., Cambridge University Press, Cambridge, 941–967, 2001.

Fink, C., Curran, G., and Sudbury, J.D., CO_2 Acceptor Process Pilot plant—1974 Rapid City, South Sakota, *Proceedings of the Sixth Synthetic Pipeline Gas Symposium*, Chicago, IL, 1974.

Freund, P. and Ormerod, W.G., Progress toward storage of carbon dioxide, *Energ. Convers. Manag.*, 38(Suppl.), S199–S204, 1997.

Gale, J., Geological storage of CO_2, What do we know, where are the gaps and what more needs to be done? *Energy*, 29, 1329–1338, 2004.

Gasior, S.J. et al., *Production of synthesis gas and hydrogen by the steam iron process—Pilot-plant study of fluidized and free-falling beds*, Bureau of Mines Report of Investigations, Pittsburgh, PA, 5911, 49, 1961.

George, R.A., Status of tubular SOFC field unit demonstrations, *J. Power Sources*, 86(1–2), 134, 2000.

Goff, F. and Lackner, K.S., Carbon dioxide sequestering using ultramafic rocks, *Environ. Geosci.*, 5(3), 89, 1998.

Goldberg, P., Mineral Sequestration Team Activities, Introduction, Issues & Plans, *Powerpoint presentation for Mineral Carbonation Workshop August 8, 2001 sponsored by NETL*. http://www.netl.doe.gov/publications/proceedings/01/minecarb/goldberg.pdf, 2001.

Griffin, K.L. and Seemann, J.R., Plants, CO_2 and photosynthesis in the 21st century, *Chem. Biol.*, 3(4), 2476, 1996.

Gupta, P. et al., Chemical Looping Reforming Process for the Production of Hydrogen from Coal. *Proceedings of the 23th Annual International Pittsburgh Coal Conference*, Pittsburgh, PA, September 2006.

Halman, M.M. and Steinberg, M., *Greenhouse Gas Carbon Dioxide Mitigation, Science and Technology*, Lewis Publishers, Boca Raton, FL, 1999.

Harvey, L.D.D., Declining temporal effectiveness of carbon sequestration, implications for compliance with the United National framework convention on climate change, *Climate Change*, 63(3), 259, 2004.

Heaven, D. et al., Synthesis gas purification in gasification to Ammonia/Urea plants, *Gasification Technologies Conference*, Washington, 2004.

Herzog, H., Drake, E., and Adams, E., *CO_2 Capture, Reuse, and Storage Technologies for Mitigating Global Climate Change—A White Paper*, Final Report, DE-AF22-96PC01257, MIT Energy Laboratory, Cambridge, MA, January 1997.

Herzog, H., Eliasson, B., and Kaarstad, O., Capturing greenhouse gases, *Sci. Am.*, 2, 72, 2000.

Herzog, H. et al., *A Research Needs Assessment for Capture Utilization, and Disposal of Carbon Dioxide from Fossil Fuel-Fired Power Plan*, DOE/ER-30194, U.S. DOE, Washington, 1993.

Herzog, H. et al., *Technology Assessment of CO_2 Ocean Disposal*, Report 95-001, MIT Energy Laboratory, Cambridge, MA, 1995.

Higman, C.A.A., Methanol production by gasification of heavy residues, Presented at the *IChemE Conference*, London, November 22–23, 1995.

Holloway, S., Storage of fossil fuel–derived carbon dioxide beneath the surface of the earth, *Ann. Rev. Energy Environ.*, 26, 145, 2001.

Houghton, R.A., Why are estimates of the terrestrial carbon balance so different? *Global Change Biol.*, 9(4), 500, 2003.

Hufton, J.R., Mayorga, S., and Sircar, S., Sorption-enhanced reaction process for hydrogen production. *AIChE J.*, 45, 248, 1999.

IPCC (Intergovernmental Panel on Climate Change), *Carbon Dioxide Capture and Storage*, Metz, B., Davidson, O., de Coninck, H., Loos, M., and Meyer, L., Eds., Cambridge University Press, Cambridge, 2005.

IPCC (Intergovernmental Panel on Climate Change), *Climate Change 2007: The Physical Science Basis—Summary for Policymakers*, Contribution of Working Group I to the Fourth Assessment Report of the Intergovernmental Panel on Climate Change, Switzerland, 2007.

Ishida, M. et al., Application of Fe_2O_3–Al_2O_3 composite particles as solid looping material of the chemical-loop combustor, *Energy & Fuels*, 19(6), 2514, 2005.

Iyer, M.V. et al., Multicyclic study on the simultaneous carbonation and sulfation of high-reactivity CaO, *I & EC Res.*, 43, 3939, 2004.

Jin, H. and Ishida, M., A new type of coal gas fueled chemical-looping combustion, *Fuel*, 83(17–18), 2411, 2004.

Johansson, T.B. et al., Options for reducing CO_2 emissions from the energy supply sector, *Energy Policy*, 24(10–11), 985, 1996.

Katell, S., Faber, J.H., and Wellman, P., *An Economic Evaluation of Hydrogen Production by the Continuous Steam-Iron Process at Seven Atmospheres*, Bureau of Mines Report of Investigations (No. 6089), Pittsburgh, PA, 13, 1962.

Keith, D.W., Sinks, Energy crops and land use, coherent climate policy demands an integrated analysis of biomass, *Climatic Change*, 49(1–2), 1, 2001.

Kikkinides, E.S., Yang, R.T., and Cho, S.H., Concentration and recovery of CO_2 from flue gas by pressure swing adsorption, *Ind. Eng. Chem. Res.*, 32, 2714–2720, 1993.

Lackner, K., Climate change, A guide to CO_2 sequestration, *Science*, 300(5626), 1677, 2003.

Lackner, K.S., Carbonate chemistry for sequestering fossil carbon, *Ann. Rev. Energy Environ.*, 27(1), 193, 2002.

Lackner, K.S., Park, A.-H.A., and Fan, L.-S., Carbon dioxide capture and disposal, carbon sequestration, in *Encyclopedia of Chemical Processing*, Lee, S.D., Ed., 305–315, 2005.

Lackner, K.S., West, A.C., and Wade, J.L., *Ion Conducting Membranes for Separation of Molecules*, U.S. Patent Publication Number: WO2006113674, 2006.

Lackner, K.S. and Ziock, H.-J., The US zero emission coal alliance, *VGB Powertech.*, 12, 57, 2001.

Lackner, K.S., Ziock, H.-J., and Grimes, P., Carbon Dioxide Extraction from Air: Is it an Option? *Proceedings of the 24th International Conference on Coal Utilization & Fuel Systems*, Clearwater, FL, 1999.

Lackner, K.S., Ziock, H.J., and Harrison, D.P., Hydrogen Production from Carbonaceous Material, U.S. Patent No. 6,790,430, 2004.

Lackner, K.S. et al., Carbon dioxide disposal in carbonate minerals, *Energy*, 20(11), 1153–1170, 1995.

Lackner, K.S. et al., *Carbon Dioxide Disposal in Mineral Form, Keeping Coal Competitive*, LA-UR-97-20941997, Los Alamos National Laboratory Report, Los Alamos, NM, 1997.

Langdon, C., Takahashi, T., Sweeney, C., Chipman, D., Goddard, J., Marubini, F., Aceves, H., Barnett, H., and Atkinson, M.J., Effect of calcium carbonate saturation state on the calcification rate of an experimental coral reef, *Global Biogeochem. Cy.*, 14, 639–654, 2000.

Leithner, R., Zero emission coal alliance process and similar cycles, *Chem. Ing. Tech.*, 76(9), 1321, 2004.

Lin, S. et al., Process analysis for hydrogen production by reaction integrated novel gasification (HyPr-RING), *Energ. Convers. Manag.*, 46, 869, 2005.

Lin, S.-Y. et al., Developing an innovative method, HyPr-RING, to produce hydrogen from hydrocarbons, *Energ. Convers. Manag.*, 43, 1283, 2002.

Lopez-Ortiz, A. and Harrison, D.P., Hydrogen production using sorption enhanced reaction, *Ind. Eng. Chem. Res.*, 40, 5102, 2001.

Lyngfelt, A., Leckner, B., and Mattisson, T., A fluidized-bed combustion process with inherent CO_2 separation, application of chemical-looping combustion, *Chem. Eng. Sci.*, 56(10), 3101, 2001.

Marchetti, C., On geoengineering and the CO_2 problem, *Climatic Change*, 1, 59, 1977.

Mattisson, T., Johansson, M., and Lyngfelt, A., Multicycle reduction and oxidation of different types of iron oxide particles-application to chemical-looping combustion, *Energy Fuels*, 18(3), 628, 2004.

Mattisson, T., Johansson, M., and Lyngfelt, A., The use of NiO as an oxygen carrier in chemical-looping combustion, *Fuel*, 85, 736, 2006.

Mattisson, T., Lyngfelt, A., and Cho, P., The use of iron oxide as an oxygen carrier in chemical-looping combustion of methane with inherent separation of CO_2, *Fuel*, 80, 1953, 2001.

McFarlan, A., Pelletier, L., and Maffei, N., An intermediate-temperature ammonia fuel cell using Gd-doped barium cerate electrolyte, *J. Electrochem. Soc.*, 151(6), A930, 2004.

Mimura, T., Simayoshi, H., Suda, T., Iijima, M., and Mituoka, S. Development of energy saving technology for flue gas carbon dioxide recovery in power plants by chemical absorption method and steam system, *Energ. Convers. and Manag.*, 38(Suppl.), S57–S62, 1997.

Möllersten, K. et al., Efficient energy systems with CO_2 capture and storage from renewable biomass in pulp and paper mills, *Renewable Energy*, 29, 1583, 2004.

National Academy of Sciences Committee (on the Science of Climate Change), *Climate Change Science, An Analysis of Some Key Questions*, National Academy Press, Washington, 2001.

Ogden, J.M., Prospects for building a hydrogen energy infrastructure, *Ann. Rev. Energy Environ.*, 24(1), 227, 1999.

Oreskes, N., The scientific consensus on climate change, *Science*, 306, 1686, 2004.
Park, A.-H.A. and Fan, L.-S., CO_2 mineral sequestration, physically activated dissolution of serpentine and pH swing process, *Chem. Eng. Sci.*, 59(22–23), 5241, 2004.
Park, A.-H.A., Jadhav, R., and Fan, L.-S., CO_2 Mineral sequestration, chemical enhanced aqueous carbonation of serpentine, *Canadian J. Chem. Eng.*, 81(3–4), 885, 2003.
Parson, E.A. and Keith, D.W., Fossil fuels without CO_2 emissions, *Science*, 282(6), 1053, 1998.
Paustian, K. et al., CO_2 mitigation by agriculture, An overview, *Climate Change*, 40(1), 135, 1998.
Penner, L.R. et al., Mineralization Strategies for Carbon Dioxide Sequestration, *Proceeding of 20th Pittsburgh Coal Conference*, Pittsburgh, PA, September 15–19, 2003.
Probstein, R.F. and Hicks, R.E., *Synthetic Fuels*, pH Press, Cambridge, MA, 1990.
Rizeq, G. et al., Advanced gasification-combustion technology for production of hydrogen, power and sequestration-ready CO_2, Presented at *Gasification Technologies 2003*, San Francisco, CA, October 12–15, 2003.
Rizeq, R.G. et al., *Fuel-Flexible Gasification-Combustion Technology for Production of H_2 and Sequestration-Ready CO_2*, Annual Technical Progress Report to DOE, 2002.
Roark, S.E., Mackay, R., and Sammells, A.F., Hydrogen Separation Membranes for Vision 21 Energy Plants, *Proceedings of the International Technical Conference on Coal Utilization & Fuel Systems*, 27(1), 101, 2002.
Ruether, J. et al., *Prospects for Early Deployment of Power Plants Employing Carbon Capture*, National Energy Technology Laboratory, Pittsburgh, PA, 2002.
Simbeck, D. and Chang, E., Hydrogen supply, cost estimate for hydrogen pathways—scoping analysis, USDOE NREL/SR-540-32525, July 2002.
Stiegel, G.J. and Ramezan, M., Hydrogen from coal gasification, An economical pathway to a sustainable energy future, *Int. J. Coal Geol.*, 65, 173, 2006.
Stix, G., A climate repair manual, *Sci. Am.*, 295(3), 46–49, September 2006.
Stocker, J., Whysall, M., and Miller, G.Q., 30 years technology for H_2 purification, *UOP LLC*, 2818, 1998.
Stultz, S.C. and Kitto, J.B., *Steam—Its Generation and Use*, 40th Ed. Babcock and Wilcox Company, Barberton, OH, 1992.
Thomas, T., Fan, L.-S., and Gupta, P., *Combustion Looping Using Composite Oxygen Carriers*, U.S. Patent Application Serial No. 11/010, 648, 2004.
United Nations, Kyoto Protocol to the United Nations Framework Convention on Climate Change, http://unfccc.int/resource/docs/convkp/kpeng.html, 1997.
U.S. DOE, *Carbon Sequestration R & D*, http://www.fe.doe.gov/programs/sequestration/, 2004.
U.S. DOE, *Enhanced Oil Recovery/CO_2 Injection*, http://www.fossil.energy.gov/programs/oilgas/eor/index.html, 2007.
Veder, A., *Coral Carbonic*, http://www.anthonyveder.nl/fleet/ccarbonic.html, accessed June 2004.
Velazquez-Vargas, L.G. et al., Hydrogen Production from Coal derived Syn Gas Using Novel Metal Oxide Particles, *Proceedings of the 23th Annual International Pittsburgh Coal Conference*, Pittsburgh, PA, September 2006.
Wallace, D., Capture and Storage of CO_2—What Needs to Be Done, *Proceedings of the 6th Conference of the Parties, COP 6, to the United Nations Framework Convention on Climate Change*, The Hague, The Netherlands, November 13–24, 2000, www.iea.org/envissu/index.htm, 2000.
White, C.M. et al., Separation and capture of CO_2 from large stationary sources and sequestration in geological formations—coalbeds and deep saline aquifers, *J. Air Waste Manage. Assoc.*, 53, 645, 2003.
Wofsy, S.C. et al., Net exchange of CO_2 in a mid-latitude forest, *Science*, 260(5112), 1314–1317, 1993.
Zeman, F.S. and Lackner, K.S., Capturing carbon dioxide directly from the atmosphere, *World Resource Rev.*, 16(2), 157–172, 2004.
Ziock, H.-J., Lackner, K.S., and Harrison, D.P., Zero Emission Coal Power, a New Concept, *Proceedings of the First National Conference on Carbon Sequestration*, http://www.netl.doe.gov/publications/proceedings/01/carbon_seq/2b2.pdf, May 14–17, 2001.

Index

D

O

P

Q

R

S

T

U

V

W

X

Y

Z

HYDROGEN FUEL

Production, Transport, and Storage